Machine Learning for Text

Charu C. Aggarwal

Machine Learning for Text

Second Edition

 Springer

Charu C. Aggarwal
Mohegan Lake
NY, USA

A Solution Manual to this book can be downloaded from https://link.springer.com/book/10.1007/978-3-030-96623-2

ISBN 978-3-030-96622-5 .ISBN 978-3-030-96623-2 (eBook)
https://doi.org/10.1007/978-3-030-96623-2

This Springer imprint is published by the registered company Springer Nature Switzerland AG
The registered company address is: Gewerbestrasse 11, 6330 Cham, Switzerland

To my wife Lata, my daughter Sayani,
and my late parents Dr. Prem Sarup and Mrs. Pushplata Aggarwal.

Preface

"If it is true that there is always more than one way of construing a text, it is
not true that all interpretations are equal."– Paul Ricoeur

The rich area of text analytics draws ideas from information retrieval, machine learning,
and natural language processing. Each of these areas is an active and vibrant field in its
own right, and numerous books have been written in each of these different areas. As a
result, many of these books have covered some aspects of text analytics, but they have not
covered all the areas that a book on learning from text is expected to cover.

At this point, a need exists for a focussed book on machine learning from text. This
book is a first attempt to integrate all the complexities in the areas of machine learning,
information retrieval, and natural language processing in a holistic way, in order to create
a coherent and integrated book in the area. Therefore, the chapters are divided into three
categories:

1. *Fundamental algorithms and models:* Many fundamental applications in text analytics, such as matrix factorization, clustering, and classification, have uses in domains
   beyond text. Nevertheless, these methods need to be tailored to the specialized characteristics of text. Chapters 1 through 8 will discuss core analytical methods in the
   context of machine learning from text.

2. *Information retrieval and ranking:* Many aspects of information retrieval and ranking are closely related to text analytics. For example, ranking SVMs and link-based
   ranking are often used for learning from text. Chapter 9 will provide an overview of
   information retrieval methods from the point of view of text mining.

3. *Sequence- and natural language-centric models:* Although multidimensional representations can be used for basic applications in text analytics, the true richness of the text
   representation can be leveraged by treating text as sequences. Chapters 10 through 16
   will discuss these advanced topics like sequence embedding, deep learning, transformers, pre-trained language models, information extraction, knowledge graphs, summarization, question-answering, opinion mining, text segmentation, and event extraction.

Because of the diversity of topics covered in this book, some careful decisions have been made
on the scope of coverage. A complicating factor is that many machine learning techniques
depend on the use of basic natural language processing and information retrieval methodologies. This is particularly true of the sequence-centric approaches discussed in Chapters 10

through 16 that are more closely related to natural language processing. Examples of analytical methods that rely on natural language processing include information extraction, event extraction, opinion mining, and text summarization, which frequently leverage basic natural language processing tools like linguistic parsing or part-of-speech tagging. Needless to say, natural language processing is a full fledged field in its own right (with excellent books dedicated to it). Therefore, a question arises on how much discussion should be provided on techniques that lie on the interface of natural language processing and text mining without deviating from the primary scope of this book. Our general principle in making these choices has been to focus on *mining* and *machine learning* aspects. If a specific natural language or information retrieval method (e.g., part-of-speech tagging) is not *directly* about text analytics, we have illustrated how to *use* such techniques (as black-boxes) rather than discussing the internal algorithmic details of these methods. Basic techniques like part-of-speech tagging have matured in algorithmic development, and have been commoditized to the extent that many open-source tools are available with little difference in relative performance. Therefore, we only provide working definitions of such concepts in the book, and the primary focus will be on their utility as off-the-shelf tools in mining-centric settings. The book provides pointers to the relevant books and open-source software in each chapter in order to enable additional help to the student and practitioner.

The book is written for graduate students, researchers, and practitioners. The exposition has been simplified to a large extent, so that a graduate student with a reasonable understanding of linear algebra and probability theory can understand the book easily. Numerous exercises are available along with a solution manual to aid in classroom teaching.

Throughout this book, a vector or a multidimensional data point is annotated with a bar, such as $\overline{X}$ or $\overline{y}$. A vector or multidimensional point may be denoted by either small letters or capital letters, as long as it has a bar. Vector dot products are denoted by centered dots, such as $\overline{X} \cdot \overline{Y}$. A matrix is denoted in capital letters without a bar, such as R. Throughout the book, the $n \times d$ document-term matrix is denoted by D, with n documents and d dimensions. The individual documents in D are therefore represented as d-dimensional row vectors, which are the bag-of-words representations. On the other hand, vectors with one component for each data point are usually n-dimensional column vectors. An example is the n-dimensional column vector $\overline{y}$ of class variables of n data points.

What Is New in the Second Edition

The second edition of the book emphasizes deep learning and natural language processing. Chapter 10 on deep learning has been significantly enhanced with discussions on different types of neural networks as well as language models like ELMo. Chapter 11 is entirely new, and it discusses transformers and pre-trained language models. Deep learning methods have been added to the chapter on text summarization with a special focus on abstractive summarization (Chapter 12). The information extraction chapter has now been updated to an integrated chapter on information extraction and knowledge graphs. The addition of knowledge graphs also lays the ground for a completely new chapter on question-answering (Chapter 14). Deep learning methods for sentiment analysis are also introduced in the book.

Mohegan Lake, NY, USA   Charu C. Aggarwal

The original version of the book has been revised. A correction to this book can be found at https://doi.org/10.1007/978-3-030-96623-2_17

Acknowledgments

Acknowledgements for the First Edition

I would like to thank my family including my wife, daughter, and my parents for their love and support. I would also like to thank my manager Nagui Halim for his support during the writing of this book.

This book has benefitted from significant feedback and several collaborations that i have had with numerous colleagues over the years. I would like to thank Quoc Le, Chih-Jen Lin, Chandan Reddy, Saket Sathe, Shai Shalev-Shwartz, Jiliang Tang, Suhang Wang, and ChengXiang Zhai for their feedback on various portions of this book and for answering specific queries on technical matters. I would particularly like to thank Saket Sathe for commenting on several portions, and also for providing some sample output from a neural network to use in the book. For their collaborations, I would like to thank Tarek F. Abdelzaher, Jing Gao, Quanquan Gu, Manish Gupta, Jiawei Han, Alexander Hinneburg, Thomas Huang, Nan Li, Huan Liu, Ruoming Jin, Daniel Keim, Arijit Khan, Latifur Khan, Mohammad M. Masud, Jian Pei, Magda Procopiuc, Guojun Qi, Chandan Reddy, Saket Sathe, Jaideep Srivastava, Karthik Subbian, Yizhou Sun, Jiliang Tang, Min-Hsuan Tsai, Haixun Wang, Jianyong Wang, Min Wang, Suhang Wang, Joel Wolf, Xifeng Yan, Mohammed Zaki, ChengXiang Zhai, and Peixiang Zhao. I would particularly like to thank Professor ChengXiang Zhai for my earlier collaborations with him in text mining. I would also like to thank my advisor James B. Orlin for his guidance during my early years as a researcher.

Finally, I would like to thank Lata Aggarwal for helping me with some of the figures created using PowerPoint graphics in this book.

Acknowledgements for the Second Edition

I would like to thank Roy Lee, Chandan Reddy, and Jiliang Tang for their feedback on the second edition of the book.

Contents

Author Biography

Charu C. Aggarwal is a Distinguished Research Staff Member (DRSM) at the IBM T. J. Watson Research Center in Yorktown Heights, New York. He completed his undergraduate degree in Computer Science from the Indian Institute of Technology at Kanpur in 1993 and his Ph.D. from the Massachusetts Institute of Technology in 1996.

 He has worked extensively in the field of data mining. He has published more than 400 papers in refereed conferences and journals and authored over 80 patents. He is the author or editor of 20 books, including textbooks on data mining, recommender systems, and outlier analysis. Because of the commercial value of his patents, he has thrice been designated a Master Inventor at IBM. He is a recipient of an IBM Corporate Award (2003) for his work on bio-terrorist threat detection in data streams, a recipient of the IBM Outstanding Innovation Award (2008) for his scientific contributions to privacy technology, and a recipient of two IBM Outstanding Technical Achievement Awards (2009, 2015) for his work on data streams/high-dimensional data. He received the EDBT 2014 Test of Time Award for his work on condensation-based privacy-preserving data mining. He is a recipient of the IEEE ICDM Research Contributions Award (2015) and ACM SIGKDD Innovation Award, which are the two most prestigious awards for influential research contributions in the field of data mining. He is also a recipient of the W. Wallace McDowell Award, which is the highest award given solely by the IEEE Computer Society across the field of Computer Science.

He has served as the general co-chair of the IEEE Big Data Conference (2014) and as the program co-chair of the ACM CIKM Conference (2015), the IEEE ICDM Conference (2015), and the ACM KDD Conference (2016). He served as an associate editor of the IEEE Transactions on Knowledge and Data Engineering from 2004 to 2008. He is an associate editor of the IEEE Transactions on Big Data, an action editor of the Data Mining and Knowledge Discovery Journal, and an associate editor of the Knowledge and Information Systems Journal. He has served or currently serves as the editor-in-chief of the ACM Transactions on Knowledge Discovery from Data as well as the ACM SIGKDD Explorations. He is also an editor-in-chief of ACM Books. He serves on the advisory board of the Lecture Notes on Social Networks, a publication by Springer. He has served as the vice-president of the SIAM Activity Group on Data Mining and is a member of the SIAM industry committee. He is a fellow of the SIAM, ACM, and the IEEE, for "contributions to knowledge discovery and data mining algorithms."

Chapter 1

An Introduction to Text Analytics

"The first forty years of life give us the text; the next thirty supply the commentary on it."– Arthur Schopenhauer

1.1 Introduction

The extraction of useful insights from text with various types of statistical algorithms is referred to as *text mining, text analytics,* or *machine learning from text.* The choice of terminology largely depends on the base community of the practitioner. This book will use these terms interchangeably. Text analytics has become increasingly popular in recent years because of the ubiquity of text data on the Web, social networks, emails, digital libraries, and chat sites. Some common examples of sources of text are as follows:

1. *Digital libraries:* Electronic content has outstripped the production of printed books and research papers in recent years. This phenomenon has led to the proliferation of digital libraries, which can be mined for useful insights. Some areas of research such as biomedical text mining specifically leverage the content of such libraries.

2. *Electronic news:* An increasing trend in recent years has been the de-emphasis of printed newspapers and a move towards electronic news dissemination. This trend creates a massive stream of news documents that can be analyzed for important events and insights. In some cases, such as Google news, the articles are indexed by topic and recommended to readers based on past behavior or specified interests.

3. *Web and Web-enabled applications:* The Web is a vast repository of documents that is further enriched with links and other types of side information. Web documents are also referred to as *hypertext.* The additional side information available with hypertext can be useful in the knowledge discovery process. In addition, many Web-enabled applications, such as social networks, chat boards, and bulletin boards, are a significant source of text for analysis.

C. C. Aggarwal, *Machine Learning for Text,*
https://doi.org/10.1007/978-3-030-96623-2_1

- *Social media:* Social media is a particularly prolific source of text because of the open nature of the platform in which any user can contribute. Social media posts are unique in that they often contain short and non-standard acronyms, which merit specialized mining techniques.

Numerous applications exist in the context of the types of insights one of trying to discover from a text collection. Some examples are as follows:

- Search engines are used to index the Web and enable users to discover Web pages of interest. A significant amount of work has been done on crawling, indexing, and ranking tools for text data.

- Text mining tools are often used to filter spam or identify interests of users in particular topics. In some cases, email providers might use the information mined from text data for advertising purposes.

- Text mining is used by news portals to organize news items into relevant categories. Large collections of documents are often analyzed to discover relevant topics of interest. These learned categories are then used to categorize incoming streams of documents into relevant categories.

- Recommender systems use text mining techniques to infer interests of users in specific items, news articles, or other content. These learned interests are used to recommend news articles or other content to users.

- The Web enables users to express their interests, opinions, and sentiments in various ways. This has led to the important area of opinion mining and sentiment analysis. Such opinion mining and sentiment analysis techniques are used by marketing companies to make business decisions.

The area of text mining is closely related to that of *information retrieval,* although the latter topic focuses on the database management issues rather than the mining issues. Because of the close relationship between the two areas, this book will also discuss some of the information retrieval aspects that are either considered seminal or are closely related to text mining.

The ordering of words in a document provides a semantic meaning that cannot be inferred from a representation based on only the frequencies of words in that document. Nevertheless, it is still possible to make many types of useful predictions without inferring the semantic meaning. There are two feature representations that are popularly used in mining applications:

1. *Text as a bag-of-words:* This is the most commonly used representation for text mining. In this case, the ordering of the words is not used in the mining process. The set of words in a document is converted into a *sparse multidimensional representation,* which is leveraged for mining purposes. Therefore, the universe of words (or *terms*) corresponds to the dimensions (or *features*) in this representation. For many applications such as classification, topic-modeling, and recommender systems, this type of representation is sufficient.

2. *Text as a set of sequences:* In this case, the individual sentences in a document are extracted as strings or sequences. Therefore, the ordering of words matters in this representation, although the ordering is often localized within sentence or paragraph

boundaries. A document is often treated as a set of independent and smaller units (e.g., sentences or paragraphs). This approach is used by applications that require greater semantic interpretation of the document content. This area is closely related to that of *language modeling* and *natural language processing*. The latter is often treated as a distinct field in its own right.

Text mining has traditionally focused on the first type of representation, although recent years have seen an increasing amount of attention on the second representation. This is primarily because of the increasing importance of artificial intelligence applications in which the language semantics, reasoning, and understanding are required. For example, question-answering systems have become increasingly popular in recent years, which require a greater degree of understanding and reasoning.

It is important to be cognizant of the sparse and high-dimensional characteristics of text when treating it as a multidimensional data set. This is because the dimensionality of the data depends on the number of words which is typically large. Furthermore, most of the word frequencies (i.e., feature values) are zero because documents contain small subsets of the vocabulary. Therefore, multidimensional mining methods need to be cognizant of the sparse and high-dimensional nature of the text representation for best results. The sparsity is not always a disadvantage. In fact, some models, such as the linear support vector machines discussed in Chapter 6, are inherently suited to sparse and high-dimensional data.

This book will cover a wide variety of text mining algorithms, such as latent factor modeling, clustering, classification, retrieval, and various Web applications. The discussion in most of the chapters is self-sufficient, and it does not assume a background in data mining or machine learning other than a basic understanding of linear algebra and probability. In this chapter, we will provide an overview of the various topics covered in this book, and also provide a mapping of these topics to the different chapters.

Chapter Organization

This chapter is organized as follows. In the next section, we will discuss the special properties of text data that are relevant to the design of text mining applications. Section 1.3 discusses various applications for text mining. The conclusions are discussed in Section 1.4.

1.2 What Is Special About Learning from Text?

Most machine learning applications in the text domain work with the bag-of-words representation in which the words are treated as dimensions with values corresponding to word frequencies. A data set corresponds to a collection of documents, which is also referred to as a *corpus*. The complete and distinct set of words used to define the corpus is also referred to as the *lexicon*. Dimensions are also referred to as terms or features. Some applications of text work with a binary representation in which the presence of a term in a document corresponds to a value of 1, and 0, otherwise. Other applications use a normalized function of the word frequencies as the values of the dimensions. In each of these cases, the dimensionality of data is very large, and may be of the order of 10^5 or even 10^6. Furthermore, most values of the dimensions are 0s, and only a few dimensions take on positive values. In other words, text is a *high-dimensional, sparse, and non-negative* representation.

These properties of text create both challenges and opportunities. The sparsity of text implies that the positive word frequencies are more informative than the zeros. There is also wide variation in the relative frequencies of words, which leads to differential importance of the different words in mining applications. For example, a commonly occurring word like "*the*" is often less significant and needs to be down-weighted (or completely removed) with normalization. In other words, it is often more important to statistically normalize the relative importance of the dimensions (based on frequency of presence) compared to traditional multidimensional data. One also needs to normalize for the varying lengths of different documents while computing distances between them. Furthermore, although most multidimensional mining methods can be generalized to text, the sparsity of the representation has an impact on the relative effectiveness of different types of mining and learning methods. For example, linear support-vector machines are relatively effective on sparse representations, whereas methods like decision trees need to be designed and tuned with some caution to enable their accurate use. All these observations suggest that the sparsity of text can either be a blessing or a curse depending on the methodology at hand. In fact, some techniques such as *sparse coding* sometimes convert non-textual data to text-like representations in order to enable efficient and effective learning methods like support-vector machines [405].

The nonnegativity of text is also used explicitly and implicitly by many applications. Nonnegative feature representations often lead to more interpretable mining techniques, an example of which is *nonnegative matrix factorization* (see Chapter 3). Furthermore, many *topic modeling* and *clustering* techniques implicitly use nonnegativity in one form or the other. Such methods enable intuitive and highly interpretable "sum-of-parts" decompositions of text data, which are not possible with other types of data matrices.

In the case where text documents are treated as sequences, a data-driven *language model* is used to create a probabilistic representation of the text. The rudimentary special case of a language model is the *unigram* model, which defaults to the bag-of-words representation. However, higher-order language models like *bigram* or *trigram* models are able to capture sequential properties of text. In other words, a language model is a data-driven approach to representing text, which is more general than the traditional bag-of-words model. Such methods share many similarities with other sequential data types like biological data. There are significant methodological parallels in the algorithms used for clustering and dimensionality reduction of (sequential) text and biological data. For example, just as *Markovian models* are used to create probabilistic models of sequences, they can also be used to create language models.

Text requires a lot of preprocessing because it is extracted from platforms such as the Web that contain many misspellings, nonstandard words, anchor text, or other meta-attributes. The simplest representation of cleaned text is a multidimensional bag-of-words representation, but complex structural representations are able to create fields for different types of *entities* and *events* in the text. This book will therefore discuss several aspects of text mining, including preprocessing, representation, similarity computation, and the different types of learning algorithms or applications.

1.3 Analytical Models for Text

The section will provide a comprehensive overview of text mining algorithms and applications. The next chapter of this book primarily focuses on data preparation and similarity computation. Issues related to preprocessing issues of data representation are also discussed in this chapter. Aside from the first two introductory chapters, the topics covered in this book fall into three primary categories:

1. *Fundamental mining applications:* Many data mining applications like matrix factorization, clustering, and classification, can be used for any type of multidimensional data. Nevertheless, the uses of these methods in the text domain has specialized characteristics. These represent the core building blocks of the vast majority of text mining applications. Chapters 3 through 8 will discuss core data mining methods. The interaction of text with other data types will be covered in Chapter 8.

2. *Information retrieval and ranking:* Many aspects of information retrieval and ranking are closely related to text mining. For example, ranking methods like ranking SVM and link-based ranking are often used in text mining applications. Chapter 9 will provide an overview of information retrieval methods from the point of view of text mining.

3. *Sequence- and natural language-centric text mining:* Although multidimensional mining methods can be used for basic applications, the true power of mining text can be leveraged in more complex applications by treating text as sequences. Chapters 10 through 16 will discuss these advanced topics like sequence embedding, neural learning, information extraction, summarization, opinion mining, text segmentation, and event extraction. Many of these methods are closely related to natural language processing. Although this book is not focused on natural language processing, the basic building blocks of natural language processing will be used as off-the-shelf tools for text mining applications.

In the following, we will provide an overview of the different text mining models covered in this book. In cases where the multidimensional representation of text is used for mining purposes, it is relatively easy to use a consistent notation. In such cases, we assume that a document corpus with n documents and d different terms can be represented as a sparse $n \times d$ *document-term matrix*, which is typically very sparse. The ith row of D is represented by the d-dimensional row vector $\overline{X_i}$. One can also represent a document corpus as a *set* of these d-dimensional vectors, which is denoted by $\mathcal{D} = \{\overline{X}_1 \ldots \overline{X}_n\}$. This terminology will be used consistently throughout the book. Many information retrieval books prefer the use of a *term-document* matrix, which is the transpose of the document-term matrix and the rows correspond to the frequencies of terms. However, using a document-term matrix, in which data instances are rows, is consistent with the notations used in books on multidimensional data mining and machine learning. Therefore, we have chosen to use a document-term matrix in order to consistent with the broader literature on machine learning.

Much of the book will be devoted to *data mining* and *machine learning* rather than the database management issues of *information retrieval*. Nevertheless, there is some overlap between the two areas, as they are both related to problems of ranking and search engines. Therefore, a comprehensive chapter is devoted to information retrieval and search engines. Throughout this book, we will use the term "learning algorithm" as a broad umbrella term to describe any algorithm that discovers patterns from the data or discovers how such patterns may be used for predicting specific values in the data.

1.3.1 Text Preprocessing and Similarity Computation

Text preprocessing is required to convert the unstructured format into a structured and multidimensional representation. Text often co-occurs with a lot of extraneous data such as tags, anchor text, and other irrelevant features. Furthermore, different words have different significance in the text domain. For example, commonly occurring words such as "*a*," "*an*," and "*the*," have little significance for text mining purposes. In many cases, words are variants of one another because of the choice of tense or plurality. Some words are simply misspellings. The process of converting a character sequence into a sequence of words (or *tokens*) is referred to as *tokenization*. Note that each occurrence of a word in a document is a token, even if it occurs more than once in the document. Therefore, the occurrence of the same word three times will create three corresponding tokens. The process of tokenization often requires a substantial amount of domain knowledge about the specific language at hand, because the word boundaries have ambiguities caused by vagaries of punctuation in different languages.

Some common steps for preprocessing raw text are as follows:

1. *Text extraction:* In cases where the source of the text is the Web, it occurs in combination with various other types of data such as anchors, tags, and so on. Furthermore, in the Web-centric setting, a specific page may contain a (useful) primary block and other blocks that contain advertisements or unrelated content. Extracting the useful text from the primary block is important for high-quality mining. These types of settings require specialized parsing and extraction techniques.

2. *Stop-word removal:* Stop words are commonly occurring words that have little discriminative power for the mining process. Common pronouns, articles, and prepositions are considered stop words. Such words need to be removed to improve the mining process.

3. *Stemming, case-folding, and punctuation:* Words with common roots are consolidated into a single representative. For example, words like "sinking" and "sank" are consolidated into the single token "sink." The case (i.e., capitalization) of the first alphabet of a word may or may not be important to its semantic interpretation. For example, the word "Rose" might either be a flower or the name of a person depending on the case. In other settings, the case may not be important to the semantic interpretation of the word because it is caused by grammar-specific constraints like the beginning of a sentence. Therefore, language-specific heuristics are required in order to make decisions on how the case is treated. Punctuation marks such as hyphens need to be parsed carefully in order to ensure proper tokenization.

4. *Frequency-based normalization:* Low-frequency words are often more discriminative than high-frequency words. Frequency-based normalization therefore weights words by the logarithm of the inverse relative-frequency of their presence in the collection. Specifically, if n_i is the number of documents in which the ith word occurs in the corpus, and n is the number of documents in the corpus, then the frequency of a word in a document is multiplied by $\log(n/n_i)$. This type of normalization is also referred to as *inverse-document frequency (idf)* normalization. The final normalized representation multiplies the term frequencies with the inverse document frequencies to create a tf-idf representation.

When computing similarities between documents, one must perform an additional normalization associated with the *length of a document*. For example, Euclidean distances are commonly used for distance computation in multidimensional data, but they would not work very well in a text corpus containing documents of varying lengths. The distance between two short documents will always be very small, whereas the distance between two long documents will typically be much larger. It is undesirable for pairwise similarities to be dominated so completely by the lengths of the documents. This type of length-wise bias also occurs in the case of the dot-product similarity function. Therefore, it is important to use a similarity computation process that is appropriately normalized. A normalized measure is the cosine measure, which normalizes the dot product with the product of the L_2-norms of the two documents. The cosine between a pair of d-dimensional document vectors $\overline{X} = (x_1 \ldots x_d)$ and $\overline{Y} = (y_1 \ldots y_d)$ is defined as follows:

$$\text{cosine}(\overline{X}, \overline{Y}) = \frac{\sum_{i=1}^{d} x_i y_i}{\sqrt{\sum_{i=1}^{d} x_i^2} \sqrt{\sum_{i=1}^{d} y_i^2}} \tag{1.1}$$

Note the presence of document norms in the denominator for normalization purposes. The cosine between a pair of documents always lies in the range $(0, 1)$. More details on document preparation and similarity computation are provided in Chapter 2.

1.3.2 Dimensionality Reduction and Matrix Factorization

Dimensionality reduction and matrix factorization fall in the general category of methods that are also referred to as *latent factor models*. Sparse and high-dimensional representations like text work well with some learning methods but not with others. Therefore, a natural question arises as whether one can somehow compress the data representation to express it in a smaller number of features. Since these features are not observed in the original data but represent hidden properties of the data, they are also referred to as *latent features*.

Dimensionality reduction is intimately related to matrix factorization. Most types of dimensionality reduction transform the data matrices into *factorized form*. In other words, the original data matrix D can be approximately represented as a product of two or more matrices, so that the total number of entries in the factorized matrices is far fewer than the number of entries in the original data matrix. A common way of representing an $n \times d$ document-term matrix as the product of an $n \times k$ matrix U and a $d \times k$ matrix V is as follows:

$$D \approx UV^T \tag{1.2}$$

The value of k is typically much smaller than n and d. The total number of entries in D is $n \cdot d$, whereas the total number of entries in U and V is only $(n + d) \cdot k$. For small values of k, the representation of D in terms of U and V is much more compact. The $n \times k$ matrix U contains the k-dimensional reduced representation of each document in its rows, and the $d \times k$ matrix V contains the k basis vectors in its columns. In other words, matrix factorization methods create reduced representations of the data with (approximate) linear transforms. Note that Equation 1.2 is represented as an approximate equality. In fact, all forms of dimensionality reduction and matrix factorization are expressed as optimization models in which the error of this approximation is minimized. Therefore, dimensionality reduction effectively compresses the large number of entries in a data matrix into a smaller number of entries with the lowest possible error.

Popular methods for dimensionality reduction in text include *latent semantic analysis*, *non-negative matrix factorization*, *probabilistic latent semantic analysis*, and *latent Dirichlet allocation*. We will address most of these methods for dimensionality reduction and matrix factorization in Chapter 3. Latent semantic analysis is the text-centric avatar of *singular value decomposition*.

Dimensionality reduction and matrix factorization are extremely important because they are intimately connected to the *representational issues* associated with text data. In data mining and machine learning applications, the *representation* of the data is the key in designing an effective learning method. In this sense, singular value decomposition methods enable high-quality retrieval, whereas certain types of non-negative matrix factorization methods enable high-quality clustering. In fact, clustering is an important application of dimensionality reduction, and some of its probabilistic variants are also referred to as *topic models*. Similarly, certain types of *decision trees* for classification show better performance with reduced representations. Furthermore, one can use dimensionality reduction and matrix factorization to convert a heterogeneous combination of text and another data type into multidimensional format (cf. Chapter 8).

1.3.3   Text Clustering

Text clustering methods partition the corpus into groups of related documents belonging to particular topics or categories. However, these categories are not known a priori, because specific examples of desired categories (e.g., politics) of documents are not provided up front. Such learning problems are also referred to as *unsupervised*, because no guidance is provided to the learning problem. In *supervised* applications, one might provide examples of news articles belonging to several natural categories like sports, politics, and so on. In the unsupervised setting, the documents are partitioned into similar groups, which is sometimes achieved with a domain-specific similarity function like the cosine measure. In most cases, an optimization model can be formulated, so that some direct or indirect measure of similarity within a cluster is maximized. A detailed discussion of clustering methods is provided in Chapter 4.

Many matrix factorization methods like probabilistic latent semantic analysis and latent Dirichlet allocation also achieve a similar goal of assigning documents to topics, albeit in a soft and probabilistic way. A *soft* assignment refers to the fact that the probability of assignment of each document to a cluster is determined rather than a hard partitioning of the data into clusters. Such methods not only assign documents to topics but also infer the significance of the words to various topics. In the following, we provide a brief overview of various clustering methods.

1.3.3.1   Deterministic and Probabilistic Matrix Factorization Methods

Most forms of non-negative matrix factorization methods can be used for clustering text data. Therefore, certain types of matrix factorization methods play the dual role of clustering and dimensionality reduction, although this is not true across every matrix factorization method. Many forms of non-negative matrix factorization are *probabilistic mixture models*, in which the entries of the document-term matrix are *assumed to be generated* by a probabilistic process. The parameters of this random process can then be estimated in order to create a factorization of the data, which has a natural probabilistic interpretation. This type of model is also referred to as a *generative* model because it assumes that the document-term matrix is created by a hidden generative process, and the data are used to estimate the parameters of this process.

1.3.3.2   Probabilistic Mixture Models of Documents

Probabilistic matrix factorization methods use generative models over the *entries* of the document-term matrix, whereas probabilistic models of documents generate the *rows* (documents) from a generative process. The basic idea is that the rows are generated by a *mixture* of different probability distributions. In each iteration, one of the mixture components is selected with a certain *a priori* probability and the word vector is generated based on the distribution of that mixture component. Each mixture component is therefore analogous to a cluster. The goal of the clustering process is to estimate the parameters of this generative process. Once the parameters have been estimated, one can then estimate the *a posteriori* probability that the point was generated by a particular mixture component. We refer to this probability as "posterior" because it can only be estimated after observing the attribute values in the data point (e.g., word frequencies). For example, a document containing the word "basketball" will be more likely to belong to the mixture component (cluster) that is generating many sports documents. The resulting clustering is a soft assignment in which the probability of assignment of each document to a cluster is determined. Probabilistic mixture models of documents are often simpler to understand than probabilistic matrix factorization methods, and are the text analogs of Gaussian mixture models for clustering numerical data.

1.3.3.3 Similarity-Based Algorithms

Similarity-based algorithms are typically either representative-based methods or hierarchical methods, In all these cases, a distance or similarity function between points is used to partition them into clusters in a deterministic way. Representative-based algorithms use representatives in combination with similarity functions in order to perform the clustering. The basic idea is that each cluster is represented by a multi-dimensional vector, which represents the "typical" frequency of words in that cluster. For example, the centroid of a set of documents can be used as its representative. Similarly, clusters can be created by assigning documents to their closest representatives such as the cosine similarity. Such algorithms often use iterative techniques in which the cluster representatives are extracted as central points of clusters, whereas the clusters are created from these representatives by using cosine similarity-based assignment. This two-step process is repeated to convergence, and the corresponding algorithm is also referred to as the k-means algorithm. There are many variations of representative-based algorithms although only a small subset of them work with the sparse and high-dimensional representation of text. Nevertheless, one can use a broader variety of methods if one is willing to transform the text data to a reduced representation with dimensionality reduction techniques.

In hierarchical clustering algorithms, similar pairs of clusters are aggregated into larger clusters using an iterative approach. The approach starts by assigning each document to its own cluster and then merges the closest pair of clusters together. There are many variations in terms of how the pairwise similarity between clusters is computed, which has a direct impact on the type of clusters discovered by the algorithm. In many cases, hierarchical clustering algorithms can be combined with representative clustering methods to create more robust methods.

1.3.3.4 Advanced Methods

All text clustering methods can be transformed into graph partitioning methods by using a variety of transformations. One can transform a document corpus into node-node similarity graphs or node-word occurrence graphs. The latter type of graph is bipartite and clustering it is very similar to the process of nonnegative matrix factorization.

There are several ways in which the accuracy of clustering methods can be enhanced with the use of either external information or with *ensembles*. In the former case, external information in the form of labels is leveraged in order to guide the clustering process towards specific categories that are known to the expert. However, the guidance is not too strict, as a result of which the clustering algorithm has the flexibility to learn good clusters that are not indicated solely by the supervision. Because of this flexible approach, such an approach is referred to as *semi-supervised* clustering, because there are a small number of examples of representatives from different clusters that are *labeled* with their topic. However, it is still not a full supervision because there is considerable flexibility in how the clusters might be created using a combination of these labeled examples and other unlabeled documents.

A second technique is to use *ensemble methods* in order to improve clustering quality. Ensemble methods combine the results from multiple executions of one or more learning algorithms to improve prediction quality. Clustering methods are often unstable because the results may vary significantly from one run to the next by making small algorithmic changes or even changing the initialization. This type of variability is an indicator of a suboptimal learning algorithm *in expectation* over the different runs, because many of these runs are often poor clusterings of the data. Nevertheless, most of these runs do contains some useful information about the clustering structure. Therefore, by repeating the clustering in multiple ways and combining the results from the different executions, more robust results can be obtained.

1.3.4 Text Classification and Regression Modeling

Text classification is closely related to text clustering. One can view the problem of text classification as that of partitioning the data into *pre-defined* groups. These pre-defined groups are identified by their *labels*. For example, in an email classification application, the two groups might correspond to *"spam"* and *"not spam."* In general, we might have k different categories, and there is no inherent ordering among these categories. Unlike clustering, a *training data set* is provided with examples of emails belonging to both categories. Then, for an unlabeled *test data set*, it is desired to categorize them into one of these two *pre-defined* groups.

Note that both classification and clustering partition the data into groups; however, the partitioning in the former case is *highly controlled* with a pre-conceived notion of partitioning defined by the training data. The training data provides the algorithm *guidance*, just as a teacher supervises her student towards a specific goal. This is the reason that classification is referred to as *supervised learning*.

One can also view the prediction of the categorical label y_i for data instance $\overline{X_i}$ as that of learning a *function* $f(\cdot)$:

$$y_i = f(\overline{X_i}) \tag{1.3}$$

In classification, the range of the function $f(\cdot)$ is a discrete set of values like $\{spam, not\ spam\ \}$. Often the labels are assumed to be drawn from the *discrete and unordered* set of values $\{1, 2, \ldots, k\}$. In the specific case of binary classification, the value of y_i can be assumed to be drawn from $\{-1, +1\}$, although some algorithms find it more convenient to use the notation $\{0, 1\}$. Binary classification is slightly easier than the case of multilabel classification because it is possible to order the two classes unlike multi-label classes such as $\{Blue, Red, Green\}$. Nevertheless, multilabel classification can be reduced to multiple applications of binary classification with simple meta-algorithms.

It is noteworthy that the function $f(\cdot)$ need not always map to the categorical domain, but it can also map to a numerical value. In other words, we can generally refer to y_i as the *dependent variable*, which may be numerical in some settings. This problem is referred to as *regression modeling*, and it no longer partitions the data into discrete groups like classification. Regression modeling occurs commonly in many settings such as sales forecasting where the dependent variables of interest are numerical. Note that the terminology "dependent variable" applies to both classification and regression, whereas the term "label" is generally used only in classification. The dependent variable in regression modeling is also referred to as a *regressand*. The values of the features in $\overline{X_i}$ are referred to as *feature variables*, or *independent variables* in both classification and regression modeling. In the specific case of regression modeling, they are also referred to as *regressors*. Many algorithms for regression modeling can be generalized to classification and vice versa. Various classification algorithms are discussed in Chapters 5, 6, and 7. In the following, we will provide an overview of the classification and regression modeling algorithms that are discussed in these chapters.

1.3.4.1 Decision Trees

Decision trees partition the training data hierarchically by imposing conditions over attributes so that documents belonging to each class are predominantly placed in a single node. In a *univariate* split, this condition is imposed over a single attribute, whereas a *multivariate* split imposes this split condition over multiple attributes. For example, a univariate split could correspond to the presence or absence of a particular word in the document. In a binary decision tree, a training instance is assigned to one or two children nodes depending on whether it satisfies the split condition. The process of splitting the training data is repeated recursively in tree-like fashion until most of the training instances in that node belong to the same class. Such a node is treated as the leaf node. These split conditions are then used to assign test instances with unknown labels to leaf nodes. The majority class of the leaf node is used to predict the label of the test instance. Combinations of multiple decision trees can be used to create *random forests*, which are among the best-performing classifiers in the literature.

1.3.4.2 Rule-Based Classifiers

Rule-based classifiers relate conditions on subsets of attributes to specific class labels. Thus, the antecedent of a rule contains a set of conditions, which typically correspond to the presence of a subset of words in the document. The consequent of the rule contains a class label. For a given test instance, the rules whose antecedents match the test instance are discovered. The (possibly conflicting) predictions of the discovered rules are used to predict the labels of test instances.

1.3.4.3 Naïve Bayes Classifier

The naïve Bayes classifier can be viewed as the supervised analog of mixture models in clustering. The basic idea here is that the data is generated by a mixture of k components, where k is the number of classes in the data. The words in each class are defined by a specific distribution. Therefore, the parameters of each mixture component-specific distribution need to be estimated in order to maximize the likelihood of these training instances being generated by the component. These probabilities can then be used to estimate the probability of a test instance belonging to a particular class. This classifier is referred to as "naïve" because it makes some simplifying assumptions about the independence of attribute values in test instances.

1.3.4.4   Nearest Neighbor Classifiers

Nearest neighbor classifiers are also referred to as *instance-based learners*, *lazy learners*, or *memory-based learners*. The basic idea in a nearest neighbor classifier is to retrieve the k-nearest training examples to a test instance and report the dominant label of these examples. In other words, it works by *memorizing* training *instances*, and leaves all the work of classification to the very end (in a *lazy* way) without doing any training up front. Nearest neighbor classifiers have some interesting properties, in that they show probabilistically optimal behavior if an infinite amount of data is available. However, in practice, we rarely have infinite data. For finite data sets, nearest neighbor classifiers are usually outperformed by a variety of *eager learning* methods that perform training up front. Nevertheless, these theoretical aspects of nearest-neighbor classifiers are important because some of the best-performing classifiers such as random forests and support-vector machines can be shown to be eager variants of nearest-neighbor classifiers under the covers.

1.3.4.5   Linear Classifiers

Linear classifiers are among the most popular methods for text classification. This is partially because linear methods work particularly well for high-dimensional and sparse data domains.

First, we will discuss the natural case of regression modeling in which the dependent variable is numeric. The basic idea is to assume that the prediction function of Equation 1.3 is in the following *linear* form:

$$y_i \approx \overline{W} \cdot \overline{X_i} + b \tag{1.4}$$

Here, $\overline{W}$ is a d-dimensional vector of coefficients and b is a scalar value, which is also referred to as the *bias*. The coefficients and the bias need to learned from the training examples, so that the error in Equation 1.4 is minimized. Therefore, most linear classifiers can be expressed in as the following optimization model:

$$\text{Minimize } \sum_i \text{Loss}[y_i - \overline{W} \cdot \overline{X_i} - b] + \text{Regularizer} \tag{1.5}$$

The function $\text{Loss}[y_i - \overline{W} \cdot \overline{X_i} - b]$ quantifies the error of the prediction, whereas the *regularizer* is a term that is added to prevent overfitting for smaller data sets. The former is also referred to as the *loss function*. A wide variety of combinations of error functions and regularizers are available in literature, which result in methods like *Tikhonov regularization* and *LASSO*. Tikhonov regularization uses the squared norm of the vector $\overline{W}$ to discourage large coefficients. Such problems are often solved with gradient-descent methods, which are well-known tools in optimization.

For the classification problem with a binary dependent variable $y_i \in \{-1, +1\}$, the classification function is often of the following form:

$$y_i = \text{sign}\{\overline{W} \cdot \overline{X_i} + b\} \tag{1.6}$$

Interestingly, the objective function is still in the same form as Equation 1.5, except that the loss function now needs to be designed for a categorical variable rather than a numerical one. A variety of loss functions such as hinge loss function, the logistic loss function, and the quadratic loss function are used. The first of these loss functions leads to a method known as the *support vector machine*, whereas the second one leads to a method referred to as *logistic regression*. These methods can be generalized to the nonlinear case with the use of *kernel methods*. Linear models are discussed in Chapter 6.

1.3.4.6 Broader Topics in Classification

Chapter 7 discusses topics such as the theory of supervised learning, classifier evaluation, and classification ensembles. These topics are important because they illustrate the use of methods that can enhance a wide variety of classification applications.

1.3.5 Joint Analysis of Text with Heterogeneous Data

Much of text mining occurs in network-centric, Web-centric, social media, and other settings in which heterogenous types of data such as hyperlinks, images, and multimedia are present. These types of data can often be mined for rich insights. Chapter 8 provides a study of the typical methods that are used for mining text in combination with other data types such as multimedia and Web linkages. Some common tricks will be studied such as the use of *shared matrix factorization* and *factorization machines* for representation learning.

Many forms of text in social media are short in nature because of the fact that these forums are naturally suited to short snippets. For example, Twitter imposes an explicit constraint on the length of a tweet, which naturally leads to shorter snippets of documents. Similarly, the comments on Web forums are naturally short. When mining short documents, the problems of sparsity are often extraordinarily high. These settings necessitate specialized mining methods for such documents. For example, such methods need to be able to effectively address the overfitting caused by sparsity when the vector-space representation is used. The factorization machines discussed in Chapter 8 are useful for short text mining. In many cases, it is desirable to use sequential and linguistic models for short-text mining because the vector-space representation is not sufficient to capture the complexity required for the mining process. Several methods discussed in Chapter 10 can be used to create multidimensional representations from sequential snippets of short text.

1.3.6 Information Retrieval and Web Search

Text data has found increasing interest in recent years because of the greater importance of Web-enabled applications. One of the most important applications is that of search in which it is desired to retrieve Web pages of interest based on specified keywords. The problem is an extension of the notion of search used in traditional information retrieval applications. In search applications, data structures such as inverted indices are very useful. Therefore, significant discussion will be devoted in Chapter 9 to traditional aspects of document retrieval.

In the Web context, several unique factors such as the citation structure of the Web also play an important role in enabling effective retrieval. For example, the well-known *PageRank* algorithm uses the citation structure of the Web in order to make judgements about the importance of Web pages. The importance of *Web crawlers* at the back-end is also significant for the discovery of relevant resources. Web crawlers collect and store documents from the Web at a centralized location to enable effective search. Chapter 9 will provide an integrated discussion of information retrieval and search engines. The chapter will also discuss recent methods for search that leverage learning techniques like *ranking support vector machines*.

1.3.7 Sequential Language Modeling and Embeddings

Although the vector space representation of text is useful for solving many problems, there are applications in which the sequential representation of text is very important. In particular, any application that requires a semantic understanding of text requires the treatment of text as a sequence rather than as a bag of words. One useful approach in such cases is to transform the sequential representation of text to a multidimensional representation. Therefore, numerous methods have been designed to transform documents and words into a multidimensional representation. In particular, kernel methods and neural network methods like *word2vec* are very popular. These methods leverage sequential language models in order to *engineer multidimensional features* which are also referred to as *embeddings*. This type of feature engineering is very useful because it can be used in conjunction with any type of mining application. Chapter 10 will provide an overview of the different types of sequence-centric models for text data, with a primary focus on feature engineering.

This chapter also introduces several deep learning models for sequence data, including *recurrent neural networks, long-short term memory*, and *gated recurrent units*. Several applications of these models, such as the generation of large-scale language models, machine translation, and sequence-centric classification are discussed. The challenges associated with such models are also discussed briefly, along with corresponding solutions. Chapter 10 also marks the beginning of the segment of the book, which is focussed more closely on natural language processing. The chapter also briefly introduces the use of *convolutional neural networks* for natural language processing.

1.3.8   Transformers and Pretrained Language Models

Although recurrent neural networks have been the method of choice for natural language processing over the past twenty years, they have been gradually replaced by transformers [560] over the past two to three years. The reason for this phenomenon is that transformers are much more accurate on longer sequences. Furthermore, they can be parallelized more easily, which enables the development of large-scale training methods. Transformers are based on the notion of the concept of *attention mechanisms*, which are introduced in Chapter 11.

Transformers have been used to build large-scale *pretrained language models*, which essentially create large-scale models of natural language using huge corpora. These models can be used for a variety of tasks in downstream processing by fine tuning them in specific natural language processing tasks like machine translation. Examples of large-scale pretrained language models include GPT [75], BERT [149], and T5 [460]. These models are discussed in detail in Chapter 11, along with key applications to natural language processing. It is noteworthy that recurrent neural networks have also been used to create pretrained language models lie ELMo (see Section 10.8.1 of Chapter 10). However, the transformer-based language models tend to exhibit significantly better performance.

1.3.9   Text Summarization

In many applications, it is useful to create short summaries of text in order to enable users to get an idea of the primary subject matter of a document without having to read it in its entirety. Such summarization methods are often used in search engines in which an abstract of the returned result is included along the title and link to the relevant document. Chapter 12 provides an overview of various text summarization techniques. Deep learning methods for text summarization are also explored in this chapter.

1.3.10   Information Extraction

The problem of information extraction discovers different types of entities from text such as names, places, and organizations. It also discovers the relations between entities. An example of a relation is that the person entity *John Doe* works for the organization entity *IBM*. Information extraction is a very key step in converting unstructured text into a structured representation that is far more informative than a bag of words. As a result, more powerful applications can be built on top of this type of extracted data. Information extraction is sometimes considered a first step towards truly intelligent applications like question-answering systems and entity-oriented search. For example, searching for a pizza location near a particular place on the Google search engines usually returns organization entities. Search engines have become powerful enough today to recognize entity-oriented search from keyword phrases. Furthermore, many other applications of text mining such as opinion mining and event detection use information extraction techniques. Methods for information extraction are discussed in Chapter 13.

Information extraction methods are often used to construct knowledge graphs. Chapter 13 provides an introduction to knowledge graphs and their use for basic natural language processing applications like search. The utility of knowledge graphs is also explored in subsequent chapters like question-answering.

1.3.11   Question Answering

One of the most fundamental problems in natural language processing is that of question answering, because it is one of the most reliable ways of assessing the level of comprehension of a machine reading system. This general principle of also true of humans, where reading comprehension tasks are used to assess the level of understanding of a particular natural language. Most of the question answering methods in the literature are focussed on *factoid question answering*, in which the answer to a question can often be restricted to a short span of words within a passage. When the underlying passage is also provided, the problem is referred to as *reading comprehension*. In more challenging settings, one must integrate the problems of search and comprehension, wherein the relevant passage must be extracted from a large collection of documents. In some cases, this collection of documents may be made available only during training time and not during prediction time — this setting is referred to as *closed-book question answering*. In the *long-form question answering task*, the answers to questions are descriptive in nature. There are several variations of the question answering task, beginning with the reading comprehension problem at the simplest end and with long-form question answering at the most difficult end. An overview of the key methods for question answering is provided in Chapter 14.

1.3.12   Opinion Mining and Sentiment Analysis

The Web provides a forum to individuals to express their opinions and sentiments. For example, the product reviews in a Web site might contain text beyond the numerical ratings provided by the user. The textual content of these reviews provides useful information that is not available in numerical ratings. From this point of view, opinion mining can be viewed as the text-centric analog of the rating-centric techniques used in recommender systems. For example, product reviews are often used by both types of methods. Whereas recommender systems analyze the numerical ratings for prediction, opinion mining methods analyze the text of the opinions. It is noteworthy that opinions are often mined from information settings like social media and blogs where ratings are not available. Chapter 15 will discuss the problem of opinion mining and sentiment analysis of text data. The use of information extraction methods for opinion mining is also discussed.

1.3.13 Text Segmentation and Event Detection

Text segmentation and event detection are very different topics from an application-centric point of view; yet, they share many similarities in terms of the basic principle of detecting *sequential change* either within a document, or across multiple documents. Many long documents contain multiple topics, and it is desirable to detect changes in topic from one part of the document to another. This problem is referred to as text segmentation. In unsupervised text segmentation, one is only looking for topical change in the context. In supervised segmentation, one is looking for specific types of segments (e.g., politics and sports segments in a news article). Both types of methods are discussed in Chapter 16. The problem of text segmentation is closely related to stream mining and event detection. In event detection, one is looking for topical changes across multiple documents in streaming fashion. These topics are also discussed in Chapter 16.

1.4 Summary

Text mining has become increasingly important in recent years because of the preponderance of text on the Web, social media, and other network-centric platforms. Text requires a significant amount of preprocessing in order to clean it, remove irrelevant words, and perform the normalization. Numerous text applications such as dimensionality reduction and topic modeling form key building blocks of other text applications. In fact, various dimensionality reduction methods are used to enable methods for clustering and classification. Methods for querying and retrieving documents form the key building blocks of search engines. The Web also enables a wide variety of more complex mining scenarios containing links, images, and heterogeneous data.

More challenging applications with text can be solved only be treating text as sequences rather than as multidimensional bags of words. From this point of view, sequence embedding and information extraction are key building blocks. Such methods are often used in specialized applications like event detection, opinion mining, and sentiment analysis. Other sequence-centric applications of text mining include text summarization and segmentation.

1.5 Bibliographic Notes

Text mining can be viewed as a specialized offshoot of the broader field of data mining [2, 233, 543] and machine learning [57, 235, 399]. Numerous books have been written on the topic of information retrieval [35, 82, 137, 367, 490] although the focus of these books is primarily on the search engines, database management, and retrieval aspect. The book by Manning *et al.* [367] does discuss several mining aspects, although this is not the primary focus. An edited collection on text mining, which contains several surveys on many topics, may be found in [15]. A number of books covering various aspects of text mining are also available [193, 573]. The most recent book by Zhai and Massung [619] provides an application-oriented overview of text management and mining applications. The natural language focus on text understanding is covered in some recent books [283, 368]. A discussion of text mining, as it relates to Web data, may be found in [92, 345].

1.5.1 Software Resources

The Bow toolkit is a classical library available for classification, clustering, and information retrieval [371]. The library is written in C, and supports several popular classification and clustering tools. Furthermore, it also supports a lot of software for text preprocessing, such as finding document boundaries and tokenization. Several useful data sets for text mining may be found in the "text" section of the UCI Machine Learning Repository [645]. The *scikit-learn* library also supports several off-the-shelf tools for mining text data in Python [646], and is freely usable. Another Python library that is more focused towards natural language processing is the NLTK toolkit [652]. The **tm** package in R [647] is publicly available and it supports significant text mining functionality. Furthermore, significant functionality for text mining is also supported in the MATLAB programming language [41]. *Weka* provides a Java-based platform for text mining [649]. Stanford NLP [650] is a somewhat more academically-oriented system, but it provides many advanced tools that are not available elsewhere.

1.6 Exercises

1. Consider a text corpus with 10^6 documents, a lexicon of size 10^5, and 100 distinct words per document, which is represented as a bag of words with frequencies.

   (a) What is the amount of space required to store the entire data matrix without any optimization?

   (b) Suggest a sparse data format to store the matrix and compute the space required.

2. In Exercise 1, let us represent the documents in 0-1 format depending on whether or not a word is present in the document. Compute the expected dot product between a pair of documents in each of which 100 words are included completely at random. What is the expected dot product between a pair with 50000 words each? What does this tell you about the effect of document length on the computation of the dot product?

3. Suppose that a news portal has a stream of incoming news and they asked you to organize the news into about 10 reasonable categories of your choice. Which problem discussed in this chapter would you use to accomplish this goal?

4. In Exercise 3, consider the case in which examples of 10 pre-defined categories are available. Which problem discussed in this chapter would you use to determine the category of an incoming news article.

5. Suppose that you have popularity data on the number of clicks (per hour) associated with each news article in Exercise 3. Which problem discussed in this chapter would you use to decide the article that is likely to be the most popular among a group of 100 incoming articles (not included in the group with associated click data).

6. Suppose that you want to find the articles that are strongly critical of some issue in Exercise 3. Which problem discussed in this chapter would you use?

7. Consider a news article that discusses multiple topics. You want to obtain the portions of contiguous text associated with each topic. Which problem discussed in this chapter would you use in order to identify these segments?

Chapter 2

Text Preparation and Similarity Computation

"Life is a long preparation for something that never happens."–William B. Yeats

2.1 Introduction

Text data is often found in highly unstructured environments, and is frequently created by human participants. In many cases, text is embedded within Web documents, which is contaminated with elements such as HyperText Markup Language (HTML) tags, misspellings, ambiguous words, and so on. Furthermore, a single Web page may contain multiple blocks, most of which might be advertisements or other unrelated content. These effects can be ameliorated with proper preprocessing. Common preprocessing methods are as follows:

1. *Platform-centric extraction and parsing:* Text can contain platform-specific content such as HTML tags. Such documents need to cleansed of platform-centric content and parsed. The parsing of the text extracts the individual *tokens* from the documents. A token is a sequence of characters from a text that is treated as an indivisible unit for processing. Each mention of the same word in a document is treated as a separate token.

2. *Preprocessing of tokens:* The parsed text contains tokens that are further processed to convert them into the terms that will be used in the collection. Words such as "*a*," "*an*," and "*the*" that occur very frequently in the collection can be removed. These words are typically not discriminative for most mining applications, and they only add a large amount of noise. Such words are also referred to as *stop words*. Common prepositions, conjunctions, pronouns, and articles are considered stop words. In general, language-specific dictionaries of stop words are often available. The words are *stemmed* so that words with the same root (e.g., different tenses of a word) are consolidated. Issues involving punctuation and capitalization are addressed. At this

C. C. Aggarwal, *Machine Learning for Text*,
https://doi.org/10.1007/978-3-030-96623-2_2

point, one can create a *vector space representation*, which is a sparse, multidimensional representation containing the frequencies of the individual words.

3. *Normalization:* As our discussion above shows, not all words are equally important in analytical tasks. Stop words represent a rather extreme case of very frequent words at one end of the spectrum that must be removed from consideration. What does one do about the varying frequencies of the remaining words? It turns out that one can weight them a little differently by modifying their document-specific term frequencies based on their corpus-specific frequencies. Terms with greater corpus-specific frequencies are down-weighted. This technique is referred to as *inverse document frequency normalization.*

Pre-processing creates a sparse, multidimensional representation. Let D be the $n \times d$ document-term matrix. The number of documents is denoted by n and the number of terms is denoted by d. This notation will be used consistently in this chapter and the book.

Most text mining and retrieval methods require similarity computation between pairs of documents. This computation is sensitive to the underlying document representation. For example, when the binary representation is used, the *Jaccard coefficient* is an effective way of computing similarities. On the other hand, the *cosine* similarity is appropriate for cases in which term frequencies are explicitly tracked.

Chapter Organization

This chapter is organized as follows. The next section discusses the conversion of a character sequence into a set of tokens. The postprocessing of the tokens into terms is discussed in Section 2.3. Issues related to document normalization and representation are introduced in Section 2.4. Similarity computation is discussed in Section 2.5. Section 2.6 presents the summary.

2.2   Raw Text Extraction and Tokenization

The first step is to convert the raw text into a character sequence. The plain text representation of the English language is already a character sequence, although text sometimes occurs in *binary* formats such as Microsoft Word or Adobe portable document format (PDF). In other words, we need to convert a set of bytes into a sequence of characters based on the following factors:

1. The specific text document may be represented in a particular type of encoding, depending on the type of format such as a Microsoft Word file, an Adobe portable document format, or a zip file.

2. The language of the document defines its character set and encoding.

When a document is written in a particular language such as Chinese, it will use a different *character set* than in the case where it is written in English. English and many other European languages are based on the Latin character set. This character set can be represented easily in the *American Standard Code for Information Interchange*, which is short for ASCII. This set of characters roughly corresponds to the symbols you will see on the keyboard of a modern computer sold in an English speaking country. The specific encoding system is highly sensitive to the character set at hand. Not all encoding systems can handle all character sets equally well.

A standard code created by the *Unicode Consortium* is the *Unicode*. In this case, each character is represented by a unique identifier. Furthermore, almost all symbols known to us from various languages (including mathematical symbols and many ancient characters) can be represented in Unicode. This is the reason that the Unicode is the default standard for representing all languages. The different variations of Unicode use different numbers of bytes for representation. For example UTF-8 uses one byte, UTF-16 uses two bytes and so on. UTF-8 is particularly suitable for ASCII, and is often the default representation on many systems. Although it is possible to use UTF-8 encoding for virtually any language (and is a dominant standard), many languages are represented in other codes. For example, it is common to use UTF-16 for various Asian languages. Similarly, other codes like ASMO 708 are used for Arabic, GBK for Chinese, and ISCII for various Indian languages, although one can represent any of these languages in the Unicode. The nature of the code used therefore depends on the language, the whims of the creator of the document, and the platform on which it is found. In some cases, where the documents are represented in other formats like Microsoft Word, the underlying binary representation has to be converted into a character sequence. In many cases, the document meta-data provides useful information about the nature of its encoding up front without having to infer it by examining the document content. In some cases, it might make sense to separately store the meta-data about the encoding because it can be useful for some machine learning applications. The key takeaway from the above discussion is that irrespective of how the text is originally available, it is always converted into a character sequence.

In many cases, the character sequence contains a significant amount of meta-information depending on its source. For example, an HTML document will contain various tags and anchor text, and an XML document will contain meta-information about various fields. Here, the analyst has to make a judgement about the importance of the text in various fields to the specific application at hand, and remove all the irrelevant meta-information. As discussed in Section 2.2.1 on Web-specific processing, some types of fields such as the headers of an HTML document may be even more relevant than the body of the text. Therefore, there is a cleaning phase is often required for the character sequence. This character sequence needs to be expressed in terms of the distinct *terms* in the vocabulary, which comprise the base dictionary of words. These terms are often created by consolidating multiple occurrences and tenses of the same word. However, before finding the base terms, the character sequence needs to be parsed into *tokens*.

A token is a contiguous sequence of characters with a semantic meaning, and is very similar to a "term," except that it allows repetitions, and no additional processing (such as stemming and stop word removal) has been done. For example, consider the following sentence:

> After sleeping for four hours, he decided to sleep for another four.

In this case, the tokens are as follows:

> { *"After" "sleeping" "for" "four" "hours" "he" "decided" "to" "sleep" "for" "another" "four"* }.

Note that the words "for" and "four" are repeated twice, and the words "sleep" and "sleeping" are also not consolidated. Furthermore, the word "After" is capitalized. These aspects are addressed in the process of converting tokens into *terms* with specific frequencies. In some situations, the capitalization is retained, and in others, it is not.

Tokenization presents some challenging issues from the perspective of deciding word boundaries. A very simple and primitive rule for tokenization is that white spaces can be used as separators after removing punctuation. White spaces refer to the character space, the tab, and the newline. However, this primitive rule is rather inadequate to address many language-specific issues. For example, how do we deal with a pair of words like "Las Vegas" that describe a city? Separating them out completely loses the semantic meaning. Should "Abraham Lincoln" be one token or two tokens? Some pairs of words like "a priori" occur together naturally, and therefore they cannot be separated on the basis of white spaces. In many cases, dictionaries of semantically co-occurring words can be used. Furthermore, common phrases can be stored and extracted from the character sequence. It is possible to not create a strict segmentation of the character sequence, but also extract overlapping character sequences. It is noteworthy that one must distinguish between *low-level tokenization* and *high-level tokenization* in this respect. Recognizing linguistically coherent phrases is an example of high-level tokenization and it requires a minimum level of linguistic processing. In many cases, the low-level phase of basic tokenization is followed up by a high-level phase of recreating semantically more meaningful tokens from the initial tokenization.

Removing punctuation marks and treating white spaces as separators will not work if a document creator has forgotten to leave a white space after a punctuation mark such as a comma. Commas, colons, and periods are therefore treated as separators, although there are some exceptions. For example, a comma or period often occurs within a number (e.g., decimal), and a colon appears between numbers when time is being represented (e.g., "8:20 PM"). Therefore, they are not treated as separators when they appear between numbers. A similar rule applies to the character '/' because it can be a separator between two words, but might be a part of a date (e.g., "06/20/2003") when it occurs between two numbers. A period has many other uses such as within an acronym, and therefore it requires special handling. Typically a list of acronyms such as "Dr." or "M.D." is stored up front by the preprocessor and compared to the character sequence as it is processed. A sequence of two dashes is treated as a separator, although a single occurrence might be a hyphen and is treated differently as discussed in a later section. Hyphens can also occur within phone numbers or social security numbers, and therefore the tokenizer should be trained to recognize them. In general, the tokenizer should be trained to recognize email addresses, Uniform Resource Locators (URLs), telephone numbers, dates, times, measures, vehicle license plate numbers, paper citations, and so on. As we can see, the process is rather tedious in the sense that we have to take care of lots of little details.

Apostrophes need to treated specially during tokenization, although some aspects are handled during the stemming phase as well. An apostrophe at the beginning of a word, at the end of a word, or ending in 's' is removed. This is because these apostrophes are often present for grammatical reasons such as a quotation or the expression of a possessive noun. Other apostrophes within the middle of the word such as "o'clock" have semantic significance, and are therefore retained within the token. In such cases, the apostrophe is simply treated as a letter within the integrated token.

In some cases, there is no unique way of performing the best tokenization. As humans, we tokenize accurately without much thought, but the task turns out to be far more ambiguous to a computer program. Therefore, different tokenizers will create a slightly different segmentation. The main rule is to use the tokenization consistently across the application at hand, when it is used at different places. An excellent off-the-shelf tokenizer is available from the Apache OpenNLP effort [644].

2.2.1   Web-Specific Issues in Text Extraction

Several aspects of text extraction are highly platform-specific. Since the Web is the most common source of text that is used in various applications, it is worthwhile examining the specific issues that arise in extracting text from the Web.

HTML documents have numerous fields in them, such as the title, the meta-data, and the body of the document. Typically, analytical algorithms treat these fields with different levels of importance, and therefore weight them differently. For example, the title of a document is considered more important than the body and is weighted more heavily. Another example is the anchor text in Web documents. Anchor text contains a description of the Web page pointed to by a link. Because of its descriptive nature, it is considered important, but it is sometimes not relevant to the topic of the page itself. Therefore, it is often removed from the text of the document. In some cases, where possible, anchor text could even be added to the text of the document *to which it points.* This is because anchor text is often a summary description of the document to which it points.

A Web page may often be organized into content blocks that are not related to the primary subject matter of the page. A typical Web page will have many irrelevant blocks, such as advertisements, disclaimers, or notices, that are not very helpful for mining. It has been shown that the quality of mining results improve when only the text in the main block is used. However, the (automated) determination of main blocks from Web-scale collections is itself a data mining problem of interest. While it is relatively easy to decompose the Web page into blocks, it is sometimes difficult to identify the main block. Most automated methods for determining main blocks rely on the fact that a *particular* site will typically utilize a similar layout for the documents on the site. Therefore, if a collection of documents is available from the site, two types of automated methods can be used:

1. *Block labeling as a classification problem:* The idea in this case is to create a new training data set that extracts visual rendering features for each block in the training data. This can be achieved using Web browsers such as Internet Explorer. Many browsers provide an API that can be used to extract the coordinates for each block. The main block is then manually labeled for some examples. This results in a training data set. The resulting training data set is used to build a classification model. This model is used to identify the main block in the remaining (unlabeled) documents of the site.

2. *Tree matching approach:* Most Web sites generate the documents using a fixed template. Therefore, if the template can be extracted, then the main block can be identified relatively easily. The first step is to extract *tag trees* from the HTML pages. These represent the frequent tree patterns in the Website. The tree-matching algorithm, discussed in the bibliographic section, can be used to determine such templates from these tag trees. After the templates have been found, the main block in each Web page is found using the extracted template. Many of the peripheral blocks often have similar content in different pages and can therefore be eliminated.

The tree-matching algorithm is discussed in [345, 620].

2.3   Extracting Terms from Tokens

Once the tokens have been extracted from the document collection, they are transformed into *terms* with specific frequencies. Note that a document may have many repetitions of a token, and these repetitions are consolidated into a single occurrence with an appropriate frequency. Furthermore, highly frequent tokens are often not discriminative, and variants of the same token need to be consolidated. We discuss these aspects in the following subsections.

2.3.1 Stop-Word Removal

Stop words are common words of a language that do not carry much discriminative content. For example, in a classification task of news articles, we would expect a word such as "the" to occur at roughly the same frequency in a sports-related article, as it would in a politics-related article. Therefore, it makes sense to remove such poorly discriminating words. The following strategies are commonly used:

1. All articles, prepositions, and conjunctions are stop words. Pronouns are sometimes considered stop words.

2. Language-specific dictionaries of stop words are available.

3. The frequent tokens in any particular collection can be identified, and a threshold on the frequency can be set in order to remove the stop words.

Stop-word removal is a *hard variant* of the softer approach of down-weighting frequent words with *inverse document frequency normalization*. In some cases, there is some loss of information associated with the hard removal of stop words. Therefore, many search and mining systems do not remove stop words, but simply rely on the approach of reducing the weight of frequent words.

2.3.2 Hyphens

Dealing with hyphens can sometimes be tricky, because in some cases they can define word boundaries, whereas in other cases they should be considered individual words. For example, compound adjectives such as "state-of-the-art" are always hyphenated, irrespective of their position in a sentence. In such a case, we can create a single term for this token. Some systems may represent this term as "stateoftheart". In other cases, two or more words might modify a noun, and therefore they might get hyphenated as a compound adjective. Depending on the usage and semantic intent, it may or may not be desirable to break it up for mining purposes. For example, consider the sentence:

> He has a dead-end job.

In such a case, the word "dead-end" naturally defines a single semantic idea, and it should probably be retained as a single term. On the other hand, consider the sentence:

> The five-year-old girl was playing with the cat.

In this case, the word "five" should probably be separated from "year-old." One can see that these decisions seem to be harder than they seem at first sight. Dictionaries of commonly hyphenated words are often available and it is possible to create automated, language-specific rules about deciding when hyphenated words should be broken up. The default rule is to retain the hyphenated word as a single term, because breaking it up leads to a change in the semantic meaning in most cases.

The other issue is that of consistency. Some writers may choose to use a hyphen between one or more words, whereas other writers might not. For example, consider the sentence:

> This road leads to a dead end.

In this usage, "dead end" is not a compound adjective and therefore it is not hyphenated. However, it might still make sense to be consistent within the semantic representation to treat "dead-end" as a single hyphenated word, because it refers to the same basic idea. In such cases, dictionaries of commonly adjacent words that (i) should be hyphenated, and (ii) should not be hyphenated, can be used in order to decide whether a pair of adjacent words should be treated as a unit. This step can be implemented in the same way as the usage-based consolidation step discussed in Section 2.3.4.

2.3.3 Case Folding

The case of a term often defines its semantic interpretation, which is relevant to the mining task at hand. Words get capitalized for various reasons, such as for beginning a sentence, for being part of a title, or for being proper nouns. In some cases, the same word could get capitalized for different reasons. For example, the word "Bob" could be a person name or a verb. In the latter case, it might be capitalized for beginning a sentence, and therefore it should be converted to lower case. On the other hand, if "Bob" is a person, then the upper case should be retained. Therefore, "Bob" and "bob" will be different terms in the lexicon.

How to decide on the specific usage of a particular term? The entire process of converting to the proper case is referred to as *truecasing* [342], and it is a machine learning problem. However, there are limits to what a machine learning model can achieve in such cases because of the ambiguities in usage and various other factors. In many cases, it is possible to use simplified heuristics. Although these are not perfect rules, their simplicity enables efficient processing. For example, words at the beginning of a sentence can always be converted to lower case, and words in titles or section headers can also be converted to lower case. The case of all other words is retained.

2.3.4 Usage-Based Consolidation

The notion of usage-based consolidation is quite similar to that of stemming, except that it is a much simpler process and is done up front during tokenization using lookup tables. The basic idea is that small variations of the same token often refer to the same word. For example, the words "color" and "colour" are just different spellings of the same word in American and British English, respectively. Similarly, usage of accents, hyphens, and white spaces may vary not only across geographical regions but also over individual writers. Different writers might use "naive" and "naïve" to refer to the same concept. In all such cases, it is important to consolidate these variations into a single term. For example, one could maintain a hash table (or other) data structure of all the possible variations of the tokens with their standardized forms. For example, hashing on either "naive" or "naïve" might return the same standardized form in both cases.

2.3.5 Stemming

Stemming is the process of consolidating related words with the same root. For example, a text document might contain the singular or plural form of the same word, various tenses, and other variations. In such cases, it makes sense to consolidate these words into a single one. After all, changing the tense of a word does not change its semantic interpretation from a mining point of view. For example, words such as "eat," "eats," "eating," and "ate" all belong to the same stem corresponding to "eat" and should therefore be consolidated into a single term.

More generally, stemming refers to the process of extracting the *morphological root* of a word, and various crude heuristics are used to achieve this goal. The common techniques are as follows:

1. *Semi-automatic lookup tables:* The lookup table of a stemmer is created up front in a semi-automatic way with various heuristics. For example, in the case of the token "eat", the variants "eats," "eated," "eatly," and "eating" may be stored in the table. Therefore, if the token "eating" is encountered at the time of text extraction, one can proceed to replace it with the word "eat." Note that not all of these are valid words, and in some cases, the constructed word could easily have a different semantic interpretation.

2. *Suffix stripping:* A small list of rules is stored in order to find the root form of a given word. For example, common suffixes such as "ing," "ed," and "ly," should be removed. Rules can also strip prefixes, although it is more common to strip suffixes.

   Sometimes, suffix stripping leads to changes in the semantic meaning. For example, the word "hoping" might get chopped to "hop," which has a completely different meaning. Similarly, this type of approach would not work with word pairs like "eat" and "ate."

3. *Lemmatization:* Lemmatization is a more sophisticated approach because it uses the specific part of speech in order to determine the root form of a word. The normalization rules depend on the part of speech and therefore, they are highly language specific.

Lemmatization is sometimes considered different from stemming, in that it goes beyond the simple stripping rules and uses the morphological roots of the words. Such an approach yields the dictionary form of the word, known as the *lemma*. For example, when the word "ate" is encountered, the approach would be able to discover that the proper root is "eat." A lemmatizer needs a significant amount of vocabulary and language-specific domain knowledge to carry out its task compared to other stemmers. The classical algorithm for lemmatization is the Porter's algorithm [557]. The latest version of Porter's algorithm is also referred to as *Snowball*. We omit the specific details of this method as it is outside the scope of this book, and packages to perform this task are readily available to the practitioner [557, 643].

2.4   Vector Space Representation and Normalization

This section will describe the *vector space representation*, which is the sparse, multidimensional representation of text used in most applications. Once the terms have been extracted we have a *dictionary* or *lexicon* as the base set of dimensions. For most mining applications, a sparse, multidimensional representation is preferred. This representation contains one dimension (feature) for each word and the value of the dimension is strictly positive only when the word is present in the document. Otherwise the value is set to 0. The positive value could either be a normalized term frequency or a binary indicator value of 1. Since a given document contains a tiny subset of the lexicon, this representation is extremely sparse. It is not uncommon for document collections of have lexicons significantly greater than a hundred-thousand words, and the average number of words in each document may only be a few hundred. Note that the entire process of conversion into this representation loses all the ordering information among words. Therefore, this model is also referred to as the *bag-of-words model*. There are two commonly used multidimensional representations of text data, corresponding to the *binary model* and the *tf-idf model*.

In some applications, it is sufficient to use a 0-1 representation corresponding to whether or not a word is present in the document. Certain types of machine learning applications such as the Bernoulli variant of the Bayes classifier only need the binary representation. However, the binary representation does lose a lot of information because it does not contain the frequencies of the individual terms, and it is also not normalized for the relative importance of words. However, the main advantages of the binary representation are that it is compact and it enables the use of many applications that would otherwise be hard to use on a representation containing the frequencies of words. For example, consider a setting in which we wish to find frequently co-occurring groups of k words, irrespective of their placement in the document. In such a case, one can leverage the binary representation and apply an off-the-shelf *frequent pattern mining* algorithm on the multidimensional representation. Another interesting aspect of text data is that the presence or absence of a particular word in a document is more informative than its precise frequency. Therefore, reasonable results can be achieved with the binary representation in some cases. It is certainly worthwhile to use the binary representation in cases where the application at hand allows only binary input data. The binary model is also sometimes referred to as the *Bernoulli* or the *boolean* model.

Most representations of text do not work with the boolean model. Rather, they use normalized frequencies of the terms. This model is referred to as the tf-idf, where *tf* stands for the *term frequency* and *idf* stands for the *inverse document frequency*. During the term extraction phase, the additional task of keeping track of the consolidated and stemmed terms is also accomplished.

Consider a document collection containing n documents in d dimensions. Let $\overline{X} = (x_1 \ldots x_d)$ be the d-dimensional representation of a document after the term extraction phase. Note that x_i represents the unnormalized frequency of a document. Therefore, all the values of x_i are nonnegative and most are zero. Since word frequencies in a long document can sometimes vary significantly, it makes sense to use *damping functions* on these frequencies. The square-root or the logarithm function may be applied to the frequencies to reduce the effect of spam. In other words, one might replace each x_i with either $\sqrt{x_i}$ or $\log(1 + x_i)$. Although the use of such damping functions is not universal, there is significant evidence to suggest that the wide variation in word frequencies makes damping extremely important in at some applications. Damping also reduces the effect of (repeated) spam words.

It is also common to normalize term frequencies based on their presence in the entire collection. The first step in normalization is to compute the inverse document frequency of each term. The inverse document frequency id_i of the ith term is a decreasing function of the number of documents n_i in which it occurs:

$$id_i = \log(n/n_i) \tag{2.1}$$

Note that the value of id_i is always nonnegative. In the limiting cases in which a term occurs in every document of the collection, the value of id_i is 0. The term frequency is normalized by multiplying it with the inverse document frequency:

$$x_i \Leftarrow x_i \cdot id_i \tag{2.2}$$

Although the use of inverse document frequency normalization is almost ubiquitous in commercial implementations of *search* applications, it is noteworthy that some *mining* algorithms have reported the use of better results with the use of raw frequencies. For example, the work in [504] reported that higher quality of clustering was obtained by *not* using the inverse document frequency normalization. One issue with inverse document frequency normalization is that even though it might help by de-emphasizing stop words, it might occasionally hurt in an inadvertent way by increasing the frequencies of misspellings and other errors that were not properly handled at preprocessing time. Therefore, the effect can be corpus-sensitive and application-sensitive. If one chooses not to use inverse document frequency normalization in a particular application, it becomes more important to be aggressive about removing stop words.

2.5   Similarity Computation in Text

Many multidimensional data mining applications use the Euclidean distance to measure the distances between pairs of points. The Euclidean distance between $\overline{X} = (x_1 \ldots x_d)$ and $\overline{Y} = (y_1 \ldots y_d)$ is defined as follows:

$$\text{Distance}(\overline{X}, \overline{Y}) = \sqrt{\sum_{i=1}^{d}(x_i - y_i)^2} \qquad (2.3)$$

It would seem at first sight that one should simply use the Euclidean distances to compute distances between pairs of points, since text is a special case of the multidimensional representation. However, the Euclidean distance is not good in computing distances in multidimensional representations that are very sparse and the number of zero values vary significantly over different points. This occurs frequently in the case of text because of the varying lengths of different documents.

In order to understand this point, consider the following four sentences:

1. She sat down.
2. She drank coffee.
3. She spent much time in learning text mining.
4. She invested significant efforts in learning text mining.

For simplicity in discussion, assume that stop words are not removed, and the text is represented in boolean form without normalization. Note that the first pair of sentences is virtually unrelated, but the two sentences are very short. Therefore, only five distinct words in the sentence have nonzero frequencies. The Euclidean distance is only $\sqrt{4} = 2$. In the case of the third and fourth sentences, there are many words in common. However, these sentences are also longer, and therefore they also have many words that are present in only one of the two sentences. As a result, the Euclidean distance between the second pair is $\sqrt{6}$, which is *larger* than the first case. This clearly does not seem to be correct because the second pair of sentences is obviously related in a semantic way, and they even share a larger *fraction* of their sentences in common.

This problem was caused by the varying lengths of the documents. The Euclidean distance will consistently report higher values for distances between longer pairs of documents even if large fractions of those documents are in common. For example, if exactly half of the terms in a pair of documents containing more than a thousand distinct words each are exactly identical, the Euclidean distance will still be more than $\sqrt{1000}$ when the documents are represented in boolean form. This distance will always be more than that between any pair of documents with less than 500 distinct words each, even if they do not share a single word in common. This type of distance function can lead to poor mining results in which longer and shorter documents are not treated with an even hand.

This suggests that we need distance (or similarity) functions that strongly normalize for the varying lengths of documents. A natural solution to this problem is to use the cosine of the *angle* between the multidimensional vectors representing the two documents. Note that the cosine between a pair of vectors does not depend on the length of the vectors but only on the angle between them. In other words, the cosine similarity between a pair of vectors, denoted by $\overline{X} = (x_1 \ldots x_d)$ and $\overline{Y} = (y_1 \ldots y_d)$, is defined as follows:

$$\text{cosine}(\overline{X}, \overline{Y}) = \frac{\sum_{i=1}^{d} x_i y_i}{\sqrt{\sum_{i=1}^{d} x_i^2} \sqrt{\sum_{i=1}^{d} y_i^2}} \qquad (2.4)$$

We can already see why this representation normalizes so well for the document length– the denominator contains the norms of the documents and therefore the effect of the varying length is blunted. The normalization also ensures that the cosine always lies in the range $(0, 1)$. Although we did not perform any idf normalization here, it is often (but not always) performed in text mining applications.

A more intuitive interpretation of the cosine can be obtained in the special case when each $\overline{X}$ and $\overline{Y}$ is a binary vectors (rather than a vector of tf-idf values). Let S_x and S_y be the indices of the words that take on the value of 1 in $\overline{X}$ and $\overline{Y}$, respectively. In such a case, the set-based variant of the cosine can be computed as follows:

$$\text{cosine}(S_x, S_y) = \frac{|S_x \cap S_y|}{\sqrt{|S_x|} \cdot \sqrt{|S_y|}}$$

$$= \text{GEOMETRIC-MEAN} \left\{ \frac{|S_x \cap S_y|}{|S_x|}, \frac{|S_x \cap S_y|}{|S_y|} \right\}$$

In other words, the cosine is the geometric mean of the fraction of shared words contained in each of the pair of documents (for the case of the binary representation of text). Even in the case where tf-idf values are used instead of the binary values, this factor plays a dominant role in the cosine computation. Since the cosine computation is so largely dependent on the *fraction of* common words in each of the documents, it is largely impervious to the lengths of the documents.

As an example, consider a pair of documents with representations $\overline{X} = (2, 3, 0, 5, 0, \ldots, 0)$ and $\overline{Y} = (0, 1, 2, 2, 0, \ldots, 0)$. Then, the cosine between the two is as follows:

$$\text{cosine}(\overline{X}, \overline{Y}) = \frac{2 \cdot 0 + 3 \cdot 1 + 0 \cdot 2 + 5 \cdot 2}{\sqrt{2^2 + 3^2 + 5^2} \cdot \sqrt{1^2 + 2^2 + 2^2}} = \frac{13}{\sqrt{38} \cdot \sqrt{9}} = \frac{13}{3 \cdot \sqrt{38}} \qquad (2.5)$$

The cosine can also be viewed as a *normalized* dot product; in other words, it is the dot product obtained after normalizing each vector to unit norm. Consider a collection in which we have normalized each vector. Therefore, for any vector $\overline{X}$, we have $\sum_{i=1}^{d} x_i^2 = 1$. Then, the cosine can be expressed as the dot product:

$$\text{cosine}(\overline{X}, \overline{Y}) = \frac{\sum_{i=1}^{d} x_i y_i}{\sqrt{\sum_{i=1}^{d} x_i^2} \sqrt{\sum_{i=1}^{d} y_i^2}} = \frac{\sum_{i=1}^{d} x_i y_i}{\sqrt{1} \sqrt{1}} = \overline{X} \cdot \overline{Y} \qquad (2.6)$$

Interestingly, if we normalize each document in the corpus to unit norm, the Euclidean distance is not very different from the cosine except that it is a distance function instead of a similarity function. The two can be shown to be related as follows:

$$\|\overline{X} - \overline{Y}\|^2 = \|\overline{X}\|^2 + \|\overline{Y}\|^2 - 2\,\overline{X} \cdot \overline{Y} \tag{2.7}$$

$$= 1 + 1 - 2\,\overline{X} \cdot \overline{Y} \tag{2.8}$$

$$= 2(1 - \overline{X} \cdot \overline{Y}) \tag{2.9}$$

The normalized Euclidean distance always lies in the range $(0, 2)$ because of the up front, length-wise normalization. Therefore, if we normalize each vector in the corpus to unit norm, we could easily use the Euclidean distance for various mining applications instead of the cosine similarity. In other words, one could obtain the same length-wise normalization advantage of the cosine by normalizing the documents up front and using the Euclidean distance. In fact, there is no difference between the use of the Euclidean distance, the dot product, or the cosine similarity for retrieval applications, once a normalization has been performed.

In the special case where the boolean representation of text is used, another commonly used measure is the Jaccard similarity. Let S_x and S_y be the set of words in a pair of documents that are represented in boolean form. Then, the Jaccard similarity is defined as follows:

$$\text{Jaccard}(S_x, S_y) = \frac{|S_x \cap S_y|}{|S_x \cup S_y|}$$

$$= \frac{\#\text{Common terms in } S_x \text{ and } S_y}{\#\text{Distinct terms in union of } S_x \text{ and } S_y}$$

The Jaccard coefficient always lies in the range $(0, 1)$ just like the cosine coefficient. It is also possible to define the Jaccard coefficient for the case where the documents $\overline{X} = (x_1 \ldots x_d)$ and $\overline{Y} = (y_1, \ldots, y_d)$ are represented in tf-idf form:

$$\text{Jaccard}(\overline{X}, \overline{Y}) = \frac{\sum_{i=1}^{d} x_i \cdot y_i}{\sum_{i=1}^{d} x_i^2 + \sum_{i=1}^{d} y_i^2 - \sum_{i=1}^{d} x_i \cdot y_i} \tag{2.10}$$

The Jaccard coefficient is especially useful for the case where the boolean representation of text is used. For the tf-idf representation, it is more common to use the cosine measure, although similar results are obtained with the Jaccard and cosine coefficients [262, 531].

2.5.1   Is idf Normalization and Stemming Always Useful?

The use of idf normalization owes its origin to information retrieval applications in which stop-words have an obviously confounding effect on the quality of the results. However, in several text *mining* applications, it is been observed that idf normalization actually has a detrimental effect. For example, in text segmentation (cf. Section 16.2 of Chapter 16), it was observed in implementations of the TextTiling algorithm [242] that the use of idf normalization in similarity computation worsened the results. Similarly, several implementations and variations of the k-means clustering algorithm have been shown to work better without idf normalization [504]. Furthermore, in many probabilistic methods for topic modeling, clustering, and classification, the underlying generative assumption implies that one should use raw term frequencies rather than idf-normalized frequencies. Methods like k-means are deterministic avatars of such probabilistic models. In linear models for classification, idf normalization is almost[1] equivalent to the use of raw term frequencies.

---

[1]Small differences are caused by regularization effects. Without regularization, the same results will be obtained in a method like linear regression, no matter how one scales the attributes.

Issues such as stemming also have similar effects in mining applications. While the effect of stemming is significant[2] in IR applications (because users specify a small number of keywords), the issue is not quite as critical when mining larger collections containing documents of reasonable length. Stemming may still be useful when mining very small documents like discussion board posts or tweets. In fact, it has been stated in [367] that techniques like stemming can sometimes degrade classification accuracy when working with larger documents. All these observations suggest that one should be careful when using different types of normalization and preprocessing methods, because they are legacy methods inherited from traditional information retrieval settings. The constraints of typical settings in mining are not always the same as those in information retrieval.

2.6 Summary

Text data requires a significant amount of preprocessing because of the unstructured nature of the environments in which it is often found. The most important phases of text processing include tokenization, term extraction, and normalization. The phases of tokenization and term extraction are highly language-specific and may often require some domain knowledge about the language at hand. After extracting the term frequencies from a collection, they are normalized so that very frequent terms receive lower weights. This type of normalization is referred to as the inverse document frequency normalization.

Similarity computation in text is highly sensitive to the length of the documents. Using the Euclidean distance on a length-unnormalized text collection can lead to disastrous results. Therefore, the common approach is to use the cosine similarity between pairs of documents. The implicit effect of the cosine similarity is to normalize each document in the corpus to unit Euclidean norm before computing the dot product. Several other similarity functions such as the Jaccard coefficient are often used, when the text is represented in boolean form.

2.7 Bibliographic Notes

A discussion of several aspects of text preprocessing may be found in several textbooks [35, 345, 367, 573]. Web-specific issues to text extraction may be found in [92, 345]. Several aspects of text preprocessing are related to information extraction of text and part-of-speech tagging. A discussion of information extraction for text data may be found in [496]. Some aspects of text preprocessing are also related to language modeling methods that are discussed in [368].

A discussion on character encoding methods may be found in [362]. Some practical suggestions on tokenization may be found in [648]. Discussions on conversions of tokens to terms may be found in [345, 367]. A variety of recent stemming algorithms are discussed in [557, 643]. Issues related to text representation and frequency-based normalization are discussed in [35, 345, 367, 490, 573]. Experimental results and theoretical justifications of various weighting schemes may be found in [136, 476, 489, 522]. A discussion of how search engines implement similarity measures efficiently is provided in [639]. An interesting method, referred to as *pivoted document length normalization* was proposed by Singhal

---

[2]When a user queries for "*eat*", documents containing "*eating*" are also useful. The main issue here is that a set of query keywords is an extremely small document, and stemming helps in reducing the effect of sparsity.

et al. [519]. Similarity measures for short segments of text are discussed in Metzler *et al.* [383]. A Web-based kernel similarity function was studied in [484], in which queries to a search engine are used to evaluate the similarities between short text snippets. The works in [262, 531] compared several similarity measures in the context of text clustering, such as the Euclidean, cosine, Jaccard, and the Pearson correlation coefficient. The Euclidean distance performed poorly because it did not normalize for the lengths of the documents. On the other hand, comparable results were obtained with the cosine, Jaccard, and Pearson correlation coefficients.

2.7.1   Software Resources

The Bow toolkit contains a tokenizer [371] written in C, which is distributed under the GNU public license. A high-quality tokenizer may also be found from the Apache OpenNLP effort [644]. The Stanford NLP [650] and NLTK site [652] also contain several natural language processing tools that can be used for tokenizing and other term extraction operations. The latest version of the Porter stemmer may be available at [643]. The *scikit-learn* [646] and R-based **tm** library [647] also have preprocessing and tokenization functionalities built into them. A Java-based tokenizer and preprocessor may be found at the *Weka* library [649].

2.8   Exercises

1. Tokenize the following sentence:

   After sleeping for two hours, he decided to sleep for another two.

2. Assume that all article, pronouns, and prepositions are stop words. Perform a sensible stemming and case folding in the example of Exercise 1, and convert to a vector-space representation. Express your representation as a set of words with associated frequencies but no normalization.

3. Consider a collection in which the words "after," "decided," and "another," each occur in 16% of the documents. All other words occur in 4% of the documents. Create an idf-normalized representation of your answer in Exercise 2.

4. Show that the Jaccard similarity between a pair of documents can never be larger than the cosine similarity between them. What are the special cases in which the Jaccard similarity is exactly equal to the cosine similarity?

5. Compute the cosine similarity between the vector pair $(1, 2, 3, 4, 0, 1, 0)$ and $(4, 3, 2, 1, 1, 0, 0)$. Repeat the same computation with the Jaccard coefficient.

6. Normalize each of the vectors in Exercise 5 to unit norm. Compute the Euclidean distance between the pair of normalized vectors. What is the relationship between this Euclidean distance and the cosine similarity computed in Exercise 5?

7. Repeat Exercise 5 with the boolean representations of the two documents.

8. Write a computer program to evaluate the cosine similarity between a pair of vectors.

Chapter 3

Matrix Factorization and Topic Modeling

"Nobody can be told what the matrix is–you have to see it for yourself."–The fictional character Morpheus in the movie *Matrix*

3.1 Introduction

Most document collections are defined by document-term matrices in which the rows (or columns) are highly correlated with one another. These correlations can be leveraged to create a low-dimensional representation of the data, and this process is referred to as *dimensionality reduction*. Almost all dimensionality reductions of this type can be expressed as *low-rank factorizations of the document-term matrix*. In order to understand this point, consider a toy corpus defined on a lexicon of seven words:

lion, lioness, cheetah, jaguar, porsche, ferrari, maserati

The first three words in the lexicon are related to the topic of cats and the last three are related to cars. The (middle) word, which is "jaguar," could be related to either topic. This because the word "jaguar" is *polysemous*, and its meaning might depend on its usage and context.

The words in a document will often be predominantly related to a particular topic, which will cause inter-attribute correlations. Therefore, consider a case where most documents contain a majority of their words from either the set { lion, lioness, cheetah, jaguar } or they contain a majority of the words from the set { jaguar, porsche, ferrari, maserati}. Intuitively speaking, these two sets define new features in terms of which the entire collection is expressed. In other words, a document containing most words from the first set can be expressed as $(a, 0)$, a document containing most words from the second set can be approximately expressed as $(0, b)$, and a document containing many words from both sets can be expressed as (c, d). One can view this new set of coordinates as a *reduced representation* of the data. Although it might seem that dimensionality reduction loses information,

© Springer Nature Switzerland AG 2022
C. C. Aggarwal, *Machine Learning for Text*,
https://doi.org/10.1007/978-3-030-96623-2_3

it is often possible to choose a representation dimensionality in which most of the semantic knowledge in the corpus is retained, and only noise is lost.

The reduction in noise can even improve representation quality. For example, in the original collection the words "lion" and "lioness" are (almost) synonymous but will not be recognized as similar words in a cosine similarity computation on the original representation. The different usages of "jaguar" to refer to either cats or to cars will also not be properly disambiguated. On the other hand, a reduced representation is often able to improve the semantic closeness of related words and disambiguate multiple uses of the same word. As a result, many retrieval and mining algorithms show improved accuracy when the reduced representation is used in lieu of the original representation. When a feature transformation improves the accuracy of an algorithm, it can be viewed as a *feature engineering* method. The goals of feature engineering are subtly different from those of dimensionality reduction. Feature engineering is focussed on improving performance accuracy of a particular algorithm by changing the data representation, and it might sometimes even *increase* the dimensionality of the representation to achieve these goals. This chapter primarily discusses dimensionality reduction methods, but it also discusses some feature engineering methods.

It is noteworthy that the new representation in the aforementioned example of cats and cars is able to pull out the hidden semantic concepts in the data, and express any document in the collection as a combination of these hidden (or *latent*) concepts. One will often see the use of the word "latent" to describe many of these techniques, which refers to the fact that these concepts are hidden in the aggregate statistics of the data. It is not difficult to observe that the notions of semantic concepts, topics, and clusters are closely related. In fact, some forms of *nonnegative* dimensionality reduction are also referred to as *topic modeling*, and they have dual use in clustering applications.

How do the notions of dimensionality reduction and latent semantic analysis relate to matrix factorization? The basic idea is that any $n \times d$ document-term matrix can be expressed in terms of $k \ll \min\{n, d\}$ d-dimensional basis vectors. The value of k defines the number of semantic concepts in the data. In our previous example of cats and cars, the value of k is 2, whereas the value of d is 7. Typically, one expresses the basis vector as a $d \times k$ matrix $V = [v_{ij}]$, in which the columns represent the basis vectors. In the example of cats and cars, one column of V (i.e., basis vector) corresponds to cats and the other corresponds to cars. If we assume that the features are ordered in the same way as shown on page 33, the basis vector for the cat concept might[1] have strongly positive components on the first four (out of seven) word components, and the car concept might have strongly positive values on the last four. Other values might be nearly zero. Furthermore, the k-dimensional reduced representations of the n documents can be expressed as the rows of an $n \times k$ matrix $U = [u_{ij}]$, which is also the reduced representation of the corpus. The rows in U provide the document coordinates (i.e., transformed representation) with respect to the basis system in V. In our previous example, the rows of U will contain the two coordinates corresponding to the strength of association of the document with cats and/or cars. Therefore, the document-term matrix can be represented in the following *factorized* form:

$$D \approx UV^T \tag{3.1}$$

The right-hand side is simply a matrix multiplication of a *embedding* matrix U with the (transpose of the) *basis* matrix V in order to transform the reduced representation into the original feature space. This type of matrix multiplication is used in all types of basis

---

[1] Here, we are assuming a specific type of factorization, referred to as non-negative matrix factorization, because of its interpretability. Other factorizations might not obey these properties.

transformations in linear algebra. However, one needs to find the best basis representation V (and corresponding reduction U) in which the error of the approximate equality "$\approx$" in Equation 3.1 is low. Therefore, one can also view this problem as that of approximate factorization of the $n \times d$ document-term matrix D into two low-rank matrices of size $n \times k$ and $d \times k$, respectively. The value of k defines the rank of the factorization. This factorization is referred to as *low-rank* because the ranks of each of U, V, and UV^T are at most $k \ll d$, whereas the rank of D might be d. The remaining $(d-k)$-dimensional subspace does not have significant representation in the corpus at hand, and it can be captured by the approximate equality "$\approx$" in Equation 3.1. Note that there will always be some *residual error* $(D-UV^T)$ from the factorization. In fact, the entries in U and V are often discovered by solving an optimization problem in which the sum of squares (or other aggregate function) of the residual errors in $(D-UV^T)$ are minimized. A low-error factorization is possible only when the underlying matrix exhibits high correlations among its different columns.

Almost all forms of dimensionality reduction and matrix factorization are special cases of the following optimization model over matrices U and V:

$$\text{Maximize similarity between entries of } D \text{ and } UV^T$$

$$\text{subject to:}$$

$$\text{Constraints on } U \text{ and } V$$

By varying the objective function and constraints, dimensionality reductions with different properties are obtained. The most commonly used objective function is the sum of the squares of the entries in $(D - UV^T)$, which is also defined as the (squared) *Frobenius norm* of the matrix $(D - UV^T)$. The (squared) Frobenius norm of a matrix is also referred to as its *energy*, because it is the sum of the second moments of all data points about the origin. However, some forms of factorizations with probabilistic interpretations use a *maximum-likelihood* objective function. Similarly, the constraints imposed on U and V enable different properties of the factorization. For example, if we impose orthogonality constraints on the columns of U and V, this leads to a model known as *singular value decomposition (SVD)* or *latent semantic analysis (LSA)*. The orthogonality of the basis vectors is particularly helpful in mapping new documents to the transformed space in a simple way. On the other hand, better semantic interpretability can be obtained by imposing nonnegativity constraints on U and V. In this chapter, we will discuss different types of reductions and their relative advantages.

Chapter Organization

This chapter is organized as follows. The remainder of this section discusses some conventions for representing the reduced representation. Section 3.2 introduces the singular value decomposition model, which is also referred to as latent semantic analysis. Nonnegative matrix factorization is introduced in Section 3.3. Probabilistic latent semantic analysis is discussed in Section 3.4. Latent Dirichlet Allocation is introduced in Section 3.5. Nonlinear dimensionality reduction methods are introduced in Section 3.6. A summary is given in Section 3.7.

3.1.1 Normalizing a Two-Way Factorization into a Standardized Three-Way Factorization

The aforementioned optimization model factorizes D into *two* matrices U and V. One can immediately notice that the factorization is not unique. For example, if we multiply each entry of U by 2, then we can divide each entry of V by 2 to get the same product UV^T. Furthermore, we can apply this trick to just a particular (say, rth) column of each of U and V to get the same result. In other words, different *normalization factors* for the columns of U and V lead to the same product.

Therefore, some forms of dimensionality reduction convert the two-way matrix factorization into a three-way matrix factorization in which each of the matrices satisfies certain normalization conventions. This additional matrix is typically a $k \times k$ diagonal matrix of nonnegative entries, in which the (r, r)th entry contains a scaling factor for the rth column. Specifically, for any two-way matrix factorization $D \approx UV^T$ into $n \times k$ and $d \times k$ matrices U and V, respectively, we can convert it into a *unique*[2] three-way matrix factorization of the following form:

$$D \approx Q\Sigma P^T \tag{3.2}$$

Here, Q is a *normalized* $n \times k$ matrix (derived from U), P is a *normalized* $d \times k$ matrix (derived from V), and Σ is a $k \times k$ diagonal matrix in which the diagonal entries contain the nonnegative normalization factors for the k concepts. Each of the columns of Q and P satisfy the constraint that its L_2-norm (or L_1-norm) is one unit. It is common to use L_2-normalization in methods like singular value decomposition and L_1-normalization in methods like probabilistic latent semantic analysis. For the purpose of discussion, let us assume that we use L_2-normalization. Then, the conversion from two-way factorization to three-way factorization can be achieved as follows:

1. For each $r \in \{1 \ldots k\}$, divide the rth column $\overline{U_r}$ of U with its L_2-norm $\|\overline{U_r}\|$. The resulting matrix is denoted by Q.

2. For each $r \in \{1 \ldots k\}$, divide the rth column $\overline{V_r}$ of V with its L_2-norm $\|\overline{V_r}\|$. The resulting matrix is denoted by P.

3. Create a $k \times k$ diagonal matrix Σ, in which the (r, r)th diagonal entry is the nonnegative value $\|\overline{U_r}\| \cdot \|\overline{V_r}\|$.

It is easy to show that the newly created matrices Q, Σ, and P satisfy the following relationship:

$$Q\Sigma P^T = UV^T \tag{3.3}$$

It is noteworthy that all diagonal entries of Σ are always nonnegative because of how the normalization is done. The three-way factorized representation is used by many dimensionality reduction methods because of its normalized properties. An example of L_1-normalization is shown later in this chapter (cf. Figure 3.2). The entries in the diagonal matrix intuitively represent the relative dominance of the different latent concepts. For example, in our previous example with the car- and cat-related documents, if car documents are more copious than cat documents and also have higher term frequencies, this will be reflected in a higher diagonal value of Σ_{rr} for the car-related entry. In a sense, Σ_{rr} reflects the relative frequency of the rth latent concept in the collection. The varying frequencies of different concepts also provide a rationale for dimensionality reduction. If we use $k = d$, then many of the values of Σ_{rr} of the infrequent latent concepts would be very small. Such concepts can be dropped without affecting the accuracy of the approximation inherent in matrix factorization. This is the reason that one can typically use values of the rank k that are much less than the dimensionality d. In text collections, it is possible for the value of d (i.e., number of terms)

---

[2]The normalization is unique up to multiplication by -1 of any particular column of P and Q for any given pair (U, V).

to be of the order of a few hundred thousand, whereas the value of k is only of the order of a few hundred.

In some of the following discussions, we will pose the optimization problem for dimensionality reduction in terms of a two-way factorization, whereas in others we will pose it as a three-way factorization. This is because different choices are provide better interpretability in different settings, although they are mathematically equivalent.

3.2 Singular Value Decomposition

Singular value decomposition (SVD) is used in all forms of multidimensional data, and its instantiation in the text domain is referred to as *latent semantic analysis (LSA)*. Consider the simplest possible factorization of the $n \times d$ matrix D into an $n \times k$ matrix $U = [u_{ij}]$ and the $d \times k$ matrix $V = [v_{ij}]$ as an *unconstrained matrix factorization problem*:

$$\text{Minimize}_{U,V} \|D - UV^T\|_F^2$$

subject to:

No constraints on U and V

Here $\|\cdot\|_F^2$ refers to the (squared) Frobenius norm of a matrix, which is the sum of squares of its entries. The matrix $(D - UV^T)$ is also referred to as the *residual matrix*, because its entries contain the residual errors obtained from a low-rank factorization of the original matrix D. This optimization problem is the most basic form of matrix factorization with a popular objective function and no constraints. This formulation has infinitely many alternative optimal solutions (see Exercises 2 and 3). However, one[3] of them is such that the columns of V are orthonormal, which allows transformations of new documents (not included in D) with simple axis rotations (i.e., matrix multiplication). A remarkable property of the unconstrained optimization problem above is that imposing orthogonality constraints does not worsen the optimal solution. *The following constrained optimization problem shares at least one optimal solution as the unconstrained version* [171, 530]:

$$\text{Minimize}_{U,V} \|D - UV^T\|_F^2$$

subject to:

Columns of U are mutually orthogonal

Columns of V are mutually orthonormal

In other words, one of the alternative optima to the unconstrained problem also satisfies orthogonality constraints. It is noteworthy that *only* the solution satisfying the orthogonality constraint is considered SVD because of its interesting properties, even though other optima do exist (see Exercises 2 and 3).

Another remarkable property of the solution (satisfying orthogonality) is that it can be computed using *eigen-decomposition* of either of the positive semi-definite matrices $D^T D$ or DD^T. The following properties of the solution can be shown (see Exercises 3[a], 5, and 6):

1. The columns of V are defined by the top-k unit eigenvectors of the $d \times d$ positive semi-definite and symmetric matrix $D^T D$. The diagonalization of a symmetric and positive semi-definite matrix results in orthonormal eigenvectors with non-negative

---

[3]This solution is unique up to multiplication of any column of U or V with -1 and rotation within subspaces of equal variance.

eigenvalues. After V has been determined, we can also compute the reduced representation U as DV, which is simply an axis rotation operation on the rows (documents) in the original data matrix. This is caused by the orthogonality of the columns of V, which results in $DV \approx U(V^T V) = U$. One can also use this approach to compute the reduced representation $\overline{X}V$ of any row-vector $\overline{X}$ that was not included in D.

2. The columns of U are also defined by the top-k *scaled* eigenvectors of the $n \times n$ *dot-product matrix* DD^T in which the (i,j)th entry is the dot-product similarity between the ith and jth documents. The scaling factor is defined so that each eigenvector is multiplied with the square-root of its eigenvalue. In other words, the *scaled eigenvectors of the dot-product matrix can be used to directly generate the reduced representation*. This fact has some interesting consequences for the nonlinear dimensionality reduction methods, which replace the dot product matrix with another similarity matrix (cf. Section 3.6). This approach is also efficient for linear SVD when $n \ll d$, and therefore the $n \times n$ matrix DD^T is relatively small. In such cases, U is extracted first by eigen-decomposition of DD^T, and then V is extracted as $D^T U$.

3. Even though the n eigenvectors of DD^T and d eigenvectors of $D^T D$ are different, the top $\min\{n, d\}$ eigenvalues of DD^T and $D^T D$ are the same values. All other eigenvalues are zero.

4. The total squared error of the approximate matrix factorization of SVD is equal to the sum of the eigenvalues of $D^T D$ that are *not* included among the top-k eigenvectors. If we set the rank of the factorization k to $\min\{n, d\}$, we can obtain an *exact* factorization into orthogonal basis spaces with zero error.

This factorization of rank $k = \min\{n, d\}$ with zero error is of particular interest. We convert the two-way factorization (of zero error) into a three-way factorization according to the methodology of Section 3.1.1, which results in a standard form of SVD:

$$D = Q\Sigma P^T = \underbrace{(Q\Sigma)}_{U} \underbrace{P^T}_{V^T} \tag{3.4}$$

Here, Q is an $n \times k$ matrix containing all the $k = \min\{n, d\}$ *non-zero* eigenvectors of DD^T, and P is a $d \times k$ matrix containing all the $k = \min\{n, d\}$ non-zero eigenvectors of $D^T D$. The columns of Q are referred to as the *left singular vectors*, whereas the columns of P are referred to as the *right singular vectors*. Furthermore, Σ is a (nonnegative) diagonal matrix in which the (r, r)th value is equal to the square-root of the rth largest eigenvalue of $D^T D$ (which is the same as the rth largest eigenvalue of DD^T). The diagonal entries of Σ are also referred to as *singular values*. Note that the singular values are always nonnegative by convention. The sets of columns of P and Q are each orthonormal because they are the unit eigenvectors of symmetric matrices. vt easy to verify (using Equation 3.4) that $D^T D = P\Sigma^2 P^T$ and that $DD^T = Q\Sigma^2 Q^T$, where Σ^2 is a diagonal matrix containing the top-k non-negative eigenvalues of $D^T D$ and DD^T (which are the same).

SVD is formally defined as the exact decomposition with zero error. What about the *approximate* variant of SVD, which is the primary goal of matrix factorization? In practice, one always uses values of $k \ll \min\{n, d\}$ to obtain *approximate* or *truncated* SVD:

$$D \approx Q\Sigma P^T \tag{3.5}$$

Using truncated SVD is the standard use-case in practical settings. Throughout this book, our use of the term "SVD" always refers to truncated SVD.

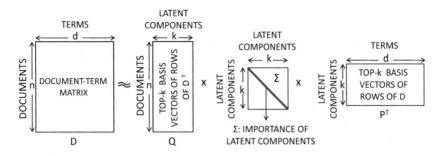

Figure 3.1: Dual interpretation of SVD in terms of the basis vectors of both D and D^T

Just as the matrix P contains the d-dimensional basis vectors of D in its columns, the matrix Q contains the n-dimensional basis vectors of D^T in its columns. In other words, *SVD simultaneously finds approximate bases of both documents and terms.* This ability of SVD to simultaneously find approximate bases for the row space and column space is shown in Figure 3.1. Furthermore, the diagonal entries of the matrix Σ provide a quantification of the relative dominance of the different semantic concepts.

One can express SVD as a weighted sum of rank-1 matrices. Let Q_i be the $n \times 1$ matrix corresponding to the ith column of Q and P_i be the $d \times 1$ matrix corresponding to the ith column of P. Then, the SVD product can be decomposed in *spectral form* using simple matrix-multiplication laws as follows:

$$Q\Sigma P^T = \sum_{i=1}^{k} \Sigma_{ii} Q_i P_i^T \tag{3.6}$$

Note that each $Q_i P_i$ is a rank-1 matrix of size $n \times d$ and a Frobenius norm of 1. Furthermore, it is possible to show that the Frobenius norm of $Q\Sigma P^T$ is given by $\sum_{i=1}^{k} \Sigma_{ii}^2$, which is the amount of *energy retained* in the representation. Maximizing the retained energy is the same as minimizing the loss defined by the sum of squares of the truncated singular values (which are small), because the sum of the two is always equal to $\|D\|_F^2$. The energy retained in the approximated matrix is the same as that in the transformed representation, because squared distances do not change with axis rotation. Therefore, the sum of the squares of the retained singular values provides the energy in the transformed representation DP. An important consequence of this observation is that the projection $D\overline{p}$ of D on any column $\overline{p}$ of P has an L_2-norm, which is equal to the corresponding singular value. In other words, SVD naturally selects the orthogonal directions along which the transformed data exhibits the largest scatter.

3.2.1 Example of SVD

An example of SVD helps in illustrating its inner workings. Consider a 6×6 matrix D defined over the lexicon of size 6 as follows:

lion, tiger, cheetah, jaguar, porsche, ferrari

The data matrix D is illustrated below:

$$D = \begin{pmatrix} & \text{lion} & \text{tiger} & \text{cheetah} & \text{jaguar} & \text{porsche} & \text{ferrari} \\ \text{Document-1} & 2 & 2 & 1 & 2 & 0 & 0 \\ \text{Document-2} & 2 & 3 & 3 & 3 & 0 & 0 \\ \text{Document-3} & 1 & 1 & 1 & 1 & 0 & 0 \\ \text{Document-4} & 2 & 2 & 2 & 3 & 1 & 1 \\ \text{Document-5} & 0 & 0 & 0 & 1 & 1 & 1 \\ \text{Document-6} & 0 & 0 & 0 & 2 & 1 & 2 \end{pmatrix}$$

Note that this matrix represents topics related to both cars and cats. The first three documents are primarily related to cats, the fourth is related to both, and the last two are primarily related to cars. The word "jaguar" is polysemous because it could correspond to either a car or a cat. Therefore, it is often present in documents of both categories and presents itself as a confounding word. We would like to perform an SVD of rank-2 to capture the two dominant concepts corresponding to cats and cars, respectively. Then, on performing the SVD of this matrix, we obtain the following decomposition:

$$D \approx Q\Sigma P^T$$

$$\approx \begin{pmatrix} -0.41 & 0.17 \\ -0.65 & 0.31 \\ -0.23 & 0.13 \\ -0.56 & -0.20 \\ -0.10 & -0.46 \\ -0.19 & -0.78 \end{pmatrix} \begin{pmatrix} 8.4 & 0 \\ 0 & 3.3 \end{pmatrix} \begin{pmatrix} -0.41 & -0.49 & -0.44 & -0.61 & -0.10 & -0.12 \\ 0.21 & 0.31 & 0.26 & -0.37 & -0.44 & -0.68 \end{pmatrix}$$

$$= \begin{pmatrix} 1.55 & 1.87 & 1.67 & 1.91 & 0.10 & 0.04 \\ 2.46 & 2.98 & 2.66 & 2.95 & 0.10 & -0.03 \\ 0.89 & 1.08 & 0.96 & 1.04 & 0.01 & -0.04 \\ 1.81 & 2.11 & 1.91 & 3.14 & 0.77 & 1.03 \\ 0.02 & -0.05 & -0.02 & 1.06 & 0.74 & 1.11 \\ 0.10 & -0.02 & 0.04 & 1.89 & 1.28 & 1.92 \end{pmatrix}$$

The reconstructed matrix is a very good approximation of the original document-term matrix. Furthermore, each point gets a 2-dimensional embedding corresponding to the rows of $Q\Sigma$. It is clear that the reduced representations of the first three documents are quite similar, and so are the reduced representations of the last two. The reduced representation of the fourth document seems to be somewhere in the middle of the representations of the other documents. This is logical because the fourth document corresponds to both cars and cats. From this point of view, the reduced representation seems to satisfy the basic intuitions one would expect in terms of *relative* coordinates. However, one annoying characteristic of this representation is that it is hard to get any *absolute* semantic interpretation from the embedding. For example, it is difficult to match up the two latent vectors in P with the original concepts of cats and cars. The dominant latent vector in P is $[-0.41, -0.49, -0.44, -0.61, -0.10, -0.12]$, in which all components are negative. The second latent vector contains both positive and negative components. Therefore, the correspondence between the topics and the latent vectors is not very clear. A part of the problem is that the vectors have both positive and negative components, which reduces their interpretability. The lack of interpretability of singular value decomposition is its primary weakness, as a result of which other nonnegative forms of factorization are sometimes preferred. Furthermore, forcing orthogonality of the vectors in P is not very natural, especially when the two topics have overlapping and confounding words like "jaguar." As a result, SVD does a relatively poor job at handling polysemy compared to many other forms of matrix factorization. However, it is not completely unsuccessful either, and it can handle the problem of synonymy relatively well.

3.2.2 The Power Method of Implementing SVD

The power method is an efficient way of finding the $d \times k$ basis matrix P. Note that the reduced representation $Q\Sigma$ can be obtained by post-multiplying the document-term matrix D with P, because we have $DP \approx Q\Sigma$. The power method can find the dominant eigenvector of any matrix (like $D^T D$) by first initializing it to a random d-dimensional column vector $\overline{p}$ and then repeatedly pre-multiplying with $D^T D$ and scaling to unit norm. To reduce the number of operations, it makes sense to compute the operations in the order dictated by the brackets in $[D^T(D\overline{p})]$. Therefore, we repeat the following step to convergence:

$$\overline{p} \Leftarrow \frac{[D^T(D\overline{p})]}{\|[D^T(D\overline{p})]\|}$$

The projection of the data matrix D on the vector $\overline{p}$ has an energy that is equal to the square of the first singular value. Therefore, the first singular value σ is obtained by using the L_2-norm of the vector $D\overline{p}$. Since $D\overline{p}$ contains the n different coordinates of the n documents along the first latent direction $\overline{p}$, the first column $\overline{q}$ of Q is obtained by a single execution of the following step:

$$\overline{q} \Leftarrow \frac{D\overline{p}}{\sigma} \tag{3.7}$$

This completes the determination of the first set of singular vectors and singular values. The next eigenvector and eigenvalue pair is obtained by making use of the spectral decomposition of Equation 3.6. First, we remove the rank-1 component contributed by the first set of singular vectors by adjusting the data matrix as follows:

$$D \Leftarrow D - \sigma\overline{q}\,\overline{p}^T \tag{3.8}$$

Note that even though $\overline{q}$ and $\overline{p}$ are vectors, we treat them as $n \times 1$ and $d \times 1$ matrices in the expression $\sigma\overline{q}\,\overline{p}^T$ to obtain a rank-1 matrix of size $n \times d$. Once the impact of the first component has been removed, we repeat the process to obtain the second set of singular vectors. The entire process is repeated k times to obtain the rank-k singular value decomposition. It is noteworthy that the matrix D is sparse in the case of document data, and therefore other efficient implementations such as the Lanczos algorithm are used [166, 167].

3.2.3 Applications of SVD/LSA

Singular value decomposition (also known as latent semantic indexing in its text-centric implementations) is used for dimensionality reduction of sparse and extremely high-dimensional text into a more traditional multidimensional format of a few hundred dimensions. A side effect of the reduction is to reduce the noise effects of synonymy and polysemy. As in the case of document-term matrix, one can use the cosine similarity to compute similarity with the reduced representations of the documents. If the rank k is chosen carefully, it is possible to improve *both* precision and recall, although the former often degrades if the value of k is chosen incorrectly. The improvement in both precision and recall occurs because of the reduction in noise effects of synonymy and polysemy. Typically, for collections containing a few hundred thousand terms, it is often sufficient to use values of k between 200 and 400. Therefore, the reduction in dimensionality is very significant, but the new representation is no longer sparse.

SVD can also be used to enable other data mining applications that do not work well with sparsity. For example, univariate decision trees work poorly with the original sparse representation of the document-term matrix. However, the transformed representation works somewhat better, especially if combinations of multiple decision trees (i.e., *random forests*) are used. Refer to Section 5.5 of Chapter 5 for a detailed discussion of decision trees.

Aside from providing a k-dimensional representation of the documents, SVD also provides a k-dimensional representation of the words. This k-dimensional representation may be extracted as the rows on the matrix $P\Sigma$. Words that are semantically similar will tend to be closer to one another in this multidimensional space. For example, the words "movie" and "film" are likely to be closer to another than the words "movie" and "song." Furthermore, the words "movie" and "song" will typically be closer to one another than "movie" and "carrot." This type of dual embedding is not exclusive to SVD, but it can be achieved in any form of matrix factorization.

The orthogonal basis representation of SVD has numerous other applications in solving systems of linear equations and other matrix operations. It also provides the mathematical framework required for generalizing SVD to nonlinear dimensionality reduction, as discussed in Section 3.6.

3.2.4   Advantages and Disadvantages of SVD/LSA

Singular value decomposition has several advantages and disadvantages compared to other matrix factorization methods. These advantages and disadvantages are as follows:

1. The orthogonal basis representation of SVD is useful for folding in the reduced representation of new documents not included in the data matrix D. For example, if $\overline{X}$ is a row vector of a new document, then its reduced representation is given by the k-dimensional vector $\overline{X}V$. This type of out-of-sample embedding is harder (albeit possible) with other forms of matrix factorization.

2. The SVD solution provides the same error as *unconstrained* matrix factorization problem. Since most other forms of dimensionality reduction are constrained matrix factorization problems, one can typically achieve a lower residual error with SVD at the same value of the rank k.

3. The topics of a text collection are often highly overlapping in terms of their vocabulary. As a result, the directions represented by the various topics are naturally not orthogonal, which matches poorly with orthogonal basis vectors. SVD does a poor job at revealing the actual semantic topics (or *clusters*) in the underlying data. Most forms of nonnegative matrix factorization that do not use orthogonal basis vectors are more adept at representing the clustering structure in the underlying data.

4. The representation provided by SVD is not very interpretable and it is hard to match with the semantic concepts in the collection. A key part of the problem is that the eigenvectors contain both positive and negative components that are hard to interpret.

The specific use of a particular method depends on the scenario or the application at hand.

3.3   Nonnegative Matrix Factorization

Nonnegative matrix factorization is a *highly interpretable* type of matrix factorization in which nonnegativity constraints are imposed on U and V. Therefore, this optimization problem is defined as follows:

$$\text{Minimize }_{U,V} \|D - UV^T\|_F^2$$

$$\text{subject to:}$$

$$U \geq 0,\ V \geq 0$$

As in the case of SVD, $U = [u_{ij}]$ is an $n \times k$ matrix and $V = [v_{ij}]$ is a $d \times k$ matrix of optimization parameters. Note that the optimization objective is the same but the constraints are different.

This type of constrained problem is often solved using Lagrangian relaxation. For the (i, s)th entry u_{is} in U, we introduce the Lagrange multiplier $\alpha_{is} \leq 0$, whereas for the (j, s)th entry v_{js} in V, we introduce the Lagrange multiplier $\beta_{js} \leq 0$. One can create a vector $(\overline{\alpha}, \overline{\beta})$ of dimensionality $(n + d) \cdot k$ by putting together all the Lagrangian parameters into a vector. Instead of using hard constraints on nonnegativity, Lagrangian relaxation uses penalties in order to relax the constraints into a softer version of the problem, which is defined by the augmented objective function L:

$$L = \|D - UV^T\|_F^2 + \sum_{i=1}^{n}\sum_{r=1}^{k} u_{ir}\alpha_{ir} + \sum_{j=1}^{d}\sum_{r=1}^{k} v_{jr}\beta_{jr} \qquad (3.9)$$

Note that violation of the nonnegativity constraints always lead to a positive penalty because the Lagrangian parameters cannot be positive. According to the methodology of Lagrangian optimization, this augmented problem is really a minimax problem because we need to minimize L over all U and V at any particular value of the (vector of) Lagrangian parameters, but we then need to maximize these solutions over all valid values of the Lagrangian parameters α_{is} and β_{js}. In other words, we have:

$$\text{Max}_{\overline{\alpha}\leq 0, \overline{\beta}\leq 0}\text{Min}_{U,V}\ L \qquad (3.10)$$

Here, $\overline{\alpha}$ and $\overline{\beta}$ represent the vectors of optimization parameters in α_{is} and β_{js}, respectively. This is a tricky optimization problem because of the way in which it is formulated with simultaneous maximization and minimization over different sets of parameters. The first step is to compute the gradient of the Lagrangian relaxation with respect to the (minimization) optimization variables u_{is} and v_{js}. Therefore, we have:

$$\frac{\partial L}{\partial u_{is}} = -(DV)_{is} + (UV^TV)_{is} + \alpha_{is} \qquad \forall i \in \{1,\ldots,n\}, s \in \{1,\ldots,k\} \qquad (3.11)$$

$$\frac{\partial L}{\partial v_{js}} = -(D^TU)_{js} + (VU^TU)_{js} + \beta_{js} \qquad \forall j \in \{1,\ldots,d\}, s \in \{1,\ldots,k\} \qquad (3.12)$$

The optimal value of the (relaxed) objective function at any particular value of the Lagrangian parameters is obtained by setting these partial derivatives to 0. As a result, we obtain the following conditions:

$$-(DV)_{is} + (UV^TV)_{is} + \alpha_{is} = 0 \qquad \forall i \in \{1,\ldots,n\}, s \in \{1,\ldots,k\} \qquad (3.13)$$

$$-(D^TU)_{js} + (VU^TU)_{js} + \beta_{js} = 0 \qquad \forall j \in \{1,\ldots,d\}, s \in \{1,\ldots,k\} \qquad (3.14)$$

We would like to eliminate the Lagrangian parameters and set up the optimization conditions purely in terms of U and V. It turns out the *Kuhn-Tucker optimality conditions* [55] are very helpful. These conditions are $u_{is}\alpha_{is} = 0$ and $v_{js}\beta_{js} = 0$ over all parameters. By multiplying Equation 3.13 with u_{is} and multiplying Equation 3.14 with v_{js}, we can use the Kuhn-Tucker conditions to get rid of these pesky Lagrangian parameters from the aforementioned equations. In other words, we have:

$$-(DV)_{is}u_{is} + (UV^TV)_{is}u_{is} + \underbrace{\alpha_{is}u_{is}}_{0} = 0 \qquad \forall i \in \{1,\ldots,n\}, s \in \{1,\ldots,k\} \qquad (3.15)$$

$$-(D^TU)_{js}v_{js} + (VU^TU)_{js}v_{js} + \underbrace{\beta_{js}v_{js}}_{0} = 0 \qquad \forall j \in \{1,\ldots,d\}, s \in \{1,\ldots,k\} \qquad (3.16)$$

One can rewrite these optimality conditions, so that a single parameter occurs on one side of the condition:

$$u_{is} = \frac{(DV)_{is}u_{is}}{(UV^TV)_{is}} \qquad \forall i \in \{1,\ldots,n\}, s \in \{1,\ldots,k\} \qquad (3.17)$$

$$v_{js} = \frac{(D^TU)_{js}v_{js}}{(VU^TU)_{js}} \qquad \forall j \in \{1,\ldots,d\}, s \in \{1,\ldots,k\} \qquad (3.18)$$

Even though these conditions are circular in nature (because the optimization parameters occur on both sides), they are natural candidates for iterative updates.

Therefore, the iterative approach starts by initializing the parameters in U and V to nonnegative random values in $(0,1)$ and then uses the following updates derived from the aforementioned optimality conditions:

$$u_{is} \Leftarrow \frac{(DV)_{is}u_{is}}{(UV^TV)_{is}} \qquad \forall i \in \{1,\ldots,n\}, s \in \{1,\ldots,k\} \qquad (3.19)$$

$$v_{js} \Leftarrow \frac{(D^TU)_{js}v_{js}}{(VU^TU)_{js}} \qquad \forall j \in \{1,\ldots,d\}, s \in \{1,\ldots,k\} \qquad (3.20)$$

These iterations are then repeated to convergence. Improved initialization provides significant advantages, and the reader is referred to [310] for such methods. Numerical stability can be improved by adding a small value $\epsilon > 0$ to the denominator during the updates:

$$u_{is} \Leftarrow \frac{(DV)_{is}u_{is}}{(UV^TV)_{is} + \epsilon} \qquad \forall i \in \{1,\ldots,n\}, s \in \{1,\ldots,k\} \qquad (3.21)$$

$$v_{js} \Leftarrow \frac{(D^TU)_{js}v_{js}}{(VU^TU)_{js} + \epsilon} \qquad \forall j \in \{1,\ldots,d\}, s \in \{1,\ldots,k\} \qquad (3.22)$$

One can also view ϵ as a type of *regularization parameter* whose primary goal is to avoid overfitting. Regularization is particularly helpful in small document collections.

As in all other forms of matrix factorization, it is possible to convert the factorization UV^T into the three-way factorization $Q\Sigma P^T$ by using the approach discussed in Section 3.1.1. It is common to use L_1-normalization on each column of U and V, or that the columns of the resulting matrices Q and P each sum to 1. Interestingly, this type of normalization makes nonnegative factorization similar to a closely related factorization known as *Probabilistic Semantic Analysis (PLSA)*. The main difference between PLSA and nonnegative matrix factorization is that the former uses a maximum likelihood optimization function whereas nonnegative matrix factorization (typically) uses the Frobenius norm. However, some forms of nonnegative matrix factorization use the I-divergence objective, which has been shown to be identical to PLSA [156, 212, 315].

3.3.1 Interpretability of Nonnegative Matrix Factorization

An important property of nonnegative matrix factorization is that it is highly interpretable in terms of the clusters in the underlying data. The rth columns U_r and V_r of each of U and V respectively contain document- and word-membership information about the rth topic (or cluster) in the data. The n entries in U_r correspond to the nonnegative components (coordinates) of the n documents along the rth topic. If a document strongly belongs to topic r, then it will have a very positive coordinate in U_r. Otherwise, its coordinate will be zero or mildly positive (representing noise). Similarly, the rth column V_r of V provides the frequent vocabulary of the rth cluster. Terms that are highly related to a particular topic will have large components in V_r. The k-dimensional representation of each document is provided by the corresponding row of U. This approach allows a document to belong to multiple clusters, because a given row in U might have multiple positive coordinates. For example, if a document discusses both science and history, it will have components along latent components with science-related and history-related vocabularies. This provides a more realistic "sum-of-parts" decomposition of the corpus along various topics, which is primarily enabled by the nonnegativity of U and V. In fact, one can create a decomposition of the document-term matrix into k different rank-1 document-term matrices corresponding to the k topics captured by the decomposition. Let us treat U_r as an $n \times 1$ matrix and V_r as a $d \times 1$ matrix. If the rth component is related to science, then $U_r V_r^T$ is an $n \times d$ document-term matrix containing the science-related portion of the original corpus. Then the decomposition of the document-term matrix is defined as the sum of the following components:

$$D \approx \sum_{r=1}^{k} U_r V_r^T \tag{3.23}$$

This decomposition is analogous to the spectral decomposition of SVD, except that its nonnegativity often gives it much better correspondence to semantically related topics.

3.3.2 Example of Nonnegative Matrix Factorization

In order to illustrate the semantic interpretability of nonnegative matrix factorization, let us revisit the same example used in Section 3.2.1, and create a decomposition in terms of nonnegative matrix factorization:

$$D = \begin{pmatrix} & \text{lion} & \text{tiger} & \text{cheetah} & \text{jaguar} & \text{porsche} & \text{ferrari} \\ \text{Document-1} & 2 & 2 & 1 & 2 & 0 & 0 \\ \text{Document-2} & 2 & 3 & 3 & 3 & 0 & 0 \\ \text{Document-3} & 1 & 1 & 1 & 1 & 0 & 0 \\ \text{Document-4} & 2 & 2 & 2 & 3 & 1 & 1 \\ \text{Document-5} & 0 & 0 & 0 & 1 & 1 & 1 \\ \text{Document-6} & 0 & 0 & 0 & 2 & 1 & 2 \end{pmatrix}$$

This matrix represents topics related to both cars and cats. The first three documents are primarily related to cats, the fourth is related to both, and the last two are primarily related to cars. The word "jaguar" is polysemous because it could correspond to either a car or a cat and is present in documents of both topics.

A highly interpretable nonnegative factorization of rank-2 is shown in Figure 3.2(a). We have shown an approximate decomposition containing only integers for simplicity, although the optimal solution would (almost always) be dominated by floating point numbers in practice. It is clear that the first latent concept is related to cats and the second latent concept

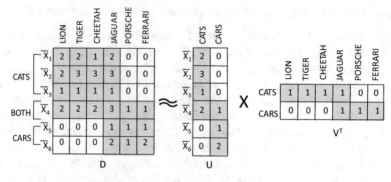

(a) Two-way factorization

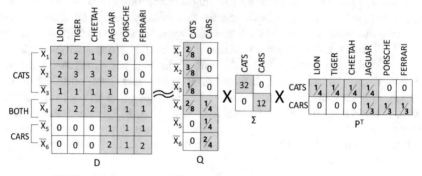

(b) Three-way factorization by applying L_1-normalization to (a) above

Figure 3.2: The highly interpretable decomposition of nonnegative matrix factorization

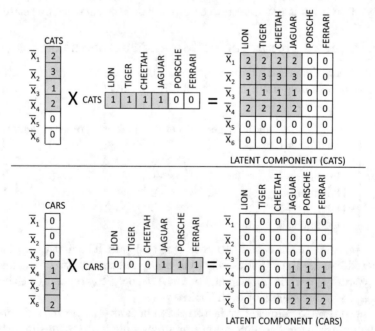

Figure 3.3: The highly interpretable "sum-of-parts" decomposition of the document-term matrix into rank-1 matrices representing different topics

is related to cars. Furthermore, documents are represented by two non-negative coordinates indicating their affinity to the two topics. Correspondingly, the first three documents have strong positive coordinates for cats, the fourth has strong positive coordinates in both, and the last two belong only to cars. The matrix V tells us that the vocabularies of the various topics are as follows:

> **Cats:** lion, tiger, cheetah, jaguar
> **Cars:** jaguar, porsche, ferrari

It is noteworthy that the polysemous word "jaguar" is included in the vocabulary of both topics, and its usage is automatically inferred from its context (i.e., other words in document) during the factorization process. This fact becomes especially evident when we decompose the original matrix into two rank-1 matrices according to Equation 3.23. This decomposition is shown in Figure 3.3 in which the rank-1 matrices for cats and cars are shown. It is particularly interesting that the occurrences of the polysemous word "jaguar" are nicely divided up into the two topics, which roughly correspond with their usage in these topics.

As discussed in Section 3.1.1, any two-way matrix factorization can be converted into a standardized three-way factorization. The three-way normalized representation is shown in Figure 3.2(b), and it tells us a little bit more about the relative frequencies of the two topics. Since the diagonal entry in Σ is 32 for cats in comparison with 12 for cars, it indicates that the topic of cats is more dominant than cars. This is consistent with the observation that more documents and terms in the collection are associated with cats as compared to cars.

3.3.3 Folding in New Documents

The process of *folding in* refers to the fact that one wants to represent the out-of-sample documents using the same basis system as the in-sample documents. It is not as easy as SVD to fold in new documents with nonnegative matrix factorization. Let D_t be a new $n_t \times d$ test data matrix with rows not included in the original matrix D. Let U_t be the $n_t \times k$ matrix containing the k-dimensional representations of the new documents. Since the basis has rank $k < d$, it is possible to determine only an approximate representation of the d-dimensional data matrix D_t in a k-dimensional basis. This can be achieved by minimizing the objective function $\|D_t - U_t V\|_F^2$ over *fixed* V and varying U_t. The matrix V is fixed because it was already estimated using the in-sample matrix D. This optimization problem can be decomposed into n_t least-squares regression problems for each of the n_t rows (documents) in U_t. As discussed in Section 6.2 of Chapter 6, the optimal solution is given by the following:

$$U_t = D_t V (V^T V)^{-1} \tag{3.24}$$

This approach can be used for any basis system, whether it is orthogonal or not. For orthonormal basis systems like SVD, we have $V^T V = I$, and therefore Equation 3.24 simplifies to $U_t = D_t V$. The main problem with this solution in the specific context of nonnegative matrix factorization is that U_t might have negative components. Nonnegativity can be forced only by fixing V after in-sample learning on D, and then learning U_t by performing the same gradient-descent updates (cf. Equation 3.19) on D_t. Note that V is not updated using out-of-sample data. This process is, of course, not as simple as the straightforward fold-in of SVD using matrix multiplication.

3.3.4 Advantages and Disadvantages of Nonnegative Matrix Factorization

Nonnegative matrix factorization has several advantages and disadvantages:

1. Nonnegativity enables a highly interpretable decomposition because of ability to represent the factorization as a sum of parts.

2. The semantic clusters (or topics) are often captured more accurately by allowing non-orthogonality in the basis vectors. This is because semantic topics are often related.

3. Nonnegative matrix factorization can better address polysemy than SVD.

4. One disadvantage of nonnegative matrix factorization is that it is harder (than SVD) to compute the reduced representations of documents that were not included in the original data matrix D. SVD is able to fold in such documents more easily as a simple projection because of its orthogonal basis system.

The advantages and disadvantages of this approach are exactly shared by PLSA, because the latter is simply a different form of nonnegative matrix factorization.

3.4 Probabilistic Latent Semantic Analysis

Probabilistic latent semantic analysis creates a normalized three-way factorization of the document-term matrix of the following form:

$$D \propto Q \Sigma P^T \tag{3.25}$$

Here, Q is an $n \times k$ matrix, Σ is a $k \times k$ diagonal matrix, and P is a $d \times k$ matrix. Furthermore, each of the columns of Q and P sum to 1, the entries in Σ sum to 1, and the individual entries are interpreted as probabilities. The use of proportionality (instead of equality) in Equation 3.25 is necessitated by the strict probability-centric scaling of Q, P, and Σ, although scaling down the entries of D to sum to 1 yields an equality relationship. The matrices P, Q, and Σ define the parameters of a generative process that is used to create the observed matrix D. These parameters are learned in order to maximize the likelihood of the observed data for this generative process. What is this generative process?

The basic idea is to assume that the frequencies in the document-term matrix are generated by a mixture of latent components $\mathcal{G}_1 \ldots \mathcal{G}_k$ sequentially incrementing entries of the document-term matrix. These mixture components are *hidden variables*, also known as *latent variables*, because they are not observed in the data, but have an explanatory role in modeling the data. A mixture component is also referred to as an *aspect* or *topic*, which leads to it being considered a *topic modeling* method. Therefore, if a given mixture component is selected, it is likely to increment topic-relevant entries. As we will see later, the number of mixture components k defines the rank of the factorization. The basic generative process may be described in terms of repeatedly selecting a position from the document-term matrix and incrementing its frequency:

1. Select a mixture component (topic) $\mathcal{G}_r$ with probability Σ_{rr}, where $r \in \{1 \ldots k\}$.

2. Select the index i of a document $\overline{X}_i$ with probability $Q_{ir} = P(\overline{X}_i | \mathcal{G}_r)$ and the index j of a term t_j with probability $P_{jr} = P(t_j | \mathcal{G}_r)$. It is assumed that the two selections are conditionally independent. Increment the (i, j)th entry of D by 1.

The generative process of incrementing matrix entries will need to be repeated as many times as the number of *tokens* in the corpus (including document-specific repetitions of term occurrences). A *plate diagram* (see explanation in Figure 3.4(a)) of this *symmetric* generative process is described in Figure 3.4(b).

One must formulate an optimization problem that maximizes the *log-likelihood* of the document-term matrix being generated by this model. In other words, the optimization problem for PLSA may be stated as follows:

Maximize$_{(P,Q,\Sigma)}$ [Log likelihood of generating D using parameters in matrices (P,Q,Σ)]

$$= \log \left(\prod_{i,j} P(\text{Adding one occurrence of term } j \text{ in document } i)^{D_{ij}} \right)$$

$$= \sum_{i=1}^{n} \sum_{j=1}^{d} D_{ij} \underbrace{\log \left(P(\overline{X_i}, t_j) \right)}_{\text{Parametrized by } P,Q,\Sigma}$$

subject to:

$P, Q, \Sigma \geq 0$

Entries in each column of P sum to 1

Entries in each column of Q sum to 1

Σ is a diagonal matrix that sums to 1

A key point here is that the entries in P, Q, and Σ are interpreted as probabilities and the generative process creates the observed matrix D on this basis. This is the reason for the normalization constraints on P, Q, and Σ.

The conditional probability $P(\overline{X_i}, t_j | \mathcal{G}_r)$ of selecting a particular document-term pair $(\overline{X_i}, t_j)$ in the generative process follows the conditional independence assumption:

$$P(\overline{X_i}, t_j | \mathcal{G}_r) = P(\overline{X_i} | \mathcal{G}_r) \cdot P(t_j | \mathcal{G}_r) \tag{3.26}$$

The main challenge in solving this optimization problem is that we do not know which mixture component generated which token. The problem would have been easy to solve, had there been only one mixture component (i.e., $k = 1$). Therefore, we need to *simultaneously* compute the mixture memberships and optimization parameters. This is achieved by using the *expectation-maximization (EM)* algorithm, which optimizes parameters and probabilistic assignments alternately in iterative fashion. The algorithm starts with random nonnegative parameters in Q, Σ, and P, which are normalized[4] so that they can be interpreted as probabilities. In the E-step, we compute the posterior probability $P(\mathcal{G}_r | \overline{X_i}, t_j)$ that each observed document-term pair $(\overline{X_i}, t_j)$ (i.e., token) was generated by a particular mixture component. Therefore, the E-step determines memberships in *expectation*. These probabilities are treated as "membership weights" of that token for the various mixture components. The M-step uses these membership weights to compute the maximum-likelihood values of all parameters in each mixture component. The M-step is referred to as the *maximization* step, because it is really solving a simplified optimization problem in which the membership weights of the tokens for various mixture components have been fixed. The specific details of the E- and M-steps are as follows:

---

[4]In other words, the columns of P, the columns of Q, and the diagonal of Σ each sum to 1.

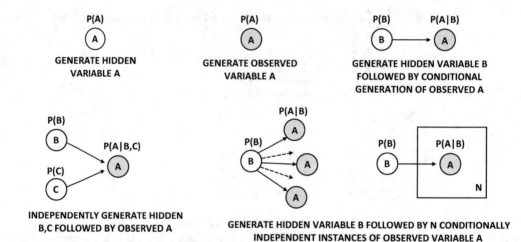

(a) Examples of plate diagrams showing generative dependencies

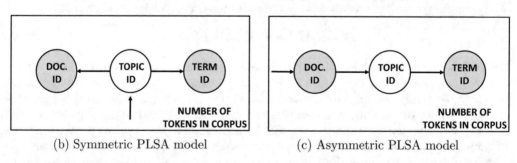

(b) Symmetric PLSA model   (c) Asymmetric PLSA model

Figure 3.4: Examples of plate diagrams and two equivalent generative models for PLSA

1. **(E-step):** Estimate the posterior probabilities $P(\mathcal{G}_r|\overline{X}_i, t_j)$ for each document-term pair $(\overline{X}_i, t_j)$ occurring in the corpus. The Bayes rule is used with the current state of the parameters:

$$P(\mathcal{G}_r|\overline{X}_i, t_j) = \frac{P(\mathcal{G}_r) \cdot P(\overline{X}_i|\mathcal{G}_r) \cdot P(t_j|\mathcal{G}_r)}{\sum_{s=1}^{k} P(\mathcal{G}_s) \cdot P(\overline{X}_i|\mathcal{G}_s) \cdot P(t_j|\mathcal{G}_s)} = \frac{(\Sigma_{rr}) \cdot (Q_{ir}) \cdot (P_{jr})}{\sum_{s=1}^{k}(\Sigma_{ss}) \cdot (Q_{is}) \cdot (P_{js})} \quad \forall i, j, r \tag{3.27}$$

2. **(M-step):** Estimate the current parameters in Q, P and Σ by using the conditional probabilities in the first step as weights for entries belonging to each generative component. This is achieved as follows:

$$Q_{ir} = P(\overline{X}_i|\mathcal{G}_r) = \frac{\sum_j P(\overline{X}_i, t_j) \cdot P(\mathcal{G}_r|\overline{X}_i, t_j)}{P(\mathcal{G}_r)} \propto \sum_j D_{ij} P(\mathcal{G}_r|\overline{X}_i, t_j) \quad \forall i, r$$

$$P_{jr} = P(t_j|\mathcal{G}_r) = \frac{\sum_i P(\overline{X}_i, t_j) \cdot P(\mathcal{G}_r|\overline{X}_i, t_j)}{P(\mathcal{G}_r)} \propto \sum_i D_{ij} P(\mathcal{G}_r|\overline{X}_i, t_j) \quad \forall j, r$$

$$\Sigma_{rr} = P(\mathcal{G}_r) = \sum_{i,j} P(\overline{X}_i, t_j) \cdot P(\mathcal{G}_r|\overline{X}_i, t_j) \propto \sum_{i,j} D_{ij} P(\mathcal{G}_r|\overline{X}_i, t_j) \quad \forall r$$

The constants of proportionality are set by ensuring that the probabilities in the columns of P, Q and the diagonal of Σ each sum to 1.

As in all applications of the expectation-maximization algorithm, these steps are iterated to convergence. Convergence can be checked by computing the likelihood function at the end of each iteration, and checking if it has improved by a minimum amount over its average value in the last few iterations.

Why can we express the estimated parameters in the factorized form of $D \propto Q\Sigma P^T$? The reasoning for this follows directly from the probabilistic interpretation of the parameters:

$$D_{ij} \propto P(\overline{X}_i, t_j) = \sum_{r=1}^{k} \underbrace{P(\mathcal{G}_r)}_{\text{Select } r} \cdot \underbrace{P(\overline{X}_i, t_j|\mathcal{G}_r)}_{\text{Select } \overline{X}_i, t_j} \quad \text{[Generative probability of incrementing } (i,j)\text{]}$$

$$= \sum_{r=1}^{k} P(\mathcal{G}_r) \cdot P(\overline{X}_i|\mathcal{G}_r) \cdot P(t_j|\mathcal{G}_r) \quad \text{[Conditional independence]}$$

$$= \sum_{r=1}^{k} P(\overline{X}_i|\mathcal{G}_r) \cdot P(\mathcal{G}_r) \cdot P(t_j|\mathcal{G}_r) \quad \text{[Rearranging product]}$$

$$= \sum_{r=1}^{k} Q_{ir} \cdot \Sigma_{rr} \cdot P_{jr} = (Q\Sigma P^T)_{ij} \quad \text{[The factorized form we are familiar with]}$$

PLSA is very similar to nonnegative matrix factorization except that we are optimizing a maximum likelihood model (equivalent to I-divergence objective in non-negative matrix factorization) rather than the Frobenius norm.

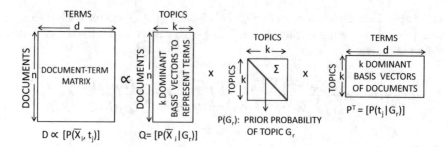

Figure 3.5: The decomposition of PLSA is similar to that of SVD (see Figure 3.1) except that the basis vectors are nonorthogonal and have a probabilistic interpretation

3.4.1   Connections with Nonnegative Matrix Factorization

The original paper on nonnegative matrix factorization [315] proposed an alternative formulation that uses an I-divergence objective rather than the Frobenius norm:

$$\text{Minimize }_{U,V} \sum_{i=1}^{n} \sum_{j=1}^{d} \left(D_{ij} \log \left\{ \frac{D_{ij}}{(UV^T)_{ij}} \right\} - D_{ij} + (UV^T)_{ij} \right)$$

$$\text{subject to:}$$

$$U \geq 0,\ V \geq 0$$

This formulation is *identical to PLSA* and requires the following iterative solution for $U = [u_{is}]$ and $V = [v_{js}]$:

$$u_{is} \Leftarrow u_{is} \frac{\sum_{j=1}^{d}[D_{ij}v_{js}/(UV^T)_{ij}]}{\sum_{j=1}^{d} v_{js}} \quad \forall i, s; \qquad v_{js} \Leftarrow v_{js} \frac{\sum_{i=1}^{n}[D_{ij}u_{is}/(UV^T)_{ij}]}{\sum_{i=1}^{n} u_{is}} \quad \forall j, s$$

The two-way factorization can be converted into a normalized three-way factorization like PLSA using the normalization approach discussed in Section 3.1.1. The aforementioned gradient-descent steps provide an alternative way of solving PLSA. Since a different computational algorithm is used, the resulting solution may not exactly be the same as that obtained with the expectation-maximization method. However, the quality of the solutions will be quite similar in the two cases since the same objective function is used.

3.4.2   Comparison with SVD

The three-way factorization of PLSA is shown in Figure 3.5, and it is similar to the corresponding factorization of SVD (cf. Figure 3.1). However, unlike SVD, the basis vectors in P and Q are not mutually orthogonal but have a probabilistic interpretation. Just as the matrix Σ in SVD contains the singular values indicating dominance of different latent concepts, the matrix Σ in PLSA contains the prior probabilities. As in SVD, the matrix Q provides a reduced representation of the documents and the matrix P provides the reduced representations of the words. The decomposition is highly interpretable. Highly positive entries in each column of P provide the lexicon for a specific topic, whereas highly positive entries in each row of P provide the most relevant topics for a particular word.

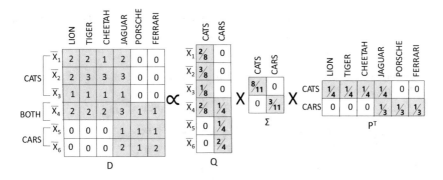

Figure 3.6: Example of PLSA (compare with Figure 3.2)

3.4.3 Example of PLSA

Let us revisit the same example used in Sections 3.2.1 and 3.3.2, respectively, to create a decomposition:

$$D = \begin{pmatrix} & \text{lion} & \text{tiger} & \text{cheetah} & \text{jaguar} & \text{porsche} & \text{ferrari} \\ \text{Document-1} & 2 & 2 & 1 & 2 & 0 & 0 \\ \text{Document-2} & 2 & 3 & 3 & 3 & 0 & 0 \\ \text{Document-3} & 1 & 1 & 1 & 1 & 0 & 0 \\ \text{Document-4} & 2 & 2 & 2 & 3 & 1 & 1 \\ \text{Document-5} & 0 & 0 & 0 & 1 & 1 & 1 \\ \text{Document-6} & 0 & 0 & 0 & 2 & 1 & 2 \end{pmatrix}$$

A possible factorization is shown in Figure 3.6. We have intentionally used the same factorization as Figure 3.2(b) to show the analogy, although they might be slightly different in practice because of the difference in objective functions. The main difference between Figures 3.2(b) and 3.6 is that the diagonal matrix Σ has been scaled down to a (prior) probability in the latter, and therefore the factorization is observed to within a constant of proportionality in PLSA. If we scale down the document-term matrix so that its entries sum to 1, then the factorization will be observed to approximate equality in PLSA.

3.4.4 Advantages and Disadvantages of PLSA

Since PLSA is a form of nonnegative matrix factorization, it inherits all the advantages and disadvantages discussed in Section 3.3.4. However, one can also view PLSA as a probabilistic model rather than a factorization model. From the probabilistic point of view, it has the following advantages and disadvantages:

1. The parameter estimation process is simple, intuitive and easy to understand. The parameters have multiple interpretations from a probabilistic or factorization point of view. This type of interpretability is often helpful to a practitioner.

2. The number of parameters estimated in PLSA grows linearly with the size of the collection, because the matrix Q has $O(n \cdot k)$ parameters. As a result, there is inability to take sufficient advantage of increasing corpus size. However, since it does not make any assumption on the distribution of topics in a document, it has the advantage of greater generality of modeling for large collections.

3. PLSA is not a fully generative model and it faces the same challenges as nonnegative matrix factorization in folding in new documents. Typically, one re-estimates $P(\mathcal{G}_r|\overline{X})$ for a new document $\overline{X}$.

Some of these challenges are addressed using a different model that is referred to as *Latent Dirichlet Allocation.*

3.5   A Bird's Eye View of Latent Dirichlet Allocation

We use this section to provide an understanding of the basic principles underlying LDA and also an understanding of its advantages and pitfalls over its cousin, PLSA, from the point of view of the practitioner. In the following, we first describe a simplified LDA model with a single Dirichlet assumption on the topic distribution of documents. Subsequently, we smooth the model with a second Dirichlet distribution on the term occurrences.

3.5.1   Simplified LDA Model

The parameter space increases proportionally with corpus size in PLSA, because the matrix Q contains $n \cdot k$ parameters and the matrix P contains $d \cdot k$ parameters. The matrix Q is particularly troublesome because it blows up the parameter space with increasing corpus size and we somehow need to find a way to get rid of it by changing the generative mechanism. Furthermore, it is not a fully generative model because new documents are difficult to fold in after parameter estimation (although heuristic fixes are possible).

A part of the problem is that PLSA tries to independently generate the different *tokens* of the document-term matrix rather than generating one *document* at a time (as is common with most mixture models in clustering). Latent Dirichlet allocation solves this problem by deciding the composition of topics in a document up front with the Dirichlet distribution, and then generating all the entries in a row of the document-term matrix in one shot. Therefore, a *prior* structure is imposed on each document with the Dirichlet distribution. Before discussing the generative process of LDA, we first discuss a slightly different *asymmetric* generative process of PLSA. This generative process is mathematical identical to the *symmetric* generative process of Section 3.4, but it is useful in relating PLSA to LDA. The asymmetric generative process of PLSA is as follows:

1. Select the ith document, $\overline{X}_i$, with probability $P(\overline{X}_i) = \sum_s P(\mathcal{G}_s)P(\overline{X}_i|\mathcal{G}_s) = \sum_s(\Sigma_{ss})(Q_{is})$.

2. Select the topic r with probability $P(\mathcal{G}_r|\overline{X}_i) = \frac{P(\mathcal{G}_r \cap \overline{X}_i)}{P(\overline{X}_i)} = \frac{(\Sigma_{rr})(Q_{ir})}{\sum_s(\Sigma_{ss})(Q_{is})}$.

3. Select the jth term, t_j, with probability $P(t_j|\mathcal{G}_r) = P_{jr}$.

Once the document-term pair has been selected, the corresponding entry in the document-term matrix is incremented by 1. The plate diagram for this asymmetric model is shown in Figure 3.4(c). This process increments *entries* of the document-term matrix. How can we generate the entire row (document) at a time? In order to do so, we need to make some kind of assumption on how the ith row of D is defined as a mixture of different topic distributions. This is achieved by using the *Dirichlet distribution* (with only k parameters) to implicitly *generate* $P(\mathcal{G}_r|\overline{X}_i)$ for the ith document. In a sense, we are imposing a Dirichlet *prior* on the topic distribution in order to generate the relative topic frequencies in a document. The relative topic frequencies in each document are different because they are defined by drawing a different instantiation of the k relative frequencies from the Dirichlet distribution. *Therefore, the document-specific parameters of the generative process are themselves generated by using another set of (compact) Dirichlet parameters.* This reduces the parameter space. Subsequently, all the terms in the ith document are generated. We still need the matrix $P_{jr} = P(t_j|\mathcal{G}_r)$ to decide the word distribution of different topics. Therefore, the fully generative process of LDA for the ith document is as follows:

1. Generate the number n_i of tokens (counting repetitions) in the ith document from a Poisson distribution.

2. Generate the relative frequencies $\overline{\Theta} = (\theta_1, \theta_2, \ldots, \theta_k)$ of different topics in the ith document from a Dirichlet[5] distribution. This step is like generating $\theta_r = P(\mathcal{G}_r|\overline{X_i})$ from the Dirichlet distribution for all topics r in the document in order to generate the document in one shot.

3. For each of the n_i tokens in the ith document, first select the rth latent component with probability $P(\mathcal{G}_r|\overline{X_i})$ and then generate the jth term with probability $P(t_j|\mathcal{G}_r)$. As in PLSA, we still need the $d \times k$ matrix of parameters P, which retain the same interpretation of containing the values $P(t_j|\mathcal{G}_r)$.

The plate diagram for the simplified LDA model is shown in Figure 3.7(a). This generative process requires only $O(d \cdot k + k)$ parameters, which reduces overfitting. Furthermore, the process is fully generative because of its document-at-a-time generative mechanism that is fully described by document-independent parameters. The probabilities of the terms of a new document being generated by any particular topic can therefore be estimated in a natural way, and can be used to create its reduced representation.

We need to use an order-k Dirichlet distribution to generate the k relative frequencies of the topics in each document in each sample. The *multivariate* probability density $f(x_1, \ldots, x_k)$ of the order-k Dirichlet distribution uses k positive *concentration* parameters denoted by $\alpha_1 \ldots \alpha_k$:

$$f(x_1 \ldots x_k) = \underbrace{\frac{\Gamma(\sum_{r=1}^{k} \alpha_r)}{\prod_{r=1}^{k} \Gamma(\alpha_r)} \prod_{r=1}^{k} (x_r)^{\alpha_r - 1}}_{\text{Multivariate density function in } k\text{-dimensional topic space}} \quad (3.28)$$

Here, Γ denotes[6] the Gamma function, which is the natural extension of the factorial function on integer numbers to the domain of real numbers. The Dirichlet distribution takes on positive probability densities only for *positive* variables x_i that sum to 1 (i.e., $\sum_{r=1}^{k} x_r = 1$). This is quite convenient because each generated tuple of k values can be interpreted as the probabilities of the k topics. Values of $0 < \alpha_i \ll 1$ lead to sparse outcomes of the random

---

[5]The Dirichlet is selected because it is the posterior distribution of *multinomial* parameters, if the prior distribution of these parameters is a Dirichlet (although the parameters of the prior and posterior Dirichlet may be different). If we throw a loaded dice repeatedly with its faces showing various topics, the resulting observations are referred to as multinomial. In LDA, the selection of the latent components of the different tokens in a document is achieved by throwing such a dice repeatedly. Formally, the Dirichlet distribution is a *conjugate prior* to the multinomial distribution. The use of conjugate priors is widespread in Bayesian statistics because of this property.

[6] For a positive integer n, the value of $\Gamma(n)$ is $(n-1)!$. For a positive real value x, the value of $\Gamma(x)$ is defined by interpolating the values at integer points with a smooth curve, which works out to an interpolated value of $\Gamma(x) = \int_0^\infty y^{x-1} e^{-y} dy$. More details of an exact definition and a specific functional form may be found at http://mathworld.wolfram.com/GammaFunction.html.

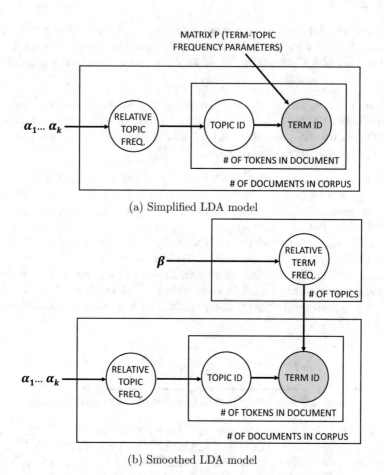

(a) Simplified LDA model

(b) Smoothed LDA model

Figure 3.7: Plate diagrams for simplified and smoothed LDA models

process in which only a small number of topics will have large probabilities. This type of sparsity is natural in real settings because a given document might contain only a couple of topics out of hundreds of topics. Furthermore, the relative presence of the ith topic will be proportional to α_i. Fixing each $\alpha_i = 1$ leads to a uniform and rather non-informative prior, which gives solutions similar to PLSA [217]. Therefore, learning appropriate values of these priors will lead to more natural models of higher quality. The topic-word parameters in matrix P need to be estimated in a data-driven manner like any other generative model. The prior parameters $\alpha_1 \dots \alpha_k$ can either be fixed up front, or they can be tuned/estimated in a data-driven manner. The default approach is to treat the prior parameters as inputs provided by the user. One can also view the priors in LDA as a clever form of regularization that reduces overfitting.

The process of parameter estimation in Latent Dirichlet Allocation is quite complex. The EM algorithm is used for parameter estimation (as in all generative models) along with techniques from variational inference for computing posterior probabilities in the E-step. In PLSA, it is a simple matter to compute the posterior probabilities in the E-step. In LDA, these posterior probabilities take on the form $P(\mathcal{G}_r, \overline{\Theta} | \alpha_1 \dots \alpha_k, t_j)$. This type of estimation is far more difficult, and it requires the use of methods from variational inference. Interested readers are referred to [61] for details. Several excellent off-the-shelf softwares are available for LDA, which are introduced in the software section of the bibliographic notes.

3.5.2 Smoothed LDA Model

Although the simplified LDA model reduces the number of parameters significantly, this can still be a problem when some of the terms are contained in only a small number of documents. When a new document contains a term that was not seen earlier, it would end up getting assigned zero probability in the simplified model. This is a common problem of sparsity in all types of probabilistic parameter estimations, and we will see several examples of this phenomenon in probabilistic classification/clustering models. A natural solution in such settings is to use *Laplacian smoothing*. The LDA model uses an additional Dirichlet distribution to perform the smoothing.

What does smoothing mean? Implicitly, smoothing is a prior assumption on the distribution of parameters in a mixture model to reduce the overfitting caused by sparsity. In this case, we have $O(d \cdot k)$ parameters of the form $P(t_j | \mathcal{G}_r)$, which we are treating as the $d \times k$ matrix P. It is therefore assumed that each of the k columns of P is an instantiation that is generated by the same order-d *exchangeable* Dirichlet distribution. Unlike the case of the Dirichlet distribution in Equation 3.28, which uses as many parameters as the order of the distribution, the exchangeable Dirichlet distribution uses a *single* parameter β to generate all the d-dimensions of the multivariate instantiation. The use of this special case of the Dirichlet distribution is important because using d parameters to describe the Dirichlet would defeat the purpose of smoothing in the first place. Therefore, the d terms in each topic are assumed to be generated according to the following order-d Dirichlet distribution, which is parameterized by a single value β:

$$f(x_1 \dots x_d) = \frac{\Gamma(d \cdot \beta)}{(\Gamma(\beta))^d} \prod_{j=1}^{d} (x_j)^{\beta - 1} \qquad (3.29)$$

$\underbrace{\qquad\qquad\qquad\qquad\qquad\qquad}$
Multivariate density function in d-dimensional term space

The generative process of the simplified LDA model is now modified in only one respect. As a very first step, before generating any of the documents, it is assumed that the d-dimensional columns of the $d \times k$ matrix $P = [P(t_j|\mathcal{G}_r)]$ are generated using the exchangeable Dirichlet distribution. After an *up front* generation of the matrix P, the individual documents are generated according to the same approach discussed in the previous section. The plate diagram for the smoothed LDA model is shown in Figure 3.7(b). It is noteworthy that the two Dirichlet distributions are used in somewhat different ways. The k parameters of the *asymmetric* topic-specific Dirichlet distribution (Equation 3.28) control the relative frequencies of various topics in documents as well as topic-specific smoothing effects, whereas the single parameter of the *symmetric* term-specific Dirichlet distribution (Equation 3.29) only controls the term-specific smoothing. Since the second Dirichlet distribution uses only a single parameter, all term-topic interactions are treated identically by it, and it does not regulate any detailed variabilities in term-topic distributions beyond smoothing. It is also possible [33] to use a single parameter, α, for the document-topic distributions, although this is not recommended for getting the most out of LDA [566]. It is noteworthy that many off-the-shelf software packages do use symmetric choices for both distributions as the default setting. In such cases, the main purpose of LDA is to use values of $\alpha, \beta \ll 1$ to encourage individual documents to each have a small number of topics and the vocabulary of each topic to be compact (i.e., sparse outcomes).

The changes in the generative process also lead to some changes in the parameter estimation process. In particular, the inference procedures are changed to treat the entries of the matrix P as random variables that are endowed with a posterior distribution. Note that this type of approach is used commonly in many probabilistic algorithms that use Laplacian smoothing.

3.6   Nonlinear Transformations and Feature Engineering

SVD provides an interesting relationship between document-document similarity matrices and dimensionality reduction. As discussed in Section 3.2, one of the ways of *directly* generating a reduced representation $Q\Sigma$ of the $n \times d$ data matrix D is to extract the eigenvectors of the $n \times n$ *dot-product similarity matrix* DD^T without generating a basis representation P in the word space. The d-dimensional columns of P, which correspond to the basis representation, are usually obtained by diagonalizing the $d \times d$ matrix $D^T D$ instead of DD^T. Note that the matrix $S = DD^T$ contains all n^2 pairwise dot products between documents. We can generate an embedding from this similarity matrix by using SVD of D:

$$S = DD^T = (Q\Sigma P^T)(Q\Sigma P^T)^T = Q\Sigma \underbrace{(P^T P)}_{I} \Sigma Q^T = Q\Sigma^2 Q^T = (Q\Sigma)(Q\Sigma)^T \qquad (3.30)$$

The columns of matrix Q contain the eigenvectors of the similarity matrix S, and the diagonal matrix Σ contains the square-root of the eigenvalues of S. In other words, if we generate the $n \times n$ dot-product similarities $S = DD^T$ between the n documents of the corpus, then we can construct the reduced representation $Q\Sigma$ from its scaled eigenvectors. This approach is an unusual way of performing SVD, because we generally use the $d \times d$ matrix $D^T D$ to generate the basis matrix P and then derive the reduced representation by the projection $Q\Sigma \approx DP$. Although the alternative similarity-matrix approach to SVD is computationally challenging for large values of n, its advantage is that we no longer have to care about the basis representation P. By using similarity matrices, we are able to escape from the need to generate these non-existent basis representations in a case where we use something other than the dot product as the similarity. This general principle forms the motivating idea of *nonlinear* dimensionality reduction in which we replace the dot-product similarity matrix with a different and more cleverly chosen similarity matrix of possibly higher quality. This principle is so important that we highlight it below:

> The large eigenvectors of high-quality similarity matrices can be used to generate useful multidimensional representations of the corpus that encode the knowledge inside the similarity matrix.

The basic assumption is that the similarity matrix represents the dot products $\Phi(\overline{X_i}) \cdot \Phi(\overline{X_j})$ in some (unknown) transformation $\Phi(\cdot)$ of the data that is more informative for particular data mining applications. We want to find this transformed representation. Virtually all nonlinear dimensionality reduction methods such as spectral methods [360], kernel SVD [502], and ISOMAP [549] use this broad approach to generate the reduced representations. Such dimensionality "reductions" are implicit transformations of the original data representation because a linear basis system for this new representation no longer exists in the original input space. Therefore, they are also referred to as *embeddings*. In fact, in some cases, embeddings do not *reduce* the input dimensionality at all because the final transformed and reduced representation might have a higher dimensionality than the original input space. Note that more than d eigenvectors of the $n \times n$ similarity matrix S can have nonzero eigenvalues; it is only in the case of the dot-product similarity that we are guaranteed at most d nonzero eigenvalues. Furthermore, if the similarity function encodes the details of a highly complex distribution, it is possible for more than d eigenvalues to be sufficiently large so that they cannot be dropped. The goal of the embedding in such cases is often to leverage a *better* similarity function (than the dot product) and obtain a more *expressive* feature representation of the complex data distribution than the overly simple dot product will provide. In a sense, nonlinear dimensionality reduction is inherently an exercise in unsupervised *feature engineering*.

The power of such embeddings is significantly greater than linear SVD and matrix factorization methods that are married to the original input space. For example, consider a setting in which we have three clusters of related topics corresponding to *Arts*, *Crafts*, and *Music*. A conceptual rendering of these clusters in two dimensions is shown in Figure 3.8. It is evident that dot-product similarity (which is similar to using the Euclidean distance) will have a hard time distinguishing between different clusters because of the non-convex shapes of the clusters. Euclidean distances and dot products implicitly favor spherical clusters. If we applied any simple clustering algorithm like k-means on the representation, it would not work well because such clustering algorithms are biased towards discovering spherical clusters. Similarly, a supervised learning method that uses linear separators to distinguish between the classes would perform very poorly.

Now imagine that we could somehow define a similarity matrix in which most of the similarities between documents of different topics are close to zero, whereas most of the similarities between documents of the same topic are nearly 1s. This similarity matrix S is shown in Figure 3.8 with a natural block structure. What type of embedding U will yield the factorization $S \approx UU^T$? First let us consider the absolutely perfect similarity function in which the entries in all the shaded blocks are 1s and all the entries outside shaded blocks are 0s. In such a case, it can be shown (after ignoring zero eigenvalues) that every document in *Arts* will receive an embedding of $(1, 0, 0)$, every document in *Music* will receive an embedding of $(0, 1, 0)$, and every document in *Crafts* will receive an embedding of $(0, 0, 1)$. Of course, in practice, we will never have a precise block structure of 1s and 0s, and there will be significant noise/finer trends within the block structure. These variations will be

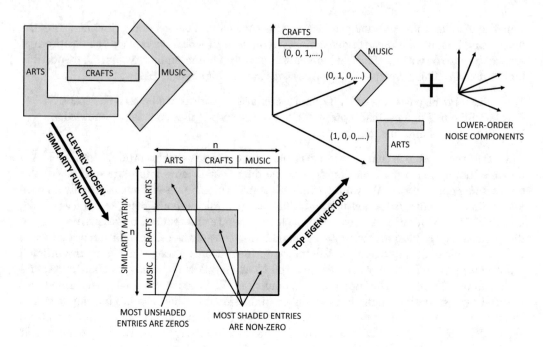

Figure 3.8: Explaining the rationale for nonlinear dimensionality reduction

captured by the lower-order eigenvectors shown in Figure 3.8. Even with these additional noise dimensions, this new representation will perform much better with many learning algorithms for clustering and classification. Where is the magic? The key idea here is that dot product similarities are sometimes not very good at capturing the *detailed* structure of the data, which other similarity functions with sharper locality-centric variations can sometimes capture. In a later section, we will also provide an intuitive illustration of how distance-exponentiated similarity functions can sometimes capture more detailed trends because of the sharper drop off in similarity values with distances.

One can even use this approach to work with richer representations of text than the multidimensional representation, without losing the convenience of a multidimensional representation. For example, imagine a setting in which we want a multidimensional embedding that preserves information about the ordering of words in documents. For example, consider the following pair of sentences:

The cat chased the mouse.
The mouse chased the cat.

From a semantic point of view, the second sentence is very different from the first, but this fact is not reflected in the bag-of-words representation. Only the sequence representation can distinguish between these two sentences. However, it is more challenging to design data mining algorithms with sequence representations because of the implicit constraints between data items (i.e., sequential ordering of tokens). A multidimensional embedding has the advantage that one does not have to worry about constraints between the individual dimensions while designing algorithms. Furthermore, the simplest and most generic setting for off-the-shelf machine learning and data mining algorithms is multidimensional data.

In such a case, we can use sequence-based similarity functions to generate the similarity matrix S. Such a similarity matrix will encode the fact that the two sentences above are different. The large eigenvectors of S will therefore also encode information about the ordering of words. As a result, mining algorithms that use this embedding will also be able to distinguish between documents based on the ordering of the words without losing the convenience of working with a multidimensional representation. The power of the embedding is limited only by how clever we can be in designing a good similarity function. As we will see in Chapter 10, there are other methods like neural networks to perform feature engineering, which have powerful applications like machine translation and image captioning.

Are there any restrictions on the types of similarity matrices one can use? As you might have noticed, it is necessary to diagonalize S with nonnegative eigenvalues:

$$S = Q\Sigma^2 Q^T = Q \underbrace{\Delta}_{\geq 0} Q^T \tag{3.31}$$

Since $\Delta = \Sigma^2$ contains only nonnegative eigenvalues, it implies that the similarity matrix must be positive semi-definite. As a practical matter, however, one can make any similarity matrix positive semi-definite by adding a sufficient amount $\lambda > 0$ to each diagonal entry.

$$S + \lambda I = Q(\Delta + \lambda I)Q^T \tag{3.32}$$

For large enough λ, the entries of the diagonal matrix $\Delta + \lambda I$ will be nonnegative as well. Therefore, $S + \lambda I$ will be positive semi-definite. Note that we are only perturbing the (less important) self-similarity values on the diagonal to achieve this goal, and all the other (critical) pairwise similarity information is preserved.

3.6.1 Choosing a Similarity Function

The choice of a proper similarity function is critical in generating an insightful embedding. In the following, commonly used similarity functions are reviewed along with their suitability to the text domain.

3.6.1.1 Traditional Kernel Similarity Functions

Traditional kernel similarity functions are positive semi-definite similarity functions that (typically) improve the performance of data mining applications by implicitly transforming the data to a higher-dimensional space before applying SVD on it. A kernel function is denoted by $K(\overline{X_i}, \overline{X_j})$ indicating the similarity between the multidimensional vectors $\overline{X_i}$ and $\overline{X_j}$. We list the commonly used kernel functions in the table below:

Function	Form
Linear Kernel	$K(\overline{X_i}, \overline{X_j}) = \overline{X_i} \cdot \overline{X_j}$ (Defaults to SVD)
Gaussian Radial Basis Kernel	$K(\overline{X_i}, \overline{X_j}) = e^{-\|\overline{X_i} - \overline{X_j}\|^2/(2 \cdot \sigma^2)}$
Polynomial Kernel	$K(\overline{X_i}, \overline{X_j}) = (\overline{X_i} \cdot \overline{X_j} + c)^h$
Sigmoid Kernel	$K(\overline{X_i}, \overline{X_j}) = \tanh(\kappa \overline{X_i} \cdot \overline{X_j} - \delta)$

Many of these kernel function have parameters associated with them, which have a critical effect on the type of feature transformation and reduction achieved by the approach.

With some commonly used kernels like the Gaussian kernel, each dimension often represents a small, densely populated locality of the input space. For example, consider a situation in which the *bandwidth* parameter σ of the Gaussian kernel is relatively small. In such a situation, two points that are located at a distance more than $4 \cdot \sigma$ will have a similarity value

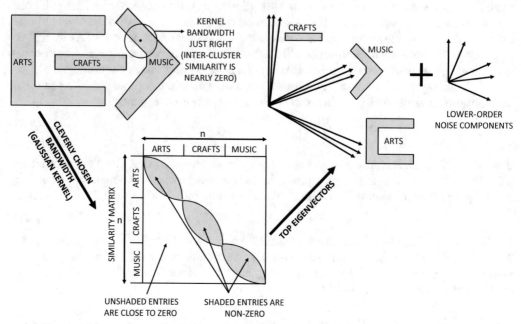

(a) Bandwidth much smaller than inter-cluster distance but not inter-point distance

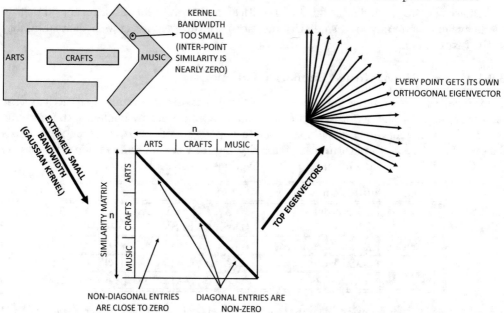

(b) Bandwidth much smaller than inter-point and inter-cluster distance

Figure 3.9: The effects of using different bandwidths of the Gaussian kernel

of virtually 0. Therefore, if σ is chosen so that the similarities between points of different clusters are close to zero, but a sufficient number of pairwise similarities within each cluster is nonzero, then each cluster will dominate a subset of the features of the embedding. This situation is shown in Figure 3.9(a) in which pairwise similarities do not have a precise block structure like Figure 3.8. Nevertheless. most of the non-zero entries reside inside the block structure with some residual variations. As a result, the embedding created by the similarity matrix of Figure 3.9(a) will be at least somewhat similar to that of Figure 3.8. The main difference is that each cluster will be represented by multiple eigenvectors, and the lower-order features will capture the residual variations from the block structure. However, the overall embedding will still be quite useful. The expression of each cluster with a subset of eigenvectors can sometimes *expand* the dimensionality of the transformed representation for low-dimensional input data even after dropping low-eigenvalue features (although text usually does not fall in this category). The basic idea here is that the (highly sensitive) distance-exponentiated similarity function creates a transformed representation that is able to better able to capture key *local* characteristics of the data distribution *within smaller subsets of dimensions* than the original representation (which locks up this information in data localities of complex shapes). Such an unlocking also makes the different dense regions of the space more clearly separable. For example, if we apply a simple clustering algorithm like k-means on the new representation in Figure 3.9(a), it will be able to nicely separate out the different high-level topics. This advantage comes at the price that the dimensionality of the transformed representation typically expands to accommodate detailed local information in individual features. For a corpus with n documents, the dimensionality of the transformed data can be as large as n when all eigenvectors are significantly nonzero. For example, if we reduce the bandwidth σ too much, each n-dimensional point will have a positive coordinate in only one dimension that is different from the dimensions chosen by any of the other $(n-1)$ points. This situation is shown in Figure 3.9(b). Such a transformation is completely useless and is a manifestation of overfitting. Therefore, the choice of the parameters of the kernel is crucial. Kernel transformations like the Gaussian radial basis function (RBF) are particularly useful in supervised settings like classification in which one can measure the algorithm performance on the labeled data to tune parameters like the bandwidth. For unsupervised settings, the rule of thumb is to set the kernel bandwidth to the median pairwise distance between points, although the exact value also depends on the data distribution and size. The bandwidth should be set to larger values for small data sets and smaller values for large data sets.

In the high-dimensional domain of text data, kernel functions like the Gaussian work poorly unless one chooses large values of σ. Text data has too many irrelevant features (terms), as a result of which the Gaussian similarity computations are noisy. Note that the irrelevant input features will be included in the exponent of the Gaussian kernel, and will become so tightly integrated with all the different transformed features of the embedding that it becomes difficult to remove their detrimental effect on data mining applications with *feature selection* and *regularization* tricks (see Chapters 5 and 6). This effect is particularly pronounced at smaller values of the bandwidth σ. Using large values of σ is similar to using the linear kernel, and the additional accuracy gains[7] over the linear kernel in such cases are small compared to the significant increase in computational complexity. Although the

---

[7]There does not seem to be a clear consensus on this issue. For the classification problem, slightly better results have been claimed in [605] for the linear kernel. On the other hand, the work in [101] shows that slightly better results are obtained with the Gaussian kernel method with proper tuning. Theoretically, the latter claim seems to be a better justified because linear kernels can be roughly simulated by the Gaussian by using a large bandwidth.

Gaussian kernel is one of the most successful kernels with other types of multidimensional data, the linear kernel often provides almost equally accurate results in the specific case of text, while retaining its computational efficiency in most application-centric settings. Some (mildly) encouraging results have been shown [101] for text data with the use of the second-order polynomial kernel $K(\overline{X_i}, \overline{X_j}) = (\overline{X_i} \cdot \overline{X_j} + c)^2$. In general, the success of traditional multidimensional kernels has been quite limited in the text domain. The main use-case of nonlinear dimensionality reduction methods in the text domain occurs in cases in which one wants to use the positioning information between words rather than using the bag-of-words approach. These methods will be discussed in the following sections.

3.6.1.2   Generalizing Bag-of-Words to N-Grams

A *bag-of-words kernel* is the same as using linear SVD, because it uses the dot product on the tf-idf representation. However, it is possible to enrich this approach by adding the N-grams to the representation of the document. The N-grams represent the groups of N words, corresponding to the sequence of N consecutively occurring words in a document. In an N-gram, this sequence of N words is treated as an indivisible entity, and becomes a pseudo-term in its own right. It is possible to discover N-grams at the time of tokenizing a text collection by allowing up to $(N - 1)$ white spaces within a token. The N-grams can often discriminate to some extent between different semantic ideas that are defined on the same bag of words. For example, consider the following three short documents:

> Document-1: The cat chased a mouse.
> Document-2: The mouse chased a cat.
> Document-3: The cat chased a rat.

Clearly, the first and third documents convey similar ideas, which can be captured only by the sequence information. A bag-of-words kernel gives a perfect similarity score of 1 between the first and second documents, even though the second document sounds quite different compared to the first. However, when we examine[8] the 2-grams of the first document, we obtain "the cat," "cat chased," "chased a," and "a mouse." The majority of the 2-grams are different from those of the second sentence. On the other hand, the first and third documents will share many 2-grams in common, which is what we want. Adding 3-grams will further enrich the representation.

One could address this situation by simply adding N-grams to the feature representation and using linear SVD. However, in such a case, the enriched dimensionality might expand significantly enough to exceed the number of documents in the collection. In such a case, the complexity of SVD becomes prohibitive. Therefore, it is more efficient to first compute the similarity matrix on the extended representation and then extract the reduced representation directly. The maximum number of nonzero eigenvectors of such a matrix is equal to the number of documents, although one can also drop very small eigenvectors. Other than N-grams, one can also use *skip-grams*, which are generalizations of N-grams. Both N-grams and skip-grams are discussed in detail in Section 10.2 of Chapter 10.

3.6.1.3   String Subsequence Kernels

String subsequence kernels [353] generalize the notion of k-grams to allow gaps inside them. We can view them as k-subsequences in which all subsequences of length k in the document are considered. A decay parameter $\lambda < 1$ is used to weight the importance of gaps in the

---

[8]For simplicity, we are including stop words in the 2-grams.

subsequence. If the first and last words are r units apart, the weight of that k-subsequence is $\lambda^r < 1$. For example, consider the following sentence:

The hungry lion ran after the rabbit, who was too clever for the lion.

Just like k-grams, we can extract k-subsequences and add them to the representation, but with appropriate weights. In this case, the weight of "the hungry" is λ, whereas that of "the lion" is $\lambda^2 + \lambda$. Note that "the lion" has two occurrences with different gaps, which accounts for the two terms in its weight. Let the document containing the single sentence above be denoted by A. Consider a different document B that has a weight of λ^2 for "the lion" in its subsequence-based representation. Then, the kernel similarity between A and B contributed by this particular subsequence will be $\lambda^2(\lambda^2 + \lambda) = \lambda^4 + \lambda^3$. This contribution is aggregated over all the subsequences in the pair of documents at hand to create an *unnormalized* kernel similarity value. The unnormalized kernel similarity is divided by the geometric mean of the self-similarities (computed in the same manner) of the pairs of documents being concerned. This type of normalization is similar to the cosine similarity, and it yields a similarity value in $(0,1)$.

As in the case of N-grams, *explicit* feature engineering can cause an explosion in the dimensionality of the representation. In fact, in the case of subsequence-based representations, the problem is so much more severe than in the case N-grams, that it is impractical to create an engineered representation even as an intermediate step for similarity computation. In the case of k-subsequences, the interesting cases are those in which the value of k is at least 4. Interestingly, it can be shown that this type of similarity can be computed using a dynamic programming approach between the pair of documents without explicitly computing the engineered features. Before introducing this dynamic programming approach, we will formalize the definition of the engineered representation.

Let Σ represent the set of all d terms in the lexicon. Then, this type of feature engineering implicitly creates a representation over the feature space Σ^k, which has d^k possible values. Let $\overline{x} = x_1 x_2 \ldots x_m$ be a sequence corresponding to a sentence or a full document in which each x_i is a token from the lexicon. Let $\overline{u} = u_1 \ldots u_k \in \Sigma^k$ be a k-dimensional sequence of words. Note that each possible k-dimensional sequence $\overline{u}$ has a single dimension (and corresponding coordinate value) in the engineered representation. Then, $\Phi_{\overline{u}}(\overline{x})$ represents the coordinate value of the dimension corresponding to $\overline{u}$ in the engineered representation. The value of $\Phi_{\overline{u}}(\overline{x})$ is obtained by determining all occurrences of the subsequence $\overline{u}$ in $\overline{x}$ and adding the credit of this subsequence over these occurrences. The credit of a particular occurrence of this subsequence is λ^l, where $l \geq k$ is the length of the substring of $\overline{x}$ that matches $\overline{u}$ as a subsequence. Let $i(1) < i(2) < \ldots < i(k)$ represent the indices of the tokens in S so that $u_r = x_{i(r)}$.

$$\Phi_{\overline{u}}(\overline{x}) = \sum_{i(1) < i(2) < \ldots < i(k) : u_r = x_{i(r)}} \lambda^{i(k) - i(1) + 1} \qquad (3.33)$$

Note that if one were to compute the engineered representation explicitly, then one would have to compute $\Phi_{\overline{u}}(\overline{x})$ for each $\overline{u} \in \Sigma^k$. This is computationally infeasible even from a storage point of view. However, one can use this definition in order to define the kernel similarity between two sequences $\overline{x} = x_1 x_2 \ldots x_m$ and $\overline{y} = y_1 y_2 \ldots y_p$. Note that $\overline{x}$ and $\overline{y}$

need not be of the same length (i.e., $m \neq p$). The kernel similarity $K(\overline{x}, \overline{y})$ is computed as follows:

$$K(\overline{x}, \overline{y}) = \sum_{\overline{u} \in \Sigma^k} \Phi_{\overline{u}}(\overline{x}) \Phi_{\overline{u}}(\overline{y})$$

$$= \sum_{[\overline{u} \in \Sigma^k]} \sum_{[i(1)<...<i(k):u_r=x_{i(r)}]} \sum_{[j(1)<...<j(k):u_r=y_{j(r)}]} \lambda^{i(k)+j(k)-i(1)-j(1)+2}$$

One would also need to normalize the above with the geometric mean of $K(\overline{x}, \overline{x})$ and $K(\overline{y}, \overline{y})$ in order to map the similarity to a value in $(0, 1)$. Since these values can be computed in a similar way, we will focus only on the computation of $K(\overline{x}, \overline{y})$. The aforementioned summation has an exponential number of terms. Therefore, it would seem at first sight that one has gained nothing over explicit feature engineering by directly computing the kernel similarity rather than creating the engineered features. However, it turns out that this similarity function can be computed efficiently using dynamic programming. To aid a proper description of the dynamic programming computation, we subscript the kernel function with the length of the matching subsequence. In other words, let $K_h(\overline{x}, \overline{y})$ represents the kernel similarity between $\overline{x}$ and $\overline{y}$ using matching subsequences of length h. In order to compute the kernel similarity over subsequences of length k, our goal is to compute $K_k(\overline{x}, \overline{y})$. Therefore, we have:

$$K_h(\overline{x}, \overline{y}) = \sum_{\overline{u} \in \Sigma^h} \Phi_{\overline{u}}(\overline{x}) \Phi_{\overline{u}}(\overline{y})$$

$$= \sum_{[\overline{u} \in \Sigma^h]} \sum_{[i(1)<...<i(h):u_r=x_{i(r)}]} \sum_{[j(1)<...<j(h):u_r=y_{j(r)}]} \lambda^{i(h)+j(h)-i(1)-j(1)+2}$$

An additional function $K'_h(\overline{x}, \overline{y})$ is defined that aids the recursive computation of the kernel for all $h \in \{1, 2, \ldots, k-1\}$:

$$K'_h(\overline{x}, \overline{y}) = \sum_{[\overline{u} \in \Sigma^h]} \sum_{[i(1)<...<i(h):u_r=x_{i(r)}]} \sum_{[j(1)<...<j(h):u_r=y_{j(r)}]} \lambda^{m+p-i(1)-j(1)+2}$$

The main difference between $K_h(\overline{x}, \overline{y})$ and $K'_h(\overline{x}, \overline{y})$ is that $i(h)+j(h)$ is replaced with $m+p$ in the exponent of λ. In other words, the latter replaces the indices of the last matching elements of $\overline{x}$ and $\overline{y}$ with the lengths of the two strings.

To facilitate a more general discussion in which different types of matchings between the tokens of the two strings are allowed, we define a match function between a pair of tokens. In the simplest definition, the match function, $M(w, v)$, is 1 when w and v are the same, and 0, otherwise:

$$M(w, v) = \begin{cases} 1 & \text{if } w = v \\ 0 & \text{if } w \neq v \end{cases} \tag{3.34}$$

Although we have defined the match function in a rudimentary way here (to be consistent with our earlier definition of $\Phi(\cdot)$ [353]), it is possible to define more general match functions in which we have features associated with tokens (e.g., part-of-speech tag). In such cases, the match function can be defined to be the similarity between the corresponding features. Such methods are used in more complex applications like information extraction [79].

The dynamic programming approach uses recursive computation in which the kernel similarity function is computed for subsequences of increasing length h from 0 to k. The similarity functions over subsequences of length $(h-1)$ are helpful in computing the similarities over subsequences of length h. Let $\overline{x} \oplus v$ denote the sequence obtained by concatenating the token v at the end of sequence $\overline{x}$. The boundary initialization is as follows:

$$K_0'(\overline{x}, \overline{y}) = 1 \quad \forall \overline{x}, \overline{y}$$
$$K_h'(\overline{x}, \overline{y}) = K_h(\overline{x}, \overline{y}) = 0 \quad \text{[if either } \overline{x} \text{ or } \overline{y} \text{ has less than } h \text{ tokens]}$$

Let $\overline{y}_a^b$ denote the substring of $\overline{y}$ from position a to position b. The recursive computations based on this initialization are as follows:

$$K_h'(\overline{x} \oplus w, \overline{y}) = \lambda K_h'(\overline{x}, \overline{y}) + \underbrace{\sum_{j=2}^{l(\overline{y})} K_{h-1}'(\overline{x}, \overline{y}_1^{j-1}) \lambda^{l(\overline{y})-j+2} M(w, y_j)}_{\text{Denote by } K_h''(\overline{x} \oplus w, \overline{y})} \quad \forall h = 1, 2 \ldots k-1$$

$$K_k(\overline{x} \oplus w, \overline{y}) = K_k(\overline{x}, \overline{y}) + \sum_{j=2}^{l(\overline{y})} K_{k-1}'(\overline{x}, \overline{y}_1^{j-1}) \lambda^2 M(w, y_j)$$

Here, $l(\overline{y})$ denotes the number of tokens in $\overline{y}$. Furthermore, we have defined an additional notation $K_h''(\cdot, \cdot)$ in the equation above, which we will use later to improve the efficiency of this recursion. An immediate observation about this recursion is that the kernel similarity computation over subsequences of length k can also be easily used to compute all the similarities over subsequences of length $1 \ldots k-1$ as byproducts. Therefore, it is relatively easy to create a composite kernel over subsequences over all lengths up to k by adding them without much additional effort. This recursion requires $O(kmp^2)$ time.

3.6.1.4 Speeding Up the Recursion

One can reduce the running time further by defining an additional function $K_h''(\overline{x} \oplus w, \overline{y})$, which is one of the terms on the right-hand side of the above recursion:

$$K_h''(\overline{x} \oplus w, \overline{y}) = \sum_{j=2}^{l(\overline{y})} K_{h-1}'(\overline{x}, \overline{y}_1^{j-1}) \lambda^{l(\overline{y})-j+2} M(w, y_j) \tag{3.35}$$

By defining this function, one can modify the aforementioned recursive equations as follows:

$$K_h''(\overline{x} \oplus w, \overline{y} \oplus v) = \lambda K_h''(\overline{x} \oplus w, \overline{y}) + \lambda^2 K_{h-1}'(\overline{x}, \overline{y}) \cdot M(w, v) \quad \forall h = 1, 2 \ldots k-1$$
$$K_h'(\overline{x} \oplus w, \overline{y}) = \lambda K_h'(\overline{x}, \overline{y}) + K_h''(\overline{x} \oplus w, \overline{y}) \quad \forall h = 1, 2 \ldots k-1$$

$$K_k(\overline{x} \oplus w, \overline{y}) = K_k(\overline{x}, \overline{y}) + \sum_{j=2}^{l(\overline{y})} K_{k-1}'(\overline{x}, \overline{y}_1^{j-1}) \lambda^2 \cdot M(w, y_j)$$

This variant of the computation requires $O(kmp)$ time.

One nice characteristic of this kernel function is that it is possible to change the match function in order to incorporate complex linguistic features associated with tokens. For example, consider a situation in which each token in $\overline{x}$ and $\overline{y}$ is associated with discrete features like the token value itself, the part-of-speech, whether the token is an *entity* (cf. Chapter 13), and so on. In such a case, one can change $M(w, v)$ to be the number of features in which the discrete feature value is the same. In fact, such an approach is used in the *relation extraction* problem (cf. Section 13.3.3.2 of Chapter 13).

3.6.1.5   Language-Dependent Kernels

It is possible to encode the rules of the grammar of the specific language into the kernel function by using the notion of *probabilistic context free grammars*. A context-free grammar is a set of rules that encodes the rules of a specific language such as the following:

$$Sentence \rightarrow NounPhrase\ VerbPhrase$$
$$NounPhrase \rightarrow Determiner\ Noun$$
$$VerbPhrase \rightarrow Verb\ NounPhrase$$
$$Noun \rightarrow \text{``}lion\text{''}$$

Typically, thousands of rules may be required to encode a specific language. Given a sentence, it is possible to *parse* the sentence into a hierarchical tree-like structure with the above rules. This results in a *constituency-based parse tree*. Given two sentences, one can compute the similarity between their parse trees with the use of *convolution tree kernels*. Since the discussion of this kernel requires a deeper understanding of parse trees, it will be deferred to Section 13.3.3.3 of Chapter 13.

3.6.2   Nyström Approximation

One of the main problems with nonlinear dimensionality reduction is that the eigenvectors of an $n \times n$ similarity matrix need to be determined. The space requirement is $O(n^2)$ and the running time requirement is $O(n^3)$, which can be computationally prohibitive even for modestly large values of n such as $1,000,000$. A corpus containing $1,000,000$ documents is not considered extraordinarily large by modern standards.

It is possible to greatly speed up the dimensionality reduction process by subsampling the rows of the document-term matrix, and then approximating the reduced kernel representation with the Nyström technique [584]. The basic idea is to first estimate the reduced representation of the in-sample points and then fold-in the out-of-sample points on the learned embedding. Although this will lead to inaccuracy in the randomized approximation, the randomization can be turned into an advantage by using ensembles. The approach can be extremely effective in a predictive setting where the predictive learning is repeated multiple times on different subsamples, and the predictions are averaged in an ensemble-centric manner [10]. Repeated engineering of features with different samples actually improves the averaged results of a predictive modeling algorithm because of the ensemble-centric effect of *variance reduction* (see Section 7.2 of Chapter 7).

The first step is to sample a set of s rows from the corpus. The value of s is typically determined by computational and space constraints. However, it is generally dependent on the corpus distribution and is independent of the size of the corpus. In other words, one can view s as a constant, although it is usually a large one like 2000. The dimensionality k of the embedding (selected by the user) can be no larger than s. An in-sample similarity matrix S_{in} of size $s \times s$ is constructed in which the $(i, j)$the entry is the similarity between the ith and jth in-sample points. Similarly, an $n \times s$ similarity matrix S_a is constructed in which the (i, j)th entry is the similarity between the ith point with the jth *in-sample* point. Then, the following pair of steps is used to first generate the embeddings of the in-sample points and then generalize the in-sample embeddings to all points (including out-of-sample points):

- **(In-sample embedding):** Diagonalize $S_{in} = Q\Sigma^2 Q^T$. Retain the top-k eigenvectors to create the matrices Q_k and Σ_k. The resulting k-dimensional representation of the

s in-sample points is available in the rows of $Q_k \Sigma_k$. If there are fewer than k nonzero eigenvectors, then reduce the value of k to the number of nonzero eigenvectors. This step requires $O(s^2 \cdot k)$ time and $O(s^2)$ space. Since s is a constant, this step requires constant time and space.

- **(Universal embedding):** Let U_k denote the unknown $n \times k$ matrix containing the k-dimensional representation of the all n points in its rows. Although we already know the embeddings of the in-sample points, we will use the properties of the similarity matrix in transformed space to derive all rows in a uniform way. Since the dot products of the n points in U_k and in-sample points in $Q_k \Sigma_k$ are (approximately) contained in the matrix S_a, we have the following:

$$S_a \approx \underbrace{U_k (Q_k \Sigma_k)^T}_{\text{Transformed Dot Products}} \qquad (3.36)$$

By postmultiplying each side with $Q_k \Sigma_k^{-1}$ and using $Q_k^T Q_k = I$, we obtain the following:

$$U_k \approx S_a Q_k \Sigma_k^{-1} \qquad (3.37)$$

Therefore, we have an embedding of all n points in k-dimensional space. This step requires a simple matrix multiplication in time $O(n \cdot s \cdot k)$, which is linear in the size of the corpus.

It is noteworthy that the s in-sample rows in U_k are approximately the same as the s rows in $Q_k \Sigma_k$ but not quite the same because of the approximation inherent in the dimensionality reduction process. Therefore, it is preferable to use the in-sample rows from U_k (rather than $Q_k \Sigma_k$) so that out-of-sample and in-sample rows are approximated in a similar way.

This approach can even be used for *linear* SVD. In linear SVD, the conventional approach (see Section 3.2.2) is to use the $d \times d$ matrix $D^T D$ to discover the basis vectors, rather than using the similarity matrix DD^T to directly extract the embedding. However, the similarity matrices are quite small when we use subsampling. The reason is that the sample size s can be selected to around 20 times the *target* dimensionality of the reduced representation rather than the *input* dimensionality of the lexicon. The typical target dimensionality of the reduced representation in linear SVD for text is often of the order of 200. This means that we can work with a sample size of about 4000 in many cases. Text can have a dimensionality of a few hundred thousand words, which makes it costly to diagonalize the $d \times d$ matrix $D^T D$ in comparison with diagonalizing the 4000×4000 similarity matrix.

Note that the entire reduction requires *linear* time in the size of the corpus, and it will execute reasonably fast at sample sizes of the order of 4000 even for large collections. Typically, this type of dimensionality reduction is coupled with an ensemble-centric setting to make repeated predictions with different transformations and then averaging the results [10]. High-quality predictive results can be obtained with such methods in both supervised and unsupervised settings because of the ensemble-centric approach. In many cases, these results are not only more accurate but also more efficient in spite of the repeated executions of an ensemble-centric approach. This is because each ensemble-centric run is often several orders of magnitude faster than using a single run on a very large corpus, and averaging the results over 20 to 25 runs still retains a computational advantage. An example of its use in the unsupervised setting is provided in Section 4.8 of Chapter 4, and a discussion in the case of the supervised setting is provided at the end of Section 6.5.1 of Chapter 6.

3.6.3   Partial Availability of the Similarity Matrix

Nonlinear dimensionality reduction methods can be viewed as clever ways of converting high-quality similarity functions into engineered features that are friendly to learning algorithms. In many cases, such similarity functions are challenging to compute, which makes their availability limited on a *de facto* basis. For example, the string subsequence kernels require dynamic programming methods to compute similarities between pairs of strings. Such methods are computationally expensive. In such cases, it is not realistic to assume that the entire similarity matrix can be computed. For a corpus containing 10^6 documents, one cannot expect to compute 10^{12} pairwise similarities, which might require a few days. However, it is possible to learn the embedding from only a subset of the entries. In cases, where there is wide variation in the similarities across different parts of the matrix, it might make sense to spread out the similarity computations randomly over the similarity matrix S in order to learn as much as possible about the structure of the embedding. In other cases, a domain expert might provide pre-specified similarities between pairs of documents, and one has no control over which pairs were selected. In such cases, it is desired to engineer a multidimensional feature representation that leverages partial information about the similarity matrix. This is a more challenging setting than the Nyström approximation of the previous section, because *entries* of the similarity matrix have been subsampled, rather than specific *rows or columns*.

Let $S = [s_{ij}]$ be an $n \times n$ similarity matrix, in which only a subset O of entries are observed:

$$O = \{(i, j) : s_{ij} \text{ is observed}\} \tag{3.38}$$

One can assume that the matrix S is symmetric, and therefore the observed set of similarities O can be grouped into symmetric pairs of entries satisfying $s_{ij} = s_{ji}$. It is desired to learn an $n \times k$ embedding U for user-specified rank k, so that for any observed entry (i, j) the dot product of the ith row of U and the jth row of U is as close as possible to the (i, j)th entry, s_{ij}, of S. In other words, the value of $\|S - UU^T\|_F^2$ should be as small as possible for the observed entries in S. This problem can be formulated only over the observed entries in O as follows:

$$\text{Minimize } J = \sum_{(i,j)\in O} \left(s_{ij} - \sum_{p=1}^{k} u_{ip}u_{jp}\right)^2$$

This problem is similar to the determination of factors in recommendation problems, and is a natural candidate for gradient-descent methods. Let $e_{ij} = s_{ij} - \sum_{p=1}^{k} u_{ip}u_{jp}$ be the error of any observed entry (i, j) from set O at a particular value of the parameter matrix U. On computing the partial derivative with respect to u_{im}, one obtains the following:

$$\frac{\partial J}{\partial u_{im}} = 2 \sum_{j:(i,j)\in O} \left(s_{ij} + s_{ji} - 2 \cdot \sum_{p=1}^{k} u_{ip}u_{jp}\right)(-u_{jm}) \qquad \forall i \in \{1\ldots n\}, m \in \{1\ldots k\}$$

$$= 2 \sum_{j:(i,j)\in O} (e_{ij} + e_{ji})(-u_{jm}) \qquad \forall i \in \{1\ldots n\}, m \in \{1\ldots k\}$$

$$= -4 \sum_{j:(i,j)\in O} e_{ij}u_{jm} \qquad \forall i \in \{1\ldots n\}, m \in \{1\ldots k\}$$

Note that s_{ij} and s_{ji} are either both present or both absent from the observed entries because of the symmetric assumption. It is possible to express these partial derivatives in matrix form. Let $E = [e_{ij}]$ be an error matrix, in which (i, j)th entry is set to the error for any observed entry (i, j) in O, and 0, otherwise. When a small number of entries are observed, this matrix is a sparse matrix. It is not difficult to see that the entire $n \times k$ matrix of partial derivatives $\left[\frac{\partial J}{\partial u_{im}}\right]_{n \times k}$ is given by $-4EU$. This suggests that one should randomly initialize the matrix U of parameters, and use the following gradient-descent steps:

$$U \Leftarrow U + \alpha EU \qquad (3.39)$$

Here, $\alpha > 0$ is the step size, which one can follow through to convergence or another stopping criterion (discussed later). Note that the error matrix E is sparse, and therefore it makes sense to compute only those entries that are present in O before converting to a sparse data structure. To improve stability of the learner, a small amount of regularization can also be used.

$$U \Leftarrow U(1 - \lambda\alpha) + \alpha EU \qquad (3.40)$$

Here, $\lambda > 0$ is a small regularization parameter.

When working with a sparsely specified similarity matrix, it is possible to determine only the most dominant features accurately. In general, the use of any rank $k > |O|/n$ will cause overfitting. For example, if we have a corpus in which the number of specified similarities is 15 times the number of documents, we can realistically learn an embedding of (much) less than 15 dimensions. To determine the optimal rank k of the factorization, one can hold out a small subset $O_1 \subset O$ of the observed entries, which are not used for learning U. These entries are used to test the squared error $\sum_{(i,j)\in O_1} e_{ij}^2$ of the matrix U learned using various values of k. The value of k at which the error of the held out entries is minimized is used. Furthermore, one can also use the held out entries to determine the stopping criterion for the gradient-descent approach. The gradient-descent is terminated when the error on the held out entries begins to rise. The recovered matrix U provides a k-dimensional embedding of the data, which can be used in conjunction with machine learning algorithms.

3.7 Summary

Many forms of dimensionality reduction can be viewed as matrix factorization methods. Singular value decomposition, nonnegative matrix factorization and PLSA fall in the category of low-rank approximation methods. Singular value decomposition has the geometric advantage of orthogonal eigenvectors, which enables out-of-sample embeddings more effectively. It can also address the problem of synonymy well and that of polysemy to a limited extent. On the other hand, it is not semantically interpretable. Nonnegative matrix factorization and PLSA, which are almost equivalent, are semantically interpretable and can handle both synonymy and polysemy very well. On the other hand, they cannot fold-in out-of-sample documents. Latent Dirichlet Allocation is a generalization of PLSA that uses a Dirichlet prior on the topic distribution of documents in order to create a fully generative model that can fold-in new documents quite as effectively.

Nonlinear dimensionality reduction methods can be viewed as generalizations of SVD that use similarity functions other than the dot product to embed the points in a transformed space. By choosing the right type of similarity function, one can often engineer more expressive features such as those that incorporate sequential word ordering information in the documents. In this sense, nonlinear dimensionality reduction is often an exercise in feature engineering. Although nonlinear dimensionality reduction methods are computationally inefficient, one can often speed them up with the use of subsampling methods.

3.8   Bibliographic Notes

Singular value decomposition has been in use in various forms since the 1800s, although some of the key proofs of the underlying results are contained in the seminal work of Eckart and Young [171]. The linear algebra book by Strang [530] is an excellent resource on the topic. The effectiveness of SVD in removing noise from high-dimensional similarity search was discussed in [6]. In the text domain, singular value decomposition is referred to as Latent Semantic Analysis (LSA). The use of LSA in text data was pioneered in the work by Deerwester et al. [170]. Subsequent experiments on TREC data sets were reported by Dumais [166, 167].

Nonnegative matrix factorization was proposed in [315]. Projected gradient-descent methods for nonnegative matrix factorization are proposed in [335]. An excellent exposition on the interpretability of nonnegative matrix factorization is provided in [316]. There are several generalizations of nonnegative matrix factorization, such as the use of orthogonal factors [157], semi-nonnegativity [155], and convexity [155]. Probabilistic latent semantic analysis is discussed in [255, 256]. The relationship of PLSA to nonnegative matrix factorization was shown in [156, 212, 315]. Latent Dirichlet Allocation was proposed independently in the fields of population genetics [447] and text mining [61]. The approach was also generalized to the dynamic setting [62]. Detailed evaluations of Latent Dirichlet Allocation may be found in [33, 575]. The work in [33] investigates the effect of hyper-parameters, when two symmetric Dirichlet distributions are used to model the document-topic and topic-term distributions. The work in [566] provides insights on the effects of using either a symmetric or asymmetric Dirichlet distribution for document-topic and topic-term distributions. An edited book on text mining [15] contains a dedicated chapter on dimensionality reduction and topic modeling techniques. A review of probabilistic topic models may be found in [59].

Nonlinear dimensionality reduction methods have a rich history in multidimensional data and include methods like Kernel PCA [502], ISOMAP [549], Local Linear Embedding (LLE) [482] and spectral clustering [360]. The Nyström technique for kernel dimensionality reduction was proposed in [584]. Local linear embedding has been used to learn semantic representations of words in text [482]. In recent years, neural networks and autoencoders have also seen an increased amount of interest for nonlinear dimensionality reduction [248]. The *word2vec* [389] and *doc2vec* [314] techniques are specific neural network-based embedding methods that retain the linguistic context of words in the embedding. In text applications, structured kernels are very useful when text is interpreted as a sequence. Details of the dynamic programming algorithm for string subsequence kernels may be found in [353]. A survey of structured kernels may be found in [207]. Similarity measures for short segments of text are discussed in [383]. A Web-based kernel similarity function was studied in [484], in which queries to a search engine are used to evaluate the similarities between short text snippets. The Nyström method was proposed in [584], and its use in the ensemble-centric setting is advocated in [10].

3.8.1   Software Resources

An R package for LSA may be found in [653], whereas a Python implementation from *scikit-learn* [646] may be found at [654]. Both implementations can handle sparse representations of text. A Java implementation of LSA may be found at *Weka* [655]. Several efficient implementations of SVD/LSA with the Lanczos algorithm in ANSI Fortran-77 and ANSI C may be found in the SVDPACK library [663]. Python implementations of various types of matrix factorization methods may be found at *scikit-learn* [656]. A Python implementation of Latent Dirichlet Allocation may also be found at that site [657]. Another free Python library of topic modeling techniques is **gensim** [464], which includes representation learning methods like *word2vec* and *doc2vec*. CRAN [658] also contains several packages for topic modeling. In particular, the packages **topicmodels** and **lda** are noteworthy. Many of these packages build on the text mining package **tm** at CRAN. A detailed discussion of the **topicmodels** package may be found in [257]. The C code from the original authors of the LDA paper is also available [659]. The **MALLET** toolkit [701] provides several fast implementations of topic models. The **kernlab** package in R [289] from CRAN provides the ability to perform nonlinear dimensionality reduction. Numerous manifold learning packages in Python are also available from *scikit-learn* [660]. The Nyström method of kernel approximation is also available [664]. However, the kernel functions available are designed for multidimensional data rather than sequence data. Since mining of sequential data is the primary use-case of kernels in the text domain, one would need to augment and modify this (open-source) Nyström implementation with a separate implementation of substring kernels in order to use it. The *word2vec* tool is available [661] under the terms of the Apache license. The **TensorFlow** version of the software is available at [662].

3.9 Exercises

1. Consider the following matrix:

$$D = \begin{pmatrix} & \text{car} & \text{truck} & \text{carrot} & \text{apple} \\ \text{Document-1} & 1 & 1 & 1 & 1 \\ \text{Document-2} & 1 & 1 & 1 & 1 \\ \text{Document-3} & 0 & 0 & 1 & 1 \\ \text{Document-4} & 0 & 0 & 1 & 1 \end{pmatrix}$$

   (a) Construct a rank-2 SVD of this matrix. You can use any off-the-shelf software you like. What is the error of the decomposition?

   (b) Perform a rank-2 nonnegative matrix factorization of this matrix with the Frobenius norm. What is the error of the decomposition?

   (c) Which of the factorizations is more easily interpretable? Can you put names to the topics of the two latent components in the case of SVD? How about nonnegative matrix factorization?

2. Let U and V be $n \times k$ and $d \times k$ matrices, respectively. Consider the unconstrained optimization problem of minimizing the Frobenius norm $\|D - UV^T\|_F^2$, which is equivalent to SVD. Show that an infinite number of alternative optimal solutions for U and V exist in which the columns of U and V are mutually non-orthogonal.

3. Consider the unconstrained optimization problem of minimizing the Frobenius norm $\|D - UV^T\|_F^2$, which is equivalent to SVD. Here, D is an $n \times d$ data matrix, U is an $n \times k$ matrix, and V is a $d \times k$ matrix.

   (a) Use differential calculus to show that the optimal solution satisfies the following conditions:

$$DV = UV^TV$$
$$D^TU = VU^TU$$

(b) Let $E = D - UV^T$ be a matrix of errors from the current solutions U and V. Show that an alternative way to solve this optimization problem is by using the following gradient-descent updates:

$$U \Leftarrow U + \alpha EV$$
$$V \Leftarrow V + \alpha E^T U$$

Here, $\alpha > 0$ is the step-size.

(c) Will the resulting solution necessarily contain mutually orthogonal columns in U and V?

4. Suppose that you change the objective function of SVD in Exercise 3 to add penalties on large values of the parameters. This is often done to reduce overfitting and improve generalization power of the solution. The new objective function to be minimized is as follows:

$$J = \|D - UV^T\|_F^2 + \lambda(\|U\|_F^2 + \|V\|_F^2)$$

Here, $\lambda > 0$ defines the penalty. How would your answers to Exercise 3 change?

5. Without using SVD, show that the nonzero eigenvalues of $D^T D$ and DD^T are the same for any matrix D. [Hint: The proof is no more than three or four lines.]

6. Suppose that you are allowed to assume that at least one of the optimal solutions of the objective function in Exercise 3 must have mutually orthogonal columns in each of U and V, and in which each column of V is normalized to unit norm.

(a) Use the optimality conditions of Exercise 3(a) to show that U must contain the largest eigenvectors of DD^T in its columns and V must contain the largest eigenvectors of $D^T D$ in its columns. What is the value of the optimal objective function?

(b) Show that the (length-normalized) optimal value for V that maximizes $\|DV^T\|_F^2$ also contains the largest eigenvectors of $D^T D$ like (a) above. You are allowed to use the same assumption of orthonormal columns in V as above. What is the value of this optimal objective function? What does this tell you about the energy preserved by the SVD projection?

(c) Show that the sum of the optimal objective function values in (a) and (b) is a constant that is independent of the rank k of the factorization but dependent only on D. How would you (most simply) describe this constant in terms of the data matrix D?

7. Suppose that you are given an $n \times n$ matrix containing the squared Euclidean distances between n data points rather than the similarities. However, you do not know the coordinates of these data points. How would you use this matrix to generate an embedding of these n data points into multidimensional space?

8. Implement the algorithms for SVD and nonnegative matrix factorization introduced in the chapter.

9. Convert the solution to Exercise 1(b) into a three-way factorization with L_1-normalization. What is the significance of the diagonal matrix?

10. Suppose you have a string kernel in which objects i and j have similarity s_{ij}. Show that the Euclidean distance between embedded objects i and j is $\sqrt{s_{ii} + s_{jj} - 2s_{ij}}$.

Chapter 4

Text Clustering

"Taxonomy is described sometimes as a science and sometimes as an art, but really it's a battleground." – Bill Bryson in *A Short History of Nearly Everything*

4.1 Introduction

The problem of text clustering is that of partitioning a corpus into groups of similar documents. Clustering is an *unsupervised learning* application because no data-driven guidance is provided about specific types of groups (e.g., sports, politics, and so on) with the use of training data. Clustering has numerous applications because of its ability to organize large collections of documents into topical groups:

1. *Web portals:* Web portals often organize documents into clusters based on content similarity, which helps the users in navigating Web pages of interest. In many cases, this organization is hierarchical, in which the higher-level clusters cover broader topics, whereas the lower-level clusters cover fine-grained topics. Such hierarchical organizations are also referred to as *taxonomies.*

2. *News portals:* Many providers of news content need to organize the documents by topic, so that users are able to find news articles of their interest. As in the case of Web portals, the organization is often hierarchical.

3. *Intermediary for other applications:* Clustering is often used as an intermediate step in other applications like outlier analysis and classification. Clustering is a type of summarization that helps in building compact predictive models for various problems.

In many settings such as that of news wire services, examples of specific groups (e.g., sports or politics) may be available. This data is then used to categorize other documents into these *pre-defined* groups. This setting is referred to as *supervised learning,* and the examples of categorized documents are collectively referred to as training data. These methods are introduced in Chapter 5. The terminology "supervised" refers to the fact that one can use

© Springer Nature Switzerland AG 2022
C. C. Aggarwal, *Machine Learning for Text,*
https://doi.org/10.1007/978-3-030-96623-2_4

the training examples to guide the grouping process, just as a teacher guides her students towards a specific goal. However, clustering is useful in applications in which no prior training examples are available, and is therefore an unsupervised method.

Clustering methods are either *flat* or *hierarchical*. In flat clustering, documents are partitioned into a set of clusters in one shot, and no hierarchical relationships exist between clusters. In hierarchical clustering, the clusters are organized in tree-like fashion as a taxonomy. For example, the sports-related documents could be at a higher-level cluster and the basketball/baseball clusters could be among the many children of the sports-related cluster. The basketball cluster could have further children containing documents related to basketball items, tournaments, clubs, and so on. Hierarchical clustering is of special importance in the text domain because of its ability to enable intuitive browsing in Web applications.

It is often useful to perform various types of feature selection and feature engineering tricks to improve the clustering process. Feature selection refers to removal of irrelevant words, whereas feature engineering refers to the transformation of text into a representation that is more amenable for clustering. This chapter will discuss several such techniques.

Clustering methods are closely related to dimensionality reduction. In particular, most nonnegative matrix factorization methods and topic models can be leveraged for clustering both words and documents. Many of these models are *mixed membership models* in which the documents are assumed to be generated by multiple mixture components containing the various topics. The basic assumption is that the corpus is defined by certain core topics (e.g., *Arts*, *Politics*, *Sports*), and a document may contain components associated with multiple topics. Many matrix factorization methods (like PLSA) exhibit these characteristics. However, if an application demands a hard partitioning of the documents into clusters, this creates some additional requirements for disambiguation of cluster membership during post-processing. The natural solution in such cases is to modify topic models with more constraints that force this type of disambiguation early on in the modeling. Such methods are also referred to as co-clustering. This chapter will discuss matrix factorization methods, the k-means method, hierarchical methods, and probabilistic methods for clustering. Clustering is closely related to the design of similarity functions, because most deterministic methods like k-means leverage similarity functions to construct clusters.

The effectiveness of clustering algorithms can be significantly improved with the use of ensemble methods. In some cases, it may be useful to cluster the text as sequences, particularly when the documents are short. In such cases, kernel methods can be used for feature engineering in an implicit or explicit way. Such sequence-centric methods can also be combined with subsampled ensembles to improve clustering quality. Furthermore, the clustering and classification problems are closely related, and one can use this fact in order to leverage classification algorithms for clustering.

Chapter Organization

This chapter is organized as follows. The next section studies several feature selection and feature engineering methods for text clustering. Section 4.3 studies the use of topic models for text clustering. Section 4.4 introduces the traditional mixture model for clustering. The k-means algorithm is discussed in Section 4.5. Hierarchical clustering algorithms are discussed in Section 4.6. Clustering ensemble methods are discussed in Section 4.7. The clustering of text as sequences is discussed in Section 4.8. The use of classification algorithms for clustering is explored in Section 4.9. Section 4.10 introduces techniques for clustering evaluation. Section 4.11 gives a summary.

4.2 Feature Selection and Engineering

Text data is high dimensional, and many words are irrelevant for clustering. The removal of such words is referred to as feature selection. Furthermore, the vector-space representation does not incorporate information about the sequential ordering of words. Such information is incorporated by using feature engineering techniques. We distinguish between these two classes of techniques as follows:

1. In feature selection techniques, the irrelevant features are dropped before applying the clustering algorithm. Such an approach improves clustering quality because a significant amount of noise is removed. Feature selection methods always reduce the dimensionality of the data.

2. In feature engineering techniques, the features are *transformed* to a new representation in which simple clustering algorithms like the k-means method can work much more effectively. In some cases, the representation of the data might change in a fundamental way (e.g., sequences to vector-space representation).

Methods like singular value decomposition (SVD) and latent semantic analysis (LSA) are somewhere in the middle, because they use only linear transformations on the features, and the primary advantage of the representation is obtained by dropping the lower-order (transformed) features. In the following, we will provide an overview of the different feature selection and engineering techniques.

4.2.1 Feature Selection

Feature selection methods remove the irrelevant words in the document collection in order to improve the effectiveness of the clustering process. Note that the removal of stop words and the inverse document weighting frequency (idf) of words is also a form of feature selection. However, these are rather rudimentary techniques that only use the raw frequencies of words in order to make judgements about their relevance. It is possible to improve on these methods by carefully evaluating the consistency of the features with intra-document similarities. One can also use any of the following feature selection methods as feature weighting methods.

4.2.1.1 Term Strength

The basic idea underlying term strength [581] is that a term is semantically relevant when *it has a higher probability of co-occurrence in similar pairs of documents.* Two documents are considered to be sufficiently similar, if the cosine similarity between them is greater than a pre-defined similarity threshold δ. Then, the term strength of t_j is defined as the fraction of such pairs in which the term occurs in the second member of the pair, given that it occurs in the first member. Therefore, the term strength $S(t_j)$ for a term t_j between documents $\overline{X}$ and $\overline{Y}$ is defined as follows:

$$S(t_j) = P(t_j \in \overline{X} | t_j \in \overline{Y}, \text{cosine}(\overline{X}, \overline{Y}) \geq \delta) \tag{4.1}$$

It is relatively straightforward to compute the term strength by sampling pairs of documents, selecting those that satisfy the similarity threshold, and then estimating the conditional probability in a data-driven manner. Features with low term strength are removed. The aforementioned computation includes the impact of the (evaluated) term t_j in the similarity calculations, which favors high-frequency terms. However, some minor changes [581] can remove the impact of the evaluated term.

4.2.1.2 Supervised Modeling for Unsupervised Feature Selection

Supervised modeling techniques are sometimes used for feature selection in unsupervised problems like clustering and outlier detection. Although the following technique was proposed [435] for unsupervised feature selection and weighting in traditional multidimensional data, it can also be used for text. We present a slight adaptation of the original idea [435] to account for the sparse representation of text.

The basic idea is to decompose the feature selection into d different prediction problems, where d is the dimensionality of the data. A feature that is largely unrelated to the remaining data set cannot be meaningfully predicted by the other $(d-1)$ features. Therefore, we create a classification problem (see Chapter 5) in which the presence of absence of the jth term t_j in a document is a binary class variable. The performance of an off-the-shelf classifier can be used to compute the relevance of this feature because relevant features can be predicted more accurately from other features. The vector space representation of the *remaining* terms is used to represent the document. This type of classification problem is typically an imbalanced learning problem because the term t_j is unlikely to be present in most of the documents. In imbalanced settings, many classifiers rank instances based on their propensity to belong to the minority class, and the ranking quality is evaluated using a measure referred to as the *Area under Curve (AUC)* of the *Receiver Operating Characteristic (ROC)* (see Chapter 7). The AUC lies in $(0,1)$, and a classifier that ranks instances randomly would be expected to receive an AUC of 0.5. Such a scenario would occur for a feature that is poorly related to the remaining features and is unlikely to help in creating coherent clusters. Therefore, the feature relevance $R(t_j)$ is given by the additional AUC beyond the random performance of 0.5:

$$R(t_j) = \max\{\text{AUC}(t_j) - 0.5, 0\} \qquad (4.2)$$

Here, $\text{AUC}(t_j)$ is the AUC of the classifier that uses term t_j as the target class. Features with low relevance are removed.

The main problem with this technique is that it requires the training of one classification model *for each feature*. Therefore, if the data contains a large number of features, as in the text domain, it becomes rather difficult to use this approach. A more efficient alternative is to divide the data into K random subsets of features, where K is a user parameter. We use $(K-1)$ subsets of features for training, and the remaining subset for prediction. Such an approach requires only K applications of the classifier.

4.2.1.3 Unsupervised Wrappers with Supervised Feature Selection

The aforementioned methods are *filter* methods because the quality of a feature is evaluated independently of the specific clustering algorithm being used. In *wrapper* methods, we wrap a clustering algorithm around the feature selection process. Therefore, the feature selection is tightly integrated with the clustering method at hand. Furthermore, the approach transforms unsupervised feature selection to supervised feature selection (cf. Section 5.2 of Chapter 5). The basic approach combines unsupervised clustering and supervised feature selection method with the following two steps:

1. Use clustering algorithm with current feature set F to partition corpus into k clusters.

2. Treat the cluster label of a document as its class. Apply any of the supervised feature selection algorithms discussed in Chapter 5 on feature set F and prune it if needed.

It is also possible to iterate on these steps, although a single application of these steps is often sufficient. A specific example of such an algorithm that combines the expectation-maximization (EM) clustering algorithm with the supervised χ^2-statistic is provided in [332]. This work also shows how such methods can also be used for feature weighting.

4.2.2 Feature Engineering

In contrast to feature *selection*, feature engineering methods transform the data to get the most out of clustering methods. Matrix factorization methods are *linear* feature engineering methods that work well in practice. In the unsupervised setting, the primary goal of such methods is to reduce the noise effects of synonymy and polysemy. In cases where it is desired to also add linguistic knowledge (i.e., word ordering knowledge) into the engineered features, nonlinear dimensionality reduction methods are required.

Many feature engineering methods exhibit a duality in the sense that they can be used to either transform documents into multidimensional space, or they can be used to transform words in the same way. For example, all matrix factorization methods simultaneously produce an embedding of documents and an embedding of words with the factor matrices. In nonlinear embedding methods, the ability to create a word embedding or document embedding depends on the specific model that is used. Nonlinear dimensionality reduction can be achieved either with kernels or with neural networks. Examples of the latter include *word2vec* [389] and *doc2vec* [314]. Such methods are discussed in Chapter 10.

4.2.2.1 Matrix Factorization Methods

Matrix factorization methods perform an approximate decomposition of an $n \times d$ document-term matrix D into two $n \times k$ and $d \times k$ matrices U and V, respectively, so that the following condition is satisfied:

$$D \approx UV^T \tag{4.3}$$

The rank k is typically chosen to be much smaller than both n and d. Numerous matrix factorization methods are discussed in detail in Chapter 3. The matrix factorization problem is typically posed as an optimization problem over minimizing the aggregate squared error of the entries in $(D - UV^T)$. In addition, different types of constraints on U and V can be used to regulate the properties of these matrices. Examples of various forms of matrix factorization include singular value decomposition, nonnegative matrix factorization, and probabilistic latent semantic analysis. The rows of the matrix U can be used as the document embeddings and the rows of the matrix V can be treated as the word embeddings. When using the factorization to create document embeddings, one scales the columns of V to unit norm and adjusts the columns of U accordingly (see Section 3.1.1 of Chapter 3). Similarly, when using the factorization to create word embeddings, one scales the columns of U to unit norm and adjusts the columns of V accordingly. Using this approach makes the corresponding embeddings sensitive to their frequencies in the collection.

The primary gain from these feature engineering methods is achieved by using a rank k that is much less than $\min\{n, d\}$. The typical value of k used in most matrix factorization methods is in the low hundreds, whereas the value of d is often in the range of hundreds of thousands. Furthermore, the value of k may vary with the collection at hand; for small collections or lexicon sizes, the value of k will be smaller. The low-rank factorization creates a residual error in Equation 4.3, which is observed only as an *approximate* equality. In such cases, the noise effects of synonymy and polysemy are removed from the collection, and the clustering tendency of the corpus improves. In other words, the approximation actually *improves* the representation quality. A detailed discussion of this phenomenon is provided in Chapter 3. One can apply a k-means algorithm on the rows of U to cluster documents and a k-means algorithm on the rows of V to cluster words.

Different types of matrix factorizations have different advantages from a feature engineering point of view. SVD is good at efficiently representing out-of-sample documents by a simple projection operation. Nonnegative matrix factorization and PLSA provide semantically interpretable representations, which can be directly used for *soft* clustering in which each document is associated with a set of probabilities of belonging to various clusters. Such methods are useful in collections with highly overlapping clusters. These issues are discussed in Section 4.3.

4.2.2.2   Nonlinear Dimensionality Reduction

Nonlinear dimensionality reduction methods are particularly well suited to creating embeddings from short texts in which linguistic/sequence knowledge is incorporated in the embedding. It is particularly important to use knowledge about the sequential ordering of words when working with short texts because the data is too sparse to be used effectively with a vector-space representation. However, unlike linear dimensionality reduction methods, which simultaneously provide word embeddings and document embeddings, these methods are optimized for either document embeddings or word embeddings. Kernel methods are best suited to the creation of document embeddings although these methods can also be generalized to create word embeddings, if suitable similarity functions can be defined between words with the use of sequence information. In all these cases, the key is to create a high-quality similarity matrix between the objects.

Nonlinear dimensionality reduction methods are discussed in Section 3.6 of Chapter 3. The basic idea is to perform an approximate rank-k diagonalization of the $n \times n$ similarity matrix S as $S = Q_k \Sigma_k^2 Q_k^T$ and then extract $Q_k \Sigma_k$ as the embedding. When the size n of the corpus is large, such an approach can be space- and time-prohibitive. In such cases, The Nyström sampling method of Section 3.6.2 can be used. Although this type of sampling method loses some accuracy, these methods become extremely powerful when they are combined with sampling-based clustering ensembles. A specific example of such an approach in the clustering context is provided in Section 4.8.

Word Embeddings

For word-clustering applications, word embeddings are required that require some knowledge of the positioning information between words. The simplest approach is to use an 2-gram embedding. For each pair of terms t_i and t_j the probability $P(t_j|t_i)$ that term t_j occurs just after t_i is computed. A matrix S is created in which S_{ij} is equal to $[P(t_i|t_j)+P(t_j|t_i)]/2$. Values of S_{ij} below a certain threshold are removed. The diagonal entries are set to be equal to the sum of the remaining entries in that row. This is done in order to ensure that the matrix is positive semi-definite. The top-k eigenvectors of this matrix can be used to generate a word embedding. Since the word space is large, the sampling technique (discussed above) may need to be leveraged. One can generalize this approach by using skip-grams with varying gaps in the modeling process. The linguistic power in the embedding depends almost completely on the type of word-word similarity function that is leveraged. The generality of the approach arises from the fact that one can even incorporate linguistic prior knowledge by using semantic databases like *WordNet* [396] to further refine the similarity matrix S. In recent years, neural network methods [53, 314, 389] like *word2vec* and *doc2vec* have also become increasingly popular for creating word embeddings (cf. Chapter 10).

4.3 Topic Modeling and Matrix Factorization

Chapter 3 introduces topic models from the perspective of matrix factorization and latent semantic analysis. In this section, we introduce the relationship between these models and clustering. All forms of *nonnegative* matrix factorization and Latent Dirichlet Allocation (LDA) can be used to generate overlapping clusters from the collection.

4.3.1 Mixed Membership Models and Overlapping Clusters

Topic models are inherently mixed membership models in which each document is generated by one or more topics or *aspects*. Although one can treat an aspect as a cluster, this point of view leads to clusters that are highly overlapping both in terms of document membership and vocabulary. One possible solution is to use the matrix U from the factorization $D \approx UV^T$ to assign each document (row of U) to the topic that has the largest positive coordinate value in U. However, documents (rows of U) that contain multiple topics (strictly positive coordinates) *do* logically belong to multiple clusters and it is not fair to force a disambiguation under such circumstances. For example, consider a document collection containing 100 documents about cats, another 100 about cars, and 30 documents discussing both cats and cars. From the perspective of topic models, this collection naturally contains two topics and the final set of 30 documents can be expressed as a combination of two topics. However, if each document is forced to be in a single cluster, a larger number of clusters are required to express the same collection, because the final set of 30 documents can be viewed as a completely distinct cluster. Therefore, there are two options in leveraging methods like nonnegative matrix factorization for clustering:

1. One can use the highly positive coordinates in U to report overlapping membership of documents in clusters as well as the topical vocabulary from the columns of V. This point of view inherently accepts mixed membership of documents in clusters.

2. If a single-membership clustering is required, can treat the matrix factorization process as a feature engineering step and apply a k-means algorithm on the engineered representation. The rationale for this view is explained in Section 4.2.2.1. Typically, the number of clusters would be larger than the rank of the factorization because of the additional clusters created by combinations of topics.

One good property of topic modeling techniques is that they allow the simultaneous discovery of word clusters and document clusters, even if they are highly overlapping.

4.3.2 Non-overlapping Clusters and Co-clustering: A Matrix Factorization View

Even though most clustering methods assign each document to one cluster, they do allow heavy term overlap across clusters. For example, a k-means algorithm assigns each document to only one cluster, but the frequent terms in the centroids might contain heavy overlaps

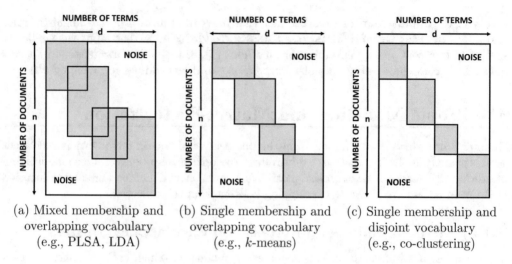

(a) Mixed membership and overlapping vocabulary (e.g., PLSA, LDA)

(b) Single membership and overlapping vocabulary (e.g., k-means)

(c) Single membership and disjoint vocabulary (e.g., co-clustering)

Figure 4.1: Mixed-membership versus single-membership models

across multiple clusters. An extreme view at the other end of the spectrum is that of co-clustering in which each cluster of a matrix is defined as a subset of rows and columns. Furthermore, no overlap is allowed among the different row sets and column sets. One can view co-clustering as the process of re-arranging the rows and columns of a matrix so that most of the positive entries lie on blocks around the diagonal. In fact, the overlaps among documents/terms in different types of clustering methods can be understood in terms of this re-arranged structure. In Figure 4.1, we have shown three common cases of clustering with varying levels of overlap between document clusters and word clusters. Even though most clustering methods strictly partition the data, they do allow overlap in the cluster vocabulary. In most of these cases of Figure 4.1(b), the cluster vocabulary is derived as a secondary output of the clustering process. However, in cases of Figures 4.1(a) and (c), the cluster vocabularies are recovered as first-class citizens along with clusters.

One can also modify nonnegative matrix factorization to also handle cases shown in Figures 4.1(b) and (c). Let D be an $n \times d$ document-term matrix, and let U and V be the $n \times k$ document factors and $d \times k$ term factors, respectively. In order to force the clusters to be non-overlapping but not the terms (Figure 4.1(b)), the non-negative matrix factorization formulation of Section 3.3 can be modified as follows:

$$\text{Minimize } _{U,V}||D - UV^T||_F^2$$
$$\text{subject to:}$$
$$U, V \geq 0$$
$$U^T U = I$$

Most clustering algorithms implicitly try to optimize an objective function of this type. In fact, it has been shown [157] that this objective function is *roughly* equivalent to that used by the k-means algorithm! One can view each of the factors in an analogous way to a cluster. The rows of matrix U contain the approximations to the cluster memberships and the columns of matrix V contain the cluster representatives (centroids). By forcing orthonormality and non-negativity at the same time, we are ensuring that only a single coordinate in each row of U has a positive scaling value in $(0, 1]$, which corresponds to the membership of that point in the corresponding cluster. Furthermore, the k columns of V contain the k nonnegative cluster representatives. Therefore, the factorization represents each point by a fractional scaling of its closest cluster representative, and the objective function minimizes the sum of squared errors of this approximation. As discussed in Section 4.5, this objective function is equivalent to that of the k-means algorithm. The main difference from the k-means algorithm is that the fractional positive value in each row scales the cluster representatives, whereas the k-means algorithm does not perform such scaling. One can obtain the precise optimization model for the k-means algorithm by further imposing the constraint that each entry from U is drawn from $\{0, 1\}$. In such a case, each row will contain exactly one value of 1, and the remaining values will be 0s. However, imposing this additional constraint results in an *integer program*, which is hard to solve from an optimization perspective. The aforementioned model for the k-means algorithm is also different from nonnegative matrix factorization to the extent that it does not impose a nonnegativity constraint on the matrix V of cluster representatives; however, this constraint is redundant, because the *optimal* cluster representative can never have a negative component for a nonnegative data matrix. Therefore, adding a nonnegativity constraint on V will result in the same optimal solution.

If one wishes to force the terms of each cluster to be non-overlapping, then orthogonality can also be forced on the columns of V:

$$\text{Minimize } _{U,V} ||D - UV^T||_F^2$$

$$\text{subject to:}$$

$$U, V \geq 0$$

$$\text{Columns of } U \text{ are mutually orthogonal}$$

$$\text{Columns of } V \text{ are mutually orthogonal}$$

SVD also forces orthogonality on the factors. However, this optimization problem is different from SVD because of the nonnegativity of the factors, which makes it more difficult. The combination of orthogonality and nonnegativity implies that each row of *both* U and V has at most a single positive value. This type of mutually exclusive membership of both documents and words in clusters is shown in Figure 4.1(c).

One can equivalently formulate this problem in standardized three-way factorization by normalizing U and V according to the approach discussed in Section 3.1.1 of Chapter 3:

$$UV^T = Q\Sigma P^T \ (Q \text{ and } P \text{ have normalized columns and } \Sigma \text{ is diagonal})$$

The *equivalent* optimization problem (with normalized matrices) is as follows:

$$\text{Minimize } _{Q,P,\Sigma} ||D - Q\Sigma P^T||_F^2$$

$$\text{subject to:}$$

$$P, Q, \Sigma \geq 0$$

$$Q^T Q = I$$

$$P^T P = I$$

$$\Sigma \text{ is diagonal}$$

Here, the diagonal matrix plays the role of pulling out the scaling factors from the columns of U and V according to the approach discussed in Section 3.1.1.

This problem can be solved [157] even in the case where Σ is not diagonal or is not a square matrix, and is referred to as *tri-factorization*. This generalization allows a different number k_q of document clusters (captured by the $n \times k_q$ matrix Q) and word clusters (captured by the $d \times k_p$ matrix P). The interactions among the document clusters and word clusters are captured by the $k_q \times k_p$ matrix Σ. Tri-factorization is a variation of topic modeling, and is more general than strict co-clustering. In tri-factorization, an exact one-to-one correspondence does not exist between the document clusters and word clusters because it is regulated by Σ.

We will first provide a solution for the (more general) optimization problem of tri-factorization in which Σ is neither diagonal nor square. Later, we will see that the special case of diagonal factors can be solved simply by using a different initialization. In such a case, Q is an $n \times k_q$ matrix, Σ is a $k_q \times k_p$ matrix, and P is a $d \times k_p$ matrix. The optimization parameters of this problem can be obtained by using the following iterative steps:

$$Q_{iq} \Leftarrow Q_{iq}\sqrt{\frac{(DP\Sigma^T)_{iq}}{(QQ^TDP\Sigma^T)_{iq}}} \quad \forall i \in \{1\ldots n\}, \forall q \in \{1\ldots k_q\}$$

$$P_{jp} \Leftarrow P_{jp}\sqrt{\frac{(D^TQ\Sigma)_{jp}}{(PP^TD^TQ\Sigma)_{jp}}} \quad \forall j \in \{1\ldots d\}, \forall p \in \{1\ldots k_p\}$$

$$\Sigma_{qp} \Leftarrow \Sigma_{qp}\sqrt{\frac{(Q^TDP)_{qp}}{(Q^TQ\Sigma P^TP)_{qp}}} \quad \forall p \in \{1\ldots k_p\}, \forall q \in \{1\ldots k_q\}$$

These steps are iterated to convergence. The matrices are all initialized to random nonnegative values in $(0, 1)$, although it is possible for a particular initialization point to arrive at a local minimum. For example, it is possible to solve the constrained problem in which Σ is square and diagonal by choosing an initialization point in which Σ is diagonal. The orthogonality constraints tend to make this approach sensitive to the presence of local minima. Better results can be achieved by relaxing the orthogonality constraints slightly [216] or by incorporating them as constraints in the objective function. Another approach, which is discussed in the next section, is to transform the problem into that of graph partitioning.

4.3.2.1 Co-clustering by Bipartite Graph Partitioning

One can pose the problem of co-clustering as a bipartite graph partitioning problem. The basic idea is to create a bipartite document-term graph in which a node exists for each document and also for each term. Edges exist only between document nodes and term nodes. An undirected edge is added to the graph between a term node and document node if and only if that term occurs between the document. The weight of the edge is equal to the normalized frequency of the term. Then, it is easy to see that a partitioning of this graph simultaneously yields both document clusters and word clusters. This situation is shown in Figure 4.2 in which each demarcated community contains both document clusters and word clusters. Therefore, this approach transforms the problem of co-clustering to that of community detection in graphs. The reader is referred to [2] for several community detection methods in graphs. A commonly used method is that of spectral graph partitioning [150], which turns out to be closely related to singular value decomposition in the special case of bipartite graphs. The following description of spectral clustering is atypical and is applicable only to the case of bipartite graphs. A more general description for all types of graphs may be found in [415].

The basic idea is to treat the $n \times d$ document-term matrix D as the adjacency matrix of the bipartite graph in which the relevant $n \times d$ portion with edges is used rather than the full $(n + d) \times (n + d)$ adjacency matrix. Let the sum of the term frequencies in document i

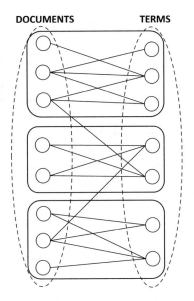

Figure 4.2: Transforming co-clustering to graph partitioning

be d_i, and let the aggregate frequency of term j be f_j. Then, the matrix D is normalized by dividing its (i,j)th entry with $\sqrt{d_i \cdot f_j}$, and the normalized matrix be denoted by D_0. This matrix is approximately decomposed with rank-p SVD as follows:

$$D_0 \approx Q\Sigma P^T \qquad (4.4)$$

Here, Q is an $n \times p$ matrix and P is a $d \times p$ matrix. The value of p is much less than $\min\{n,d\}$. The original work [150] recommends using $p = \log_2(k)+1$ for a k-way clustering, although this can turn out to be too conservative in practice. The two matrices Q and P are stacked[1] to create a single $(n+d) \times p$ matrix as follows:

$$Z = \begin{pmatrix} Q \\ P \end{pmatrix} \qquad (4.5)$$

Note that the matrix Z has one p-dimensional row for each document or term. Furthermore, it can be shown that the rows of Z are not comparable in terms of their scaling because of the varying frequencies of words and documents. In order to remedy this issue, each row in Z is divided by the square-root of the aggregate frequency[2] of the corresponding document/term in the corpus. These frequencies are the same as d_i or f_j computed above. Subsequently, a k-means algorithm is applied to cluster the rows of Z in order to extract the simultaneous partitioning of documents and words. Note that each cluster will typically contain a subset of documents *and* words, depending on which rows of Z are grouped into that cluster. Therefore, this partitioning turns out to be a co-clustering.

---

[1]The first eigenvector is not discriminative in terms of the clustering structure and can be dropped. Its value can be shown to depend only on the square-root of the frequency of the corresponding term or document.

[2]The general form of symmetric spectral clustering [415] (cf. Section 4.8.2), which is applicable to all types of bipartite and non-bipartite graphs, normalizes each row to unit norm. This choice is a worthy alternative.

4.4   Generative Mixture Models for Clustering

Generative models assume that the corpus is generated by a mixture of distributions, and estimate the parameters of these distributions based on the observed corpus. The k clusters in the mixture are denoted by $\mathcal{G}_1 \ldots \mathcal{G}_k$, where k is an input parameter. The terms in each document of a mixture component are modeled by a distribution specific to that mixture component. These assumptions provide the analyst the ability to incorporate some domain knowledge into the modeling process by selecting a particular type of distribution. The most commonly used assumptions correspond to the *Bernoulli* and the *multinomial* models. The Bernoulli model is appropriate when the text documents are represented as vectors of 0–1 values, corresponding to the presence or absence of terms. The multinomial model is used to model arbitrary word frequencies. The generative process of mixture modeling is as follows:

1. Select the rth mixture component $\mathcal{G}_r$ with prior probability $\alpha_r = P(\mathcal{G}_r)$.

2. Generate the vector space representation of a document using the probability distribution of $\mathcal{G}_r$. The common choices are Bernoulli or multinomial distributions.

For a given corpus, the goal of the expectation-maximization algorithm is to estimate the parameters of the distributions, so that the observed data has the maximum *likelihood* of being generated by this model. One can compactly denote the entire vector of mixture distribution parameters and prior probabilities $\alpha_1 \ldots \alpha_k$ by $\overline{\Theta}$. The probability of a single document $\overline{X_i}$ being generated by the model is $\sum_{m=1}^{k} P(\mathcal{G}_m) \cdot P(\overline{X_i}|\mathcal{G}_m) = \sum_{m=1}^{k} \alpha_m P(\overline{X_i}|\mathcal{G}_m)$. We want to learn the entire vector $\overline{\Theta}$ of parameters, so as to maximize the *product* of these probabilities over *all* documents in the corpus:

$$\text{Maximize}_{\overline{\Theta}} \left\{ P(Corpus|\overline{\Theta}) = \prod_{i=1}^{n} \left(\sum_{m=1}^{k} \alpha_m P(\overline{X_i}|\mathcal{G}_m) \right) \right\} \qquad (4.6)$$

In practice, one uses the logarithm of this value to create a *log-likelihood* objective function, which is then maximized.

The main challenge in estimating these parameters is that it is not known which mixture component generated which document; if we knew which mixture component generated which document, then parameter estimation would be a very simple matter by fitting the relevant subset of documents to that mixture component in an optimal way. The expectation-maximization algorithm therefore uses an iterative approach, in which the expected probability of membership is estimated based on the current state of the parameters (i.e., *expectation* step). Then, the parameters are optimized by holding this membership probability fixed. This step is simplified because the membership probabilities can be viewed as (fixed) weights on points and we can optimally estimate the parameters of each mixture component without worrying about the other components (i.e., *maximization* step). The two-step iterative approach is then executed to convergence. In the following, we will describe the steps of the expectation-maximization algorithm for the Bernoulli and multinomial models.

4.4.1   The Bernoulli Model

In the Bernoulli model, it is assumed that the jth term, t_j, in the lexicon is present in a document generated from the rth mixture component with probability $p_j^{(r)}$. Then, the probability $P(\overline{X_i}|\mathcal{G}_r)$ of the generation of the document $\overline{X_i}$ from mixture component $\mathcal{G}_r$ is given[3] by the product of the d different Bernoulli probabilities corresponding to presence or absence of various terms:

$$P(\overline{X_i}|\mathcal{G}_r) = \prod_{t_j \in \overline{X_i}} p_j^{(r)} \prod_{t_j \notin \overline{X_i}} (1 - p_j^{(r)}) \tag{4.7}$$

An important assumption here is that the presence or absence of the various terms are conditionally independent with respect to the choice of mixture component. Therefore, one can express the joint probability of the attributes in $\overline{X_i}$ as the product of the corresponding values on individual attributes. This assumption is also referred to as the *naïve Bayes assumption*, and is commonly used for clustering and classification with the Bernoulli model.

Then, the expectation-maximization algorithm starts by randomly assigning documents to clusters, and estimates the initial parameters by applying the M-step (see below) with respect to this random assignment. Subsequently, it uses the following two steps iteratively:

1. **(E-step):** In the expectation step, the probabilistic assignments of documents to clusters are computed using the Bayes rule of *posterior probabilities*. The probability of a document $\overline{X_i}$ belonging to the rth cluster can be viewed as the posterior probability that the rth mixture component, $\mathcal{G}_r$, was used to generate it. This posterior probability is computed as follows:

$$P(\mathcal{G}_r|\overline{X_i}) = \frac{P(\mathcal{G}_r) \cdot P(\overline{X_i}|\mathcal{G}_r)}{\sum_{m=1}^{k} P(\mathcal{G}_m) \cdot P(\overline{X_i}|\mathcal{G}_m)} = \frac{\alpha_r \cdot \prod_{t_j \in \overline{X_i}} p_j^{(r)} \prod_{t_j \notin \overline{X_i}}(1 - p_j^{(r)})}{\sum_{m=1}^{k} \alpha_m \cdot \prod_{t_j \in \overline{X_i}} p_j^{(m)} \prod_{t_j \notin \overline{X_i}}(1 - p_j^{(m)})} \tag{4.8}$$

The right-most expression above is a result of substitution of $P(\overline{X_i}|\mathcal{G}_r)$ from Equation 4.7 in the above equation.

2. **(M-step):** The soft assignment probability $w_{ir} = P(\mathcal{G}_r|\overline{X_i})$ above is used to enable the estimation of parameters by treating it as a "membership weight." The value of $\alpha_r = P(\mathcal{G}_r)$ is estimated as the fraction of membership weights assigned to cluster r. One can estimate this value as $\sum_{i=1}^{n} w_{ir}/n$. One also needs to estimate the parameters of the Bernoulli distribution for various mixture components. One can estimate $p_j^{(r)}$ as the weighted fraction of documents in component r containing term t_j:

$$p_j^{(r)} = \frac{\sum_{i:t_j \in \overline{X_i}} w_{ir}}{\sum_{i=1}^{n} w_{ir}} \tag{4.9}$$

The two aforementioned steps are iterated to convergence. Convergence is checked by evaluating whether the (log-likelihood) objective function has improved by a minimum amount over its average value in the previous few iterations. Laplacian smoothing is used in the estimation of the M-step. Let d_a be the average number of 1s in each (binary representation of a) document and d be the size of the lexicon. The basic idea is to add a Laplacian smoothing parameter $\gamma > 0$ to the numerator of Equation 4.9 and $d \cdot \gamma/d_a$ to the denominator. Similarly, one can smooth the estimation of α_r by adding $\beta > 0$ to the numerator and $k \cdot \beta$ to the denominator.

It is noteworthy that the posterior probability $P(\mathcal{G}_r|\overline{X_i})$ provides the probability of assignment of document $\overline{X_i}$ to cluster $\mathcal{G}_r$. The k posterior probabilities specific to a document

---

[3] Although $\overline{X_i}$ is a binary vector, we are treating it like a set when we use a set-membership notation like $t_j \in \overline{X_i}$. Any binary vector can also be viewed as a set of the 1s in it.

will always sum to 1, as is expected in a probabilistic assignment of a document to clusters. If desired, one can also convert this *soft* assignment to a hard assignment by assigning each document to the cluster to which it has the largest posterior probability. For any particular cluster r, the terms with large values of $p_j^{(r)}$ are assumed to be the topical vocabulary of the cluster. Therefore, the approach returns (overlapping) word clusters along with document clusters. The initialization can be performed randomly, although improved results can be obtained by using another simple clustering algorithm in lieu of the first E-step to assign documents to clusters (corresponding to 0–1 posteriors).

4.4.2   The Multinomial Model

The multinomial model is designed to handle arbitrary term frequencies. The parameters of the k mixture components are defined by a $d \times k$ matrix of multinomial probability parameters $Q = [q_{jr}]$, in which $(q_{1r}, q_{2r}, \ldots q_{dr})$ represent the d parameters of a multinomial distribution of terms for the rth mixture component. The different values of q_{jr} sum to 1 for a particular mixture component, $\mathcal{G}_r$, over all terms (i.e., $\sum_{j=1}^{d} q_{jr} = 1$).

The generative process first selects the rth mixture component $\mathcal{G}_r$ with probability $\alpha_r = P(\mathcal{G}_r)$ and then throws a loaded die (owned by the rth component) L times to generate a document with L *tokens* (counting repetitions). The loaded die has as many faces as the number of terms d, and the probability of the jth face showing up is given by q_{jr} for the die owned by the rth mixture component. Therefore, if the die is thrown L times, then the number of times each face shows up provides the number of times each term shows up in the document. If we assume that the frequency vector of the document $\overline{X_i}$ is given by $(x_{i1} \ldots x_{id})$, then the generative probability of the ith document is defined by the multinomial distribution:

$$P(\overline{X_i}|\mathcal{G}_r) = \frac{(\sum_{j=1}^{d} x_{ij})!}{x_{i1}! x_{i2}! \ldots x_{id}!} \prod_{j=1}^{d} (q_{jr})^{x_{ij}} \propto \prod_{j=1}^{d} (q_{jr})^{x_{ij}} \tag{4.10}$$

The constant of proportionality holds for fixed $\overline{X_i}$ and varying mixture component, because it depends only on $\overline{X_i}$ and is independent of the mixture component $\mathcal{G}_r$.

One can now perform the E-step by using this new probability instead:

1. **(E-step):** Compute the posterior probability of document $\overline{X_i}$ using Equation 4.10 as follows:

$$P(\mathcal{G}_r|\overline{X_i}) = \frac{P(\mathcal{G}_r) \cdot P(\overline{X_i}|\mathcal{G}_r)}{\sum_{m=1}^{k} P(\mathcal{G}_m) \cdot P(\overline{X_i}|\mathcal{G}_m}) = \frac{\alpha_r \cdot \prod_{j=1}^{d} (q_{jr})^{x_{ij}}}{\sum_{m=1}^{k} \alpha_m \cdot \prod_{j=1}^{d} (q_{jm})^{x_{ij}}} \tag{4.11}$$

2. **(M-step):** The soft assignment probability $w_{ir} = P(\mathcal{G}_r|\overline{X_i})$ above is used to enable the estimation of parameters by treating it as a "membership weight." As in the Bernoulli model, the value of $\alpha_r = P(\mathcal{G}_r)$ is estimated as $\sum_{i=1}^{n} w_{ir}/n$. One also needs to estimate the parameters of the multinomial distribution for various mixture components. One can estimate q_{jr} as the weighted fraction of tokens in mixture component r that correspond to term t_j:

$$q_{jr} = \frac{\sum_{i=1}^{n} w_{ir} \cdot x_{ij}}{\sum_{i=1}^{n} \sum_{v=1}^{d} w_{ir} \cdot x_{iv}} \tag{4.12}$$

It is also possible to use Laplacian smoothing to improve the estimation of the parameters for sparse data. In such a case, we add γ to the numerator and $\gamma \cdot d$ to the denominator for a small value of $\gamma > 0$.

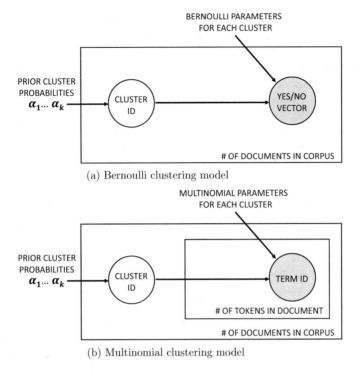

(a) Bernoulli clustering model

(b) Multinomial clustering model

Figure 4.3: Plate diagrams for Bernoulli and multinomial clustering models

As in the case of the Bernoulli model, these steps are iterated to convergence. One can use the estimated value of $P(\mathcal{G}_r|\overline{X_i})$ as the probability of assignment of document $\overline{X_i}$ to cluster r. The soft assignment can also be converted to a hard assignment by selecting the value of r for which this probability is as large as possible. For any particular cluster r, the terms with large values of q_{jr} are assumed to be the topical vocabulary of the cluster.

4.4.3 Comparison with Mixed Membership Topic Models

The topic models of Chapter 3 are referred to as *mixed membership models*, whereas the clustering models of this section are *single membership models*. Here, it is important to understand that the generative process in (mixed-membership) topic models is quite different from (single-membership) clustering models. Although both the PLSA model and the clustering models discussed above yield a cluster assignment probability for each document (i.e., $P(\mathcal{G}_r|\overline{X_i})$), this value should be interpreted differently in the two cases:

Clustering: $P(\mathcal{G}_r|\overline{X_i}) = \mathrm{P}(\ \mathcal{G}_r$ given *entire document* $\overline{X_i}\ )$
Topic Models: $P(\mathcal{G}_r|\overline{X_i}) = \mathrm{P}(\ \mathcal{G}_r$ given *a randomly chosen token* from $\overline{X_i})$

This difference is crucial because a single-membership model will always generate a document about cars and cats from a single mixture component, whereas topic models might generate different tokens of that document from different mixture components.

In this context, we present the plate diagrams for the single membership clustering models in Figure 4.3. Note that the cluster identifier is always generated exactly once for each *document*. However, in the plate diagrams for topic models (cf. Figures 3.4 and 3.7 of

Chapter 3), it is evident that the topic identifiers are generated once for every *token*. These differences are crucial while trying to convert soft probabilities into a hard assignment. In a single membership model, it is theoretically justified to assign each document to the cluster with the highest probability of assignment because it was *assumed* to be generated from a single component. The soft nature of the assignment is simply caused by the uncertainty of the statistical estimation process. In a topic model, the actual generation could be truly overlapping across multiple topics even after accounting for the estimation uncertainty. Therefore, in the case of topic models, it makes more sense to either accept the overlapping nature of the clustering, or to treat the soft probabilities as engineered features on top of which a k-means algorithm is applied.

4.4.4   Connections with Naïve Bayes Model for Classification

The naïve Bayes model[4] for classification (cf. Section 5.3 of Chapter 5) is a rudimentary special case of the expectation-maximization (EM) algorithm in which a single iteration of the E-step and M-step is sufficient. Imagine that you were given *labeled* training data in which the labels indicate which mixture component generated which point. How could you use the EM algorithm to create probabilistic assignments for *unlabeled* test points? The basic idea is to apply the M-step *only to the labeled training data* and estimate all the parameters. This step is greatly simplified because the posterior probabilities of the labeled points are all pre-defined to be either 0 or 1 rather than soft "membership weights." Furthermore, the M-step does not need to be iteratively repeated because the labeling is assumed to be an unquestioned ground truth, which cannot be improved upon. Subsequently, these estimated parameters are used with the *unlabeled* points to assign probabilities in one execution of the E-step. This process we have just described is the same as the naïve Bayes algorithm.

> The naïve Bayes classification algorithm is a rudimentary special case of the expectation-maximization algorithm in which the M-step is applied once to the labeled training data and the E-step is applied once to the unlabeled test data.

This connection between clustering and classification can be extended to any type of classifier (cf. Section 4.9) and not just naïve Bayes.

Is it also possible to use the unlabeled data in the M-step? If we choose to do so, the algorithm remains iterative, and the resulting algorithm [419] is referred to as *semi-supervised* classification. This type of algorithm can sometimes perform more accurate classification than the naïve Bayes algorithm, particularly if the amount of available labeled data is small. This algorithm is discussed in Section 5.3.6 of Chapter 5.

4.5   The k-Means Algorithm

The k-means algorithm is a very simple clustering algorithm that identifies a strict partitioning of the data into k clusters. The value of k is an input parameter to the algorithm. Consider an $n \times d$ data matrix in which the ith row vector (document) is denoted by $\overline{X_i}$. The k-means problem is that of finding the k d-dimensional representatives $\overline{Y_1} \ldots \overline{Y_k}$, such that the sum of squared distances of each document to its closest centroid is as low as possible. In other words, we wish to determine $\overline{Y_1} \ldots \overline{Y_k}$, so that the following objective function is

---

[4]This model is discussed only in later chapters. The uninitiated reader may choose to skip over this section in the first reading.

minimized:

$$J = \sum_{i=1}^{n} \min_{j=1}^{d} ||\overline{X_i} - \overline{Y_j}||^2 \tag{4.13}$$

Note that this objective function uses the Euclidean distance, which is unusual for text data. However, for the purpose of the following discussion, assume that the document-term matrix is normalized as a preprocessing step, so that the L_2-norm $||\overline{X_i}||$ of each document is one unit. As discussed in Section 2.5 of Chapter 2 (cf. Equation 2.9), there is no difference between the use of the Euclidean distance, cosine similarity, or the dot product similarity, after such a normalization has been performed. We will first discuss a simple k-means algorithm with this *length-wise* normalization assumption, and then discuss the heuristic variations utilized for text data.

The main obstacle to solving Equation 4.13 is that the optimal assignments of data points to representatives depend on the values of the representatives, and the representatives themselves depend on these assignments in a circular way. This circularity naturally suggests an iterative approach, in which we alternately determine the best assignments (while fixing the representatives) and determine the best representatives (while fixing the assignments) until convergence is achieved. Therefore, the k-means algorithm starts by initializing a set of k seed representatives $\overline{Y_1} \ldots \overline{Y_k}$ as k randomly chosen documents, and improves them by using the following pair of iterative steps:

1. **Optimal assignments with fixed representatives:** Each document is assigned to the representative to which it has the largest cosine measure. For normalized data, maximization of the cosine is the same as the minimization of the Euclidean distance, and therefore this assignment provides the lowest objective function for Equation 4.13. Assume that the n points are partitioned into k clusters denoted by $\mathcal{C}_1 \ldots \mathcal{C}_k$, where each cluster $\mathcal{C}_i$ contains a subset of points in $\{\overline{X_1} \ldots \overline{X_n}\}$.

2. **Optimal representatives with fixed assignments:** If the assignments are fixed then one can separately determine the optimal value of $\overline{Y_r}$ for the rth cluster $\mathcal{C}_r$, which turns out to be the centroid of that cluster:

$$\overline{Y_r} = \frac{\sum_{\overline{X_i} \in \mathcal{C}_r} \overline{X_i}}{|\mathcal{C}_r|} \tag{4.14}$$

The proof of this result is provided in Lemma 4.5.1 and its intuitive explanation is that a cluster is best represented by its most central point for minimizing error.

The two steps are then iterated to convergence. Typically, the convergence criterion is that the objective function does not change by a certain minimum amount. Furthermore, there is typically also a bound on the maximum number of iterations in order to prevent very long running times.

We now show that the contribution of the rth cluster to objective function J with assigned points in $\mathcal{C}_r$ if the representative $\overline{Y_j}$ is chosen to be the centroid of $\mathcal{C}_r$.

Lemma 4.5.1 *Let the contribution J_r of the rth cluster $\mathcal{C}_r$ to the objective function value of Equation 4.13 be defined as follows:*

$$J_r = \sum_{\overline{X_i} \in \mathcal{C}_r} ||\overline{X_i} - \overline{Y_r}||^2 \tag{4.15}$$

Then, the value of J_r is minimized when $\overline{Y_r}$ is chosen to be the centroid of $\mathcal{C}_r$.

Algorithm *KMeans*(Documents: $\overline{X_1} \ldots \overline{X_n}$, Number of clusters: k)
begin
    Initialize each of $\overline{Y_1} \ldots \overline{Y_k}$ to random points from $\overline{X_1} \ldots \overline{X_n}$;
    repeat
        Create partitioning $\mathcal{C}_1 \ldots \mathcal{C}_k$ by assigning each $\overline{X_i}$ to
            its closest representative (i.e., largest cosine) from $\overline{Y_1} \ldots \overline{Y_k}$;
        for each cluster $\mathcal{C}_r$ set $\overline{Y_r}$ to the centroid of $\mathcal{C}_r$;
    until convergence;
    return $\mathcal{C}_1 \ldots \mathcal{C}_k$;
end

Figure 4.4: The k-means algorithm

Proof: The gradient of the objective function with respect to $\overline{Y_r}$ needs to be set to 0 as the optimality condition. Setting the gradient to 0 leads to the following optimality condition:

$$\sum_{\overline{X_i} \in \mathcal{C}_r} 2 \cdot (\overline{Y_r} - \overline{X_i}) = 0 \qquad (4.16)$$

This condition simplifies to the following:

$$\overline{Y_r} = \frac{\sum_{\overline{X_i} \in \mathcal{C}_r} \overline{X_i}}{|\mathcal{C}_r|} \qquad (4.17)$$

In other words, the gradient-based optimality condition implies that $\overline{Y_r}$ is the centroid of $\mathcal{C}_r$. ∎

The aforementioned discussion provides a mathematical description of using *normalized* vectors that is *theoretically* optimal. However, for text data, a few practical changes are made that deviate from this presentation in the following ways:

1. We do not normalize the documents length-wise to unit norm up front. Note that the cosine function is insensitive to the length-wise normalization step and therefore the similarity computation is not affected. Furthermore, the use of cosine ensures that we can drop the factor of $|\mathcal{C}_r|$ in the denominator of Equation 4.14 without changing the similarity computation. However, if the documents are not normalized to unit norm up front, the solution will no longer be exactly the same because longer documents will have more influence on the centroids. The practical effect of this change is, however, not significant in most reasonable settings.

2. The infrequent terms in the centroid of a cluster can be dropped [504]. Dropping infrequent terms has the dual advantage of improving the quality of computation (by removing noise) and efficiency (by reducing the number of computations). It was suggested in [504] that as few as 200 to 400 of the most frequent words can be retained in the centroid of each cluster.

The frequent words in the centroid of each cluster provide a *cluster digest* that summarizes the topical content of the cluster. A partial example [7] of a cluster digest containing a cluster of documents related to American history is as follows:

history (183), lincoln (122), washington (23), abolition (38), constitution (95), bill (124), independence (165), columbus (63), settlers (44), civil (91), president (105), war (83), treaty (36), jefferson (23), confederate (43), union (29), british (61), ...

The numbers in the brackets represent the term weights in the truncated centroid. Because of truncation, only large term weights are retained. Thus, by examining the frequent words in each cluster, it is often possible to get an idea of the semantic content of the cluster. A pesudo-code of the k-means algorithm is provided in Figure 4.4.

4.5.1   Convergence and Initialization

It is noteworthy that each execution of the aforementioned steps is guaranteed to not worsen the objective function J of Equation 4.13, and therefore the objective function changes *monotonically* with algorithm progression. Since the number of possible clusterings is finite, a monotonically changing objective function is usually a recipe for convergence after a finite number of iterations. One needs to be careful that ties in assignment are broken using a consistent rule (e.g., using the lowest cluster index), so that the algorithm will never cycle to the same solution unless it has converged to a fixed point. Although convergence of the k-means algorithm is guaranteed, it is not guaranteed to converge to a global optimum solution. In particular, the algorithm can be sensitive to the choice of seeds that are selected up front. If outliers are selected as the initial seeds, the quality of the approach can be poor. In fact, it is often better to use completely randomly generated vectors as initialization points, rather than the use of a document from the collection. The k-means algorithm is often combined with other hierarchical algorithms [141] to provide a high-quality starting point. Such an approach is described in Section 4.6.2.

4.5.2   Computational Complexity

The k-means typically requires a relatively small number of iterations. It is not uncommon to require less than 10 iterations for the algorithm to give high-quality solutions, provided that a reasonable starting point is used. In this sense, the use of a proper starting point with other algorithms becomes even more important. For all practical purposes, the number of iterations is assumed to be a constant.

In each iteration, the similarities of n documents to k clusters is computed. This process requires $O(n \cdot k)$ similarity computations. If the number of words in each centroid is restricted to maximum value $d_t \ll d$, the time complexity of each similarity computation is $O(d_t)$. Therefore, the overall computational complexity is given by $O(n \cdot k \cdot d_t)$.

4.5.3   Connection with Probabilistic Models

The k-means algorithm can be viewed as the deterministic avatar of the expectation-maximization (EM) algorithm. Just as EM algorithms determine a probabilistic assignment of documents to clusters (E-step), the k-means algorithm computes a deterministic assignment in each iteration. The EM algorithm optimizes the parameters of its mixture component in the M-step, whereas the k-means algorithm determines the optimal representative (i.e., centroid) in each iteration. Just as the EM-algorithm optimizes the mean-squared error, the EM algorithm maximizes a log-likelihood criterion. In fact, with mixture distributions like the Gaussian, the log-likelihood of a Gaussian simplifies to the Euclidean distance! Of course, one rarely uses the Gaussian modeling assumption in the text domain. Nevertheless, the connections between the two methods are useful to keep in mind. The expectation-maximization algorithm flexibility of incorporating domain-specific knowledge about the corpus by choosing a particular mixture distribution (e.g., Bernoulli or multinomial). On the other hand, the k-means algorithm has the advantage of greater simplicity.

The benefits of simplicity should not be underestimated, because it makes the k-means approach more robust and less likely to get stuck in local minima.

4.6 Hierarchical Clustering Algorithms

Hierarchical clustering algorithms naturally create a tree-like structure (or *taxonomy*) of the documents. The creation of the taxonomies has a special place in document clustering because of its ability to enable intuitive browsing of large collections. The taxonomy can be created in either top-down or bottom-up fashion. The bottom-up technique can be viewed as a standalone clustering method, whereas the top-down approach can be viewed as a meta-algorithm that uses another clustering algorithm as a subroutine. This makes the bottom-up approach inherently more interesting, and it is the primary focus of this section.

Bottom-up methods start with individual documents in each cluster and successively agglomerate them into higher-level clusters by merging similar pairs of clusters. This successive merging leads to a natural hierarchical relationship among the clusters, where clusters in later stages are supersets of clusters in earlier stages. The main differences among different hierarchical methods arise because of the differences in the criteria with which clusters are successively merged. The basic framework for a hierarchical clustering algorithm is to always work with a *current cluster set* (or model) $\mathcal{M} = \{\mathcal{C}_1 \ldots \mathcal{C}_m\}$ of clusters and reduce the size of this set by 1 by using the following iterative step:

> Determine the most similar pair of clusters $(\mathcal{C}_i, \mathcal{C}_j)$ from $\mathcal{M} = \{\mathcal{C}_1 \ldots \mathcal{C}_m\}$ and replace them with a single larger cluster containing the points in both clusters as follows:

$$\mathcal{M} \Leftarrow \mathcal{M} - \underbrace{\{\mathcal{C}_i, \mathcal{C}_j\}}_{\text{Remove similar pair}} \cup \underbrace{\{\mathcal{C}_i \cup \mathcal{C}_j\}}_{\text{Add larger cluster}} \qquad (4.18)$$

Therefore, each step of this clustering process reduces the number of clusters by 1, until one arrives at the desired number of clusters. In the initialization step, each document lies in its own cluster and therefore there are n clusters. After the first merge of the most similar pair of singleton clusters, a single cluster will contain two documents, and therefore there will be $(n - 1)$ clusters. In general, the hierarchical clustering process will require us to determine the similarity between *sets* of documents, $\mathcal{C}_i$ and $\mathcal{C}_j$, in order to determine which pair to select for merging.

How can one determine the similarity between two sets, $\mathcal{C}_i$ and $\mathcal{C}_j$, of documents? It turns out that there is no single way to do this. There are $|\mathcal{C}_i| \times |\mathcal{C}_j|$ possible pairwise similarity computations between the documents in these sets and one must somehow combine these similarity values to create a global measure of the similarity between these sets. For example, there are $4 \times 2 = 8$ possible similarity computations between the sets of points shown in Figure 4.5. There are several natural criteria for combining the similarities, which lead to the different variations of this *family* of bottom-up clustering algorithms:

1. *Single-linkage clustering:* In single-linkage clustering, the similarity between the closest pair of documents from $\mathcal{C}_i$ and $\mathcal{C}_j$ is used to quantify the similarity between $\mathcal{C}_i$ and $\mathcal{C}_j$. Therefore, if s_{ij} is the similarity between clusters $\mathcal{C}_i$ and $\mathcal{C}_j$, then we have:

$$s_{ij} = \text{MAX}_{\overline{X} \in \mathcal{C}_i, \overline{Y} \in \mathcal{C}_j} \text{cosine}(\overline{X}, \overline{Y}) \qquad (4.19)$$

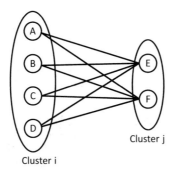

Figure 4.5: The similarity between the two sets of documents is expressed as a function of the similarities between individual document pairs

2. *Group-average linkage clustering:* In group-average linkage the average of the similarities between all documents in $\mathcal{C}_i$ and all documents in $\mathcal{C}_j$ is computed. Therefore, if s_{ij} is the similarity between clusters $\mathcal{C}_i$ and $\mathcal{C}_j$, we have:

$$s_{ij} = \text{MEAN}_{\overline{X} \in \mathcal{C}_i, \overline{Y} \in \mathcal{C}_j} \text{cosine}(\overline{X}, \overline{Y}) \qquad (4.20)$$

3. *Complete linkage clustering:* In complete linkage clustering, the similarity between the most dissimilar pair is used as the relevant criterion. Therefore, the similarity between the cluster pairs $\mathcal{C}_i$ and $\mathcal{C}_j$ is defined as follows:

$$s_{ij} = \text{MIN}_{\overline{X} \in \mathcal{C}_i, \overline{Y} \in \mathcal{C}_j} \text{cosine}(\overline{X}, \overline{Y}) \qquad (4.21)$$

4. *Centroid similarity:* In centroid similarity, the cosine similarity between the centroids of $\mathcal{C}_i$ and $\mathcal{C}_j$ is used as the merging criterion.

Both the single-linkage and the complete-linkage criteria have weaknesses that are caused by the fact that they depend on a *single* pair of documents in order to compute the similarity criterion. In the case of single-linkage clustering, the main problem is that of *chaining* where a sequence of successive merges caused by individual pairs of documents eventually leads to the merging of unrelated groups of documents. For example, consider a set of four clusters, which contain the following representative documents:

Cluster 1 contains: "The sergeant looked at the platoon."
Cluster 2 contains: "The sergeant looked at the moon."
Cluster 3 contains: "The dog looked at the moon."
Cluster 4 contains: "The dog howled at the moon."

It is easy to see that successive clusters contain very similar documents although there is no relationship between the documents in clusters 1 and 4. It is quite conceivable that successive merges might eventually lead to the merging of these clusters because of the presence of a *chain* of similar documents between the pair. An example of such a chain of undesirable merges is shown in Figure 4.6(a). Indeed, this situation occurs annoyingly often with single-linkage clustering algorithms. The complete-linkage method also performs poorly at later stages of the merging. When the clusters are large, they will often contain pairs of outlier documents between which the similarity is quite low. In general, complete-linkage similarity computation is often dominated by the outliers in these clusters. Clearly,

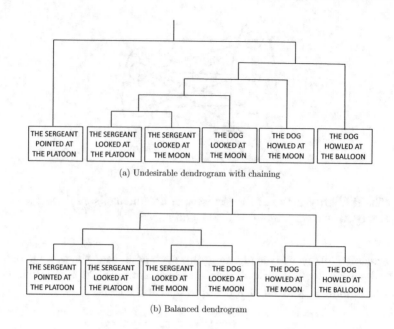

(a) Undesirable dendrogram with chaining

(b) Balanced dendrogram

Figure 4.6: Different dendrograms of the same set of six documents

making merging decisions about clusters based on the properties of outliers (i.e., *atypical* points) inside them does not seem to be a wise choice. Therefore, group-average linkage and centroid clustering are more desirable choices.

This successive clustering process leads to a natural hierarchy of the clusters, which is referred to as a *dendrogram*. A dendrogram is a binary tree in which each merge is an internal node of the tree, and its two children nodes correspond to the sets of clusters that have been merged. Therefore, lower nodes of the tree are more fine grained, and the leaf nodes contain individual documents. An example of two possible dendrograms based on a different sequence of merges from the same set of documents is shown in Figure 4.6. In one case, the dendrogram is well balanced. whereas in another it is not. In general, one has little control on the shape of the dendrogram in bottom-up clustering methods, and exercising good judgement in the choice of merging function is crucial. For example, a poorly structured dendrogram like that in Figure 4.6(a) can be caused by single linkage clustering. One can obtain a flat clustering from this dendrogram by cutting it at a higher level of the tree. The default approach of hierarchical clustering cuts the dendrogram in a specific way, which depends on the order in which the merges are performed. To create a flat clustering with k clusters, one can omit the last set of $(k-1)$ merges. However, it is also possible to construct the dendrogram up to the root and then use the hindsight gained from the structure in the dendrogram. In such a case, the dendrogram can be cut so as to obtain a clustering that seems semantically appealing to a domain expert (on manual inspection) or to obtain a more balanced clustering structure.

4.6.1   Efficient Implementation and Computational Complexity

It is important to implement the approach properly in order to obtain fast performance. For example, a naive implementation might compute $m \times m$ similarities in each step, when

m clusters remain in the data. This is obviously not optimal because most of the pairwise similarities between clusters can be carried over from one step to the next without re-computation. At any given moment in time, when m clusters remain in the data, an $m \times m$ similarity matrix S is maintained. This similarity matrix is updated (and shrinks in size) as the clusters are successively merged over the course of algorithm progression.

At the very beginning of the algorithm, an $n \times n$ similarity matrix $S = [s_{ij}]$ is computed between all pairs of documents, in which the (i, j)th entry s_{ij} corresponds to the similarity between the ith and jth documents. As the algorithm progresses and clusters are merged, the indices of the clusters are updated, and the entry s_{ij} corresponds to the similarity between the ith and jth cluster in the data. During a merge of $\mathcal{C}_i$ and $\mathcal{C}_j$, the rows/columns for the ith and jth clusters need to be removed and a new row/column needs to be added to the similarity matrix for the merged cluster. Therefore, we need a way to compute the similarity between this new cluster and every other cluster in the data. For the case of centroid similarity, this re-computation is a simple matter of just recomputing the centroid of the new cluster, and computing its similarity with respect to the centroids of the remaining clusters. However, even for the other cases, this re-computation is generally quite simple. Let $Sim(\mathcal{C}_i \cup \mathcal{C}_j, \mathcal{C}_k)$ be the similarity of any other cluster $\mathcal{C}_k$ with the merged cluster $\mathcal{C}_i \cup \mathcal{C}_j$. Then, one can compute the new similarities in terms of the current entries of the similarity matrix S as follows:

$$
Sim(\mathcal{C}_i \cup \mathcal{C}_j, \mathcal{C}_k) = \begin{cases} \max\{s_{ik}, s_{jk}\} & \text{(Single-Linkage Clustering)} \\ \frac{s_{ik} \cdot |\mathcal{C}_i| + s_{jk} \cdot |\mathcal{C}_j|}{|\mathcal{C}_i| + |\mathcal{C}_j|} & \text{(Group-Average Linkage Clustering)} \\ \min\{s_{ik}, s_{jk}\} & \text{(Complete Linkage Clustering)} \end{cases} \quad (4.22)
$$

Therefore, when a cluster is merged, one only has to drop the rows/columns for clusters $\mathcal{C}_i$ and $\mathcal{C}_j$ from the similarity matrix, and add a single row/column for the merged cluster. Therefore, the number of rows and number of columns both reduce by 1.

For a corpus containing n documents, the space complexity of the approach is $O(n^2)$, which is the size of the similarity matrix at the very beginning of the algorithm. The computation of the similarity matrix requires $O(n^2)$ cosine similarity computations at the very beginning of the algorithm. Let d_a be the average document size. Since the cosine similarity computation is linearly related to average document size, the initialization of the similarity matrix requires $O(n^2 d_a)$ time. In addition, the algorithm contains $O(n)$ merges. However, other than in the case of centroid merging (see Exercise 7), this step is independent of document size. This is because each similarity re-computation in Equation 4.22 requires only $O(1)$ time rather than $O(d_a)$ time and there are $O(n)$ such computations for the various clusters. Therefore, the total time for similarity re-computation is $O(n^2)$. In addition, one must determine the similarity of the highest quality among $O(n)$ possible values, which requires $O(n^2 \cdot \log(n))$ time over the course of the algorithm is a heap data structure is maintained. Therefore, the overall computational complexity is $O(n^2 \cdot d_a + n^2 \log(n))$, of which $O(n^2 \cdot d_a)$ turns out to be running time of the initialization step. For values of d_a of the order of a couple of hundred words, it is possible for the initialization time to become both the running time and space bottleneck for the algorithm.

The space complexity is particularly problematic even for data sets of modest size. For example, if the corpus contains a million documents, the space complexity is of the order of 10^{12} bytes, which is about a terabyte. In the modern age, it is not uncommon to encounter collections of such sizes. This space complexity increases by a factor of 100 for every ten-fold increase in corpus size. In cases, where the similarities cannot be maintained in main memory, they may need to be recomputed from scratch in each iteration. This would dramatically increase the time-complexity to $O(n^3)$, which is unmanageable even for small data sets containing a few thousand documents. Luckily, one good property of hierarchical methods is that they provide excellent clusterings even on small samples of the data. In such cases, they can be combined with k-means methods to obtain the best of both worlds. This approach is described in the next section.

4.6.2   The Natural Marriage with k-Means

Hierarchical algorithms and k-means algorithms have strengths and weaknesses that are complementary in terms of running time and accuracy. The k-means algorithm is efficient and generally accurate on large data sets, unless the seed set is very poor. On the other hand, a hierarchical clustering algorithm is expensive, but it is quite robust even when applied to a small sample of the data. This observation suggests that a hierarchical method can be used to merge a relatively small sample of documents to a robust set of k clusters, whose centroids can be used to create an excellent seed set for the k-means algorithm. This results in a two-phase approach in which the first phase uses hierarchical clustering and the second phase uses k-means.

The size of the sample used in the first phase should be such that the running times of the two phases are balanced. The running time of the k-means algorithm is $O(k{\cdot}n{\cdot}d_t)$, where d_t is the average lexicon size retained in each centroid. The running time of the hierarchical approach for a sample of size s is roughly given by $O(s^2 d_t + s^2 \log(s)) \approx O(s^2 d_t)$, if we assume that d_t is larger than $\log(s)$. These are reasonable assumptions to make in practical settings. In order for the running time to be balanced between the phases of k-means and hierarchical clustering, the following condition must hold:

$$s^2 \cdot d_t = k \cdot n \cdot d_t \qquad\qquad (4.23)$$

Therefore, we have $s = \sqrt{k \cdot n}$. In such a case, the running time of the two-phase approach is given by $O(s^2 d_t) = O(k \cdot n \cdot d_t)$, which is linear in the corpus size and the number of clusters. This is generally the best running time that one can hope to achieve with a clustering algorithm.

The above description of the hierarchical phase is (roughly) that of a technique, referred to as *buckshot* [141]. Another alternative for the hierarchical phase is referred to as *fractionation* [141]. The fractionation method is the more robust one, but the buckshot method is faster in many practical settings. Unlike the buckshot method, which uses a sample of $\sqrt{k \cdot n}$ documents, the fractionation method works with all the documents in the corpus. The fractionation algorithm initially breaks up the corpus into n/m buckets, each of size $m > k$ documents. An agglomerative algorithm is applied to each of these buckets to reduce them by a factor $\nu \in (0,1)$. This step creates $\nu \cdot m$ agglomerated documents in each bucket, and therefore $\nu \cdot n$ agglomerated documents over all buckets. An "agglomerated document" is defined as the concatenation of the documents in a cluster. The entire process (including the creation of m buckets) is repeated by treating each of these agglomerated documents as a single document. The approach terminates when a total of k seeds remains.

It remains to be explained how the documents are partitioned into buckets. One possibility is to use a random partitioning of the documents. However, a more carefully designed procedure sorts the documents by the index of the jth most common word in the document. Here, j is chosen to be a small number, such as 3, that corresponds to medium frequency words in the documents. Contiguous groups of m documents in this sort order are mapped to clusters. This approach ensures that the resulting groups have at least a few common words in them and are therefore not completely random.

The agglomerative clustering of m documents in the first iteration of the fractionation algorithm requires $O(m^2)$ time for each group, and sums to $O(n \cdot m)$ over the n/m different groups. Because the number of individuals reduces geometrically by a factor of ν in each iteration, the total running time over all iterations is $O(n \cdot m \cdot (1 + \nu + \nu^2 + \ldots))$. For $\nu < 1$, the running time over all iterations is still $O(n \cdot m)$. By selecting $m = O(k)$, one still ensure a running time of $O(n \cdot k)$ for the initialization procedure.

4.7 Clustering Ensembles

Clustering is an unsupervised problem and therefore it is possible for specific parameter or algorithmic choices to perform poorly in an individual run. However, by combining the results from multiple runs, the results are more robust. Such methods are referred to as *ensembles*. It is common for the combined result to be better than the results obtained from *most* of the individual runs. This effect is observed because of the impact of *variance reduction* (cf. Section 7.2 of Chapter 7). The basic idea of a clustering ensemble is to use the following two steps:

1. Apply clustering on the data set D for m times to obtain m different partitions by either using a randomized variant of the algorithm, or by using different clustering algorithms/parameter choices in each run. Each such run is referred to as an *ensemble component* or *base method*.

2. Merge the results from the m runs in the first step to obtain a single (more robust) clustering. This step is referred to as the *consensus step*, and it usually requires the application of a (simple) clustering algorithm on the point-to-cluster assignment information obtained in the first phase. The basic idea of the consensus phase is that pairs of points that repeatedly get assigned to the same cluster over different ensemble components should be grouped in the same cluster at the very end.

The following sections will describe each of these steps in detail.

4.7.1 Choosing the Ensemble Component

The specific ensemble component chosen will have different effects on the overall accuracy and efficiency of the clustering approach. The most common ways of choosing the different ensemble components are as follows:

1. One can choose to use different clustering algorithms such as the k-means, hierarchical, and the EM-methods in different ensemble components.

2. One can use a randomized approach such as k-means with different choices of initialized seeds. This will result in different outputs from different ensemble components.

3. One can run a clustering algorithm on a *derived data set* containing a subset of features. This approach is referred to as *feature bagging* or *multiview clustering*.

4. One can apply the computationally intensive parts of the clustering algorithm to only a subsample of the data set. A useful example is the pairing of Nyström sampling with ensemble methods. Kernel methods are often used in the text domain to incorporate the sequential ordering information between the words, although the primary impediment in their use is their high computational complexity. Subsampling methods

have great power when paired with base methods of super-linear complexity, because they can improve *both* the accuracy *and* the efficiency of the approach. In such cases, running many ensemble components on subsamples is faster than a single application of the base method on the full data set. The leveraging of the Nyström ensemble for incorporating sequence knowledge into text clustering methods is described in Section 4.8.

In general, it is advisable to use base methods with high diversity in order to get the most out of an ensemble method.

4.7.2   Combining the Results from Different Components

The final step in an ensemble method is to combine the results from different components. It turns out that the final step of combining the results is also a clustering problem on the outputs of the results. However, this new clustering problem is much simpler because the outputs have a more natural tendency to cluster. Therefore, very simple clustering methods like k-means can be used effectively in these cases. In the final phase, each point is represented in m keywords, where m is the number of ensemble components. Each keyword in this new pseudo-document corresponds to an ensemble component, and it contains a concatenation of its cluster identifier and ensemble component identifier. For example, if a document was assigned to cluster identifier 23 in the 45th ensemble component, then the keyword "23#45" is created and added to the new representation of that document. Therefore, each document will have exactly as many keywords as the number of ensemble components. Two documents that co-occur in the same cluster frequently will have many keywords in common. One can view this new representation as a kind of *feature engineering* like the *stacking ensemble* in classification [2]. The new features will have an extremely high tendency to cluster because of the fact that they are cluster identifiers from obtained from various base clustering methods. Therefore, a simple application of a k-means approach on this new representation will provide high-quality results.

4.8   Clustering Text as Sequences

Most mining methods use the bag-of-words model when working with text. However, a lot of knowledge in the collection is hidden in the sequential and positioning information between words. For short documents, the use of sequential information becomes more critical, because a bag of words is often too sparse to be used robustly with conventional mining techniques.

   Almost all the successful learning methods that treat text as sequences use representation learning and feature engineering in one form or the other. Typically, the sequential representation of text is converted into a multidimensional representation that encodes information about sequential word ordering. This multidimensional representation may often be quite high-dimensional, because sequence information is inherently complex. The common feature engineering methods are as follows:

1. *Enriching with phrases and k-grams:* Frequent phrases in the corpus can be used as features to enrich the representation. It has been shown [615] that using frequent phrases improves the quality of clustering of Web documents. One can also add frequent k-grams to the vector-space representation to use the bag-of-words model.

2. *Kernel methods:* Kernel methods provide a natural approach for incorporating sequence information into text mining applications by using string-based kernel func-

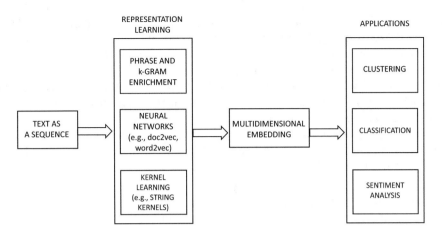

Figure 4.7: Representation learning for converting sequences to semantically knowledgeable embeddings

tions in order to perform nonlinear dimensionality reduction. Refer to Section 3.6 of Chapter 3.

3. *Neural networks:* In recent years, a number of neural network techniques such as *word2vec* [389] and *doc2vec* [314] have been proposed that incorporate sequence information in creating word and document embeddings. These methods are discussed in detail in Chapter 10.

All these methods achieve the same task of creating a *multidimensional* representation that can be leveraged with existing off-the-shelf methods for clustering. This overall framework is shown in Figure 4.7.

4.8.1 Kernel Methods for Clustering

As discussed in Section 3.6 of Chapter 3, string kernels can be used to incorporate sequence information into kernel representations. The basic idea is that these methods work with similarity matrices (defined on the string representations) rather than the document-term representations. The most common algorithms for kernel-based clustering include kernel k-means and spectral clustering. The former performs implicit feature engineering with the kernel trick, whereas the latter performs explicit feature transformation with only a small number of eigenvectors. In the following, we describe both these methods.

4.8.1.1 Kernel k-Means

The kernel k-means approach can be implemented in various ways, the most convenient of which is with the use of a normalized similarity matrix. Let $S = [s_{ij}]$ be an $n \times n$ similarity matrix, which contains the pairwise similarity information between the string representations of the documents. The matrix is normalized as follows:

$$s_{ij} \Leftarrow \frac{s_{ij}}{\sqrt{s_{ii}} \cdot \sqrt{s_{jj}}} \qquad (4.24)$$

The effect of the normalization is that every data point lies on the unit sphere in (transformed) kernel space. As discussed in Section 2.5 of Chapter 2, it is equivalent to use the dot product, the Euclidean distance, or the cosine with such a normalization. Therefore, one can use the dot product for simplicity. Assume that the transformation implied by the kernel similarity matrix is denoted by $\Phi(\cdot)$ so that $s_{ij} = \Phi(\overline{X_i}) \cdot \Phi(\overline{X_j})$.

The kernel k-means algorithm proceeds as follows. We start with a random assignment of points to the k clusters, denoted by $\mathcal{C}_1 \ldots \mathcal{C}_k$. The usual implementation of the k-means algorithm determines the centroids of the clusters as the representatives of the next iteration. The kernel k-means algorithm computes the dot product of each point to the various clusters *in transformed space* and re-assigns each point to its closest centroid in the next iteration. How can one compute the dot product between a transformed point $\Phi(\overline{X_i})$ and the centroid $\overline{Y_j}$ of $\mathcal{C}_j$ (in transformed space)? This can be achieved as follows:

$$\Phi(\overline{X_i}) \cdot \overline{Y_j} = \Phi(\overline{X_i}) \cdot \frac{(\sum_{q \in \mathcal{C}_j} \Phi(\overline{X_q}))}{|\mathcal{C}_j|} = \frac{\sum_{q \in \mathcal{C}_j} \Phi(\overline{X_i}) \cdot \Phi(\overline{X_q})}{|\mathcal{C}_j|} = \sum_{q \in \mathcal{C}_j} \frac{s_{iq}}{|\mathcal{C}_j|}$$

Therefore, for any given point $\overline{X_i}$, we only need to compute its average kernel similarity to all points in that cluster. The basic idea of being able to perform such operations in transformed space without explicitly performing the transformation is referred to as the *kernel trick*.

Instead of the centroids, the approach does require the explicit maintenance of assignments of each point to various clusters in order to recompute the assignments for the next iteration. As in all k-means algorithms, the approach is iterated to convergence. For a data set containing n points, the approach requires $O(n^2)$ time in each iteration of the k-means algorithm, which can be quite costly for large data sets. The approach also requires the computation of the entire kernel matrix, which might require $O(n^2)$ storage. However, if the similarity function can be computed efficiently, then one does not need to store the kernel matrix a priori, but simply recompute individual entries on the fly when they are needed. The main problem is that many substring similarity functions require dynamic programming, which are not particularly efficient to compute. In such cases, one must consider the fact that the approach will require as many as $O(n^2)$ similarity computations *in each iteration* of the k-means method. Another disadvantage of the kernel trick is that it can be paired with only a restricted subset of clustering algorithms (e.g., k-means) that use similarity functions between points. Not all clustering algorithms are equally friendly to the use of the kernel trick. Furthermore, one can perform no further engineering or normalization of the extracted features, if they are being used only indirectly via the kernel trick.

4.8.1.2   Explicit Feature Engineering

Explicit feature engineering works by actually materializing the kernel SVD transformation and applying an off-the-shelf algorithm on the transformed data. This is a more flexible approach of enabling arbitrary clustering algorithms to work with kernel transformations, rather than simply algorithms (like k-means) that can be expressed in terms of pairwise similarities. The broader approach of explicit feature engineering works by diagonalizing an $n \times n$ similarity matrix $S = Q\Sigma^2 Q^T$ as follows:

Diagonalize $S = Q\Sigma^2 Q^T$;
Extract the n-dimensional embeddings in rows of $Q\Sigma$;
Drop any zero eigenvectors from $Q\Sigma$ to create $Q_0\Sigma_0$;
Apply any existing clustering algorithm on rows of $Q_0\Sigma_0$;

The columns of Q_0 contain the non-zero eigenvectors, and the n rows of $Q_0\Sigma_0$ contain the embeddings of the n points. It is noteworthy that all n eigenvectors are extracted and only the zero eigenvectors are dropped. Such zero eigenvectors show up as zero columns in $Q\Sigma$. Explicit feature engineering is exactly equivalent to the use of the kernel trick only when all non-zero eigenvectors (no matter how small) are retained. The embedding dimensionality can be as large as the number of points n, if no dimensions are dropped. The *space requirements* of such an approach can therefore be $O(n^2)$. Furthermore, the running time requirement for extracting all n eigenvectors is $O(n^3)$, which can be prohibitive.

A natural question arises as to whether one can drop the lower-order eigenvectors to improve the space requirements and computational efficiency. Indeed, many implementations of kernel methods such as spectral clustering do drop lower-order eigenvectors. However, dropping lower-order eigenvectors is not without its pitfalls. Often, a complex data set will require a large number of dimensions to express the complex variations in the local shapes of the data distribution. An example is provided in Section 3.6, in which a large number of eigenvectors are required to express non-convex clusters (cf. Figure 3.9(a)). In this sense, nonlinear dimensionality reduction methods should be viewed as feature engineering techniques (unlike linear dimensionality reduction methods where the primary goal is feature space compression). The main problem in unsupervised settings is that it is hard to know the correct number of dimensions to use, and even extremely low-order eigenvectors are sometimes informative in complex distributions. The safest solution is to keep all nonzero eigenvectors (or drop only a small percentage of them). However, this might result in an n-dimensional data set with n points, which requires $O(n^2)$ space.

A solution to this computational dilemma is the use of Nyström sampling, which subsamples a set of s documents in order to create an s-dimensional representation. Typically, the value of s is independent of the corpus size, although it depends on the complexity of the underlying data distribution (e.g., number of clusters). Then, the approach proceeds as follows:

> Draw a subsample of s documents from the corpus;
> Use the Nyström method (cf. Section 3.6.2) to create an s-dimensional
>   representation of all documents denoted by the $n \times s$ matrix U_s;
> Apply any existing clustering algorithm on U_s;

In order to improve robustness, the approach can be used in the ensemble-centric setting discussed in Section 4.7. The clustering is repeated m times, and the results are integrated into a single robust clustering using the methodology discussed in Section 4.7.2.

4.8.1.3 Kernel Trick or Explicit Feature Engineering?

A natural question arises as whether one should use the kernel trick or explicit feature engineering. When using explicit feature engineering with the full data, the kernel trick provides equivalent results (with k-means), but will require $O(n^2)$ *similarity computations*, whereas feature engineering will require not only these similarity computations, but also an additional $O(n^3)$ time required for eigenvector extraction. The kernel trick will also require only $O(n)$ space, if one is willing to recompute kernel similarities every time they are required. Therefore, it would make sense to use the kernel trick.

On the other hand, the choice is not quite as simple, if one uses Nyström sampling with m ensemble components. In terms of accuracy, the ensembles will almost always provide higher quality results because of variance reduction effects. The comparisons in computational time and space requirements are more interesting. The Nyström method requires $O(n \cdot s)$ similarity computations and $O(n \cdot s^2)$ time to extract the eigenvectors for each ensemble component. The time required for k-means clustering in each ensemble component is $O(n \cdot k \cdot s)$ with the s-dimensional engineered representation. Since there are m ensemble components, the overall time required is $O(n \cdot s \cdot m(k + T + s))$, which strictly dominates the $O(n \cdot k \cdot m)$ time required for the post-processing phase of the ensemble-centric method. Here, T is the time required for each similarity computation. The running time of kernel k-means is always $O(n^2 \cdot T)$.

   Which is larger? For substring kernels that use dynamic programming, the value of T can be quite large. However, even if we ignore this factor and set it to 1, it seems that the sampling approach has an advantage. If the corpus has a very large number of documents, it is possible for $s^2 \cdot m$ to be less than n. For example, if the corpus contains 100 million documents, then one would do better with Nyström sampling at $m = 20$ and $s = 2000$. This does not yet include the effect of expensive kernel computation. Increasing T to 1000 results in a break-even corpus size of a few hundred thousand documents. The only advantage of the kernel k-means approach is that the Nyström method requires $O(n \cdot s)$ space, whereas one can choose to compute all similarities on the fly with kernel k-means and reduce the space requirements to $O(n)$. However, this saving comes at the expense of repeated computation of the same kernel similarity value across different iterations of k-means. Explicit feature engineering also provides the opportunity to further enhance the extracted feature representation by normalization, or by using any of the feature selection methods discussed earlier in this chapter. These methods are not available by using the kernel trick. One can also use an arbitrary clustering algorithm, and not be restricted to the use of k-means. Therefore, explicit feature engineering has significant advantages, which are often not recognized when the kernel trick is used.

4.8.2   Data-Dependent Kernels: Spectral Clustering

Explicit feature engineering is useful in cases where *data-dependent kernels* are used. A data-dependent kernel adjusts the similarity matrix with local or global data statistics, and therefore the computation of any particular similarity value requires knowledge of the entire data distribution rather than just a pair of points. Spectral clustering is an instantiation of kernel k-means in which one is compelled to use explicit feature engineering rather than the kernel trick. This compulsion is caused by the data-dependent nature of the kernel and subsequent feature selection/normalization. Spectral clustering uses the following steps, which are refinements of those used in kernel k-means with explicit feature engineering:

1. **(Breaking inter-cluster links):** Let $S = [s_{ij}]$ be a symmetric $n \times n$ similarity matrix defined over n documents, in which s_{ij} is the similarity between documents i and j. The similarity matrix might be created with the use of a domain-specific similarity function such as a string subsequence kernel (cf. Section 3.6.1.3). The diagonal entries of S are set to 0. All pairs (i, j) are identified such that documents i and j are *mutual* κ-nearest neighbors of each other according to the similarity matrix S. Such similarity values, s_{ij}, are retained in S. Otherwise, the value of s_{ij} is set to 0. This step sparsifies the similarity matrix, and intuitively tries to "break" the inter-cluster links, so that the resulting points are less likely to be close to one another in the engineered representation. The number of nearest neighbors, κ, regulates the sparsity of the similarity matrix.

2. **(Normalizing for dense and sparse regions):** For each row i, the sum of each row in the symmetric matrix S is computed as follows:

$$S_i = \sum_j s_{ij}$$

Intuitively, the value of S_i quantifies the "density" in the locality of document i. Then, each similarity value is normalized using the following relation:

$$s_{ij} \Leftarrow \frac{s_{ij}}{\sqrt{S_i \cdot S_j}} = \frac{s_{ij}}{\text{GEOMETRIC-MEAN}(S_i, S_j)}$$

The basic idea is to normalize the similarities between documents with the geometric mean of the "densities" at their end points. Therefore, the similarity is *relative* to the *local* data distribution. For example, the similarity between two modestly similar documents in a local region belonging to a rare topic (e.g., beetle fighting) becomes magnified, whereas the similarity between two documents on a popular topic (e.g., stock market) is de-emphasized. This type of adjustment makes the similarity function more adaptive to the statistics of data locality. For example, if a document is in a very dense region, it facilitates the creation of a larger number of fine-grained clusters in that region. At the same time, it becomes possible to create fewer clusters with more widely separated points in sparse regions. An intuitive way of understanding this (in the context of a spatial application) is that population clusters in sparsely-populated Alaska would be geographically larger than those in densely-populated California.

3. **(Explicit feature engineering):** The resulting similarity matrix S is diagonalized to $S = Q \Delta Q^T$, where the columns of Q contain the eigenvectors, and Δ is a diagonal matrix containing the eigenvalues. Only the largest $r \ll n$ eigenvectors (columns) of Q need to be computed to create a smaller $n \times r$ matrix Q_0. Furthermore, each row of Q_0 is scaled to unit norm, so that all engineered points (i.e., rows of Q_0) lie on the unit sphere. This type of normalization ensures that the use of Euclidean distance between points is identical to the use of cosine similarity (cf. Equation 2.9 of Chapter 2). At this point, the k-means algorithm is applied on the normalized and engineered points with the Euclidean distance.

The first two steps change the kernel matrix in a data-dependent way because aggregated statistics from multiple points are used to change the entries. As in the case of spectral clustering, a data-dependent kernel often cannot be computed without materializing the similarity matrix first. Materializing the similarity matrix loses the space-efficiency advantage of the kernel trick over explicit feature engineering, which is one of the reasons that the kernel trick is not used in this case. Furthermore, the various adjustments to the engineered representation such as the dropping of lower-order eigenvectors cannot be exactly replicated with the kernel trick. Therefore, spectral clustering is a good example of the numerous advantages of explicit feature engineering over the kernel trick.

One quirk with spectral clustering is that the diagonal entries of S are set to 0, which will always[5] allow negative eigenvalues in $Q \Delta Q^T$. From this point of view, the spectral kernel S is not positive semi-definite, as is required in kernel methods. However, increasing all diagonal entries of S by an amount equal to the most negative eigenvalue does not change the eigenvectors (embedding), and also makes the matrix positive semi-definite. It is much easier to interpret spectral clustering as a kernel technique with this cosmetic change. The main difference is that one uses Q as the embedding in spectral clustering (rather than $Q\sqrt{\Delta}$), and the former is invariant to translation of the diagonal entries of S. Because of these types of minor quirks, it is often forgotten that the spectral method is an approximate instantiation of kernel k-means after data-dependent modification of the kernel matrix.

---

[5]Each of the respective sums of the diagonal entries of S and Δ are the same because the trace of a matrix is invariant under similarity transformation [530]. Therefore, the eigenvalues sum to 0. Unless all eigenvalues are 0 (i.e., $S = 0$), at least one negative eigenvalue will exist.

4.9   Transforming Clustering into Supervised Learning

A neat connection exists between unsupervised and supervised[6] learning, because of which clustering problems can be solved by repeated execution of any classification algorithm. This is a useful result, because it unlocks the use of hundreds of off-the-shelf classifiers for clustering. The expectation-maximization algorithm discussed in this chapter is a special case of this approach, which uses a repeated application of the naïve Bayes classifier. Similarly, the k-means algorithm can be viewed as a repeated application of the centroid classifier. In spite of the wide popularity of expectation-maximization and k-means, it is surprising that the notion of using an *arbitrary* classifier for clustering is rarely explored. Such an approach offers modeling choices with more interesting properties than the naïve Bayes or the centroid classifier. For example, numerous deep learning methods like long short-term memory (cf. Section 10.8.4 of Chapter 10) are designed for text classification. These classifiers treat text as sequences and their use as subroutines also results in the incorporation of semantic properties of text in clustering. In general, any classifier that works at the sequence level is potentially interesting.

The basic idea is to assume that each of the k clusters corresponds to a "class" in the data. We have a classification algorithm $\mathcal{A}$ available, which can be trained on a *labeled* data set, and it returns scores associated with each of the k different classes when applied to a *test instance* (i.e., unlabeled instance). Without loss[7] of generality, we can assume that the scores are non-negative and sum to 1. Furthermore, it is assumed that classifier $\mathcal{A}$ can be used in cases where the training instances are weighted, and the classifier gives proportional importance to instances in accordance with their weight. One can always convert an unweighted classifier to a weighted classifier by repeatedly sampling training instances in proportion to their weights, training the classifiers, and averaging the predictions over these models.

The approach starts by randomly assigning the n documents in the corpus to the k clusters (or, preferably, by using a simple algorithm like k-means), and estimates the initial parameters of algorithm $\mathcal{A}$ by applying the training step (see below) with respect to this "class" labeling. Subsequently, it uses the following two steps iteratively:

1. **(Prediction step):** Use currently trained algorithm $\mathcal{A}$ to predict scores of each class (i.e., cluster) for each document in the data set. For each document-class pair, create a training instance with weight equal to the corresponding score. This process will create a training data set with $O(n \cdot k)$ instances in which each document takes on all the labels, albeit with different weights.

2. **(Training step):** Train the algorithm $\mathcal{A}$ on the weighted training data set from the prediction step to create an updated model.

These steps are repeated to convergence. The reader is strongly encouraged to compare the iterative approach above with the expectation-maximization method discussed in Section 4.4. One issue in the training step is that the training data set contains k copies of the same instance, albeit with different weights and class labels. Some classifiers can cause problems with such data sets. One way of avoiding this problem is to sample only one of the duplicate instances in proportion to its weight while creating the training model.

---

[6]This section requires an understanding of the classification problem. We recommend the uninitiated reader to skip this section at the first reading of the book, and return to it only after covering the material in the next chapter. The notations and terminologies used in this section assume such an understanding.

[7]The sigmoid function $1/(1 + e^{-\lambda s})$ can be used to convert an arbitrary score s to the range $(0, 1)$, which is followed by normalizing the scores to sum to 1 over all classes.

Practical Issues

As in the case of expectation-maximization, a pervasive risk with the use of this approach is that it can get stuck in local minima. Local minima become particularly likely when a complex classifier with an *over-fitting* tendency is used. The approach depends crucially on generalizing and smoothing out the (initial) random variations in cluster identifier distribution into a more coherent distribution over many iterations. This is not possible with complex classifiers that are too ready to fit to any non-smooth class distribution. For example, one of the reasons that k-means gets stuck in local minima less often than expectation-maximization is because it uses a relatively simple centroid classifier. Larger data sets allow the use of more complex classifiers including neural networks. Furthermore, it might sometimes be advisable to use simpler classifiers in the first few iterations and gradually increase the complexity of the classifier. For example, one can use a neural network with a smaller number of parameters in the initial iterations, and increase the number of parameters in later iterations. Creating the model on sampled data (in proportion to weights) in each training iteration is also helpful in avoiding overfitting and local minima. Other options include the use of different classification algorithms in different iterations, or the use of different random subsets of features for training/prediction in different iterations.

4.10 Clustering Evaluation

Clustering algorithms can be evaluated using either *internal validity measures*, or by using *external validity measures*.

4.10.1 The Pitfalls of Internal Validity Measures

Internal validity measures use a criterion, such as the average cosine similarity to the nearest cluster centroid, to evaluate a clustering. It is not difficult to see that this criterion is used within the objective function optimized by clustering algorithms like k-means. In fact, most internal validity measures use the criteria derived from the objective functions of various clustering algorithms or are at least related to these criteria in some way. This creates a problem in using internal validity measures to fairly compare two clustering algorithms with very different objective functions. For example, if we use the average cosine similarity to the nearest cluster centroid as an internal validity criterion, it is virtually impossible for any other clustering algorithm to outperform k-means for the same number of clusters. The main problem is that the measure does not tell us anything about the inherent goodness of a particular clustering, but more about how well the criterion of a particular clustering algorithm matches with the evaluation criterion. In other words, internal validity measures are often inherently biased towards specific algorithms or specific parameter settings of the same algorithm, and are dangerous to use because they can lead to misleading views on the accuracy of particular clustering algorithms. This book will, therefore, pointedly, omit the discussion of internal validity measures.

4.10.2 External Validity Measures

External validity measures use the dependent variables (or *labels*) from supervised learning problems to evaluate the clustering. The dependent variable is not used by the clustering algorithm, and therefore the criterion is inherently external both to the algorithm and the data set used for clustering. For example, consider a classification problem from the domain of earth science in which the features describe characteristics of trees, and the class labels correspond to the forest cover type. In such a case, the clustering algorithm would only use the features to create the clusters without using the feature cover type. Subsequently, it is measured whether the class labels are spread out randomly over the different clusters or whether each cluster is dominated by a single class label. It is generally desirable for the individual clusters to be dominated by particular class labels.

The main assumption in external validity measures is that the external class labels respect the inherent clustering structure of the data to a large degree. Although this might not be a perfect assumption, it is still a better choice than the use of internal validity measures. After all, class labels are often selected on the basis of natural semantic groupings in the corpus. Over a large number of data sets, any particular external validity measure can provide a very good indicator of the quality of the results in real settings.

The following will provide an overview of the key validity measures that are used frequently. Therefore, we introduce the notation that will be used consistently in this section. Consider a situation in which a particular clustering algorithm finds k_d clusters from a corpus of n documents containing $n_1 \ldots n_{k_d}$ documents. Furthermore, the number of class labels (or *ground-truth* clusters) in the underlying data set is denoted by k_t, and the numbers of documents belonging to the different ground-truth clusters are denoted by $g_1 \ldots g_{k_t}$. The number of algorithm-determined clusters, k_d, may not be the same as the number, k_t, of class labels/ground-truth clusters. The number of documents in the ith algorithm-determined cluster that belong to the jth class label is denoted by m_{ij}. Then, the following relationships become immediately evident:

$$\sum_{j=1}^{k_t} m_{ij} = n_i \quad \forall i \in \{1 \ldots k_d\}$$

$$\sum_{i=1}^{k_d} m_{ij} = g_j \quad \forall j \in \{1 \ldots k_t\}$$

$$\sum_{i=1}^{k_d} n_i = \sum_{j=1}^{k_t} g_j = n$$

One of the simplest validity measures used is the *cluster purity*. The basic idea in cluster purity is to determine the level of dominance of the class labels in the algorithm-determined clusters. This purity, P, can be computed as follows:

$$P = \frac{\sum_{i=1}^{k_d} \max_j \{m_{ij}\}}{\sum_{i=1}^{k_d} n_i} = \frac{\sum_{i=1}^{k_d} \max_j \{m_{ij}\}}{n} \tag{4.25}$$

An equivalent way of computing the cluster purity is to compute the purity of the ith cluster as $(\max_j \{m_{ij}\}/n_i)$ and compute its weighted average over all clusters. The weight of the ith cluster is proportional to the number of documents in it. A different way of understanding purity is by viewing the clustering method as a classifier. Each document is labeled with the dominant label of the algorithm-determined cluster it belongs to. The accuracy of such a prediction with respect to the external ground-truth is the cluster purity. Therefore, the cluster purity always lies between 0 and 1. It is noteworthy that most external validity measures are related to various quantifications used in supervised learning in one form or the other. This fact is not a co-incidence because external validity measures use a supervised setting to test the effectiveness of an unsupervised algorithm.

The main advantage of the cluster purity measure is that it is simple and easy to understand in an intuitive way. However, it pays too much attention to only the most dominant label in a particular cluster, and it ignores the relative distribution of other labels. Consider a setting in which we have three algorithm-determined clusters and ten class labels. The data set is clustered in two different ways, which are referred to as partitioning A and partitioning B, respectively. Suppose that each of the three clusters have 70% presence of a unique label in both A and B in an identical way. However, in partitioning A, the remaining seven labels are randomly distributed across the different clusters. In partitioning B, the remaining seven labels are neatly segmented across the three clusters in a mutually exclusive way. Clearly, one would prefer partitioning B over partitioning A. However, cluster purity is unable to distinguish between partitioning A and partitioning B because it ignores the non-dominant labels in the clusters. Such clusterings can be distinguished using two other measures, which are referred to as Gini index and entropy. As in the case of cluster purity, these measures are also borrowed from the supervised learning domain.

Let $p_{ij} = m_{ij}/n_i$ be the fraction of the points in cluster i that belong to class (ground-truth cluster) j. Therefore, we have $\sum_{j=1}^{k_t} p_{ij} = 1$. Both the Gini index and entropy are defined in terms of p_{ij}. For Gini index, we first define the Gini index $G(i)$ that is specific to cluster i as follows:

$$G(i) = 1 - \sum_{j=1}^{k_t} p_{ij}^2 \tag{4.26}$$

The Gini index lies between 0 and $1 - 1/k_t$, in which a perfectly homogeneous cluster receives a Gini index of 0, whereas a cluster with equally distributed class labels receives a value of $1 - 1/k_t$. Therefore, smaller values of the Gini index are indicative of superior clustering. The overall Gini index G is the weighted average of the Gini index of the individual clusters:

$$G = \frac{\sum_{i=1}^{k_d} n_i \cdot G(i)}{n} \tag{4.27}$$

A second measure that is commonly used is that of conditional entropy, which is very similar to the Gini index. Let $E(i)$ be the conditional entropy specific to cluster i. Then, the value of $E(i)$ is defined as follows:

$$E(i) = - \sum_{j=1}^{k_t} p_{ij} \log(p_{ij}) \tag{4.28}$$

As in the case of the Gini index, lower values of the conditional entropy are indicative of a clustering of higher quality, and it always lies in the range $(0, \log(k_t))$. The overall conditional entropy E is obtained by computing the weighted average of the cluster-specific values:

$$E = \frac{\sum_{i=1}^{k_d} n_i \cdot E(i)}{n} \tag{4.29}$$

The conditional entropy measures how much uncertainty remains in predicting the class labels, if one were given the clustering. For example, if one sets $k_d = n$, the clusters would be singleton points containing only one class, and there would be no uncertainty in predicting the class label. This is consistent with the fact that the conditional entropy of this case can be shown to be 0. It is noteworthy that all of the aforementioned measures will generally give better values of clustering quality when the number of algorithm-determined clusters k_d is increased. Therefore, they cannot be used to compare clusterings of varying granularity.

A measure related to conditional entropy is the *normalized mutual information*, which is better normalized for the number of clusters in the data. First, we define the notion of *mutual information* between the algorithm-determined clusters and class labels:

$$MI = \sum_{i=1}^{k_d} \sum_{j=1}^{k_t} \frac{m_{ij}}{n} \log \left(\frac{n \cdot m_{ij}}{n_i \cdot g_j} \right) \tag{4.30}$$

The mutual information is always nonnegative, and higher values are desirable. For a particular data set and class labeling, the sum of the conditional entropy and mutual information can be shown to be a constant that depends only on the entropy of the class labeling, irrespective of the algorithm used for clustering. Therefore, the mutual information can be viewed as an information gain over the original class labeling, and it conveys almost the same information about the quality of clustering as conditional entropy. The only difference is that larger values are more desirable in this case, with a value of 0 indicating independence between clustering and class labels. These relationships are also discussed in Section 5.2.4 of Chapter 5 in the context of feature selection. However, one advantage of the mutual information is that it can be normalized to a value in $(0, 1)$ that is less sensitive to the number of algorithm-determined clusters. The normalized mutual information, NMI, is defined as follows:

$$NMI = \frac{2 \cdot MI}{-\sum_{i=1}^{k_d} \frac{n_i}{n} \log \left(\frac{n_i}{n} \right) - \sum_{j=1}^{k_t} \frac{g_j}{n} \log \left(\frac{g_j}{n} \right)} \tag{4.31}$$

The denominator is simply the sum of the entropies in the ground-truth labels and algorithm-determined clusters, and is at least equal to twice the mutual information. The normalized mutual information can take on a value of 1, when $k_d = k_t$ and the clusters match up exactly with the class labels. Therefore, the measure is biased in favor of clusterings where the number of algorithm-determined clusters is close to the number of classes in the data. This can sometimes be a problem if the number of natural clusters in the data is not equal to the number of classes in the data. However, if we assume that the ground-truth classes reflect the true number of clusters in the data, then the measure is a reasonable one to use to compare between clusters of varying granularities. After all, not picking the correct number of clusters should also be considered a mistake made by the algorithm-determined clustering in such a case.

Finally, a number of measures sample pairs at objects and quantify the agreement between the algorithm-determined cluster indices and class labels. The *Rand Index* samples pairs of documents and computes the fraction of pairs in which the algorithm-determined cluster indices and class labels come to the same conclusion about whether or not they should belong to the same cluster. Therefore, the Rand Index lies in the range $(0, 1)$, and higher values are better. The *Fowlkes-Mallows measure* computes the geometric mean between the precision and recall. The precision is defined as the average fraction of pairs in an algorithm-determined cluster that belong to the same ground-truth label. The recall is defined as the average fraction of the pairs in a ground-truth cluster that belong to the same algorithm-determined cluster. The geometric mean of these two quantities is the Fowlkes-Mallows measure. For large data sets, the precision and recall must be estimated by sampling because the total number of pairs is large. One advantage of measures like the Rand Index and Fowlkes-Mallows is that it is possible (to a limited extent) to compare two clusterings with a varying number of algorithm-determined clusters k_d, which is not possible with other measures like purity that improve with increasing k_d. This is because the precision and recall are affected in opposite directions by varying k_d. However, these measures would still be biased towards values of k_d that are close to k_t, because the assumption is that the class labeling reflects the true number of clusters in the data. Note that the best possible value of 1 can only be achieved by these measures when $k_d = k_t$.

4.10.2.1 Relationship of Clustering Evaluation to Supervised Learning

Clustering evaluation measures are closely related to supervised learning in two ways:

1. *Supervised accuracy measures:* The cluster purity measure can be viewed as a generalization of the accuracy measure in classification, where the clustering is used to perform classification. Similarly, the Fowlkes-Mallows measure generalizes the precision/recall measures used in supervised learning. A discussion of classification evaluation is provided in Section 7.5 of Chapter 7.

2. *Categorical feature selection measures:* All feature selection measures used in supervised learning for categorical attributes can be generalized easily to clustering evaluation. This is because feature selection methods effectively use the discrete values of a categorical attribute in an analogous way to a cluster of repeated values, when measuring the discriminative power of that categorical attribute. As a result, measures like the Gini index and entropy are also used for feature selection in supervised learning. Therefore, many supervised feature selection measures like the χ^2-statistic can also be used for clustering evaluation. A discussion of feature selection measures in supervised learning is provided in Section 5.2 of Chapter 5.

This relationship between clustering and supervised evaluation measures is useful, because one can use it to design many high-quality evaluation measures for clustering.

4.10.2.2 Common Mistakes in Evaluation

There are several common mistakes made by practitioners while bench-marking clustering algorithms:

1. Most clustering measures cannot evaluate the relative quality of clusterings of different granularities in an unbiased way. For example, increasing the value of k will usually improve the cluster purity, Gini index, and the entropy. When each point is in its own cluster, a perfect value of the measure will be achieved. Although the Fowlkes-Mallows measure is less sensitive, it is biased in favor of clusterings in which the number of algorithm-determined and ground-truth clusters match.

2. A common temptation for a practitioner is to evaluate different variations of a clustering algorithm using some external validity measure and then select the best option. However, by using an external validity measure for tuning, the analyst has unwittingly incorporated supervision in the algorithm. To ensure that the clustering is truly unsupervised, one must assume that the ground-truth labels do not exist while setting the algorithm parameters.

Clustering is a hard problem to evaluate because of its unsupervised nature. Often, the only true evaluation of a clustering is its utility in an application-centric setting.

4.11   Summary

The problem of clustering is that of unsupervised learning in which no guidance is provided to a learner about the natural groupings in the data. Feature selection methods typically evaluate the consistency in the similarities over individual features with those over other features. Most matrix factorization and topic modeling methods can be used to discover overlapping document clusters and word clusters from the corpus. Traditional mixture models use a specific model of generation of a document from each mixture component. Similarity-based methods such as the k-means algorithm are closely related to mixture models, but are less likely to be stuck in local minima because of their simplicity. Hierarchical methods are more expensive than k-means methods but they often provide clusterings of better quality. Therefore, it sometimes makes sense to combine hierarchical and k-means methods to obtain high-quality results.

Clustering ensembles are useful for combining the results of different clustering algorithms and in obtaining a single more robust clustering. Such methods also have utility in improving the efficiency of clustering methods when they are combined with methods like subsampling. Clustering methods that use the sequence information inside text are almost always feature engineering methods. Methods like kernel k-means and explicit feature engineering can both prove useful when combined with string-based kernels. Feature engineering has the advantage that one can obtain high-quality results with ensembles, and also use algorithms other than the k-means approach. Clustering evaluation measures are either internal or external. Internal evaluation measures are often misleading and generally not recommended. External measures use class labels as the ground-truth and typically adapt classification accuracy measures in order to quantify the quality of a clustering.

4.12   Bibliographic Notes

Surveys on text clustering may be found in [9, 15]. A feature selection survey for clustering may also be found in [9], and some of the these methods are also applicable to text data. Term strength was one of the earliest unsupervised methods [581] proposed for text feature selection. The use of unsupervised models for supervised feature selection was proposed in [435] in the context of outlier detection. A wrapper method that combines the χ^2-statistic with probabilistic clustering is proposed in [332]. Most of the matrix factorization and dimensionality reduction techniques discussed in Chapter 3 can be used as feature engineering methods for improving clustering applications, because they reduce the effects of synonymy and polysemy and bring out the key latent concepts in the data. Interestingly, it is also possible to use clustering to engineer such types of concept decompositions for other applications like similarity search [13, 151].

Nonnegative matrix factorization was proposed in [315], and PLSA was proposed in [255, 256]. The equivalence between the two was shown in [156]. The use of nonnegative matrix factorization for clustering is advocated in several works [154, 157, 511, 591]. Among these works, the work in [157] discusses how different types of constraints within the nonnegative matrix factorization lead to clusterings with varying levels of overlaps among the rows and the columns. A chapter on nonnegative matrix factorization methods for clustering may be found in [9]. A survey on co-clustering for biological data may be found in [363].

The earliest works on probabilistic clustering [330, 437] were focused on distributional clustering of words based on co-occurrence. These ideas were generalized to the supervised setting in [38]. The multinomial version of unsupervised clustering may be found in [104]. The work in [419] is a semi-supervised variant of the expectation-maximization algorithm. This work illuminates the entire spectrum of possibilities between the unsupervised expectation-maximization algorithm and the fully supervised naïve Bayes algorithm.

The k-means algorithm has been explored extensively by several researchers. The projection-based approach discussed in this book is based on [504]. The basic ideas in k-means clustering have been generalized to the streaming setting [14, 629]. Numerous hierarchical methods have also been proposed for clustering, and a comprehensive overview may be found in [9]. In the text domain, a single-linkage implementations are discussed in [21, 135], and the centroid clustering method is discussed in [563]. The combination of hierarchical and k-means clustering (cf. Section 4.6.2) is discussed in the Scatter/Gather work in [141]. This approach also discusses alternatives to using hierarchical methods, such as buckshot and fractionation in order to make the algorithms more efficient. A semi-supervised variant of k-means is found in [7], and this paper also illuminates the connections between semi-supervised clustering and classification. A detailed comparison of various clustering algorithms may be found in [628].

A survey on ensemble methods for clustering may be found in [214]. Sequential knowledge can be incorporated into text clustering methods by using frequent phrases [615]. Representation learning methods have found much interest in the text community for embedding sequential relationships among words into multidimensional representations. In particular, neural network methods like *word2vec* [389] and *doc2vec* [314] are used to embed text sequences into multidimensional methods. The Nyström sampling method is discussed in [584], and its use for unsupervised learning is discussed in [10]. Numerous clustering validity measures are discussed in detail in [9, 614].

4.12.1 Software Resources

One of the earliest libraries for clustering is the Bow toolkit [371], which is written in C. The Python library *scikit-learn* [646] contains several text clustering tools [665]. The R-based **tm** library [647] can be used for preprocessing the documents. Most R distributions contain the **stats** package, which contains the **kmeans** and **hclust** functions by default. These functions perform k-means and hierarchical clustering, respectively. However, since these implementations use the Euclidean distance function rather than the cosine function, it is important to normalize each vector-space representation *up front* to unit norm, so that using the cosine, dot product, or the Euclidean distances create the same result (cf. Section 2.5 of Chapter 2). The *Weka* library also contains several Java implementations of clustering algorithms [649]. The statistics and machine learning toolbox in MATLAB has functions [666] for k-means and hierarchical clustering. It also provides the ability to automatically compute the dendrogram from a data set. In many of these packages, it is important to normalize the documents up front to unit length, because they use the Euclidean distance under the covers.

4.13 Exercises

1. The Gini index criterion is discussed in this chapter (for cluster validity). Show how you can pair this criterion with the k-means algorithm to perform unsupervised feature selection. Which other cluster validity criterion (or criteria) can you use for unsupervised feature selection in this manner?

2. Implement the feature selection criterion for term strength.

3. Consider the nonnegative tri-factorization $D = Q\Sigma P^T$ of rank-k, in which Σ is constrained to be diagonal/nonnegative. Furthermore P and Q are constrained to satisfy $Q^T Q = P^T P = I$. The optimization formulation of this problem is discussed in the chapter. Show how you can use the k-means algorithm to create an initialization point for the gradient-descent steps of this optimization formulation.

4. Suppose your text documents have a representation in which you only know about the presence or absence of words in half the lexicon and you know the exact frequencies of words in the remaining half. Show how you can combine the Bernoulli and multivariate models to perform text clustering.

5. Implement the k-means algorithm for clustering.

6. Suppose that you represent your corpus as a graph in which each document is a node, and the weight of the edge between a pair of nodes is equal to the cosine similarity between them. Interpret the single-linkage clustering algorithm in terms of this similarity graph.

7. Suppose you were given only the similarity graph of Exercise 5 and not the actual documents. How would you perform k-means clustering with this input?

8. For the case of hierarchical clustering algorithms, what is the complexity of centroid merging? How would you make it efficient?

9. What is the number of possible clusterings of a data set of n points into k groups? What does this imply about the convergence behavior of algorithms whose objective function is guaranteed not to worsen from one iteration to the next?

10. Implement the group-average linkage clustering algorithm.

11. As discussed in the chapter, explicit feature engineering methods can be made faster and more accurate with Nyström sampling. Spectral clustering has also been presented as a special case of kernel methods with explicit feature engineering in this chapter. Discuss the difficulties in using Nyström sampling with spectral clustering. Can you think of any way of providing a reasonable approximation? [The second part of the question is open-ended without a crisp answer.]

Chapter 5

Text Classification: Basic Models

"Science is the systematic classification of experience."–George Henry Lewes

5.1 Introduction

In classification, the corpus is partitioned into *classes* that are typically defined by application-specific criteria. Therefore, training *examples* are provided that associate data points with *labels* indicating their class membership. For example, the training examples extracted from a news portal on political matters might attach one of three labels associated with each of the documents, such as *"senate," "congress,"* and *"legislation."* Then, for a given set of *test examples* in which labels are not available, the goal is to place them in one of these categories with the use of a *supervised model* that was constructed using the training examples. The process of learning a categorization model from the training data, and then applying it to the test data is referred to as *generalization.* The basic principle here is that we are generalizing our experiences from (specific) training examples with known labels to arbitrary test data with unknown labels.

Text classification and clustering are closely related problems. One can view each class in an analogous way to a cluster. Unlike clustering, the problem of classification distinguishes between training examples and test examples, and labels are observed only for training examples. Therefore, the supervised model from the training data is used to predict the labels of the test examples. For example, the model might learn that the word *"representative"* is related to the label *"congress"* and it might use this fact to assign test documents containing this word to the label *"congress."* A key observation is that the training instances inherently fix the nature of these "clusters" (i.e., classes) with the use of labels. Therefore, the test examples are always assigned to one of the pre-defined training labels (groupings) in classification, whereas clustering has a more open-ended view in which it uses the similarity structure of the data to define its own groupings (which can eventually be manually labeled by a domain expert). This is the reason that classification is referred to as *supervised learning,* because the training examples play the role of a teacher who guides the students towards a specific goal of finding a particular type of grouping. This type of guided grouping provides significant control in many application-centric settings:

© Springer Nature Switzerland AG 2022
C. C. Aggarwal, *Machine Learning for Text,*
https://doi.org/10.1007/978-3-030-96623-2_5

1. *News portals:* News portals often organize incoming documents on the basis of a specific topic such as politics, sports, entertainment, and so on. In many cases, the topical categorization needs to be done in real time, as new articles are received continuously. This process is also referred to as *news filtering.* A similar principle applies to the organization of large groups of document collections such as digital libraries or scientific literature.

2. *Email and spam filtering:* Many email providers allow the ability to filter spam in an automated way. This is a classification application in which each email is labeled as either "spam" or "not spam."

3. *Opinion mining and sentiment analysis:* In opinion mining and sentiment analysis, the basic idea is to use the text of reviews, blogs, or social posts in order to make judgements about the opinions and sentiments of users. As discussed in Chapter 15, this problem is a direct application of classification.

The problem of text classification is formally defined as follows. Consider an $n \times d$ training data matrix D, whose n rows are the tf-idf representations of the n documents. These rows contain the d-dimensional row vectors $\overline{X_1} \ldots \overline{X_n}$. In addition, the ith document $\overline{X_i}$ is associated with the class label y_i. We can assume that the column vector $\overline{y} = [y_1 \ldots y_n]^T$ contains all the class labels associated with the n training instances. It is assumed that the class label of each training instance is drawn from the set of k label values denoted by $\mathcal{L} = \{1, \ldots k\}$, although there are some special conventions for binary classes (which will be discussed later). Therefore, the pair $(D, \overline{y})$ represents the training data, and a one-to-one correspondence exists between rows of D and entries of $\overline{y}$. This data matrix is used to create a model for classifying each unlabeled test instance $\overline{Z}$:

Definition 5.1.1 (Data Classification) *Given an $n \times d$ document-term matrix D associated with the n-dimensional vector of class labels $\overline{y}$, predict the class label for an unlabeled test document $\overline{Z}$.*

More generally, one can create a *test matrix* D_t of size $n_t \times d$. Therefore, there are n_t test instances, $\overline{Z_1} \ldots \overline{Z_{n_t}}$, which are rows of this matrix. Each such test instance needs to be independently classified using the above model. The aforementioned definition is the simplest model formulation. In some cases, instead of predicting the labels of each instance independently, one might want to sort all the test instances in D_t in order of their propensity to belong to a particularly important class. For example, in a spam-detection application, one might want to rank all the emails in order of their propensity to be spam.

5.1.1   Types of Labels and Regression Modeling

For k-way classification, it is assumed that the label set is denoted by $\mathcal{L} = \{1 \ldots k\}$. Note that the values $1 \ldots k$ represent only discrete identifiers without any ordering among them. For example, the semantic interpretation of the labels in a color-prediction application could correspond to $\mathcal{L} = \{Blue, Red, Green\}$. The only case in which one might impose an arbitrary ordering between labels (and use them as numeric quantities) is the binary case in which the value of k is 2. Many binary classification algorithms use either the convention $\mathcal{L} = \{0, 1\}$, or they work with the convention $\mathcal{L} = \{-1, +1\}$. The binary classification problem is particularly common in practical settings, and some classification models (cf. Chapter 6) are naturally designed to solve only the binary case. Nevertheless, these classifiers can also be used for k-way classification with some algorithmic tricks.

So far, we have viewed the class label only from the point of view of partitioning the data, and therefore it is defined as a categorical label. A more general view is that this label could be an arbitrary numerical quantity, such as a decimal value drawn from the real domain. In such a case, we use the term *dependent variable* to refer to this quantity rather than as a class variable. The entries in each row (document) of D are referred to as the *independent variables* of that instance. The problem in which the dependent variable is numerical is also referred to as *regression modeling*. It is easy to see that binary classification can be considered a rudimentary special case of regression modeling. The dependent variable is also referred to as the *response variable* or *regressand*. The independent variables are also referred to as the *explanatory variables*, *input variables*, *feature variables*, *predictor variables*, or *regressors*. All the models discussed in this chapter can be used for both classification and regression, although our primary focus will be on classification.

5.1.2 Training and Testing

Most classifiers have an up front training phase in which only the labeled training data is used to build a *summarized* model that relates the characteristics of the documents (e.g., term distributions) to the classes. This phase is referred to as *training* or learning. The summarized model essentially *generalizes* the knowledge gained from the training data to *unseen* test instances. This prediction of the labels of unseen test instances is referred to as the *testing* or *prediction* phase. It is noteworthy that the accuracy of a trained classifier will typically be much higher if it is used to "predict" the (known) labels of the (seen) instances that it was trained on than on the unseen test instances. This is because the trained model "remembers" some of the specific and unimportant nuances about the training instances within the summarized model, which improves the accuracy only on these specific instances. Classifiers that have small gaps in their training and test data accuracy are said to have good *generalization* power. It is easy[1] to construct classifiers in which the accuracy on the training data is very high, but that on the test data is extremely poor. This phenomenon is referred to as *overfitting*. Overfitting is undesirable, because the only utility of a classifier arises from correctly predicting instances for which labels are not already available (i.e., test data). As a general rule, classifiers with a *concise* summary model will have better generalization power, although the overall accuracy depends on several other factors. These issues will be discussed in Section 7.2 of Chapter 7.

The training phase might include a phase of *model selection*, which corresponds to the tuning of parameters or other design choices in the algorithm. A very simple way to implement model selection is by hiding (i.e., *holding out*) a part of the labeled training data during model construction and then evaluating the accuracy of using various values of the parameters (or training design choices) on the held out data. This set is referred to as the *validation set*, and it is distinct from the test set on which the predictions are finally applied. After the phase of model selection, the unlabeled instances are predicted with the optimized design choices.

Finally, a *decision boundary* is a hyperplane or hyper-surface in the data space that partitions the data into various classes. All classification algorithms attempt to model this

---

[1] Consider a classifier that memorizes the training examples as follows. For any test instance, it is determined whether a training instance has zero distance to it (which is guaranteed when the test instance is drawn from the training data). If such an instance is found, the label of that training instance is returned. Otherwise a random label is returned. Such a classifier will have 100% accuracy on the training data, but will perform randomly on unseen test instances. The key point is that *generalization* is about extrapolating predictions from known *instances* of the data space (i.e., training points) to all *regions* of the data space. Memorizing only the known instances is the worst possible way to achieve this.

decision boundary directly or indirectly using the training data. It is noteworthy that the hyper-surface might not be contiguous when a class is not contiguously located in the data space. Furthermore, a real data set may not contain a sharply defined decision boundary because there might be regions where the classes are overlapping. As a result, the decision boundary is sometimes viewed as a *region* of the data space in which the classification is ambiguous, and the predicted decision boundary by a particular model is often chosen somewhere in this region to provide the best performance on unseen data. Simplified modeling assumptions (e.g., linear shape of boundary) are often made during learning, and therefore it is common for classifiers to make mistakes near the modeled decision boundary.

5.1.3 Inductive, Transductive, and Deductive Learners

Not all classifiers have a clear separation between the training and testing phase. Classifiers that do create a summarized model up front from the training data are said to be *inductive* learners, and their primary goal is to *generalize* the observation from training data to unseen instances. These classifiers generalize easily to *any* unlabeled test instance. However, if more training data is received, then it could invalidate the model because of the presence of additional data that conflicts with the currently available summarized model. After all, the currently available model is only a *hypothesis* about unseen instances. Most of the classifiers discussed in this chapter are inductive learners.

In *transductive* learners, the (unlabeled) test data is included with the labeled training data in the training phase, and the predictions can be specific only to that particular set of unlabeled data. Therefore, the generalization achieved with a transductive learner is less than that achieved with an inductive learner because the resulting models may not generalize to unseen test instances. However, this specificity also (often) provides the advantage that the predictions for those specific test instances are more accurate. Such methods are closely related to *semi-supervised* methods, because they use both labeled and unlabeled data.

Finally, a fundamentally different way of classifying data is by using deductive learners. Deductive learners use rules of logic to capture fundamental properties of the instances. These rules are often obtained using knowledge of the world or other domain characteristics. In a sense, these rules are considered absolute truth that cannot be invalidated by future observations. For example, consider the following pair of rules: *"Bald men do not have hair. Only people with hair need combs."* Now if you had a feature in your instance containing information about whether or not someone was bald, you could use it to predict whether they will need a comb. A human-centric analogy would be that deductive learning comprises the lessons you learned from your parents, whereas inductive learning comprises the lessons you learned from your own life experiences. The latter is known to more effective both in real life and in machine learning, although one should not discount the guidance provided by deductive learning where it is available. Inductive learners that encode domain knowledge about the data within the classification process can be viewed to have *some* characteristics of deductive learners, and are therefore hybrid models [434]. The power of deductive learning is often incorporated indirectly in inductive models by incorporating mild constraints (or *bias*) into the model with domain-specific insights. One needs to be careful when using such methods because strong levels of bias suffocate the ability of the learner to benefit from more examples. Most of machine learning focuses on inductive learning, because of its emphasis on observation-driven inference. This chapter will primarily focus on inductive learners.

5.1.4 The Basic Models

This chapter will discuss the four basic models for text classification, which are the naïve Bayes classifier, the nearest-neighbor classifier, decision trees, and rule-based classifiers. These four classifiers are selected because they are among the oldest methods in the literature, and are related in fundamental ways to other learning models. For example, the naïve Bayes classifier can be shown to be a supervised variant of the probabilistic clustering model discussed in Chapter 4. Similarly, some of the most powerful classifiers in the supervised domain like random forests and support-vector machines can be shown to be *adaptive* variants of nearest-neighbor classifiers (cf. Sections 5.5.6 and 6.3.6).

5.1.5 Text-Specific Challenges in Classifiers

Text is extremely sparse and high-dimensional, which causes off-the-shelf, multidimensional models to behave in unexpected ways. The frequency of a single term often contains little predictive power, and it is only by using combinations of many features that robust classification can be achieved. If a classifier uses *sequential* decisions that prioritizes one feature strictly before another, this can affect the accuracy of classification negatively because of overfitting. This observation has implications in the design of classifiers like *univariate decision trees* that use sequential decisions over individual attributes. In fact, if all features are used simultaneously, then some simple models like *linear classification* work better than in other domains without the need for sophisticated nonlinear transformations of the data (cf. Section 6.5.3 of Chapter 6).

Another consequence of sparsity is that the presence of a particular term in a document is much more informative than its absence for inferring the class label. This is because the presence of a term is statistically rare in a sparse document and thereby more informative. Some classifiers that use excessive information about absent terms perform poorly because of overfitting. Furthermore, the precise frequency of a term contains much less *incremental* information compared to that obtained by knowing that the term is present in the document. This asymmetry in the relative importance of different values of the term frequencies is important to keep in mind while attempting to adapt classifiers from the traditional multidimensional domain (which tend to treat all values in a symmetric way).

Chapter Organization

This chapter is organized as follows. The next section introduces feature selection methods for classification. The naïve Bayes model is introduced in Section 5.3. Section 5.4 discusses nearest-neighbor methods. Decision trees are discussed in Section 5.5. Rule-based classifiers are introduced in Section 5.6. A summary is given in Section 5.7.

5.2 Feature Selection and Engineering

Text data is often extracted from sources like the Web in which the authorship varies widely, with many misspellings and use of non-standard vocabulary and acronyms. Many features are irrelevant, and including them leads to overfitting, particularly when labeled data are limited. The discriminative features can be identified by examining the co-occurrence statistics of the various terms with respect to the classes. For example, an undiscriminating term will be randomly distributed across all classes. On the other hand, a highly relevant term will be concentrated in a smaller subset of the classes. A number of measures such as the

Gini index, conditional entropy, and the χ^2-*statistic* are used to measure this type of association. Such models are referred to as *filter* models because a single quantification is used up front to filter features. All the models discussed in this section, other than those in Section 5.2.6, are filter models. In *wrapper* models, an iterative feature selection process is tied to a particular classification model, and the effect of a particular term on the accuracy of that model is used for feature selection. We will omit a detailed discussion of wrapper models because they are rarely used in the text domain. Finally, in embedded models (cf. Section 5.2.6), the feature discrimination can be quantified using the intermediate outputs of a particular classification algorithm.

5.2.1   Gini Index

The Gini index measures the imbalance in the class distribution of a set of instances that include a particular term. The basic idea is that discriminative features tend to increase this imbalance. From all instances that contain the term t_j, let $P(c_r|t_j)$ be the fraction (i.e., observed probability) that belong to the class r. Therefore, for a k-class problem we have:

$$\sum_{r=1}^{k} P(c_r|t_j) = 1 \quad \forall t_j \tag{5.1}$$

When the term t_j is poorly discriminative of the class label, all the values of $P(c_r|t_j)$ for varying r and fixed j will be similar and close to $1/k$. On the other hand, if the feature is extremely discriminative, then all documents containing that term will belong to a single class. As a result, only one of these fractions will be 1, and others will be 0s. How can we provide a single measure of goodness that captures the desirability of greater skew? A simple measure is the Gini index $G(t_j)$, which is defined as follows:

$$G(t_j) = 1 - \sum_{r=1}^{k} [P(c_r|t_j)]^2 \tag{5.2}$$

When all documents containing a term belong to a single class, the Gini index takes its minimum value of 0. On the other hand, if documents containing the term are evenly distributed across different classes, the Gini index takes on its maximum value of $1 - 1/k$. In other words, the Gini index always lies in the range $(0, 1 - 1/k)$, and smaller values are desirable. Features with large values of the Gini index can be removed. One issue with this measure is that it does not work very well when the class distributions are imbalanced in the original data [7]. Therefore, one has to compute a re-normalized value of $P(c_r|t_j)$ with respect to the numbers of instances $n_1 \ldots n_k$ in various classes:

$$f_r(t_j) = \frac{P(c_r|t_j)/n_r}{\sum_{s=1}^{k} P(c_s|t_j)/n_s} \tag{5.3}$$

Then, the normalized value of the Gini index may be computed as follows:

$$G_n(t_j) = 1 - \sum_{r=1}^{k} f_r(t_j)^2 \tag{5.4}$$

The re-normalization is a way of forcing the original class distribution to be an even distribution, and examining how much the addition of term t_j changes the class distribution. Unlike most other measures, the absence of the term in not used in the Gini coefficient computation. In the text domain, it is sometimes desirable to not use the absence of terms too strongly, because it is noisy information.

5.2.2 Conditional Entropy

Let $n(t_j)$ be the number of documents containing term t_j out of a corpus of size $n \geq n(t_j)$. Among all these instances that contain the term t_j, let $P(c_r|t_j)$ be the fraction (i.e., observed probability) of documents belonging to the class r. Furthermore, among the $(n - n(t_j))$ documents that do not contain t_j, let $P(c_r|\neg t_j)$ be the fraction that belong to class r. Then, the conditional entropy $E(t_j)$ is defined as follows:

$$E(t_j) = -\sum_{r=1}^{k} \left\{ \left[\frac{n(t_j)}{n} \right] P(c_r|t_j) \cdot \log[P(c_r|t_j)] + \left[\frac{n - n(t_j)}{n} \right] P(c_r|\neg t_j) \cdot \log[P(c_r|\neg t_j)] \right\}$$

$$(5.5)$$

The conditional entropy lies between $(0, \log(k))$ and it measures how much the presence or absence of a term affects our certainty of being able to determine the class label. For example, if all documents containing a term belong to one class, and all documents not containing that term belong to another class, then the conditional entropy will be 0. Lower values are indicative of more discriminative features. The features can be ranked in order of conditional entropy and the ones with the largest values can be pruned.

5.2.3 Pointwise Mutual Information

First, the point-wise mutual information with respect to a single class is defined. Subsequently, the idea is generalized to multiple classes. The point-wise mutual information $PMI_r(t_j)$ with respect to class r is defined as follows:

$$PMI_r(t_j) = \log \left[\frac{P(c_r|t_j)}{P(c_r)} \right]$$

The notions used in this section are the same as those used above in the discussions on the Gini index and conditional entropy. The overall point-wise mutual information across all classes can be defined in two different ways:

$$PMI_{avg}(t_j) = \sum_{r=1}^{k} \frac{n_r}{n} PMI_r(t_j)$$

$$PMI_{max}(t_j) = \max_r PMI_r(t_j)$$

The point-wise mutual information is positive when the presence of the term is positively correlated with respect to a particular class. Larger values of the point-wise mutual information are more desirable.

5.2.4 Closely Related Measures

Many authors and practitioners use closely related measures like *mutual information* (different from *pointwise* mutual information) and *information gain*, which turn out to give identical results to conditional entropy. Therefore, it is useful to know these relationships to avoid redundancy in usage. Note that point-wise mutual information uses only information about the presence of terms but not about the absence of terms. A different measure is the mutual information, which uses both the presence and absence of terms to compute pointwise mutual information values such as $\{PMI_r(t_j), PMI_r(\neg t_j)\}$, and then computes a weighted average over all possibilities. Let $P(c_r \cap t_j)$ represent the fraction of all documents from the corpus that both belong to class r and contain term t_j. Then, the mutual information $MI(t_j)$ is computed as follows:

$$MI(t_j) = \sum_{r=1}^{k} [P(c_r \cap t_j)PMI(t_j) + P(c_r \cap \neg t_j)PMI(\neg t_j)] \qquad (5.6)$$

The mutual information measures the amount of information that the term t_j has with respect to the class distribution. The mutual information is always nonnegative and takes on the minimum value of 0 when the two terms are statistically independent. Either positive or negative correlation between the term and a particular class increases mutual information. As discussed in Section 4.10.2 of Chapter 4, a normalized variant of mutual information is also used in measuring clustering validity. The mutual information is also referred to the *information gain*. Interestingly, one can compute the mutual information (i.e., information gain) $I(t_j)$ in terms of the aforementioned measure of conditional entropy and the entropy of the original class frequencies $n_1 \ldots n_k$:

$$\underbrace{\text{Information Gain } I(t_j) \text{ of } t_j}_{\text{Same as Mutual Information}} = \underbrace{- \sum_{r=1}^{k} \frac{n_r}{n} \log\left(\frac{n_r}{n}\right)}_{\text{Entropy in class distribution}} - \underbrace{E(t_j)}_{\text{Conditional Entropy}}$$

$$(5.7)$$

In other words, information gain tells us the gain in conditional entropy (after knowing occurrence data of term t_j) with respect to base entropy of class distribution. Since the first term in the RHS above is independent of t_j, the use of information gain only flips the ordering of the different features in relation to conditional entropy. The information gain is always a nonnegative value with higher values indicating a greater degree of discrimination. The fact that the information gain $I(t_j)$ is the same as the mutual information $MI(t_j)$ is left as an exercise for the reader (see Exercise 2). It makes sense to use only one of the three measures of conditional entropy, mutual information, and information gain, while performing feature selection, because they will provide the same results. However, pointwise mutual information will provide different results, because it does not use the absence of terms. The normalized Gini index also does not use the absence of terms.

5.2.5 The χ^2-Statistic

The basic idea of the χ^2-statistic is to treat the co-occurrence between the term and class as a *contingency table*. For example, consider a scenario where we are trying to determine whether the term *"elections"* is relevant to the class *Politics*. Consider a collection of 1000 documents in which 10% of the documents belong to the *Politics* category, and the term *"elections"* occurs in about 20% of the documents. Then, the *expected* number of occurrences of each possible combination of word occurrence and class contingency is as follows:

	Term *"elections"* $\in$ document	Term *"elections"* $\notin$ document
Document $\in$ *Politics*	$1000 * 0.1 * 0.2 = 20$	$1000 * 0.1 * 0.8 = 80$
Document $\notin$ *Politics*	$1000 * 0.9 * 0.2 = 180$	$1000 * 0.9 * 0.8 = 720$

The aforementioned expected values are computed under the assumption that the occurrence of the term in the document and the occurrence of a document in the *Politics* class are independent events. If these two events are truly independent, then clearly the term will be irrelevant to the learning process. Therefore, the goal of the χ^2-computation is to evaluate how far the *observed* quantities in the contingency table different from the aforementioned *expected* quantities. For example, consider a scenario where the contingency table deviates from expected values and the term *"elections"* and class label *Politics* are related. In such a case, the *observed* contingency table may appear as follows:

	Term *"elections"* $\in$ document	Term *"elections"* $\notin$ document
Document $\in$ *Politics*	$O_1 = 60$	$O_2 = 40$
Document $\notin$ *Politics*	$O_3 = 140$	$O_4 = 760$

The χ^2-statistic measures the normalized deviation between observed and expected values across the various cells of the contingency table. In this case, the contingency table contains $p = 2 \times 2 = 4$ cells. Let O_i be the observed value of the ith cell and E_i be the expected value of the ith cell. Then, the χ^2-statistic is computed as follows:

$$\chi^2 = \sum_{i=1}^{p} \frac{(O_i - E_i)^2}{E_i} \tag{5.8}$$

Therefore, in the particular example of this table, the χ^2-statistic evaluates to the following:

$$\chi^2 = \frac{(60 - 20)^2}{20} + \frac{(40 - 80)^2}{80} + \frac{(140 - 180)^2}{180} + \frac{(760 - 720)^2}{720}$$
$$= 80 + 20 + 8.89 + 2.22 = 111.11$$

It is also possible to compute the χ^2-statistic as a function of the observed values in the contingency table without explicitly computing expected values. This is possible because the expected values are also functions of these observed values. The arithmetic formula to compute the χ^2-statistic in a 2×2 contingency table is as follows:

$$\chi^2 = \frac{(O_1 + O_2 + O_3 + O_4) \cdot (O_1 O_4 - O_2 O_3)^2}{(O_1 + O_2) \cdot (O_3 + O_4) \cdot (O_1 + O_3) \cdot (O_2 + O_4)} \tag{5.9}$$

Here, $O_1 \ldots O_4$ are the observed frequencies according to the table above. It is easy to verify that this formula yields the same χ^2-statistic of 111.11. Note that if the observed values are exactly equal to the expected values, then it implies that the corresponding term is irrelevant to the class at hand. In such a case, the χ^2-statistic will evaluate to its least possible value of 0. Therefore, the top-k features with the largest χ^2-statistic are retained. The χ^2-test can also be probabilistically interpreted in terms of a χ^2 distribution.

One can extend the χ^2-statistic for binary classification (as discussed above) to the k-way setting by combining the class-wise results [606]. Then, if $\chi_r^2(t_j)$ represents the χ^2-statistic for term t_j and occurrence/non-occurrence of class r, the integrated values are as follows:

$$\chi_{avg}^2(t_j) = \sum_{r=1}^{k} \frac{n_r}{n} \chi_r^2(t_j)$$
$$\chi_{max}^2(t_j) = \max_r \chi_r^2(t_j)$$

Here, $n_1 \ldots n_k$ represent the number of documents in the k classes, and n is the total number of documents.

5.2.6   Embedded Feature Selection Models

Many classification and regression models provide the ability to perform embedded feature selection by leveraging the output of intermediate steps. Feature selection is accomplished with the use of *regularization* in order to reduce overfitting, which is similar in principle to the goals of feature selection. As a result, the intermediate outputs of these regularized algorithms provide useful insights for feature selection. For example, consider the following linear regression model (see Section 6.2.2 of Chapter 6), in which the numerical dependent variable y_i is predicted using the following linear relationship to the feature variables $\overline{X_i}$:

$$y_i \approx \overline{W} \cdot \overline{X_i} \quad \forall i \in \{1 \ldots n\} \tag{5.10}$$

The notation $\overline{W}$ represents a d-dimensional vector of coefficients that is learned by the training model. This vector is computed by solving the following optimization model:

$$\text{Minimize} \underbrace{\sum_{i=1}^{n} (\overline{W} \cdot \overline{X_i} - y_i)^2}_{\text{Prediction Error}} + \underbrace{\lambda \sum_{i=1}^{d} |w_i|}_{\text{Penalty for using features}}$$

Here, $\lambda > 0$ is a regularization parameter, which controls the severity of the penalty. Such a penalty ensures that the optimization will not assign a large non-zero coefficient for that feature, unless the feature conveys important and irreplaceable information about the dependent variable. Feature penalization is referred to as regularization. The type of penalty discussed above is referred to as the L_1-penalty, and it has the remarkable property of favoring a coefficient vector $\overline{W}$ in which many values of w_i are zero. Such features are effectively dropped because they will have no influence on prediction of test instances according to Equation 5.10. The natural idea in embedded feature selection is that it leverages on built-in (regularization) mechanisms by many algorithms to avoid overfitting. After all, the main goal of feature selection is also the prevention of overfitting. A detailed discussion of L_1-regularization is provided in Section 6.2.2 of Chapter 6.

5.2.7   Feature Engineering Tricks

Two types of feature engineering tricks are commonly used in the text domain. The first trick is done to get rid of sparsity, which can be a problem for some classifiers such as decision trees. The second technique uses representation mining techniques to embed sequential representations of text to multidimensional representations. The latter approach is able to leverage the sequential ordering information among words to incorporate greater semantic knowledge in learning. Since the second approach will be discussed in Chapter 10, the following will discuss only the feature engineering methods used to address sparsity.

Sparsity can cause challenges with certain types of classifiers like decision trees, which use attributes *one at a time* in the modeling process. Since each term contains information relevant to only a small subset of documents in which it is present, and the absence of terms is noisy information, it often causes overfitting when classifiers make important decisions with individual attributes. Therefore, in such cases, methods like latent semantic analysis (LSA) are not just useful for dimensionality reduction, but they can be viewed as feature engineering methods that enable the use of certain types of classifiers. A particular variant of LSA, known as a *Rotation Ensemble* is particularly useful for ensemble-centric implementations. The basic idea is to use the following approach:

Randomly split the d terms into K disjoint subsets of size d/K to
        create K projected data sets;
Perform LSA on *each* projected data set to extract $r \ll d/K$ features;
Pool all extracted features to create a $(K \cdot r)$-dimensional data set;
Apply a classifier on the new representation;

This approach can be applied multiple times, and the prediction of a test instance can be averaged over multiple such transformations. A particularly common classifier that is used with this approach is the *decision tree*, and the resulting classifier is referred to as the *Rotation Forest* [478]. Another feature engineering method is the Fisher's linear discriminant (cf. Section 6.2.3 of Chapter 6), which provides *discriminative* directions in the space. Such methods have also been used in conjunction with decision trees [95].

5.3 The Naïve Bayes Model

The naïve Bayes classifier uses a probabilistic generative model that is identical to the mixture model used for clustering (cf. Section 4.4 of Chapter 4). The model assumes that the corpus is generated from a mixture of different classes. The generative process, which is applied once for each observed document, is as follows:

1. Select the rth class (mixture component) C_r with prior probability $\alpha_r = P(C_r)$.

2. Generate the next document from the probability distribution for C_r. The most common choices are the Bernoulli and multinomial distributions.

The observed (training and test) data are assumed to be outcomes of this generative process, and the parameters of this generating process are estimated so that the log-likelihood of this data set being created by the generative process is maximized. Generally, only the training data is used to estimate the parameters, because the training data contains additional information about the identity of the mixture component that generated each document. Subsequently, these parameters are used to estimate the probability of the generation of each unlabeled test document from each mixture component (class). This results in a probabilistic classification of unlabeled documents.

Each cluster $\mathcal{G}_r$ in the expectation-maximization algorithm of Section 4.4 is analogous to a class C_r in this setting. One can view naïve Bayes as a simplification of the iterative expectation-maximization algorithm in which the presence of labels allows the execution of the approach in a single iteration. Unlike clustering, the training process in classification uses *a single* application of the M-step (on labeled data), and the probabilistic prediction of test instances is a single application of the E-step on the unlabeled test instances (to estimate posterior probabilities). Furthermore, the naïve Bayes classifier has analogous Bernoulli and multinomial models to those used in clustering.

5.3.1 The Bernoulli Model

In the Bernoulli model, it is assumed that only the presence or absence of each term in the document is observed. Therefore, the frequencies of the terms are ignored, and the vector-space representation of a document is a sparse binary vector. The Bernoulli model assumes that the jth term, t_j, in the lexicon is present in a document generated from the rth class (mixture component) with probability $p_j^{(r)}$. Then, the probability $P(\overline{Z}|C_r)$ of the generation of the document $\overline{Z}$ from mixture component $\mathcal{C}_r$ is given[2] by the product of the d different Bernoulli probabilities corresponding to presence of absence of various terms:

$$P(\overline{Z}|\mathcal{C}_r) = \prod_{t_j \in \overline{Z}} p_j^{(r)} \prod_{t_j \notin \overline{Z}} (1 - p_j^{(r)}) \tag{5.11}$$

An important assumption here is that the presence or absence of the various terms are conditionally independent with respect to the choice of class. Therefore, one can express the joint probability of the attributes in $\overline{Z}$ as the product of the corresponding values on individual attributes. This assumption is also referred to as the *naïve Bayes assumption*, which is also the reason that the method is referred to as a *naïve Bayes classifier*. The term "naïve" is used because this type of approximation is generally not true in real settings.

The main task in the *training phase* of the Bayes classifier is to estimate the (maximum likelihood) values of the prior probabilities α_r and class-specific generative probabilities $p_j^{(r)}$. These parameters are estimated so that the observed data has the maximum likelihood of being generated by the model, and are then used to perform the prediction of the labels of unseen test instances. One can summarize this process as follows:

- **Training phase:** Estimate the maximum-likelihood values of the parameters $p_j^{(r)}$ and α_r using only the training data.

- **Prediction phase:** Use the estimated values of the parameters to predict the class of each unlabeled test instance.

The training phase is executed first, which is followed by the prediction phase. However, since the prediction phase of a naïve Bayes classifier is the key to understanding it, we will present the prediction phase before the training phase. Therefore, the following section will assume that the model parameters have already been learned in the training phase.

Prediction Phase

The prediction phase uses the Bayes rule of posterior probabilities to predict an instance. The basic idea is that the learner uses the aggregate frequency of each class in the training data to learn a *prior* probability $\alpha_r = P(\mathcal{C}_r)$, of each class. Subsequently, it needs to estimate the *posterior* probability $P(\mathcal{C}_r|\overline{Z})$ after observing a *specific* document (with binary representation $\overline{Z} = (z_1 \ldots z_d)$) for which the label is not known. This estimation provides a probabilistic prediction for the test instance $\overline{Z}$ of belonging to a particular class.

According to the Bayes rule of posterior probabilities, the posterior probability of $\overline{Z}$ being generated by the mixture component $\mathcal{C}_r$ of the rth class can be estimated as follows:

$$P(\mathcal{C}_r|\overline{Z}) = \frac{P(\mathcal{C}_r) \cdot P(\overline{Z}|\mathcal{C}_r)}{P(\overline{Z})} \propto P(\mathcal{C}_r) \cdot P(\overline{Z}|\mathcal{C}_r) \tag{5.12}$$

A constant of proportionality[3] is used instead of the $P(\overline{Z})$ in the denominator, because the estimated probability is only compared between multiple classes to determine the predicted class, and $P(\overline{Z})$ is independent of the class.

---

[2] Although $\overline{X_i}$ is a binary vector, we are treating it like a set when we use a set-membership notation like $t_j \in \overline{X_i}$. Any binary vector can also be viewed as a set of the 1s in it.

[3] The constant of proportionality can be easily inferred by ensuring that the sum of the posterior probabilities across all classes is 1. As we will see later, there are scenarios associated with ranking instances to belong to specific classes, where the constant of proportionality does matter.

An important observation here is that all the parameters on the right-hand side of the conditional can be estimated using the Bernoulli model. We further expand the relationship in Equation 5.12 using the Bernoulli distribution of Equation 5.11 as follows:

$$P(\mathcal{C}_r|\overline{Z}) \propto P(\mathcal{C}_r) \cdot P(\overline{Z}|\mathcal{C}_r) = \alpha_r \prod_{t_j \in \overline{Z}} p_j^{(r)} \prod_{t_j \notin \overline{Z}} (1 - p_j^{(r)}) \tag{5.13}$$

Note that all the parameters on the right-hand side are estimated during the training phase discussed below. Therefore, one now has an estimated probability of each class being predicted up to a constant factor of proportionality. The class with the highest posterior probability is predicted as the relevant one, although the output is sometimes provided in the form of probabilities. It is noteworthy that this step is identical to the E-step used for mixture modeling in clustering (cf. Section 4.4.1), except that it is applied only to the unlabeled test instances.

Training Phase

The training phase of the Bayes classifier uses the labeled training data to estimate the maximum likelihood values of the parameters in Equation 5.13. It is evident that we need to estimate two sets of parameters, which are the prior probabilities α_r and the Bernoulli generative parameters, $p_j^{(r)}$, for each mixture component. The statistics available for parameter estimation include the number of labeled documents n_r belonging to the rth class $\mathcal{C}_r$, and the number, $m_j^{(r)}$, of the documents belonging to class $\mathcal{C}_r$ that contain term t_j. The maximum likelihood estimates of these parameters can be shown to be the following:

1. *Estimation of prior probabilities:* Since the training data contains n_r documents for the rth class in a corpus size of n, the natural estimate for the prior probability of the class is as follows:

$$\alpha_r = \frac{n_r}{n} \tag{5.14}$$

If the corpus size is small, it is helpful to perform Laplacian smoothing by adding a small value $\beta > 0$ to the numerator and $\beta \cdot k$ to the denominator:

$$\alpha_r = \frac{n_r + \beta}{n + k \cdot \beta} \tag{5.15}$$

The precise value of β contains the amount of smoothing, and it is often set to 1 in practice. When the amount of data is very small, this results in the prior probabilities being estimated closer to $1/k$, which is a sensible assumption in the absence of sufficient data.

2. *Estimation of class-conditioned mixture parameters:* The class-conditioned mixture parameters, $p_j^{(r)}$, are estimated as follows:

$$p_j^{(r)} = \frac{m_j^{(r)}}{n_r} \tag{5.16}$$

It is particularly important to use Laplacian smoothing on the class-conditioned probabilities because a particular term t_j might not even be present in the training documents of the rth class, particularly when the corpus is small. In such a case, one would estimate the corresponding value of $p_j^{(r)}$ to 0. As a result of the multiplicative nature of Equation 5.13, the presence of term t_j in an unseen document will always lead to an estimated probability of 0 for the rth class. Such predictions are often erroneous, and are caused by overfitting to the small training data size.

Laplacian smoothing of class-conditioned probability estimation is performed as follows. Let d_a be the average number of 1s in the binary representation of each training document and d be the size of the lexicon. The basic idea is to add a Laplacian smoothing parameter $\gamma > 0$ to the numerator of Equation 5.16 and $d\gamma/d_a$ to the denominator:

$$p_j^{(r)} = \frac{m_j^{(r)} + \gamma}{n_r + d\gamma/d_a} \tag{5.17}$$

The value of γ is often set to 1 in practice. When the amount of training data is very small, this choice leads to a default value of d_a/d for $p_j^{(r)}$, which reflects the level of sparsity in the document collection.

It is noteworthy that the training phase in the Bayes classifier is a simplified variant of the M-step used in the mixture model for clustering (cf. Section 4.4.1). This simplification is because *labeled* training data is available to infer the membership of documents in mixture components.

5.3.2 Multinomial Model

While the Bernoulli model uses only the presence of absence of terms in documents, the multinomial model explicitly uses their term frequencies. Just as the parameter $p_j^{(r)}$ in the Bernoulli model denotes the probability whether a term is observed in a particular component, the parameter q_{jr} in the multinomial model denotes the fractional presence of term t_j in the rth mixture component, including the effect of repetitions. The values of q_{jr} sum to 1 for a particular mixture component r over all terms (i.e., $\sum_{j=1}^{d} q_{jr} = 1$).

The generative process for the multinomial mixture model first selects the rth class (mixture component) with probability $\alpha_r = P(\mathcal{C}_r)$. Then, it throws a loaded die (owned by the rth class) L times to generate a document with L *tokens* (counting repetitions). The loaded die has as many faces as the number of terms d, and the probability of the jth face showing up is given by q_{jr} for the die owned by the rth class. Therefore, if the die is thrown L times, then the number of times each face shows up provides the number of times each term shows up in the observed document. If we assume that the frequency vector of the document $\overline{Z}$ is given by $(z_1 \ldots z_d)$, then the generative probability of the ith document is given by the following multinomial distribution:

$$P(\overline{Z}|\mathcal{C}_r) = \frac{(\sum_{j=1}^{d} z_j)!}{z_1! z_2! \ldots z_d!} \prod_{j=1}^{d} (q_{jr})^{z_j} \propto \prod_{j=1}^{d} (q_{jr})^{z_j} \tag{5.18}$$

The constant of proportionality holds for fixed $\overline{Z}$ and varying class, because it depends only on $\overline{Z}$ and is independent of the class $\mathcal{C}_r$.

The overall process of both prediction and training in the multinomial model is very similar to that of the Bernoulli model. As in the case of the Bernoulli model, one can use the Bayes rule and Equation 5.18 to derive the following values for the estimated posterior probability that the test instance $\overline{Z}$ belongs to class $\mathcal{C}_r$:

$$P(\mathcal{C}_r|\overline{Z}) \propto P(\mathcal{C}_r) \cdot P(\overline{Z}|\mathcal{C}_r) \propto \alpha_r \prod_{j=1}^{d} (q_{jr})^{z_j} \tag{5.19}$$

If needed, the constant of proportionality can be inferred by ensuring the posterior probabilities over all classes sum to 1. The class with the largest posterior probability can be predicted as the relevant one for the test instance $\overline{Z}$.

In order to compute the values on the right-hand side of Equation 5.19, one only needs to estimate the parameters α_r and q_{jr} during the training phase. The fractional presence of each class in the training data is used as the estimate of α_r. Laplacian smoothing can be used if needed. Furthermore, if $\nu(j, r)$ is the number of times that the term t_j shows up in the documents belonging to class r (with proportionate credit given to repetitions in a single document), then the estimate q_{jr} can be computed as follows:

$$q_{jr} = \frac{\nu(j, r)}{\sum_{j=1}^{d} \nu(j, r)} \qquad (5.20)$$

One can also view this estimate as the fraction of the number of *tokens* (i.e., positions) in a class that correspond to a particular term. This is different from the Bernoulli model that estimates the class-conditioned probabilities as the fraction of class-specific documents containing a particular term. It is also possible to use Laplacian smoothing in order to smooth the estimation. In this case, we add a small value $\gamma > 0$ to the numerator, and $\gamma \cdot d$ to the denominator. This results in the following estimation:

$$q_{jr} = \frac{\nu(j, r) + \gamma}{\sum_{j=1}^{d} \nu(j, r) + \gamma \cdot d} \qquad (5.21)$$

It is common to set γ to 1. This type of smoothing biases the estimation of the probability of each of the d faces in the multinomial die roll towards $1/d$, which implies that all terms are equally favored. This is a reasonable assumption in the absence of sufficient data.

5.3.3 Practical Observations

The naïve assumption of conditional independence is never really true in practical settings. In spite of this fact, the actual predictions are surprisingly robust. Using more complicated assumptions often end up overfitting the data. Several insights are provided in [159] about why the naïve assumption works so well in practice.

A natural question arises as to when it is preferable to use either the Bernoulli or the multinomial models. Note that the Bernoulli model uses both the presence and the absence of terms in a document, but it does not use the term frequencies. The two main factors are (i) the typical length of each document and, (ii) the size of the lexicon from which the terms are drawn. For short documents that have a non-sparse representation with respect to a small lexicon, it makes sense to use the Bernoulli model. In short documents, there are a limited number of repetitions of terms, which reduces the gain obtained from including frequency information. Furthermore, if the lexicon size is very small and the vector-space representation is non-sparse, then even the absence of a term in a document is informative. When the document representation is sparse, information about absence of terms is noisy, which hurts the Bernoulli model. Furthermore, the ignoring of frequency information will also increase the inaccuracy of the Bernoulli model. Therefore, it makes sense to use the multinomial model in such cases.

5.3.4 Ranking Outputs with Naïve Bayes

The prediction problem of classification is not always posed in terms of selecting the class of a single test instance. In many cases, a set of test instances $\overline{Z_1} \ldots \overline{Z_{n_t}}$ is provided, and it is desired to rank them in order of their propensity to belong to a particularly valuable class of interest. This problem is closely related to that of ranking in search engines.

Consider a situation where an aficionado in automobiles is interested in the rth class for which the label is *Cars*. How would one use the trained Bayes model to rank the test documents $\overline{Z}_1 \ldots \overline{Z}_{n_t}$ for this user? The aforementioned discussion already shows how one can estimate $P(\mathcal{C}_r|\overline{Z}_i)$ for each test instance $\overline{Z}_i$ *up to a constant of proportionality.* This scaling factor is not relevant when comparing the probabilities across different *classes*, but it is relevant when comparing the prediction across different *instances* because it varies across instances. The scaling factor for each test instance can be easily estimated by using the fact that the posterior probabilities of all classes must always sum to 1:

$$\sum_{r=1}^{k} P(\mathcal{C}_r|\overline{Z}_i) = 1 \tag{5.22}$$

After scaling, the normalized value of the posterior probabilities of the rth class are compared across different instances, and the documents are ranked in order of decreasing probability.

5.3.5   Example of Naïve Bayes

In the following, we will provide a numerical example of the naïve Bayes model. A similar example will be provided for both the Bernoulli and the multinomial model, in which documents are categorized either as *Cars* or as *Cats*.

5.3.5.1   Bernoulli Model

Consider the following corpus containing four training documents and two test documents. The corpus is represented in binary form in which the frequencies of the terms are ignored:

$$\begin{pmatrix} & \text{lion} & \text{tiger} & \text{cheetah} & \text{jaguar} & \text{porsche} & \text{ferrari} & Label \\ \text{Train1} & 1 & 1 & 1 & 1 & 0 & 0 & Cats \\ \text{Train2} & 1 & 1 & 1 & 1 & 0 & 0 & Cats \\ \text{Train3} & 0 & 0 & 0 & 1 & 1 & 1 & Cars \\ \text{Train4} & 0 & 0 & 0 & 1 & 1 & 1 & Cars \\ \text{Test1} & 1 & 1 & 1 & 1 & 1 & 1 & - \\ \text{Test2} & 1 & 1 & 1 & 1 & 0 & 0 & - \end{pmatrix}$$

For illustrative purposes, the lexicon contains only six terms. The class label of each instance is shown in the final column. The first four documents are the training documents, and the labels shown for them in the final column are *Cats* and *Cars*. However, the last two rows correspond to test instances, and therefore their labels are missing.

In the following, it is only shown how to use the training data to predict the probability of the two labels for the document *Test1*. The prediction of *Test2* is left as an exercise for the reader (see Exercise 4). The steps for the training and prediction phase are as follows.
Training: In order to perform the training, the prior probabilities and the class conditioned probabilities need to be estimated. Laplacian smoothing is used with $\beta = \gamma = 1$. The prior probabilities are estimated as:

$$P(Car) = \frac{2+\beta}{4+2\beta} = \frac{1}{2}, \qquad P(Cat) = \frac{2+\beta}{4+2\beta} = \frac{1}{2}$$

Next, we need to estimate the parameters for the Bernoulli distribution. We first show how to estimate $P(lion|Cats)$. The average number d_a of terms in the four training documents is $14/4$, and the total size of lexicon is $d = 6$. Therefore, the sparsity factor required for Laplacian smoothing is $6 \times 4/14 = 12/7$. In order to estimate $P(lion|Cats)$ note that the term is present in both of the two training documents on cats. Therefore, the estimation of this Bernoulli parameter is as follows:

$$P(lion|Cats) = \frac{2+\gamma}{2+\frac{12\gamma}{7}} = \frac{2+1}{2+\frac{12}{7}}$$

$$= \frac{21}{26}$$

By using an identical argument, we can show the following:

$$P(lion|Cats) = \frac{21}{26}, \ P(tiger|Cats) = \frac{21}{26}, \ P(cheetah|Cats) = \frac{21}{26}, \ P(jaguar|Cats) = \frac{21}{26}$$

$$P(porsche|Cats) = \frac{7}{26}, \ P(ferrari|Cats) = \frac{7}{26}$$

Similarly, one can compute the parameters of the Bernoulli distribution for *Cars* as follows:

$$P(lion|Cars) = \frac{7}{26}, \ P(tiger|Cars) = \frac{7}{26}, \ P(cheetah|Cars) = \frac{7}{26}, \ P(jaguar|Cars) = \frac{21}{26}$$

$$P(porsche|Cars) = \frac{21}{26}, \ P(ferrari|Cars) = \frac{21}{26}$$

Note that *"jaguar"* is the only term to get a high probability for both classes. These estimated probabilities represent the entire training model used by a naïve Bayes classifier. Next, we show how these estimated probabilities can be used for prediction of *Test1*.

Prediction: The prediction phase of *Test1* is particularly simple because it contains all the terms of the lexicon. Therefore, the class conditional probabilities may be computed as follows:

$$P(Cats|Test1) \propto P(Cats) \cdot P(lion|Cats) \cdot P(tiger|Cats) \cdot P(cheetah|Cats) \cdot$$
$$P(jaguar|Cats) \cdot P(porsche|Cats) \cdot P(ferrari|Cats)$$
$$= \frac{1}{2} \left(\frac{21}{26}\right)^4 \left(\frac{7}{26}\right)^2$$

$$P(Cars|Test1) \propto P(Cars) \cdot P(lion|Cars) \cdot P(tiger|Cars) \cdot P(cheetah|Cars) \cdot$$
$$P(jaguar|Cars) \cdot P(porsche|Cars) \cdot P(ferrari|Cars)$$
$$= \frac{1}{2} \left(\frac{21}{26}\right)^3 \left(\frac{7}{26}\right)^3$$

These computations only provide the inference to a constant of proportionality. One can also compute the exact probabilities of each class by ensuring that the corresponding probabilities sum to 1. Using that relationship, we obtain the fact that $P(Cats|Test1) = \frac{3}{4}$ and $P(Cars|Test1) = \frac{1}{4}$. Therefore, the test instance is more likely to belong to the *Cat* category. This is a logical conclusion because a larger number of terms in the document belong to the category of *Cats*. It is noteworthy that Laplacian smoothing is essential for obtaining reasonable results. If Laplacian smoothing had not been used, then one would have arrived at a probability of 0 for both outcomes, which would have lead to an indefinite prediction.

5.3.5.2 Multinomial Model

In the case of the multinomial model, the document-term matrix is assumed to contain frequencies. Therefore a very similar matrix is used as in the previous case, except that it also contains frequencies. The corresponding matrix is shown below:

$$
\begin{pmatrix}
 & \text{lion} & \text{tiger} & \text{cheetah} & \text{jaguar} & \text{porsche} & \text{ferrari} & \textit{Label} \\
\text{Train1} & 2 & 2 & 1 & 2 & 0 & 0 & \textit{Cats} \\
\text{Train2} & 2 & 3 & 3 & 3 & 0 & 0 & \textit{Cats} \\
\text{Train3} & 0 & 0 & 0 & 1 & 1 & 1 & \textit{Cars} \\
\text{Train4} & 0 & 0 & 0 & 2 & 1 & 2 & \textit{Cars} \\
\text{Test1} & 2 & 2 & 2 & 3 & 1 & 1 & - \\
\text{Test2} & 1 & 1 & 1 & 1 & 0 & 0 & -
\end{pmatrix}
$$

The prior probabilities are computed in exactly the same way as before. Therefore, the prior probabilities can be estimated to $(1/2)$ for each class. In order to compute the multinomial parameters, the number of occurrences of each term in the various classes are computed (including the effect of repetitions in the same document). This is summarized in the table below:

$$
\begin{pmatrix}
 & \text{lion} & \text{tiger} & \text{cheetah} & \text{jaguar} & \text{porsche} & \text{ferrari} & \text{Total} \\
\textit{Cats} & 4 & 5 & 4 & 5 & 0 & 0 & 18 \\
\textit{Cars} & 0 & 0 & 0 & 3 & 2 & 3 & 8
\end{pmatrix}
$$

The last column contains the total number of tokens in that class over all its documents. Now we need to compute the probabilities of each multinomial parameter q_{jr}. Without Laplacian smoothing, one can derive these parameters from the above counts by simply dividing each row with the total at the very end. However, with smoothing, we need to add 1 to each numerator and 6 to each denominator, since there are six terms in the lexicon. The corresponding values of q_{jr} are provided in the matrix below:

$$
\begin{pmatrix}
 & \text{lion} & \text{tiger} & \text{cheetah} & \text{jaguar} & \text{porsche} & \text{ferrari} \\
\textit{Cats} & \frac{5}{24} & \frac{6}{24} & \frac{5}{24} & \frac{6}{24} & \frac{1}{24} & \frac{1}{24} \\
\textit{Cars} & \frac{1}{14} & \frac{1}{14} & \frac{1}{14} & \frac{4}{14} & \frac{3}{14} & \frac{4}{14}
\end{pmatrix}
$$

Note that each row sums to 1, because it represents the probabilities of the different faces of the die in a multinomial event of selecting a word at a particular position.

One can use these estimated parameters to perform the prediction. Since, the frequency vector of *Test1* is $(2, 2, 2, 3, 1, 1)$, these frequencies are the exponents of the probabilistic parameters for each term:

$$
P(Cats|Test1) \propto \frac{1}{2} \left(\frac{5}{24} \right)^2 \left(\frac{6}{24} \right)^2 \left(\frac{5}{24} \right)^2 \left(\frac{6}{24} \right)^3 \left(\frac{1}{24} \right) \left(\frac{1}{24} \right)
$$

$$
P(Cars|Test1) \propto \frac{1}{2} \left(\frac{1}{14} \right)^2 \left(\frac{1}{14} \right)^2 \left(\frac{1}{14} \right)^2 \left(\frac{4}{14} \right)^3 \left(\frac{3}{14} \right) \left(\frac{4}{14} \right)
$$

On simplification and normalization, it can be shown that the probabilities of *Cats* and *Cars* are around 0.94 and 0.06, respectively. Therefore, one arrives at the same conclusion, except that the predictions are more definitive in this case. This is because of the greater frequency of the cat-related words in the test document. It is noteworthy that Laplacian smoothing is essential for obtaining reasonable results. If Laplacian smoothing had not been used, then one would have arrived at a probability of 0 for both outcomes, which would have lead to an indefinite prediction. The multinomial model also does not use terms absent from the test document. Although *Test1* contains all the terms, the document *Test2* does not. If the multinomial model is used to classify *Test2*, both $P(Cats|Test2)$ and $P(Cars|Test2)$ can be expressed in terms of only the conditional probability estimates of "*lion*," "*tiger*," "*cheetah*," and "*jaguar*." The conditional estimates of "*porsche*" and "*ferrari*" will be ignored (see Exercise 5).

5.3.6 Semi-Supervised Naïve Bayes

The Bayes model provides a remarkably clear picture of the connections between supervised and unsupervised models of learning. It is noteworthy that the mixture-modeling algorithm for clustering in Section 4.4 of Chapter 4 uses *exactly* the same generative model as the naïve Bayes model. A mixture component represents a cluster in unsupervised learning, whereas a mixture component represents a class in supervised learning. The differences in their computational procedures are explained by the fact that unsupervised mixture modeling is handicapped by the absence of labels. Labels are useful in identifying the mixture component that generates each training point so that the parameters of each mixture component can be estimated easily. In the absence of labels, one is forced to use an *iterative* approach of probabilistically predicting the mixture component associated with each data point (E-step) and estimating mixture parameters (M-step). The presence of labels simplifies the learning process *to a single M-step* in naïve Bayes classification, because *the unlabeled data is not used in parameter estimation*. Furthermore, the unlabeled instances are classified with a single application of the E-step using the learned parameters.

Semi-supervised learning is useful when the amount of labeled data is limited, and therefore the unlabeled data is incorporated in the parameter estimation process in order to improve classification accuracy. The use of unlabeled data in parameter estimation [419] causes semi-supervised methods to be iterative like the expectation-maximization algorithm of Section 4.4. The semi-supervised approach assumes that each mixture component is associated with a class. The labeled and unlabeled points of each class are generated by its mixture component. At initialization, the parameters of each mixture component and the prior probabilities are estimated with an application of the naïve Bayes algorithm on the labeled instances. Subsequently, the following pair of steps is iteratively used:

1. **(E-step):** The E-step estimates the probabilities of the unlabeled instances using the Bayes rule of posterior probabilities. Therefore, the first iteration of the E-step would yield exactly the same probabilities as computed by the naïve Bayes algorithm. Therefore, the E-step remains the same but it is applied only to the unlabeled data during the iterations in order to predict their class memberships. As in the EM-algorithm of Section 4.4, we use the soft membership probabilities derived in the E-step to associate membership weights with unlabeled instances. The membership weights of a point across different clusters sum to 1 because they represent posterior probabilities. *An important modification to the E-step in the semi-supervised setting is that the labeled instances are also associated with a membership weight $\lambda > 0$ to the class/cluster it belongs to and 0 to all other classes.* The value of λ is a user-driven parameter in $(0, \infty)$, which regulates the level of supervision.

2. **(M-step):** The M-step remains identical to what is discussed in the mixture-modeling algorithm of Section 4.4, except that it is executed with the help of the modified

membership weights in which labeled instances are given the user-defined weight of λ.

The two steps are iterated to convergence. The probabilistic predictions of the E-step in the final iteration can be used to predict the class labels. Therefore, the modifications to the expectation-maximization algorithm of Section 4.4 are relatively minor, and involve the incorporation of labeled data within the parameter estimation step. The degree of impact of this change depends on the value of λ.

The parameter λ controls the trade-off between the importance of labeled and unlabeled data. Setting $\lambda = 0$ results in the EM-algorithm of Section 4.4, and setting $\lambda = \infty$ results in the naïve Bayes algorithm of this section. All other positive values of λ provide varying levels of supervision in which the iterative approach is still needed. It is generally sensible to choose $\lambda > 1$ in semi-supervised classification applications, because one should weight each labeled point to a greater degree than each unlabeled point.

Note that such intermediate values of λ can often outperform the naïve Bayes method in cases where the amount of labeled data is very small. With limited labeled data, the conditional probabilities of absent terms from the labeled data will be estimated poorly by the fully supervised naïve Bayes method. Such probabilities will be estimated far more robustly in the semi-supervised setting because the unlabeled documents in the relevant mixture component can be leveraged for robust estimation. A different way of understanding this is that unlabeled data can learn the shape of the underlying data distribution and ensure that the labeled data is required only to map the learned clusters of this data distribution to the different labels. Therefore, most of the "heavy-lifting" of learning the shape of the data distribution is done with unlabeled data, and only a small amount of data is needed to map the dense segments (mixture components) of this data distribution to different classes. The natural assumption here is that class labels do not change abruptly within contiguous, dense, and clustered regions of the data. This situation occurs often in real data sets due to the natural *smoothness* and *clustered* properties of real-world class distributions [103]. It is also possible to construct semi-supervised models in which the number of mixture components is larger than the number of labeled classes to learn class distributions that are locally contiguous in specific regions (see Exercises 6 and 7).

Another advantage of semi-supervision is that the learning process is specific to the test instances we are interested in. Purely supervised methods build models that are more general than what we really need. This provides semi-supervision an advantage based on *Vapnik's principle* [103]:

> "When trying to solve some problem, one should not solve a more difficult problem as an intermediate step."

One can get better results by solving the narrower problem and tuning the learning process to the specific test instances at hand. For example, if a small training data set contains only a couple of instances of each class, the number of instances is too small to robustly estimate the prior probabilities. On the other hand, if a large test data contains these classes in the proportion of 9:1, then the semi-supervised parameter estimation process will use this additional information to assign more robust prior probabilities. If a different test data set contains these classes in the reverse proportion of 1:9, it will assign different prior probabilities.

This approach can be used for both semi-supervised clustering and semi-supervised classification. Semi-supervised clustering has slightly different applications from semi-supervised classification, because the supervision is gentler in the former and the goal is to create a semantically meaningful partition with external input rather than to label instances. For semi-supervised clustering applications, it makes sense to use smaller values of λ to give more importance to the clustering structure inherent in the unlabeled data.

5.4 Nearest Neighbor Classifier

Nearest-neighbor classifiers use the following principle:

> Similar instances have similar labels.

A natural way of implementing this principle is to use a κ-nearest-neighbor classifier. The basic idea is to identify[4] the κ-nearest neighbors of a test point, and compute the number of points that belong to each class. The class with the largest number of points is reported as the relevant one. The cosine similarity is used to compute the nearest neighbors, although one can use advanced methods like the substring kernel in order to incorporate sequence information in the classification process. Nearest-neighbor classification can be used for both binary classes and multi-way classes, as long as the class with the largest vote is used. If the dependent variable is numeric, the average value of the dependent variable among the nearest neighbors can be reported.

Nearest-neighbor classifiers are also referred to as *lazy learners*, *memory-based learners*, and *instance-based learners*. They are referred to as lazy learners because most of the work of classification is postponed to the very end. In a sense, these methods *memorize* all the training examples, and use the best matching ones to the *instance* at hand. Unlike model-based methods, less generalization and learning is done up front, and most of the work of classification is left to the very end in a *lazy* way. However, there are many natural variations of nearest-neighbor classifiers in which some of the work of learning is brought up front. Such classifiers are referred to as *adaptive nearest-neighbor classifiers*.

A straightforward implementation of the nearest-neighbor method requires no training, but it requires $O(n)$ similarity computations for classifying *each test instance*. One can speed this process up using a data structure called an *inverted index*. This data structure is discussed in detail in Section 9.2.2 of Chapter 9. An inverted index contains a list of document identifiers associated with each term. For a given test document, one needs to access as many inverted lists as the number of terms in it, and access only those documents whose identifiers are included in one of these inverted lists.

The number of nearest neighbors, κ, is a parameter for the algorithm. Its value can be set by trying different values of κ on the training data. The value of κ at which the highest accuracy is achieved on the training data is used. While computing accuracy on the training data, a *leave-one-out* approach is used, in which the point to which the κ-nearest neighbors are computed is not included among the nearest neighbors. For example, if we did not take this precaution, every point with be its own nearest neighbor, and a value of $\kappa = 1$ would always be deemed as optimal. This is a manifestation of overfitting, which is avoided with the leave-one-out approach. The classification accuracy is computed by using a *validation sample* of size s. For each point in the sample, the similarities with respect to the entire data are computed in a leave-one-out manner. These computed similarities are used to rank the $n - 1$ training points for each sample, and test various values of κ. This process requires $O(n \cdot s)$ similarity computations and $O(n \cdot s \cdot \log(n))$ time for sorting the

---

[4]Most of the literature uses the notation of k instead of κ to denote the number of nearest neighbors. We use κ instead of k for notational disambiguation, since the latter variable has been used consistently in this chapter to denote the number of classes. Using k to denote both the number of classes and the number of neighbors would cause confusion.

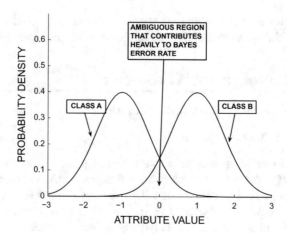

Figure 5.1: Example of how the noise in a data set affects error

points. For a validation sample size of s, the time required is $O(s \cdot n \cdot (T + \log(n)))$ for tuning the parameter κ. Here, T is the time required for each similarity computation. One can reduce this running time with an inverted index.

5.4.1   Properties of 1-Nearest Neighbor Classifiers

A special case of nearest-neighbor classifiers is one in which the value of κ is set to 1. Such classifiers are not very robust in practice because they are sensitive to the specific data set at hand. This lack of robustness is caused by the fact that the predictions can overfit the vagaries of the particular training sample at hand. Whenever the classification of the same test instance varies significantly with the choice of training sample, it contributes to increased classifier error, and this portion of the error is referred to as the *variance* (cf. Section 7.2 of Chapter 7). As the size of the training sample increases, the accuracy of the 1-nearest neighbor classifier increases as well. In fact, it can be shown that with an infinite amount of data, the error of a 1-nearest-neighbor classifier is at most twice the *Bayes optimal error rate*. The Bayes optimal error rate refers to the minimum achievable error rate of a particular data distribution. In order to understand this point, consider a 1-dimensional data set with two normally distributed classes as shown in Figure 5.1. It is noteworthy that the class distribution is overlapping in a particular region of the data. Even if a learner were given the extraordinary advantage of being told the (true) generative distribution of the two classes, it would still make mistakes on some of these ambiguous instances in this overlapping region. The Bayes error rate quantifies this intrinsic error from a probabilistic point of view. This notion is closely related to that of *intrinsic noise* in a data set, which is a fundamental component of the error in any classifier (cf. Section 7.2 of Chapter 7).

The boundary between various classes is also referred to as the decision boundary in classification. In general, boundaries of complex nonlinear shapes are considered more challenging for classification. The aforementioned observation of the error rate of a 1-nearest-neighbor classifier implies that it can approximate any nonlinear boundary very well, given a "sufficient" amount of data. This point can be better understood with the *Voronoi diagram* induced by the training data.

Given a set of training points, one can divide the data space into a set of *Voronoi regions* or *cells* induced by these points. A Voronoi region or cell is a portion of the data

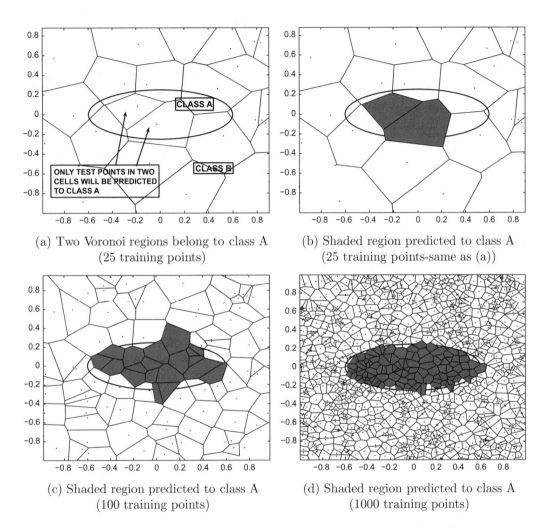

(a) Two Voronoi regions belong to class A (25 training points)

(b) Shaded region predicted to class A (25 training points-same as (a))

(c) Shaded region predicted to class A (100 training points)

(d) Shaded region predicted to class A (1000 training points)

Figure 5.2: Increasing the number of training points improves the accuracy of 1-nearest neighbor classification. The shaded region approximates the (true) elliptical boundary between classes A and B more closely with increasing number of training points. With an infinite amount of data, only the error caused by intrinsic noise (e.g., overlapping regions and mislabeled training points) will remain. The cumulative effect of the noise contributed by both training and test points is equal to twice the Bayes error rate.

space that is closest to the single point inside it as compared to all the other training points. From a 1-nearest neighbor classifier point of view, each Voronoi region "belongs" to a single training point, and all test instances within that cell will take on the same class label as that training point. This situation is shown in Figure 5.2(a), in which the class A is enclosed by an elliptical decision boundary. However, only 25 training points are used, and therefore test points in only two Voronoi regions will be assigned class A. Note that the shapes of the Voronoi cells are jagged, and therefore if the decision boundary between the two classes is smooth, the 1-nearest-neighbor classifier will try to approximate this boundary with jagged edges, which increases its error. As shown in Figure 5.2(b), the 1-nearest neighbor classifier tries to approximate the elliptical region for class A with the shaded region, which causes a rather poor decision boundary. This approximation varies with random choice of training data, which increases classification error *in expectation*. However, if the number of training points is increased, the size of each Voronoi region reduces, and therefore the jagged approximation of the 1-nearest-neighbor classifier improves, as shown in Figures 5.2(c), and (d) in which 100 and 1000 points are respectively used. With an infinite amount of data, any arbitrary boundary can be approximated very well, and only the ambigous/overlapping regions of the decision boundary will be incorrectly classified. This portion of the error is a result of the specific noise or mislabeling in the data set. There is little that most classifiers to do to handle such instances, and they contribute to the portion of the error referred to as the Bayes error rate.

The main problem with the 1-nearest neighbor classifier is that the amount of data required to achieve this error rate depends *exponentially* on the intrinsic dimensionality of the data set. Text data may have hundreds of thousands of terms, and the intrinsic dimensionality may often be on the order of hundreds. As a result, the required amount of data is too large for a 1-nearest-neighbor classifier to achieve an error anywhere close the Bayes error rate. Using a κ-nearest neighbor classifier with larger values of κ is a way of smoothing the aforementioned jagged boundary to improve the error rate with a limited amount of data. There are several other ways of smoothing this boundary, such as the use of clustering, the use of weighted nearest neighbors, or the use of some level of supervision in determining the nearest neighbors. The last of these is also referred to as *adaptive* nearest-neighbor classification, and it provides a family of the most powerful classifiers in machine learning. Two of the most powerful classifiers in machine learning, which are *random forests* and *kernel support vector machines*, can be shown to be adaptive nearest-neighbor classifiers. These points will be discussed in Sections 5.5.6 and 6.3.6. This section will provide an overview of methods for smoothing the predicted decision boundary, such as the Rocchio method, the weighted nearest-neighbor method, and adaptive nearest-neighbor method.

5.4.2   Rocchio and Nearest Centroid Classification

The Rocchio classifier can be viewed as a modification of the nearest-neighbor classifier. In Rocchio classification, the centroids of each of the classes is computed up front. For a given test instance, the nearest class centroid is computed with cosine similarity. The label of the closest centroid is reported as the classification of the test instance. The Rocchio classifier is extremely efficient in both training and prediction. The training step requires only the computation of the centroid of each class, which scales linearly with training data size. The testing step requires only the computation of k cosine similarities for a k-class problem.

The Rocchio method provides stable predictions over different choices of training data sets. However, it shows significant *bias* in the predictions. For example, Rocchio's method

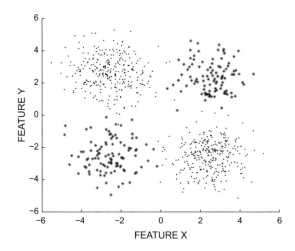

Figure 5.3: A bad case for the Rocchio method

would not work very well if documents of the same class were separated into distinct clusters. In such cases, the centroid of a class of documents may not be representative of that class. A bad case for Rocchio's method is illustrated in Figure 5.3, in which each class is associated with two distinct clusters. Furthermore, the centroid of each class is similar, and therefore, the Rocchio method would have difficulty in distinguishing between the classes. On the other hand, a 1-nearest-neighbor classifier would perform quite well in this case. The Rocchio method does not adjust well to the varying frequencies of different classes. By using one centroid for each class, it effectively sets an equal prior probability of each class.

A natural trade-off between the two extremes of a 1-nearest neighbor classifier and the Rocchio method is to use centroid-based classification. The basic idea is to use an off-the-shelf clustering algorithm to partition the documents of each class into clusters. Class labels are associated with clusters rather than documents. The number of clusters in each class is proportional to the number of documents in that class. This ensures that the clusters in each class are of approximately the same granularity.

The *cluster digests* from the centroids are extracted by retaining only the most frequent words in that centroid. Typically, about 200 to 400 words are retained in each centroid. The lexicon in each of these centroids provides a stable and topical representation of the subjects in each class. An example of the (weighted) word vectors for two classes corresponding to the labels "*Business schools*" and "*Law schools*" could be as follows:

1. *Business schools:* business (35), management (31), school (22), university (11), campus (15), presentation (12), student (17), market (11), ...

2. *Law schools:* law (22), university (11), school (13), examination (15), justice (17), campus (10), courts (15), prosecutor (22), student (15), ...

Typically, most of the noisy words have been truncated from the cluster digest. Similar words are represented in the same centroid, and words with multiple meanings can be represented in contextually different centroids. Therefore, this approach also indirectly addresses the issues of synonymy and polysemy, with the *additional* advantage that the nearest-neighbor classification can be performed more efficiently with a smaller number of centroids. The dominant label from the top-κ matching centroids, based on cosine similarity, is reported.

Such an approach can provide comparable or better accuracy than the vanilla κ-nearest neighbor classifier in many cases.

5.4.3   Weighted Nearest Neighbors

A κ-nearest neighbor classifier can be viewed through the lens of a similarity weighted classifier. Such a view helps in generalizing the κ-nearest neighbor classifier to a surprisingly powerful family of methods (e.g., *adaptive* nearest neighbor methods), and also illustrates how a proper choice of weight balances robustness (resistance to overfitting) and reduction in bias. Consider a training data set with documents $\overline{X_1} \ldots \overline{X_n}$ with labels $y_1 \ldots y_n$. Although we assume binary labels $y_i \in \{-1, +1\}$ for notational simplicity and closed-form expressions, the basic ideas underlying these arguments can be generalized to multi-way classification and regression modeling with minor modifications. Then, for any test instance $\overline{Z}$, one can view a κ-nearest neighbor classifier as a *similarity weighted classifier*, where the similarity between test instance $\overline{Z}$ and training instance $\overline{X_i}$ is denoted[5] by $K(\overline{Z}, \overline{X_i})$. The prediction $F(\overline{Z})$ of the test instance $\overline{Z}$ can be expressed as a similarity weighted classifier as follows:

$$F(\overline{Z}) = \text{sign} \left\{ \sum_{i=1}^{n} K(\overline{Z}, \overline{X_i}) y_i \right\} \tag{5.23}$$

Here, the function "sign" returns either -1 or $+1$, depending on the sign of its argument. One can view the κ-nearest neighbor classifier as a weighted nearest-neighbor classifier in which the value of $K(\overline{Z}, \overline{X_i})$ is defined as follows:

$$K(\overline{Z}, \overline{X_i}) = \begin{cases} 1 & \overline{X_i} \text{ is among the } \kappa\text{-nearest neighbors of } \overline{Z} \\ 0 & \text{otherwise} \end{cases} \tag{5.24}$$

The similarity function $K(\overline{Z}, \overline{X_i})$ can be viewed as a weight that decays with reducing similarity of $\overline{X_i}$ to the test point $\overline{Z}$. For infinitely large data sets, it is desirable to choose the sharpest possible decay in weight, which is achieved by the 1-nearest neighbor classifier. Such a classifier yields an error of at most twice the Bayes optimal rate for infinite data but very poor results for small data sets. Choosing $\kappa = n$ results in a (relatively stable) majority-vote classifier even for minuscule data sets, but the predictions are unable to take advantage of more data. In particular, the predictions are not very discriminating in different regions of the space because every test point gets the same prediction. This is a manifestation of excessive bias in predictions. Clearly, a trade-off needs to be selected that works well for the data set at hand.

Setting $K(\overline{Z}, \overline{X_i})$ to the dot product $\overline{Z} \cdot \overline{X_i}$ (after scaling the training and test vectors to unit norm) results in the use of the cosine similarity as the weight. One can also use Gaussian kernel similarity (with normalized documents), which exponentiates the negative (squared) distances $D(\overline{Z}, \overline{X_i})$ to create similarity values:

$$K(\overline{Z}, \overline{X_i}) = e^{-D(\overline{Z}, \overline{X_i})^2 / (2 \cdot \sigma^2)} = e^{-||\overline{Z} - \overline{X_i}||^2 / (2 \cdot \sigma^2)} \tag{5.25}$$

The choice of the bandwidth σ controls the rate of decay of the weight with increasing distance of training points to the test point. If we have a small data set, we should use a large value of σ to encourage slow decay. On the other hand, for a larger data set, we can

---

[5]We intentionally use the seemingly unusual notation $K(\cdot, \cdot)$ for a similarity function, as we will later connect this principle with the kernel similarity function used by support vector machines.

use a smaller value of σ to encourage sharper decay. The value of σ can be tuned by using a leave-one-out validation approach. The weighted nearest-neighbor method can also be used for regression. The only difference is that one does not need to use the sign function in Equation 5.23, and one must normalize the similarities to sum to 1 over all points. In other words, the values of $K(\overline{Z}, \overline{X_i})$ should be proportionally scaled to sum to 1 for fixed $\overline{Z}$ and all $\overline{X_i}$.

5.4.3.1 Bagged and Subsampled 1-Nearest Neighbors as Weighted Nearest Neighbor Classifiers

A 1-nearest neighbor classifier makes unstable predictions over different choices of the training data. It stands to reason that the classifier is making mistakes in at least some of the training data instantiations, and therefore the instability contributes to higher *expected* error. A weighted nearest-neighbor classifier has less variability than a 1-nearest neighbor classifier. Interestingly, one can show that combining some *ensemble* methods like bagging or subsampling with the 1-nearest neighbor classifier can simulate the effect of a weighted nearest-neighbor classifier, and reduce the variability of the base predictor.

Bagging works as follows. In each iteration, a sample of size $s \leq n$ is selected from the training data of size n. The sample is selected *with replacement* so that it might contain duplicates. The 1-nearest neighbor classifier is used on each test point to make a prediction of that point in each ensemble component. The predictions of that point over all ensemble components are averaged. For a regression model, the average prediction is returned. For a binary classifier model with class labels in $\{-1, +1\}$, the sign of the average (or aggregate) prediction is returned. Subsampling is similar to bagging, except that the sampling is done without replacement.

A bagged nearest-neighbor classifier is a weighted nearest-neighbor classifier in which the weight of each training point is the probability that it is the 1-nearest neighbor of the test instance in a sample of size s. Let $P(\overline{X_i}|\overline{Z})$ be the probability that $\overline{X_i}$ is the 1-nearest neighbor of $\overline{Z}$ in the bagged sample of size s. Furthermore, let $R(\overline{Z}, \overline{X_i}) \in \{1 \dots n\}$ represent the *rank* of the nearest neighbor distance of $\overline{X_i}$ to $\overline{Z}$. Then, the probability that the point $\overline{X_i}$ is the nearest neighbor of $\overline{Z}$ in a bagged sample of size s is as follows:

$$P(\overline{X_i}|\overline{Z}) = P[\text{Not sampling nearest } (R(\overline{Z}, \overline{X_i}) - 1)] - P[\text{Not sampling nearest } R(\overline{Z}, \overline{X_i})]$$

$$(5.26)$$

$$= \left(1 - \frac{R(\overline{Z}, \overline{X_i}) - 1}{n}\right)^s - \left(1 - \frac{R(\overline{Z}, \overline{X_i})}{n}\right)^s \tag{5.27}$$

Then, the effect of the bagged 1-nearest neighbor classifier is to create a weighted prediction of the form of Equation 5.23, where $K(\overline{Z}, \overline{X_i})$ is set to $P(\overline{X_i}|\overline{Z})$. The sample size s regulates the rate of decay. Using a sample size of $s = 1$ is equivalent to using the κ-nearest neighbor classifier with $\kappa = n$, and using a sample size of $s = n$ is equivalent to the 1-nearest neighbor classifier on all the points with a single ensemble component. In general, increasing the sample size makes the weight decay sharper. Another observation is that one does not need to implement a bagged 1-nearest neighbor classifier with Monte Carlo sampling. One can directly use Equation 5.23 and set $K(\overline{Z}, \overline{X_i}) = P(\overline{X_i}|\overline{Z})$ according to Equation 5.27. One can also derive a similar result for the case of subsampling without replacement:

$$P(\overline{X_i}|\overline{Z}) = \begin{cases} \binom{n - R(\overline{Z}, \overline{X_i})}{s-1} / \binom{n}{s} & \text{if } R(\overline{Z}, \overline{X_i}) \leq n - s + 1 \\ 0 & \text{if } R(\overline{Z}, \overline{X_i}) > n - s + 1 \end{cases} \tag{5.28}$$

We leave the proof of this result as an exercise for the reader (see Exercise 8).

These results show that weighted nearest-neighbor classifiers are connected to well-known techniques in ensemble-learning, and their use can provide robust results. It is particularly noteworthy that the weights decay exponentially with the *rank* of the distances of the training points to the test point in the bagged and subsampled 1-nearest neighbor methods. This is similar to using Gaussian decay (cf. Equation 5.25), except that the weights decay exponentially with the *raw* distances in Equation 5.25 (rather than the rank). Bagged/subsampled 1-nearest neighbors are known to give good results, and this also suggests that one can get good results with Gaussian decay. In fact, combining Gaussian decay with supervised importance weighting of points can be shown to be equivalent to *kernel support vector machines* (cf. Chapter 6). Such methods are referred to as *adaptive nearest neighbor methods*.

5.4.4   Adaptive Nearest Neighbors: A Powerful Family

Nearest-neighbor methods are sensitive to several factors, which add to the error:

1. Noisy and irrelevant points add to the error of the nearest-neighbor classifier.

2. Irrelevant features add to the instability of the computations, which can further increase the variability in predictions.

Is there any way to modify the nearest-neighbor classifier to make it less sensitive to these effects? It turns out that this is indeed possible by using one of the following two strategies either in isolation or in combination:

1. It can be learned up front which *points* are more important for improving classification accuracy. Such points can be weighted to a greater degree.

2. It can learned up front, which dimensions (or directions) are more important, and the similarity function $K(\overline{Z}, \overline{X_i})$ can be modified to give greater importance to the discriminative directions.

One can now augment the weighted nearest-neighbor classification prediction function with an additional weight λ_i with point $\overline{X_i}$:

$$F(\overline{Z}) = \text{sign} \left\{ \sum_{i=1}^{n} \lambda_i \, K(\overline{Z}, \overline{X_i}) \, y_i \right\} \qquad (5.29)$$

The value of λ_i needs to be learned up front in a data-driven manner. Furthermore, the similarity function $K(\overline{Z}, \overline{X_i})$ might be data-driven and learned in a supervised manner based on the labeled training data. It is often overlooked that some of the most powerful classifiers in all of machine learning are adaptive nearest-neighbor classifiers:

> The kernel support vector machine and the random forest, which are known to be extremely powerful classifiers [194], are special cases of adaptive nearest-neighbor classifiers. Specifically, their prediction function can be reduced to the form of Equation 5.29.

In the case of the support vector machine, the prediction function is almost identically of the form of Equation 5.29, whereas a random forest uses $\lambda_i = 1$ but uses a data-driven similarity function, which is defined algorithmically (i.e., not in closed form) [72]. These

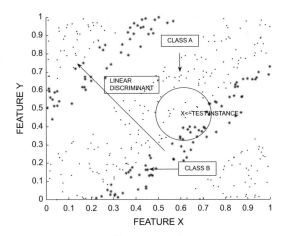

Figure 5.4: Increasing sensitivity of distance function towards discriminating directions reduces impact of noise

points will be explained in greater detail in the sections on random forests and support vector machines (cf. Sections 5.5.6 and 6.3.6).

In this section, we will provide a specific example of an adaptive method in which $K(\overline{Z}, \overline{X_i})$ is designed in a supervised way. We describe the discriminant adaptive nearest-neighbor classifier [236], which weights specific directions in the data in order to make the distance function more sensitive to the distribution of classes. This makes the classifier less sensitive to noise, and able to perform better classification with a small amount of data. In order to understand this point, consider a two-class data distribution shown in Figure 5.4, which contains two classes denoted by A and B. For the purpose of the following discussion, assume that the documents have been normalized to unit norm, so that using the Euclidean is equivalent to using the cosine similarity (cf. Section 2.5 of Chapter 2). Although the test instance belongs to class A, the spherical distance contour of the Euclidean distance finds a larger number of points belonging to class B. This is caused by the fact that the data set is very small, and only one of the two directions in the data is discriminating (see direction of arrow shown in Figure 5.4). As a result, the noisy direction contributes to the error caused by the vagaries inherent in a small data set.

One can improve the Euclidean distance function by incorporating information about the class distribution. Consider the Euclidean distance $D(\overline{Z}, \overline{X_i})$, which is defined as follows:

$$D(\overline{Z}, \overline{X_i}) = ||\overline{Z} - \overline{X_i}||^2 = (\overline{Z} - \overline{X_i})(\overline{Z} - \overline{X_i})^T$$

One can augment this distance function with a $d \times d$ *distortion matrix* A, which contains all the useful knowledge in the training data about the discriminating directions:

$$D(\overline{Z}, \overline{X_i}) = (\overline{Z} - \overline{X_i})A(\overline{Z} - \overline{X_i})^T \qquad (5.30)$$

How is A learned from the training data? Basically, the matrix A is set to the *linear discriminant analysis metric* that implicitly scales the directions in the data, so that less discriminating directions are given less importance.

Let Σ_i be the covariance matrix of the ith class, so that the (j, k)th entry of Σ_i is equal to the covariance between the jth and kth dimensions in the ith class. Let n_i be the number of points in the ith class. Let $\overline{\mu}$ be the d-dimensional row vector representing the mean of the entire data set, and $\overline{\mu_i}$ be the d-dimensional row vector representing the mean of the ith class. Then, the $d \times d$ within-class scatter matrix is defined as follows:

$$S_w = \sum_{i=1}^{k} n_i \Sigma_i \tag{5.31}$$

The $d \times d$ between-class scatter matrix is defined as the sum of the following rank-1 matrices:

$$S_b = \sum_{i=1}^{k} n_i (\overline{\mu_i} - \overline{\mu})^T (\overline{\mu_i} - \overline{\mu}) \tag{5.32}$$

Note that each term in the above summation is a $d \times d$ matrix, because it is the product of a $d \times 1$ matrix with a $1 \times d$ matrix. Then, the distortion matrix A is defined as follows:

$$A = S_w^{-1} S_b S_w^{-1} \tag{5.33}$$

This matrix A is used to compute the distance function of Equation 5.30 and the corresponding nearest neighbors. The dominant class among the κ nearest neighbors is reported as the relevant one.

It is also possible to exponentiate the (negative of the) squared distance function to create a similarity value like Equation 5.25, and then substitute in the prediction function of Equation 5.29. The value of λ_i in Equation 5.29 is set to 1. This is an adaptive approach because the similarity function has been learned up front with the use of labeling information.

This approach requires the inversion of matrices of size $d \times d$ in the original space, which might be computationally onerous. Therefore, latent semantic analysis can be used to transform all the training and test documents to a space of less than 500 dimensions. The operations in the transformed space are far more efficient and computationally tractable.

5.5   Decision Trees and Random Forests

As the name implies, a decision tree is a tree-like (i.e., hierarchical) partitioning of the data space, in which the partitioning is achieved with a series of split conditions (i.e., decisions) on the attributes. The idea is to partition the data space into attribute regions that are heavily biased towards a particular class during the training phase. Therefore, partitions are associated with their favored class labels. During the testing phase, the relevant partition of the data space is identified for the test instance, and the label of the partition is returned. Note that each *node* in the decision tree corresponds to a region of the data space defined by the split conditions at its ancestor nodes, and the root node corresponds to the entire data space. Random forests are ensemble-centric implementations of decision trees, which are known to be highly robust and accurate.

5.5.1   Basic Procedure for Decision Tree Construction

Decision trees partition the data space recursively in top-down fashion using *split conditions* or *predicates*. The basic idea is to choose the split conditions in such a way that the subdivided portions are dominated by one or more classes. The evaluation criteria for such split predicates are often similar to feature selection criteria in classification. The split criteria typically correspond to constraints on the frequencies of one or more words. A split that uses a single attribute is referred to as a *univariate split*, whereas a split using multiple attributes is referred to as a *multivariate split*. It is common for each node in the decision tree to have only two children. For example, if the split predicate corresponds to the presence of absence of a word, then all documents containing the word will be contained in one child and the remaining documents will be in the other child. The splits are applied recursively in top-down fashion, until each node in the tree contains a single class. These nodes are the leaf nodes, and are labeled with the classes of their instances. In order to classify a test instance for which the label is unknown, the split predicates are used in top-down fashion over various nodes of the tree in order to identify the branch to follow down the tree until the leaf node is reached. For example, if the split predicates correspond to presence or absence of words, it is checked whether the test document contains the word or not to determine the relevant branch to follow. This process is repeated until the relevant leaf node is identified, and its label is reported as the prediction of the test instance.

This type of extreme way of creating a tree until each leaf contains instances of only a single class is referred to as growing a tree to full height. Such a fully-grown tree will provide 100% accuracy on the *training data* even for a data set in which class labels are generated randomly and independently of the features in the training instances. This is clearly the result of overfitting, because one cannot expect to learn anything from a data set with random labels. A fully-grown tree will often misinterpret random nuances in the training data as indicative of discriminative power, and these types of overfitted choices will cause the predictions of the same test instance to vary significantly between trees constructed on different training samples. This type of variability is usually a sign of a poor classifier *in expectation*, because at least some of these diverse predictions are bound to be incorrect. As a result, the performance on the test data of such a tree will be poor even for those data sets in which the feature values are related to the class labels. This problem is addressed by *pruning* the nodes at the lower levels of the tree that do not contribute in a positive way to the generalization power on unseen test instances. As a result, the leaves of the pruned tree may no longer contain a single class, and are therefore labeled with the majority class (or *dominant* class for k-way classification).

Pruning is accomplished by holding out a part of the training data, which is not used in the (initial) decision-tree construction. For each internal node, it is tested whether or not the accuracy improves on the held out data by removing the subtree rooted at that node (and converting that internal node to a leaf). Depending on whether or not the accuracy improves, the pruning is performed. Internal nodes are selected for testing in bottom-up order, until all of them have been tested once. It is also noteworthy that pruning is not required in ensemble-centric implementations of decision trees like random forests, because such implementations avoid overfitting by other mechanisms. The overall procedure of decision-tree construction is shown in Figure 5.5. Note that the specific split criterion is not spelled out in these generic pseudo-code. This is an issue that will be discussed in the next section. The notion of eligibility of a node to be split is also not specified in the pseudo-code. Since bottom nodes are pruned anyway, it is possible to stop early using other criteria than growing the tree to full height. Various stopping criteria make nodes ineligible for splitting, such as a maximum threshold on the number of instances, or a minimum percentage threshold on the dominant class.

5.5.2   Splitting a Node

The split criteria can use any of the feature selection criteria discussed in Section 5.2. Common choices are the Gini index (cf. Equation 5.2) and conditional entropy (cf. Equation 5.5).

Algorithm *ConstructDecisionTree* (Labeled Training Document Set: D_y)
begin
　Hold out a document subset H from D_y to create $D'_y = D_y - H$;
　Initialize decision tree T to a single root node containing D'_y;
　{ **Tree Construction Phase** }
　repeat
　　Select any eligible leaf node from T with data set L;
　　Use split criteria of section 5.5.2 to partition L into subsets L_1 and L_2;
　　Store split condition at L and make $\{L_1, L_2\}$ children of L in T;
　until no more eligible nodes in T;
　{ **Tree Pruning Phase** }
　repeat
　　Select an untested internal node N in T in bottom-up order;
　　Create T_n obtained by pruning subtree of T at N;
　　Compare accuracy of T and T_n on held out set H;
　　if T_n has better accuracy **then** replace T with T_n;
　until no untested internal nodes remain in T;
　Label each leaf node of T with its dominant class;
　return T;
end

Figure 5.5: Training process in a decision tree

In the following, conditional entropy is used as an example because of its popularity.

Consider the case of univariate splits, in which only the presence or absence of a term in a document is used as the split criterion. In other words, the frequency of the term is ignored. For a given node L, let $L_1(j)$ and $L_2(j)$ be the respective sets of documents that contain or do not contain the jth term t_j. Then, the conditional entropy values, $E_1(t_j)$ and $E_2(t_j)$, are computed for $L_1(j)$ and $L_2(j)$, respectively, using Equation 5.5. The overall entropy O_j of the split with term t_j is defined as the weighted average of these two values, where the weight is defined by the number of data points in $L_1(j)$ and $L_2(j)$, respectively:

$$O_j = \frac{|L_1(j)|}{|L_1(j)| + |L_2(j)|} E_1(t_j) + \frac{|L_2(j)|}{|L_1(j)| + |L_2(j)|} E_2(t_j) \qquad (5.34)$$

The split is tested for each term t_j, and the one providing the lowest conditional entropy is selected. The identity of the term t_j is also stored at node L, so that it can be used at prediction time, when a test instance is classified with the decision tree.

Prediction

Once the decision tree has been set up, it is relatively easy to use it for prediction. The split criterion associated with each node is always stored with that node during decision tree construction. For a test instance, the split criterion at the root node is tested (e.g., presence or absence of a word) to decide which branch to follow. This process is repeated recursively until the leaf node is reached. The label of the leaf node is returned as the prediction. A confidence is associated with the prediction, corresponding to the fraction of the labels belonging to the predicted class in the relevant leaf node.

5.5.3   Multivariate Splits

In the case of multivariate splits, more than one attribute is used for making splitting decisions. The vector-space representation of documents is used for implementing the split.

The basic idea is to sample r directions $\overline{Y_1} \ldots \overline{Y_r}$ in the d-dimensional vector space and project all the documents in L along each of these r directions. Here, r is user-defined parameter. The projection of the document $\overline{X_i}$ along the qth direction is given by $\overline{X_i} \cdot \overline{Y_q}$. The projection of each document (contained in node L) along the qth direction creates an ordering among these documents, which is used to test $|L| - 1$ possible split points. Furthermore, since there are r directions, one can test a total of $r(|L|-1)$ possible split points by repeating the process along each of the directions. The quantification of Equation 5.34 is used to evaluate the quality of each split, and the best one is selected. How are the directions $\overline{Y_1} \ldots \overline{Y_r}$ selected? One possibility is to choose random directions in the space. However, it is often helpful to use biased directions [498], where $\overline{Y_q}$ is a vector joining the centroids of *random samples* drawn from documents in L, and each of the centroids is defined by documents of only a single randomly chosen class. Such a direction is more likely to yield a good split that discriminates well between two classes.

A special case of this setting [498] is one in which a pair of documents $(\overline{X_u}, \overline{X_v})$ belonging to different classes is chosen to define $\overline{Y_q} = \overline{X_u} - \overline{X_v}$. In such a case, the projection of data point $\overline{X_i}$ on the direction $\overline{Y_q}$ is given by $\overline{X_u} \cdot \overline{X_i} - \overline{X_v} \cdot \overline{X_i} = s_{ui} - s_{vi}$. Here, s_{ui} and s_{vi} represent dot-product similarities between corresponding training pairs. Instead of the dot product, we can use any type of similarity function, even if the multidimensional representation of the document is not used. For example, we can use string kernel similarities in cases where we want to use the sequence information. In other words, it is possible to build multivariate decision trees only with similarities to make decision trees sensitive to word ordering. This notion is referred to as *similarity forests*, when used with an ensemble-centric implementation [498], and is an adaptation of kernel methods to decision trees (see Chapter 6 for kernel methods).

5.5.4 Problematic Issues with Decision Trees in Text Classification

Because of the high-dimensional and sparse nature of text, off-the-shelf implementations of decision trees do not always work well. However, with the proper implementation, it is possible to obtain high-quality results with decision trees. In the following, some practical guidance is provided.

Like nearest-neighbor classifiers, decision trees have the capability to approximate arbitrary decision boundaries, *given an infinite amount of data*. This is because the successive localization of small regions of the data with splits is similar[6] to the implicit Voronoi-based partitioning of the data space in nearest-neighbor classifiers. However, with a finite amount of data, the predictions of a decision tree are not only inaccurate, but they are also heavily biased in favor of specific classes in particular regions of the data. In other words, if training data samples of small size are drawn from a large base data set, the predictions will all be biased towards favoring particular classes in specific regions of the data. This bias is caused by the split criteria at the top levels of the tree, which have a disproportionately large effect on the final prediction. Often, the split criteria at the top levels of the tree are relatively stable with choice of training sample. Note that this correlated behavior is quite different from a 1-nearest neighbor classifier in which the predictions of different training samples are quite diverse. It is easy to make mistakes in the split criteria at the top levels because they are made in myopic way without an understanding of the interactions between various attributes. The problem starts becoming particularly severe with increasing dimensionality of the data set. Text collections often contain hundreds of thousands of dimensions.

---

[6]In Section 5.5.6, we show further connections between nearest-neighbor classifiers and randomized variants of decision trees.

One observation about univariate splits is that they lead to imbalanced decision trees in which the paths dominated by absence of terms are much longer than paths dominated by presence of terms. Univariate splits are generally best for classification of short text, or text documents drawn from a smaller lexicon. For longer documents, multivariate split criteria can often provide better results. This is because univariate split criteria give too much importance to the absence of terms in many long paths of the tree. Such paths might lead to noisy decisions. When working with long documents, it is particularly important to use models that use many terms *simultaneously* at key decision points in the learning process, and also to give greater importance to presence of terms (rather than absence).

Another issue is that multi-way classification tends to work poorly in text if one constructs a single tree to explain all classes. This is because the terms relevant to various classes are largely disjoint, which increases the size of the relevant vocabulary. Since decision trees use sequential decisions on individual attributes, the small number of terms used for splitting at the higher levels of the tree assume a disproportionately high importance. Therefore, a multi-way classification problem is usually decomposed into multiple binary, one-against-all classification problems, and the results from these different classifiers are integrated by reporting the most confident prediction.

5.5.5 Random Forests

Even though decision trees can capture arbitrary decision boundaries with an infinite amount of data, they can capture only piecewise linear approximations of these boundaries with a finite amount of data. These approximations are particularly inaccurate in smaller data sets. Another problem with decision trees is that the bagging and subsampling tricks used for 1-nearest neighbors do not work quite as well because the splits at the higher levels of the tree are highly correlated. In other words, the expected prediction of a decision tree with randomly chosen training data sets *of small size* has a bias in terms of *consistently* classifying certain test examples incorrectly. One cannot correct the predictions of such test instances by using bagging or subsampling.

A more effective approach is to randomize the tree construction process by allowing the splits at the higher levels of the tree to use the best feature selected out of a restricted subset of features. In other words, r features are randomly selected at each node, and the best splitting feature is selected only out of these features. Furthermore, different nodes use different subsets of randomly selected features. Using smaller values of r results in an increasing amount of randomization in tree construction. At first sight, it would seem that using such a randomized tree construction should impact the prediction in a detrimental way. However, the key is that multiple such randomized trees are grown, and the predictions of each test point over different trees are averaged to yield the final result. By averaging, we mean that the number of times a class is predicted by a randomized tree for a test instance is counted. The class receiving the most number of votes is predicted for the test instance. This averaging process improves the quality of the predictions significantly over a single tree by effectively using diverse terms at higher levels of the different trees in various ensemble components. This results in more robust predictions. The individual trees are grown to full height without pruning because the averaged predictions do not have the overfitting problem of the predictions of individual trees.

The approach can be generalized easily to the multivariate case, which is already randomized to some extent. The multivariate case uses r randomized directions in the data $\overline{Y_1} \ldots \overline{Y_r}$, in which each direction $\overline{Y_q}$ is defined as the vector joining documents belonging to two randomly chosen classes. It is helpful to use a small value of r in order to optimize the split with respect to a smaller number of directions (thereby increasing randomization). This approach can even be made to work when only *similarities* between documents are available, such as with the use of string kernels [498]. A pair of documents $(\overline{X_u}, \overline{X_v})$ belonging to different classes is chosen to define $\overline{Y_q} = \overline{X_u} - \overline{X_v}$. In such a case, the projection of data point $\overline{X_i}$ on the direction $\overline{Y_q}$ is given by $\overline{X_u} \cdot \overline{X_i} - \overline{X_v} \cdot \overline{X_i} = s_{ui} - s_{vi}$. Here, s_{ui} and s_{vi} represent dot-product similarities between corresponding training pairs (which can be replaced with string kernel similarities during the split). This notion is referred to as *similarity forests* [498], and is an adaptation of kernel methods to decision trees (see Chapter 6 for kernel methods).

The random forest can also be constructed by building a conventional (deterministic) decision tree on a randomized feature engineering of the data set. This is a slightly different approach to randomization than the one obtained by using a bag of features at a node. In particular, the LSA-based feature extraction trick discussed in Section 5.2.7 is used to build each decision tree. The resulting forest is referred to as a *Rotation Forest* [478], and it is particularly well suited to text because the new representation is able to get rid of the sparsity in the original representation.

5.5.6 Random Forests as Adaptive Nearest Neighbor Methods

Random forests are adaptive nearest neighbor methods. The intuitive similarity between a decision tree and a 1-nearest neighbor method is easy to see by treating a 1-nearest neighbor method as a technique that performs a Voronoi partitioning of the space with singleton training points (cf. Figure 5.2). Each Voronoi region is labeled with the class of its training instance. A decision tree also partitions the space into hypercubes (in the case of univariate splits), but the hypercubes are constructed more carefully by the hierarchical tree construction process. This supervision in hypercube-based partitioning is what gives the decision tree its adaptivity. The forest adds robust weights to the neighbors, just as an ensemble of 1-nearest neighbors results in weighted nearest neighbors (cf. Section 5.4.3.1).

In the following, we will show this result more formally for a random forest. It is assumed that each decision tree in the random forest is grown to full height without pruning (which is common in the random forest setting). Let $I_t(\overline{X}, \overline{Y})$ be a binary 0–1 indicator function that takes on the value of 1 when $\overline{X}$ and $\overline{Y}$ are mapped to the same node in the tth randomized decision tree from a forest containing $m \geq t$ trees. Let $N(i, t)$ be number of training instances in the node containing $\overline{X_i}$ for the tth randomized decision tree. Consider the following similarity function between the test instance $\overline{Z}$ and training instance $\overline{X_i}$.

$$K(\overline{Z}, \overline{X_i}) = \sum_{t=1}^{m} \frac{I_t(\overline{Z}, \overline{X_i})}{N(i, t)} \tag{5.35}$$

Then, it can be shown (see Exercise 12) that the prediction $F(\overline{Z})$ of a random forest for binary classification of test instance $\overline{Z}$ with labels $\overline{y_i} \in \{-1, +1\}$ takes on the following form of weighted nearest-neighbor classification over all n training instances:

$$F(\overline{Z}) = \text{sign} \left\{ \sum_{i=1}^{n} K(\overline{Z}, \overline{X_i}) \, y_i \right\} \tag{5.36}$$

Note that this form is exactly the same of that of adaptive nearest-neighbor prediction in Equation 5.29, except that the value of λ_i has been set to 1. The classification is still *adaptive* because the similarity function $K(\overline{Z}, \overline{X_i})$ needs to be learned up front in a supervised way with the construction of the random forest.

5.6   Rule-Based Classifiers

Rule-based classifiers use a set of "if then" rules $\mathcal{R} = \{R_1 \ldots R_m\}$ to match conditions on features on the left-hand side of the rule to the class labels on the right-hand side. The expression on the left-hand side of the rule is referred to as the *antecedent* and that on the right-hand side of the rule is referred to as the *consequent*. A rule is typically expressed in the following form:

IF Condition THEN Conclusion

The condition on the left-hand side of the rule, also referred to as the antecedent, often contains conditions of the form $(t_j \in \overline{X})$ AND $(t_l \in \overline{X})$ AND $(\ldots)$. In other words, all terms included in the antecedent, such as t_j and t_l, must be present in the document for the rule to be triggered. Each condition $(t_j \in \overline{X})$ is referred to as a *conjunct*. The right-hand side of the rule is referred to as the consequent, and it contains the class variable. Therefore, a rule R_i is of the form $Q_i \Rightarrow c$ where Q_i is the antecedent, and c is the class variable. The "$\Rightarrow$" symbol denotes the "THEN" condition. In other words, the rules relate the presence of terms like t_j and t_l in the document to the class variable c. Although it is possible for more general conditions to be used on the left-hand side, this is often not done in practice. For example, it is possible to include conditions like $(t_j \notin \overline{X})$ corresponding to *absence* of terms, although this is not recommended in sparse domains like text because such conditions are noisy and could lead to overfitting [121]. Therefore, throughout this section, it will be assumed that only rules corresponding to the presence of terms are generated.

As in all inductive classifiers, rule-based methods have a training phase and a prediction phase. The training phase of a rule-based algorithm creates a set of rules. The prediction phase for a test instance discovers some or all rules that are *triggered* or *fired* by the test instance. A rule is said to be triggered by a training or test instance when the logical condition in the antecedent is satisfied by the features in the instance. Alternatively, for the specific case of training instances, it is said that such a rule *covers* the training instance. In some algorithms, the rules are *ordered* by priority and therefore, the first rule fired by the test instance is used to predict the class label in the consequent. In some algorithms, the rules are unordered, and multiple rules with (possibly) conflicting consequent values are triggered by the test instance. In such cases, methods are required to resolve the conflicts in class label prediction. Rules generated from *sequential covering algorithms* are ordered. On the other hand, rules that are generated from association pattern mining are unordered.

5.6.1   Sequential Covering Algorithms

The basic idea in sequential covering algorithms is to generate the rules for each class at one time, by treating the class of interest as the positive class, and the union of all other classes as the negative class. Each generated rule always contains the positive class as the consequent. In each iteration, a single rule is generated using a *Learn-One-Rule* procedure and training examples that are covered by the class are removed. The generated rule is added to the bottom of the rule list. This procedure is continued until at least a certain minimum fraction of the instances of that class have been covered. Other termination criteria are often used. For example, the procedure can be terminated when the error of the next generated rule exceeds a certain pre-determined threshold on a separate validation set. A minimum description length (MDL) criterion is sometimes used when further addition of a rule increases the minimum description length of the model by more than a certain amount. The procedure is repeated for all classes. Note that less prioritized classes start with a smaller training data set because many instances have already been removed in the rule generation of higher priority classes. The *RIPPER* algorithm orders the rules belonging to the rare classes before those of more frequent classes, although other criteria are used by other algorithms, whereas *C4.5rules* uses various accuracy and information-theoretic measures to order the classes. The broad framework of the sequential covering algorithm is as follows:

> **for** each class c in a particular order **do**
>> **repeat**
>>> Extract the next rule $R \Rightarrow c$ using *Learn-One-Rule* on training data V;
>>> Remove examples covered by $R \Rightarrow c$ from training data V;
>>> Add extracted rule to bottom of rule list;
>> **until** class c has been sufficiently covered

The procedure for learning a single rule is described in Section 5.6.1.1. Only rules for $(k-1)$ classes are grown, and the final class is assumed to be a default catch-all class. One can also view the final rule for the remaining class c_l as the catch-all rule $\{\} \Rightarrow c_l$. This rule is added to the very bottom of the entire rule list. This type of ordered approach to rule generation makes the prediction process a relatively simple matter. For any test instance, the first triggered rule is identified. The consequent of that rule is reported as the class label. Note that the catch-all rule is guaranteed to be triggered when no other rule is triggered. One criticism of this approach is that the ordered rule generation mechanism might favor some classes more than others. However, since multiple criteria exist to order the different classes, it is possible to repeat the entire learning process with these different orderings, and report an averaged prediction.

5.6.1.1 Learn-One-Rule

It remains to be explained how the rule for a single class is generated. Although the original *RIPPER* algorithm allows antecedent conditions corresponding to both presence and absence of terms in documents, the absence of terms is a noisy indicator. As a result, their inclusion often causes overfitting [121]. Therefore, the following description will only consider the case in which rules corresponding to presence of terms in a document are used in the antecedent. For brevity, we will concisely denote a rule such as $(t_j \in \overline{X})$ AND $(t_l \in \overline{X}) \Rightarrow c$ by $\{t_j, t_l\} \Rightarrow c$. When the rules for class c are generated, each term is sequentially added to the antecedent. The approach starts with the empty rule $\{\} \Rightarrow c$ for the class c, and then adds terms one by one to the antecedent. What should be the criterion for adding a term to the antecedent of the current rule $R \Rightarrow c$?

1. The simplest criterion is to add the term to the antecedent that increases the accuracy of the rule as much as possible. In other words, if n_* is the number of training examples covered by the rule (after addition of a candidate term t_j to antecedent), and n_+ is the number of positive examples among these instances, then the accuracy of the rule is given by n_+/n_*. However, such an approach can sometimes favor rare terms or misspellings if the small number of training examples covered by the corresponding rule all belong to the positive category (by random chance). This is a manifestation of overfitting. To address this issue, the accuracy of adding the term t_j to the antecedent of the current rule $R \Rightarrow c$ is computed as follows:

$$A(R \Rightarrow c, t_j) = \frac{n_+ + 1}{n_* + k} \tag{5.37}$$

Here, k is the total number of classes.

2. Another criterion is *FOIL's information gain.* The term "FOIL" stands for *First Order Inductive Learner.* Consider the case where a rule covers n_1^+ positive examples and n_1^- negative examples, where positive examples are defined as training examples matching the class in the consequent. Furthermore, assume that the addition of a term to the antecedent changes the number of positive examples and negative examples to n_2^+ and n_2^-, respectively. Then, FOIL's information gain FG is defined as follows:

$$FG = n_2^+ \left(\log_2 \frac{n_2^+}{n_2^+ + n_2^-} - \log_2 \frac{n_1^+}{n_1^+ + n_1^-} \right) \tag{5.38}$$

This measure tends to select rules with high coverage because n_2^+ is a multiplicative factor in FG. At the same time, the information gain increases with higher accuracy because of the term inside the parentheses. This particular measure is used by the *RIPPER* algorithm.

Several other measures are often used, such as the *likelihood ratio* and *entropy.* Terms can be successively added to the antecedent of the rule, until 100% accuracy is achieved by the rule on the training data or when the addition of a term cannot improve the accuracy of a rule. In many cases, this point of termination leads to overfitting. Just as node pruning is done in a decision tree, antecedent pruning is necessary in rule-based learners to avoid overfitting. Another modification to improve generalization power is to grow the r best rules simultaneously at a given time, and only select one of them at the very end based on the performance on a held-out set. This approach is also referred to as *beam search.*

Rule Pruning

Overfitting may result from the presence of too many conjuncts. As in decision-tree pruning, the Minimum Description Length principle can be used for pruning. For example, for each conjunct in the rule, one can add a penalty term δ to the quality criterion in the rule-growth phase. This will result in a *pessimistic error rate.* Rules with many conjuncts will therefore have larger aggregate penalties to account for their greater model complexity. A simpler approach for computing pessimistic error rates is to use a separate holdout validation set that is used for computing the error rate (without a penalty). However, this type of approach is not used by Learn-One-Rule.

The conjuncts successively added during rule growth (in sequential covering) are then tested for pruning in reverse order. If pruning reduces the pessimistic error rate on the training examples covered by the rule, then the generalized rule is used. While some algorithms such as *RIPPER* test the most recently added conjunct first for rule pruning, it is not a strict requirement to do so. It is possible to test the conjuncts for removal in any order, or in greedy fashion, to reduce the pessimistic error rate as much as possible. Rule pruning may result in some of the rules becoming identical. Duplicate rules are removed from the rule set before classification.

5.6.2   Generating Rules from Decision Trees

Decision trees can also be used to generate rules because each path in a decision tree corresponds to a rule. Generally, rules are generated from univariate decision trees because of their interpretability although it is possible, in principle, possible to also generate rules from multivariate decision trees. For each path in a decision tree, a rule can be generated corresponding to the conjuncts of the conditions required to reach a leaf. One difference between the rules generated from decision trees and other methods is that many paths in a (univariate) decision tree correspond to absence of attributes. Therefore, the rules may contain conjuncts corresponding to absence of attributes. Since all the paths in a decision tree represent non-overlapping regions of the space, the *initial* set of rules generated from a decision tree are mutually exclusive in terms of coverage. However, this situation changes with further processing of this initial set of rules with rule pruning.

Rules are processed one by one, and conjuncts are pruned from them in greedy fashion to improve the accuracy as much as possible on the covered examples in a separate holdout validation set. This approach is similar to decision-tree pruning except that one is no longer restricted to pruning the conjuncts at the lower levels of the decision tree. Therefore, the pruning process is more flexible than that of a decision tree, because it is not restricted by an underlying tree structure. Duplicate rules may result from pruning of conjuncts. These rules are removed. The rule-pruning phase increases the coverage of the individual rules and, therefore, the mutually exclusive nature of the rules is lost. A single test instance might fire multiple rules. As a result, it again becomes necessary to order the rules.

In *C4.5rules* [454], all rules that belong to the class whose rule set has the smallest description length are prioritized over other rules. The total description length of a rule set is a weighted sum of the number of bits required to encode the size of the model (rule set) and the number of examples covered by the class-specific rule set in the training data, which belong to a different class. Typically, classes with a smaller number of training examples are favored by this approach. A second approach is to order the class first whose rule set has the least number of false-positive errors on a separate holdout set. A rule-based version of a decision tree generally allows the construction of a more flexible decision boundary with limited training data than the base tree from which the rules are generated. This is primarily because of the greater flexibility in the model, which is no longer restrained by the straitjacket of an exhaustive and mutually exclusive rule set. As a result, the approach generalizes better to unseen test instances.

5.6.3 Associative Classifiers

Associative classifiers [348] leverage association rule mining techniques [1, 2] in order to perform text classification. Such methods are particularly well suited to the text domain because associative classifiers were originally designed for sparse domains like market basket data, which are similar to text. The basic idea of such classifiers is to relate a bag of terms in the antecedent of the rules to a class label in the consequent of the rule. Therefore, a rule is of the following form:

$$S \Rightarrow c$$

Here, S is a set of terms, and c is a class label (identifier) drawn from $\{1 \ldots k\}$ The bags of terms, S, in the antecedent of the rule always correspond to the presence of all terms in S in a document. Therefore, absence of terms in a document is never used. Furthermore, this approach borrows ideas from association rule mining to define the rule set. A rule $S \Rightarrow c$ is mined from the training data, if it satisfies two conditions:

1. At least a minimum fraction *minsup* of the training documents both contain S and belong to class c. The value of *minsup* is a user-defined parameter, which is referred to as the *minimum support*. In general, the support of the rule is defined as the fraction of documents that contain S and belong to class c.

2. Among all documents that contain S, at least a minimum fraction *minconf* of the documents belong to class c. The value of *minconf* is a user-defined parameter, which is referred to as the *minimum confidence*. In general, the confidence of the rule is the conditional probability of a document belonging to class c, given that it contains S.

Imposing a minimum support requirement prevents overfitting by only selecting rules with significant presence, whereas imposing a minimum confidence requirement ensures that predictive rules are selected.

Associative classifiers are easy to implement in an efficient way because many off-the-shelf association pattern mining techniques are available to mine the rules from the underlying data in an efficient way. We refer the reader to [1, 2] for a review of various rule mining techniques. Significant amount of rule pruning is often used in order to reduce redundancy. Rule pruning is a heuristic process in which the different factors are integrated to create the final rule set [30]. For example, for two rules $S_1 \Rightarrow c$ and $S_2 \Rightarrow c$, in which $S_1 \subseteq S_2$, the second rule is redundant with respect to the first. However, if the confidence of the second rule is significantly higher, then it does convey additional information. Therefore, the rule $S_2 \Rightarrow c$ can be pruned, only if it has lower confidence. All such rules are removed. Subsequently, the rules are ordered by decreasing confidence, with ties broken by successive criteria of decreasing support and increasing number of terms in antecedent. The rules are processed in this order and documents that fire a rule are marked, if they have not already been marked. A rule is considered non-redundant only if it is fired by at least one unmarked document during the aforementioned processing. At the end of the process, the majority class of those training instances that do not fire any rule is treated as the default class.

Prediction: For any given test instance, those rules that are fired by the test instance are identified. If no rules are fired, then the default class is predicted. If all fired rules predict the same class, then the corresponding class is reported. The main challenge arises in cases where the fired rules conflict with one another. The simplest approach is to use the sum of the confidences of all the fired rules for a particular class as its prediction propensity. The class with the highest propensity is reported. However, more complex prediction mechanisms are used by various rule-based methods. Refer to the bibliographic notes.

5.7   Summary

The sparsity of text causes a number of unique challenges for classification. For example, the absence of words conveys noisier information as compared to the presence of words. The chapter studies numerous feature selection methods such as the Gini index, conditional entropy, mutual information, and the χ^2-statistic. The four most fundamental classification methods, which are the naïve Bayes classifier, the κ-nearest neighbor method, the decision tree, and rule-based methods are studied in this chapter. The naïve Bayes classifier is closely related to mixture models for clustering. Nearest-neighbor classifiers are theoretically very accurate if an infinite amount of data is available, although their accuracy is limited by the finiteness of the data and the noise in the features. A powerful family of nearest-neighbor classifiers is that of adaptive methods, of which random forests and support vector machines are special cases. Decision trees face many challenges for effective implementation in text data; however, with the proper implementation, they can provide good results. Rule-based classifiers are closely connected to decision trees, because rules can also be extracted from decision trees. Numerous methods like sequential covering and association pattern mining have been developed for extracting rules from text documents.

5.8 Bibliographic Notes

Surveys on text classification may be found in [1, 15, 505]. A comparative study of several feature selection methods for text categorization may be found in [606]. The use of word clusters for dimensionality reduction is explored in [13, 38, 293, 326, 520]. Many classifiers in the text domain have been compared in [168, 272, 605].

The basic ideas of the naïve Bayes classifier for text are discussed in [275, 327, 373]. The differences between the Bernoulli and multinomial models are discussed in [373]. Discussions on the independence assumption in naïve Bayes are provided in [130, 159]. A hierarchical classifier with the naïve Bayes method is discussed in [93]. The semi-supervised method for probabilistic classification is based on the ideas in [419]. The work in [330] also uses supervised clustering with a mixture model, which is then leveraged for categorization. A variety of semi-supervised methods for learning are discussed in [63, 64, 400]. A book on semi-supervised learning [103] provides an excellent overview.

The κ-nearest neighbor classifier and its variants have been studied extensively in the literature [133]. Early studies on the effectiveness of κ-nearest neighbor methods are provided in [602, 603, 605]. A nearest-neighbor classifier that weights words is discussed in [231]. The Rocchio classifier is based on the relevance feedback ideas developed in [479]. A probabilistic variant of the Rocchio algorithm for text classification is provided in [275]. Several centroid classifiers are studied in [7, 66, 232, 293]. The work in [309] uses ideas from linear classifiers to create generalized instance sets for nearest-neighbor classification. The ideas of bagging were presented in [71] and those of subsampling are presented in [76]. The proof of Bayes optimality of the 1-nearest neighbor classifier is available in [165]. The connections between bagged/subsampled 1-nearest neighbors and weighted nearest-neighbor classifiers are explored in [494]. The discriminant metric-based distance function is presented in [236]. The work in [236] introduces a local variation of this approach as well.

The well-known C4.5 decision tree classifier was proposed in [454] and ID3 was proposed in [455]. Decision trees were generalized to random forests in [70, 72]. DT-min10 [328] was an early decision-tree algorithm for building the tree for each category. The algorithm derives its name from the fact that the tree construction was stopped when fewer than 10 examples were mapped to a leaf node. No pruning was done. Several recommendations on the construction of decision trees on sparse data like text are provided in [278]. Early studies on the advantages of ensemble-centric implementations of decision trees for text categorization are provided in [32, 574]. The work in [108] provides some of the earliest proposals on decision tree construction, and also suggested that a separate decision tree should be grown for each category. The use of Fisher's linear discriminant for constructing decision trees is presented in [95], and rotation forests are presented in [478]. The work in [232] highlights some of the problems associated with decision-tree classification in sparse domains like text. The work in [331] also does not show encouraging results of decision trees in comparison with methods like naïve Bayes. There is, however, not a clear consensus on this issue. For example, the work in [168] reports relatively good results with the decision tree proposed in [108], especially in comparison with the naïve Bayes classifier. Independent results in [272] suggest that the decision tree does not work as well as the support vector machine, but it works approximately as well as κ-nearest neighbor and Rocchio, and (much) better than the naïve Bayes classifier. Indeed, the repeated under-performance of the (widely revered) naïve Bayes classifier in independent experiments [168, 272, 605] would make this classifier a more questionable choice. It is also noteworthy that none of the compared results use the random forest implementation of a decision tree, which should provide better results. Therefore, it would seem that even though decision trees have sparsity-centric challenges in the text domain, their true potential might be widely underestimated and that of the naïve Bayes might be widely overestimated. In fact, most of the sparsity-based drawbacks of random forests vanish, when splits in arbitrary directions are used. In such cases, one can even use kernelized variants of random forests [498].

The earliest methods for rule-based text classification were proposed in [31]. Many key ideas for sequential covering algorithms were laid by Fürkranz and Widmer [205] in the *IREP* algorithm. The *RIPPER* method [119] for rule-induction, which is a popular method for text classification, is closely related to *IREP*. Its use in email classification was studied in [120]. It was shown in [121] that it is inadvisable to use the absence of words in rule-based classification for text. The use of context-sensitive methods for improving the classification accuracy of rule-based methods is studied in [123]. The foundation for classification based on associations was laid in the seminal work of [348]. The description of the associative classifier in this chapter is roughly based on [30], although several portions have been simplified.

5.8.1   Software Resources

The Bow toolkit [371], which is written in C, contains many basic algorithms for text classification. Several text classification tools are included in the Python library *scikit-learn* [646]. The most well known library for classification in R is the **caret** package [305]. Although this package is not specifically designed for text data, many of the core predictive modeling techniques can be adapted easily to the text domain with appropriate text preprocessing tools. The R-based **tm** library [647] can be used for preprocessing and tokenization in combination with the **caret** package. The package **RTextTools** [667] in R also has numerous categorization methods, which are specifically designed for text. The primary focus on this package is on ensemble methods, although it also contains standalone classifiers like decision trees. In addition, the **rotationForest** package [668] in R, which is available from CRAN, can be used to address the sparsity challenges associated with text. An implementation of this method is also available in *Weka* [649]. The *Weka* library [649] in Java contains numerous text classification tools, and it is particularly rich in conventional tools like decision trees and rule-based methods. The **MALLET** toolkit [701] supports many classifiers like naïve Bayes and decision trees.

5.9   Exercises

1. Consider the term "elections" which is present in only 50 documents in a corpus of 1000 documents. Furthermore, assume that the corpus contains 100 documents belonging to the *Politics* category, and 900 documents belonging to the *Not-Politics* category. The term "election" is contained in 25 documents belonging to the *Politics* category.

   (a) Compute the unnormalized Gini index and the normalized Gini index $G_n(\cdot)$ of the term "elections."

   (b) Compute the entropy of the class distribution with respect to the entire data set.

   (c) Compute the conditional entropy of the class distribution with respect to the term "elections."

   (d) Compute the mutual information of the term "elections" according to Equation 5.6. How are your answers to (b), (c), and (d) related?

(e) Compute the information gain of the term "elections" according to Equation 5.7. How are your answers to (d) and (e) related?

2. Show that the sum of (i) the mutual information between a class and term (Equation 5.6), and (ii) the conditional entropy of the class distribution with respect to the same term, is equal to the total entropy of the class distribution.

3. The χ^2 distribution is defined by the following formula, as discussed in the chapter:

$$\chi^2 = \sum_{i=1}^{p} \frac{(O_i - E_i)^2}{E_i}$$

Show that for a 2×2 contingency table, the aforementioned formula can be rewritten as follows:

$$\chi^2 = \frac{(O_1 + O_2 + O_3 + O_4) \cdot (O_1 O_4 - O_2 O_3)^2}{(O_1 + O_2) \cdot (O_3 + O_4) \cdot (O_1 + O_3) \cdot (O_2 + O_4)}$$

Here, $O_1 \ldots O_4$ are defined in the same way as in the tabular example in the text.

4. Predict the probabilities of categories *Cat* and *Car* of *Test2* on the toy corpus example in Section 5.3.5.1. You can use the Bernoulli naïve Bayes model with the same level of smoothing as used in the example in the book. Return normalized probabilities that sum to 1 over the two categories.

5. Predict the probabilities of categories *Cat* and *Car* of *Test2* on the toy corpus example in Section 5.3.5.2. You can use the multinomial naïve Bayes model with the same level of smoothing as used in the example in the book. Return normalized probabilities that sum to 1 over the two categories.

6. Naïve Bayes is a generative model in which each class corresponds to one mixture component. Design a *fully supervised* generalization of the naïive Bayes model in which each of the k classes contains exactly $b > 1$ mixture components for a total of $b \cdot k$ mixture components. How would you perform parameter estimation in this model?

7. Naïve Bayes is a generative model in which each class corresponds to one mixture component. Design a *semi-supervised* generalization of the naïive Bayes model in which each of the k classes contains exactly $b > 1$ mixture components for a total of $b \cdot k$ mixture components. How would you perform parameter estimation in this model?

8. Provide a proof of Equation 5.28 on subsampling. Specifically, show that if the 1-nearest neighbor algorithm is used with a subsample of size s out of n points, then the prediction $F(\overline{Z})$ is equivalent to that of a weighted nearest-neighbor classifier of the following form:

$$F(\overline{Z}) = \text{sign} \left\{ \sum_{i=1}^{n} P(\overline{X_i}|\overline{Z}) y_i \right\}$$

Here, the ith training point and its label are denoted by $(\overline{X_i}, y_i)$ in a training data set of n points. The value of $P(\overline{X_i}|\overline{Z})$ is defined as follows:

$$P(\overline{X_i}|\overline{Z}) = \begin{cases} \binom{n - R(\overline{Z}, \overline{X_i})}{s-1} / \binom{n}{s} & \text{if } R(\overline{Z}, \overline{X_i}) \leq n - s + 1 \\ 0 & \text{if } R(\overline{Z}, \overline{X_i}) > n - s + 1 \end{cases}$$

The notation $R(\overline{Z}, \overline{X_i}) \in \{1, \ldots, n\}$ denotes the rank of the distance of the ith training point $\overline{X_i}$ in sorted order from $\overline{Z}$.

9. The adaptive nearest-neighbor method discussed in the chapter uses a single distortion metric over the entire data space in order to compute the nearest neighbor of a point. Propose a training algorithm to make this metric locally adaptive, so that an optimized distortion metric is used for each test instance based on the local class patterns in the data. What are the possible advantages and disadvantages of using such an approach?

10. Consider a bagged 1-nearest neighbor classifier in which a bagged sample of size s is selected out of 1000 training points repeatedly in order to create a prediction. You have two different data sets with the following types of (binary) class distributions:

   - Distribution A: Class 1 is linearly separable from class 2 with a hyperplane although there might be some mixing of the classes near the boundary. Both classes have 50% presence in the data.

   - Distribution B: There are 10 spherical clusters of each of class A and class B containing exactly 50 points each, and there is also some overlap of the clusters of different classes.

   How would you choose the optimal size of the sample s in each training data set. Would this optimal sample size be the same in the two data sets? In which data set do you think that the optimal sample size will be larger?

11. Imagine a document data set in which the class label is generated by the following hidden function (which is unknown to the analyst and therefore has to be learned by a supervised learner):

   If a term has an odd number of consonants, then the term is of type 1. Otherwise the term is of type 2. The class label of a document is of type 1, if the majority of the tokens in it are of type 1. Otherwise, the class label is of type 2.

   For a document collection of this type, would you prefer to use (i) a Bernoulli naïve Bayes classifier, (ii) a multinomial naïve Bayes classifier, (iii) a nearest-neighbor classifier, or (iv) a univariate decision tree? What is the impact of the lexicon size and average document size on various classifiers?

12. Show that the prediction $F(\overline{Z})$ of the test instance $\overline{Z}$ by a random forest with m ensemble components is given by the following:

$$F(\overline{Z}) = \text{sign} \left\{ \sum_{i=1}^{n} \sum_{t=1}^{m} \frac{I_t(\overline{Z}, \overline{X_i})}{N(i,t)} y_i \right\}$$

   Here, $\overline{X_1} \ldots \overline{X_n}$ are the training instances, $y_1 \ldots y_n$ are the binary labels drawn from $\{-1, +1\}$, $N(i,t)$ is the number of instances in the same leaf as $\overline{X_i}$ in the tth ensemble component, and $I(i,t)$ is an indicator function that tells us whether or not $\overline{Z}$ and $\overline{X_i}$ end up in the same leaf in the tth ensemble component.

13. Discuss the advantages of rule-based learners over decision trees, when the amount of data is limited.

14. Discuss how one might integrate domain knowledge with rule-based learners.

Chapter 6

Linear Models for Classification and Regression

"When the solution is simple, God is answering."– Albert Einstein

6.1   Introduction

Linear models for classification and regression express the dependent variable (or class variable) as a linear function of the independent variables (or feature variables). Specifically, consider the case in which y_i is the dependent variable of the ith document, and $\overline{X}_i = (x_{i1} \ldots x_{id})$ are the d-dimensional feature variables of this document. In the case of text, these feature variables correspond to the term frequencies of a lexicon with d terms. The value of y_i is a numerical quantity in the case of regression, and it is a binary value drawn from $\{-1, +1\}$ in the case of classification. Then, linear models for both classification and regression assume that the dependence between y_i and $\overline{X}_i$ is expressed in terms of d linear coefficients $\overline{W} = (w_1 \ldots w_d)$, and a *bias term b* as follows:

$$y_i = \sum_{j=1}^{d} w_j x_{ij} + b = \overline{W} \cdot \overline{X}_i + b \qquad \text{Linear Regression (Numerical Dependent Variable)}$$

$$y_i = \text{sign}\{\sum_{j=1}^{d} w_j x_{ij} + b\} = \text{sign}\{\overline{W} \cdot \overline{X}_i + b\} \quad \text{Classification (Binary Dependent Variable)}$$

If the coefficients $\overline{W}$ and bias b can be learned, so that they are satisfied for the training data, then the aforementioned prediction function can be used to predict the dependent variable of any (unlabeled) test document. An important point here is that it may not be possible to find a coefficient vector $\overline{W}$ and bias b, so that these conditions are *exactly* satisfied for *all* training data points. After all, the modeling assumption is only a rough *hypothesis* about the true function relating $\overline{X}_i$ and y_i. In particular, for any data set in which the

© Springer Nature Switzerland AG 2022
C. C. Aggarwal, *Machine Learning for Text*,
https://doi.org/10.1007/978-3-030-96623-2_6

number of training records n is greater than d, the aforementioned system of equations is *over-determined*. Therefore, a set of coefficients $w_1 \ldots w_d$ and bias b will usually not exist in which the aforementioned equations are *exactly* satisfied. In such a case, it makes sense to learn $\overline{W}$ and b, so that the equations are satisfied with the least possible cumulative error. For the case of numeric dependent variables, one could set up an optimization objective function of the following form:

$$\text{Minimize } J = \frac{1}{2} \sum_{i=1}^{n} (y_i - \sum_{j=1}^{d} w_j x_{ij} - b)^2 = \frac{1}{2} \sum_{i=1}^{n} (y_i - \overline{W} \cdot \overline{X_i} - b)^2$$

Note that the objective function punishes violations of the linear condition $y_i = \sum_{j=1}^{d} w_j x_{ij} + b$ with a squared penalty. Of course, the penalty for binary dependent variables is different from that used for numeric dependent variables. Furthermore, there are many other ways in which one can penalize errors, which will lead to subtle variations in the properties of the model. Therefore, it is important to view all these choices as being part of a larger family of linear models with many subtle distinguishing characteristics. This chapter will discuss many such variations.

6.1.1   Geometric Interpretation of Linear Models

Both classification and regression have a neat geometric interpretation in terms of linear hyperplanes. In the case of classification, one can view the hyperplane $\overline{W} \cdot \overline{X} + b = 0$ as a $(d-1)$-dimensional *separating hyperplane* between the two classes in d-dimensional *feature* space. A two-class example is shown in Figure 6.1(a), in which the first class is marked by 'o' and the second class is denoted by '*'. The best linear separator $x_1 + x_2 = 1$ between the two classes is shown, although four points do occur on the wrong side of the separator. Such training instances are penalized by the optimization model used to learn $\overline{W}$ and b. The aggregate errors of such points are minimized by the linear model. Points are typically penalized as a function of how far they are from the separating hyperplane on the "wrong" side. Typically, only training points on the wrong side of the separator are penalized by the linear model, although some linear models also penalize points for being "close enough" to the separator even when they are on the correct side. The specific choice of the penalty is a key distinguishing characteristic between different members of the family of linear models. The bias is proportional to the distance of the hyperplane from the origin, and the proportionality factor is equal to 1 when $\overline{W}$ and b are proportionately scaled so that the former is normalized to unit length.

Linear *regression* models can also be interpreted in terms of linear hyperplanes. In the case of regression, a d-dimensional hyperplane is constructed in $(d+1)$-dimensional space including the dependent variable. The corresponding hyperplane is $y = \overline{W} \cdot \overline{X} + b$, which is shown in Figure 6.1(b). As in the case of Figure 6.1(a), there are two independent variables, although the hyperplane needs to be drawn in three dimensions to include the dependent variable. For any given training point $(\overline{X_i}, y_i)$, deviations in the *observed value* of y_i from the predicted value $\overline{W} \cdot \overline{X_i} + b$ are penalized. Unlike the case of classification, most training points will be penalized to some extent (and not just the misclassified points) unless they lie *exactly* on the linear hyperplane that is learned by the algorithm. There are numerous relationships between the models used for linear classification and regression. For example, the linear regression model can also be used to directly solve linear classification by treating the binary class variables as numerical response values. This particular special case is referred to as the *regularized least-squares classification*.

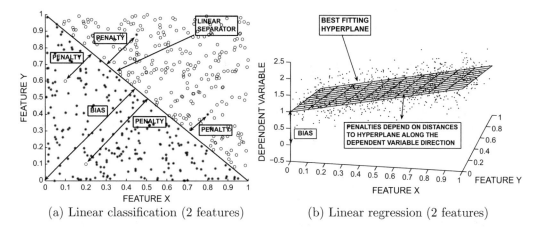

(a) Linear classification (2 features)   (b) Linear regression (2 features)

Figure 6.1: Geometric interpretation of linear classification and regression

For both classification and regression, the bias is shown in Figure 6.1 as a kind of offset of the learned hyperplane from the origin. Although the effect of bias is important, it can be incorporated indirectly without explicitly introducing a bias variable. The following section discusses such an *algebraic* simplification to reduce notational complexity.

6.1.2 Do We Need the Bias Variable?

The bias variable b captures the invariant portion of the prediction y_i that is unrelated to the values of the feature variables. For example, consider a regression setting in which all dependent variables y_i are drawn from $[99.99, 100.01]$, and all feature values correspond to relatively modest term frequencies, which are less than 10. Since the dependent variable does not vary much over different training instances, one can simply set the coefficient vector $\overline{W}$ to a d-dimensional vector of 0s, and set the value of b to 100 to obtain a reasonable prediction. Note that it would be hard to obtain an accurate prediction without a bias variable b, especially if different documents contain very different terms. In other words, the bias variable captures the *invariant* portion of the prediction over different documents, which is difficult to model using highly varying features.

It is also possible to model the bias variable as one of the coefficients of a feature variable by using a simple feature engineering trick of incorporating an *invariant* feature. The basic idea is to add a single dummy feature to each training record with a value of 1. The coefficient of the newly added dummy variable is the bias. This approach is equivalent to inventing a single dummy pseudo-word, and adding it to every document in the corpus. Therefore, if we change the notation to assume (without loss of generality) that the lexicon contains $d - 1$ terms (instead of d terms), and the dth term is the dummy term, then one can set $b = w_d$. Therefore, one can write the aforementioned linear model as follows:

$$y_i = \sum_{j=1}^{d-1} w_j x_{ij} + w_d = \overline{W} \cdot \overline{X_i} \qquad \text{Linear Regression (Numerical Dependent Variable)}$$

$$y_i = \text{sign}\{\sum_{j=1}^{d-1} w_j x_{ij} + w_d\} = \text{sign}\{\overline{W} \cdot \overline{X_i}\} \qquad \text{Classification (Binary Dependent Variable)}$$

The addition of a single dummy term with an invariant frequency naturally captures the invariant part of the dependent variable within its coefficient (which is the bias). This way of modeling the bias is helpful in promoting algebraic simplicity without losing anything in terms of modeling generality. Depending on algebraic convenience, some of the models in this chapter will use the bias term, whereas others may not. In cases where the bias term is not used, it is important to keep in mind that the derived algorithms are based on the assumption of adding a dummy feature, and one must actually perform this preprocessing on the data set when using these algorithms. The corresponding optimization model for numeric dependent variables now becomes the following:

$$\text{Minimize } J = \frac{1}{2} \sum_{i=1}^{n} (y_i - \overline{W} \cdot \overline{X_i})^2$$

This model is the classical objective function for linear regression, although other optimization criteria are used for binary dependent variables.

Another heuristic way of getting rid of the bias in the case of numeric dependent variables (i.e., regression) is by mean centering all independent and dependent variables. The intuition is that the bias is caused by the offset of the data distribution from the origin along the dependent variable (cf. Figure 6.1(b)). In the special case of the *least-squares* objective function for regression, mean-centering all variables and then using a bias-free model can be shown to be mathematically identical to the use of a bias-inclusive model on the original data (see Exercise 2).

6.1.3   A General Definition of Linear Models with Regularization

All forms of supervised learning emphasize the ability to *generalize* a learned model from (seen) training data to (unseen) test data. Unfortunately, the coefficients learned from the training data may not always generalize very well to making predictions on the test data, particularly if the size of the training data is small. In order to understand this point, consider a situation in which number of features d is greater than the number of training instances n. Furthermore, the dependent variable y_i is always twice the value of the first feature in document $\overline{X_i}$, and the remaining $(d-1)$ feature values are completely unrelated to the dependent variable. In such a case, it is evident that the optimal coefficient vector would be $w_1 = 2$, and $w_2 = w_3 = \ldots = w_d = 0$. Setting a coefficient value of a feature to 0 has the same effect as discarding the feature, and it may be viewed as a type of feature selection. Unfortunately, however, when the number of features d is greater than n, the system of equations above is an *under-determined* system with infinitely many solutions with zero error. In other words, an optimization model might easily discover solutions in which irrelevant features are used. This will inevitably lead to overfitting, and therefore poor generalization on the unseen test data. Note that the problem of overfitting occurs not only in the case when $d > n$, but also in cases where the number of instances is larger than the number of features only to a modest degree. This is because every data set is bound to have random nuances, which can cause freak correlations between features and the dependent variable. The optimization model is bound to assign non-zero coefficients to such features as well. In general, the larger the amount of data available, the better the coefficients that can be learned by optimizing the squared error of the prediction in terms of their generalization power to unseen test instances.

How can we encourage the linear model to use only relevant features and discard the irrelevant ones by setting zero (or small) coefficient values for such features? One possibility is to use feature selection up front. Although such a solution does help to some extent, it creates several problems in terms of accounting for the specific effect of redundant features and predicting the precise effect of removing features on the optimization process. A more natural solution is to try to impose a budget on the number of features that are used. In other words, one might try to optimize the following problem:

$$\text{Minimize } J = \frac{1}{2} \sum_{i=1}^{n} (y_i - \overline{W} \cdot \overline{X_i})^2$$

$$\text{subject to:}$$

$$\text{At most } r \text{ coefficients from } \overline{W} \text{ have non-zero values}$$

Such an optimization problem is hard to solve in practice. A more natural solution is to impose a penalty for using large values of coefficients. This is a *soft* form of feature selection, because it encourages the absolute values of coefficients to be small (thereby de-emphasizing the impact of weakly correlated features). Therefore, one uses a *regularization* parameter $\lambda > 0$ to create the following regularized linear regression model:

$$\text{Minimize } J = \underbrace{\frac{1}{2} \sum_{i=1}^{n} (y_i - \overline{W} \cdot \overline{X_i})^2}_{\text{Prediction Error}} + \underbrace{\frac{\lambda}{2} ||\overline{W}||^2}_{\text{Penalty for using features}}$$

The penalty term on the coefficients, which is $||\overline{W}||^2 = \sum_{j=1}^{d} w_j^2$, is also referred to as the regularizer. This particular form of regularization is also referred to as L_2-*regularization*, since the L_2-norm of the coefficients is used. Other types of regularization, such as L_1-regularization, are commonly used. The various linear models differ in terms of the choice of the objective function quantifying the prediction error (which is different between classification and regression) and the choice of the regularizer. More generally, almost all linear models for both classification and regression can be shown to be special cases of the following optimization problem:

$$\text{Minimize } J = \underbrace{\sum_{i=1}^{n} L(y_i, \overline{W} \cdot \overline{X_i})}_{\text{Loss Function}} + \underbrace{\lambda \Omega(\overline{W})}_{\text{Regularization}}$$

The function $L(\cdot, \cdot)$ denotes the *loss function* that tries to quantify the error penalty of trying to predict y_i with the linear function $\overline{W} \cdot \overline{X_i}$, and the function $\Omega(\cdot)$ is the regularization term to prevent overfitting. The properties of the learned model depend in several interesting ways on these choices. This chapter will study the most common choices, which result in models such as *linear regression, linear least-squares fit (LLSF), Fisher's linear discriminant, support vector machines,* and *logistic regression.*

6.1.4 Generalizing Binary Predictions to Multiple Classes

You might have noticed that the class label y_i is often assumed to be binary in the case of linear models. What do we do when the data contains $k > 2$ classes? It is common to use generic meta-algorithms, which can take a binary classification algorithm $\mathcal{A}$ as input and use it to make multilabel predictions. Several strategies are possible to convert binary classifiers into multilabel classifiers.

The first strategy is the *one-against-rest* approach, which is also referred to as the *one-against-all* approach. In this approach, k different binary classification problems are created, such that one problem corresponds to each class. In the ith problem, the ith class is considered the set of positive examples whereas all the remaining examples are considered negative examples. The binary classifier $\mathcal{A}$ is applied to each of these training data sets. This creates a total of k models. If the positive class is predicted in the ith problem, then the ith class is rewarded with a vote that is proportional to the confidence of prediction. One may also use the numeric output of a classifier (e.g., a function of the distance of instance from separator) to weight the corresponding vote. The highest numeric score for a particular class is selected to predict the label. Note that the choice of the numeric score for weighting the votes depends on the classifier at hand.

The second strategy is the *one-against-one* approach. In this strategy, a training data set is constructed for each of the $\binom{k}{2}$ pairs of classes. The algorithm $\mathcal{A}$ is applied to each training data set. This results in a total of $k(k-1)/2$ models. For each model, the prediction provides a vote to the winner. The class label with the most votes is declared as the winner at the end. At first sight, it seems that this approach is computationally more expensive, because it requires us to train $k(k-1)/2$ classifiers, rather than training k classifiers, as in the one-against-rest approach. However, the computational cost is ameliorated by the smaller size of the training data in the one-against-one approach. Specifically, the training data size in the latter case is approximately $2/k$ of the training data size used in the one-against-rest approach on the average. If the running time of each individual classifier scales super-linearly with the number of training points, then the overall running time of this approach may actually be lower than the first approach that requires us to train only k classifiers. This can be the case with many nonlinear classifiers. The one-against-one approach may also result in ties between different classes that receive the same number of votes. In such cases, the numeric scores output by the classifier may be used to weight the votes for the different classes. As in the previous case, the choice of the numeric score depends on the choice of the base classifier model.

There are also some optimized methods for converting the binary classifier into a multilabel classifier by changing the problem formulation. The change in the formulation is specific to the linear model at hand, and it is not designed as a meta-algorithm (like one-against-rest). Examples include multinomial logistic regression (cf. Section 6.4.4) and the *Weston-Watkins multi-class SVM* [579].

6.1.5   Characteristics of Linear Models for Text

The sparse and high-dimensional representation of text is particularly suitable for linear models. Several independent evaluations have shown that linear models are among the best performing classifiers in the text domain. Furthermore, linear models are very efficient, and can be implemented in time that is linear to corpus size.

Linear models can be extended to modeling nonlinear relationships by using feature engineering tricks in which the documents are implicitly transformed into a new space before applying a linear model. In fact, it is even possible to use these models in cases where one only has access to similarities between pairs of documents rather than the actual feature representation. Such methods are particularly helpful when one wants to use the sequence representation of text in mining algorithms by leveraging sequence-centric similarity functions. Therefore, linear models present a broad class of highly flexible algorithms that are considered state-of-the-art in the text domain.

Chapter Notations

The following notations will be used in this chapter. The training data is defined by the $n \times d$ data matrix D, whose rows are the documents denoted by the d-dimensional row vectors $\overline{X_1} \ldots \overline{X_n}$. Let $\overline{X_i}$ be a d-dimensional tuple denoted by $(x_{i1}, x_{i2}, \ldots x_{id})$ corresponding to the d term frequencies in the document. In addition, the ith document $\overline{X_i}$ is associated with the class label y_i. We can assume that the column vector $\overline{y} = [y_1 \ldots y_n]^T$ contains the class labels (or dependent variables in regression) for the n training instances. In the context of classification, this chapter only considers binary class labels, which are drawn from $\{-1, +1\}$. Therefore, the pair $(D, \overline{y})$ represents the training data, and a one-to-one correspondence exists between the n rows (documents) of D and the n entries of $\overline{y}$. In addition to the training data, we have a *test matrix* D_t of size $n_t \times d$. Therefore, there are n_t test instances, denoted by $\overline{Z_1} \ldots \overline{Z_{n_t}}$, and the class label (or numeric dependent variable) is not observed for these instances.

Chapter Organization

This chapter is organized as follows. The next section will introduce the least-squares family of regression and classification models, which includes the Fisher discriminant. The support vector machine is introduced in Section 6.3. Logistic regression is introduced in Section 6.4. Nonlinear models are discussed in Section 6.5. The summary is given in Section 6.6.

6.2 Least-Squares Regression and Classification

The least-squares family is one of the most fundamental ones for classification and regression. Although this particular family is inherently designed for regression, it can also be adapted to classification. In fact, many important classification models used in the text domain, such as the linear least-squares fit and the Fisher discriminant are applications of this basic model to categorical dependent variables. There are also some important properties of models based on the type of regularization that is used. So far, we have only considered a regularization penalty in which the sum of squares of the coefficients is penalized. This type of regularization is referred to as L_2-*regularization* or *Tikhonov regularization*.

6.2.1 Least-Squares Regression with L_2-Regularization

Consider an $n \times d$ document-term matrix D (training data), for which the n-dimensional column vector containing the numeric dependent variables is denoted by $\overline{y} = [y_1, y_2, \ldots y_n]^T$. Then, the ith row (i.e., document) of D, denoted by $\overline{X_i}$, is approximately related to the class variable using a d-dimensional row vector $\overline{W}$ of coefficients as follows:

$$y_i \approx \overline{W} \cdot \overline{X_i} \quad \forall i \in \{1 \ldots n\} \tag{6.1}$$

Therefore, the least-squares formulation, which includes a regularization term (i.e., penalty on coefficients) is as follows:

$$\text{Minimize } J = \underbrace{\frac{1}{2} \sum_{i=1}^{n} (y_i - \overline{W} \cdot \overline{X_i})^2}_{\text{Prediction Error}} + \underbrace{\frac{\lambda}{2} \sum_{j=1}^{d} w_j^2}_{L_2\text{-Regularization}}$$

We have already seen this form of the objective function in the previous section. One can express this objective function in terms of the $n \times d$ document-term matrix D, dependent variable vector $\overline{y}$, and coefficient vector $\overline{W}$ as follows:

$$\text{Minimize } J = \frac{1}{2}||D\overline{W}^T - \overline{y}||^2 + \frac{\lambda}{2}||\overline{W}||^2$$

The optimality condition for this problem is obtained by setting the partial derivative of J with respect to each element of the vector $\overline{W}$ to 0. The partial derivative across all elements of the vector $\overline{W}$ can be expressed in an integrated way using matrix calculus notation:

$$\frac{\partial J}{\partial \overline{W}} = D^T(D\overline{W}^T - \overline{y}) + \lambda\overline{W}^T = 0 \tag{6.2}$$

By re-arranging the aforementioned condition, we obtain the following:

$$(D^T D + \lambda I)\overline{W}^T = D^T\overline{y} \tag{6.3}$$

The matrix $D^T D$ is positive semi-definite, and the regularization with $\lambda > 0$ makes the matrix $D^T D + \lambda I$ positive definite. Any positive definite matrix is always invertible, and therefore the coefficient vector $\overline{W}$ can be obtained as follows:

$$\overline{W}^T = (D^T D + \lambda I)^{-1} D^T\overline{y} \tag{6.4}$$

Here, I is the $d \times d$ identity matrix. Once the coefficient vector has been determined, the dependent variable of an unseen test instance $\overline{Z}_i$ can be predicted as the dot product of the coefficient vector and the test instance:

$$F(\overline{Z}_i) = \overline{W} \cdot \overline{Z}_i \tag{6.5}$$

In fact, one can predict the labels of the entire $n_t \times d$ test data matrix D_t in one shot as $\overline{y_t} = D_t\overline{W}^T$. Note that $\overline{y_t}$ is a column vector of n_t entries containing the predicted values of the dependent variable of each of the n_t rows in D_t. The value of λ can be tuned by holding out of a part of the data, and testing the accuracy of using various values of λ for training.

6.2.1.1   Efficient Implementation

The solution of Equation 6.4 requires the inversion of a $d \times d$ matrix. The value of d can be greater than 10^5 for the text domain, which can create significant challenges. One possibility is to use a gradient-descent method for more efficient prediction.

Instead of setting the gradient vector $\frac{\partial J}{\partial \overline{W}}$ to 0 to obtain the solution in closed form, one can choose to use it for gradient descent. The approach initializes the vector $\overline{W}$ randomly. In each iteration, the coefficient vector $\overline{W}$ can be updated using a step-size $\alpha > 0$ as follows:

$$\overline{W}^T \Leftarrow \overline{W}^T - \alpha \left[\frac{\partial J}{\partial \overline{W}}\right]$$

$$= \overline{W}^T(1 - \alpha\lambda) - \alpha D^T \underbrace{(D\overline{W}^T - \overline{y})}_{\text{Current Errors}}$$

The gradient-descent steps are repeated to convergence. Note that the last term contains an error vector $(D\overline{W}^T - \overline{y})$, which is computed first, and then D^T is pre-multiplied with it to create the update. Such an ordering of matrix/vector computations ensures efficiency.

6.2.1.2   Approximate Estimation with Singular Value Decomposition

Singular value decomposition provides an efficient way to perform approximate matrix inversion of $(D^T D + \lambda I)$. Furthermore, this approximation actually *helps* the prediction because it is an indirect form of regularization. In other words, we can set λ to 0 and use this alternative form of regularization instead. Therefore, we will discuss the following by assuming that λ is set to 0. Truncated singular value decomposition of rank-k approximately decomposes the $n \times d$ document-term matrix D into a $n \times k$ matrix Q, a $k \times k$ diagonal matrix Σ, and a $d \times k$ matrix P as follows:

$$D \approx Q\Sigma P^T \tag{6.6}$$

The rank k should always be chosen small enough that each entry of Σ is strictly positive. A key observation is that it is often possible to set $k \ll \min\{d, n\}$, and the "loss" resulting from such a truncation actually improves the representation by reducing the noise effects of synonymy and polysemy. This is what improves the generalizability of the learned coefficients to unseen test instances. By substituting Equation 6.6 in Equation 6.4 and setting $\lambda = 0$, one obtains the following:

$$
\begin{aligned}
\overline{W}^T &= (P\Sigma^2 P^T)^{-1}(Q\Sigma P^T)^T \overline{y} \\
&= (P\Sigma^{-2}P^T)P\Sigma Q^T \overline{y} && [\text{Using } P^{-1} = P^T] \\
&= P\Sigma^{-2}\Sigma Q^T \overline{y} && [\text{Using } P^T P = I] \\
&= P\Sigma^{-1} Q^T \overline{y}
\end{aligned}
$$

The key point here is that one only needs to compute the top-k singular vectors/values of D using Lanczos algorithm [166, 167]. Even the power method of Section 3.2.2 can be used for document collections of modest size. Since the value of $k \sim [200, 500]$ is often much smaller than the dimensionality $d > 100,000$ of a typical collection, such an approach turns out to be very efficient. The value of k now serves the same role as the regularization parameter λ, in which *small* values of k indicate a *higher* level of regularization. The benefits of such a noise reduction approach have also been shown [601] in the context of the linear least-squares fit method for classification (cf. Section 6.2.3.3).

In order to understand why the approach improves generalizability of the approach to unseen test instances, one can view this approach as a way of building concise models with fewer parameters. It is noteworthy that one could transform the training matrix D to k-dimensional space using the transformation $D_k = DP$. Each test instance $\overline{Z}$ can be transformed to k-dimensional space using $\overline{Z}^{(k)} = \overline{Z}P$. Then, one can perform the linear regression in this new lower-dimensional space (without any need for regularization) and predict the dependent variable. It can be shown that the SVD-truncated prediction in the original space is exactly equivalent to the prediction in this transformed problem. Note that the transformed problem needs to compute only k coefficients on a non-redundant set of variables, and is therefore far more concise. Furthermore, much of the noise in the lower-order singular vectors (which is a source of overfitting) has been removed by the transformation to a semantically coherent space. As a result, the overall accuracy of the results improves using this approach on unseen test instances.

Relationship with Principal Components Regression

The aforementioned use of truncated singular value decomposition is closely related to *principal components regression* [280]. In principal components regression, the data is transformed to a lower-dimensional space using principal component analysis (PCA) [279]. The data is then regressed in this new space by treating the transformed attributes as the explanatory variables. This approach is very similar to the truncated singular value decomposition discussed above, except that PCA is used instead of SVD for the transformation. The SVD of a data matrix after centering it to zero mean will result in the same solution as that obtained with the use of PCA. For sparse data matrices like text, the means of the attributes (term frequencies) are close to zero anyway, and centering the data does not make a large difference to the final predictions. It is better to not center the data matrix, because doing so destroys the sparsity of the data matrix. Sparsity is very desirable from a computational and space-efficiency point of view with such decompositions.

6.2.1.3   The Path to Kernel Regression

The prediction of test point $\overline{Z}$ of linear regression (cf. Equation 6.5) can be *equivalently* expressed purely in terms of dot products between training instances, as well the dot products between $\overline{Z}$ and the training instances. Let $K(\overline{Z}, \overline{X_i}) = \overline{Z} \cdot \overline{X_i}$ represent the dot product between the test instance $\overline{Z}$ and the training instance $\overline{X_i}$. Let S be the $n \times n$ matrix representing the dot-product similarities between the n training points and $\overline{y}$ represent the n-dimensional column vector of response variables. Then, the prediction $F(\overline{Z})$ of test point $\overline{Z}$ can be expressed in terms of an n-dimensional row vector $[K(\overline{Z}, \overline{X_1}), \dots K(\overline{Z}, \overline{X_n})]$ of similarities between training points and the test instance:

$$F(\overline{Z}) = \underbrace{[K(\overline{Z}, \overline{X_1}), K(\overline{Z}, \overline{X_2}) \dots K(\overline{Z}, \overline{X_n})]}_{\text{Test-Training Similarities}} (S + \lambda I)^{-1} \overline{y} \qquad (6.7)$$

We leave the proof of this result as an exercise to the reader (see Exercise 3). Note that we cannot derive the coefficient vector $\overline{W}$ without using the features of the training points, but we can still express the predictions of test instances purely in terms of the dot products in matrix S. If one chooses to use similarity functions other than the dot product for $K(\overline{Z}, \overline{X_i})$, then the approach becomes *least-squares kernel regression*, which is able to capture non-linear relationships[1] between the regressors and regressand. In fact, a multidimensional representation of the data is not even needed, and one can use string kernels (cf. Chapter 3) as similarities. There are two useful applications of this form of the prediction:

1. If the number of documents n is far less than the size of the lexicon d, then it is easier to invert an $n \times n$ matrix $S + \lambda I$ than to invert a $d \times d$ matrix. Therefore, the aforementioned solution may be preferable.

2. Consider a setting in which one wishes to use the sequential relationships among the terms in the documents as salient information for predicting the regressand. In such a case, one can use string kernels in conjunction with Equation 6.7. A string kernel effectively incorporates the linguistic and semantic information in sentences, which is not available from the bag-of-words representation. As a result, this approach leverages deeper semantic knowledge embedded in the documents.

In general, it is hard to invert an $n \times n$ similarity matrix in large collections. Therefore, it makes sense to use the SVD-based low-rank trick discussed in Section 6.2.1.2.

---

[1] When using kernel methods, it is customary to add a small constant amount to every entry in the similarity matrix between points to account for the effect of the dummy variable representing the bias term [365] (see Exercise 5).

6.2.2   LASSO: Least-Squares Regression with L_1-Regularization

The acronym LASSO stands for *Least Absolute Shrinkage and Selection Operator*, and it uses L_1-regularization instead of L_2-regularization for least-squares regression. As in the case of all least-squares problems, it is assumed that the ith training instance $\overline{X_i}$ is related to the dependent variable as follows:

$$y_i \approx \overline{W} \cdot \overline{X_i} \tag{6.8}$$

In order to learn the regression coefficients, the least-squares error of the prediction needs to be minimized as follows:

$$\text{Minimize } J = \underbrace{\frac{1}{2}\sum_{i=1}^{n}(y_i - \overline{W} \cdot \overline{X_i})^2}_{\text{Prediction Error}} + \underbrace{\lambda \sum_{j=1}^{d}|w_j|}_{L_1\text{-Regularization}}$$

Note that the regularization term now uses the L_1-norm of the coefficient vector rather than the L_2-norm. One can write this objective function in terms of the $n \times d$ training data matrix D and the n-dimensional column vector $\overline{y}$ of dependent variables as follows:

$$\text{Minimize } J = \frac{1}{2}||D\overline{W}^T - \overline{y}||^2 + \lambda||\overline{W}||_1$$

Here, $||W||_1$ represents the L_1-norm of the vector $\overline{W}$. This optimization problem cannot be solved in closed form like the case of L_2-regularization. An important point here is that the function J is non-differentiable for any $\overline{W}$ in which even a single component w_j is 0. Specifically, if w_j is infinitesimally larger than 0, then the partial derivative of $|w_j|$ is $+1$, whereas if w_j is infinitesimally smaller than 0, then the partial derivative of $|w_j|$ is -1. This makes the derivative of w_j undefined exactly at 0. For such objective functions with specific points of non-differentiability, *subgradients* are often used. In these methods, the partial derivative of w_j at 0 is selected uniformly at random from $[-1, +1]$, whereas the derivative at values different from 0 is computed in the same way as the gradient. Let the subgradient of w_j be denoted by s_j. Then, for step-size $\alpha > 0$, the update is as follows:

$$\overline{W}^T \Leftarrow \overline{W}^T - \alpha\,\lambda\,[s_1, s_2, \ldots, s_d]^T - \alpha D^T \underbrace{(D\overline{W}^T - \overline{y})}_{\text{Error}}$$

Here, each s_j is the subgradient of w_j and is defined as follows:

$$s_j = \begin{cases} -1 & w_j < 0 \\ +1 & w_j > 0 \\ \text{Sample uniformly from } [-1, +1] & w_j = 0 \end{cases} \tag{6.9}$$

One issue here is that the random choice of s_j from $[-1, +1]$ can sometimes cause the objective function to worsen. Therefore, this method is not a gradient-descent method because some iterations will cause the objective function value to worsen. Nevertheless, it can be shown that the approach has guaranteed convergence properties for convex objective functions. However, the fact that the objective function can worsen is important from the point of view of using the best possible value of $\overline{W}_{best}$ that was obtained in any iteration. At the beginning of the process, both $\overline{W}$ and $\overline{W}_{best}$ are initialized to the same random vector.

After each update of $\overline{W}$, the objective function value is evaluated with respect to $\overline{W}$, and $\overline{W}_{best}$ is set to the recently updated $\overline{W}$ if the objective function value provided by $\overline{W}$ is better than that obtained by the stored value of $\overline{W}_{best}$. At the end of the process, the vector $\overline{W}_{best}$ is returned by the algorithm as the final solution. One issue with this solution is that it can be slow in practice, and therefore another technique called least-angle regression [173] is often used. Another option is to use $s_j = 0$ at $w_j = 0$, which often turns out to be a more practical choice.

6.2.2.1   Interpreting LASSO as a Feature Selector

Almost all L_1-regularization methods, including LASSO, always lead to sparse solutions in which most values of w_j are exactly zero. This is different from L_2-regularization, in which the penalty reduces the size of the coefficients but most of them are non-zero. From a prediction point of view, a zero coefficient has no influence on the prediction, and therefore such a feature can be dropped. In L_2-regularization, the feature selection is softer in the sense that the influence of each feature is reduced by shrinking its coefficient, but most of them still have non-zero influence on the prediction. This observation provides LASSO a very nice interpretability and also dual use as a feature selector. In fact, such feature selectors are referred to as *embedded models* because they embed the feature selection within the modeling process. From a semantic point of view, one gets to learn which terms are relevant or irrelevant for the modeling process. LASSO is particularly useful in very high-dimensional domains like text in which a small number of features can have a high level of explanatory power.

A natural question arises as to when one should choose L_1- or L_2-regularization. In terms of prediction accuracy, L_2-regularization almost always performs better than L_1-regularization, and is the safe option over arbitrary data sets. For sparse and high-dimensional domains like text, L_1-regularization can sometimes provide comparable performance, but is almost always outperformed by combining L_1- and L_2-regularization with the *elastic net* [640]. The real utility of pure L_1-regularization is in providing highly interpretable feature selection, and that should also be viewed as its primary use case. Combining with L_2-regularization provides high-quality solution with good interpretability. However, if prediction accuracy is the primary goal and one does not want to use the more complicated optimization algorithms to combine L_1- and L_2-regularization, the simple and safe choice is to use only L_2-regularization. Although this chapter primarily focuses on L_2-regularization for classification, it is noteworthy that all the linear classification models in this chapter have L_1-variants, which have similar sparsity properties to the LASSO for regression. A detailed discussion of many of these generalizations is provided in [237].

6.2.3   Fisher's Linear Discriminant and Least-Squares Classification

The Fisher's linear discriminant can be shown to be a special case of least-squares regression on appropriately coded response variables, although this is not how the discriminant is defined. Rather, the linear discriminant is defined as the direction that maximizes the ratio of the inter-class variance to the intra-class variance, if all points were to be projected along that direction.

One can view the Fisher's discriminant as a supervised cousin of principal component analysis (PCA). The latter finds a direction in the data space that maximizes the variance of all points along that direction irrespective of class. On the other hand, the Fisher's discrim-

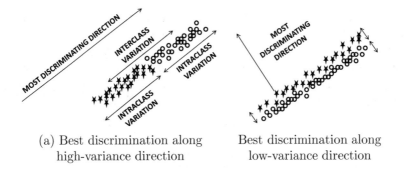

(a) Best discrimination along   Best discrimination along
high-variance direction   low-variance direction

Figure 6.2: Sensitivity of Fisher's discriminant to class distribution

inant focuses on maximizing the ratio of inter-class to intra-class variance, and therefore discovers very different solutions. A two-class example is illustrated in Figure 6.2, in which the effect of using different labeling of classes on the same data set is shown. The directions in Figures 6.2(a) and 6.2(b) are very different because the ratio of inter-class to intra-class variance is maximized in different directions in the two cases. Let the discovered direction be $\overline{W}$. Then, the projection of any data point $\overline{X}_i$ along this direction is given by $\overline{W} \cdot \overline{X}_i$. In each case, it is a relatively simple matter to classify the data set by using an appropriately chosen threshold with respect to the 1-dimensional coordinate from the projection. The negative of this threshold can be used to define the bias b. The value of b can be estimated on a held-out set using cross-validation. Therefore, the prediction $\hat{y}_i \in \{-1, +1\}$ of the ith data point can be computed as follows:

$$\hat{y}_i = \text{sign}\{\overline{W} \cdot \overline{X}_i + b\}$$

This is the well-known prediction function used in linear classification. However, as we will discuss later, the Fisher's method is also used as a feature engineering method, and particularly so in multiclass settings.

Next, we discuss the derivation of the Fisher direction $\overline{W}$. For a d-dimensional data set, let $\overline{\mu_0}$ be the row vector denoting the d-dimensional mean of the negative class (i.e., class with label -1), and $\overline{\mu_1}$ be the row vector denoting the d-dimensional mean of the positive class (i.e., class with label $+1$). Similarly, let Σ_0 be the $d \times d$ covariance matrix of only the points belonging to the negative class in which the (j, k)th entry is the covariance between the jth and kth attributes of the points in this class. The corresponding covariance matrix for the positive class is Σ_1. Furthermore, let n_0 and n_1 be the number of training examples, respectively, belonging to the negative and positive class, so that the total number of training examples n is given by $n_0 + n_1$.

The squared distance between the means of the two classes along $\overline{W}$ is given by $(\overline{W} \cdot \overline{\mu_1} - \overline{W} \cdot \overline{\mu_0})^2$. This quantity is proportional to the inter-class variance[2] (or between-class scatter) $B(\overline{W})$:

$$B(\overline{W}) \propto n(\overline{W} \cdot (\overline{\mu_1} - \overline{\mu_0}))^2 = \overline{W} \underbrace{\left[n(\overline{\mu_1} - \overline{\mu_0})^T (\overline{\mu_1} - \overline{\mu_0}) \right]}_{d \times d \text{ matrix } S_b \text{ of rank-1}} \overline{W}^T = \overline{W} S_b \overline{W}^T$$

---

[2]The notions of scatter and variance are different only in terms of scaling. The scatter of a set of n values is equal to n times their variance. Therefore, it does not matter whether the scatter or variance is used within a constant of proportionality.

The above relationships introduce an additional notation S_b by replacing the $d \times d$ rank-1 matrix $\left[n(\overline{\mu}_1 - \overline{\mu}_0)^T(\overline{\mu}_1 - \overline{\mu}_0)\right]$ with a between-class scatter matrix[3] S_b.

In order to compute the scatter *within* each class along direction $\overline{W}$, we make use of the well-known fact [279] that the scatter of a set of n points along a direction $\overline{W}$ can be expressed in terms of the covariance matrix Σ as $n\overline{W}\Sigma\overline{W}^T$. Then, we compute the scatter within each class along $\overline{W}$ and compute their sum in $I(\overline{W})$ as follows:

$$I(\overline{W}) = n_1(\overline{W}\Sigma_1\overline{W}^T) + n_0(\overline{W}\Sigma_0\overline{W}^T)$$
$$= \overline{W}\ \underbrace{(n_1\Sigma_1 + n_0\Sigma_0)}_{d \times d \text{ matrix } S_w}\ \overline{W}^T$$
$$= \overline{W}S_w\overline{W}^T$$

An additional notation, S_w, corresponding to the within-class scatter matrix is introduced above. Then, the objective function of the Fisher discriminant maximizes the ratio of the interclass to intra-class scatter along $\overline{W}$ as follows:

$$\text{Maximize } J = \frac{B(\overline{W})}{I(\overline{W})} = \frac{\overline{W}S_b\overline{W}^T}{\overline{W}S_w\overline{W}^T}$$

Note that only the direction of $\overline{W}$ matters in the above solution, and its scaling (i.e., norm) does not affect J. Therefore, in order to make the optimal solution unique, one can choose a scaling in which the denominator is 1. This creates a constrained optimization problem:

$$\text{Maximize } J = \overline{W}S_b\overline{W}^T$$
$$\text{subject to:}$$
$$\overline{W}S_w\overline{W}^T = 1$$

Setting the gradient of the Lagrangian relaxation $\overline{W}S_b\overline{W}^T - \alpha(\overline{W}S_w\overline{W}^T - 1)$ to 0 yields the generalized eigenvector condition $S_b\overline{W}^T = \alpha S_w\overline{W}^T$. Therefore, $\overline{W}^T$ is the only nonzero eigenvector of the rank-1 matrix $S_w^{-1}S_b$. Because $S_b\overline{W}^T = (\overline{\mu}_1^T - \overline{\mu}_0^T)\left[n(\overline{\mu}_1 - \overline{\mu}_0)\overline{W}^T\right]$ always points in the direction of $(\overline{\mu}_1^T - \overline{\mu}_0^T)$, it follows that $S_w\overline{W}^T \propto \overline{\mu}_1^T - \overline{\mu}_0^T$. Therefore, we have the following:

$$\overline{W}^T \propto S_w^{-1}(\overline{\mu}_1 - \overline{\mu}_0)^T \tag{6.10}$$
$$= (n_1\Sigma_1 + n_0\Sigma_0)^{-1}(\overline{\mu}_1 - \overline{\mu}_0)^T \tag{6.11}$$

It is also common to use a variant of this methodology in which a parameter γ is introduced to give differential weight to the various classes:

$$\overline{W}^T \propto (\Sigma_1 + \gamma\Sigma_0)^{-1}(\overline{\mu}_1 - \overline{\mu}_0)^T \tag{6.12}$$

One can choose γ by optimizing a desired cost function on held out portion of the data. Equal weighting to the classes irrespective of their relative population is achieved by setting $\gamma = 1$. However, the "official" Fisher discriminant is defined only by Equation 6.11, which is what will be used in this chapter.

---

[3]This two-class variant of the scatter matrix S_b is not exactly the same as defined in the multi-class version S_b of Section 6.2.3.1. Nevertheless, all entries in the two matrices are related with the proportionality factor of $\frac{n_1 \cdot n_0}{n^2}$ which turns out to be inconsequential to the *direction* of the Fisher discriminant. In other words, the use of the multi-class formulas in Section 6.2.3.1 will yield the same result in the binary case.

6.2.3.1 Linear Discriminant with Multiple Classes

The aforementioned solution can be generalized to multiple classes in two ways. One can perform the classification using a one-against-all approach in which one class is selected as the positive class and the remaining classes are selected as the negative classes. This process is repeated k times, and the most confident prediction is returned for a test instance. This approach is used frequently in the text domain [95, 601, 604]. Although the approach can be reasonably used for prediction, a more powerful approach is to use all the classes simultaneously to derive the $k - 1$ directions.

First, we need to compute the scatter matrices S_w and S_b for the multi-class setting. The scatter matrices are computed in a similar way to the linear discriminant metric of Section 5.4.4 in Chapter 5. Let Σ_i be the covariance matrix of the ith class, so that the (j, k)th entry of Σ_i is equal to the covariance between the jth and kth dimensions in the ith class. Let n_i be the number of points in the ith class, and $n = \sum_i n_i$ be the total number of points. Let $\overline{\mu}$ be the d-dimensional row vector representing the mean of the entire data set, and $\overline{\mu_i}$ be the d-dimensional row vector representing the mean of the ith class. Then, the $d \times d$ within-class scatter matrix is defined as follows:

$$S_w = \sum_{i=1}^{k} n_i \Sigma_i \tag{6.13}$$

The $d \times d$ between-class scatter matrix[4] is defined as the sum of the following rank-1 matrices:

$$S_b = \sum_{i=1}^{k} n_i (\overline{\mu_i} - \overline{\mu})^T (\overline{\mu_i} - \overline{\mu}) \tag{6.14}$$

Note that each product above is the product of a $d \times 1$ matrix with a $1 \times d$ matrix. The matrix S_b is n times the covariance matrix of a data set containing the means of the classes in which the mean of the ith class is repeated n_i times. Then, the top-$(k - 1)$ eigenvectors of the rank-$(k - 1)$ matrix $S_w^{-1} S_b$ provides a low-dimensional space of data representation. Other classifiers like decision trees can be constructed in this space. The multiclass variant of linear discriminant analysis is often used to perform feature engineering for other classifiers. It is sometimes also used for soft feature scaling as shown in Section 5.4.4.

An immediate observation is that the approach is computationally expensive for high dimensional data sets. It requires $O(n \cdot d^2)$ time to compute each scatter matrix, and $O(d^3)$ time to invert the within-class scatter matrix. The value of d is usually greater than 10^5 in text. To improve efficiency, one can first preprocess the feature variables of both the training and test data with latent semantic analysis, and reduce the dimensionality to less than 500. It is much easier to compute a 500×500 matrix and invert it.

6.2.3.2 Equivalence of Fisher Discriminant and Least-Squares Regression

The binary Fisher's discriminant classifier *is the same as* least-squares regression with respect to the (binary) class indicator variable [57]. This is an important result, because it enables the use of many of the efficient techniques for least-squares regression, such as gradient-descent and SVD-based approximation. Furthermore, this equivalence also enables the use of kernel methods presented for the case of least-squares regression.

---

[4]Note that this matrix is different from the one introduced for the two-class case only by a proportionality factor, which does not affect the final solution.

This result is algebraically easiest to show by mean-centering both the data matrix and the response variable in a particular way and setting[5] the bias to 0. Therefore, the following will not assume the use of a dummy column to adjust for bias. Consider the following case in which the $n \times d$ document-term matrix D and the n-dimensional column vector $\overline{y}$ containing the class variables have been preprocessed as follows. The matrix D is mean-centered by simply subtracting the mean of each column from the corresponding variable. Similarly, the column vector $\overline{y}$ of class variables in $\{-1, +1\}$ has been mean-centered by setting its positive entries to n_0/n and negative entries to $-n_1/n$. Then, the coefficient vector $\overline{W}^T$ of least-squares regression without regularization satisfies the following;

$$(D^T D)\overline{W}^T = D^T \overline{y} \tag{6.15}$$

Because of the special way in which the response variable has been coded, the right-hand side of the above expression simplifies as follows (convince yourself why this is true):

$$(D^T D)\overline{W}^T \propto (\overline{\mu}_1 - \overline{\mu}_0)^T \tag{6.16}$$

A key relationship between the within-class scatter matrix S_w, the between-class scatter matrix S_b, and the full scatter matrix $D^T D$ is as follows:

$$D^T D = S_w + \frac{n_1 \cdot n_0}{n^2} S_b \tag{6.17}$$

$$= S_w + K \left[(\overline{\mu}_1 - \overline{\mu}_0)^T (\overline{\mu}_1 - \overline{\mu}_0) \right] \tag{6.18}$$

Here, K is a suitably chosen scalar. Note that this relationship holds when the matrix D is mean-centered. Here, we simply assume this relationship, and leave it as an exercise for the reader to show its correctness (see Exercise 4).

By substituting Equation 6.18 in Equation 6.16, one obtains the following:

$$\left(S_w + K \left[(\overline{\mu}_1 - \overline{\mu}_0)^T (\overline{\mu}_1 - \overline{\mu}_0) \right] \right) \overline{W}^T \propto (\overline{\mu}_1 - \overline{\mu}_0)^T \tag{6.19}$$

Now, a key point here is that the vector $\left[(\overline{\mu}_1 - \overline{\mu}_0)^T (\overline{\mu}_1 - \overline{\mu}_0) \right] \overline{W}^T$ always points in the direction of $(\overline{\mu}_1 - \overline{\mu}_0)^T$ because we can write this vector as $(\overline{\mu}_1 - \overline{\mu}_0)^T \left[(\overline{\mu}_1 - \overline{\mu}_0)\overline{W}^T \right]$. This means that the second term on the left-hand side of Equation 6.19 can be dropped without affecting the proportionality relationship of vectors:

$$S_w \overline{W}^T \propto (\overline{\mu}_1 - \overline{\mu}_0)^T$$

$$\overline{W}^T \propto S_w^{-1} (\overline{\mu}_1 - \overline{\mu}_0)^T$$

Note that the vector on the right-hand side is the same as that provided by the Fisher discriminant. In other words, with the proper preprocessing of the data matrix and response variable, one obtains the same result with least-squares regression as the Fisher discriminant.

The aforementioned result uses mean-centered matrices to obtain the equivalence with algebraic simplicity. Centering both the data matrix and the response variable is simply a way of ensuring that the bias is 0 in the optimal solution of least-squares regression, and one does not have to worry about an (uncentered) dummy column of 1s in D. One can also show more general equivalence by allowing for a bias variable and adding a dummy column

---

[5]One can also show more general equivalence by allowing for bias.

of 1s to the data matrix to absorb the bias coefficient. This result has considerable practical significance because it shows that one can use any of the efficient solution methods discussed earlier in this section for least-squares regression in the case of Fisher discriminant with two classes. The equivalence between least-squares regression and the Fisher discriminant also means that one can extend the kernel regression methods discussed in Section 6.2.1.3 to the Fisher discriminant.

Although the Fisher discriminant can be simulated with least-squares regression, this does not mean that the entire family of discriminant methods is subsumed by the least-squares family. Fisher's discriminant is only one member of a larger family of linear discriminators. The objective functions of linear discriminators and least-squares regression try to capture geometrically different notions but turn out to be equivalent in special cases like the Fisher discriminant with binary data. Furthermore, the multi-class treatment is different in the two cases.

6.2.3.3 Regularized Least-Squares Classification and LLSF

When regularization is combined with linear regression on binary class variables drawn from $\{-1, +1\}$, the formulation is referred to as *regularized least-squares classification.* The formulation for least-squares classification can be written as follows:

$$\text{Minimize } J = \frac{1}{2} \sum_{i=1}^{n} [y_i - (\overline{W} \cdot \overline{X_i})]^2 + \frac{\lambda}{2} ||\overline{W}||^2 \tag{6.20}$$

$$= \frac{1}{2} \sum_{i=1}^{n} [1 - y_i(\overline{W} \cdot \overline{X_i})]^2 + \frac{\lambda}{2} ||\overline{W}||^2 \tag{6.21}$$

Note that the second relationship of Equation 6.21 is only true when the class variable is coded to $\{-1, +1\}$ because the value of y_i^2 is always 1. As we will see later, this form of the objective function is very closely related to that of a support-vector machine. A test instance $\overline{Z}$ is classified using the following prediction function with the learned value of $\overline{W}$:

$$F(\overline{Z}) = \text{sign}\{\overline{W} \cdot \overline{Z}\} \tag{6.22}$$

At a learning rate of η, the stochastic gradient-descent update of least-squares classification is exactly the same as the one shown earlier for least-squares regression with numeric responses (cf. Section 6.2.1.1):

$$\overline{W} \Leftarrow \overline{W}(1 - \eta\lambda) + \eta y(1 - y(\overline{W} \cdot \overline{X}))\overline{X}$$

The equivalence to the updates in Section 6.2.1.1 follow from using $y^2 = 1$. Furthermore, the above updates represent *stochastic* gradient descent because the gradients are computed with respect to a single training point $(\overline{X}, y)$ that is randomly sampled from the training data. We use this form of the updates to relate them better with other types of linear classification models.

This formulation is also referred to as the linear least-squares fit (LLSF) method in the text domain [601, 604]. However, the original formulation in [601, 604] does not use L_2-regularization, and it uses truncated singular value decomposition instead (cf. Section 6.2.1.2). A formulation was also proposed for the multiclass case [601], although it can be shown that it is equivalently decomposable into a one-against-all approach applied to the binary formulation.

The LLSF and least-squares classification methods are equivalent to the Fisher discriminant, when regularization is not used. The LLSF method does not center the document-term matrix and uses binary variables as the responses to learn the regressors. In contrast, the results in Section 6.2.3.2 show that Fisher's discriminant performs the same regression on *centered* variables. Is this difference significant? It turns out that these differences are not significant because they can be adjusted for by simply adding a bias variable in the form of a dummy column of 1s to D when running LLSF. Note that a binary indicator response variable can be obtained from the response variable of Section 6.2.3.2 by adding n_1/n to each response value. Furthermore, each column of a mean-centered data matrix D is different from the uncentered matrix only in terms of translation of each column by its mean. These differences in translation can be fully absorbed with the use of different values of the bias variable (dummy-column coefficient) without changing the non-trivial regression coefficients (i.e., those belonging to observed variables). The LLSF implementation does have the advantage of working with the original sparse data matrices, which is particularly useful in the text domain.

As a historical note, it should be pointed out that the regularized least-squares family has been re-invented several times. The Fisher discriminant was proposed in 1936 as a method for finding class-sensitive directions. Least-squares classification and regression date back to Widrow-Hoff learning in the sixties [580] and Tikhonov-Arsenin's seminal work [550] in the seventies. The remarkable relationship between the Fisher discriminant and these methods was eventually discovered [57]. Another closely related variation is the *perceptron* algorithm (cf. Section 10.6.1.1 of Chapter 10), which (also remarkably) is a shifted version of the support vector machine loss function (pp. 322). As discussed in the next section, the support vector machine is itself a repaired version of the least-squares classification loss function. In fact, Hinton [247] repaired the Widrow-Hoff avatar of the least-squares classification loss function to create the L_2-loss of the support vector machine, three years before Cortes and Vapnik's seminal work [132] on support vector machines. The first application of least-squares methods to text categorization was proposed in [601, 604].

6.2.3.4 The Achilles Heel of Least-Squares Classification

The least-squares classification family (including the Fisher discriminant), has an important weakness in the nature of its loss function. By directly penalizing the squared difference between the indicator variable y_i and the prediction $\overline{W} \cdot \overline{X}_i$, one not only penalizes misclassified points but also the "easy" points that are correctly classified by $\overline{W} \cdot \overline{X}_i$ in a very strong way. For example, consider an instance $\overline{X}_i$ belonging to the positive class for which the value of $\overline{W} \cdot \overline{X}_i$ turns out to be 10. Even though this prediction is correct in a very confident way, this confidence will be penalized by the least-squares objective function, in which the coded value of y_i is 1. Such points typically correspond to well-separated points from the decision boundary, and their influence on the learned value of $\overline{W}$ often has a detrimental effect on the classification of points that are close to the decision boundary.

In order to illustrate this point, a two-class distribution is illustrated in Figure 6.3. It is noteworthy that the points that are far away from the decision boundary (on the correct side) skew the direction of the Fisher discriminant, which results in two misclassified regions near the true decision boundary. If the well-separated points in Figure 6.3 were to be thrown away, the Fisher discriminant does much better in approximating the true boundary. This observation is intriguing in the sense that one expects a classification model to be punished by the presence of "delinquent" (i.e., mislabeled) training points on the wrong side of the decision boundary, but one rarely expects to be penalized for having outstanding citizens

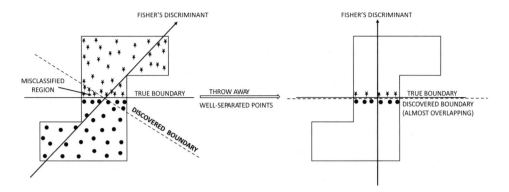

Figure 6.3: Well-separated points have a detrimental effect on the Fisher discriminant

in the training data!

The Fisher discriminant often lags behind another linear classifier, referred to as the support vector machine. The support vector machine *removes* the well-separated points, and keeps only the points near the decision boundary, which are referred to as "support vectors" for learning. Interestingly, it has been shown [513] that the Fisher discriminant is similar[6] to the support vector machine, if the well-separated points were to be discarded.

> The difference in accuracy performance of the support-vector machine and the Fisher discriminant/least-squares classification can be primarily explained by the differences in their treatment of the well-separated points.

Of course, since the problem of finding the well-separated points is the most difficult part of a support-vector machine, this observation does not help us much from an algorithmic point of view. This observation is, nevertheless, helpful from a heuristic point of view because one can discard well-separated points by using various heuristic tricks [95, 129]. Such heuristics can often boost the accuracy of the Fisher discriminant significantly.

While the superiority of SVMs over least-squares classification is generally accepted, some researchers have also pointed out that the differences are not large enough to be considered significant [471, 605]. Furthermore, points near the decision boundary can also be noisy points on the wrong side of the boundary, and therefore their primacy over the well-separated points is not guaranteed. Not all real-world settings are as neat as that shown in Figure 6.3. One can easily construct examples of toy data sets to make the counter-argument that well-separated examples are more informative than points near the boundary. Support-vector machines also require greater care and computational effort in parameter tuning. In particular, the work in [471] shows several examples in which least-squares methods are less sensitive to parameter choice (such as the regularization parameter) as compared to support vector machines.

6.3 Support Vector Machines

A support vector machine (SVM) has a special geometric interpretation of its regularizer, which leads to the notion of *margin-based separation* of the points belonging to the two

---

[6]The SVM generally uses the hinge loss rather than the quadratic loss. The use of quadratic loss is possible in an SVM but it is less common. This is another key difference between the Fisher discriminant and the (most common implementation of the) SVM.

classes. The basic idea here is that an SVM creates two parallel hyperplanes symmetrically on each side of the decision boundary, so that most points lie on either side of these two *margin hyperplanes* on the correct side. Although most textbooks introduce SVMs with this geometric interpretation, we believe that the regularized optimization view of a support vector machine is more helpful in understanding its true origins, and relating it to other linear models like least-squares classification. Therefore, we will first start with the regularized optimization view, and introduce the geometric interpretation later.

Some expositions of SVMs explicitly use a bias variable b, whereas others do not. The bias variable can be absorbed by addition of a single columns of 1s to the document-term matrix D. The coefficient of this dummy term is the bias variable (cf. Section 6.1.2). This does lead to a small change in the final predictions when regularization is used. This is because only the coefficients of the feature variables are regularized but an explicit bias variable is not. However, when the bias variable is treated as a coefficient of a dummy feature, it is regularized as well. Although the use of a dummy variable changes the optimization model slightly, the effect on the final predictions is quite small. The following exposition will work with the assumption of a dummy column like the other models of this chapter.

6.3.1   The Regularized Optimization Interpretation

Consider a data set with n training point-class variable pairs $(\overline{X_1}, y_1) \ldots (\overline{X_n}, y_n)$, in which the class variable y_i is always drawn from $\{-1, +1\}$. We start with the optimization formulation of least-squares classification in Equation 6.21, which is treated as the "parent problem" throughout this chapter:

$$\text{Minimize } J = \frac{1}{2} \sum_{i=1}^{n} [1 - y_i(\overline{W} \cdot \overline{X_i})]^2 + \frac{\lambda}{2} ||\overline{W}||^2 \quad \text{[Regularized Least-Squares Classification]}$$

The primary criticism of the least-squares classification model (cf. Section 6.2.3.4) is the fact that it not only penalizes the points for being on the incorrect side of the decision boundary, but it also penalizes them for being too far on the correct side. In particular, any point $\overline{X_i}$ for which $y_i(\overline{W} \cdot \overline{X_i}) > 1$ is actually being classified in a comfortable way on the correct side, and it should not be penalized. How can we remove this weakness of the least-squares classification model? The simplest way is to modify the aforementioned objective function so that points with $y_i(\overline{W} \cdot \overline{X_i}) > 1$ are not penalized. We present two such modifications below corresponding to different variations of the SVM objective function:

$$\text{Minimize } J = \frac{1}{2} \sum_{i=1}^{n} \max\{0, [1 - y_i(\overline{W} \cdot \overline{X_i})]\}^2 + \frac{\lambda}{2} ||\overline{W}||^2 \quad \text{[Quadratic-Loss SVM]}$$

$$\text{Minimize } J = \sum_{i=1}^{n} \max\{0, [1 - y_i(\overline{W} \cdot \overline{X_i})]\} + \frac{\lambda}{2} ||\overline{W}||^2 \quad \text{[Hinge-Loss SVM]}$$

As in the case of regularized least-squares regression, the prediction $F(\overline{Z})$ for test point $\overline{Z}$ is as follows:

$$F(\overline{Z}) = \text{sign}\{\overline{W} \cdot \overline{Z}\} \tag{6.23}$$

The linear separator $\overline{W} \cdot \overline{X} = 0$ therefore defines the decision boundary between the positive and negative classes. Therefore, *the support vector machine is a modification of the least-squares classification model, which addresses the latter's weakness in handling well-separated training points.*

The quadratic-loss SVM is more closely related to the regularized least-squares classification as compared to hinge loss. However, since the hinge-loss SVM is more common, the following description will primarily focus on this setting. One notational quirk used by the SVM community is that the optimization formulation is (equivalently) parameterized with the *slack penalty* $C = 1/\lambda$ rather than the regularization parameter λ. Therefore, for greater consistency with widely accepted notations, we use a similar form:

$$\text{Minimize } J = \frac{1}{2}||\overline{W}||^2 + C \cdot \sum_{i=1}^{n} \max\{0, [1 - y_i(\overline{W} \cdot \overline{X_i})]\} \qquad \text{[Hinge-Loss SVM]}$$

From an intuitive point of view, the slack penalty C quantifies the amount by which each point is penalized for "slacking off" from its target value of y_i in a one-sided way. For example, a positive point (i.e., $y_i = 1$) with $\overline{W} \cdot \overline{X_i} = 0.7$ will be penalized by $0.3C$, whereas a point with $\overline{W} \cdot \overline{X_i} = 1.3$ will not be penalized. Note that the former point will be classified correctly by Equation 6.23, but it is still penalized for being "too close" to the decision boundary. After all, such a point could be on the correct side of the decision boundary simply by virtue of overfitting. One can immediately see that the optimization formulation of the support vector machine is naturally designed to discourage overfitting.

6.3.2 The Maximum Margin Interpretation

Support vector machines also have an interesting geometric interpretation, which often helps in visualizing their solutions and motivating several solution methodologies. Note that the decision surface $\overline{W} \cdot \overline{X} = 0$ lies in the middle of the two hyperplanes $W \cdot X = 1$ and $\overline{W} \cdot \overline{X} = -1$. The two parallel hyperplanes to the decision boundary are shown in Figure 6.4(a). These hyperplanes are key because the distance between them is referred to as the *margin*, and the region between them reflects the zone of "uncertainty" near the decision boundary. It is undesirable to have too many points in this region, and therefore a training point $\overline{X_i}$ lying in this region is always penalized, even when it is correctly classified by virtue of satisfying $y_i = \text{sign}\{\overline{W} \cdot \overline{X_i}\}$ [or, equivalently $y_i(\overline{W} \cdot \overline{X_i}) > 0$]. Such correctly classified training points in the uncertain margin region satisfy $y_i(\overline{W} \cdot \overline{X_i}) \in (0, 1)$, and the corresponding penalty will be at most C. Other points on the incorrect side of the decision boundary can have arbitrarily large values of the penalty depending on their distance to the (relevant) margin hyperplane. The quantity $(1 - y_i(\overline{W} \cdot \overline{X_i})) > 0$ captures this "slack," and will be explicitly represented as a slack variable ξ_i later in this section. Four examples of penalized points are shown in Figure 6.4(a), all of which are circled. Note that the point A will be penalized even though it lies on the correct side of the decision boundary.

The contribution of the regularizer has a more interesting interpretation. The distance between the two hyperplanes $\overline{W} \cdot \overline{X} = 1$ and $\overline{W} \cdot \overline{X} = -1$ can be shown[7] to be $2/||\overline{W}||$ using elementary rules of coordinate geometry. Note that the regularizer is the squared inverse of this quantity, and therefore minimizing the regularizer is equivalent to increasing the distance between the two hyperplanes. Increasing the distance between the two hyperplanes is a natural way of achieving the goals of a regularizer because it discourages correctly classified training points from being too close to the decision boundary, which might be a result of overfitting. Therefore, one can recast the goals of the regularization and prediction

---

[7]http://mathworld.wolfram.com/Point-PlaneDistance.html

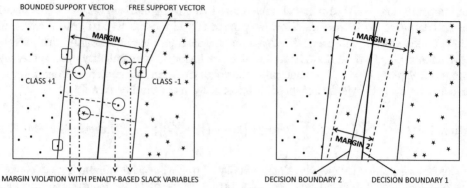

(a) Support vectors and slack penalties   (b) Trade-off between margin and slack penalties

Figure 6.4: Illustrating the notions of support vectors and margins in SVMs

errors in terms of the margin maximization principle as follows:

$$\text{Minimize } J = \underbrace{\frac{1}{2}||\overline{W}||^2}_{\text{Encourage Greater Margin}} + \underbrace{C \cdot \sum_{i=1}^{n} \max\{0, [1 - y_i(\overline{W} \cdot \overline{X_i})]\}}_{\text{Discourage Margin Violation}}$$

As in all regularized problems, there is a trade-off between the loss function and the regularizer. For example, in Figure 6.4(b), two possible sets of decision boundaries are shown. In one of them, the margin is undesirably thin, but there are only two penalties for margin violation. In the other case, the margins are thicker, but there are four penalties for margin violation. Which of these would the SVM optimization formulation select? If C is small, then it would go for the thick margin with increased regularization. If C is large, then the SVM would go for the thin margin with less regularization. In practice, parameters like C are chosen in a data-driven way by holding out a portion of the training data and selecting them in order to maximize the accuracy.

A key concept in SVM optimization is the notion of *support vectors*, based on which the SVM derives its name. An important point about SVM optimization is that each of the two separating hyperplanes on either side of the decision boundary could touch one or more training points *at optimality*. Such training data points are referred to as *free support vectors*. There are three free support vectors in the example of Figure 6.4(a). The notion of support vector naturally conveys the geometric interpretation of these training points "supporting" the hyperplanes on either side of the decision boundary. The training data points that are explicitly penalized for margin violation are also considered support vectors, but they are considered *bounded support vectors*. Since four points are penalized in Figure 6.4(a), there are four bounded support vectors. Note that a bounded support vector could either be a correctly classified training point within the margin region, or it could be a misclassified point inside/outside the margin region.

6.3.3   Pegasos: Solving SVMs in the Primal

Although dual formulations of SVMs are common, linear SVMs can be solved quite efficiently in the primal. As in the least-squares models, the first line of attack should be to examine if gradient-descent methods can be used on the original (i.e., primal) objective function. Unfortunately, the hinge-loss objective is not differentiable at a particular value of the vector $\overline{W}$, because of the presence of points satisfying the condition $y_i(\overline{W} \cdot \overline{X_i}) = 1$ in the training data. This problem is caused by the maximization function inside the loss term, $\max\{0, [1 - y_i(\overline{W} \cdot \overline{X_i})]\}$, of each point. For margin-violating points satisfying $y_i(\overline{W} \cdot \overline{X_i}) < 1$, the portion of the gradient contributed by these points is $-y\overline{X_i}$. For points satisfying $y_i(\overline{W} \cdot \overline{X_i}) > 1$ the contribution to the gradient is 0. The main uncertainty arises for points where the condition is exactly satisfied with equality, where the gradient is non-differentiable. In spite of this fact, a particular form of *mini-batch stochastic gradient descent* works very well, in which such non-differentiable points are dropped from the sampled set.

One such solution is *Pegasos* [512], which also has a sub-gradient interpretation. The approach randomly samples training points of *batch-size* s, and retains only those points in the batch violating the margin (i.e., satisfying $y_i(\overline{W} \cdot \overline{X_i}) < 1$). The gradient is updated with respect to only these retained points in each iteration. Since the points are selected based on margin violation, the differentiability of the objective function with respect to these points is guaranteed. The learning rate η_t in the tth iteration is set to $1/t$. The *Pegasos* algorithm starts by initializing $\overline{W}$ to a vector of 0s and then uses the following steps:

for $t = 1$ to T **do begin**
$\quad \eta_t = 1/t; \overline{W} \Leftarrow \overline{W}(1 - \eta_t);$
$\quad A_t = $ Random sample of s training pairs $(\overline{X_i}, y_i);$
$\quad A_t^+ = \{(\overline{X}, y) \in A_t : y(\overline{W} \cdot \overline{X}) < 1\};$
$\quad \overline{W} \Leftarrow \overline{W} + \frac{\eta_t \cdot n \cdot C}{s} \sum_{(\overline{X}, y) \in A_t^+} y\overline{X};$
$\quad \overline{W} \Leftarrow \min \left\{ 1, \frac{\sqrt{n \cdot C}}{\|\overline{W}\|} \right\} \overline{W}; \{ \text{Optional} \}$
endfor

Aside from the stochastic gradient update[8] step, the approach has an additional parameter shrinking step before the end of the iterative loop, which is optional. Another notable characteristic of *Pegasos* is in the bold nature of the step sizes, which are shown to converge fast. Aside from the step-size and shrinking innovations, these updates are *almost identical to those of a regularized perceptron* (cf. Equation 10.23 of Chapter 10), except that a perceptron defines A_t^+ as the set of all misclassified points satisfying $y(\overline{W} \cdot \overline{X}) < 0$ (without including the marginally correct points near the decision boundary). It has been shown in [512] that the number of iterations required depends on $O(C_0/\epsilon)$, where ϵ is the desired accuracy and $C_0 = n \cdot C$ is the *relative* weight of the slack penalty term compared to the regularization term after accounting for the effect of training data size. With careful handling of sparsity in the update process, the complexity of each update is $O(s \cdot q)$ where s is the (typically small) mini-batch size, and q is the average number of terms with non-zero frequency in each training example. In other words, the running time of the approach is independent of the training sample size, because one can assume that the relative weight C_0 is chosen in an insensitive way to training data size. The implementation of each update requires some care in handling sparsity.

Sparsity-Friendly Updates

This method is also particularly suitable for sparse domains like text, in which most entries of each $\overline{X_i}$ are 0s. Note that $\overline{W}$ might be a dense vector, whereas the vector added to it in

---

[8]On the surface, these steps look different from [512]. However, they are mathematically the same, except that the objective function uses different parametrizations and notations. The parameter λ in [512] is equivalent to $1/(n \cdot C)$ in this book.

each iteration from a small batch of s entries might be sparse. One wants the update time to be proportional to the number of non-zero entries in the sparse vector rather than the dense vector. A part of the problem is that some of the updates on $\overline{W}$ are multiplicative with respect to all the entries, which might require $O(d)$ time at first sight. One does not want to perform the multiplicative updates on each of the d elements of $\overline{W}$ explicitly because the value of d could easily be greater than 10^5. An important point is that multiplicative updates only affect a proportional scaling of the vector, which can be maintained separately from the relative values of its entries. In other words, one maintains two scalars θ and γ, and an unnormalized vector $\overline{V}$. The vector $\overline{W}$ is equal to $\theta\overline{V}$, and the norm of $\overline{W}$ is maintained in $\gamma = ||\overline{W}||$. Note that this is a redundant representation of $\overline{W}$ (because one is using $d+2$ values instead of d values to represent $\overline{W}$), but it helps in performing the additive and multiplicative portions of the update on different parts of the representation. An update is implemented as follows. First, θ and γ are multiplied with $(1 - \eta_t)$ to account for the multiplicative part of the update. Then, the relevant entries of $\overline{V}$ are updated with the additive quantity $\frac{\eta_t n \cdot C}{s \cdot \theta} \sum_{(\overline{X},y) \in A_t^+} y\overline{X}_i$. Note the use of θ in the denominator of the additive quantity so that $\overline{W} = \theta \cdot \overline{V}$ is appropriately updated. This additive update changes the value of γ, which can be updated[9] in time proportional to the sparsity level in the added quantity. Then, the multiplicative updates caused by the final shrinking step are used to update θ and γ. The final shrinking step is able to avoid the expensive computation of the norm of $\overline{W}$ because it is readily available in γ.

6.3.4 Dual SVM Formulation

The dual formulation of SVMs has been the dominant methodology for solving SVMs by historical accident [102], although there is no special reason to prefer the dual over the primal. In order to formulate the dual SVM, one first needs to explicitly introduce slack variables ξ_i in order to get rid of the maximization function in the objective. Such a restatement of the objective function results in the following constrained optimization problem:

$$\text{Minimize } J = \frac{1}{2}||\overline{W}||^2 + C \cdot \sum_{i=1}^{n} \xi_i$$

subject to:

$$\xi_i \geq 1 - y_i(\overline{W} \cdot \overline{X}_i) \quad \forall i \in \{1 \ldots n\} \quad \text{[Satisfied tightly for poorly separated points]}$$
$$\xi_i \geq 0 \quad \forall i \in \{1 \ldots n\} \quad \text{[Satisfied tightly for well-separated points]}$$

Intuitively, the slack variables ξ_i represent the amount by which the margin rules are violated, and they are penalized with C. The objective function therefore naturally tries to minimize each ξ_i. As a result, at least one of the two constraints involving ξ_i will be satisfied to equality (at optimality) depending on whether the training point is poorly separated (i.e., a support vector) or well separated (i.e., correctly classified outside margin hyperplanes).

A Lagrangian relaxation methodology is commonly used to solve such constrained optimization problems. We introduce two sets of Lagrangian parameters corresponding to the two sets of constraints. The margin violation constraints are assigned the Lagrangian parameters α_i, whereas the nonnegativity constraints are assigned the Lagrangian parameters

---

[9]When a sparse vector $\overline{a}$ is added to a dense vector $\overline{b}$, the change in the squared norm of $\overline{b}$ is $||\overline{a}||^2 + 2\overline{a} \cdot \overline{b}$. This can be computed in time proportional to the number of nonzero entries in the sparse vector $\overline{a}$.

γ_i. The Lagrangian relaxation J_L is as follows:

$$L_D = \text{Minimize } J_L = \frac{1}{2}||\overline{W}||^2 + \{C\sum_{i=1}^{n}\xi_i\} - \underbrace{\sum_{i=1}^{n}\alpha_i(\xi_i - 1 + y_i(\overline{W}\cdot\overline{X_i}))}_{\text{Relax margin rule}} - \underbrace{\sum_{i=1}^{n}\gamma_i\xi_i}_{\text{Relax }\xi_i \geq 0}$$

subject to:

$$\alpha_i \geq 0, \gamma_i \geq 0 \quad \forall i \in \{1\ldots n\} \quad [\text{ Since relaxed constraints are inequalities}]$$

In Lagrangian optimization, one wants to minimize the optimization problem at fixed values of the Lagrangian parameters, and then maximize this objective function with respect to all values of the Lagrangian parameters. Such a problem is referred to as the dual problem of the Lagrangian. In other words, we have:

$$L_D^* = \max_{\alpha_i,\gamma_i \geq 0} L_D = \max_{\alpha_i,\gamma_i \geq 0} \min_{\overline{W},\xi_i} J_L$$

For convex optimization problems like support vector machines, the solution of this rather odd optimization problem can be shown to be the same as the optimal solution of the original problem. Such a solution is referred to as the *saddle point* of the Lagrangian. The first step in finding the saddle point is to get rid of the minimization variables, so that we are left with a pure maximization problem in terms of the Lagrangian parameters. Therefore, one must set the partial derivatives with respect to the $\overline{W} = (w_1 \ldots w_d)$ and ξ_i to 0.

$$\nabla J_L = \overline{W} - \sum_{i=1}^{n}\alpha_i y_i \overline{X_i} = 0 \quad [\text{Gradient with respect to } \overline{W} \text{ is 0}] \tag{6.24}$$

$$\frac{\partial J_L}{\partial \xi_i} = C - \alpha_i - \gamma_i = 0 \quad \forall i \in \{1\ldots n\} \tag{6.25}$$

The first of these two constraints is particularly interesting because it shows that the co-efficients of the separating hyperplane can be fully expressed in terms of the training data points. Therefore, solving for α_i is sufficient to derive the separating hyperplane. Furthermore, one can even use α_i to *directly* provide a prediction $F(\overline{Z})$ of the test instance $\overline{Z}$ in terms of pairwise dot products between points:

$$F(\overline{Z}) = \text{sign}\{\overline{W}\cdot\overline{Z}\} = \text{sign}\{\sum_{i=1}^{n}\alpha_i y_i \overline{X_i}\cdot\overline{Z}\} \tag{6.26}$$

In order to eliminate the minimization variables, we substitute for $\overline{W}$ in the objective function. As an added bonus, we can also get rid of γ_i by substituting $\gamma_i = C - \alpha_i$ (based on Equation 6.25) to derive an objective function (and constraints) purely in terms of α_i. On substituting for these variables and simplifying, one can write the dual problem in maximization form as follows:

$$\text{Maximize } L_D = \left\{\sum_{i=1}^{n}\alpha_i\right\} - \frac{1}{2}\sum_{i=1}^{n}\sum_{j=1}^{n}\alpha_i\alpha_j y_i y_j (\overline{X_i}\cdot\overline{X_j})$$

subject to:

$$0 \leq \alpha_i \leq C \quad \forall i \in \{1\ldots n\}$$

Once we solve for α_i, the prediction function of Equation 6.26 can be used to classify a test instance. For linear SVMs, one can also derive the coefficient vector $\overline{W}$ using Equation 6.24. The dual formulation has the following properties:

1. The dual objective function and the prediction of Equation 6.26 can be expressed purely in terms of dot products without knowing the feature representations of the points. As we will see later, this fact has important consequences in order to use the approach for arbitrary data types.

2. The *Kuhn-Tucker optimality conditions* of the Lagrangian dual are obtained by setting the penalty terms in the Lagrangian relaxation to 0:

$$\alpha_i(\xi_i - 1 + y_i(\overline{W} \cdot \overline{X_i})) = 0$$
$$(C - \alpha_i)\xi_i = 0$$

Based on the Kuhn-Tucker optimality conditions, one can derive the following:

- Any point satisfying $y_i(\overline{W} \cdot \overline{X_i}) > 1$ (i.e., non-support vector) must satisfy $\alpha_i = 0$ because of the first Kuhn-Tucker condition and the nonnegativity of ξ_i. Furthermore, the second Kuhn-Tucker condition $(C - 0)\xi_i = 0$ ensures that $\xi_i = 0$ for non-support vectors. Such well-separated points are not penalized in the primal.

- Any point satisfying $y_i(\overline{W} \cdot \overline{X_i}) < 1$ (i.e., bounded/margin-violating support vector) must satisfy (i) $\xi_i > 0$, (ii) $\alpha_i = C$. These points are penalized in the primal objective function, because they are either too close to the decision boundary (on the correct side), or are on the incorrect side of the decision boundary.

- Points with $0 < \alpha_i < C$ are free support vectors and satisfy (i) $\xi_i = 0$, (ii) $y_i(\overline{W} \cdot \overline{X_i}) = 1$. These points are not penalized in the primal objective function, since slacks are 0. These points lie on the margin hyperplanes.

Points that are not support vectors do not contribute to either the primal or dual objective function value at optimality. This means that the well-separated points are redundant with respect to both the optimization objective and the constraints, and can be thrown away without affecting the optimal solution. This observation is often used in SVM optimization algorithms.

6.3.5   Learning Algorithms for Dual SVMs

In the following, we provide a generalized description of the dual solution by replacing dot products $\overline{X_i} \cdot \overline{X_j}$ with kernel similarity values $K(\overline{X_i}, \overline{X_j})$. This generalized description will be helpful in using support vector machines in the context of kernel methods.

$$\text{Maximize } L_D = \left\{ \sum_{i=1}^{n} \alpha_i \right\} - \frac{1}{2} \sum_{i=1}^{n} \sum_{j=1}^{n} \alpha_i \alpha_j y_i y_j K(\overline{X_i}, \overline{X_j})$$

$$\text{subject to:}$$
$$0 \leq \alpha_i \leq C \quad \forall i \in \{1 \ldots n\}$$

A natural solution is to use gradient ascent in which the n-dimensional vector of Lagrangian parameters is updated according to a gradient direction. The partial derivative of L_D with respect to α_k is as follows:

$$\frac{\partial L_D}{\partial \alpha_k} = 1 - y_k \sum_{s=1}^{n} y_s \alpha_s K(\overline{X_k}, \overline{X_s}) \tag{6.27}$$

This direction is used to update α_k. However, an update might lead to α_k violating the feasibility constraints. One possible solution to address this problem is reset the value of α_k to 0 if it becomes negative, and to reset it to C if it exceeds C. Therefore, one starts by setting the vector of Lagrangian parameters $\overline{\alpha} = [\alpha_1 \ldots \alpha_n]$ to an n-dimensional vector of 0s and uses the following update steps with learning rate η_k for the kth component:

> **repeat**
>   **for** each $k \in \{1 \ldots n\}$ **do begin**
>   Update $\alpha_k \Leftarrow \alpha_k + \eta_k \left[1 - y_k \sum_{s=1}^{n} y_s \alpha_s K(\overline{X_k}, \overline{X_s})\right]$;
> $\left\{ \text{Update is equivalent to } \alpha_k \Leftarrow \alpha_k + \eta_k \left[\frac{\partial L_D}{\partial \alpha_k}\right] \right\}$
> $\alpha_k \Leftarrow \min\{\alpha_k, C\}$;
> $\alpha_k \Leftarrow \max\{\alpha_k, 0\}$;
>   **endfor**;
> **until** convergence

The learning rate η_k for the kth component is set to $1/K(\overline{X_k}, \overline{X_k})$, because it causes the partial derivative of the objective with respect to α_k to fall to 0 after making this step. This result can be shown by replacing α_k with $\alpha'_k = \alpha_k + \eta_k(1 - y_k \sum_{s=1}^{n} y_s \alpha_s K(\overline{X_k}, \overline{X_s}))$ in Equation 6.27 (see Exercise 19). In the pseudo-code above, the values of all the α_k are not updated simultaneously, and the updated value of α_k is allowed to influence the updates of other components of $\overline{\alpha}$. This results in faster convergence.

The aforementioned algorithm is not optimized for efficiency. Efficiency can be improved by leveraging decomposition techniques that optimize with respect to only an active subset of Lagrangian variables at any given time [273, 424]. In such cases, the ideas on *Sequential Minimal Optimization (SMO)* [189, 440] restrict the working set of variables to a minimal value of 2. Some cutting plane algorithms like *SVMPerf* [274] are focused on constructing only linear models in sparse domains like text. The algorithm scales linearly with the number of non-zero entries in the document-term matrix.

6.3.6 Adaptive Nearest Neighbor Interpretation of Dual SVMs

The dual formulation of an SVM has an adaptive nearest-nearest interpretation. Consider the prediction function $F(\overline{Z})$ of test instance $\overline{Z}$ (which is introduced in Equation 6.26 and repeated below):

$$F(\overline{Z}) = \text{sign}\{\overline{W} \cdot \overline{Z}\} = \text{sign}\{\sum_{i=1}^{n} \alpha_i y_i \overline{X_i} \cdot \overline{Z}\} \tag{6.28}$$

It is useful to compare this equation with the adaptive nearest-neighbor prediction of Equation 5.29 in Chapter 5. The two prediction functions are *identical* because the weight λ_i in Equation 5.29 is analogous to the Lagrangian parameter α_i, and the similarity function $K(\overline{Z}, \overline{X_i})$ of Equation 5.29 is the dot product $\overline{X_i} \cdot \overline{Z}$. Well-separated data points are not support vectors, and therefore have $\alpha_i = 0$. Such points have no influence on the objective function. In other words, the SVM learns the relative importance of points using the Lagrangian parameters α_i, which results in throwing away the unimportant points (i.e., well-separated points). After throwing away the unimportant points, the SVM performs a weighted nearest-neighbor prediction on the remaining points, in which the weights correspond to the learned Lagrangian parameters. This is the basic principle of *adaptive* nearest neighbors in which some of the work in identifying "important" points or dimensions is done up front, rather than in a purely lazy fashion. Is there a way in which one can interpret the nature of the adaptivity learned by the dual? To understand this point, consider the only data-dependent term $-\sum_{i=1}^{n} \sum_{j=1}^{n} \alpha_i \alpha_j y_i y_j (\overline{X_i} \cdot \overline{X_j})$ in the dual objective L_D. This term is

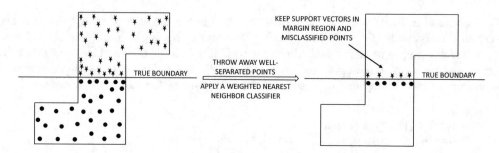

Figure 6.5: Support vector machines are adaptive nearest-neighbor methods. A support vector machine summarizes the data in a small number of support vectors, which contributes to its generalization power.

maximized when the weights of pairs of points $(\overline{X_i}, \overline{X_j})$ that belong to opposite classes (i.e., $y_i y_j = -1$) and are located close to one another (i.e., high $(\overline{X_i} \cdot \overline{X_j})$) also have large weights (α_i, α_j). In other words, points in "mixed-class regions" should have large weights, and these are precisely the uncertain points near the decision boundary. Furthermore, well-separated points have no influence at all. As we will see later, the shape of the decision boundary can be nonlinear (like a nearest-neighbor classifier) if we use something other than dot products as the similarity in the dual objective function. Consider, what happens when instead of using $\overline{X_i} \cdot \overline{X_j}$ as the similarity in the dual, we use a 0–1 similarity $K(\overline{X_i}, \overline{X_j})$ defining neighbors, which is 1 only if the similarity is greater than a threshold and 0, otherwise. In such a case, one can interpret the dual as roughly[10] solving the following problem:

$$\text{Maximize}_{\alpha_i} \sum_{\text{Opposite class neighbor pairs}} \alpha_i \cdot \alpha_j - \sum_{\text{Same class neighbor pairs}} \alpha_i \cdot \alpha_j$$

subject to:

Each nonnegative weight α_i is less than C

This optimization formulation will try to maximize the weights of points located in regions near other classes and will set of weights of points fully surrounded by same-class neighbors to 0. This will result in a subset of "uncertain" points together with point-specific weights. *The basic idea is that giving greater importance to uncertain points in the boundary region for nearest-neighbor prediction is more accurate than using a naïve implementation of nearest-neighbor classification.* This "importance weight" is learned in the dual parameters. We summarize this point below.

A support vector machine is an adaptive nearest-neighbor method.

The equivalence between a support-vector machine and an adaptive nearest-neighbor method is illustrated in Figure 6.5. The fact that most of the training points can be thrown away without changing the prediction means that SVMs have a more *concise* model compared to lazy nearest-neighbor methods. This type of *model compression* is how an adaptive nearest-neighbor classifier sometimes expresses itself. Compressed learning algorithms always have good generalization power to unseen test data because they do not have sufficient memory to remember irrelevant training data nuances.

This equivalence between the SVMs and nearest-neighbor methods also provides an intuitive explanation why one can capture nonlinear decision boundaries by changing the

---

[10]We say "roughly" because we are ignoring the data-independent term $\sum_{i=1}^{n} \alpha_i$.

dot product $\overline{X_i} \cdot \overline{Z}$ in both the optimization formulation and the prediction function to a weight that decays more sharply with distance than the dot product (e.g., Gaussian kernel). After all, weighted nearest-neighbor methods are also able to capture nonlinear boundaries when the weights are sharply decaying (cf. Section 5.4 of Chapter 5). Such kernel methods will be discussed in more detail in Section 6.5.

6.4 Logistic Regression

Logistic regression falls in a class of probabilistic models referred to as *discriminative models*. Such models assume that the dependent variable is an observed value generated from a probabilistic distribution defined by a function of the feature variables. First, we present a regularized optimization interpretation in order to relate it better to the other models discussed in this chapter.

6.4.1 The Regularized Optimization Interpretation

Consider a classification problem with training-test pairs $(\overline{X_1}, y_1) \ldots \overline{X_n}, y_n)$. Each class variables y_i is drawn from $\{-1, +1\}$. We start with the optimization formulation of least-squares classification in Equation 6.21 (which is treated as the "parent problem" throughout this chapter):

$$\text{Minimize } J = \frac{1}{2} \sum_{i=1}^{n} [1 - y_i(\overline{W} \cdot \overline{X_i})]^2 + \frac{\lambda}{2} ||\overline{W}||^2 \quad \text{[Regularized Least-Squares Classification]}$$

SVMs address an important criticism of the least-squares classification model (cf. Section 6.2.3.4), which is the fact that least-squares not only penalizes the points for being on the incorrect side of the decision boundary, but it also penalizes them for being too far on the correct side. SVMs do not penalize such points by setting negative values of the slack $[1 - y_i(\overline{W} \cdot \overline{X_i})]$ to 0. However, one unusual effect of this change is that there is no variation in the value of the objective for points that are sufficiently well separated. Logistic regression uses a smooth log loss, in which there is still some variation in the objective function value of such points. It is a debatable matter whether or not such a change will help the model; this is an issue that we will explore in Section 6.4.5.

One can write the objective function for logistic regression as follows:

$$\text{Minimize } J = \sum_{i=1}^{n} \log[1 + \exp\{-y_i(\overline{W} \cdot \overline{X_i})\}] + \frac{\lambda}{2} ||\overline{W}||^2 \qquad (6.29)$$

Here, the exponentiation function is denoted by "$\exp(\cdot)$." A key point here is that an increasing level of distance of a training point $\overline{X_i}$ from the decision boundary on the correct side, which is captured by increasingly positive values of $y_i(\overline{W} \cdot \overline{X_i})$, is penalized less by logistic regression (albeit with smoothly diminishing returns). This is the opposite of least-squares classification, where it is increasingly penalized beyond a particular point. In support-vector machines, increasing distance in the correct direction from the decision boundary beyond a particular point (i.e., margin boundary) is neither rewarded nor penalized.

To show the differences between the various loss functions, we have plotted (cf. Figure 6.6) the penalty at varying values of $\overline{W} \cdot \overline{X}$ of a positive training point $\overline{X}$ with label $y = +1$. The three loss functions of regularized least-squares classification, SVM, and logistic regression are shown. The loss functions of logistic regression and the support vector

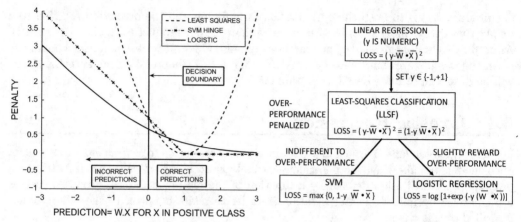

(a) Loss functions of various linear models   (b) Relationships among linear models

Figure 6.6: (a) The loss for a training instance $\overline{X}$ belonging to the positive class at varying values of $\overline{W} \cdot \overline{X}$. Logistic regression and SVM are similar except that the former is smooth, whereas the latter abruptly flattens out beyond the margin point with $\overline{W} \cdot \overline{X} \geq 1$. Least-squares classification is the only case in which the penalty *increases* in some regions with increasing $\overline{W} \cdot \overline{X}$ for the (positive) class training instance. (b) All linear models in classification derive their motivation from the parent problem of linear regression, which historically precedes the classification formulations. The modifications treat the well-separated (i.e., over-performing) points in different ways.

machine look strikingly similar, except that the former is a smooth function, and the SVM sharply bottoms at zero loss beyond $\overline{W} \cdot \overline{X} \geq 1$. This similarity in loss functions is important, because it explains why the two models seem to provide similar results in many practical cases. The regularized least-squares model, which is equivalent to the Fisher discriminant, provides a very different loss function. In fact, this is the only loss function where there is a region of the space in which increasing $\overline{W} \cdot \overline{X}$ actually increases the penalty on the point. One consequence of the smooth objective function of logistic regression is that it considers all points including well-separated points relevant to the model, albeit to a smaller degree. As a result, the model no longer throws away most of the points (like SVMs). Furthermore, unlike SVMs, logistic regression is commonly used in the linear setting. This is not a problem in the specific case of the text domain, where linear models are recommended anyway. Although it is possible to design nonlinear variants of logistic regression, SVMs are generally preferable in those settings.

As logistic regression also has a probabilistic interpretation, it turns out that one can perform the prediction $F(\overline{Z})$ of a test instance both deterministically as well as in a probabilistic sense. The deterministic prediction is identical to an SVM, but the probabilistic prediction is unique[11] to logistic regression.

$$F(\overline{Z}) = \text{sign}\{\overline{W} \cdot \overline{Z}\} \text{ [Deterministic Prediction]}$$

$$P(F(\overline{Z}) = 1) = \frac{1}{1 + \exp(-\overline{W} \cdot \overline{Z})} \text{ [Probabilistic Prediction]}$$

---

[11]It has been shown [441] how one can derive heuristic probability estimates with an SVM.

It is noteworthy that points on the decision boundary satisfying $\overline{W} \cdot \overline{Z} = 0$ will be predicted to a probability of $1/(1 + \exp(0)) = 0.5$, which is a reasonable prediction. The probabilistic predictions in logistic regression can be learned using stochastic gradient descent.

6.4.2 Training Algorithms for Logistic Regression

In order to derive the stochastic gradient-descent iterations for logistic regression, let us consider the gradient ∇J of its objective function J with respect to $\overline{W}$:

$$\nabla J = \lambda \overline{W} - \sum_{i=1}^{n} \frac{y_i \exp\{-y_i(\overline{W} \cdot \overline{X_i})\} \overline{X_i}}{1 + \exp\{-y_i(\overline{W} \cdot \overline{X_i})\}} \tag{6.30}$$

For mini-batch *stochastic* gradient descent, only the gradient with respect to a subset A of s randomly chosen training instances is considered. We can write the corresponding gradient as follows:

$$\nabla J = \frac{\lambda s}{n} \overline{W} - \sum_{(\overline{X_i}, y_i) \in A} \frac{y_i \exp\{-y_i(\overline{W} \cdot \overline{X_i})\} \overline{X_i}}{1 + \exp\{-y_i(\overline{W} \cdot \overline{X_i})\}} \tag{6.31}$$

Choosing $s = 1$ leads to pure stochastic gradient descent. One can use these updates to design the mini-batch stochastic gradient-descent algorithm for logistic regression, starting with $\overline{W} = 0$ and updating for T iterations with learning rate η as follows:

> **for** $t = 1$ to T **do begin**
> $A_t = $ Random sample of s training pairs $(\overline{X_i}, y_i)$;
> $\overline{W} \Leftarrow \overline{W}\left(1 - \frac{\eta \lambda s}{n}\right) + \eta \sum_{(\overline{X}, y) \in A_t} \frac{y \exp\{-y(\overline{W} \cdot \overline{X})\} \overline{X}}{1 + \exp\{-y(\overline{W} \cdot \overline{X})\}}$;
> **endfor**

The reader is encouraged to examine the similarity of this update process to the *Pegasos* algorithm described in Section 6.3.3. The main differences arise in the handling of well-separated points and choice of learning rate. For simplicity, we have used a constant learning rate η. There are several other techniques like the Newton method that are used for fast convergence in logistic regression.

6.4.3 Probabilistic Interpretation of Logistic Regression

Logistic regression is a member of the family of *generalized linear models*, which have a natural probabilistic interpretation. Although logistic regression is designed to deal with binary dependent variables, the family of generalized linear models can handle dependent variables of all types like ordinal data (ratings), categorical data, and count occurrence data. Both SVMs and logistic regression use different ways of modifying least-squares regression to the binary nature of the dependent variable. Logistic regression is more systematic in the sense that the ideas can be adapted to other types of target variables.

In essence, logistic regression assumes that the target variable $y_i \in \{-1, +1\}$ is the observed value generated from a hidden Bernoulli probability distribution that is parameterized by $\overline{W} \cdot \overline{X_i}$. Since $\overline{W} \cdot \overline{X_i}$ might be an arbitrary quantity (unlike the parameters of a Bernoulli distribution), we need to apply some type of function to it in order to bring it to the range $(0, 1)$. The specific function chosen is the sigmoid function. In other words, we have:

$$y_i \sim \text{Bernoulli distribution parametrized by sigmoid of } \overline{W} \cdot \overline{X_i}$$

It is this probabilistic interpretation because of which we get our prediction function $F(\overline{Z})$ for a given data point $\overline{Z}$:

$$P(F(\overline{Z}) = 1) = \frac{1}{1 + \exp(-\overline{W} \cdot \overline{Z})}$$

One can write this prediction function more generally for any target $y \in \{-1, +1\}$.

$$P(F(\overline{Z}) = y) = \frac{1}{1 + \exp(-y(\overline{W} \cdot \overline{Z}))} \tag{6.32}$$

It is easy to verify that the sum of the probabilities over both outcomes of y is 1.

The key here is that if we have another type of target variable (e.g., categorical, multinomial, or ordinal), we can choose to use a different type of distribution and a different function of $\overline{W} \cdot \overline{X_i}$ to define the parametrization of the hidden probabilistic process. The ability to handle arbitrary types of target variables is where the real power of this family of generalized linear models is derived.

Probabilistic models learn the parameters of the probabilistic process in order to maximize the likelihood of the data. The likelihood of the entire training data set with n pairs of the form $(\overline{X_i}, y_i)$ is as follows:

$$\mathcal{L}(\text{Training Data}|\overline{W}) = \prod_{i=1}^{n} P(F(\overline{X_i}) = y_i) = \prod_{i=1}^{n} \frac{1}{1 + \exp(-y_i(\overline{W} \cdot \overline{X_i}))}$$

One must maximize the likelihood and minimize the negative log-likelihood. Therefore, the minimization objective function $\mathcal{LL}$ of the log-likelihood can be expressed by using the negative logarithm of the aforementioned expression:

$$\mathcal{LL} = \sum_{i=1}^{n} \log[1 + \exp\{-y_i(\overline{W} \cdot \overline{X_i})\}] \tag{6.33}$$

After adding the regularization[12] term, this (negative) log-likelihood function is *identical* to the objective function of logistic regression in Equation 6.29. Therefore, logistic regression is essentially a (negative) log-likelihood minimization algorithm.

6.4.3.1   Probabilistic Interpretation of Stochastic Gradient Descent Steps

Most gradient-descent models are *mistake-driven methods*, in that the update step is often a function of the errors made on the training data. In order to understand this point, note that the gradient-descent steps for least-squares regression in Section 6.2.1.1 are direct functions of errors made on the training data. How do the updates in logistic regression compare to this characteristic of other methods? Let us examine an update made by stochastic gradient descent on a subset of points A_t in the tth iteration (see pseudocode on page 189):

$$\overline{W} \Leftarrow \overline{W}\left(1 - \frac{\eta\lambda s}{n}\right) + \eta \sum_{(\overline{X}, y) \in A_t} \frac{y \exp\{-y(\overline{W} \cdot \overline{X})\}\overline{X}}{1 + \exp\{-y(\overline{W} \cdot \overline{X})\}}$$

$$= \overline{W}\left(1 - \frac{\eta\lambda s}{n}\right) + \eta \sum_{(\overline{X}, y) \in A_t} y \left\{P(F(\overline{X}) = -y)\right\} \overline{X}$$

$$= \overline{W}\left(1 - \frac{\eta\lambda s}{n}\right) + \eta \sum_{(\overline{X}, y) \in A_t} y \left\{P\left[\text{Mistake on } (\overline{X}, y)\right]\right\} \overline{X}$$

Therefore, logistic regression is also a mistake-driven method, and the *probabilities* of the mistakes are used. This is in consonance with the fact that logistic regression is a probabilistic method.

---

[12]Regularization is equivalent to assuming that the parameters in $\overline{W}$ are drawn from a Gaussian prior and it results in the addition of the term $\lambda||\overline{W}||^2/2$ to the log-likelihood to incorporate this prior assumption.

6.4.3.2 Relationships among Primal Updates of Linear Models

SVMs replace $P\left[\text{Mistake on }(\overline{X}, y)\right]$ with a 0/1 value in the probabilistic update of the previous section, depending on whether or not the point $(\overline{X}, y)$ meets the margin requirement. In fact, it is possible to write a unified form of the update for least-squares classification, SVM, and logistic regression. This form of the update is as follows:

$$\overline{W} \Leftarrow \overline{W}(1 - \eta\lambda) + \eta y[\delta(\overline{X}, y)]\overline{X}$$

Here, the *mistake function* $\delta(\overline{X}, y)$ is an error value for least-squares classification, an indicator variable for SVMs, and a probability for logistic regression (see Exercise 15). The close relationships among the updates mirror the close relationships among their loss functions (cf. Figure 6.6). Remarkably, the perceptron update is identical to the SVM update, but with a different definition of the indicator variable (cf. page 322).

6.4.4 Multinomial Logistic Regression and Other Generalizations

The probabilistic interpretation of logistic regression is particularly convenient because it provides a path to modeling target variables of other types with the use of *generalized linear models*. After all, the whole point of the probabilistic process in logistic regression is to convert the continuous value $\overline{W} \cdot \overline{X}_i$ into a binary prediction y_i with a probabilistic interpretation. In the case of the k-class problem, the target variable y_i is generated as follows:

$$y_i \sim \text{Target-sensitive distribution parametrized by functions of } \overline{W}_1 \cdot \overline{X}_i \dots \overline{W}_k \cdot \overline{X}_i$$

The choice of the distribution above depends on the type of target variable (i.e., dependent variable) one is trying to learn. In the aforementioned setting, the target variable has k *categorical* values denoted by $\{1 \dots k\}$. Therefore, the classes have probability distributions defined by the following:

$$P(y_i = r|\overline{X}_i) = \frac{\exp(\overline{W}_r \cdot \overline{X}_i)}{\sum_{m=1}^{k} \exp(\overline{W}_m \cdot \overline{X}_i)} \quad \forall r \in \{1 \dots k\} \tag{6.34}$$

As in the case of logistic regression one learns the parameters in $\overline{W}_1 \dots \overline{W}_k$ by maximizing the likelihood of the observed targets on the training data. Specifically, the loss function is also referred to as the *cross-entropy loss*:

$$\mathcal{LL} = -\sum_{i=1}^{n}\sum_{r=1}^{k} I(y_i, r) \cdot \log\left[P(y_i = r|\overline{X}_i)\right] \tag{6.35}$$

Here, the indicator function $I(y_i, r)$ is 1 when the observed value of y_i is r, and 0, otherwise. Therefore, the approach for learning the multiclass parameters is different only in the specific details of the maximum-likelihood function, and the principles of the overall framework remain unchanged from logistic regression. One can use the following stochastic gradient-descent steps for *each W_r*, when trained on $(\overline{X}_i, y_i)$:

$$\overline{W}_r \Leftarrow \overline{W}_r(1 - \eta\lambda) + \eta\overline{X}_i\left[I(y_i, r) - P(y_i = r|\overline{X}_i)\right] \quad \forall r \in \{1 \dots k\} \tag{6.36}$$

Here, η is the step size and λ is the regularization parameter. The reader should convince herself that the special case of the multinomial objective function (Equation 6.35) for binary classes turns out to be identical to logistic regression (see Exercise 13).

In essence, the approach is learning k different linear separators simultaneously, and each separator tries to discriminate a particular class from the remaining data. This bears some resemblance to a one-against-all approach (see Section 6.1.4), which is often used to convert binary classifiers like support vector machines to multi-way classifiers by voting on different predictions obtained by building such models separately. However, the difference is that the separators are learned *simultaneously* in multinomial logistic regression by *jointly optimizing* training log-likelihood with respect to all k classes *at once*. This results in a more flexible model rather than a decomposable one-against-all approach, which is done *sequentially* after learning each $\overline{W_r}$ individually. This model is also referred to as *multinomial logistic regression, maximum entropy (MaxEnt)*, or the *softmax* model. One can also use appropriate distributions to model count-occurrence data (with a multinomial distribution), or ratings data (with an ordered probit model). Refer to the bibliographic notes for pointers on generalized linear models. It is noteworthy that using different linear separators simultaneously can also be achieved in other binary models like SVMs. For example, it is possible to design a *multi-class SVM loss function*, known as the *Weston-Watkins SVM* [579], that learns k different separators simultaneously (see Exercise 14). However, the SVM is not quite as flexible as the family of generalized linear models in handling different types of target variables.

6.4.5   Comments on the Performance of Logistic Regression

Logistic regression has very similar performance to that of support vector machines. This is because the loss functions of the two methods are very similar. In fact, in highly noisy data sets with overlapping class distributions, linear logistic regression may slightly outperform a linear support vector machine. Support vector machines tend to do well when the classes are well separated. One reason for this is that support vector machines always include misclassified training points among the support vectors. Therefore, if the data set contains a large number of mislabeled points or other intrinsic noise, it can affect the SVM classification to a larger extent. This is caused by the fact that the SVM throws away a lot of the correctly labeled points for not being support vectors. Therefore, the misclassified training points occupy an even larger proportion of the support vectors retained by the SVM model. In these specific cases, the smooth objective function of logistic regression might provide some protection because it gives some weight to all correctly labeled points in the loss function (albeit a small amount to the well separated ones) to balance out the noise. However, even in these cases, the performance of the SVM is often statistically comparable to logistic regression, provided that the regularization parameters are properly tuned.

A difficult case for logistic regression is that of well-separated classes in which support vector machines generally provide superior performance. In such cases, logistic regression methods tend to become unstable in terms of their probability estimates. However, they can usually be used to reasonably classify the test instances even if the probability estimates are poor. Note that well separated classes are an easy case, and many classifiers can be used to solve such cases. In summary, it is often difficult to choose between SVMs and logistic regression. Multinomial variations of logistic regression often have an advantage in multi-way classification because of the ability to build a more powerful model with multiple classes. If nonlinear models are desired, then the support vector machine is the method of choice. This will be the subject of discussion in the next section.

6.5 Nonlinear Generalizations of Linear Models

Nonlinear methods for classification use linear models on a transformation of the data that is defined by kernel singular value decomposition (SVD). Therefore, a simplistic way to implement nonlinear models is as follows:

1. Transform the n training data points in d-dimensional space to a new representation D'. For a finite data set of n points, a data-specific representation of at most n dimensions can always be found. The n-dimensional representation is contained in the n rows of the $n \times n$ matrix U by diagonalizing an appropriately chosen $n \times n$ similarity matrix S between the points to express it in the form $S = UU^T$.

2. Apply any linear model (e.g., Fisher discriminant, SVM, or logistic regression) on the transformed representation of the training data in the rows of U to create a model.

3. For any test point, transform it to the same space as the training data, and apply the learned model on the transformed representation to predict its class label.

The basic idea is that a linear separator in the transformed space maps to a nonlinear separator in the original space. Although this crude way of implementing kernel classification is not what is done in practice, it is *identical* to what is achieved using methods like the kernel trick, which will be discussed later. Before reading further, the reader is advised to revisit the material in Section 3.6 of Chapter 3 on kernel SVD. Kernel SVMs are *direct* applications of this transformation.

Singular value decomposition can recover[13] the original data representation from an $n \times n$ similarity (i.e., dot product) matrix by computing its top eigenvectors. Any data matrix D can be recovered (in a rotated and de-correlated axis system) using the eigenvectors of its $n \times n$ dot product matrix $S = DD^T$. One can diagonalize S as follows:

$$S = Q\Sigma^2 Q^T = \underbrace{(Q\Sigma)}_{U} \underbrace{(Q\Sigma)^T}_{U^T} \tag{6.37}$$

The matrix U will have at most d nonzero columns when S contains dot products, because at most $\min\{n, d\}$ entries (SVD singular values) of the diagonal matrix Σ are non-zero. The remaining $(n - d)$ dimensions of U can be dropped. In such a case, the matrix U contains the d-dimensional embedding of all n points, which are returned by traditional SVD. Now imagine that you replaced the dot product in the (i, j)th entry of $S = DD^T$ with another kernel similarity value $K(\overline{X_i}, \overline{X_j})$ such as one of the following:

Function	Form
Linear Kernel	$K(\overline{X_i}, \overline{X_j}) = \overline{X_i} \cdot \overline{X_j}$ (Defaults to rotated/reflected version of original data as in SVD)
Gaussian Radial Basis Kernel	$K(\overline{X_i}, \overline{X_j}) = \exp(-\|\overline{X_i} - \overline{X_j}\|^2/(2 \cdot \sigma^2))$
Polynomial Kernel	$K(\overline{X_i}, \overline{X_j}) = (\overline{X_i} \cdot \overline{X_j} + c)^h$
Sigmoid Kernel	$K(\overline{X_i}, \overline{X_j}) = \tanh(\kappa \overline{X_i} \cdot \overline{X_j} - \delta)$

The basic idea is that these kernel similarities represent the dot products between data points in transformed space with unknown transformation $\Phi(\cdot)$:

$$K(\overline{X_i}, \overline{X_j}) = \Phi(\overline{X_i}) \cdot \Phi(\overline{X_j}) \tag{6.38}$$

---

[13]The data will typically be rotated and reflected in particular directions.

The extraction of the nonzero eigenvectors of any of the similarity matrices above will yield an n-dimensional representation $\Phi_s(\overline{X})$ of the transformed data. Consider the case, where for any of the $n \times n$ similarity matrices above, if we extract all nonzero eigenvectors using the same approach as above:

$$S = Q\Sigma^2 Q^T = \underbrace{(Q\Sigma)}_{U} \underbrace{(Q\Sigma)^T}_{U^T} \tag{6.39}$$

In this case, the n rows of U provide the data-specific[14] transformed representation $\Phi_s(\overline{X})$, and it is possible for U to have more than d nonzero columns. In other words, the transformation can have higher dimensionality than the original data. The linear kernel is a special case in which we obtain a rotated and reflected version of the original data with at most d nonzero dimensions. For many kernels, this higher-dimensional representation unlocks the local clustering characteristics of the data along the different transformed dimensions, and the clusters (or classes) now become linearly separable. Therefore, it would make sense to use linear SVM on $\Phi_s(\overline{X})$ rather than the original data.

6.5.1   Kernel SVMs with Explicit Transformation

Even though it is uncommon to implement kernel SVMs with explicit transformation, it is possible to do so. For the purpose of discussion, assume that the eigenvectors and eigenvalues of the $n \times n$ kernel similarity matrix S are denoted by Q and Σ (see previous section). One can drop the zero eigenvectors (columns) of Σ and Q to yield the $n \times r$ matrix $U_0 = Q_0 \Sigma_0$ with $r < n$ dimensions. The rows of U_0 contain the *explicit* transformations of the training points. Any *out-of-sample* test point $\overline{Z}$ can also be projected into this r-dimensional representation by observing that its dot products with training points must evaluate to the corresponding kernel similarities between the test and training points:

$$\underbrace{\Phi_s(\overline{Z})}_{1 \times r} \underbrace{(Q_0 \Sigma_0)^T}_{r \times n} = \underbrace{[K(\overline{Z}, \overline{X_1}), K(\overline{Z}, \overline{X_2}), \dots K(\overline{Z}, \overline{X_n})]}_{1 \times n \text{ row vector of similarities}} \tag{6.40}$$

Multiplying both sides with $Q_0 \Sigma_0^{-1}$ and using $Q_0^T Q_0 = I$ on the left-hand side, we obtain:

$$\Phi_s(\overline{Z}) = [K(\overline{Z}, \overline{X_1}), K(\overline{Z}, \overline{X_2}), \dots K(\overline{Z}, \overline{X_n})]Q_0 \Sigma_0^{-1} \tag{6.41}$$

The point $\Phi_s(\overline{Z})$ contains the r-dimensional data-specific transformation of the test point in the same r-dimensional space as the training data was transformed. Therefore, we present the algorithm for kernel SVMs (with *explicit* transformation starting from the training data similarity matrix S) as follows:

---

[14]Strictly speaking, the transformation $\Phi(\overline{X})$ would need to be infinite dimensional to adequately represent the universe of *all possible* data points for Gaussian kernels. However, the relative positions of n points (and the origin) in any dimensionality can always be projected on an n-dimensional plane, just as a set of a two 3-dimensional points (with the origin) can always be projected on a 2-dimensional plane. The eigenvectors of the $n \times n$ similarity matrix of these points provide precisely this projection. This is referred to as the *data-specific* Mercer kernel map. Therefore, even though one often hears of the impossibility of extracting infinite dimensional points from a Gaussian kernel, this makes the nature of the transformation sound more abstract and impossible than it really is (as a practical matter). The reality is that we can always work with the data-specific n-dimensional transformation. As long as the similarity matrix is positive semi-definite, a finite dimensional transformation always exists for a finite data set, which is adequate for the learning algorithm. We use the notation $\Phi_s(\cdot)$ instead of $\Phi(\cdot)$ to represent the fact that this is a data-specific transformation.

Diagonalize $S = Q\Sigma^2 Q^T$;
Extract the n-dimensional embedding in rows of $Q\Sigma$;
Drop any zero eigenvectors from $Q\Sigma$ to create $Q_0\Sigma_0$;
{ The n rows of $Q_0\Sigma_0$ and their class labels constitute training data }
Apply linear SVM on $Q_0\Sigma_0$ and class labels to learn model $\mathcal{M}$;
Convert test point $\overline{Z}$ to $\Phi_s(\overline{Z})$ using Equation 6.41;
Apply $\mathcal{M}$ on $\Phi_s(\overline{Z})$ to yield prediction;

In other words, kernel SVMs *can be* implemented with *explicit* transformation. Furthermore, one can substitute the SVM with any learning algorithm like logistic regression or Fisher discriminant. Note that this approach is applicable across the entire spectrum of supervised and unsupervised learning algorithms. An unsupervised example of kernel-based k-means clustering algorithm (with explicit feature transformation) is described in Section 4.8.1.2 of Chapter 4. In the form described above, the explicit transformation approach is highly inefficient, because the extracted representation might require $O(n^2)$ space for the matrix $U_0 = Q_0\Sigma_0$. This is the (practical) reason that one resorts to the kernel trick, which is discussed later in this chapter. The kernel trick provides an *equivalent solution* to that provided by the pseudo-code above.

However, explicit transformations with kernels have a worse reputation than they deserve. It is noteworthy that one can use Nyström sampling (cf. Section 3.6.2 of Chapter 3) in combination with ensemble tricks to improve the efficiency *and* the accuracy of this approach. In fact, this type of sampling approach (explicit transformation) has many benefits, but is often underappreciated by researchers and practitioners alike. Such a sampling-based approach is described for clustering in Sections 4.7 and 4.8.1.2. We leave the implementation for classification to the reader (see Exercise 12). It is also useful to explore the explicit transformation approach, because it provides an understanding of *how* kernels improve the separability of the different classes in transformed space. This will be the topic of the discussion in the next section.

6.5.2   Why do Conventional Kernels Promote Linear Separability?

Conventional kernels like the Gaussian kernel transform the data into a higher-dimensional space in which the points of different classes become linear separable. As discussed in the previous section, these transformations actually *expand* the dimensionality of the embedded data over the representation in the input space. An expanded dimensionality leads to a greater number of ways to separate two sets of points, and therefore a linear separator is easier to find. Even better insight might be obtained by examining the *way in which* this larger number of dimensions is used. The key point is that embedded kernel can capture the local clustering (i.e., class) structure of the data in dedicated subsets of engineered features that are often disjoint from one another. In order to understand this point, consider a case in which the text documents are all normalized to unit norm. Then, the dot product between any pair of documents $(\overline{X_i}, \overline{X_j})$ can be expressed in terms of the squared Euclidean distance R^2 between them:

$$\overline{X_i} \cdot \overline{X_j} = \frac{||\overline{X_i}||^2 + ||\overline{X_j}||^2 - ||\overline{X} - \overline{Y}||^2}{2}$$

$$= \frac{1 + 1 - R^2}{2} = 1 - R^2/2$$

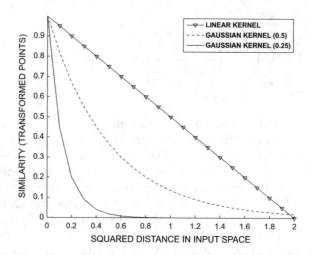

Figure 6.7: Similarities between pairs of points in the transformed space are very sensitive to the squared distances in the input space, when a small bandwidth of the Gaussian kernel is used.

Conventional kernels can therefore be expressed in terms of R^2 as follows:

$$\overline{X_i} \cdot \overline{X_j} = 1 - R^2/2 \quad \text{[Linear Kernel]}$$
$$(\overline{X_i} \cdot \overline{X_j})^2 = (1 - R^2/2)^2 \quad \text{[Quadratic Kernel]}$$
$$\exp(-\|\overline{X_i} - \overline{X_j}\|^2/(2 \cdot \sigma^2)) = \exp(-R^2/2\sigma^2) \quad \text{[Exponential Kernel]}$$

In each case, it is evident that higher-order kernel similarities in the transformed space decay much more sharply than the dot product with increasing distance in the input space. In Figure 6.7, it is shown how the similarity values of the Gaussian kernel (i.e., dot products in transformed space) vary with different values of the squared distance R^2 in the input space. For the Gaussian kernel, two different values of σ at 0.25 and 0.5 are used. It is immediately evident that the drop is much sharper with higher-order kernels and small bandwidths. In such cases, the similarity is almost zero between many pairs of points. Since nonnegative kernels like the Gaussian can always be assumed[15] to create a nonnegative embedding in a single orthant, the only way in which the similarity between a pair of transformed points is zero if they take positive components along different dimensions. In other words, kernels like the Gaussian map distant points in the input space to different dimensions, and they map closely clustered points (typically belonging to the same class) to a dedicated subset of dimensions. With the right choice of the bandwidth σ, different classes will be dominated by different subsets of dimensions in the transformed space, which promotes linear separability. However, this linear separator in the transformed space maps to a nonlinear separator in the original input space.

In order to explain this point, we revisit an example from Chapter 3 in Figure 6.8. In this case, the data is segmented into three classes, corresponding to *Arts*, *Crafts*, and *Music*. Suppose, we want to separate *Arts* from the other classes. It is evident that a linear separator cannot separate out this class, because it is tightly integrated with the other classes. Now imagine that you use a Gaussian kernel to transform the data. If a sufficiently

---

[15]When all entries in the kernel matrix are nonnegative, it means that all pairwise angles between points are less than 90°. One can always reflect the points to the nonnegative orthant without loss of generality.

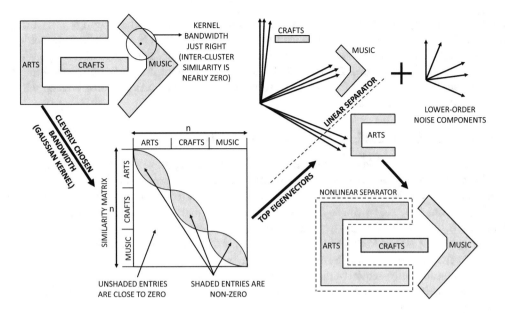

Figure 6.8: Linear separators in transformed space often serve the purpose of nonlinear separators in the input space. Refer to Figure 3.9 of Chapter 3 to relate the approach to kernel SVD.

small bandwidth is chosen, the similarity between pairs of points belonging to different classes will be close to zero, although there will always be pairs of points within the same class that have high similarity to one another. As a result, the populated entries in the similarity matrix might look like the ones shown in Figure 6.8. The only way in which such a matrix can be represented as dot products of points is the case in which different dimensions of this embedding are dominated by the different classes. In such a case, a linear separator will be able to separate the *Arts* class from the other classes. Note that this linear separator in the transformed space corresponds to a nonlinear separator in the original input space. In essence, such transformation methods are designed to unlock the local information captured by combinations of (input) dimensions into individual (transformed) dimensions. A key point is that it is crucial to tune the parameters of the kernel (e.g., bandwidth σ) appropriately in order to get the best classification performance.

6.5.3 Strengths and Weaknesses of Different Kernels

Conventional kernels like the Gaussian kernel and the polynomial kernel have had only limited success in the text domain. One issue is that text data is sparse and high-dimensional, and such data domains are often linearly separable to a large degree. Note that the Gaussian kernel and polynomial kernels will usually provide slightly better performance than linear classifiers with sufficient tuning, because choosing a large bandwidth in the Gaussian kernel is (almost) equivalent to the linear kernel. Therefore, with sufficient tuning of bandwidth an operating point can usually be found where the nonlinear kernel wins over the linear kernel. The main problem is that the nonlinear variations of most SVM algorithms are computationally expensive compared to the linear variations, and the small accuracy advantages may not be worth the additional effort. The other point to keep in mind is that one now needs to tune two parameters (corresponding to regularization and kernel parameters), which requires a more expensive grid search[16] for parameter tuning. This further increases computational costs. If the grid search is not exhaustive enough, it is possible for the nonlinear kernel method to perform worse than the linear kernel in which it is easier to tune effectively. Second-order polynomial kernels can provide modest improvements [101] in accuracy because they capture the interactions between pairs of terms, although the advantages are still quite limited.

Capturing Linguistic Knowledge with Kernels

The main potential of kernels lies in its ability to incorporate the linguistic knowledge in the corpus for classification. In this context, substring kernels [353] use the sequential positioning of words in order to capture deeper semantic concepts from the data than are available from the bag-of-words representation. A number of such kernels are discussed in Chapter 3. The ability to incorporate semantic and linguistic concepts directly into the model by using string kernels is a powerful notion. In the longer run, such settings may be the primary use case for kernel methods in text, although much research needs to be done on linguistically-sensitive similarity learning in this domain. Truly cognitive forms of artificial intelligence require the ability to integrate sequence-based learning models into the classification process.

6.5.4   The Kernel Trick

As discussed earlier, the transformation $\Phi(\overline{X})$ is obtained by using an $n \times n$ similarity matrix S containing all pair-wise similarities $\Phi(\overline{X_i}) \cdot \Phi(\overline{X_j})$ in transformed space. One way of using kernel methods is to extract the data-specific Mercer kernel map $\Phi_s(\overline{X})$ by diagonalizing the $n \times n$ similarity matrix S and then building a linear model on the extracted representation. However, in many cases, if the solutions to a linear model can be expressed in terms of dot products, it is not necessary to explicitly perform this feature engineering. In such cases, replacing dot products with similarities provide identical results to explicit feature engineering. So, the essence of the kernel trick is as follows:

> Create a closed-form solution or optimization formulation that is defined in terms of dot products. Also derive a form of the test instance prediction function in terms of the dot products of the test instance with other training instances. Now replace all dot products with entries of the similarity matrix S.

Several sections of this chapter show how the training as well as prediction of many linear models can be expressed in terms of dot products. For example, consider the dual of the support vector machine introduced earlier in this chapter. The dual can be expressed as follows:

$$\text{Maximize } L_D = \left\{ \sum_{i=1}^{n} \alpha_i \right\} - \frac{1}{2} \sum_{i=1}^{n} \sum_{j=1}^{n} \alpha_i \alpha_j y_i y_j (\overline{X_i} \cdot \overline{X_j})$$

subject to:

$$0 \leq \alpha_i \leq C \quad \forall i \in \{1 \ldots n\}$$

It is evident that this dual only contains dot products like $\overline{X_i} \cdot \overline{X_j}$ between training data pairs. We can replace this dot product with kernel similarity (e.g., the similarity obtained

---

[16]Suppose one has $p_1 \ldots p_t$ different possibilities for t different parameters. One now has to evaluate the algorithm at each combination of $p_1 \times p_2 \ldots \times p_t$ possibilities over a held out set.

from a string kernel) and solve for the various values of α_i. Note that the gradient-ascent update for the dual problem (cf. Section 6.3.5) is already expressed in terms of kernel similarities $K(\overline{X_i}, \overline{X_j})$ rather than dot products.

How can we use this similarity to return the prediction for a test document (say, in string form)? In order to understand this point, consider the prediction function of kernel SVMs for test point $\overline{Z}$:

$$F(\overline{Z}) = \text{sign}\{\overline{W} \cdot \overline{Z}\} = \text{sign}\{\sum_{i=1}^{n} \alpha_i y_i \overline{X_i} \cdot \overline{Z}\} \tag{6.42}$$

One can replace each $\overline{X_i} \cdot \overline{Z}$ with the corresponding string kernel similarity between training point $\overline{X_i}$ and test point $\overline{Z}$ in order to yield the final prediction.

6.5.5 Systematic Application of the Kernel Trick

The use of the kernel trick with the dual of an SVM almost seems like a serendipitous observation in retrospect. However, there are large numbers of possible variations of linear models, each of which might have its own objective function and its own set of constraints. The techniques of least-squares regression, Fisher's discriminant, and logistic regression are examples from a large family of possibilities. Given a linear problem, how can we kernelize it? Would the dual of an optimization problem always work for kernelization? Is there a systematic way to do it?

Although several methods have been proposed in recent years for using the kernel trick with the primal, the use of the kernel trick on the dual is more well known. Using primal methods in conjunction with the kernel trick is far more systematic, and even has several efficiency advantages. However, this (more) useful and systematic technique has always toiled in relative obscurity compared to its more famous dual cousin because of historical reasons, such as the fact that the very first paper on this topic used the dual optimization method [132]. In this context, the following observation was made in an insightful paper written about a decade back [102]:

> "The vast majority of text books and articles introducing support vector machines (SVMs) first state the primal optimization problem, and then go directly to the dual formulation. A reader could easily obtain the impression that this is the only possible way to train an SVM."

An important idea that can be used in order to solve nonlinear SVMs in the primal is the *representer theorem*. Consider the L_2-regularized form of all linear models discussed in this chapter, in which the loss function is $L(y_i, \overline{W} \cdot \overline{X_i})$:

$$\text{Minimize } J = \sum_{i=1}^{n} L(y_i, \overline{W} \cdot \overline{X_i}) + \frac{\lambda}{2}||\overline{W}||^2 \tag{6.43}$$

Consider a situation in which the training data points have dimensionality d, but all of them lie on a 2-dimensional plane. Note that the optimal linear separation of points on this plane can always be achieved with the use of a 1-dimensional line on this 2-dimensional plane. Furthermore, this separator is more concise than any higher dimensional separator and will therefore be preferred by the L_2-regularizer. A 1-dimensional separator of training points lying on a 2-dimensional plane is shown in Figure 6.9(a). Although it is also possible to get the same separation of training points using any 2-dimensional plane (e.g., Figure 6.9(b))

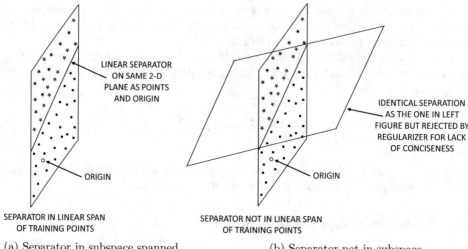

(a) Separator in subspace spanned
spanned by training points

(b) Separator not in subspace
spanned by training points

Figure 6.9: Both the linear separators in (a) and (b) provide exactly the same separation of training points, except that the one in (a) can be expressed as a linear combination of the training points. The separator in (b) will always be rejected by the regularizer. The key point of the representer theorem is that a separator $\overline{W}$ can always be found in the plane (subspace) of the training points with an identical separation to one that does not.

passing through the 1-dimensional separator of Figure 6.9(a), such a separator would not be preferred by an L_2-regularizer because of its lack of conciseness. In other words, given a set of training data points $\overline{X_1} \ldots \overline{X_n}$, the separator $\overline{W}$ always lies in the space spanned by these vectors. We state this result below, which is a very simplified version of the representer theorem, and is specific to linear models with L_2-regularizers.

Theorem 6.5.1 (Simplified Representer Theorem) *Let J be any optimization problem of the following form:*

$$Minimize \; J = \sum_{i=1}^{n} L(y_i, \overline{W} \cdot \overline{X_i}) + \frac{\lambda}{2}||\overline{W}||^2$$

Then, any optimum solution $\overline{W}^*$ to the aforementioned problem lies in the subspace spanned by the training points $\overline{X_1} \ldots \overline{X_n}$. In other words, there must exist real values $\beta_1 \ldots \beta_n$ such that the following is true:

$$\overline{W}^* = \sum_{i=1}^{n} \beta_i \overline{X_i}$$

Proof: Suppose that $\overline{W}^*$ cannot be expressed in the subspace spanned by the training points. Then, let us decompose $\overline{W}^*$ into the portion $\overline{W}_{\|} = \sum_{i=1}^{n} \beta_i \overline{X_i}$ spanned by the training points and an additional orthogonal residual $\overline{W}_{\perp}$. In other words, we have:

$$\overline{W}^* = \overline{W}_{\|} + \overline{W}_{\perp} \tag{6.44}$$

Then, it suffices to show that $\overline{W}^*$ can be optimal only when $\overline{W}_{\perp}$ is the zero vector.

Each $(\overline{W}_\perp \cdot \overline{X}_i)$ has to be 0, because $\overline{W}_\perp$ is orthogonal to the subspace spanned by the various training points. The optimal objective J^* can be written as follows:

$$J^* = \sum_{i=1}^{n} L(y_i, \overline{W}^* \cdot \overline{X}_i) + \frac{\lambda}{2}||\overline{W}^*||^2 = \sum_{i=1}^{n} L(y_i, (\overline{W}_\| + \overline{W}_\perp) \cdot \overline{X}_i) + \frac{\lambda}{2}||\overline{W}_\| + \overline{W}_\perp||^2$$

$$= \sum_{i=1}^{n} L(y_i, \overline{W}_\| \cdot \overline{X}_i + \underbrace{\overline{W}_\perp \cdot \overline{X}_i}_{0}) + \frac{\lambda}{2}||\overline{W}_\|||^2 + \frac{\lambda}{2}||\overline{W}_\perp||^2$$

$$= \sum_{i=1}^{n} L(y_i, \overline{W}_\| \cdot \overline{X}_i) + \frac{\lambda}{2}||\overline{W}_\|||^2 + \frac{\lambda}{2}||\overline{W}_\perp||^2$$

It is noteworthy that $||\overline{W}_\perp||^2$ must be 0, or else $\overline{W}_\|$ will be a better solution than $\overline{W}^*$. Therefore, $\overline{W}^* = \overline{W}_\|$ lies in the subspace spanned by the training points. ∎

Intuitively, the representer theorem states that for a particular family of loss functions, one can always find an optimal linear separator within the subspace spanned by the training points (see Figure 6.9), and the regularizer ensures that this is the concise way to do it.

The representer theorem provides a boilerplate method to create an optimization model that is expressed as a function of dot products:

For any given optimization model of the form of Equation 6.43 plug in $\overline{W} = \sum_{i=1}^{n} \beta_i \overline{X}_i$ to obtain a new optimization problem parameterized by $\beta_1 \ldots \beta_n$, and expressed only in terms of dot products between training points. Furthermore, the same approach is also used while evaluating $\overline{W} \cdot \overline{Z}$ for test instance $\overline{Z}$.

Consider what happens when one evaluates $\overline{W} \cdot \overline{X}_i$ in order to plug it into the loss function:

$$\overline{W} \cdot \overline{X}_i = \sum_{p=1}^{n} \beta_p \overline{X}_p \cdot \overline{X}_i \tag{6.45}$$

Furthermore, the regularizer $||\overline{W}||^2$ can be expressed as follows:

$$||\overline{W}||^2 = \sum_{i=1}^{n} \sum_{j=1}^{n} \beta_i \beta_j \overline{X}_i \cdot \overline{X}_j \tag{6.46}$$

In order to kernelize the problem, all we have to do is to substitute the dot product with the similarity value $s_{ij} = K(\overline{X}_i, \overline{X}_j) = \Phi(\overline{X}_i) \cdot \Phi(\overline{X}_j)$ from the $n \times n$ similarity matrix S. Therefore, one obtains the following optimization objective function:

$$J = \sum_{i=1}^{n} L(y_i, \sum_{p=1}^{n} \beta_p s_{pi}) + \frac{\lambda}{2} \sum_{i=1}^{n} \sum_{j=1}^{n} \beta_i \beta_j s_{ij} \quad \text{[General form]}$$

In other words, all we need to do is to substitute each $\overline{W} \cdot \overline{X}_i$ in the loss function with $\sum_p \beta_p s_{pi}$. Therefore, one obtains the following form for least-squares regression:

$$J = \frac{1}{2} \sum_{i=1}^{n} (y_i - \sum_{p=1}^{n} \beta_p s_{pi})^2 + \frac{\lambda}{2} \sum_{i=1}^{n} \sum_{j=1}^{n} \beta_i \beta_j s_{ij} \quad \text{[Least-squares regression]}$$

The aforementioned formulation provides an alternative way of proving the closed-form solution of kernel regression in Section 6.2.1.3 (see Exercise 18).

By substituting $\overline{W} \cdot \overline{X_i} = \sum_p \beta_p s_{pi}$ into the loss functions of linear classification, one can obtain corresponding optimization formulations:

$$J = \sum_{i=1}^{n} \max\{0, 1 - y_i \sum_{p=1}^{n} \beta_p s_{pi}\} + \frac{\lambda}{2} \sum_{i=1}^{n} \sum_{j=1}^{n} \beta_i \beta_j s_{ij} \quad [\text{SVM}]$$

$$J = \sum_{i=1}^{n} \log(1 + \exp(-y_i \sum_{p=1}^{n} \beta_p s_{pi})) + \frac{\lambda}{2} \sum_{i=1}^{n} \sum_{j=1}^{n} \beta_i \beta_j s_{ij} \quad [\text{Logistic Regression}]$$

These *unconstrained* optimization problems are conveniently expressed in terms of pairwise similarities, and parameterized by $\beta_1 \ldots \beta_n$. In order to classify a test instance $\overline{Z}$, one only needs to compute $\overline{W} \cdot \overline{Z} = \sum_i \beta_i K(\overline{X_i}, \overline{Z})$ after $\beta_1 \ldots \beta_n$ have been learned.

In SVMs, the primal variables $\beta_1 \ldots \beta_n$ can be related to the dual variables $\alpha_1 \ldots \alpha_n$ at optimality. At least one optimal solution pair $(\overline{\alpha}^*, \overline{\beta}^*)$ will exist in which we have $\beta_i^* = y_i \alpha_i^*$ because $\overline{W}^* = \sum_i \alpha_i^* y_i \Phi(\overline{X_i}) = \sum_i \beta_i^* \Phi(\overline{X_i})$. However, this relationship does not hold over all points in the solution space, and the corresponding non-optimal objective function value of the primal at $\beta_i = \alpha_i y_i$ is always larger than that of the dual at α_i. Any optimal solution to the dual can be used to derive an optimal solution $\beta_i^* = y_i \alpha_i^*$ for the primal, although the converse is not true because dual variables are bounded. Furthermore, an "almost" optimal solution for the dual problem can map to a much poorer solution for the primal (which is a potential drawback of dual optimization).

The unconstrained variables $\beta_1 \ldots \beta_n$ in the primal (in contrast to the bounded variables $\alpha_1 \ldots \alpha_n$ in the dual) allow easier optimization. Furthermore, a neat re-parametrization trick is available with the primal. One can perform stochastic gradient descent with respect to $\overline{W}$ (as in Section 6.3.3), while updating $\overline{W}$ only indirectly using $\beta_1 \ldots \beta_n$ via the representer theorem. We describe the kernelized variant of *Pegasos* for SVMs using $C = 1/\lambda$:

Initialize $\beta_1 \ldots \beta_n$ to 0;
for $t = 1$ to T **do begin**
  $\eta_t = 1/t$; $\overline{\beta} \Leftarrow (1 - \eta_t)\overline{\beta}$;
  Select $(\overline{X}_{i_t}, y_{i_t})$ randomly;
  if $\underbrace{(y_{i_t} \sum_{p=1}^{n} \beta_p K(\overline{X_{i_t}}, \overline{X_p}) < 1)}_{y_{i_t} \overline{W} \cdot \overline{X_{i_t}} < 1}$ **then** $\underbrace{\beta_{i_t} \Leftarrow \beta_{i_t} + \eta_t \cdot n \cdot C \cdot y_{i_t}}_{\text{Update } \overline{W} \text{ indirectly}}$;
endfor

Note that this algorithm is almost *identical* to that discussed in Section 6.3.3 except that we are *indirectly* updating $\overline{W} = \sum_{i=1}^{n} \beta_i \overline{X_i}$ by updating β_i instead of $\overline{W}$. The batch-size selected is 1, and the optional projection step has been omitted to simplify the updates. An algorithm like the above can be derived for many linear methods with the use of the representer theorem (see Exercise 9).

The learning rate of $\eta_t = 1/t$ is convenient because it allows some optimizations in *Pegasos*. At the tth iteration, the amount added to the coefficient β_{i_t} for a margin-violating point $\overline{X_{i_t}}$ is $n \cdot C \cdot y_{i_t}/t$, which is proportional to $1/t$. This proportionality is maintained for all $t' > t$ iterations because of successive scaling down of $\overline{\beta}$ by $(r-1)/r$ in the rth iteration for each $r \in (t, t']$. This property allows us to simply add 1 to the unnormalized value of β_{i_t} in the tth iteration, drop the regularization scaling, and multiply each (i.e., ith) coefficient at the end with $n \cdot C \cdot y_i/T$ after the final (i.e., Tth) iteration. The checking of the margin condition is modified to $y_{i_t} \sum_{p=1}^{n} \beta_p y_p K(\overline{X_{i_t}}, \overline{X_p}) < t/nC$, which is the only potentially expensive step. The time for this check depends on the number of nonzero entries in the vector $\overline{\beta}$. At most one nonzero β_i is introduced in each iteration, and $\overline{\beta}$ is sparse if there is early generalization accuracy. It is sometimes beneficial to initialize $\overline{\beta}$ to a sparse and "almost optimal" vector by deriving it from execution on a smaller data sample. The *Pegasos* algorithm is considered a state-of-the-art method because of its efficiency. What this algorithm shows is that after more than two decades of complex research in optimizing dual SVMs, one can do as well or better with primal optimization in a few lines of code.

6.6 Summary

All linear models for classification are closely related, as they optimize a loss function that is expressed in terms of a linear combination of the feature variables. Linear classification problems adapt the loss function from linear regression in various ways in order to address the binary nature of the class variable. Methods like the Fisher discriminant are straightforward adaptations of linear regression in this respect. The SVM varies on the Fisher discriminant in terms of its handling of the well-separated points in the data. Logistic regression uses a loss function that is a smooth variation on the one used in support vector machines and it provides similar results. All of these models can be generalized to the nonlinear setting by using kernel transformations.

6.7 Bibliographic Notes

Least-squares regression and classification dates back to the Widrow-Hoff algorithm [580] and Tikhonov-Arsenin's seminal work [550]. L_2-regularization is sometimes referred to as Tikhonov regularization. A detailed discussion of regression analysis may be found in [162], and regression with L_1-regularization is discussed in [237]. Neural networks like perceptrons [58] are also based on a modified version of least-squares regression, which is much closer to a support vector machine. A discussion of these methods may be found in [218]. Several independent works [204, 539] re-derived these methods in terms of their relationship with support-vector machines. The first application of least-squares methods to text categorization is provided in [601, 604]. All these methods are straightforward applications of regularized least-squares regression on the training data by treating the binary response variable as a numeric response. The Fisher discriminant was proposed by Ronald Fisher [191] in 1936, and is a specific case of the family of linear discriminant analysis methods [376]. The kernel version of Fisher discriminant is discussed in [388]. Even though the Fisher discriminant uses a different looking objective function that least-squares regression, it turns out to be a special case of least-squares regression in which the binary response variable is used as the regressand [57]. The relationship of the Fisher discriminant with the support-vector machine was shown in [513] in terms of the treatment of well-separated points. A variation [129] of the Fisher discriminant has also been proposed that removes the well-separated points in order to improve its performance.

The support-vector machine is generally credited to Cortes and Vapnik [132], although the primal method for L_2-loss SVMs was proposed several years earlier by Hinton [247]. This approach repairs the loss function in least-squares classification by keeping only one-half of the quadratic loss curve and setting the remaining to zero, so that it looks like a smooth version of hinge loss (try this on Figure 6.6(a)). The specific significance of this contribution was lost within the broader literature on neural networks. Hinton's work also does not focus on the importance of regularization in SVMs, although the general notion of adding shrinkage to gradient-descent steps in neural networks was well known. The hinge-loss SVM [132] is heavily presented from the perspective of duality and the maximum-margin interpretation, which makes its relationship to regularized least-squares classification somewhat opaque. The relationship of SVMs to least-squares classification is more evident from other related works [471, 513], where it becomes evident that quadratic and hinge-loss SVMs are natural variations of regularized L_2-loss (i.e., Fisher discriminant) and L_1-loss classification that use the binary class variables as the regression responses [218]. The main differences account for the fact that binary responses should be treated differently than numerical responses, and points with $y_i(\overline{W} \cdot \overline{X_i}) > 1$ should not be penalized because they represent correct classification of training instances (see Figure 6.6). All these variations of the objective function can be kernelized in the same way using the representer theorem [565]. The margin-centric interpretation has been used to create a different variant of linear regression for numeric targets, referred to as support-vector regression [163, 558].

The decomposition methods for the dual were pioneered by [424] and adapted in *SVM-Light* [273] and the *Sequential Minimal Optimization (SMO)* [440] algorithms. An optimized version of this algorithm [189] is implemented in LIBLINEAR [188], which is a software library for many linear learning algorithms. A cutting plane algorithm for text data was proposed in *SVMPerf* [274]. Primal optimization of kernel SVMs was advocated in [102]. The *Pegasos* algorithm was proposed in [512], and the approach was based on primal optimization. General material on support vector machines is available in [80, 134, 558]. String kernels are discussed in [353]. The logistic regression model smooths the hinge-loss in a support vector machine, and it belongs to the broader family of generalized linear models. A detailed discussion of generalized linear models is provides in [374]. The use of maximum entropy models for text classification is explored in [418]. A variety of procedures such as *generalized iterative scaling*, *iteratively reweighted least-squares*, and *gradient descent* for multinomial logistic regression are discussed in [238].

6.7.1   Software Resources

Two important libraries for large-scale SVMs and linear classification are LIBSVM [100] and LIBLINEAR [188]. Both these libraries are implemented in C++. These libraries implement many of the linear classification algorithms discussed in this chapter, and also contain specialized implementations for sparse data like text. The former library is focused more on SVMs, whereas the latter library has various linear algorithms like SVMs and logistic regression. Interfaces in several languages like Python, Java, and MATLAB have been made available by the creators of LIBSVM and LIBLINEAR. Furthermore, many other third-party platforms use LIBLINEAR or LIBSVM's implementations under the covers. Therefore, many of the tools mentioned below also use these implementations, but it is important to discuss them as they use different programming language platforms to provide the user interface. The Python library *scikit-learn* [646] contains many tools for linear classification and regression. The **kernlab** package [289] from CRAN can be used to perform linear and nonlinear classification in R. The **caret** package [305] is a good choice for those working in the R programming language, although it sources the implementations of specific algorithms from other packages and constructs a wrapper around them. The R-based **tm** library [647] can be used for preprocessing and tokenization in combination with the **caret** package. The package **RTextTools** [667] in R also has numerous categorization methods, which are specifically designed for text. The *Weka* library [649] in Java has also implemented various tools for text classification and regression. The **MALLET** toolkit [701] supports an implementation of the *MaxEnt* classifier, which uses multinomial logistic regression.

6.8   Exercises

1. The bias variable is often addressed in least-squares classification and regression by adding an additional column of 1s to the data. Discuss the differences with the use of an explicit bias term when regularized forms of the model are used.

2. Write the optimization formulation for least-squares regression of the form $y = \overline{W} \cdot \overline{X} + b$ with a bias term b. Do not use regularization. Show that the optimal value of the bias term b always evaluates to 0 when the data matrix D and response variable vector $\overline{y}$ are both mean-centered.

3. For any $n \times d$ data matrix D, use singular value decomposition to show the following for any value of $\lambda > 0$:

$$(D^T D + \lambda I)^{-1} D^T = D^T (D D^T + \lambda I)^{-1}$$

   Note that the two identity matrices on either side of the equation are of sizes $d \times d$ and $n \times n$, respectively. What you showed is a special case of the Sherman-Morrison-Woodbury identity in matrix algebra. Explain the consequences of this identity for kernel least-squares regression.

4. Suppose that the within-class scatter matrix S_w is defined as in Section 6.2.3, and the between-class scatter matrix S_b is defined as $S_b = n \left[(\overline{\mu}_1 - \overline{\mu}_0)^T (\overline{\mu}_1 - \overline{\mu}_0) \right]$. Assume that the data matrix D is mean-centered. Show that the full scatter matrix can be expressed as follows:

$$D^T D = S_w + \frac{n_1 \cdot n_0}{n^2} S_b \tag{6.47}$$

   Here, $\overline{\mu}_1$ and $\overline{\mu}_0$ are the means of the positive and negative classes in the training data. Furthermore, n_1 and n_0 are the number of positive and negative examples in the training data.

5. Show that the effect of the bias term can be accounted for by adding a constant amount to each entry of the $n \times n$ kernel similarity matrix when using kernels with linear models.

6. Formulate a variation of regularized least-squares classification in which L_1-loss is used instead of L_2-loss. How would you expect each of these methods to behave in the presence of outliers? Which of these methods is more similar to SVMs with hinge loss? Discuss the challenges of using gradient-descent with this problem as compared to the regularized least-squares formulation.

7. Derive stochastic gradient-descent steps for the variation of $L1$-loss classification introduced in Exercise 6. You can use a constant step size.

8. Derive stochastic gradient-descent steps for SVMs with quadratic loss instead of hinge loss. You can use a constant step size.

9. Consider loss functions of the following form:

$$\text{Minimize } J = \sum_{i=1}^{n} L(y_i, \overline{W} \cdot \overline{X}_i) + \frac{\lambda}{2} ||\overline{W}||^2$$

   Derive stochastic gradient-descent steps for this general loss function. You can use a constant step size.

10. Consider loss functions of the following form:

$$\text{Minimize } J = \sum_{i=1}^{n} L(y_i, \overline{W} \cdot \overline{X_i}) + \frac{\lambda}{2}||\overline{W}||^2$$

Use the representer theorem to derive stochastic gradient-descent steps for this general loss function in the kernel setting, where the gradient is computed with respect to $\overline{\beta}$. Here, $\overline{\beta}$ defines the n-dimensional vector of representer-theorem coefficients.

11. Consider loss functions of the following form:

$$\text{Minimize } J = \sum_{i=1}^{n} L(y_i, \overline{W} \cdot \overline{X_i}) + \frac{\lambda}{2}||\overline{W}||^2$$

Use the representer theorem to derive stochastic gradient-descent steps for this general loss function in the kernel setting, where the gradient is computed with respect to $\overline{W}$. Here, $\overline{W}$ defines the linear hyperplane in the transformed space of unknown dimensionality. Your gradient-descent steps should update the hyperplane $\overline{W}$ indirectly via the representer theorem. Discuss the difference from the previous exercise.

12. Provide an algorithm to perform classification with explicit kernel feature transformation and the Nyström approximation. How would you use ensembles to make the algorithm efficient and accurate?

13. **Multinomial logistic regression:** Show that the special case of Equation 6.35 for binary classes is identical to the objective function of logistic regression.

14. **Multi-class SVMs:** Consider a k-class problem for $k > 2$. An alternative to the one-against-all approach for learning multi-class SVMs is to learn the coefficient vectors $\overline{W_1} \ldots \overline{W_k}$ of the k separators simultaneously like the multinomial logistic regression model. Set up a loss function and an optimization model for multi-class SVMs. Discuss the advantages and disadvantages of this approach versus the one-against-all approach.

15. Show that the stochastic gradient-descent updates of least-squares classification, SVM, and logistic regression are all of the form $\overline{W} \Leftarrow \overline{W}(1 - \eta\lambda) + \eta y[\delta(\overline{X}, y)]\overline{X}$. Here, the mistake function $\delta(\overline{X}, y)$ is $1 - y(\overline{W} \cdot \overline{X})$ for least-squares classification, an indicator variable for SVMs, and a probability value for logistic regression. Assume that η is the learning rate, and $y \in \{-1, +1\}$. Write the specific forms of $\delta(\overline{X}, y)$ in each case.

16. Consider an SVM with properly optimized parameters. Provide an intuitive argument as to why the out-of-sample error rate of the SVM will be usually less than the fraction of support vectors in the training data.

17. Suppose that you perform least-squares regression without regularization with the loss function $\sum_{i=1}^{n}(y_i - \overline{W} \cdot \overline{X_i})^2$, but you add spherical Gaussian noise with variance λ to each feature. Show that the expected loss with the perturbed features provides a loss function that is identical to that of L_2-regularization. Use this result to provide an intuitive explanation of the connection between regularization and noise addition.

18. Show how to use the representer theorem to derive the closed-form solution of kernel least-squares regression.

19. Show that the partial derivative of the Lagrangian dual L_D falls to zero in Equation 6.27, when the step-size η_k in Section 6.3.5 is set to $1/K(\overline{X_k}, \overline{X_k})$.

Chapter 7

Classifier Performance and Evaluation

"All models are wrong, but some are useful."– George E. P. Box

7.1 Introduction

Among all machine learning problems, classification is the most well studied, and has the most number of solution methodologies. This embarrassment of riches also leads to the natural problems of *model selection* and *evaluation*. In particular, some natural questions that arise are as follows:

1. Given a classifier, what are the causes for its error? Is there a theoretical way in which one might decompose the error into intuitively interpretable components?

2. Can one use the insights from the aforementioned analysis to choose a particular classifier in a domain in general, and text in particular? Are there specific design criteria that one should be aware of while using a particular supervised learning algorithm? Are there ways in which the performance of off-the-shelf classifiers can be enhanced with these insights?

3. Given a set of learning algorithms, is there an empirical way to evaluate their performance and choose the best performer among them?

The theoretical analysis of classification models is closely related to their evaluation, model design, and selection. Therefore, this chapter will discuss these issues in an integrated way.

Classification models are often designed to maximize accuracy on the training data either directly or indirectly. Although the maximization of accuracy on the training data is desirable in general, it does not always translate to increased accuracy on the test data (i.e., better *generalizability*), particularly when the training data is small. For example, decision trees prune nodes, rule-based classifiers prune rules, and almost all optimization-based learning models use regularizers that are designed to make the model *concise* at the expense of training accuracy. Concise models have better generalizability to (unseen) test data, even though they may be unable to take sufficient advantage of an increasing amount

© Springer Nature Switzerland AG 2022
C. C. Aggarwal, *Machine Learning for Text*,
https://doi.org/10.1007/978-3-030-96623-2_7

of training data. The natural trade-off between these goals is quantified with the use of the *bias-variance trade-off*.

The theoretical analysis of classifier performance is useful because it provides some guidance about classifier design and other tricks such as the use of *ensembles*. Previous chapters have already discussed some ensemble methods like bagging and random forests. This chapter will revisit these methods and introduce other methods like *boosting*. Finally, this chapter will discuss classifier evaluation, *model selection*, and parameter tuning.

Chapter Organization

This chapter is organized as follows. The bias-variance trade-off is introduced in Section 7.2. The implications of the bias-variance trade-off on text classification performance are discussed in Section 7.3. Classification ensemble methods are introduced in Section 7.4. Methods for classifier evaluation are introduced in Section 7.5. A summary is given in Section 7.6.

7.2   The Bias-Variance Trade-Off

The bias-variance trade-off provides theoretical insights into the varying causes of modeling error. All classifiers attempt to learn the shape of the decision boundary separating different classes in one form or another. Classifiers like *linear* support vector machines impose strong prior assumptions on the shape of the decision boundary and are therefore inherently less powerful than nonlinear classifiers like kernel support vector machines that can learn an arbitrary shape of the boundary. From a conceptual point of view, a nonlinear model is more "correct" because it does not make as many assumptions (i.e., does not have predefined *biases*) about the shape of the decision boundary. However, the fact that more powerful models do not always win with a finite data set is the most important takeaway from the bias-variance trade-off. A key point is that the prediction of a model is not only dependent on the correctness of the model used but also on the specific nuances of the training data set at hand, which may cause accidental relationships between the feature and target variables from a particular training data set. A complex model may result in more opportunities for these accidental relationships to influence the final prediction, particularly if the training data set is small. This sensitivity in prediction to the specific nuances of the training data contributes to the error and makes the comparison in accuracy between different models more subtle than it seems at first sight. In particular, the error of a classifier can be decomposed into the following three components:

1. *Bias:* Loosely speaking, the bias can be viewed as an error caused by erroneous assumptions made in the model. For example, consider a situation in which the two classes are separated by a nonlinear decision boundary. However, if we choose to use a linear support vector machine (SVM) in this setting, the classifier will be consistently incorrect over different choices of training data sets. Bias often results in *consistently* incorrect classification of particular test instances. Another example of a highly biased classifier is an n-nearest neighbor classifier for a training data set of size n. This classifier is essentially a majority vote classifier over the full data set, and will (almost always) predict minority class examples incorrectly irrespective of the specific draw of the training data one receives, as long as the draw is of reasonable size.

2. *Variance:* The variance of a learning algorithm is a measure of its stability over different choices of training data sets. For example, a 1-nearest neighbor classifier is highly

unstable with respect to the choice of the specific training data set that is used. When the variance is high, the same test point might receive inconsistent predictions over different choices of the training data. This inconsistency is a result of *overfitting*, in which the classifier learns the specific nuances of the training data that do not generalize well to test instances. As a result, changing the training data set changes the prediction on the same test instance, and the classifier predictions become less stable. Clearly, variance always adds to the error *in expectation* because at least some of the training instantiations in which the same test point is predicted differently must be incorrect. Therefore, variance causes *inconsistency* in classification of the same test instance over different choices of training data sets, which naturally adds to the error.

3. *Noise:* The intrinsic noise is a property of the specific data set at hand. Any data set will have regions of the space in which the two classes overlap or in which the points are mislabeled. There is little that any classifier can do to reduce this type of noise. While bias and variance are specific to a particular learning model, the intrinsic noise is considered a property of the data, and is independent of the model at hand. Noise is considered an irreducible part of the error that cannot be addressed by a learning algorithm. For example, even if a learning algorithm were to be seeded with the extraordinary advantage of being told the distribution of each class, the noise would still be a part of the error.

As shown above with the example of the nearest neighbor classifier, different choices of parameters in the same model may lead to different levels of bias and variance, which typically (but not always) exhibit in the form of a trade-off between the two. The goal of a supervised learning algorithm is to attain an optimal point of this trade-off in which the overall error is minimized.

7.2.1 A Formal View

We assume that the base distribution from which the training data set is generated is denoted by $\mathcal{B}$. One can generate a data set $\mathcal{D}$ from this base distribution:

$$\mathcal{D} \sim \mathcal{B} \tag{7.1}$$

One could draw the training data in many different ways, such as selecting only data sets of a particular size. For now, assume that we have some well defined generative process according to which training data sets are drawn from $\mathcal{B}$. The analysis below does not rely on the specific mechanism with which training data sets are drawn from $\mathcal{B}$.

Access to the base distribution $\mathcal{B}$ is equivalent to having access to an infinite resource of training data, because one can use the base distribution an unlimited number of times to generate training data sets. In practice, such base distributions (i.e., infinite resources of data) are not available. As a practical matter, an analyst uses some data collection mechanism to collect only *one finite instance* of $\mathcal{D}$. However, the conceptual existence of a base distribution from which other training data sets can be generated is useful in theoretically quantifying the sources of error in training on this finite data set.

Now imagine that the analyst had a set of t test instances in d dimensions, denoted by $\overline{Z_1} \ldots \overline{Z_t}$. The dependent variables of these test instances are denoted by $y_1 \ldots y_t$. For clarity in discussion, let us assume that the test instances and their dependent variables were also generated from the same base distribution $\mathcal{B}$ by a third party, but the analyst was provided access only to the feature representations $\overline{Z_1} \ldots \overline{Z_t}$, and no access to the dependent variables

$y_1 \ldots y_t$. Therefore, the analyst is tasked with job of using the single finite instance of the training data set $\mathcal{D}$ in order to predict the dependent variables of $\overline{Z_1} \ldots \overline{Z_t}$.

Now assume that the relationship between the dependent variable y and its feature representation $\overline{Z_i}$ is defined by the *unknown* function $f(\cdot)$ as follows:

$$y_i = f(\overline{Z_i}) + \epsilon_i \qquad (7.2)$$

Here, the notation ϵ_i denotes the intrinsic noise, which is independent of the model being used. The value of ϵ_i might be positive or negative, although it is assumed that $E[\epsilon_i] = 0$. If the analyst knew what the function $f(\cdot)$ corresponding to this relationship was, then they could simply apply the function to each test point $\overline{Z_i}$ in order to approximate the dependent variable y_i, with the only remaining uncertainty being caused by the intrinsic noise.

The problem is that the analyst does not know what the function $f(\cdot)$ is in practice. Note that this function is used within the generative process of the base distribution $\mathcal{B}$, and the entire generating process is like an oracle that is unavailable to the analyst. The analyst only has examples of the input and output of this function. Clearly, the analyst would need to develop some type of *model* $g(\overline{Z_i}, \mathcal{D})$ using the training data in order to *approximate* this function in a data-driven way.

$$\hat{y}_i = g(\overline{Z_i}, \mathcal{D}) \qquad (7.3)$$

Note the use of the circumflex (i.e., the symbol '^') on the variable $\hat{y}_i$ to indicate that it is a *predicted* value by a specific algorithm rather than the observed (true) value of y_i.

All prediction functions of supervised learning models such as Bayes classifiers, SVMs, and decision trees are examples of the estimated function $g(\cdot, \cdot)$. Some algorithms (such as linear regression and SVMs) can even be expressed in a concise and understandable way:

$$g(\overline{Z_i}, \mathcal{D}) = \underbrace{\overline{W} \cdot \overline{Z_i}}_{\text{Learn } \overline{W} \text{ with } \mathcal{D}} \qquad \text{[Linear Regression]}$$

$$g(\overline{Z_i}, \mathcal{D}) = \underbrace{\text{sign}\{\overline{W} \cdot \overline{Z_i}\}}_{\text{Learn } \overline{W} \text{ with } \mathcal{D}} \qquad \text{[SVMs]}$$

Other models like decision trees are expressed algorithmically as computational functions. The choice of computational function includes the effect of its specific parameter setting, such as the number of nearest neighbors in a κ-nearest neighbor classifier.

The goal of the bias-variance trade-off is to quantify the expected error of the learning algorithm in terms of its bias, variance, and the (data-specific) noise. For generality in discussion, we assume a numeric form of the target variable, so that the error can be intuitively quantified by the *mean-squared error* between the predicted values $\hat{y}_i$ and the observed values y_i. This is a natural form of error quantification in regression, although one can also use it in classification by using probabilistic predictions of test instances. The mean squared error, MSE, of the learning algorithm $g(\cdot, \mathcal{D})$ is defined over the set of test instances $\overline{Z_1} \ldots \overline{Z_t}$ as follows:

$$MSE = \frac{1}{t} \sum_{i=1}^{t} (\hat{y}_i - y_i)^2 = \frac{1}{t} \sum_{i=1}^{t} (g(\overline{Z_i}, \mathcal{D}) - f(\overline{Z_i}) - \epsilon_i)^2$$

The best way to estimate the error in a way that is independent of the specific choice of training data set is to compute the *expected* error over different choices of training data sets:

$$E[MSE] = \frac{1}{t} \sum_{i=1}^{t} E[(g(\overline{Z_i}, \mathcal{D}) - f(\overline{Z_i}) - \epsilon_i)^2]$$

$$= \frac{1}{t} \sum_{i=1}^{t} E[(g(\overline{Z_i}, \mathcal{D}) - f(\overline{Z_i}))]^2 + \frac{\sum_{i=1}^{t} E[\epsilon_i^2]}{t} \quad [\text{Using } E[\epsilon_i] = 0]$$

The second relationship is obtained by expanding the quadratic expression on the right-hand side of the first equation.

The right-hand side of the above expression can be further decomposed by adding and subtracting $E[g(\overline{Z_i}, \mathcal{D})]$ within the squared term on the right-hand side:

$$E[MSE] = \frac{1}{t} \sum_{i=1}^{t} E[\{(f(\overline{Z_i}) - E[g(\overline{Z_i}, \mathcal{D})]) + (E[g(\overline{Z_i}, \mathcal{D})] - g(\overline{Z_i}, \mathcal{D}))\}^2] + \frac{\sum_{i=1}^{t} E[\epsilon_i^2]}{t}$$

One can expand the quadratic polynomial on the right-hand side to obtain the following:

$$E[MSE] = \frac{1}{t} \sum_{i=1}^{t} E[\{f(\overline{Z_i}) - E[g(\overline{Z_i}, \mathcal{D})]\}^2]$$

$$+ \frac{2}{t} \sum_{i=1}^{t} \{f(\overline{Z_i}) - E[g(\overline{Z_i}, \mathcal{D})]\}\{E[g(\overline{Z_i}, \mathcal{D})] - E[g(\overline{Z_i}, \mathcal{D})]\}$$

$$+ \frac{1}{t} \sum_{i=1}^{t} E[\{E[g(\overline{Z_i}, \mathcal{D})] - g(\overline{Z_i}, \mathcal{D})\}^2] + \frac{\sum_{i=1}^{t} E[\epsilon_i^2]}{t}$$

The second term on the right-hand side of the aforementioned expression evaluates to 0 because one of the multiplicative factors is $E[g(\overline{Z_i}, \mathcal{D})] - E[g(\overline{Z_i}, \mathcal{D})]$. On simplification, we obtain the following:

$$E[MSE] = \underbrace{\frac{1}{t} \sum_{i=1}^{t} \{f(\overline{Z_i}) - E[g(\overline{Z_i}, \mathcal{D})]\}^2}_{\text{Bias}^2} + \underbrace{\frac{1}{t} \sum_{i=1}^{t} E[\{g(\overline{Z_i}, \mathcal{D}) - E[g(\overline{Z_i}, \mathcal{D})]\}^2]}_{\text{Variance}} + \underbrace{\frac{\sum_{i=1}^{t} E[\epsilon_i^2]}{t}}_{\text{Noise}}$$

We examine each of the aforementioned terms to understand the parts of the error they represent. Consider the (squared) bias term corresponding to the expression $\frac{1}{t} \sum_{i=1}^{t} E[\{f(\overline{Z_i}) - E[g(\overline{Z_i}, \mathcal{D})]\}^2]$. This measure computes the difference between the true value of the function $f(\overline{Z_i})$, and the *expected* prediction by the model, which is denoted by $E[g(\overline{Z_i}, \mathcal{D})]$. For example, a 1-nearest neighbor classifier is known to have very low bias because the averaged prediction over a large number of training data sets will be close to the true prediction. Similarly, a nonlinear classifier will often have low bias because of the ability to model complex decision boundaries.

However, these excellent bias characteristics do not always result in low values of the expected mean-squared error (MSE). This is because of the additional variance term, which is generally irreducible to 0. The main problem is that one only has access to a single finite instance of the training data set $\mathcal{D}$, and therefore it is not possible to exactly compute

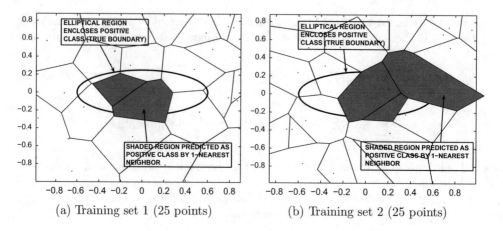

(a) Training set 1 (25 points)   (b) Training set 2 (25 points)

Figure 7.1: Illustration of high variance in prediction of 1-nearest neighbor classifier

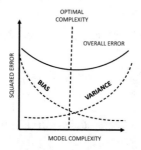

Figure 7.2: The point of optimal model complexity

$E[g(\overline{Z_i}, \mathcal{D})]$, which remains only a theoretical prediction. As we will see later, ensemble methods try to approximate this prediction using some tricks, albeit in an imperfect way.

The variance term is exacerbated by the use of powerful classifiers on small data sets. For example, the powerful 1-nearest neighbor classifier is almost Bayes optimal for infinitely large data. Now consider a tiny data set in which all points belonging to the positive class are enclosed inside the following ellipse:

$$\frac{25}{9}x_1^2 + 16x_2^2 = 1$$

Two examples of the Voronoi regions (cf. Chapter 5) induced by different samples of 25 training points are shown in Figures 7.1(a) and 7.1(b), respectively. Test instances lying in the shaded region are predicted to the positive class in each case, and this shaded region is very different from the true elliptical boundary. This high level of inaccuracy is caused by the inherent instability (i.e., variance) of the model on a small training data set.

Using a very powerful model on a tiny data set is like using a sledgehammer to swat a fly, which causes unpredictability in controlling it properly (i.e., increased variance). The optimal model complexity depends on the delicate trade-off between bias and variance. Although bias reduces with increasing model complexity, the variance increases. Therefore, the optimal error is reached at some intermediate model complexity (cf. Figure 7.2).

7.2.2 Telltale Signs of Bias and Variance

For a given data set and learning algorithm, how can an analyst tell whether the main causative factor in the error is the bias or the variance? This is a useful piece of information to have in order to make appropriate adjustments to the algorithm at hand. In general, it is impossible to exactly estimate the bias and the variance without access to an infinite resource of data. However, there are some telltale signs that an analyst can use in order to make decisions on the source of the error. High-variance algorithms are particularly easy to identify because they will often overfit the data, and there will be large gaps between the accuracy on the training data and a held-out portion of the labeled data (which is not used for training). Furthermore, it is also possible to run the algorithm on multiple samples to estimate the variance term on an out-of-sample test data set, although the estimate will only be a very approximate one.

Bias is generally harder to identify. Although algorithms with small gaps between training and test accuracy (and large error) might have high bias, one cannot be certain whether the errors are caused by intrinsic noise in the training data. One way to check if the errors are caused by intrinsic noise is to use other types of the models on the data set to check if the same test instances are being classified incorrectly by very different models. The noisy instances will cause problems for all models, and will tend to be misclassified in a more consistent way. On the other hand, since the bias of different models is different, it will be reflected in the fact that each model is consistently incorrect on its own specific set of test instances which is somewhat different from that of others. Although this approach can provide rough hints about the nature of the bias and noise, one should not view this approach as a formal methodology. An important issue to always keep in mind is that an analyst only sees the integrated form of the error on a particular data set, which is usually not possible to precisely decompose into different components with the use of a finite data resource.

7.3 Implications of Bias-Variance Trade-Off on Performance

This section will discuss the implications of the bias-variance trade-off on classifier performance. The discussion will be specifically focussed on text data, which is high-dimensional and sparse.

7.3.1 Impact of Training Data Size

Increased training data size almost always reduces the variance of a classifier because of the robustness of using a larger amount of data. It is common for classifiers to overfit the specific characteristics of a particular data distribution when a small training data set is used. The expected value $E[V]$ of the variance V in the bias-variance trade-off is as follows:

$$E[V] = \frac{1}{t} \sum_{i=1}^{t} E[\{g(\overline{Z_i}, \mathcal{D}) - E[g(\overline{Z_i}, \mathcal{D})]\}^2]$$

Note that if the expectation is computed conditionally over data sets $\mathcal{D}$ of small size, the value of $g(\overline{Z_i}, \mathcal{D})$ will vary more significantly with choice of $\mathcal{D}$ for most reasonable models. Examples of this drastic variation in the case of a 1-nearest neighbor classifier on data sets of size 25 are shown in Figure 7.1. Furthermore, if the size of the training data is increased beyond 25, then examples of predicted regions are shown in Figure 5.2 of Chapter 5. It is immediately evident that the use of larger training data sets leads to more stable predictions.

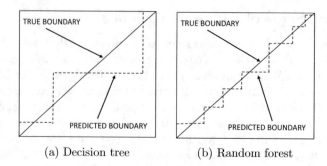

(a) Decision tree   (b) Random forest

Figure 7.3: Even though a decision tree can model an arbitrary decision boundary with an infinite amount of data, it has a consistent bias with a small amount of data. This is reflected in its piecewise linear boundaries that do not change significantly with choice in training data. Randomizing the tree construction is a way of reducing this data-centric bias by forcibly inducing model-centric diversity and averaging the predictions. A random forest can be viewed either as a bias reduction method or as a variance-reduction method, depending on the specific choice of bias-variance decomposition that one uses for analysis.

Increasing data size also reduces bias in many classifiers, although the effect is usually less pronounced and can sometimes be reversed if the parameters of the algorithm are fixed at values that are suitable to small data sets. In the case of the 1-nearest neighbor classifier, increases in data size lead to reduction in both bias and variance. With an infinite amount of data, the only remaining effect is that of intrinsic noise. In particular, the accuracy of a 1-nearest neighbor classifier is that of twice the Bayes optimal rate (cf. Section 5.4 of Chapter 5). The factor of two is because the noise in the training data and the test instance contribute equally to the error.

Another interesting example of the effect of data size on bias is that in a decision tree. A decision tree can model arbitrary decision boundaries with an infinite amount of data. However, when a small amount of data is used, piecewise linear boundaries are created. Such piecewise linear boundaries do not necessarily imply a high level of bias if they vary significantly over different choices of the training data. However, since a decision tree is heavily influenced by the splits at the top level of the tree, which do not change significantly with different choices of training data, the result is that the predictions of the decision tree could be very stable to different choices of training data sets. This situation is shown in Figure 7.3(a), in which the coarse and piece-wise linear approximations of the true decision boundary are shown. This type of coarse approximation would often result in greater bias when the bias is estimated only over training data sets of small size. In other words, the value of $E[g(\overline{Z_i}, \mathcal{D})]$ is often further away from $f(\overline{Z_i})$, when the expectation computation is restricted to training data sets of small size. A key issue is that the bias of a decision tree depends on its height. Smaller trees are more biased, and small data sets prevent the creation of deep trees. On the other hand, the random forest has better bias performance over smaller training data sets because it averages the predictions from trees constructed using different choices of splits. The averaging process leads to smoother decision boundaries like Figure 7.3(b) that approximate the true decision boundary more accurately, which results in lower bias. Although the random forest is often viewed as a variance-reduction method, that point of view needs a non-traditional definition[1] of the bias-variance trade-off

---

[1]Instead of computing the expected values of the bias-variance trade-off over different choices of training data sets, one can compute it over different randomized choices of models. This approach is referred to as the *model-centric* view of the bias-variance trade-off [10]. The traditional view of the bias-variance trade-off

in which the expected bias, variance, and error are computed using a random process defined by the choice of model at hand. The traditional view of the bias-variance trade-off is one in which the expectation is computed over randomized choices of training data sets [10]. A detailed discussion of the random forest is provided in Section 5.5.5 of Chapter 5.

7.3.2 Impact of Data Dimensionality

Increased data dimensionality almost always leads to an increase in the error because of the presence of irrelevant attributes. This issue is particularly important in the text domain because of the high dimensionality of text data. However, the increased error may be either reflected in the bias or the variance, depending on the choice of classifier. Classifiers like linear regression, in which the parameter space increases with dimensionality, tend to show increased variance with dimensionality if regularization is not used. Regularization in linear models can be viewed as an indirect form of feature selection. Feature selection improves the accuracy of a complex model by reducing variance. This is the reason that using regularization is crucial when using linear models with text data. Interestingly, even though the increased dimensionality increases the variance of linear models, it has a beneficial effect on the bias. In high-dimensional cases like text, the different classes are often (almost) linearly separable. Therefore, even though linear models can have high bias in many data sets, they seem to work well in the text domain. This is a ringing endorsement of linear SVMs for text, because linear SVMs have lower variance than nonlinear SVMs, and they also seem to have low bias in the *specific case* of the text domain.

On the other hand, classifiers in which the contributions from different dimensions are pre-aggregated before prediction tend to show increased bias. An example is the nearest neighbor classifier in which the contributions from different dimensions are aggregated in the distance function. In such cases, increased dimensionality actually tends to make the predictions of the classifier more stable (albeit with increased bias because of the aggregated impact of irrelevant dimensions). Just as a random forest works well because of randomized choices of splits, one of the tricks that is used with nearest neighbor classifiers in high dimensions is to build classifiers on random subsets of dimensions, and average the predictions from various subsets. This approach is referred to as *feature bagging*. In fact, the idea of feature bagging was a precursor to the idea of random forests in classification [250, 251].

7.3.3 Implications for Model Choice in Text

There are several implications of the high-dimensional and sparse nature of text on classifier design. Although many of these issues are discussed in Chapter 5 and 6, this chapter will also provide an analytical explanation of these behaviors in terms of the bias-variance trade-off. Such explanations also provide guidance in designing models for text.

Linear versus nonlinear models: Although linear models often have high bias because of strong prior assumptions, this is not the case in the text domain in which the sparse, high-dimensional nature of text tends to make the different classes (almost) linearly separable. As a result, linear models often have low bias in the text domain. Although nonlinear models like the Gaussian kernels can also simulate (or slightly improve) linear performance by using a large bandwidth, the additional accuracy advantages are often not worth the

---

is a data-centric view in which the randomized process to describe the bias-variance trade-off is defined by using different choices of training data sets. From the data-centric view, a random forest is really a bias reduction method over training data sets of small size.

increased computational effort. A key point is that tuning kernel parameters becomes exceedingly important with a nonlinear model, and it is easy to be less than exhaustive in searching the space of parameter choices with a nonlinear method (because of the computational cost). In such cases, it is actually possible for a nonlinear model to deliver poorer performance than a linear model as a practical matter. The use of nonlinear methods should be largely restricted to cases in which linguistic or sequencing information inside the text is used with string kernels. It makes little sense to use a nonlinear kernel with the vector space (i.e., multidimensional) representation of text.

Importance of feature selection: Text is a high-dimensional domain with many irrelevant attributes. Such attributes increase the error of the classifier in terms of either bias or variance, depending on the choice of model used. Models in which the number of parameters increase with data dimensionality tend to show increased variance with dimensionality. In such cases, feature selection is an effective way of reducing variance. The regularization of parameters in a linear model is a form of feature selection.

Presence versus absence of words: In Chapters 5 and 6, several examples have been provided in which classifiers using presence of words generally perform better than those using absence of words. For example, this is an important reason why the multinomial model often performs better than the Bernoulli model in text classification. A category can often be expressed using thousands of words, and most of the topical words of the category may be missing from a small document purely as a matter of chance. If a classifier uses the absence of these words as conclusive evidence of a particular document belonging to that class, it is likely to have poor generalization power to unseen test documents. This will lead to overfitting, which is a manifestation of high variance. In general, imbalanced frequencies of categorical features are important to account for in classification models because the presence of a feature is far more informative than its absence.

7.4   Systematic Performance Enhancement with Ensembles

From the aforementioned discussion, it is evident that key choices in the design of an algorithm can optimize the error by choosing the bias-variance trade-off appropriately. Ensembles provide a natural way to use the bias-variance theory in a judicious way to optimize performance. These methods are *meta-algorithms* that take a base method as input and improve its performance by applying it repeatedly over different modifications of the data or with different variants of the same model. The results from the different models are then combined to yield a single robust prediction. The specific choice of the model and the way in which the outputs of different models are combined regulate how an ensemble method reduces the bias or variance.

7.4.1   Bagging and Subsampling

Bagging and subsampling are two methods to reduce the variance of an ensemble method. A brief description of these methods in the context of the 1-nearest neighbor detector is provided in Section 5.4.3.1 of Chapter 5. The basic ideas in these methods are as follows:

1. In the case of bagging, the training data is sampled with replacement. The sample size s may be different from the size of the training data size n, although it is common to set s to n. In the latter case, the resampled data will contain duplicates, and about a fraction $1/e$ of the original data set will not be included at all (see Exercise 6). Here, the notation e denotes the base of the natural logarithm. A model is constructed on the resampled training data set, and each test instance is predicted with the resampled data. The entire process of resampling and model building is repeated m times. For a given test instance, each of these m models is applied to the test data. The predictions from different models are then averaged to yield a single robust prediction. Although it is customary to choose $s = n$ in bagging, the best results are often obtained by choosing values of s much less than n.

2. Subsampling is similar to bagging, except that the different models are constructed on the samples of the data created *without* replacement. The predictions from the different models are averaged. In this case, it is essential to choose $s < n$, because choosing $s = n$ yields the same training data set and identical results across different ensemble components.

Both bagging and subsampling are variance-reduction methods. In order to understand this point, consider the variance term in the bias-variance trade-off:

$$E[V] = \frac{1}{t} \sum_{i=1}^{t} E[\{g(\overline{Z_i}, \mathcal{D}) - E[g(\overline{Z_i}, \mathcal{D})]\}^2]$$

If the analyst had access to an infinite resource of data (i.e., the base distribution $\mathcal{B}$), she could go back to it as many times as she wanted in order to draw different training data sets $\mathcal{D}$, estimate the value of $E[g(\overline{Z_i}, \mathcal{D})]$ with an averaged prediction of $\overline{Z_i}$, and report it as the final result instead of $g(\overline{Z_i}, \mathcal{D})$ on a single finite instance of $\mathcal{D}$. Such an approach would result in a variance of 0, which will provide lower error.

The main problem with this approach is that the analyst does not have access to such an infinite resource of data. Bagging is an imperfect way of performing the same simulation by drawing $\mathcal{D}$ from the original instance of the finite data. Of course, such a simulation is imperfect because of two reasons:

1. The different instances of $\mathcal{D}$ drawn from the same base data are correlated with one another because of overlaps in instances. This limits the amount of variance reduction, and a part of it is *irreducible* in a way that is hidden from the analyst (without knowledge of the base distribution). This irreducible variance is a consequence of the fact that one cannot hope to determine the expected values over draws from a base distribution with a single finite instance. Nevertheless, if unstable configurations of the detector are used (e.g., a 1-nearest neighbor detector), then the variance-reduction effects are very significant.

2. The samples from the original data set do not provide as accurate results as using the original data. For example, a bagged sample contains repetitions, which are not naturally reflective of the original distribution. Similarly a subsample is smaller in size than the original instance, as a result of which some useful pattern for modeling will be irretrievably lost. All these effects will lead to a slight increase in bias.

The heuristic simulation above can either improve or worsen the accuracy, depending on the choice and configuration of the detector at hand. For example, if an extremely stable detector is used, then the variance reduction will not be sufficient to compensate for the loss in bias. However, *in practice*, the overall effect of the simulation is to improve the accuracy of most reasonable configurations of the base detector.

Methods like bagging and subsampling help unstable configurations of detectors to achieve their full potential. The unstable configuration of a detector has inherently higher potential for improvement because of fewer prior assumptions than a stable configuration and the fact that its base performance is impeded by variance to a greater degree. The overall error at very stable configurations (i.e., large number of nearest neighbors) actually increases slightly because of bagging, and therefore this choice is inappropriate for bagging.

7.4.2   Boosting

In boosting, a weight is associated with each training instance, and the different classifiers are trained with the use of these weights. The weights are modified iteratively based on classifier performance. In other words, the future models constructed are dependent on the results from previous models. Thus, each classifier in this model is constructed using a the same algorithm $\mathcal{A}$ on a weighted training data set. The basic idea is to focus on the incorrectly classified instances in future iterations by increasing the relative weight of these instances. The hypothesis is that the errors in these misclassified instances are caused by classifier bias. Therefore, increasing the instance weight of misclassified instances will result in a new classifier that corrects for the bias on these *particular* instances. By iteratively using this approach and creating a weighted combination of the various classifiers, it is possible to create a classifier with lower *overall* bias.

The most well-known approach to boosting is the *AdaBoost* algorithm. For simplicity, the following discussion will assume the binary class scenario. It is assumed that the class labels are drawn from $\{-1, +1\}$. This algorithm works by associating each training example with a *weight* that is updated in each iteration, depending on the results of the classification in the last iteration. The base classifiers therefore need to be able to work with weighted instances. Weights can be incorporated either by direct modification of training models, or by (biased) bootstrap sampling of the training data. The reader should revisit the section on rare class learning for a discussion on this topic. Instances that are misclassified are given higher weights in successive iterations. Note that this corresponds to intentionally biasing the classifier in later iterations with respect to the *global* training data, but reducing the bias in certain *local* regions that are deemed "difficult" to classify by the specific model $\mathcal{A}$.

In the tth round, the weight of the ith instance is $W_t(i)$. The algorithm starts with equal weight of $1/n$ for each of the n instances, and updates them in each iteration. In the event that the ith instance is misclassified, then its (relative) weight is increased to $W_{t+1}(i) = W_t(i)e^{\alpha_t}$, whereas in the case of a correct classification, the weight is decreased to $W_{t+1}(i) = W_t(i)e^{-\alpha_t}$. Here α_t is chosen as the function $\frac{1}{2}\log_e((1-\epsilon_t)/\epsilon_t)$, where ϵ_t is the fraction of incorrectly predicted training instances (computed after weighting with $W_t(i)$) by the model in the tth iteration. The approach terminates when the classifier achieves 100% accuracy on the training data ($\epsilon_t = 0$), or it performs worse than a random (binary) classifier ($\epsilon_t \geq 0.5$). An additional termination criterion is that the number of boosting rounds is bounded above by a user-defined parameter T. The overall training portion of the algorithm is illustrated in Figure 7.4.

It remains to be explained how a particular test instance is classified with the ensemble learner. Each of the models induced in the different rounds of boosting is applied to the test instance. The prediction $p_t \in \{-1, +1\}$ of the test instance for the tth round is weighted with α_t and these weighted predictions are aggregated. The sign of this aggregation $\sum_t p_t\alpha_t$ provides the class label prediction of the test instance. Note that less accurate components

Algorithm *AdaBoost*(Data Set: $\mathcal{D}$, Base Classifier: $\mathcal{A}$, Maximum Rounds: T)
begin
   $t = 0$;
  for each i initialize $W_1(i) = 1/n$;
  repeat
    $t = t + 1$;
    Determine weighted error rate ϵ_t on $\mathcal{D}$ when base algorithm $\mathcal{A}$
             is applied to weighted data set with weights $W_t(\cdot)$;
    $\alpha_t = \frac{1}{2}\log_e((1 - \epsilon_t)/\epsilon_t)$;
    for each misclassified $\overline{X_i} \in \mathcal{D}$ **do** $W_{t+1}(i) = W_t(i)e^{\alpha_t}$;
       else (correctly classified instance) **do** $W_{t+1}(i) = W_t(i)e^{-\alpha_t}$;
    for each instance $\overline{X_i}$ **do** normalize $W_{t+1}(i) = W_{t+1}(i)/[\sum_{j=1}^{n} W_{t+1}(j)]$;
  until $((t \geq T)$ OR $(\epsilon_t = 0)$ OR $(\epsilon_t \geq 0.5))$;
  Use ensemble components with weights α_t for test instance classification;
end

Figure 7.4: The *AdaBoost* algorithm

are weighted less by this approach.

An error rate of $\epsilon_t \geq 0.5$ is as bad or worse than the expected error rate of a random (binary) classifier. This is the reason that this case is also used as a termination criterion. In some implementations of boosting, the weights $W_t(i)$ are reset to $1/n$ whenever $\epsilon_t \geq 0.5$, and the boosting process is continued with the reset weights. In other implementations, ϵ_t is allowed to increase beyond 0.5, and therefore some of the prediction results p_t for a test instance are effectively inverted with negative values of the weight $\alpha_t = \log_e((1 - \epsilon_t)/\epsilon_t)$.

Boosting primarily focuses on reducing the bias. The bias component of the error is reduced because of the greater focus on misclassified instances. The ensemble decision boundary is a complex combination of the simpler decision boundaries, which are each optimized to specific parts of the training data. For example, if the *AdaBoost* algorithm uses a linear SVM on a data set with a nonlinear decision boundary, it will be able to learn this boundary by using different stages of the boosting to learn the classification of different portions of the data. Because of its focus on reducing the bias of classifier models, such an approach is capable of combining many weak (high bias) learners to create a strong learner. Therefore, the approach should generally be used with simpler (high bias) learners with low variance in the individual ensemble components. In spite of its focus on bias, boosting can occasionally reduce the variance slightly when re-weighting is implemented with sampling. This reduction is because of the repeated construction of models on randomly sampled, albeit re-weighted, instances. The amount of variance reduction depends on the re-weighting scheme used. Modifying the weights less aggressively between rounds will lead to better variance reduction. For example, if the weights are not modified at all between boosting rounds, then the boosting approach defaults to bagging, which only reduces variance. Therefore, it is possible to leverage variants of boosting to explore the bias-variance trade-off in various ways. However, if one attempts to use the vanilla *AdaBoost* algorithm with a high-variance learner, severe overfitting is likely to occur.

Boosting is vulnerable to data sets with significant noise in them. This is because boosting assumes that misclassification is caused by the bias component of instances near the *incorrectly modeled decision boundary*, whereas it might simply be a result of the mislabeling of the *data*. This is the noise component that is intrinsic to the *data*, rather than the *model*. In such cases, boosting inappropriately overtrains the classifier to low-quality portions of the data. Indeed, there are many noisy real-world data sets where boosting does not perform well. Its accuracy is typically superior to bagging in scenarios where the data sets are not excessively noisy.

7.5   Classifier Evaluation

Evaluation algorithms are important not only from the perspective of understanding the performance characteristics of a learning algorithm, but also from the point of view of optimizing algorithm performance via *model selection*. Given a particular data set, how can we know which algorithm to use? Should we use a support vector machine or a random forest? Therefore, the notions of model evaluation and model selection are closely intertwined.

Given a labeled data set, one cannot use all of it for model building. This is because the main goal of classification is to *generalize* a model of labeled data to unseen test instances. Therefore, using the same data set for both model building and testing grossly overestimates the accuracy. Furthermore, the portion of the data set used for *model selection* and *parameter tuning* also needs to be different from that used for model building. A common mistake is to use the same data set for both parameter tuning and final evaluation (testing). Such an approach partially mixes the training and test data, and the resulting accuracy is overly optimistic. Given a data set, it should always be divided into three parts.

1. *Training data:* This part of the data is used to build the training model such as a decision tree or a support vector machine. The training data may be used multiple times over different choices of the parameters or completely different algorithms to build the models in multiple ways. This process sets up the stage for *model selection*, in which the best algorithm is selected out of these different models. However, the actual *evaluation* of these algorithms for selecting the best model is not done on the training data but on a separate validation data set to avoid favoring overfitted models.

2. *Validation data:* This part of the data is used for model selection and parameter tuning. For example, the choice of the kernel bandwidth and the regularization parameters may be tuned by constructing the model multiple times on the first part of the data set (i.e., training data), and then using the validation set to estimate the accuracy of these different models. The best choice of the parameters is determined by using this accuracy. In a sense, validation data should be viewed as a kind of test data set to tune the parameters of the algorithm, or to select the best choice of the algorithm (e.g., decision tree versus support vector machine).

3. *Testing data:* This part of the data is used to test the accuracy of the final (tuned) model. It is important that the testing data are not even looked at during the process of parameter tuning and model selection to prevent overfitting. The testing data are *used only once at the very end of the process.* Furthermore, if the analyst uses the results on the test data to adjust the model in some way, then the results will be contaminated with knowledge from the testing data. The idea that one is allowed to look at a test data set only once is an extraordinarily strict requirement (and an important one). Yet, it is frequently violated in real-life benchmarks. The temptation to use what one has learned from the final accuracy evaluation is simply too high.

The division of the labeled data set into training data, validation data, and test data is shown in Figure 7.5. Strictly speaking, the validation data is also a part of the training data, because it influences the final model (although only the model building portion is often referred to as the training data). The division in the ratio of 2:1:1 is quite common. However, it should not be viewed as a strict rule. For very large labeled data sets, one needs only a modest number of examples to estimate accuracy. When a very large data set is available, it makes sense to use as much of it for model building as possible, because the variance induced by the validation and evaluation stage is often quite low. A constant

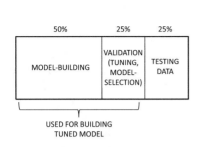

Figure 7.5: Partitioning a labeled data set for evaluation design

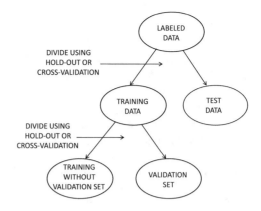

Figure 7.6: Hierarchical division into training, validation, and testing portions

number of examples (e.g., less than a few thousand) in the validation and test data sets are sufficient to provide accurate estimates.

7.5.1 Segmenting into Training and Testing Portions

The aforementioned description of partitioning the labeled data into three segments is an implicit description of a method referred to as *hold-out* for segmenting the labeled data into various portions. However, the division into *three* parts is not done in one shot. Rather, the training data is first divided into *two* parts for training and testing. The testing part is then carefully hidden away from any further analysis *until the very end where it can be used only once.* The remainder of the data set is then divided again into the training and validation portions. This type of recursive division is shown in Figure 7.6.

A key point is that the types of division at both levels of the hierarchy are conceptually identical. In the following, we will consistently use the terminology of the first level of division in Figure 7.6 into "training" and "testing" data, even though the same approach can also be used for the second-level division into model building and validation portions. This consistency in terminology allows us to provide a common description for both levels of the division.

7.5.1.1 Hold-Out

In the hold-out method, a fraction of the instances are used to build the training model. The remaining instances, which are also referred to as the *held out* instances, are used for testing. The accuracy of predicting the labels of the held out instances is then reported as the overall accuracy. Such an approach ensures that the reported accuracy is not a result of overfitting to the specific data set, because different instances are used for training and testing. The approach, however, underestimates the true accuracy. Consider the case where the held-out examples have a higher presence of a particular class than the labeled data set. This means that the held-in examples have a lower average presence of the same class, which will cause a mismatch between the training and test data. Furthermore, the class-wise frequency of the held-in examples will always be inversely related to that of the held-out examples. This will lead to a consistent pessimistic bias in the evaluation.

7.5.1.2   Cross-Validation

In the cross-validation method, the labeled data is divided into q equal segments. One of the q segments is used for testing, and the remaining $(q-1)$ segments are used for training. This process is repeated q times by using each of the q segments as the test set. The average accuracy over the q different test sets is reported. Note that this approach can closely estimate the true accuracy when the value of q is large. A special case is one where q is chosen to be equal to the number of labeled documents and therefore a single document is used for testing. Since this single document is left out from the training data, this approach is referred to as *leave-one-out cross-validation*. Although such an approach can closely approximate the accuracy, it is usually too expensive to train the model a large number of times. Nevertheless, leave-one-out cross-validation is the method of choice for lazy learning algorithms like nearest neighbor classifiers.

7.5.2   Absolute Accuracy Measures

Once the data have been segmented between training and testing, a natural question arises about the type of accuracy measure that one can use in classification and regression.

7.5.2.1   Accuracy of Classification

When the output is presented in the form of class labels, the ground-truth labels are compared to the predicted labels to yield the following measures:

1. *Accuracy:* The accuracy is the fraction of test instances in which the predicted value matches the ground-truth value.

2. *Cost-sensitive accuracy:* Not all classes are equally important in all scenarios, while comparing the accuracy. This is particularly important in imbalanced class problems, in which one of the classes is much rarer than the other. For example, consider an application in which it is desirable to classify tumors as *malignant* or *non-malignant* where the former is much rarer than the latter. In such cases, the misclassification of the former is often much less desirable than misclassification of the latter. This is frequently quantified by imposing differential costs $c_1 \ldots c_k$ on the misclassification of the different classes. Let $n_1 \ldots n_k$ be the number of test instances belonging to each class. Furthermore, let $a_1 \ldots a_k$ be the accuracies (expressed as a fraction) on the subset of test instances belonging to each class. Then, the overall accuracy A can be computed as a weighted combination of the accuracies over the individual labels.

$$A = \frac{\sum_{i=1}^{k} c_i n_i a_i}{\sum_{i=1}^{k} c_i n_i} \tag{7.4}$$

The cost sensitive accuracy is the same as the unweighted accuracy when all costs $c_1 \ldots c_k$ are the same.

Aside from the accuracy, the statistical robustness of a model is also an important issue. For example, if two classifiers are trained over a small number of test instances and compared, the difference in accuracy may be a result of random variations, rather than a truly *statistically significant* difference between the two classifiers. This measure is related to that of the variance of a classifier that was discussed earlier in this chapter. When the variance of two classifiers is high, it is often difficult to assess whether one is truly better than the other. One way of testing the robustness is to repeat the aforementioned process of cross-validation (or hold-out) in many different ways (or *trials*) by repeating the randomized process of creating the folds in many different ways. The difference δa_i in accuracy between the ith pair of classifiers (constructed on the same folds) is computed, and the standard deviation σ of this difference is computed as well. The overall difference in accuracy over s trials is computed as follows:

$$\Delta A = \frac{\sum_{i=1}^{s} \delta a_i}{s} \tag{7.5}$$

Note that ΔA might be positive or negative, depending on which classifier is winning. The standard deviation is computed as follows:

$$\sigma = \sqrt{\frac{\sum_{i=1}^{s} (\delta a_i - \Delta A)^2}{s - 1}} \tag{7.6}$$

Then, the overall statistical level of significance by which one classifier wins over the other is given by the following:

$$Z = \frac{\Delta A \sqrt{s}}{\sigma} \tag{7.7}$$

The factor $\sqrt{s}$ accounts for the fact that we are using the sample mean ΔA, which is more stable that the individual accuracy differences δa_i. The standard deviation of ΔA is a factor $1/\sqrt{s}$ of the standard deviation of individual accuracy differences. Values of Z that are significantly greater than 3, are strongly indicative of one classifier being better than the other in a statistically significant way.

7.5.2.2 Accuracy of Regression

The effectiveness of linear regression models can be evaluated with a measure known as the Mean Squared Error (MSE), or the Root Mean Squared Error, which is the RMSE. Let $y_1 \ldots y_r$ be the observed values over r test instances, and let $\hat{y}_1 \ldots \hat{y}_r$ be the predicted values. Then, the mean-squared error, denoted by MSE is defined as follows:

$$MSE = \frac{\sum_{i=1}^{r} (y_i - \hat{y}_i)^2}{r} \tag{7.8}$$

The Root-Mean-Squared Error (RMSE) is defined as the square root of this value:

$$RMSE = \sqrt{\frac{\sum_{i=1}^{r} (y_i - \hat{y}_i)^2}{r}} \tag{7.9}$$

Another measure is the R^2-*statistic*, or the *coefficient of determination*, which provides a better idea of the *relative performance* of a particular model. In order to compute the R^2-statistic, we first compute the variance σ^2 of the observed values. Let $\mu = \sum_{j=1}^{r} y_j / r$ be the mean of the dependent variable. Then, the variance σ^2 of the r observed values of the test instances is computed as follows:

$$\sigma^2 = \frac{\sum_{i=1}^{r} (y_i - \mu)^2}{r} \tag{7.10}$$

Then, the R^2-statistic is as follows:

$$R^2 = 1 - \frac{MSE}{\sigma^2} \tag{7.11}$$

Larger values of the R^2 statistic are desirable, and the maximum value of 1 corresponds to an MSE of 0. It is possible for the R^2-statistic to be negative, when it is applied on an out-of-sample test data set, or even when it is used in conjunction with a nonlinear model.

Although we have described the computation of the R^2-statistic for the test data, this measure is often used on the training data in order to compute the fraction of unexplained variance in the model. In such cases, linear regression models always return an R^2-statistic in the range $(0, 1)$. This is because the mean value μ of the dependent variable in the training data can be predicted by a linear regression model, when the coefficients of the features are set to 0 and only the bias term (or coefficient of dummy column) is set to the mean. Since the linear regression model will always provide a solution with a lower objective function value on the training data, it follows that the value of MSE is no larger than σ^2. As a result, the value of the R^2-statistic on the training data always lies in the range $(0, 1)$. In other words, a training data set can never be predicted better using its mean than by using the predictions of linear regression. However, an out-of-sample test data set *can* be modeled better by using its mean than by using the predictions of linear regression.

One can increase the R^2-statistic on the training data simply by increasing the number of regressors, as the MSE reduces with increased overfitting. When the dimensionality is large, and it is desirable to compute the R^2-statistic on the training data, the adjusted R^2-statistic provides a more accurate measure. In such cases, the use of a larger number of features for regression is penalized. The adjusted R^2-statistic for a training data set with n documents and d dimensions is computed as follows:

$$R^2 = 1 - \frac{(n - d)}{(n - 1)} \frac{MSE}{\sigma^2} \tag{7.12}$$

The R^2-statistic is generally used only for linear models. For nonlinear models, it more common to use the MSE as a measure of the error.

7.5.3   Ranking Measures for Classification and Information Retrieval

The classification problem is posed in different ways, depending on the setting in which it is used. The absolute accuracy measures discussed in the previous section are useful in cases where the labels or numerical dependent variables are predicted as the final output. However, in some settings, a particular *target* class is of special interest, and all the test instances are *ranked* in order of their propensity to belong to the target class. A particular example is that of classifying email as "*spam*" or "*not spam*." When one has a large number of documents with a high imbalance in relative proportion of classes, it makes little sense to directly return binary predictions. In such cases, only the top-ranked emails will be returned based on the probability of belonging to the "*spam*" category, which is the target class. Ranking-based evaluation measures are often used in imbalanced class settings in which one of the classes (i.e., the rare class) is considered more relevant from a detection point of view.

Ranking-based evaluations are also useful in information retrieval settings, in which a keyword query is entered by a user, and a ranked list of documents is returned based on their relevance. Such methods are also used for Web search, which will be discussed in Chapter 9. All such information retrieval problems can implicitly be considered two-class problems in which the documents belong to either the *"relevant"* class or *"not relevant"* class, and the ranking is returned based on the propensity to belong to the former. Therefore, the evaluation discussion in this section is not only relevant to classification, but is also useful from the broader point of view of information retrieval, Web search, and some other applications:

1. In outlier analysis, one often returns a ranked list of anomalies. Although outlier detection is an unsupervised problem, a binary ground truth is often available for evaluation.

2. In recommender systems with implicit feedback, binary ground truth may be available about which items have been consumed. A ranked list of recommendations can be evaluated against this ground truth.

3. In information retrieval and search, the ground-truth set of relevant documents may be available. The ranked list of retrieved documents can be evaluated against this binary ground truth.

Discussions of some of these different ranking measures is also provided in different contexts [3, 4].

7.5.3.1 Receiver Operating Characteristic

Ranking methods are used frequently in cases where a ranked list of a particular class of interest is returned. The ground-truth is assumed to be binary in which the class of interest corresponds to the positive class, and the remaining documents belong to the negative class. In most such settings, the relative frequencies of the two classes are heavily imbalanced, so that the discovery of (rare) positive class instances is more desirable. This situation is also true in information retrieval and search, in which the set of documents returned in response to a keyword search can be viewed as belonging to the *"relevant"* class.

The instances that belong to the positive class in the *observed data* are *ground-truth positives* or *true positives*. It is noteworthy that when information retrieval, search, or classification applications are used, the algorithm can *predict* any number of instances as positives, which might be different from the number of *observed* positives (i.e., true positives). When a larger number of instances are predicted as positives, one would recover a larger number of the true positives, but a smaller percentage of the predicted list would be correct. This type of trade-off can be visualized with the use of a precision-recall or a *receiver operating characteristic (ROC)* curve. Such trade-off plots are commonly used in rare class detection, outlier analysis evaluation, recommender systems, and information retrieval. In fact, such trade-off plots can be used in any application where a binary ground truth is compared to a ranked list discovered by an algorithm.

The basic assumption is that it is possible to rank all the test instances using a numerical score, which is the output of the algorithm at hand. This numerical score is often available from classification algorithms in the form of a probability of belonging to the positive class in methods like the naïve Bayes classifier or logistic regression. For methods like SVMs, one can report the (signed) distance of a point from the separating class instead of converting it into a binary prediction. A threshold on the numerical score creates a predicted list of positives. By varying the threshold (i.e., size of predicted list), one can quantify the fraction of relevant (ground-truth positive) instances in the list, and the fraction of relevant

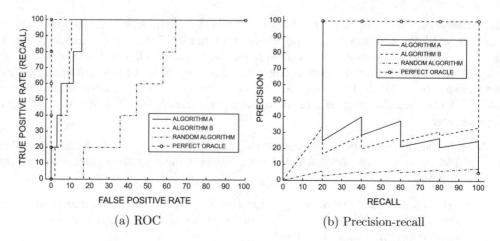

(a) ROC   (b) Precision-recall

Figure 7.7: ROC curve and precision-recall curves

Table 7.1: Rank of ground-truth positive instances

Algorithm	Rank of ground-truth positives (ground-truth positives)
Algorithm A	1, 5, 8, 15, 20
Algorithm B	3, 7, 11, 13, 15
Random Algorithm	17, 36, 45, 59, 66
Perfect Oracle	1, 2, 3, 4, 5

instances that are missed by the list. If the predicted list is too small, the algorithm will miss relevant instances (false-negatives). On the other hand, if a very large list is recommended, there will be too many spuriously predicted instances (i.e., false-positives). This leads to a trade-off between the false-positives and false-negatives, which can be visualized with the *precision-recall* curve or the *receiver operating characteristic (ROC)* curve.

Assume that one selects the top-t set of ranked instances and predicted them to belong to the positive class. For any given value t of the size of the positively predicted list, the set of instances predicted to belong to the positive class is denoted by $\mathcal{S}(t)$. Note that $|\mathcal{S}(t)| = t$. Therefore, as t changes, the size of $\mathcal{S}(t)$ changes as well. Let $\mathcal{G}$ represent the true set of relevant documents (ground-truth positives). Then, for any given size t of the predicted list, the *precision* is defined as the percentage of percentage of instances predicted to belong to the positive class that truly turn out to belong to the positive class in the predicted labels:

$$Precision(t) = 100 \cdot \frac{|\mathcal{S}(t) \cap \mathcal{G}|}{|\mathcal{S}(t)|}$$

The value of $Precision(t)$ is *not* necessarily monotonic in t because both the numerator and denominator may change with t differently. The *recall* is correspondingly defined as the percentage of *ground-truth* positives that have been recommended as positive for a list of size t.

$$Recall(t) = 100 \cdot \frac{|\mathcal{S}(t) \cap \mathcal{G}|}{|\mathcal{G}|}$$

While a natural trade-off exists between precision and recall, this trade-off is not necessarily monotonic. In other words, an increase in recall does not always lead to a reduction in precision. One way of creating a single measure that summarizes both precision and recall is the F_1-measure, which is the harmonic mean between the precision and the recall.

$$F_1(t) = \frac{2 \cdot Precision(t) \cdot Recall(t)}{Precision(t) + Recall(t)} \qquad (7.13)$$

While the $F_1(t)$ measure provides a better quantification than either precision or recall, it is still dependent on the size t of the number of instances predicted to belong to the positive class, and is therefore still not a complete representation of the trade-off between precision and recall. It is possible to visually examine the entire trade-off between precision and recall by varying the value of t and plotting the precision versus the recall. The lack of monotonicity of the precision makes the results hard to interpret.

A second way of generating the trade-off in a more intuitive way is through the use of the ROC curve. The *true-positive rate*, which is the same as the recall, is defined as the percentage of ground-truth positives that have been included in the predicted list of size t.

$$TPR(t) = Recall(t) = 100 \cdot \frac{|\mathcal{S}(t) \cap \mathcal{G}|}{|\mathcal{G}|}$$

The false-positive rate $FPR(t)$ is the percentage of the falsely reported positives in the predicted list out of the ground-truth negatives (i.e., irrelevant documents belonging to the negative class in the observed labels). Therefore, if $\mathcal{U}$ represents the universe of all test instances, the ground-truth negative set is given by $(\mathcal{U} - \mathcal{G})$, and the falsely reported part in the predicted list is $(\mathcal{S}(t) - \mathcal{G})$. Therefore, the false-positive rate is defined as follows:

$$FPR(t) = 100 \cdot \frac{|\mathcal{S}(t) - \mathcal{G}|}{|\mathcal{U} - \mathcal{G}|} \qquad (7.14)$$

The false-positive rate can be viewed as a kind of "bad" recall, in which the fraction of the ground-truth negatives (i.e., test instances with observed labels in the negative class), which are incorrectly captured in the predicted list $\mathcal{S}(t)$, is reported. The ROC curve is defined by plotting the $FPR(t)$ on the X-axis and $TPR(t)$ on the Y-axis for varying values of t. In other words, the ROC curve plots the "good" recall against the "bad" recall. Note that both forms of recall will be at 100% when $\mathcal{S}(t)$ is set to the entire universe of test documents (or entire universe of documents to return in response to a query). Therefore, the end points of the ROC curve are always at $(0,0)$ and $(100,100)$, and a random method is expected to exhibit performance along the diagonal line connecting these points. The *lift* obtained above this diagonal line provides an idea of the accuracy of the approach. The area under the ROC curve provides a concrete quantitative evaluation of the effectiveness of a particular method. Although one can directly use the area shown in Figure 7.7(a), the staircase-like ROC curve is often modified to use local linear segments which are not parallel to either the X-axis or the Y-axis. The trapezoidal rule [190] is then used to compute the area slightly more accurately. From a practical point of view, this change often makes very little difference to the final computation.

To illustrate the insights gained from these different graphical representations, consider an example of a scenario with 100 test instances, in which 5 documents truly belong to the positive class. Two algorithms A and B are applied to this data set that rank all test instances from 1 to 100 to belong to the positive class, with lower ranks being selected first in the predicted list. Thus, the true-positive rate and false-positive rate values can be generated from the ranks of the 5 test instances in the positive class. In Table 7.1, some hypothetical ranks for the 5 truly positive instances have been illustrated for the different algorithms. In addition, the ranks of the ground-truth positive instances for a random algorithm have been indicated. This algorithm ranks all the test instances randomly. Similarly, the ranks for a "perfect oracle" algorithm are such that the correct positive instances are placed as the top 5 instances in the ranked list. The resulting ROC curves are illustrated in Figure 7.7(a). The corresponding precision-recall curves are illustrated in Figure 7.7(b). Note that the ROC curves are always increasing monotonically, whereas the precision-recall curves are not monotonic. While the precision-recall curves are not quite as nicely interpretable as the ROC curves, it is easy to see that the *relative trends* between different algorithms are the same in both cases. In general, ROC curves are used more frequently because of greater ease in interpretability.

What do these curves really tell us? For cases in which one curve strictly dominates another, it is clear that the algorithm for the former curve is superior. For example, it is immediately evident that the oracle algorithm is superior to all algorithms and that the random algorithm is inferior to all the other algorithms. On the other hand, algorithms A and B show domination at different parts of the ROC curve. In such cases, it is hard to say that one algorithm is strictly superior. From Table 7.1, it is clear that Algorithm A ranks three positive instances very highly, but the remaining two positive instances are ranked poorly. In the case of Algorithm B, the highest ranked positive instances are not as well ranked as Algorithm A, though all 5 positive instances are determined much earlier in terms of rank threshold. Correspondingly, Algorithm A dominates on the earlier part of the ROC curve, whereas Algorithm B dominates on the later part. It is possible to use the area under the ROC curve as a proxy for the overall effectiveness of the algorithm. However, not all parts of the ROC curve are equally important because there are usually practical limits on the size of the predicted list.

Application to information retrieval and search: The ROC can also be used for evaluation in information retrieval and search. The only difference is that instead of a single prediction problem, we have a set Q of multiple queries. Each such query has its own ROC, and the AUCs of the different queries are averaged to provide a final result.

Intuitive interpretation of Area under Curve (AUC): The area under the curve has a natural intuitive interpretation. If one samples two random test instances, such that one of them belongs to the positive class and the other belongs to the negative class, the AUC provides the probability that the two instances are ranked correctly with respect to each other by the ranking algorithm. When the algorithm returns random rankings, each of these instances is equally likely to occur ahead of the other in the ranked list. As a result, the AUC of a random algorithm is 0.5.

Average precision and mean-average precision (MAP): The average precision is defined in the single-query setting (like classification), whereas the mean average precision is defined in a multi-query setting like information retrieval where multiple queries are used. In the context of information retrieval applications, the precision is also referred to as the *hit rate*. Let L be the maximum size of the recommended list in an information retrieval setting, and $Precision(t)$ be the precision, when the size of the predicted list is t. Then, the average precision AP is computed as follows:

$$AP = \frac{\sum_{t=1}^{L} Precision(t)}{L} \tag{7.15}$$

This defines the average precision over a single query. However, if we have a query set Q, and AP_i corresponds to the precision of the ith query, then the mean average precision is defined as the mean of these values over the $|Q|$ different queries.

$$MAP = \frac{\sum_{i=1}^{|Q|} AP_i}{|Q|} \tag{7.16}$$

It is possible to set the value of L to the size of the universe of documents, although this is often not done in practice. In practice, a maximum "reasonable" size of the recommended list is used to set the value of L.

7.5.3.2 Top-Heavy Measures for Ranked Lists

One disadvantage of the receiver operating characteristic is that it places an equal level of importance on the top-ranked instances versus the lower-ranked instances. For example, moving an instance belonging to the positive class in the ranking below 10 additional negative instances has the same incremental effect on the AUC, irrespective of whether that instance was originally at the top of the list or whether that instance was in the middle of the list. However, from an application-centric point of view, the user often pays much more attention to the top of the list. Therefore, it is useful to design performance measures that pay greater attention to the top of the ranked list. These types of measures are particularly important in information retrieval and search, in which the returned list of results is an extremely small fraction of the universe of documents, and it is not realistic to use measures such as the AUC that require the *entire* ranked list. Top-heavy measures provide decreasing importance to the instances ranked lower in the list, so that the effect of changing the list lower down the order has little effect on the performance metric. From a practical point of view, this approach truncates the ranked list of predicted positive instances because the vast majority of items that are very low down the list have little effect on the overall evaluation.

In utility-based ranking, the basic idea is that each positive instance in the recommended list contributes a utility value that depends on its position in the list. If a positive item is ranked higher in the recommended list, then it has greater utility to the user, because it is more likely to be noticed by virtue of its position. This is a somewhat different concept from the AUC, which only uses the *relative* positions of the positive and negative instances, and pays little attention to their absolute position.

Let v_j be the position of the jth test instance in the recommended list. Furthermore, let $y_j \in \{0, 1\}$ be its label corresponding to whether it is relevant or not. A value of 1 indicates that it is relevant. In the classification[2] setting, a value of 0 corresponds to the negative class, whereas a value of 1 corresponds to the positive class.

An example of such a measure is the discounted cumulative gain (DCG). In this case, the discount factor of the jth test instance is set to $\log_2(v_j + 1)$, where v_j is the rank of this instance in the recommended list. Then, the discounted cumulative gain *for a single query* is defined as follows:

$$DCG = \sum_{j : v_j \leq L} \frac{2^{rel_j} - 1}{\log_2(v_j + 1)} \tag{7.17}$$

Here, rel_j is the ground-truth relevance of test instance j. In the classification setting, the value of rel_j might simply be set to y_j. In information retrieval settings, the value of rel_j is set to the numerical score that a human evaluator gives to the document. Note that the discounted cumulative gain only gets credit for those test instances in which the

---

[2]Throughout this book, we have used $y_j \in \{-1, +1\}$ in the classification setting. However, we switch to the notation $\{0, 1\}$ here for greater conformity with the information retrieval literature.

value of v_j is at most L. In some settings like classification, the value of L is set to the total number of test instances. However, in other settings, the value of L is set to some reasonably large value beyond which it does not make sense to examine the recommended list. The above description is for the case of single query setting like classification. In the multi-query setting, the value of the DCG is averaged over the different queries. Note that each query would have its own ground truth set, and corresponding values of rel_j, and therefore the discounted cumulative gain needs to be computed independently for each query.

Then, the normalized discounted cumulative gain (NDCG) is defined as ratio of the discounted cumulative gain to its ideal value, which is also referred to as ideal discounted cumulative gain (IDCG).

$$NDCG = \frac{DCG}{IDCG} \qquad (7.18)$$

The ideal discounted cumulative gain is computed by repeating the computation for DCG, except that the ground-truth rankings are used in the computation. The basic idea in the computation is that it is assumed that the ranking system can correctly place the ground-truth positive instances at the top of the ranked list. The ideal score is computed under this assumption.

7.6 Summary

This chapter discusses the theoretical aspects of text classification performance and its applications in improve the accuracy of text classifiers. In particular, the use of ensemble methods in improving classifier accuracy was discussed. In addition, the evaluation of classification algorithms was discussed. Text classifier evaluation is closely related to that of evaluation of search engines, particularly when ranking-based measures are used. The receiver operating characteristic is commonly used for evaluating the accuracy of classifiers. In addition, a number of top-heavy measures such as the use of normalized discounted cumulative gain are introduced in this chapter.

7.7 Bibliographic Notes

A detailed discussion of the bias-variance trade-off may be found in [235]. The bias-variance trade-off was originally proposed for regression, though it was eventually generalized for binary loss functions in classification [301, 302]. The bias-variance trade-off has also been studied from the perspective of unsupervised problems such as outlier analysis [10]. A discussion of ensemble methods for classification may be found in [507, 631], and a discussion for outlier detection is found in [10]. Bagging and random forest methods for classification are discussed in [70, 71, 76]. In addition, feature begging methods for classification were introduced in [250, 251]. These methods were precursors to the random forest technique. The use of bagging for 1-nearest neighbor detectors is studied in [494]. The *AdaBoost* algorithm was introduced in [200], and a ranking variant for information retrieval, referred to as *AdaRank*, is discussed in [596]. Stochastic gradient boosting methods are proposed in [201].

Connection of Boosting to Logistic Regression

Although boosting methods may sometimes seem mysterious in their ability to consistently improve accuracy, they can be understood better when viewed from the perspective of iterative variations of linear regression that fit linear models to nonlinear data with the use of example re-weighting. Such models are referred to as *generalized additive models* [238] that attempt to fit a (simpler) linear model to a complex data distribution by applying multiple instantiations of the linear model on re-weighted or modified instances of the training data. An example of such a model with numerical data is to apply linear regression iteratively with the residuals as new response variables and also reweighting instances. This is a classical form of generalized additive models that was known long before the advent of the boosting algorithm [238].

However, iterative linear regression with residuals is suited to numerical response variables. Logistic regression is a probabilistic approach that uses the logistic function to convert the numerical response in linear regression to a binary response with a Bernoulli assumption on the response variable. For the two-class problem, boosting can be viewed as an approximation to additive modeling on the logistic scale using maximum Bernoulli likelihood as a criterion [202]. Note that the use of the logistic scale is a standard way to convert numerical regression responses to the binary case. The exponential re-weighting in boosting can be explained on the basis of this assumption. In essence, *AdaBoost* adapts the loss function from the approach used in generalized additive models in a suitable form for classification, and then uses an iterative approach to optimize it. Furthermore, *AdaBoost* is applied as an ensemble method with any classification algorithm rather than as a generalized variant of a specific classification (i.e., logistic regression) model. This historical connection is also consistent with the fact that *AdaBoost* should only be used in combination with simple models with low variance (like linear models). Like *AdaBoost*, generalized linear models are susceptible to overfitting.

Classifier Evaluation

Evaluation methods for classification, recommender systems, regression, information retrieval, and outlier analysis are closely related. In fact, many techniques like the precision-recall measures and the receiver operating characteristic are used in all these cases. A detailed discussion of evaluation methods in recommender systems may be found in [3]. Such evaluation measures have significant usefulness in the context of information retrieval applications. The receiver operating characteristic curve is discussed in detail in [190]. A detailed discussion of several evaluation methods for classification and information retrieval may also be found in [367].

7.7.1 Software Resources

Ensemble methods are available in many of the libraries such as the Python library *scikit-learn* [646], the **caret** package in R [305], and the package **RTextTools** [667]. The **rotationForest** package [668] in R, which is available from CRAN, can be used to address the sparsity challenges associated with text. The *Weka* library [649] in Java contains numerous ensemble methods for classification. Most of the aforementioned libraries contain built-in tools for classifier evaluation, such as the accuracy, precision, recall, and the receiver operating characteristic.

7.7.2 Data Sets for Evaluation

Many data sets are available for the evaluation of classification algorithms. Any discussion of data sets for text is incomplete without the discussion of the pioneering *20 Newsgroups* and *Reuters* data sets. The *20-Newsgroups* data set [672] contains about 1000 articles from twenty different Usenet groups. Therefore, this is a multiclass problem, which is used commonly in many classification settings. The *Reuters* data set has two variations corresponding to *Reuters-21578* [673] and (the larger) *Reuters Corpus Volume 1 (RCV1)* [674]. The former data set derives its name from the back that it contains 21578 news articles from the *Reuters* newswire service. The latter collection contains more than 800,000 articles. The *University of California at Irvine Machine Learning Repository* [645] contains several labeled text data sets. The *Europeana Linked Open Data* initiative [675] has a collection of text data sets, which includes other types of rich data such as links, images, videos, and other metadata. The *ICWSM 2009 data set* challenge [676] has also published a very large data set of 44 million blog posts. Although the data set is not specific to classification, it can be used for a variety of supervised applications with the appropriate annotation. In addition, Stanford University NLP [651] contains a large number of text corpora, albeit for linguistically focused applications.

A number of data sets are also available for evaluating search and retrieval relevance, which is related to but not quite the same as classification. Among them, the Text Retrieval Conference (TREC) collections are among the most well known for information retrieval evaluation [669]. The data sets also contain relevance judgements that are useful for evaluation purposes. The NII Test Collection for IR Systems (NTCIR) focuses on cross-language retrieval and may be found in [670]. The cross-language evaluation forum (CLEF) also provides a similar source of data sets [671].

7.8   Exercises

1. Discuss the effect on the bias and variance by making the following changes to a classification algorithm: (a) Increasing the regularization parameter in a support vector machine, (b) Increasing the Laplacian smoothing parameter in a nïve Bayes classifier, (c) Increasing the depth of a decision tree, (d) Increasing the number of antecedents in a rule, (e) Reducing the bandwidth σ, when using the Gaussian kernel in conjunction with a support vector machine.

2. Suppose you found the optimal setting of the Gaussian kernel in and SVM for a data set. Now you were given additional training data to use for your SVM, which is identically distributed to your previous data set. Given the larger data set, would the optimal value of the kernel bandwidth increase or decrease in most settings?

3. A key parameter while designing a subsampling ensemble is the size of the subsample. Discuss the effect of this choice from the point of view of the bias-variance trade-off.

4. Implement a subsampling ensemble in combination with a 1-nearest neighbor classifier.

5. Suppose you used a 1-nearest neighbor classifier in combination with an ensemble method, and you are guaranteed that each application gets you the correct answer for a test instance with 60% probability. What is the probability that you get the correct answer using a majority vote of three tries, each of which are guaranteed to be independent and identically distributed?

6. Suppose that a data set is sampled without replacement to create a bagged sample of the same size as the original data set. Show that he probability that a point will not be included in the re-sampled data set is given by $1/e$, where e is the base of the natural logarithm.

Chapter 8

Joint Text Mining with Heterogeneous Data

"We become not a melting pot but a beautiful mosaic. Different people, different beliefs, different yearnings, different hopes, different dreams." – Jimmy Carter

8.1   Introduction

Text documents often occur in combination with other heterogeneous data such as images, Web links, social media, ratings, and so on. Examples of such settings are as follows:

1. *Web and social media links:* In Web and social media networks, the text documents are often associated with nodes. For example, the Web can be a viewed as a graph in which each node contains a Web page and also connects to other nodes via hyperlinks. Similarly, a social network is a friendship graph of user-to-user linkages in which each node contains the textual posting activity of the user. Therefore, one can associate a node with a list of terms as well as a list of other nodes.

2. *Image data:* Many images on the Web and from other sources are associated with textual captions and other content. One can therefore conceptualize two separate collections of text and images, which are connected to one another with co-occurrence/captioning links.

3. *Cross-lingual data:* Cross-lingual data contains a separate corpus for each language, and associations might exist between the two collections. These associations might be based on either cross-lingual dictionaries or pairs of similar documents. The goal is to discover a joint representation for text mining.

The heterogeneous nature of the features creates numerous challenges for algorithm design. Although researchers and practitioners have often studied these different problem domains independently, there are surprising similarities in the underlying techniques. Therefore, this chapter will provide a unified presentation of these heterogeneous settings.

© Springer Nature Switzerland AG 2022
C. C. Aggarwal, *Machine Learning for Text*,
https://doi.org/10.1007/978-3-030-96623-2_8

The aforementioned problems arise in both supervised (e.g., classification) and unsupervised settings (e.g., clustering and topic modeling). The resulting methods are fairly general, and they cover nontraditional variations of these problems such as *transfer learning*. In transfer learning, labeled data (e.g., documents) in one domain are used to perform classification in another domain (e.g., images), when the amount of labeled data in the first domain is significantly greater than the second. Transfer learning is also used in the unsupervised context, in which the unlabeled data in the text domain is used to learn semantically coherent features for a domain like image data.

The main challenge in all these cases is that the input feature spaces of the different domains are different. Machine learning methods can be best implemented when the data from the different modalities are embedded in a single feature space. This opens the door to *latent modeling techniques*, which can take on one of three forms:

1. *Shared matrix factorization:* In these case, the various types of data are represented as matrices, and the relationships among these matrices are expressed in the form of a *factorization graph*, which expresses a set of factorization relationships among matrices. This terminology should not be confused with the notion of a factor graph used in probabilistic graphical models. Each vertex corresponds to the *latent variables* of a row (e.g., document, image, social network node) or a column (e.g., term, visual feature, social network node). Each edge corresponds to a factorization relationship by multiplying the latent variable matrices at its end points, and therefore the edge is labeled with the relevant matrix that is factorized by these latent variables. Therefore, given a problem with heterogeneous data, all that the analyst has to do is to set up an appropriate factorization graph, which is followed by setting up its corresponding optimization problem and gradient-descent steps. It is noteworthy that even though this approach has implicitly been used by many researchers repeatedly across many data domains, this book will formalize the approach as a more systematic framework. It is hoped that such a systematic framework will reduce the burden on a practitioner and researcher who is faced with a new setting involving heterogeneous data.

2. *Factorization machines:* Factorization machines were recently proposed in the context of heterogeneous data in recommender systems. The approach can be extended to all types of heterogeneous settings beyond recommender systems. Many special cases of factorization machines are identical to shared matrix factorization methods. One advantage of using factorization machines is that the problem can be reduced to the systematic process of *feature engineering*. Furthermore, off-the-shelf software for factorization machines is available. On the other hand, factorization machines are better suited to supervised problems like regression and classification, whereas shared matrix factorization methods can be applied to broader settings.

3. *Joint probabilistic modeling:* In joint probabilistic modeling, a generative process is assumed to create the documents and other data types based on a hidden variable. Each data type is drawn from its own probability distribution.

Another common approach is to convert the heterogeneous data to a *relationship graph* on which network mining algorithms are directly applied. In such cases, each data item (document) or feature corresponds to a node in the graph and the edges represent the relationships among them (e.g., presence of a term in a document). One can view the relationship graph as an expanded representation of the factorization graph in which one does not attempt to summarize the data with latent variables. This approach is different from the aforementioned methods in that it does not directly try to find to a compressed (latent) representation of the data, but it explicitly tries to model structural relationships between the individual data items and/or features of various domains, and represent them as a network. The resulting network is often quite large because each data item (e.g., document) or feature (e.g., term) corresponds to a distinct node in the resulting network. This type of modeling allows the use of off-the-shelf structural mining algorithms.

Finally, a general comment about the goals of this chapter is in order. While it is possible to create separate chapters for text-image mining, text-link mining, or cross-lingual mining, such an approach does not help one grasp the main ideas behind the *common* principles of the corresponding mathematical techniques. Therefore, if one is faced with a new setting involving a different type of data, one has to start from scratch in designing a suitable methodology for the mining process without being able to generalize the ideas already available from known settings. The goal of this chapter is not to teach the reader about a *specific* type of heterogeneous setting such as text, images, links, or cross-lingual data. Rather, it is to point out the *common threads* that run through these (seemingly independent) lines of research and educate the reader to use them in a *systematic* way. This book therefore summarizes this commonly used "bag of tricks" from various domains (and independent threads of work) in the context of a unified framework for the first time. In addition, for each specific approach, examples from various application domains will be provided to compare and contrast the different tricks.

Chapter Organization

This chapter is organized as follows. The next section introduces shared matrix factorization. Factorization machines are introduced in Section 8.3. Joint probabilistic modeling techniques are introduced in Section 8.4. The use of graph mining techniques for mining heterogeneous data domains is discussed in Section 8.5. A summary is given in Section 8.6.

8.2 The Shared Matrix Factorization Trick

The main challenge while dealing with heterogeneous data is the fact that the different data domains use completely different feature spaces to represent their data. For example, while documents are represented as document-word matrices, images are represented as image-(visual word) matrices. Furthermore, indirect relationships between documents and images such as captioning and user tagging may also be expressed in the form of matrices. Note that the document-term matrix shares the document modality with the document-image matrix. In general, *we have a set of matrices in which some of the modalities are shared*, and we wish to extract latent representations of the shared relationships implicit in these matrices. These latent representations are typically expressed in the form of low-rank matrices, which can be used for any application such as clustering, classification, and so on. The key in this entire process is to use *shared latent factors* between different modalities so that they are able to incorporate the impact of these relationships in an indirect (i.e., latent) way within the extracted embedding. A key aspect of this approach is to create a *factorization graph* that expresses the latent relationships between the different data modalities.

8.2.1 The Factorization Graph

The factorization graph is a way of expressing how different data matrices are created by different features and data items. The simplest possible factorization graph is that of

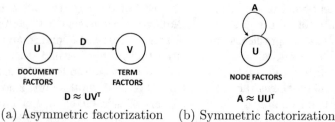

(a) Asymmetric factorization   (b) Symmetric factorization

Figure 8.1: Simplest possible examples of factorization graphs

factorizing the $n \times d$ document-term matrix D into the $n \times k$ latent factor U and the $d \times k$ latent factor V. The corresponding document-term matrix D is factorized as follows:

$$D \approx UV^T \tag{8.1}$$

Note that the matrix U contains the latent factors (i.e., embedding coordinates) of each document in its rows, and the matrix V contains the latent factors (i.e., embedding coordinates) of each term in its rows. This factorization can be expressed as a directed graph shown in Figure 8.1(a). The nodes and edges of this factorization graph are defined as follows:

1. Each node corresponds to a matrix of latent factors. These latent factors might either correspond to the rows of a data matrix or the columns of a data matrix. For example, the node on the left in Figure 8.1(a) corresponds to the document embedding U, and the node on the right corresponds to the term embedding V. Each of these latent factor matrices is defined in such a way that the number of columns is equal to the rank of the factorization.

2. A directed edge from U to V is defined for the factorization $D \approx UV^T$. The edge is labeled with the matrix D that is factorized. It is noteworthy that the direction of the edge defines the fact that U occurs first in the factorization. Furthermore, even though the node at the arrow-head of the edge is labeled with V, its *transpose* is used in the factorization.

   It is possible for the edges in the factorization graph to be self-loops. For example, a factorization of an $n \times n$ symmetric matrix A (e.g., adjacency relations of a social network) can be of the form UU^T. Such an example is shown in Figure 8.1(b). In such a case, the factorization is expressed with a single node in the factorization graph.

All the factorizations discussed in Chapter 3 are two-node or single-node factorizations. It is also possible to add other constraints to the factorization, such a orthogonality, nonnegativity, and so on. Such constraints are optional, and are not included in the factorization graph. Therefore, a given factorization graph could represent many possible factorizations, depending on the constraints that are added to the factorization process. Although the two-node factorization graph does not seem to convey much information, this graphical representation is more useful in complex types of multi-modal settings.

8.2.2   Application: Shared Factorization with Text and Web Links

Consider a setting in which one wants to factorize a Web graph using both the text content and links. In this case, we have two data matrices. The first data matrix D is an $n \times d$

document-term matrix corresponding to the textual content of all the Web pages. The other matrix A is an $n \times n$ *directed* adjacency matrix, which is not symmetric. Furthermore, the structural and textual information are intimately connected. Therefore, the latent factors of a node should be regulated by both the text and linkage structure.

The $n \times k$ document factors are denoted by the matrix $U = [u_{ij}]$ (with rows containing k-dimensional document factors) and the term factors are contained in the matrix $V = [v_{ij}]$ (with rows containing k-dimensional term factors). Since there is a one-to-one correspondence between documents and the vertices of the Web graph, one of the factors of the adjacency matrix A should be U. However, there are two modeling options as to whether to use U as the *outgoing* factors of the adjacency matrix or to use U as the *incoming* factors of the adjacency matrix. With these options the corresponding factorizations are as follows:

1. When U is used as an *outgoing* factor of the adjacency matrix, the semantic interpretation is that the low-rank representation of a Web document is closely related to the types of Web pages that *it points towards*. In such a case, one must introduce an additional $n \times k$ *incoming* factor matrix $H = [h_{ij}]$, and try to find U, V, and H, such that the following conditions are satisfied:

$$D \approx UV^T, \quad A \approx UH^T \tag{8.2}$$

The corresponding factorization graph is illustrated in Figure 8.2(a).

2. When U is used as an *incoming* factor of the adjacency matrix, the semantic interpretation is that the low-rank representation of Web document are closely related to the types of Web pages that *point to it*. In such a case, one must introduce an additional $n \times k$ *outgoing* factor matrix H, and try to find U, V, and H, such that the following conditions are satisfied:

$$D \approx UV^T, \quad A \approx HU^T \tag{8.3}$$

The corresponding factorization graph is illustrated in Figure 8.2(b).

3. An additional option is to treat the outgoing and incoming link factors in a symmetric way by using the following set of conditions:

$$D \approx UV^T, \quad D \approx HV^T, \quad A \approx UH^T \tag{8.4}$$

Note that in this case, it does not matter whether the last condition is used as $A \approx UH^T$ or whether we use $A \approx HU^T$, although only one of these two conditions is imposed for an asymmetric matrix A. The two equivalent factorizations are shown in Figure 8.2(c).

The specific choice of the shared factorization depends on which semantic interpretation is considered more likely by the analyst. For now, let us consider the case in which the content of documents is more closely related to their outgoing links (cf. Figure 8.2(a)). In such a case, the optimization model should enforce the following factorizations:

$$D \approx UV^T, \quad A \approx UH^T \tag{8.5}$$

One can, therefore, define the following optimization problem in order to learn the matrices U, V, and H as follows:

$$\text{Minimize } J = ||D - UV^T||_F^2 + \beta||A - UH^T||_F^2 + \underbrace{\lambda(||U||_F^2 + ||V||_F^2 + ||H||_F^2)}_{\text{Regularization}}$$

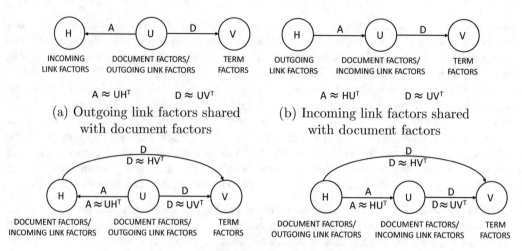

(a) Outgoing link factors shared
with document factors

(b) Incoming link factors shared
with document factors

BOTH FACTORIZATION GRAPHS ARE EQUIVALENT

(c) Symmetric treatment of link factors

Figure 8.2: Factorization graph for shared factorization of text and Web links

Here, the notation $||\cdot||_F$ denotes the Frobenius norm of a matrix, the parameter β regulates the relative importance of the structure versus the content, and λ controls the amount of regularization. It is also possible to use different regularization weights for each of U, V, and W. Such optimization problems are solved with the use of gradient-descent methods (cf. Section 8.2.2.1). It is relatively straightforward to formulate the optimization problem for the case of Figure 8.2(b) by changing the term $\beta||A - UH^T||_F^2$ in the aforementioned optimization problem to $\beta||A - HU^T||_F^2$. In the case of the factorization graph in Figure 8.2(c), it would suffice to add the term $\gamma||D - HV^T||_F^2$ to either of the aforementioned two formulations. This particular family of optimization problems is designed for the unsupervised setting. The resulting factors can be used for clustering either documents or terms by applying the k-means algorithm to the k-dimensional rows of U and V, respectively. Although these embeddings can be directly used for classification, one can also make enhancements to the optimization model for supervised settings (cf. Section 8.2.2.2).

8.2.2.1   Solving the Optimization Problem

We compute the gradient of J with respect to the entries in U, V, and H to update them in the direction of the (negative) gradient. For any current values of U, V, and H, let e_{ij}^D represent the (i,j)th entry of the error matrix $(D - UV^T)$, and e_{ij}^A represent the (i,j)th entry of the error matrix $(A - UH^T)$. The relevant partial derivatives of J can be expressed as follows:

$$\frac{\partial J}{\partial u_{iq}} = -\sum_{j=1}^{d} e_{ij}^D v_{jq} - \beta \sum_{p=1}^{n} e_{ip}^A h_{pq} + \lambda u_{iq} \quad \forall i \in \{1 \ldots n\},\ \forall q \in \{1 \ldots k\}$$

$$\frac{\partial J}{\partial v_{jq}} = -\sum_{i=1}^{n} e_{ij}^D u_{iq} + \lambda v_{jq} \quad \forall j \in \{1 \ldots d\},\ \forall q \in \{1 \ldots k\}$$

$$\frac{\partial J}{\partial h_{pq}} = -\beta \sum_{i=1}^{n} e_{ip}^A u_{iq} + \lambda h_{pq} \quad \forall j \in \{1 \ldots d\},\ \forall q \in \{1 \ldots k\}$$

For notational simplicity, a factor of 2 is omitted from the aforementioned gradients because it can be absorbed by the step-size in the gradient-descent method. These gradients can be used to update the entire set of $(2 \cdot n + d)k$ parameters with a step-size of α. However, such an approach can sometimes be slow to converge. It can also be impractical because it requires the computation of large error matrices corresponding to $(D - UV^T)$ and $(A - UH^T)$. The latter is particularly large when n is large.

A more effective approach is to use *stochastic* gradient descent, which effectively computes the gradients with respect to residual errors in randomly sampled *entries* of the matrices. One can sample any entry in *either* the document-term matrix or the adjacency matrix, and then perform the gradient-descent step with respect to the error in this single entry. In other words, one performs the following steps:

Randomly sample any entry from either D or A;
Perform a gradient-descent step with respect to entry-specific loss;

Consider a case in which the (i, j)th entry in the document-term matrix is sampled with error e_{ij}^D. Then, the following updates are executed for each $q \in \{1 \ldots k\}$ and step-size α:

$$u_{iq} \Leftarrow u_{iq}(1 - \alpha \cdot \lambda/2) + \alpha e_{ij}^D v_{jq}$$
$$v_{jq} \Leftarrow v_{jq}(1 - \alpha \cdot \lambda) + \alpha e_{ij}^D u_{iq}$$

On the other hand, if the (i, p)th entry in the adjacency matrix is sampled, then the following updates are performed for each $q \in \{1 \ldots k\}$ and step-size α:

$$u_{iq} \Leftarrow u_{iq}(1 - \alpha \cdot \lambda/2) + \alpha \beta e_{ip}^A h_{pq}$$
$$h_{pq} \Leftarrow h_{pq}(1 - \alpha \cdot \lambda) + \alpha \beta e_{ip}^A u_{iq}$$

These steps are repeated to convergence. The main advantage of stochastic gradient descent is the fast update and the ability to deal with large matrices.

The aforementioned updates are designed for the case of the factorization of Figure 8.2(a). It is also possible to derive the stochastic gradient-descent steps for the case of Figure 8.2(b) in an analogous way.

8.2.2.2   Supervised Embeddings

It is possible to extract better embeddings in supervised settings by incorporating information available from the dependent variable within the optimization model. Consider the case in which some of the rows of the $n \times d$ data matrix D are associated with class labels y_i drawn from $\{-1, +1\}$. Therefore, we define the set S of observed labels as follows:

$$S = \{i : \text{Label } y_i \text{ of } i\text{th row of } D \text{ is observed}\} \tag{8.6}$$

In such a case, one can create a k-dimensional coefficient (row) vector $\overline{W}$, and assume that the class label is related to features by regression. Let $\overline{u_i}$ be the k-dimensional row vector corresponding to the ith row of U. Then, the linear-regression condition is as follows:

$$y_i \approx \overline{u_i} \cdot \overline{W} \quad \forall i \in S \tag{8.7}$$

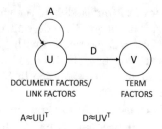

Figure 8.3: Joint factorization with undirected social networks

In such a case, the optimization problem is modified as follows:

$$\text{Minimize } J = ||D - UV^T||_F^2 + \beta||A - UH^T||_F^2 + \gamma\sum_{i\in S}(y_i - \overline{u_i} \cdot \overline{W})^2$$

$$+ \underbrace{\lambda(||U||_F^2 + ||V||_F^2 + ||H||_F^2 + ||\overline{W}||^2)}_{\text{Regularization}}$$

Note that the added term is derived from the optimization formulation of least-squares classification (cf. Chapter 6). Therefore, the approach can also be used in cases where the dependent variables are real-valued. One can then use gradient-descent methods in order to learn the supervised embeddings U, V, and H. In such a case, the vector $U\overline{W}^T$ yields the predicted values of y_i for both the training and the test data. Using supervised embeddings can sometimes yield better quality results for classification than unsupervised embeddings. The derivation of the gradient-descent steps uses the same broad approach as in the case of Section 8.2.2.1.

8.2.3   Application: Text with Undirected Social Networks

Many graphs like social networks are undirected, which raises the possibility of *symmetric* matrix factorization. We have a set of n documents, such that each document corresponds to a node (or *actor*) in the social network. Therefore, we have an $n \times d$ document-term matrix D, and an $n\times n$ *symmetric* and *undirected* adjacency matrix of the social friendship network. Furthermore, it is assumed that a one-to-one correspondence exists between documents and social actors, representing the content (e.g., summary of all Facebook wall posts) associated with that actor. In such a case, we have an $n \times k$ node-linkage factor matrix U, and a $d \times k$ term factor matrix V. The relevant factorizations are as follows:

$$D \approx UV^T, \quad A \approx UU^T \tag{8.8}$$

The corresponding factorization graph is illustrated in Figure 8.3. It is noteworthy that the adjacency matrix can be factorized with a single matrix because of the fact that it is symmetric. However, it is important to set the diagonal entries of A to the degrees of the nodes in order to ensure that the matrix A is positive semi-definite. Otherwise, the diagonal entries might increase the error of factorization UU^T (which is always positive semi-definite). The optimization formulation is as follows:

$$\text{Minimize } J = ||D - UV^T||_F^2 + \beta||A - UU^T||_F^2 + \underbrace{\lambda(||U||_F^2 + ||V||_F^2)}_{\text{Regularization}}$$

One can use similar stochastic gradient-descent steps as in Section 8.2.2.1. At any particular values of U and V, the error of the (i,j)th entry of the document-term matrix is denoted by $e_{ij}^D = (D - UV^T)_{ij}$ and the error of the (i,j)th entry of the adjacency matrix is denoted by $e_{ij}^A = (A - UU^T)_{ij}$. In stochastic gradient-descent approach, the gradients are computed with respect to the loss in a single entry. Consider a case in which the (i,j)th entry in the document-term matrix has been sampled with error $e_{ij}^D = (D - UV^T)_{ij}$. Then, the following updates are performed for each $q \in \{1 \ldots k\}$ and step-size α:

$$u_{iq} \Leftarrow u_{iq}(1 - \alpha \cdot \lambda/2) + \alpha e_{ij}^D v_{jq}$$
$$v_{jq} \Leftarrow v_{jq}(1 - \alpha \cdot \lambda) + \alpha e_{ij}^D u_{iq}$$

On the other hand, if the (i,p)th entry in the adjacency matrix is sampled, then the following updates are performed for each $q \in \{1 \ldots k\}$ and step-size α:

$$u_{iq} \Leftarrow u_{iq}(1 - \alpha \cdot \lambda/2) + 2\alpha\beta e_{ip}^A u_{pq}$$
$$u_{pq} \Leftarrow u_{pq}(1 - \alpha \cdot \lambda) + 2\alpha\beta e_{ip}^A u_{iq}$$

These steps are repeated to convergence. One can also perform a supervised factorization of the matrix by adapting the approach discussed in Section 8.2.2.2.

8.2.3.1 Application to Link Prediction with Text Content

The *link prediction* problem in social networks is that of finding pairs of actors, who are not currently connected but are likely to become connected in the future [333]. The aforementioned factorization can used easily for link prediction in social networks. In particular, consider any pair of nodes (i,j), between which a link does not currently exist. Then, the (i,j)th entry of UU^T will provide a numerical propensity of a link existing between nodes i and j. Therefore, the entries of UU^T provide link-prediction scores.

8.2.4 Application: Transfer Learning in Images with Text

Text documents have the advantages of having a semantically coherent feature space, which is often closely related to the semantic nature of clusters and classes in real-world applications. In other words, text has a data representation, which is often application-friendly because of the natural way in which its features are extracted. This is not quite the case in image data in which the features are semantically less informative. Is there any way in which one can use the higher-quality features of text data in order to *engineer* better features for image mining? This problem is referred to as that of "closing the semantic gap" in image classification. One way of achieving this goal is the use of *transfer learning*, in which knowledge is transferred from the text to the image domain. The key idea here is that images often co-occur with various types of text such as tags or captions. This co-occurrence can be used to map the images into a new latent semantic space in which the feedback from the co-occurrence information is incorporated. The new multidimensional representation is semantically more coherent, because it incorporates knowledge from the text modality. As a result, the classification is significantly improved when off-the-shelf classifiers are used on the representation. There are two distinct settings in which transfer learning is used. The first is one in which the text is unlabeled and the only purpose of transfer learning is to engineer better quality features with the use of co-occurrence information. This approach is a kind of semi-supervision, except that the unlabeled data belongs to a different (text) domain from that in which the classification is performed. The second setting is one in which

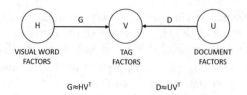

Figure 8.4: Joint factorization with document-tag and (visual feature)-tag matrix

the text data are already labeled and one also uses the labeling to perform the classification in the image domain. Unlike the first setting, the text also helps in compensating for the paucity of labels associated with the images.

8.2.4.1 Transfer Learning with Unlabeled Text

Consider the case in which we have an $m \times d'$ image-(visual feature) matrix denoted by M in d' dimensions (corresponding to visual words), and each of the m rows in this matrix is denoted by a class label. It is noteworthy that visual words correspond to image features, they are often semantically a lot less friendly than textual words. Therefore, transfer learning methods are used [633] in order to extract a semantically coherent representation of the images. One can view this approach as a kind of semi-supervised learning with unlabeled data, except that the semi-supervision is performed with data from a different (text) domain in order to classify image data [633].

Where does one obtain the semantic knowledge about the visual features? This is extracted from *tagging* data. In many social media sites like *Flickr*, images are often tagged and each tag can be viewed as a short set of keywords. Each tag typically contains less than two or three words, and rarely more than 10 words. As a practical matter, one can consider a tag set as new and informal lexicon, which is semantically very descriptive. Consider the case in which we have a vocabulary of d tags, and each tag can be applied to one or more of a set of p images. Therefore, we have a tagged set of p images, with a $p \times d'$ representation (denoted by Z) in visual-word space, and a corresponding tag matrix T, which has a $p \times d$ binary representation. In other words, the matrix T contains 0–1 entries corresponding to which tag is applied to which image. This matrix is extremely sparse. The matrix $G = Z^T T$ is then a $d' \times d$ mapping between visual words and tags. In other words, it provides knowledge about which visual word corresponds with which tag frequently, and is a kind of mapping of the (semantically obscure) visual words to the space of (semantically coherent) tags. In addition, it is assumed that we have a set of n documents that are expressed in terms of the informal vocabulary of d tags. Although a document collection might originally be expressed using a conventional lexicon of English words, it is not very difficult to express it in terms of a tag vocabulary by setting the value of the jth tag for the ith document to 1 if at least one conventional word is shared between the document and the tag. In other words, we have an $n \times d$ document-tag matrix in terms of this non-traditional tag-vocabulary of size d. This document-tag matrix is denoted by D and it provides useful co-occurrence information between the different tags, which is further useful for extracting a semantic representation of the images. One can view D as the unlabeled collection that is used for semi-supervision.

Let H be a $d' \times k$ matrix and V be a $d \times k$ matrix, where k is the rank of the factorization process. Note that the factorization of the matrix $G = HV^T$ provides a latent representation H of each visual word, although it is unable to account for the co-occurrences and relationships among different tags. This is particularly important because the matrix G is often sparse, which makes it difficult to extract reliable factor matrices. Therefore, we propose to use a shared factorization with an additional $n \times k$ factor matrix U of the n documents in D:

$$D \approx UV^T, \quad G \approx HV^T$$

The corresponding factorization graph is illustrated in Figure 8.4, and its associated optimization problem is as follows:

$$\text{Minimize } J = ||D - UV^T||_F^2 + \beta ||G - HV^T||_F^2 + \lambda(||U||_F^2 + ||V||_F^2 + ||H||_F^2)$$

Here, β is the balancing parameter that regulates the relative importance of the different terms. This optimization problem is very similar to that discussed in Section 8.2.2.1. Therefore, the gradient-descent steps of that section can be used for this problem, although other types of optimization methods are also discussed in [633].

The matrix H can be viewed as a kind of translator matrix to transform data points from visual-word space to a latent semantic space in which the representation quality is improved. Given the labeled $m \times d'$ matrix M, one can transform it to k-dimensional space by using the new representation $M' = MH$. The classification is performed on this transformed representation of the data.

8.2.4.2 Transfer Learning with Labeled Text

A second setting is one in which have a *labeled* $n \times d$ document-term matrix, and also a set of images for which very few labels are observed. In the case of the documents, the labels are available as an n-dimensional column vector $\bar{y} = [y_1 \ldots y_n]^T$. It is assumed that each y_i is drawn from $\{-1, +1\}$. In addition, we have an $m \times d'$ image-(visual word) matrix M, which is defined over a lexicon of d' visual words. The labels for a subset S of images in M is observed, and these labels are drawn from the same base set as the documents. Therefore, we have:

$$S = \{i : \text{ Label of } i\text{th row of } M \text{ is observed}\}$$

In the event that the label of the ith row of M is observed, it is denoted by z_i. Each z_i is also drawn from $\{-1, +1\}$. It is noteworthy that the size of the set S may be quite small in many real settings, which is why transfer learning is required in the first place. It is assumed that the documents and images may co-occur in various ways through either Web links, or through the use of inline placement of images within Web pages. Therefore, we assume that we have an $m \times n$ co-occurrence matrix C between the images and the Web pages.

In order to perform the factorization, both the images and the documents are mapped into a k-dimensional latent space, with corresponding factors denoted by U_M and U_D, respectively. Here U_M is an $m \times k$ matrix because there are m images, and U_D is an $n \times k$ matrix because there are n documents. In addition, the $d' \times k$ latent-factor matrix of the visual words is denoted by H and the $d \times k$ latent-factor matrix of the (textual) words is denoted by V. Then, in order to force U_M and U_D to be the relevant embeddings of the image and text domains, respectively, we have the following:

$$M \approx U_M H^T, \quad D \approx U_D V^T$$

The key point is that the matrices U_M and U_D are in the same k-dimensional space and the dot products between their rows correspond to similarities, which are also reflected in the co-occurrence matrix C. This condition can be enforced by using the following factorization:

$$C \approx U_M U_D^T \tag{8.9}$$

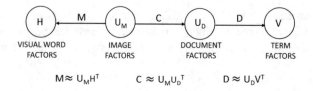

$$M \approx U_M H^T \qquad C \approx U_M U_D^T \qquad D \approx U_D V^T$$

Figure 8.5: Joint factorization with text, images, and co-occurrence matrix

The corresponding factorization graph is illustrated in Figure 8.5.

Without considering the labels, it is possible to create an unsupervised embedding for the documents and images within a joint latent space. The corresponding optimization problem is as follows:

$$\text{Minimize } J = ||D - U_D V^T||_F^2 + \beta ||M - U_M H^T||_F^2 + \gamma ||C - U_M U_D^T||_F^2 + \lambda \cdot \text{Regularizer}$$

As in all the previous cases, the regularizer is defined by the sum of the squares of the Frobenius norms of the various parameter matrices. Furthermore, β and γ are balancing parameters that regulate the relative importance of various terms.

However, when additional labels y_i are available for documents and z_j for images, it is possible to add supervision by forcing documents and images with the same labels to be somewhat similar. Because the labels are drawn from $\{-1, +1\}$, the value of $1 + y_i z_j$ will be 2 when $y_i = z_j$, and it will be 0, otherwise. Let $\overline{u}_i^D$ be the ith row of the document factor matrix U_D, and let $\overline{u}_j^M$ be the jth row of the image factor matrix U_M. Both these rows are k-dimensional row vectors and the difference between them provides the distance between the relevant embeddings of these rows. A *label-agreement term*, J_L, is defined by penalizing distances between embeddings with the same label:

$$J_L = \sum_{i=1}^{n} \sum_{j \in S} \underbrace{(1 + y_i z_j)||\overline{u}_i^D - \overline{u}_j^M||^2}_{\text{Nonzero when } y_i = z_j}$$

In order to construct a supervised embedding, an additional term of θJ_L needs to be added to the objective function J of the unsupervised embedding, where θ regulates the importance of supervision.

Once the embeddings have been learned (with a gradient-descent method), the sign of each of the m entries in the column vector $U_M U_D^T \overline{y}$ provides the label prediction of the m images (including the originally labeled ones). The basic idea is that $U_M U_D^T$ provides pair-wise similarity between image-document pairs. By post-multiplying with $\overline{y}$, one is classifying each image by using a similarity-weighted linear combination of the labels of documents. Hyper-parameters like β, γ, and θ can be tuned in order to maximize accuracy on a held-out set. This discussion is broadly based on the ideas presented in [450].

8.2.5   Application: Recommender Systems with Ratings and Text

Content-based recommender systems use the textual descriptions of items to learn user propensities about particular items. Ratings indicate the degree of like or dislike of users towards items. In such cases, the data for each user is converted into a user-specific text classification problem. The training documents for each user-specific classification problem correspond to the descriptions of items rated by that user, and the dependent variable is its

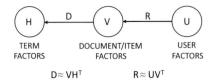

Figure 8.6: Joint factorization with users, items, and item descriptions

item-specific rating from that user. This training data can be used to learn a user-specific classification or regression model for rating prediction.

However, content-based systems do not use the *collaborative* power of like-mined users to make predictions. A different class of recommendation methods, referred to as *collaborative filtering* methods, use the similarities in rating patterns between users and items to make predictions. Let R be an $m \times n$ ratings matrix R over m users and n items. The matrix $R = [r_{ij}]$ is *massively incomplete*, and only a small subset O of the ratings in R are observed:

$$O = \{(i, j) : r_{ij} \text{ is observed}\}$$

In addition, we have an $n \times d$ document-term matrix D, in which each of the n rows contains the descriptions of the n items over a lexicon of size d.

Collaborative filtering problems are often solved using matrix factorization methods in which the ratings matrix R is decomposed into user and item factors. A key complication with recommender systems is that the ratings matrix is only partial observed and therefore, one can only define the optimization problem in matrix factorization over the observed ratings in O. Let U be the $m \times k$ matrix representing the factors of the users, V be the $n \times k$ matrix representing the factors of the items, and let $H = [h_{ij}]$ be the $d \times k$ matrix representing the factors of the textual words (terms). Then, we have the following relationships:

$$\underbrace{R \approx UV^T}_{\text{Observed entries}} \quad , \quad D \approx VH^T$$

The corresponding factorization graph is illustrated in Figure 8.6.

Note that the first factorization is defined only over the observed entries in O. Therefore, the corresponding optimization problem also needs to be formulated over the observed entries as follows:

$$\text{Minimize } J = \sum_{(i,j)\in O} \left(r_{ij} - \sum_{s=1}^{k} u_{is} v_{js}\right)^2 + \beta \|D - VH^T\|_F^2 + \lambda(\|U\|_F^2 + \|V\|_F^2 + \|H\|_F^2)$$

Here, β is the balancing parameter. The gradient-descent steps for this optimization problem are similar to those discussed in Section 8.2.2.1, except that only the observed entries are used to compute the gradients. Setting β to 0 results in traditional recommender system updates (see Exercise 5). Furthermore, almost the same optimization problem is used in the following related settings:

1. One can combine user-user trust matrices with ratings matrices instead of combining text with ratings matrices. This approach is described in Chapter 11 of [3].

2. One can combine social tagging matrices with ratings matrices. Such an approach is similar to the technique discussed above, except that the tags are used as the

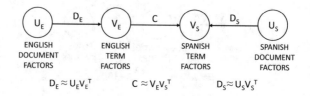

Figure 8.7: Joint factorization of English and Spanish documents with feature co-occurrence

"lexicon" to represent the items. Furthermore, since tags are related to the users, it is also possible to create a user-tag matrix, which can be factorized.

It is noteworthy that shared matrix factorization methods form the backbone of a wide variety of hybrid techniques for recommender systems.

8.2.6 Application: Cross-Lingual Text Mining

Cross-lingual text mining shares a number of similarities with the case in which images and text are mined together. However, in the case of the image/text mining, the co-occurrence matrices are created using *instances* of images and documents. In the case of cross-lingual text mining, sufficient domain knowledge is available to create cross-lingual matrices at the feature level.

Consider a setting in which we have two document collections in English and Spanish, respectively. The $n \times d$ document-term matrix for the English collection is denoted by D_E, whereas the $m \times d'$ document-term matrix for the Spanish collection is denoted by D_S. In addition, we have a $d \times d'$ feature-level co-occurrence matrix C between English and Spanish. The feature-level co-occurrence matrix between English and Spanish can be extracted in a variety of ways. For example, one can use a cross-lingual dictionary [39] or thesaurus [386] in order to create the co-occurrence matrix. Each entry (i, j) in C takes on the value of 1 if the ith English term is related to the jth Spanish term. It is relatively easy to use a dictionary to create a co-occurrence matrix. It is also possible to create co-occurring feature matrix from document pairs that are translations of one another. For example, let C_E and C_S be two $c \times d$ and $c \times d'$ document-term matrices containing c documents in English and Spanish, respectively, so that the ith rows in C_E and C_S are translations of the same sentence in English and Spanish, respectively. Then, one can derive $d \times d'$ the *feature-level* co-occurrence matrix as follows:

$$C = C_E C_S^T \tag{8.10}$$

Unlike image-text mining, the co-occurrence matrices are specified at the feature level rather than the instance level. Let U_E and V_E be the respective $n \times k$ and $d \times k$ document-factor and term-factor matrices for the English documents. Similarly, let U_S and V_S be the respective $n \times k$ and $d \times k$ document-factor and term-factor matrices for the Spanish documents. Then, the corresponding factorizations are as follows:

$$D_E \approx U_E V_E^T, \quad D_S = U_S V_S^T, \quad C \approx V_E V_S^T$$

The corresponding factorization graph is illustrated in Figure 8.7. This particular factorization is similar to the case of image/text mining except that the co-occurrence matrix is defined as the product of the term-factor matrices in the two languages (rather than the instance factors). The optimization problem can be formulated in a similar way, and the gradient-descent steps can be obtained by computing the derivative of the squared error.

8.3 Factorization Machines

Factorization machines are closely related to shared matrix factorization methods, and are particularly suitable under the following conditions:

1. Each data instance contains features from multiple domains. For example, consider an item that is tagged with particular keywords by a user and also rated by that user. In such a case, the feature set corresponds to all the item identifiers, all the possible keywords, and the user identifiers. The feature values of the user identifier, item identifier, and the relevant keywords are set to 1, whereas all other feature values are set to 0. The dependent variable is equal to the value of the rating.

2. The feature representation is often sparse, which contains a large number of 0s. Many homogeneous text domains, such as *short text snippets* can also be used in conjunction with factorization machines. For example, a tweet in *Twitter* is limited to 140 characters, which imposes natural constraints on the number of words in each such "document." Traditional classification and regression methods, such as support vector machines, work poorly in this setting. In many natural applications, the feature representation is sparse *and* binary, although this is not always necessary.

Factorization machines are polynomial regression techniques, in which strong regularization conditions are imposed on the regression coefficients in order to handle the challenges of sparsity. Sparsity is common in short-text domains, such as the social content on bulletin boards, social network datasets, and chat messengers. It is also common in recommender systems.

An example of a data set drawn from the recommendation domain is illustrated in Figure 8.8. It is evident that there are three types of attributes corresponding to user attributes, item attributes, and tagging keywords. Furthermore, the rating corresponds to the dependent variable, which is also the regressand. At first sight, this data set seems to be no different from a traditional multidimensional data set to which one might apply least-squares regression in order to model the rating as a linear function of the regressors.

Unfortunately, the sparsity of the data in Figure 8.8 ensures that a least-squares regression method does rather poorly. For example, each row might contain only three or four non-zero entries. In such cases, linear regression may not be able to model the dependent variable very well, because the presence of a small number of non-zero entries provides little information. Therefore, a second possibility is to use higher-order interactions between the attributes in which we use the simultaneous presence of multiple entries for modeling. As a practical matter, one typically chooses to use second-order interactions between attributes, which corresponds to second-order polynomial regression. One possibility is to use a second-order polynomial kernel in order to perform kernel regression with the use of the kernel trick (cf. Chapter 6). However, as we will discuss below, an attempt to do so leads to overfitting, which is exacerbated by the sparse data representation.

Let $d_1 \ldots d_r$, be the number of attributes in each of the r data modalities such as text, images, network data and so on. Therefore, the total number of attributes is given by $p = \sum_{k=1}^{r} d_k$. We represent the variables of the row by $x_1 \ldots x_p$, most of which are 0s, and a few might be nonzero. In many natural applications in the recommendation domain, the values of x_i might be binary. Furthermore, it is assumed that a target variable is available for each row. In the example of Figure 8.8, the target variable is the rating associated with each row, although it could be any type of dependent variable in principle.

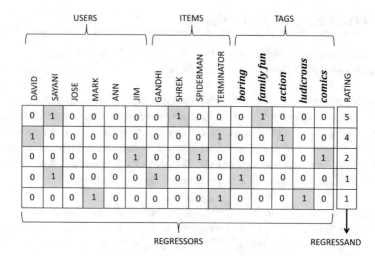

	USERS						ITEMS				TAGS					
DAVID	SAYANI	JOSE	MARK	ANN	JIM	GANDHI	SHREK	SPIDERMAN	TERMINATOR	boring	family fun	action	ludicrous	comics	RATING
0	1	0	0	0	0	0	1	0	0	0	1	0	0	0	5
1	0	0	0	0	0	0	0	0	1	0	0	1	0	0	4
0	0	0	0	0	1	0	0	1	0	0	0	0	0	1	2
0	1	0	0	0	0	1	0	0	0	1	0	0	0	0	1
0	0	0	1	0	0	0	0	0	1	0	0	0	1	0	1

REGRESSORS   REGRESSAND

Figure 8.8: An example of a sparse regression modeling problem with heterogeneous attributes

Consider the use of a regression methodology in this setting. For example, the simplest possible prediction would be use linear regression with the variables $x_1 \ldots x_p$.

$$\hat{y}(\overline{x}) = b + \sum_{i=1}^{p} w_i x_i \tag{8.11}$$

Here, b is the bias variable and w_i is the regression coefficient of the ith attribute. This is in an almost identical form to the linear regression discussed in Section 6.2 of Chapter 6, except that we have explicitly used a global bias variable b. Although this form can provide reasonable results in some cases, it is often not sufficient for sparse data in which a lot of information is captured by the correlations between various attributes. For example, in a recommender system, the co-occurrence of a user-item pair is far more informative than the separate coefficients of users and items. Therefore, the key is to use a *second-order* regression coefficient s_{ij}, which captures the coefficient of the interaction between the ith and jth attribute.

$$\hat{y}(\overline{x}) = b + \sum_{i=1}^{p} w_i x_i + \sum_{i=1}^{p} \sum_{j=i+1}^{p} s_{ij} x_i x_j \tag{8.12}$$

Note that one could also include the second-order term $\sum_{i=1}^{p} s_{ii} x_i^2$, although x_i is often drawn from sparse domains with little variation in nonzero values of x_i, and the addition of such a term is not always helpful. For example, if the value of x_i is binary (as is common), the coefficient of x_i^2 would be redundant with respect to that of x_i.

One observation is that the above model is very similar to what one would obtain with the use of kernel regression with a second-order polynomial kernel. In sparse domains like text, such kernels often overfit the data, especially when the dimensionality is large and the data is sparse. Even for an application in a single domain (e.g., short-text tweets), the value of d is greater than 10^5, and therefore the number of second-order coefficients is more than 10^{10}. With any training data set containing less than 10^{10} points, one would perform quite poorly. This problem is exacerbated by sparsity, in which pairs of attributes co-occur rarely in the training data, and may not generalize to the test data. For example, in a recommender application, a particular user-item pair may occur only once in the entire training data, and it will not occur in the test data if it occurs in the training data. In fact, all the user-item pairs that occur in the test data will not have occurred in the training data. How, then, does one learn the interaction coefficients s_{ij} for such user-item pairs? Similarly, in a short-text mining application, the words "movie" and "film" may occur together, and the words "comedy" and "film" may also occur together, but the words "comedy" and "movie" might never have occurred together in the training data. What does one do, if the last pair occurs in the test data?

A key observation is that one can use the learned values of s_{ij} for the other two pairs (i.e., "comedy"/"film" and "movie"/"film") in order to make some inferences about the interaction coefficient for the pair "comedy" and "movie." How does one achieve this goal? The key idea is to assume that the $d \times d$ matrix $S = [s_{ij}]$ of second-order coefficients has a *low-rank* structure for some $d \times k$ matrix $V = [v_{is}]$:

$$S = VV^T \qquad (8.13)$$

Here, k is the rank of the factorization. Intuitively, one can view Equation 8.13 as a kind of regularization constraint on the (massive number of) second-order coefficients in order to prevent overfitting. Therefore, if $\overline{v_i} = [v_{i1} \dots v_{ik}]$ is the k-dimensional row vector representing the ith row of V, we have:

$$s_{ij} = \overline{v_i} \cdot \overline{v_j} \qquad (8.14)$$

By substituting Equation 8.14 in the prediction function of Equation 8.12, one obtains the following:

$$\hat{y}(\overline{x}) = b + \sum_{i=1}^{p} w_i x_i + \sum_{i=1}^{p} \sum_{j=i+1}^{p} (\overline{v_i} \cdot \overline{v_j}) x_i x_j \qquad (8.15)$$

The variables to be learned are b, the different values of w_i, and each of the vectors $\overline{v_i}$. Although the number of interaction terms might seem large, most of them will evaluate to zero in sparse settings in Equation 8.15. This is one of the reasons that factorization machines are designed to be used only in sparse settings where most of the terms of Equation 8.15 evaluate to 0. A crucial point is that we only need to learn the $O(d \cdot k)$ parameters represented by $\overline{v_1} \dots \overline{v_k}$ in lieu of the $O(d^2)$ parameters in $[s_{ij}]_{d \times d}$.

A natural approach to solve this problem is to use the stochastic gradient-descent method, in which one cycles through the observed values of the dependent variable to compute the gradients with respect to the error in the observed entry. The update step with respect to any particular model parameter $\theta \in \{b, w_i, v_{is}\}$ depends on the error $e(\overline{x}) = y(\overline{x}) - \hat{y}(\overline{x})$ between the predicted and observed values:

$$\theta \Leftarrow \theta(1 - \alpha \cdot \lambda) + \alpha \cdot e(\overline{x}) \frac{\partial \hat{y}(\overline{x})}{\partial \theta} \qquad (8.16)$$

Here, $\alpha > 0$ is the learning rate, and $\lambda > 0$ is the regularization parameter. The partial derivative in the update equation is defined as follows:

$$\frac{\partial \hat{y}(\overline{x})}{\partial \theta} = \begin{cases} 1 & \text{if } \theta \text{ is } b \\ x_i & \text{if } \theta \text{ is } w_i \\ x_i \sum_{j=1}^{p} v_{js} \cdot x_j - v_{is} \cdot x_i^2 & \text{if } \theta \text{ is } v_{is} \end{cases} \qquad (8.17)$$

The term $L_s = \sum_{j=1}^{p} v_{js} \cdot x_j$ in the third case is noteworthy. To avoid redundant effort, this term can be pre-stored while evaluating $\hat{y}(\overline{x})$ for computation of the error term $e(\overline{x}) =$

$y(\overline{x}) - \hat{y}(\overline{x})$. This is because Equation 8.15 can be algebraically rearranged as follows:

$$\hat{y}(\overline{x}) = b + \sum_{i=1}^{p} w_i x_i + \frac{1}{2} \sum_{s=1}^{k} \left([\sum_{j=1}^{p} v_{js} \cdot x_j]^2 - \sum_{j=1}^{p} v_{js}^2 \cdot x_j^2 \right)$$

$$= b + \sum_{i=1}^{p} w_i x_i + \frac{1}{2} \sum_{s=1}^{k} \left(L_s^2 - \sum_{j=1}^{p} v_{js}^2 \cdot x_j^2 \right)$$

Furthermore, the parameters $\overline{v}_i$ and w_i do not need to be updated when $x_i = 0$. This allows for an efficient update process in sparse settings, which is linear in both the number of nonzero entries and the value of k.

Factorization machines can be used for any (massively sparse) classification or regression task; ratings prediction in recommender systems is only one example of a natural application. Although the model is inherently designed for regression, binary classification can be handled by applying the logistic function on the numerical predictions to derive the probability whether $\hat{y}(\overline{x})$ is $+1$ or -1. The prediction function of Equation 8.15 is modified to a form used in logistic regression:

$$P[y(\overline{x}) = 1] = \frac{1}{1 + \exp(-[b + \sum_{i=1}^{p} w_i x_i + \sum_{i=1}^{p} \sum_{j=i+1}^{p} (\overline{v}_i \cdot \overline{v}_j) x_i x_j])} \qquad (8.18)$$

Note that this form is identical to that used in Equation 6.32 of Chapter 6. The main difference is that we are also using second-order interactions within the prediction function. A log-likelihood criterion can be optimized to learn the underlying model parameters with a gradient-descent approach [199, 467, 468].

The description in this section is based on second-order factorization machines that are popularly used in practice. In third-order polynomial regression, we would have $O(p^3)$ additional regression coefficients of the form w_{ijk}, which correspond to interaction terms of the form $x_i x_j x_k$. These coefficients would define a massive third-order *tensor*, which can be compressed with tensor factorization. Although higher-order factorization machines have also been developed, they are often impractical because of greater computational complexity and overfitting. A software library, referred to as *libFM* [468], provides an excellent set of factorization machine implementations. The main task in using *libFM* is an initial feature engineering effort, and the effectiveness of the model mainly depends on the skill of the analyst in extracting the correct set of features. Other useful libraries include *fastFM* [48] and[1] *libMF* [677], which have some fast learning methods for factorization machines.

8.4   Joint Probabilistic Modeling Techniques

Probabilistic modeling techniques like expectation-maximization and naïve Bayes can be naturally used with heterogeneous data, because different data modalities are generated by different distributions. In other words, the individual data instances contain elements from all the different domains, which are generated from different domain-specific distributions. In fact, methods like *collective topic modeling* [147, 148] can be viewed as probabilistic variants of shared matrix factorization.

For ease in discussion, consider a setting in which each data instance contains attributes corresponding to the text, numerical, and categorical domains. Therefore, we will assume

---

[1]The libraries *libFM* and *libMF* are different.

that there are a total of n data instances denoted by the vectors $\overline{X_1} \ldots \overline{X_n}$. Each data instance $\overline{X_i}$ can be segmented into three parts $\overline{X_i} = (\overline{X_i}^D, \overline{X_i}^C, \overline{X_i}^N)$. Here, $\overline{X_i}^D$ contains the values of the d attributes (word frequencies) for the text portion of the data instances, $\overline{X_i}^C$ contains the values of the attributes for the categorical portion of the data instances, and $\overline{X_i}^N$ contains the values of the attributes for the numerical portion of the data instances.

8.4.1 Joint Probabilistic Models for Clustering

It is relatively easy to create generative models for clustering data instances with attributes of different types. Consider the case in which we wish to use a mixture modeling approach in order to determine the clusters. Therefore, we will discuss a generalized form of the mixture modeling approach discussed in Section 4.4 of Chapter 4.

We assume that the mixture contains k hidden components (clusters) denoted by $\mathcal{G}_1 \ldots \mathcal{G}_k$. The value of k is an input parameter to the algorithm. Each iteration of the generative process creates a particular data instance $\overline{X_i} = (\overline{X_i}^D, \overline{X_i}^C, \overline{X_i}^N)$. Therefore, the text, categorical, and numerical components of each instance need to be generated at the same time. An important assumption made in the generative process is that once the mixture component has been selected, the text, categorical, and numerical components are generated in a conditionally independent way by a distribution that is most suitable for that particular data modality. For example, the following assumptions could be made:

1. The term-frequency component $\overline{X_i}^D$ is generated from a multinomial distribution.

2. The categorial component $\overline{X_i}^C$ is generated from a categorical distribution.

3. The numerical component $\overline{X_i}^N$ is generated rom a Gaussian distribution.

It is noteworthy that the relevant parameters of each distribution are specific to the mixture component at hand. Therefore, by selecting a particular component, the shape and location of the relevant cluster is fixed across all data modalities in the form of relevant probability distributions. Therefore, by independently generating the instances from these three different probability distributions, one can generate the entire data instance. The generative process uses the following steps:

1. Select the rth mixture component $\mathcal{G}_r$ with prior probability $\alpha_r = P(\mathcal{G}_r)$.

2. Independently generate $\overline{X_i}^D$ from the multinomial distribution of the rth component, $\overline{X_i}^C$ from the categorical distribution of the rth component, and $\overline{X_i}^N$ from the Gaussian distribution of the rth component.

It is relatively easy to adapt the expectation maximization algorithm to this setting. The key differences lie in the E-step. In the E-step, the goal is to estimate $P(\mathcal{G}_r | \overline{X_i})$, which is expressed in the following way using Bayes theorem:

$$P(\mathcal{G}_r | \overline{X_i}) = \frac{P(\mathcal{G}_r) \cdot P(\overline{X_i} | \mathcal{G}_r)}{\sum_{m=1}^{k} P(\mathcal{G}_m) \cdot P(\overline{X_i} | \mathcal{G}_m)} \tag{8.19}$$

The key point here is that one can express $P(\overline{X_i} | \mathcal{G}_r)$ in terms of the product of the corresponding values over the different data modalities because of conditional independence:

$$P(\overline{X_i} | \mathcal{G}_r) = P(\overline{X_i}^D | \mathcal{G}_r) \cdot P(\overline{X_i}^C | \mathcal{G}_r) P(\overline{X_i}^M | \mathcal{G}_r) \tag{8.20}$$

Since each of these quantities has its own (discrete or continuous) probability distribution, one can compute these values using the current values of the corresponding parameters. Furthermore, the M-step remains the same as the case of homogeneous data, except that the parameters of each data modality are estimated independently for each mixture component. Methods for estimating the parameters of the multinomial distribution for the text modality are discussed in Section 4.4. The estimation of parameters for the numerical and categorical distributions are discussed in [2].

8.4.2   Naïve Bayes Classifier

It is natural to generalize the naïve Bayes classifier to heterogeneous data using the same approach as the expectation-maximization algorithm for clustering. This is because the naïve Bayes classifier can be viewed as a supervised variant of the expectation-maximization algorithm, in which a single iteration of the M-step is applied to the labeled data in order to estimate the parameters of each mixture component (class). Furthermore, these estimated parameters are used to estimate the probability of each class for unlabeled data points with the Bayes rule, as in the E-step:

$$P(\mathcal{G}_r|\overline{X_i}) = \frac{P(\mathcal{G}_r) \cdot P(\overline{X_i}|\mathcal{G}_r)}{\sum_{m=1}^{k} P(\mathcal{G}_m) \cdot P(\overline{X_i}|\mathcal{G}_m)} \tag{8.21}$$

As in the case of the expectation-maximization algorithm, the quantities on the right-hand side can be estimated using the product of the corresponding values over the different data modalities (cf. Equation 8.20).

8.5   Transformation to Graph Mining Techniques

Many of the heterogeneous text mining problems can be transformed to graph mining problems. This opens the door to the use of a vast variety of graph mining techniques like community detection and collective classification [2]. Virtually all the shared matrix factorization methods discussed in Section 8.2 can be addressed with transformation to graph mining techniques. This is because the factorization graphs discussed in Section 8.2 can be expanded into more detailed relationship graphs.

Consider an undirected social network, in which we have a set of n documents, such that each document corresponds to a node in the social network. Therefore, we have an $n \times d$ document-term matrix D, and an $n \times n$ *symmetric* and *undirected* adjacency matrix of the social friendship network. Furthermore, it is assumed that a one-to-one correspondence exists between documents and social actors, representing the content (e.g., summary of all Facebook wall posts) associated with that actor. This case is discussed in Section 8.2.3. In such a case, we have an $n \times k$ node-linkage factor matrix U, and a $d \times k$ term factor matrix V. The relevant factorizations are as follows:

$$D \approx UV^T, \quad A \approx UU^T \tag{8.22}$$

The corresponding factorization graph is illustrated in Figure 8.3, which is repeated above in Figure 8.9(a). Furthermore, the corresponding relationship graph is illustrated in Figure 8.9(b). Note that the document factor node in Figure 8.9(a) is now replaced by the actual nodes in the social network (containing the documents) in Figure 8.9(b). Similarly, the self-loop (labeled by A) in Figure 8.9 is replaced by the corresponding links in the adjacency matrix A. The term factor node in Figure 8.9(a) is replaced by the actual terms in

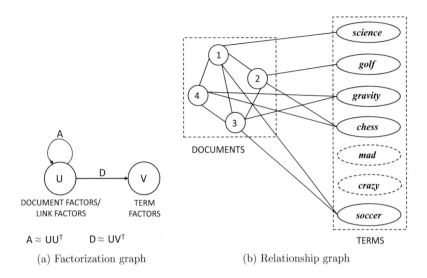

(a) Factorization graph   (b) Relationship graph

Figure 8.9: Expanding a factorization graph into a relationship graph with undirected social networks

Figure 8.9(b). The document-term matrix is now replaced by links between documents and nodes. It is possible for these edges to be weighted corresponding to the term-frequencies. It is noteworthy that the relationship graph is generally undirected, whereas the factorization graph is always directed.

The entire process creates a semi-bipartite network, which can be used in conjunction with many graph mining algorithms [2] for clustering and classification. The area of graph mining contains a rich variety of combinatorial algorithms that can be used to gain various insights. For example, the *PageRank* techniques discussed in Chapter 9 can also be used with these network mining algorithms to discover various insights about the relationships between documents and terms. The broad approach is to use the following steps:

1. Create nodes for the various data instance identifiers (e.g., document identifiers) and attribute values (e.g., terms) in the data.

2. Create undirected, weighted links depending on the available data matrices across different domains. These matrices may correspond to document-term matrices, image-(visual word) matrices, or co-occurrence matrices.

Consider the cross-lingual mining application discussed in Section 8.2.6. In this case, we assume that English and Spanish documents are available as respective document-term matrices D_E and D_S. In addition, an explicit mapping between English terms and Spanish terms is available in a co-occurrence matrix C, where the rows of C correspond to English terms and the columns correspond to Spanish terms. The factorization graph and a possible relationship graph are illustrated in Figure 8.10. In this case, we have used an exact mapping between the terms of the two languages, although it is also possible to construct the co-occurrence matrix with matching pairs of sentences in the two languages. In such a case, a nonzero entry is placed between an English and Spanish term in C when they co-occur in a matching pair of sentences. Although such an approach creates noisy co-occurrence links, it can also capture useful semantic relationships between terms without exact equivalence.

It is evident from Figure 8.10 that a one-to-one relationship exists between nodes in the factorization graph, and the various *types* of nodes in the relationship graph. Furthermore,

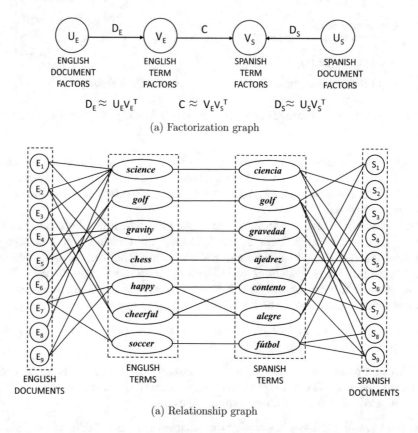

(a) Factorization graph

(a) Relationship graph

Figure 8.10: Comparing factorization and relationship graphs in cross-lingual text mining

the data matrices in the factorization graph have been expanded into explicit links in the relationship graph. Once such a network has been constructed, it can be used in the conjunction with a variety of graph mining algorithms such as the following:

1. Node clustering methods can be used to partition the nodes into disjoint groups and create a segmentation of *both* the data instances and terms in all modalities.

2. Collective classification methods can be used to leverage a subset of labeled nodes (in any modality) to propagate the labels to other nodes with *label propagation methods* [2]. A specific approach that uses label propagation with *PageRank*-like random walk methods for social networks is discussed in [8].

The main advantage of graph-based transformation techniques is that they allow the use of broader classes of *discrete* combinatorial techniques, which are inherently different from the *continuous* optimization methods used by techniques like matrix factorization. In many cases, off-the-shelf graph mining techniques can be used.

8.6 Summary

Text mining applications arise frequently in combination with various types of heterogeneous data such as Web links, social links, images, recommender systems, and cross-lingual data. Shared matrix factorization methods are among the most flexible methods for mining heterogeneous data sources, and they can be used in both supervised and unsupervised settings. Factorization machines are closely related to shared matrix factorization methods, and are particularly suitable for supervised modeling of sparse data. Many mixture models and their supervised variants like the naïve Bayes classifier can be extended easily to heterogeneous data domains by modeling a heterogeneous data instance with the use of conditionally independent data distributions. Finally, many heterogeneous data mining problems can be transformed to graph mining techniques.

8.7 Bibliographic Notes

Shared matrix factorization has been used for many heterogeneous mining applications with text data. In particular, the notion of collective matrix factorization is introduced in [517], which provides a generic view of using matrix factorization methods with shared entity types. Since topic modeling methods like PLSA are instantiations of nonnegative matrix factorization, such probabilistic models can also be generalized easily to other domains. For example, collective topic modeling methods are discussed in [147, 148], and topic-modeling methods with network regularization are discussed in [378]. The method in [378] is also referred to as *NetPLSA*. The work in [535] discusses a topic modeling approach in the context of *heterogeneous* networks. The use of matrix factorization methods for community detection with edge content are discussed in [451]. The link prediction problem was proposed in [333]. The use of matrix factorization methods for link prediction are discussed in [3, 382]. Numerous methods have also been proposed for transfer learning between text and images for clustering [599] and classification [142, 450, 633]. A survey of heterogeneous transfer learning may be found in [15]. Methods for cross-lingual text mining are discussed in [561, 570]. An overview of cross-lingual methods for text mining is provided in [15].

Factorization machines are proposed in [467, 468], and a detailed discussion is provided in [199]. Although factorization machines have primarily been used in recommender systems, they have significant potential to be used in other applications like network link prediction and heterogeneous classification. They are also useful for short-text data, although this aspect remains relatively unexplored.

An early work that uses a Bayesian approach for hypertext categorization with hyperlinks is provided in [94]. The use of a Bayesian approach for clustering and classification of text data is provided in [17]. A generative approach for community detection in nodes with content is proposed in [598]. A discriminative probabilistic approach for combining link and content in community detection is provided in [600]. A detailed discussion of several node classification methods with content and structure is provided in [506]. Probabilistic models that combine content and structure for link prediction are discussed in [12, 213].

A graph-based approach to clustering with structure and content is presented in [630]. A classification technique that uses random walks on derived graphs for classification with text in social networks is discussed in [8]. Random walks for social media settings with image, text, and links are discussed in [551]. These walks are used for applications such as search and recommendations in heterogeneous social media settings.

8.7.1 Software Resources

Numerous software resources are available for performing matrix factorization in *scikit-learn* [646] (in Python) and at *Weka* [649] (in Java). However, most of these matrix factorization methods are designed for homogeneous settings, based on the ideas in Chapter 3. Most of the shared matrix factorization methods are designed as research prototypes, and few are available as off-the-shelf software for practical use. The easiest to use software for heterogeneous data is that of factorization machines [467]. In particular, three different libraries are available in the form of *libFM* [468] (from the original author), *libMF* [677], and *fastFM* [48]. The *libMF* library also provides access to other matrix factorization methods, and is different from the similar-sounding *libFM* library. Many of these libraries have freely downloadable Python wrappers.

8.8 Exercises

1. Show how to use a factorization machine to perform undirected link prediction in a social network with content.

2. Show how to convert a link prediction problem with structure and content into a link prediction problem on a derived graph.

3. Suppose that you have a user-item ratings matrix with numerical/missing values. Furthermore, users have rated each other's trustworthiness with binary/missing values.

   (a) Show how you can use shared matrix factorization for estimating the rating of a user on an item that they have not already rated.

   (b) Show how you can use factorization machines to achieve similar goals as (a).

4. Derive the gradient update equations for using factorization machines in binary classification with logistic loss. Derive the prediction function and updates for hinge loss.

5. Derive the gradient-descent updates for the optimization problem in Section 8.2.5. Discuss the special case of $\beta = 0$.

Chapter 9

Information Retrieval and Search Engines

"Making a wrong decision is understandable. Refusing to search continually for learning is not."–Phil Crosby

9.1 Introduction

Information retrieval is the process of satisfying user information needs that are expressed as textual queries. Search engines represent a Web-specific example of the information retrieval paradigm. The problem of Web search has many additional challenges, such as the collection of Web resources, the organization of these resources, and the use of hyperlinks to aid the search. Whereas traditional information retrieval only uses the content of documents to retrieve results of queries, the Web requires stronger mechanisms for quality control because of its open nature. Furthermore, Web documents contain significant meta-information and zoned text, such as title, author, or anchor text, which can be leveraged to improve retrieval accuracy. This chapter discusses the following aspects of information retrieval:

1. What types of data structures are most suitable for retrieval applications? The classical data structure for enabling search in text is the inverted index, and it is surprisingly versatile in handling various types of queries. The discussion of the inverted index will be paired with that of query processing.

2. The additional design issues associated with Web-centric search engines will be discussed. For example, we will discuss the collection of document resources from the Web, which is referred to as *crawling*.

3. How does one decide which Web documents are of high quality? Documents that are pointed to by many other pages are often considered more reputable, and it is desirable to assign such documents higher ranks in the search results.

© Springer Nature Switzerland AG 2022
C. C. Aggarwal, *Machine Learning for Text*,
https://doi.org/10.1007/978-3-030-96623-2_9

4. Given a search query, how does one score the matching between the keywords and the document? This is achieved with the use of *information retrieval models*. In recent years, such models have been enhanced with machine learning techniques in order to account for user feedback.

From the aforementioned discussion, it is evident that the Web-centric application of information retrieval (i.e., a search engine) has several additional layers of complexity. This chapter will discuss these additional layers.

The query processing can either provide a 0-1 response (i.e., *Boolean* retrieval), or it can provide a score that indicates the relevance of the document to the query. The Boolean model is the traditional approach used in classical information retrieval in which all results satisfying a logical keyword query are returned. The scoring model is more common for queries on very large document collections like the Web, because only a tiny fraction of the top-ranked results are relevant. Although thousands of Web pages might exactly match the keywords specified by the user, it is crucial to rank the results with various relevance- and quality-centric criteria in order to ease the burden on the user. After all, a user cannot be expected to browse more than ten or twenty of the top results. In such cases, quality-scoring techniques and learning techniques on user feedback are often used to enhance the search results. Although traditional forms of information retrieval are unsupervised, a supervised variant of information retrieval has gained increasing attention in recent years. Search can be viewed as a ranking-centric variant of classification. This is because a user query to a document collection is a binary classification problem over the entire corpus in which a label of "*relevant*" indicates that the document is relevant, and a label of "*non-relevant*" indicates otherwise. This is the essence of the learning-to-rank approach, which is also discussed in this chapter.

Chapter Organization

This chapter is organized as follows. Indexing and query processing are discussed in the next section, whereas scoring models are covered in Section 9.3. Methods for Web crawling are discussed in Section 9.4. The special issues associated with query processing in search engines are discussed in Section 9.5. The different ranking algorithms such as *PageRank* and *HITS* are discussed in Section 9.6. A summary is given in Section 9.7.

9.2   Indexing and Query Processing

Queries in information retrieval are typically posed as sets of keywords. The older *boolean retrieval systems* were closer to database querying systems in which users could enter sets of keywords connected with the "AND" and "OR" clauses:

text AND *mining*
(*text* AND *mining*) OR (*recommender* AND *systems*)

Each keyword in the aforementioned expression implicitly refers to the fact that the document contains the relevant keywords. For example, the first query above can be viewed as the *conjunct* of two conditions:

(*text* $\in$ Document) AND (*mining* $\in$ Document)

Most natural keyword queries in information retrieval systems are posed as conjuncts. Because of the ease in providing keywords as sets of relevant terms, it is often implicitly assumed that a query like "*text mining*" really refers to a conjunct of two conditions without explicitly using the "AND" operator. The use of the "OR" operator is increasingly rare in modern retrieval systems both because of the complexity of using it, and the fact that too many results are returned with queries containing the "OR" operator unless the individual conjuncts are very restrictive. In general, the most common approach is to simply pose the query as a set of keywords, which implicitly uses the "AND" operator. However, search engines also use the relative ordering of the keywords when interpreting such queries. For example, the query "*text mining*" might not yield the same result as "*mining text.*" For simplicity in discussion, we will first discuss the case in which queries are posed as sets of keywords that are implicitly interpreted as conjuncts of membership conditions. Later, we will show how to extend the approach to more complex settings.

In all keyword-centric queries, two important data structures are commonly used:

1. *Dictionary:* Given a query containing a set of terms, the first step is to discover whether that term occurs in the vocabulary of the corpus. If the term does occur in this vocabulary, a pointer is returned to a second data-structure indexing the documents containing this term. The second data structure is an *inverted list*, which is a component of the *inverted index*.

2. *Inverted index:* As the name implies, the inverted index can be viewed as an "inverted" representation of the document-term matrix, and it comprises a set of inverted lists. Each inverted list contain the identifiers of documents containing a term. The inverted index is connected to the dictionary data structure in the sense that the dictionary data structure contains pointers to the heads of the inverted lists of each term. These pointers are required during query processing.

For a given query, the dictionary is first used to locate the pointers to the relevant term-specific inverted lists, and subsequently these inverted lists are used for query processing. The intersection of the different inverted lists provides the list of document identifiers that are relevant to a particular query. In practice, too many or too few documents might satisfy all query keywords. Therefore, other types of scoring criteria such as partial matches and word positions are used to rank the results. The inverted index is versatile enough to address such ranking queries, as long as the scoring function satisfies certain convenient *additivity* properties with respect to the query terms. In the following, we will describe these query processing techniques together with their supporting data structures.

9.2.1 Dictionary Data Structures

The simplest dictionary data structure is a hash table. Each entry of the hash table contains the (i) string representation of the term, (ii) a pointer to the first element of the inverted list of the term, and (iii) the number of documents in which the term occurs. Consider a hash table containing N entries. The data structure is initialized to an array of NULL values. The hash function $h(\cdot)$ uses the string representation of the term t_j to map it to a random value $v = h(t_j)$ in $[0, N - 1]$. In the event that the vth entry in the hash table is empty, the term t_j is inserted as the vth entry in the hash table. The main problem arises in cases where the vth entry is already occupied, which results in a *collision*. Collisions can be resolved in two ways, depending on the type of hash table that is used:

Chained hashing: In the case of chained hashing, one creates a linked list of multiple terms, which is pointed to by each hash table entry. When the vth entry is already occupied, it is first checked whether term t_j already exists within the linked list. If this is the case, then an insertion does not need to be performed. Otherwise, the linked list is augmented with the term t_j, and its length increases by 1. The entries of the linked list contain the string representation of the term, the number of documents in which it occurs, and a pointer to the first item on the inverted list of the term. The linked list is maintained in (lexicographically) sorted order[1] to enable faster searching of terms. When a term is to be checked against the linked list, one simply scans the linked list in sorted order until the term is found or a lexicographically larger term is reached.

Linear probing: In linear probing, a linked list is not maintained at each position in the hash table. Rather each position in the hash table contains the meta-information (e.g., string representation, inverted list pointer, and inverted list size) for a single term. For a given term t_j, the $h(t_j)$th position is checked to see if it is empty. If the position is empty, then the string for term t_j is inserted at that position along with its meta-information (document frequency and inverted list pointer). Otherwise, it is checked if the occupied position already contains term t_j. If the occupied position does not contain the term t_j, the same check is repeated with the $[h(t_j) + 1]$th position. Thus, one "probes" successive positions $h(t_j), h(t_j) + 1, \ldots h(t_j) + r$, until the term t_j is encountered or an empty position is reached. If the term t_j is encountered, then nothing needs to be done, since the hash table already contains the term t_j. Otherwise, the term t_j is inserted at the first empty position encountered during the linear probing process. This probing process is also useful during query time, when a term needs to be searched in the dictionary to obtain the pointer to its inverted list.

The hash table data structure does not provide any natural way to identify terms with closely related spellings. It is often useful to identify such terms as query suggestions to the user, when they make a mistake in entering a query. For example, if a user enters the (misspelled) query term "*recieve*," it is often desirable to suggest the alternate query term "*receive*." One can find such terms in the context of the hash table data structure by creating a separate dictionary of misspelled words (from historical queries) together with the possible spellings that might be correct. A more challenging case arises when users misspell words to their *homonyms*. For example, the term "*school principle*" is most likely intended to be "*school principal*." Such a spelling correction is referred to as a *contextual* spelling correction, and it can be detected only by using the surrounding phrase in the form of a k-gram dictionary of incorrect query phrases.

An alternative that allows the detection of closely related spellings is to implement the dictionary as a variant of the binary search tree in which terms are stored only at the leaf levels of the tree, and the internal nodes contain the meta-information in order to find the relevant leaf efficiently. In the binary search tree, the entire set of terms can be viewed as a lexicographically sorted list, which is partitioned at some intermediate letter between 'a' and 'z.' For example, all terms starting with letters between 'a' and 'h' belong to the left branch of the tree, whereas all terms starting with letters between 'i' and 'z' belong to the right branch of the tree. Similarly, the left branch may be divided into two parts, corresponding to the beginning portions between $[a, de]$, and $[df, h]$, respectively. This type of recursive division is shown in Figure 9.1. The leaf nodes of the binary search tree contain the actual terms. The process of searching for a term is a relatively simple matter. One only

---

[1]A lexicographically sorted order refers to the order in which terms occur in a dictionary.

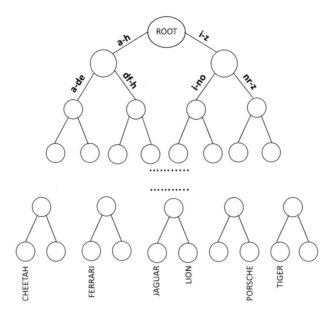

Figure 9.1: A binary tree structure for storing a searchable dictionary of terms. The leaf nodes point to the data structures indexed by the terms, which are the inverted lists

needs to traverse the path corresponding to the front portion of the query term until the appropriate leaf node is reached (or it is determined that the binary tree does not contain the search term).

If the binary tree is relatively well balanced, the search process is efficient because the depth of the tree is $O(\log(d))$ over a dictionary of d terms. It is often difficult to fully balance a binary tree in the presence of dynamic updates. One way of creating a more balanced tree structure is to use a *B-Tree* instead of a binary search tree. Interested readers are referred to [493] for details of these data structures. Although the tree-like structures do offer better search capabilities, the hash table is often the data structure of choice for dictionaries. One advantage of the hash table is that it has $O(1)$ lookup and insertion time.

9.2.2 Inverted Index

The inverted list is designed to identify all the document *identifiers* related to a particular term. Each *inverted list* or *postings list* corresponds to a particular term in the lexicon, and it contains a list of the identifiers of all documents containing the term. Each element of this list is also referred to as a *posting*. The document identifiers of the inverted list are often (but not always) maintained in sorted order to enable efficient query processing and index update operations. The relevant term frequencies are often stored with document identifiers.

An example of an inverted representation of a document-term matrix is shown in Figure 9.2. The hash-based dictionary data structure, which is tied to this index, is also included in this figure. Note that the dictionary data structure also contains the document-wise frequency of each term (i.e., number of occurrences across distinct documents), whereas each individual posting of the inverted list contains the document-specific term frequency. These additional statistics are required to compute match-based scores between queries and doc-

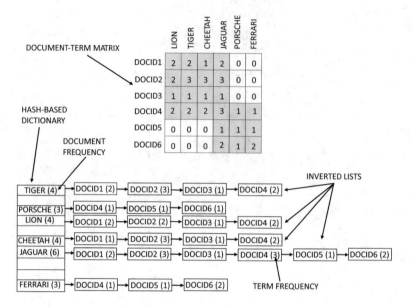

Figure 9.2: A hash-based dictionary and an inverted index together with its parent document-term matrix. The dictionary is used to retrieve the pointer to the first element of the inverted list during query processing

uments with inverse document frequency normalization. As we will see later, the postings list might also contain other meta-data about the position of that term in the document. Such meta-data can be useful for positional queries.

It is common to use linked lists to store the inverted index when it is maintained in main memory. Even when inverted lists are too long to be stored in main memory, smaller portions of them are often maintained in main memory for fast query processing. Linked lists can be used to insert a document identifier at any position in the inverted list efficiently by a single pointer deletion and two pointer additions. Therefore, such data structures are efficient from the perspective of incremental updates. The first element of each inverted list is pointed to by the entry of the relevant term in the dictionary data structure. This mapping is crucial for query processing.

One issue with the inverted list is that many of the lists of uncommon or unique terms are extremely short. Storing such lists as separate files is inefficient. In practical implementations, multiple inverted lists are consolidated into files on disk, and the dictionary data structure contains the pointer to the offset in the relevant file on disk. This pointer provides the first posting in the inverted list of the term being queried.

9.2.3   Linear Time Index Construction

Given a document corpus, how does one create the dictionary and the inverted lists? Modern computers usually have sufficient memory to maintain the dictionaries in main memory. However, the construction of inverted lists is a completely different matter. The space required by an inverted index is of the same magnitude required by a sparse representation of the document-term matrix within a constant of proportionality (see Exercise 1). A document corpus is usually too large to be maintained in main memory and so is its inverted index.

Converting one disk-resident representation (i.e., corpus) to another (i.e., inverted index) is often an inefficient task, if care is not taken to limit the reads and writes to disk. The most important algorithm design criterion is to minimize random accesses to disk and favor sequential reads as far as possible during index construction. The following will describe a linear-time method, which is referred to as *single-pass in-memory indexing*. The basic idea is to work with the available main memory and build both the dictionary and inverted index within the memory until it is exhausted. When memory is exhausted, the current dictionary and inverted index structure are both stored on disk with care being taken to store the inverted lists in sorted lexicographic order of the terms. At this point, a new dictionary and inverted index structure is started, and the entire process is repeated. Therefore, at the end of the process, one will have multiple dictionaries and inverted index structures. These are then merged in a single pass through the inverted lists. The following discussion explains both the phases of multiple index construction and merging.

An important assumption is that the document identifiers are processed in sorted order, which is easy to implement when the document identifiers are created during index construction. The practical effect of this design choice is that the elements of the inverted lists are arranged in sorted order as identifiers are appended to the end of each list. Furthermore, the document identifiers in the list of an earlier block are all strictly smaller than those in a later block, which enables easy merging of these lists. The approach starts by initializing a hash-based dictionary $\mathcal{H}$ and an inverted index $\mathcal{I}$, to empty structures and then updating them as follows:

> **while** remaining memory is sufficient to process next document **do begin**
>   Parse next document with identifier DocID;
>   Extract set S of distinct terms in DocID with term frequencies;
>   Use $\mathcal{H}$ to identify existing and new terms in S;
>   For each new term in S, create a new entry in $\mathcal{H}$ pointing to a newly created
>   singleton inverted list in $\mathcal{I}$ containing DocID and term frequency;
>   For each existing term in S, add DocID to end of corresponding inverted list
>   in $\mathcal{I}$ together with the term frequency;
> **end while**
> Sort the entries of $\mathcal{H}$ in lexicographic order of term;
> Use the sorted entries of $\mathcal{H}$ to create a single disk file containing
>   the inverted lists of $\mathcal{I}$ in lexicographic order of term;
> Store sorted dictionary $\mathcal{H}$ on disk containing file offset pointers to inverted lists;

The sorted dictionaries can be stored on disk as lists of sorted term-string/document-frequency/offset triplets rather than as hash tables. After processing the entire corpus, the (multiple) disk files containing partial inverted indexes need to be merged. Let these disk files containing the inverted lists be denoted by $\mathcal{I}_1, \mathcal{I}_2, \ldots \mathcal{I}_k$. The merging is a simple matter because (i) each inverted list in $\mathcal{I}_j$ is sorted by document identifier, (ii) the different inverted lists within each $\mathcal{I}_j$ are arranged in lexicographically sorted order of term, and (iii) all document identifiers in earlier block writes are smaller than the document identifiers in later block writes. The conditions (i) and (iii) are consequences of the fact that document identifiers are selected (or created) in sorted order for parsing. An example of two partial indexes containing three documents each is shown in Figure 9.3. Note that the first index only contains sorted lists with document identifiers between DocId1 and DocId3, whereas the second index contains sorted inverted lists with document identifiers between DocId4 and DocId6. Therefore, the merged and sorted list of any term (e.g., *Jaguar*) can be obtained by concatenating one list after the other.

In order to merge the inverted lists, one can simultaneously open all the files containing $\mathcal{I}_1 \ldots \mathcal{I}_k$ and $\mathcal{H}_1 \ldots \mathcal{H}_k$. We do not need to read these files in memory but scan them sequentially in order to process each term in sorted order. The merging can be achieved

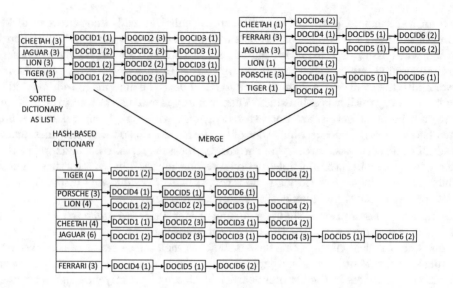

Figure 9.3: Merging a pair of partial indexes

with a simultaneous linear scan of the different files, because the inverted lists of various terms are stored in lexicographically sorted order. For any particular term, its inverted list is identified in each block (if it exists), and these lists are simply concatenated. The lists of later blocks are concatenated after those of earlier blocks to ensure that document identifiers remain in sorted order. At the same time, the dictionary of the merged index is created from scratch. For each term-specific merging, a new entry is added to the dictionary containing the pointer to the merged postings list and the number of documents containing that term (which is the length of the merged list). It is also possible to create the index without processing the document identifiers in sorted order, although doing so will increase the running time slightly (see Exercise 2).

9.2.4 Query Processing

Query processing is of two types. One of them is *Boolean retrieval*, in which documents are returned only when they *exactly* match a particular query. There is no focus on ranking the results, even when a large number of them are returned. Furthermore, in the Boolean retrieval model, one can construct Boolean expressions for queries containing "AND," "OR," and "NOT," whereas ranked retrieval generally uses free text queries. As a practical matter, however, ranked retrieval is almost always necessary in order to distinguish between the varying levels of matches between target queries and documents.

9.2.4.1 Boolean Retrieval

In Boolean retrieval, results are returned depending on whether or not they match a particular query. The query can be in forms such as the following:

(*text* AND *mining*) OR (*data* AND *mining*)
(*text* OR *data*) AND *mining*

The two queries above are actually equivalent, which also reflect the different ways in which they can be resolved. Consider the case where one wishes to use the first form of the query. The first step is to use the dictionary data structure in order to locate the terms "*text*," "*data*," and "*mining*," and their corresponding inverted lists. First, the lists of "*text*" and "*mining*" are intersected into a sorted list L_1, and then the lists of "*data*" and "*mining*" are intersected into another sorted list L_2. Subsequently, the lists L_1 and L_2 are merged with a union operation in order to implement the "OR" operator. An important point about Boolean retrieval is that the returned results are unordered, and there is a single correct result set for a given search query.

The following will describe the process of intersecting two sorted lists in linear time. Assume that each inverted list is defined as a linked list in which the document identifiers are in sorted order. Then, the algorithm uses two pointers, which are initialized to the beginning of the two lists. These pointers are used to scan through the two linked lists in order to identify common document identifiers. The query-result list, which is denoted by Q, is initialized to the empty list. If the document identifiers at the current pointer values in the two inverted lists are the same, then this document identifier is appended to the end of Q and both pointers are incremented by 1. Otherwise, it is determined which pointer corresponds to the smaller document identifier. Consider the case where the pointer of the list corresponding to the keyword "*mining*" has the smaller document identifier. The pointer to the inverted list for "*mining*" is incremented until the corresponding document identifier is either the same or larger than that of "*text*." This process of advancing the pointers to the two lists is continued (with the list Q growing continuously) until the end of at least one of the lists is reached. When the intersection of more than two lists is performed in succession, it is advisable to start with the most restrictive pair of words first to perform the intersection, so that the size of the intermediate result is as small as possible. In other words, the inverted lists are used in decreasing order of inverse document frequency in the intersection process. This is done in order to ensure that smaller documents are processed first. The process of merging two lists with the "OR" operator uses a similar approach as that of intersection (see Exercise 3).

9.2.4.2 Ranked Retrieval

Boolean retrieval is rarely used in information retrieval and search engines, because it provides no understanding of the ranking of the retrieved results. Even though the Boolean retrieval model does allow the ability to combine different logical operators to create potentially complex queries, the reality is that it is often cumbersome for the end user to effectively use this type of functionality. Most practical applications use *free text queries*, in which users specify sets of keywords. Although a free text query can be interpreted in terms of maximizing the match over the query keywords, there are often many other factors that influence the matching. In ranked retrieval, the results need to be *scored* and *ranked* in response to the query, and the system often performs this type of ranking using a variety of different factors (e.g., relative positions of terms in document or document quality) that are not always specified explicitly in the query. In this sense, ranked retrieval allows the use of a variety of different *models* for retrieving search results, and there is no single model that is considered fundamentally "correct" in a way that can be crisply defined. This is a different concept from Boolean retrieval, in which the correct set of results is exactly defined, and the returned results are unordered.

For large-scale applications like Web search, the Boolean relevance of the (possibly thousands of) documents to a set of search terms is not quite as important as ensuring that the tiny set of results at the very top of the search are relevant to the user. This is a far more difficult problem than Boolean search, and many aspects of it have a distinct machine learning flavor. Although one can restrict the search results based on relevance criteria (e.g., all query terms must be present), the ability to correctly score the large number of valid search results remains exceedingly important in these settings. Most natural scoring functions satisfy the following properties:

1. The presence of a term in the document that matches a query term increases the score, and the score increases with the frequency of the term.

2. Matching terms that are rarely present in the document collection (i.e., terms with high inverse document frequency) increase the score to a greater degree. This is because rare terms are less likely to be matched by chance.

3. The score of candidate documents with longer length is penalized because terms might be matched to the query purely by chance.

The cosine similarity function with tf-idf normalization satisfies all of the above properties, although it does not account for many factors used in modern search engines such as the ordering of the terms or their proximity. Furthermore, when multiple factors are used for computing similarity, it is helpful to be able to weight the relative importance of these factors. This problem has the flavor of supervised learning, which leads to the notion of machine learning in information retrieval. This section will provide a broad overview of the index structures, query processing, and scoring functions, whereas Section 9.3 will focus more deeply on the basic principles with which various scoring functions are designed.

There are two fundamental paradigms for query processing in ranked retrieval, which correspond to *term-at-a-time* and *document-at-a-time* query processing with the inverted index. Many nicely behaved scoring functions like the cosine can be computed using either paradigm because they can be expressed as additive functions over query terms. However, the document-at-a-time processing is more convenient for complex functions that use various factors involving multiple terms, such as the relative positions of the terms. In both cases, the document identifiers are accessed using the inverted lists and their scores are continually updated using *accumulator variables* (each of which is associated with a document identifier). In the following, we will describe each of these paradigms.

Term-at-a-time Query Processing with Accumulators

Accumulators are intermediate aggregation variables that can help in evaluating surprisingly general scoring functions between queries and documents, as long as the scoring function is computed in an additive way over the target query terms. For small subsets of query terms, even more general functions incorporating positional information between terms can be computed with accumulators. Consider a query $\overline{Q} = (q_1 \ldots q_d)$ with a small number of query terms in which most values of q_i are 0. Consider a document $\overline{X} = (x_1 \ldots x_d)$ defined over the same lexicon of size d. Now consider a simple scoring function $F(\overline{X}, \overline{Q})$ of the following form:

$$F(\overline{X}, \overline{Q}) = \sum_{j:q_j>0} g(x_j, q_j) \qquad (9.1)$$

Note that the summation is only over the small number of terms satisfying $q_j > 0$, and $g(\cdot, \cdot)$ is another function that increases with both x_j and q_j. For example, using $g(x_j, q_j) = x_j q_j$ yields the dot product, which is the unnormalized variant of the cosine function.

The inverted lists of all the terms with $q_j > 0$ are accessed one after another to perform the scoring. Every time a new document identifier is encountered on an inverted list, a new accumulator needs to be created to track the score of that document. For each document identifier encountered on the inverted list of a query term with $q_j > 0$, the value of $g(x_j, q_j)$ is added to the accumulator of that document identifier. In cases where the corpus is large, too many document identifiers might be encountered and one might run out of space to create new accumulators. There are several solutions for addressing this issue. First, the inverted lists should always be accessed in decreasing order of inverse document frequency, so that the most number of terms are used when one runs out of memory. Furthermore, since the terms with higher inverse document frequency are assumed to be more discriminative, this ordering is also helpful in ensuring that the accumulators are more likely to be assigned to relevant documents. A hash table is used to keep track of the accumulators for various documents. When one runs out of memory in the hash table, the results are returned with respect to only[2] those identifiers that have been encountered so far. New accumulators are no longer added because such documents are not assumed to be strong matches. However, the counts of existing accumulators continue to be updated.

Finally, the documents with the largest accumulators are returned. The naïve approach would be to scan the accumulators to identify the top-k values. A more efficient approach is to scan the accumulator values and maintain the top-k in a min-heap (i.e., a heap containing the minimum value at the root). The heap is initialized by inserting the first k scanned accumulators. Subsequent accumulators are compared with the value at the root of the heap, and dropped if they are less than the value at the root. Otherwise, they are inserted into the heap, and the minimum value at the root is deleted. This approach requires time that is $O(n_a \cdot \log(k))$ time, where n_a is the number of accumulators.

It is noteworthy that term-at-a-time query processing does not require the elements on the inverted list to be sorted by document identifier. In fact, for term-at-a-time query processing, it makes sense to sort the lists by decreasing order of term-frequency in the various document identifiers and use only those documents whose term-frequency is above a particular threshold. Furthermore, one can also handle more general functions than Equation 9.1, which are of the following form:

$$F(\overline{X}, \overline{Q}) = \frac{\sum_{j:q_j>0} g(x_j, q_j)}{G(\overline{X})} + \alpha \cdot Q(\overline{X}) \tag{9.2}$$

Here, $G(\overline{X})$ is some normalization function (like the length of the document), α is a parameter, and the function $Q(\overline{X})$ is some global measure of the quality of the document (such as the *PageRank* of Section 9.6.1). It is not difficult to see that the cosine is a special case[3] of this measure. It is also assumed that such global measures for document normalization or quality are pre-stored up front in a hash table indexed by document identifier. This type of scoring function can be addressed by using an additional processing step at the end in which the values of $G(\overline{X})$ and $Q(\overline{X})$ are accessed from the hash table to adjust the scores.

---

[2]If all query terms must be included in the result, then the intersection of the inverted lists can be performed up front and accumulators are assigned only to document identifiers that lie in this intersection. There are many such *index elimination* tricks that one can use to speed up the process.

[3]One can set $Q(\overline{X}) = 0$ and select $G(\overline{X})$ to be the length of document $\overline{X}$. Normalization with the query length is not necessary because it is constant across all documents and does not change the relative ranking.

Document-at-a-time Query Processing with Accumulators

Unlike the term-at-a-time query processing paradigm, the document-at-a-time approach requires each inverted list to be sorted by document identifier. The document-at-a-time approach can handle more general query functions than the term-at-a-time approach because it accesses all the inverted lists for the query terms simultaneously in order to identify all the query-specific meta-information associated with a document identifier. For a given query vector $\overline{Q} = (q_1 \ldots q_d)$, let $\overline{Z}_{X,Q}$ represent all the meta-information in the document $\overline{X}$ about the *matching terms* in the document with respect to the query. This meta-information could correspond to the position of the matching terms in the document $\overline{X}$, the portion of the document in which matching terms lie, and so on. As we will discuss later, such meta-information can often be stored along with the inverted lists. Then, consider the following scoring function, which is a generalization of Equation 9.2:

$$F(\overline{X},\overline{Q}) = \frac{H(\overline{Z}_{X,Q})}{G(\overline{X})} + \alpha \cdot Q(\overline{X}) \tag{9.3}$$

Here, $G(\overline{X})$ and $Q(\overline{X})$ are global document measures as in Equation 9.2. The function $H(\overline{Z}_{X,Q})$ is more general than the additive form of Equation 9.2 because it could include the effect of the interaction of multiple query terms. This function could, in principle, be quite complex and include factors such as the positional distance between the query terms in the document. However, to enable such a query, the inverted index needs to contain the meta-information about the positions of query terms (cf. Section 9.2.4.3).

In such a case, one simultaneously traverses the inverted list for *each* term t_j satisfying $q_j > 0$ (i.e., terms included in the query). As in the case of list intersection, one traverses each of the *sorted* lists in parallel until one reaches the same document identifier. At this point, the value of $H(\overline{Z}_{X,Q})$ is computed (using the meta-information associated with document identifiers) and added to the accumulator variable for that document identifier. The other post-processing steps in document-at-a-time querying are identical to those of term-at-a-time query processing. If the space for accumulator variables is limited, the document-at-a-time processing maintains the best scores so far, which turns out to be a more sensible approach for obtaining the best results. In such cases, it might also make sense to incorporate the impact of global document measures like $G(\overline{X})$ and $Q(\overline{X})$ at the time the document is processed rather than leaving it to the post-processing phase.

Although it is possible to enable scoring functions like Equation 9.3 with term-at-a-time querying, it increases the space overhead in impractical ways. One would need to store all the meta-information in the traversed lists along with the accumulator variables and then evaluate Equation 9.3 in the final step.

Term-at-a-Time or Document-at-a-Time?

The two schemes have different advantages and disadvantages. The document-at-a-time approach allows the maintenance of the best k results found so far dynamically. Furthermore, the types of queries that can be resolved with document-at-a-time processing are more complex, because one can use the relative positions of terms and other statistics that use the properties of multiple query terms. On the other hand, the document-at-a-time processing requires multiple disk seeks and buffers because multiple inverted lists are explored simultaneously. In term-at-a-time processing, one can read in large chunks of a single inverted list at one time in order to perform the processing efficiently.

What Types of Scores Are Common?

In many search engines, global meta-features of the document such as its provenance or its citation structure are included in the final similarity score. In fact, modern search engines often learn the importance of various meta-features (cf. Sections 9.2.4.5 and 9.2.4.6) by leveraging user click-through behavior. For example, Equations 9.2 and 9.3 contain the parameter α, which regulates the importance of page quality in ranking. Such a parameter can be learned using machine learning models from previous user click-through behavior. It needs to be pointed out that most of the popular scoring functions in information retrieval and search engines (including advanced machine learning models) can be captured using Equations 9.2 and 9.3 by instantiating the various terms in these equations appropriately. Several such models will be explored in this section and in Section 9.3.

9.2.4.3 Positional Queries

It is often desirable for query processing to account for the positions of the query terms. There are several ways in which the positioning can be taken into account. The first is to include common phrases as "terms" and created inverted lists for them. However, this approach greatly expands the term set. Furthermore, for a given query, there are multiple ways in which one can process the query using either the phrases or the individual terms.

In order to resolve queries with the positional index, the same inverted list is maintained, except that all the positions of a term in the document are maintained as meta-information along with a document identifier in the inverted list. Specifically, in the inverted list for any particular term, the following meta-information is retained along with document identifiers:

$$\text{DocId}, \; freq, \; (Pos_1, \; Pos_2, \; \ldots, \; Pos_{freq})$$

Here, $freq$ denotes the number of times the term occurs in the document with identifier DocId. For example, if the term "*text*" occurs at position 7 and 16 in DocId, and the term "*mining*" occurs at positions 3, 8, and 23 of DocId, then all these positions are stored with the document identifiers in the inverted list. Therefore, in the first case, the meta-information DocId,2, (7, 16) is maintained as one of the entries in the inverted list of "*text*," whereas in the second case, the meta-information DocId,3, (3, 8, 23) is maintained in one of the entries of the inverted list of "*mining*." In this particular case, it is evident that the term "*text*" occurs at position 7 in the document with identifier DocId, whereas the term "*mining*" occurs at position 8 in the same document. Therefore, it is evident that the phase "*text mining*" is present in DocId. For Boolean queries, one will need to check these positions at the time of intersecting the inverted lists of "*text* and "*mining*."

In ranking queries (like search engines), the relative positions of terms in a document can affect the scoring function used to quantify the degree to which a document matches a specific set of keywords in a particular order. Therefore, the queries "*text mining*" and "*mining text*" will not return the same ranking of the results, when using a search engine like Google. It turns out that this type of query processing can be performed with accumulator variables, because the effect of relative positioning can be captured by Equation 9.3. In particular, the function $H(\overline{Z}_{X,Q})$ of Equation 9.3 should be *defined* by the search engine architect in order to capture the impact of term positioning. The natural approach in these cases is to use document-at-a-time query processing (see page 268).

It is common to combine phrase-based indexes with positional indexes. The basic idea is to keep track of the commonly queried phrases, and maintain inverted lists for these frequent phases in addition to the lists of the individual terms. For a given query, the frequent phrases in it are combined with positional indexes in order to use the positioning information. How is such a combination achieved? An important point here is that search engines typically use *free text queries*, which are often mapped to an internal representation by the system with the use of a *query parser*. In many cases, the query parser can issue multiple queries, in which phrase indexes are used in combination with positional indexes to yield an efficient query result. For example, the following approach might be used:

1. The inverted lists of frequent phrases might be used in order to provide a first response to the query. Note that it is not necessary that the query phrase is frequent and is available in the index. In such a case, one might try different 2-word subsets of the query phrase to check if it is available in the inverted index.

2. In the event that sufficient query results cannot be generated using the aforementioned approach, one might try to use the positional index in order to generate a query result that scores the documents based on relative proximity of query terms.

The specific heuristic used to resolve a query depends on the goals of the search system at hand. Modern search engines use a number of query optimizations that include all types of meta-data about the document to score and rank results. Examples include *zoned scoring* and *machine learned scoring*, which are discussed in Sections 9.2.4.4 and 9.2.4.5, respectively. Furthermore, qualitative judgements about the document are inferred based on the co-citation structure, and incorporated in the final ranking. These issues are discussed in a later section (cf. Sections 9.5 and 9.6).

9.2.4.4 Zoned Scoring

In zoned scoring, different parts of a document, such as the author, title, keywords, and other meta-data are given varying amounts of weight. These different parts are referred to as *zones*. Although zones seem similar to fields at first sight, they are different in the sense that they might contain arbitrary and free-form text. For example, in search engines, the title of a document is quite important as compared to the body of the document. In some cases, the zoning can be implemented by simply adding a more important zone to the vector space representation with a higher weight. For example, the title can be given a higher weight than the body of the document. However, in most cases, zoning is implemented by storing the information about the zoning within the inverted list. Specifically, consider the inverted list of each term that contains the frequency as well as the positioning information. Along with each positioning information, we also maintain the zone in which the term occurs. In other words, consider a position-based inverted index in which one of the entries in the inverted list of "*text*" is of the form DocId, 2, (7, 16). Therefore there are two occurrences of the term "*text*" in DocId with positions 7 and 16, respectively. This type of entry, however, assumes that the positioning is defined with respect to a document with a single zone. More generally, the entry can be of the form DocId, 2, (2-Title, 9-Body). In this case, the term "*text*" occurs as the second token of the title and the ninth position in the body. This type of meta-information can be used easily for *weighted zone scoring*, in which the matched document identifiers during the intersection of inverted lists are scored based on the specific zones in which they reside. An important point here is in deciding how much weight to give each zone. While it is clear that some zones such as the title are more important, the process of finding specific weights for zones has the flavor of a machine learning algorithm.

9.2.4.5 Machine Learning in Information Retrieval

How can one find the appropriate weights for each zone in an information retrieval setting? Consider a situation in which the documents of a corpus have r weights $w_1 \ldots w_r$ over the r zones. Furthermore, the frequencies of a particular document-term combination over these zones are $x_1 \ldots x_r$. For simplicity, one can also assume that each $x_i \in \{0, 1\}$ depending on whether the term is present in the zone. However, it is also possible to have non-binary values of x_i. For example, if a term occurs more than once in a ith zone, the value of x_i might be larger than one. In practical settings, many values of x_i might be 0. Then, the contribution of that document-term combination to the scoring is given by $\sum_{j=1}^{r} w_j \cdot x_j$, where w_j is an unknown weight. It is relatively easy to compute this type of additive score with accumulators at query processing time, if one knew the values $w_1 \ldots w_r$ of the weights. Therefore, the weights are learned up front in offline fashion.

In order to learn the appropriate values of w_i, one can use the *relevance feedback values from the user* over a set of training queries. The training data contains a set of engineered features that are extracted from each document in response to a query (such as the zones in which query words lie) together with a user relevance judgement of whether the document is relevant to the corresponding query. The relevance judgement might be a binary quantity (i.e., relevant/not relevant), a numerical quantification of the relevance judgement, or a ranking-based judgement between pairs of documents. The Web-centric approach of collecting relevance feedback leverages user click-through behavior, which is discussed later.

Learning the importance of zones is not the only application of such weight-learning techniques. There is significant meta-information associated with both terms and documents, whose importance can be learned for query processing. Examples include the following:

1. *Document-specific features:* The meta-information associated with a document on the Web, such as its geographical location, creation date, Web linkage-based co-citation measures (cf. Section 9.6.1), number of words in pointing anchor text, or Web domain can be used in the scoring process. Document-specific meta-data is often independent of the query at hand, and has been shown to be effective for improving retrieval performance with machine learning techniques [470].

2. *Impact features:* The impact of several terms in a scoring function is often regulated with parameters. For example, the scoring functions of Equations 9.2 and 9.3 contain the parameter α, which regulates the importance of document quality. Sometimes multiple scoring functions like the cosine, binary-independence model, and the BM25 model (see Section 9.3) can be combined with weights. The importance of these weights can be learned with user feedback.

3. *Query-document pair-specific meta-data:* The zones of the document in which query terms occur and query-term ordering/positioning within the candidate document can be used in the scoring process.

The scoring function in modern search engines is quite complex and is often tuned using machine learning. In general, one might have any arbitrary set of parameters $w_i \ldots w_m$. In other words, these weight parameters include the zoning weights, and they might represent the importance of different *features* $x_1 \ldots x_m$ extracted from a document-query pair. Note that the same set of features is extracted from any document-query pair, which allows the learning to be generalized from one query to another. In the zoning example, the features correspond to the different zones of the documents in which the query words lie, and their corresponding frequencies. Therefore, if the user feedback data consistently shows the user preferentially clicking on search results with query words in the title (over the body), then this fact will be learned by the algorithm irrespective of the query at hand. This is achieved with a relevance function $R(w_1 x_1, \ldots w_m x_m)$, which is defined in terms of the weighted features. For example, a possible relevance function could be as follows:

$$R(w_1 x_1, \ldots, w_m x_m) = \sum_{j=1}^{m} w_j x_j \qquad (9.4)$$

The values of $w_1 \ldots w_m$ are learned from user feedback. The choice of the relevance function is a part of search engine design, and virtually all functions that are defined in terms of the meta-data about matching terms between the query and the target can be modeled with scoring functions like Equation 9.3. Such scores can be efficiently computed at query processing time with accumulators.

It is noteworthy that relevance judgements can also be inferred using *implicit feedback* based on user actions rather than their explicit judgements. For example, search engines provide a large amount of implicit feedback based on user clicks on returned query results. Such feedback should, however, be used carefully because top-ranked items are more likely to be clicked by a user, and therefore one must adjust for the rank of the returned items during the learning process. Consider a situation in which a search engine ranks document $\overline{X_j}$ above document $\overline{X_i}$, but the user clicks on the document $\overline{X_i}$ but not $\overline{X_j}$. Because of the preferential clicking pattern of the user, there is evidence that document $\overline{X_i}$ might be more relevant than document $\overline{X_j}$ to the user. In such a case, the training data is defined in terms of *ranked pairs* like $(\overline{X_i}, \overline{X_j})$, which indicate relative preference. This type of data is highly noisy but the saving grace is that copious amounts of it can be collected easily. Machine learning methods are particularly good at learning from large amounts of noisy data.

For the purpose of this section and the next, each $\overline{X_i}$ refers to the query-specific features extracted from the document (e.g., impact features), rather than text vectors. This choice requires an understanding of the importance of various characteristics of the document-query pair, such as the impact features, including the use of zones, physical proximities, ordering of matched words, document authorship, domain, creation date, page citation structure, and so on. *The extraction of the features depending on the match between the query and the document is an important modeling and feature engineering process, which depends on the search application at hand.* Each of these features either need to be pre-stored (e.g., document-specific *PageRank*), or they need to be computed on-the-fly using accumulators. At query time, they need to be combined using the linear condition in Equation 9.4.

9.2.4.6 Ranking Support Vector Machines

The previous section discusses the importance of learning methods by extracting m query-specific features and learning their associated parameters $w_1 \ldots w_m$ in order to quantify their relevance to a new query. How can one use such pairwise judgements by the end user in order to learn key parameters such as $w_1 \ldots w_m$ that are used in ranking the results? This is typically achieved by *learning-to-rank* algorithms. A classical example of such an algorithm is the *ranking support vector machine*, which is also referred to as the *ranking SVM*. The ranking SVM uses previous queries to create training data for the extracted features. The features for a document contain attributes corresponding to the various zones in which the query terms occur, meta-information about the document such as its geographical location, and so on. The training data contains pairs of documents (in this query-centric representation), denoted by $(\overline{X_i}, \overline{X_j})$, which signifies the fact that $\overline{X_i}$ should occur earlier than $\overline{X_j}$. We would like to learn $\overline{W} = (w_1 \ldots w_m)$, so that $\overline{W} \cdot \overline{X_i} > \overline{W} \cdot \overline{X_j}$ for the training documents $\overline{X_i}$ and $\overline{X_j}$, which *contain the query-specific "match" features* (see Section 9.2.4.5 for examples). Once such weights have been learned from a training corpus (created by past queries and user feedback), they can be used in real time to rank the different documents.

We will now formalize the optimization model for the ranking SVM. The training data $\mathcal{D}_R$ contains the following set of ranked pairs:

$$\mathcal{D}_R = \{(\overline{X_i}, \overline{X_j}) : \overline{X_i} \text{ should be ranked above } \overline{X_j}\}$$

For each such pair in the ranking support vector machine, the goal is learn $\overline{W}$, so that $\overline{W} \cdot \overline{X_i} > \overline{W} \cdot \overline{X_j}$. However, we impose an additional *margin requirement* to penalize pairs where the difference between $\overline{W} \cdot \overline{X_i}$ and $\overline{W} \cdot \overline{X_j}$ is not sufficiently large. Therefore, we would like to impose the following stronger requirement:

$$\overline{W} \cdot (\overline{X_i} - \overline{X_j}) > 1$$

Any violations of this condition are penalized by $1 - \overline{W} \cdot (\overline{X_i} - \overline{X_j})$ in the objective function. Therefore, one can formulate the problem as follows:

$$\text{Minimize } J = \sum_{(\overline{X_i}, \overline{X_j}) \in \mathcal{D}_R} \max\{0, [1 - (\overline{W} \cdot [\overline{X_i} - \overline{X_j}])]\} + \frac{\lambda}{2}||\overline{W}||^2$$

Here, $\lambda > 0$ is the regularization parameter. Note that one can replace each pair $(\overline{X_i}, \overline{X_j})$ with the new set of features $\overline{X_i} - \overline{X_j}$. Therefore, one can now assume that the training data simply contains n instances of the m-dimensional *difference features* denoted by $\overline{U_1} \ldots \overline{U_n}$, where n is the number of ranked pairs in the training data. In other words, each $\overline{U_p}$ is of the form $U_p = \overline{X_i} - \overline{X_j}$ for a ranked pair $(\overline{X_i}, \overline{X_j})$ in the training data. Then, the ranking SVM formulates the following optimization problem:

$$\text{Minimize } J = \sum_{i=1}^{n} \max\{0, [1 - \overline{W} \cdot \overline{U_i}]\} + \frac{\lambda}{2}||\overline{W}||^2$$

One can also write this optimization formulation in terms of the *slack penalty* $C = 1/\lambda$ in order to make it look cosmetically more similar to a traditional SVM:

$$\text{Minimize } J = \frac{1}{2}||\overline{W}||^2 + C \sum_{i=1}^{n} \max\{0, [1 - \overline{W} \cdot \overline{U_i}]\}$$

Note that the only difference from a traditional support-vector machine is that the class variable y_i is missing in this optimization formulation. However, this change is extremely easy to incorporate in all the optimization techniques discussed in Section 6.3 of Chapter 6. In each case, the class variable y_i is replaced by 1 in the corresponding gradient-descent steps of various methods discussed in Section 6.3. The linear case is particularly easy to extend using the aforementioned approach, although one can also extend the techniques to kernel SVMs with some minor modifications. The main point to keep in mind is that kernel SVMs work with dot products between training instances. In this case, the training instances are of the form $\overline{U_p} = \Phi(\overline{X_i}) - \Phi(\overline{X_j})$, where $\Phi(\cdot)$ is the nonlinear transformation that is (implicitly) used by a particular kernel similarity function. Let the kernel similarity function define a similarity matrix $S = [s_{ij}]$ over the training instances so that we have the following:

$$s_{ij} = \Phi(\overline{X_i}) \cdot \Phi(\overline{X_j}) \tag{9.5}$$

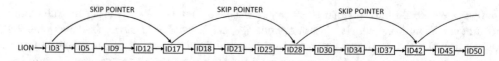

Figure 9.4: Skip pointers with skip values of 4

In kernel methods, only the similarity values s_{ij} are available (as a practical matter) rather than the explicit transformation.

Consider two training instances $\overline{U_p}$ and $\overline{U_q}$ in the following form:

$$\overline{U_p} = \Phi(\overline{X_i}) - \Phi(\overline{X_j}) \quad [\overline{X_i} \text{ ranked higher than } \overline{X_j}]$$
$$\overline{U_q} = \Phi(\overline{X_k}) - \Phi(\overline{X_l}) \quad [\overline{X_k} \text{ ranked higher than } \overline{X_l}]$$

Then, the dot product between $\overline{U_p}$ and $\overline{U_q}$ can be computed as follows:

$$\overline{U_p} \cdot \overline{U_q} = (\Phi(\overline{X_i}) - \Phi(\overline{X_j})) \cdot (\Phi(\overline{X_k}) - \Phi(\overline{X_l}))$$
$$= \{\Phi(\overline{X_i}) \cdot \Phi(\overline{X_k}) + \Phi(\overline{X_j}) \cdot \Phi(\overline{X_l})\} - \{\Phi(\overline{X_i}) \cdot \Phi(\overline{X_l}) + \Phi(\overline{X_j}) \cdot \Phi(\overline{X_k})\}$$
$$= \underbrace{\{s_{ik} + s_{jl}\}}_{\text{Similarly ranked}} - \underbrace{\{s_{il} + s_{jk}\}}_{\text{Differently ranked}}$$

One can use these pairwise similarity values to adapt the kernel methods discussed in Section 6.3 to the case of the ranking SVM. As a practical matter, however, linear models are preferable because they can be efficiently used in conjunction with an inverted index with the use of accumulators. The basic idea here is that the inverted index can be used in conjunction with all the meta-data available in it to efficiently compute $\overline{W} \cdot \overline{Z}$ for a test (candidate) document $\overline{Z}$, once the weights in $\overline{W}$ have been learned during the (offline) scoring phase.

9.2.5   Efficiency Optimizations

There are several other optimizations associated with query processing. Some of these optimizations are particularly important in the context of Web retrieval in which the inverted lists are long and lead to many disk space accesses.

9.2.5.1   Skip Pointers

Skip pointers are like shortcuts in the inverted lists at various positions in order to be able to skip over irrelevant portions of the lists in the intersection process. Skip pointers are useful for intersecting lists of unequal size. In such cases, the skip pointers in the longer list can be useful in performing efficient intersection, because the longer inverted list will have large segments that are irrelevant to the intersection. Consider a term t_j with an inverted list of length n_j. We assume that the inverted list is sorted with respect to the document identifiers. Skip pointers are placed only at positions in the inverted list of the form $s \cdot k + 1$ for fixed skip value s and $k = 0, 1, 2, \ldots, \lfloor n_j/s \rfloor - 1$. An example of skip pointers with $s = 4$ is shown in Figure 9.4.

Now consider the simple problem of intersecting a long list with an extremely short list of length 1 containing a single document identifier. In order to determine whether or not the long list contains this document identifier, we simply traverse its skip pointers, until we identify the segment in which the identifier lies. Subsequently, only this segment is scanned in order to determine whether or not the document identifier lies inside it. In this case, if we use $s = \sqrt{n_j}$, then it can be shown that at most $2\sqrt{n_j}$ traversals will be required. Now, if we need to intersect a short list of length n_t with a longer list of length n_j, then we can repeat this process one by one with elements of the shorter list in sorted order. For best efficiency, care must be taken to use the starting point in the longer list in each case where the search for the previous element of the shorter list was concluded. In the worst case, this approach might incur a small overhead, whereas one will generally do extremely well in cases where the lists have asymmetric lengths. In general, the use of the square-root of the length of the inverted list is a good heuristic for setting the skip values. The main drawback of skip pointers is that they are best suited to static lists that do not change frequently. For a dynamically changing list, it is impossible to maintain the structured pattern of the skip pointers without incurring large update overheads.

9.2.5.2 Champion Lists and Tiered Indexes

One problem with the solutions in all of the above cases is that the query processing can be quite slow for larger collections in which the inverted lists are very long. In such cases, one typically does not even need all the responses to the query, as long as the top-ranked results can be identified reasonably accurately.

Since large term frequencies often have a favorable impact on the scores, they can be used to identify the portions of the inverted lists that are most likely to yield good matches. A natural approach is to use *champion lists*, in which only the subset of document identifiers in which only the top-p documents with highest frequency with respect to a term are maintained in a *truncated* inverted list. Any additional meta-information such as term frequency and term positions can also be maintained along with the document identifiers. In order to resolve a query, the first step is to determine if a "sufficient" number (say, q) of documents is returned by using only the champion lists. If a sufficient number of documents is returned, then one does not need to use the entire inverted index. Otherwise, the query has to be resolved using the full inverted index. The values of p and q are therefore parameters in this process, which need to be chosen in an application-specific way. Champion lists are particularly useful in document-at-a-time querying in which inverted lists are sorted by document identifier. In the event that the inverted lists are sorted in decreasing order of term frequency (for term-at-a-time querying), the effect of champion lists can be realized by using only the initial portions of the inverted lists. Therefore, champion lists need not be explicitly maintained in such cases.

A generalization of the notion of champion lists is the use of *tiered indexes*. In tiered indexes, the idea is that the inverted list only contains the subset of document identifiers with frequency more than a particular threshold. Therefore, the highest threshold corresponds to tier 1, which has the shortest inverted lists. The next higher threshold corresponds to tier 2, and so on. If a query can be resolved using only tier 1 lists, then the results are accepted. Otherwise, the query is processed using the next tier. This approach is continued, until a sufficient number of results can be returned.

9.2.5.3 Caching Tricks

In query processing systems, large numbers of users might be simultaneously making queries, as a result of which many inverted lists will be accessed repeatedly. In such cases, it makes sense to store the inverted lists of frequently queried terms in fast caches for quick retrieval.

The query processing system first checks the cache to retrieve the inverted list for the term. If the inverted list is not available in the cache, then the pointer to the disk (available in the dictionary) is used.

Caches are expensive and therefore only a small fraction of the inverted lists can be stored. Therefore, one needs an *admission control* mechanism to decide which inverted lists to store in the cache. The admission control mechanism must be sufficiently adaptive that the inverted lists stored in the cache are statistically likely to have been accessed frequently in the recent past. The time-tested method for achieving this goal is to use a least recently used (LRU) cache. The cache maintains the last time that each inverted list in it was accessed. When an inverted list for a term is requested, the cache is checked to see if its is available. If the inverted list is found in the cache, its time stamp is updated to the current time. On the other hand, if the inverted list is not found in the cache, it needs to be accessed from disk. Furthermore, it is now inserted in the cache at the expense of one or more existing lists in the cache. This is achieved by removing a sufficient number of least recently used inverted lists from the cache to make room for the newly inserted list. Refer to the bibliographic notes for pointers to multilevel caching methods that are used in information retrieval applications.

9.2.5.4 Compression Tricks

Both the dictionary and the inverted index are often stored in compressed form. Although compression obviously saves on storage, a more important motivation for compression is that it improves efficiency. This is because smaller files improve the caching behavior of the system. Furthermore, it takes less time to read a file from disk and load it to main memory.

Dictionaries are often stored in main memory because they require much less space than the inverted index. However, for some systems even the memory requirements of a dictionary become a burden. Therefore, one needs to reduce its memory footprint as much as possible to ensure that it fits in main memory and possibly free up storage for other parts of the index. How does one allocate memory for the terms in the dictionary? One approach is to allocate *fixed-width* for the string representation of the term in a hash-based dictionary. For example, if 25 characters are allocated for each term in the hash table, but a term like *"golf"* requires only four characters. Therefore, 21 characters are being wasted. Furthermore, the fixed-length approach would cause problems in storage of long terms or phrases with more than 25 characters. One approach is to allocate space only for pointers within the hash table to the string representation of each term. Such pointers are referred to as *term pointers*. Dictionaries are compressed by concatenating all the terms in the lexicon into a single string, in which the terms occur in lexicographically sorted order. The delimiters between two terms can be obtained by using term pointers. Therefore, instead of the hash table, one now maintains a lexicographically sorted array of entries containing these term pointers. A term pointer points to the position on the string at which the term starts. Because of the sorting of both the string dictionary and the array in the same way, the next pointer in the array also provides the end delimiter of the current term in the string. Aside from the term pointers, the array also contains a numerical entry with the number of documents containing the term and the pointer to the first element of the inverted list of the term. Therefore each entry in the sorted table contains 4 bytes each for two pointers (to terms and postings), and 4 bytes to store the document frequency. An example of a compressed dictionary for the example of Figure 9.2 is shown in Figure 9.5. When a query is entered by the user, one needs to efficiently locate the pointer to the relevant inverted lists. To achieve this goal, binary search can be used on this dictionary with the help of term pointers. Once

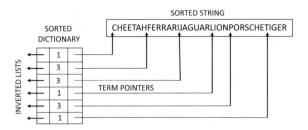

Figure 9.5: Compressing a dictionary by avoiding fixed width allocation of terms

the relevant entry of the array has been isolated, the pointer to the inverted list can be returned. For a dictionary containing a million terms of 8 characters per term, the size required by the string is 8 MB and the size required by the array is 12 MB. Although these requirements might seem tiny (and unimportant to compress), they do enable the use of very fast caches or severely constrained hardware settings. One can also use hash tables with dynamically allocated memory for terms, if space is not at a premium.

Inverted lists can also be compressed. The most common approach is to use *variable byte codes*, in which each number is encoded using as many bytes as needed. Only 7 bits within the byte are used for encoding, and the last bit is a continuation indicator telling us whether or not the next byte is part of the same number. Therefore any number less than $2^7 = 128$ requires a single byte, and any number less than 128^2 requires at most two bytes. Most term frequencies in a document are small values less than 128, and they can be stored in a single byte. However, document identifiers can be arbitrarily long integers. In the case of inverted lists, which are sorted by document identifier, one can use the idea of *delta encoding*.

When the document identifiers are in sorted order, one can store the *differences* between consecutive document identifiers using variable byte codes (or any other compression scheme that favors small numbers). For example, consider the following sequence of document identifiers:

23671, 23693, 23701, 23722, 23755, 23812

One does have to store the first document identifier, which is rather large. However, subsequent document identifiers can be stored as successive offsets, which are the differences between consecutive values in the aforementioned sequence:

22, 8, 21, 33, 57

These values are also referred to as *d-gaps*. Each of these values is small enough to be stored in a single byte in this particular example. Another important point to keep in mind is that these differences between successive document identifiers will be small for more frequent terms (with larger inverted lists). This means that larger inverted lists will be compressed to a greater degree, which is desirable for storage efficiency. This is a recurring idea in many compression methods where frequently occurring items are represented using codes of smaller length, whereas rarer items are allowed codes of longer length. We refer the reader to the bibliographic notes for pointers to various compression schemes.

9.3   Scoring with Information Retrieval Models

The previous section provides a broad idea of the scoring process in information retrieval with the use of different types of indexes. A broad picture is provided about the various types of factors that are used for scoring and ranking documents (e.g., aggregate matches or proximity of keywords). However, it does not discuss the specific types of models that are used in information retrieval applications for scoring and ranking documents. Such models can often be used in combination with relevance judgements by combining them with weights. It is noteworthy that most of the models in this section can be captured using scoring functions of the form discussed in Equations 9.2 and 9.3. This fact enables the use of efficient term-at-a-time or document-at-a-time query processing methods for these models (see page 266).

9.3.1   Vector Space Models with tf-idf

The simplest approach is to use the tf-idf representation discussed in Section 2.4 of Chapter 2. We briefly recap some of the concepts in using the tf-idf representation.

Consider a document collection containing n documents in d dimensions. Let $\overline{X} = (x_1 \ldots x_d)$ be the d-dimensional representation of a document after the term extraction phase. The square-root or the logarithm function may be applied to the frequencies to reduce the effect of terms that occur too often in a document. In other words, one might replace each x_i with either $\sqrt{x_i}$, $\log(1 + x_i)$, or $1 + \log(x_i)$.

It is also common to normalize term frequencies based on their presence in the entire collection. The first step in normalization is compute the inverse document frequency of each term. The inverse document frequency id_i of the ith term is a decreasing function of the number of documents n_i in which it occurs:

$$id_i = \log(n/n_i) \tag{9.6}$$

Note that the value of id_i is always nonnegative. In the limiting cases in which a term occurs in every document of the collection, the value of id_i is 0. The term frequency is normalized by multiplying it with the inverse document frequency:

$$x_i \Leftarrow x_i \cdot id_i \tag{9.7}$$

Once the normalized representation of each document in the corpus is computed, it is used to respond to similarity-based queries. The most common similarity function is the cosine function, which is introduced in Section 2.5 of Chapter 2.

Consider a target document $\overline{X} = (x_1 \ldots x_d)$ and the query vector $\overline{Q} = (q_1 \ldots q_d)$. The query vector might either be binary or it might be based on term frequencies. The cosine function is defined as follows:

$$\text{cosine}(\overline{X}, \overline{Q}) = \frac{\sum_{i=1}^{d} x_i q_i}{\sqrt{\sum_{i=1}^{d} x_i^2}\sqrt{\sum_{i=1}^{d} q_i^2}}$$

$$\propto \frac{\sum_{i=1}^{d} x_i q_i}{\sqrt{\sum_{i=1}^{d} x_i^2}} = \frac{\sum_{i=1}^{d} x_i q_i}{||\overline{X}||}$$

It is sufficient to compute the cosine to a constant of proportionality because we only need to rank the different instances for a particular query. It is easy to see that this scoring function is of the form captured by Equations 9.2 and 9.3, which can be computed with either term-at-a-time or document-at-a-time query processing (see pages 266 and 268).

9.3.2 The Binary Independence Model

The binary independence model uses binary relevance judgements about training documents in order to score previously unseen documents with the use of a naïve Bayes classifier. In particular, the Bernoulli classifier of Section 5.3.1 in Chapter 5 is used. Let $R \in \{0,1\}$ indicate whether or not a the document is relevant to be particular query. As discussed in Section 5.3.1, the Bernoulli model implicitly assumes that each document $\overline{X}$ is represented in a vector space representation with Boolean attributes containing information about whether or not each term t_j is present in $\overline{X}$.

Assume that we have some training data available, which tells us whether or not a document is relevant to a particular query. Note that relevance judgement data that is *query-specific*[4] is often hard to come by, although one can allow the user to provide feedback to query results to collect data about document relevance or non-relevance. Note that the collected data is useful only for that specific query, which is different from the machine learning approach of previous sections in which the training data for importance of specific types of meta-features are learned over *multiple* queries. In order to use these models without human intervention or training data, we will eventually make a number of simplifying assumptions. In that sense, these models also provide the intuitions necessary for query processing without these (query-specific) relevance judgements.

Let $p_j^{(0)}$ be the fraction of non-relevant documents in the training data (i.e., user relevance judgements) that do not contain term t_j, and $p_j^{(1)}$ be the fraction of documents containing term t_j. Similarly, let α_0 be the fraction of non-relevant documents and α_1 be the fraction of relevant documents in the training data. The scoring function for a given query is represented in terms of the Bayes classification probabilities. We wish to find the ratio of the probability $P(R=1|\overline{X})$ to that of $P(R=0|\overline{X})$. Based on the results from the Bernoulli model in Section 5.3.1, we can state the following:

$$P(R=1|\overline{X}) = \frac{P(R=1) \cdot P(\overline{X}|R=1)}{P(\overline{X})} = \frac{\alpha_1 \prod_{t_j \in \overline{X}} p_j^{(1)} \prod_{t_j \notin \overline{X}} (1 - p_j^{(1)})}{P(\overline{X})}$$

$$P(R=0|\overline{X}) = \frac{P(R=0) \cdot P(\overline{X}|R=0)}{P(\overline{X})} = \frac{\alpha_0 \prod_{t_j \in \overline{X}} p_j^{(0)} \prod_{t_j \notin \overline{X}} (1 - p_j^{(0)})}{P(\overline{X})}$$

Then, the ratio of the two quantities may be computed after ignoring the prior probabilities (because they are not document-specific and do not affect the ranking):

$$\frac{P(R=1|\overline{X})}{P(R=0|\overline{X})} \propto \frac{\prod_{t_j \in \overline{X}} p_j^{(1)} \prod_{t_j \notin \overline{X}} (1 - p_j^{(1)})}{\prod_{t_j \in \overline{X}} p_j^{(0)} \prod_{t_j \notin \overline{X}} (1 - p_j^{(0)})} \tag{9.8}$$

The constant of proportionality is used here because the document-independent ratio α_1/α_0 is ignored in the above expression. We can rearrange the above expression to within a constant of proportionality and make it dependent only on the terms occurring in $\overline{X}$. This is achieved by multiplying both sides of Equation 9.8 with a document-independent term, and then dropping it only from the left-hand side (which retains the proportionality relationship

---

[4]In all the previous discussions on machine learned information retrieval, the training data is not specific to a particular query. However, each set of values of the extracted features is query-specific and multiple queries are represented in the same training data. The importance of the query-specific values of the meta-features (e.g., zones, authorship, location) of the document is learned with feedback data.

because of document-independence):

$$\frac{P(R=1|\overline{X})}{P(R=0|\overline{X})} \cdot \underbrace{\frac{\prod_{t_j}(1-p_j^{(0)})}{\prod_{t_j}(1-p_j^{(1)})}}_{\text{Document Independent}} \propto \frac{\prod_{t_j\in\overline{X}}p_j^{(1)}(1-p_j^{(0)})}{\prod_{t_j\in\overline{X}}p_j^{(0)}(1-p_j^{(1)})} \quad \text{[Multiplying both sides]}$$

$$\frac{P(R=1|\overline{X})}{P(R=0|\overline{X})} \propto \prod_{t_j\in\overline{X}} \frac{p_j^{(1)}(1-p_j^{(0)})}{p_j^{(0)}(1-p_j^{(1)})} \quad \text{[Dropping document-independent term on LHS]}$$

The logarithm of the above quantity is used in order to compute an additive form of the relevance score. The additive form of the ranking score is always desirable because it can be computed using inverted indexes and accumulators, as discussed on page 266. In the binary independence model, the *retrieval status value* $RSV_{bi}(\overline{X},\overline{Q})$ of document $\overline{X}$ with respect to query vector $\overline{Q}$ can be expressed in terms of a summation only over the terms present in the document $\overline{X}$ as follows:

$$RSV_{bi}(X,Q) = \sum_{t_j\in\overline{X}} \log\left(\frac{p_j^{(1)}(1-p_j^{(0)})}{p_j^{(0)}(1-p_j^{(1)})}\right) \tag{9.9}$$

For any term t_j that is not included in the binary query vector $\overline{Q}$, it is assumed that the term is distributed in a similar way across relevant and non-relevant documents. This is equivalent to assuming that $p_j^{(0)} = p_j^{(1)}$ for terms not in $\overline{Q}$, which results in the dropping of non-query terms from the right-hand side of Equation 9.9. This results in a retrieval status value that is expressed only as a summation over the *matching* terms in the query and the document:

$$RSV_{bi}(\overline{X},\overline{Q}) = \sum_{t_j\in\overline{X}, t_j\in\overline{Q}} \log\left(\frac{p_j^{(1)}(1-p_j^{(0)})}{p_j^{(0)}(1-p_j^{(1)})}\right) \tag{9.10}$$

An important point here is that the quantities such as $p_j^{(0)}$ and $p_j^{(1)}$ are not usually available on a query-specific basis, unless the human is actively involved in providing relevance feedback to the results of queries. Some systems do allow the user to actively enter relevance feedback values. In these cases, a list of results is presented to the user based on some matching model for retrieval. The user then indicates which results are relevant. As a result, one can now label documents as *relevant* or *non-relevant*, which is a straightforward classification setting. In such cases, the problem of parameter estimation of $p_j^{(0)}$ and $p_j^{(1)}$ becomes *identical* to the way in which parameters are estimated in the Bernoulli model in Section 5.3.1. As in Section 5.3.1, Laplacian smoothing is used in order to provide more robust estimates of the probabilities. Note that such an approach has to perform the parameter estimation in real time after receiving the feedback, so that the next round of ranking can be presented to the user on the basis of computed values of $RSV_{bi}(\overline{X},\overline{Q})$. If R is the number of relevant documents out of N documents, and r_j is the number of relevant documents containing term t_j out of n_j such documents, then the values of $p_j^{(0)}$ and $p_j^{(1)}$ are set as follows:

$$p_j^{(1)} = \frac{r_j + 0.5}{R+1}, \qquad p_j^{(0)} = \frac{n_j - r_j + 0.5}{N-R+1} \tag{9.11}$$

The constant values of 0.5 and 1 are respectively added to the numerator and denominator for smoothing.

However, not all systems are able to use the feedback to specific queries in real time. In such cases, a number of simplifying assumptions are used in order to estimate quantities such as $p_j^{(0)}$ and $p_j^{(1)}$. For any term t_j that is included in the binary query vector $\overline{Q}$, the value of $p_j^{(1)}$ is assumed to be a large constant[5] such as 0.5 (relative to fractional occurrences of random terms in documents) because the terms in the queries are generally extremely relevant. However, such terms can also be present in non-relevant documents at a statistical frequency that is similar to that of the remaining collection. The value of $p_j^{(0)}$ is computed based on the statistical frequency of the term t_j across the whole collection. The frequency of the term across the whole collection is n_j/n, where n_j is the number of documents in the whole collection in which the term t_j occurs and n be the total number of documents. Then, the value of $p_j^{(0)}$ is set to n_j/n.

Therefore, if relevance feedback is not used, the retrieval status value with the binary independence model, specific to query $\overline{Q}$ and candidate document $\overline{X}$ is given by substituting $p_j^{(1)} = 0.5$ and $p_j^{(0)} = n_j/n$ in Equation 9.10:

$$RSV_{bi}(\overline{X}, \overline{Q}) = \sum_{t_j \in \overline{X}, t_j \in \overline{Q}} \log\left(\frac{n - n_j}{n_j}\right) \approx \sum_{t_j \in \overline{X}, t_j \in \overline{Q}} \log\left(\frac{n}{n_j}\right) \qquad (9.12)$$

An alternative version of the expression above is also used, when Laplacian smoothing is desired:

$$RSV_{bi}(\overline{X}, \overline{Q}) = \sum_{t_j \in \overline{X}, t_j \in \overline{Q}} \log\left(\frac{n - n_j + 0.5}{n_j + 0.5}\right) \qquad (9.13)$$

It is easy to see that the expression on the right-hand side of Equation 9.12 is equal to the sum of the inverse document frequency (idf) weights over matching terms. Therefore, one can even view this probabilistic model as a theoretical confirmation of the soundness of using inverse document frequencies in other similarity functions (e.g., cosine similarity) for computing matching scores. There are, however, several key differences of this model from the cosine similarity. First, the term frequencies of documents are not used, and secondly, document length normalization is missing. The fact that the term frequency is missing is a consequence of the fact that the documents are treated as binary vectors. Unfortunately, the missing term frequencies and document-length normalization do hurt the retrieval performance. Nevertheless, the binary independence model provides an initial template for constructing a more refined probabilistic model using the term frequencies that accounts for the same factors as the cosine similarity, but is better grounded in terms of probabilistic interpretation. This model is referred to as the *BM25* model, which is discussed in the next section.

9.3.3 The BM25 Model with Term Frequencies

The BM25 model, which is also referred to as the *Okapi* model, augments the binary independence model with term frequencies and document length normalization in order to improve the retrieved results. Let $(x_1 \ldots x_d)$ represent the raw term frequencies in document

---

[5]This was one of the earliest ideas proposed by Croft and Harper [136]. However, other alternatives are possible. Sometimes, a few relevant documents may be available, which can be used to estimate $p_j^{(1)}$. The other idea is to allow $p_j^{(1)}$ to rise with the number of documents n_j containing term t_j. For example, one can use $p_j^{(1)} = \frac{1}{3} + \frac{2 \cdot n_j}{3 \cdot n}$ [211].

$\overline{X}$ without any form of frequency damping[6] or inverse document frequency (idf) normalization. Similarly, let $\overline{Q} = (q_1 \ldots q_d)$ represent the term frequencies in the query Q. Then, the retrieval status value $RSV_{bm25}(\overline{X}, \overline{Q})$ is closely related to that of the binary independence model:

$$RSV_{bm25} = \sum_{t_j \in \overline{Q}} \underbrace{\left(\log \frac{p_j^{(1)}(1 - p_j^{(0)})}{p_j^{(0)}(1 - p_j^{(1)})} \right)}_{\approx idf_j} \cdot \underbrace{\frac{(k_1 + 1)x_j}{k_1(1 - b) + b \cdot L(\overline{X}) + x_j}}_{\text{doc. frequency/length impact}} \cdot \underbrace{\frac{(k_2 + 1)q_j}{k_2 + q_j}}_{\text{query impact}} \quad (9.14)$$

The values k_1, k_2, and b are parameters, which respectively regulate impact of document term frequency, query term frequency, and document length normalization, respectively. The first term is identical to that in the binary independence model and can be simplified in a similar way to Equation 9.13. The second term incorporates the impact of term frequencies and the document lengths. Small values of k_1 lead to the frequencies of the term being ignored, and large values of k_1 lead to linear weighting with the term frequency x_j. Intermediate values[7] of $k_1 \in (1, 1.5)$ have the same effect as that of applying the square-root or logarithm to the term frequency in order to reduce excessive impact of repeated term occurrences. The expression $L(\overline{X})$ is the normalized length of document $\overline{X}$, which is the ratio of its length to the average length of a document in the collection. Note that $L(\overline{X})$ will be larger than 1 for long documents. The parameter b is helpful for document length normalization. Setting the value of b to 0 results in no document length normalization, whereas setting the value of $b = 1$ leads to maximum normalization. A typical value of $b = 0.75$ is used. The parameter k_2 serves the same purpose as k_1, except that it does so for the query document frequencies. The choice of k_2 is, however, not quite as critical because the query documents are typically short with few repeated occurrences of terms. In such cases, almost any choice of $k_2 \in (1, 10)$ will yield similar results, and in some cases the entire term for query frequency normalization is dropped. Query length normalization is unnecessary because it is a proportionality factor that does not affect ranking of documents. Unlike the binary independence model, the summation is over all the terms in the query $\overline{Q}$ rather than only the matching terms between $\overline{Q}$ and $\overline{X}$. However, since the value of x_j is multiplicatively included in the expression, the absence of the query term in a document will automatically set its retrieval status value to 0. This is important because it means that only documents with matching terms contribute to the ranking score, and it is possible to perform query processing with an inverted index. One can express the first term in a data-driven manner, which is similar to Equation 9.13:

$$RSV_{bm25} = \sum_{t_j \in \overline{Q}} \underbrace{\left(\log \frac{N - n_j + 0.5}{n_j + 0.5} \right)}_{\approx idf_j} \cdot \underbrace{\frac{(k_1 + 1)x_j}{k_1(1 - b) + b \cdot L(\overline{X}) + x_j}}_{\text{doc. frequency/length impact}} \cdot \underbrace{\frac{(k_2 + 1)q_j}{k_2 + q_j}}_{\text{query impact}} \quad (9.15)$$

The aforementioned expression is for cases where relevance feedback is not available. If relevance feedback is available, then the values of $p_j^{(0)}$ and $p_j^{(1)}$ in the first term are set using Equation 9.11. Since the retrieval status value is computed in an additive way over query terms, and only matching documents are relevant, one can use the document-at-a-time (page 268) query processing technique in order to evaluate the score.

---

[6]As discussed earlier, the square root or logarithm is frequently applied to term frequencies to reduce the impact of repeated words.

[7]Such values of k_1 are recommended in TREC experiments.

9.3.4 Statistical Language Models in Information Retrieval

A statistical language model assigns a probability to a sequence of words in a given language in a data-driven manner. In other words, given a corpus of documents, the language model estimates the probability that it was generated using this model. The use of language models in information retrieval is based on the intuition that users often formulate queries based on terms that are likely to appear in the returned documents. In some cases, even the ordering of the terms in the query might be chosen on the basis of the expected sequence of terms in the document. Therefore, if the user creates a *language model* for each document, it effectively provides a language model for the query. In other words, the assumption is that the document and query were generated from the same model. Documents can then be scored by computing the posterior probability of generating the document from the same model as the query. This is a fundamentally different notion from the concept of relevance that is used in the binary independence and BM25 models for ranking documents.

A language model for a document provides a generative process of constructing the document. The most *primitive* language model is the *unigram* language model in which no sequence information is used and only the frequencies of terms are used. The basic assumption is that each *token* in a document is generated by rolling a die independently from the previous tokens in the document, where each face of the die shows a particular term. Note that the unigram language model creates a multinomial distribution of terms in a document, as discussed in Chapters 4 and 5. Therefore, the unigram language model can be fully captured by using the probabilities of the different terms in the documents and no information about the sequence of the terms in the collection.

More complex language models such as bigram and n-gram language models use sequence information. A bigram language model uses only the previous term to predict the term at a particular position, and a trigram model uses the previous two words. In general, an n-gram model uses the previous $(n - 1)$ terms to predict a term at a particular position. In this case, the parameters of the model correspond to the conditional probabilities of tokens, given a fixed set of previous $(n - 1)$ tokens. An n-gram model falls in the broad category of Markovian models, which refers to a *short-memory assumption*. In this particular case, only a history of $(n - 1)$ terms in the sequence is used to predict the current term, and therefore the amount of memory used for modeling is limited by the parameter n. Large values of n result in theoretically more accurate models (i.e., lower bias), but sufficient data is often not available to estimate the exponentially increasing number of parameters of the model (i.e., higher variance). As a practical matter, only small values of n can be used in a realistic way because of the rapid increase in the amount of data needed to estimate the parameters at large values of n. A broader discussion of language models is provided in Section 10.2 of Chapter 10, although this section will restrict the discussion to unigram language models. In general, unigram language models are used frequently because of their simplicity and the ease in estimation of the parameters with a limited amount of data.

9.3.4.1 Query Likelihood Models

How are language models used for information retrieval? Given a document $\overline{X}$, one can estimate the parameters of the language model, and then compute the posterior probabilities of $\overline{X}$, given the additional knowledge about the query $\overline{Q}$. Therefore, the overall approach may be described as follows:

1. Estimate the parameter vector $\overline{\Theta}_X$ of the language model $\mathcal{M}$ being used with the use of each candidate document $\overline{X}$ in the corpus. For example, if a unigram language

model is used, the parameter vector $\overline{\Theta}_X$ will contain the probabilities of the different faces of the die that generated $\overline{X}$. The value of $\overline{\Theta}_X$ can therefore be estimated as the fractional presence of various terms in $\overline{X}$. Note that each parameter vector $\overline{\Theta}_X$ is specific only to a particular document $\overline{X}$.

2. For a given query $\overline{Q}$, estimate the posterior probability $P(\overline{X}|\overline{Q})$. The documents are ranked on the basis of this posterior probability.

In order to compute the posterior probability, the Bayes rule is used:

$$P(\overline{X}|\overline{Q}) = \frac{P(\overline{X}) \cdot P(\overline{Q}|\overline{X})}{P(\overline{Q})} \propto P(\overline{X}) \cdot P(\overline{Q}|\overline{X})$$

The constant of proportionality above is identified as the document-independent term, which does not affect relative ranking. A further assumption is that the prior probability $P(\overline{X})$ is uniform over all documents. Therefore, we have the following:

$$P(\overline{X}|\overline{Q}) \propto P(\overline{Q}|\overline{X})$$

Finally, the value of $P(\overline{Q}|\overline{X})$ is computed by using the underlying language model, whose parameters were estimated using $\overline{X}$. This is the same as estimating $P(\overline{Q}|\overline{\Theta}_X)$.

In the context of a unigram model, this estimation takes on a particularly simple form. Let $\overline{\Theta}_X = (\theta_1 \ldots \theta_d)$ be the probabilities of the different terms in the collection, which were estimated using $\overline{X} = (x_1 \ldots x_d)$. The parameter θ_j can be estimated as follows:

$$\theta_j = \frac{x_j}{\sum_{j=1}^{d} x_j} \tag{9.16}$$

Then, the estimation of $P(\overline{Q}|\overline{\Theta}_X)$ can be accomplished using the multinomial distribution:

$$P(\overline{Q}|\overline{\Theta}_X) = \prod_{j=1}^{d} \theta_j^{q_j} \tag{9.17}$$

Note that the logarithm of Equation 9.17 is additive in nature, and it can be computed efficiently with the use of an inverted index and accumulator variables (cf. page 266). Furthermore, the use of the logarithm avoids the multiplication of very small probabilities.

One way of understanding the query likelihood probability in the context of the unigram language model is that it estimates the probability that the query is generated as a sample of terms from the document. Of course, this interpretation would not be true for more complex language models like a bigram model.

One issue with this estimation is that it would give nonzero scores for a document only if it contained all the terms in that document. This is because the parameter vector $\overline{\Theta}_X$ is computed using a single document $\overline{X}$, which inevitably leads to a lot of zero values in the parameter vector. One can use the Laplacian smoothing methods that are commonly used for multinomial distributions, as discussed in Chapters 4 and 5. Another option is to use *Jelinek-Mercer smoothing*, in which the value of θ_j is estimated using the statistics of both the document $\overline{X}$ and the whole collection. Let θ_j^X and θ_j^{All} be these two estimated values. The parameter θ_j^X is estimated as before, whereas the estimation of θ_j^{All} is simply he fraction of the tokens in the whole collection that are t_j. Then, the estimated value of θ_j is a convex combination of these two values with the use of the parameter $\lambda \in (0, 1)$:

$$\theta_j = \lambda \theta_j^X + (1 - \lambda)\theta_j^{All} \tag{9.18}$$

Using $\lambda = 1$ reverts to the aforementioned model without smoothing, whereas using $\lambda = 0$ causes so much smoothing that all documents tie with the same ranking score.

9.4 Web Crawling and Resource Discovery

Web crawlers are also referred to as *spiders* or *robots*. The primary motivation for Web crawling is that the resources on the Web are dispensed widely across globally distributed sites. While the Web browser provides a graphical user interface to access these pages in an interactive way, the full power of the available resources cannot be leveraged with the use of only a browser. In many applications, such as search and knowledge discovery, it is necessary to download all the relevant pages *at a central location* (or a modest number of distributed locations), to allow search engines and machine learning algorithms to use these resources efficiently. In this sense, search engines are somewhat different from information retrieval applications; even the compilation of the corpus for querying is a difficult task because of the open and vast nature of the Web.

Web crawlers have numerous applications. The most important and well-known application is search, in which the downloaded Web pages are indexed to provide responses to user keyword queries. All the well-known search engines, such as Google and Bing, employ crawlers to periodically refresh the downloaded Web resources at their servers. Such crawlers are also referred to as *universal crawlers* because they are intended to crawl all pages on the Web irrespective of their subject matter or location. Web crawlers are also used for business intelligence, in which the Websites related to a particular subject are crawled or the sites of a competitor are monitored and incrementally crawled as they change. Such crawlers are also referred to as *preferential crawlers* because they discriminate between the relevance of different pages for the application at hand.

9.4.1 A Basic Crawler Algorithm

While the design of a crawler is quite complex, with a distributed architecture and many processes or threads, the following describes a simple sequential and universal crawler that captures the essence of how crawlers are constructed.

A crawler uses the same mechanism used by browsers to fetch Web pages based on the *Hypertext Transfer Protocol* (HTTP). The main difference is that the fetching is now done by an automated program using automated selection decisions, rather than by the manual specification of a *Uniform Resource Locator (URL)* by a user with a Web browser. In all cases, a particular URL is fetched by the system. Both browsers and crawlers typically[8] use GET requests to fetch Web pages, which is a functionality provided by the HTTP protocol. The difference is that the GET request is invoked in a browser when a user clicks a link or enters a URL, whereas the GET request is invoked in an automated way by the crawler. In both cases, a *domain name system (DNS) server* is used to translate the URL into an internet protocol (IP) address. The program then connects to the server using that IP address and sends a GET request. In most cases, servers listen to requests at multiple *ports*, and port 80 is typically used for Web requests.

The basic crawler algorithm, described in a very general way, uses a seed set of Universal Resource Locators (URLs) S, and a selection algorithm $\mathcal{A}$ as the input. The algorithm $\mathcal{A}$ decides which document to crawl next from a current *frontier list* of URLs. The frontier list represents URLs extracted from the Web pages. These are the candidates for pages that can eventually be fetched by the crawler. The selection algorithm $\mathcal{A}$ is important because it regulates the basic strategy used by the crawler to discover the resources. For example, if new

---

[8]Browsers also use POST requests, when additional information is needed by the Web server. For example, an item is usually bought on the POST request. However, such requests are not used by crawlers because they might inadvertently causes actions (such as buying), which were not desired by the crawler.

Algorithm *BasicCrawler*(Seed URLs: S, Selection Algorithm: $\mathcal{A}$)
begin
  $FrontierList = S$;
  repeat
    Use algorithm $\mathcal{A}$ to select URL $X \in FrontierSet$;
    $FrontierList = FrontierList - \{X\}$;
    Fetch URL X and add to repository;
    Add all relevant URLs in fetched document X to
      end of $FrontierList$;
  until termination criterion;
end

Figure 9.6: The basic crawler algorithm

URLs are appended to the end of the frontier list, and the algorithm $\mathcal{A}$ selects documents from the beginning of the list, then this corresponds to a breadth-first algorithm.

The basic crawler algorithm proceeds as follows. First, the seed set of URLs is added to the frontier list. In each iteration, the selection algorithm $\mathcal{A}$ picks one of the URLs from the frontier list. This URL is deleted from the frontier list and then fetched using the GET request of the HTTP protocol. The fetched page is stored in a local repository, and the URLs inside it are extracted. These URLs are then added to the frontier list, provided that they have not already been visited. Therefore, a separate data structure, in the form of a hash table, needs to be maintained to store all visited URLs. In practical implementations of crawlers, not all unvisited URLs are added to the frontier list due to Web spam, spider traps, topical preference, or simply a practical limit on the size of the frontier list. After the relevant URLs have been added to the frontier list, the next iteration repeats the process with the next URL on the list. The process terminates when the frontier list is empty. If the frontier list is empty, it does not necessarily imply that the entire Web has been crawled. This is because the Web is not strongly connected, and many pages are unreachable from most randomly chosen seed sets. Because most practical crawlers such as search engines are *incremental* crawlers that refresh pages over previous crawls, it is usually easy to identify unvisited seeds from previous crawls and add them to the frontier list, if needed. With large seed sets, such as a previously crawled repository of the Web, it is possible to robustly crawl most pages. The basic crawler algorithm is described in Figure 9.6.

Thus, the crawler is a graph-search algorithm that discovers the outgoing links from nodes by parsing Web pages and extracting the URLs. The choice of the selection algorithm $\mathcal{A}$ will typically result in a bias in the crawling algorithm, especially in cases where it is impossible to crawl all the relevant pages due to resource limitations. For example, a breadth-first crawler is more likely to crawl a page with many links pointing to it. Interestingly, such biases are sometimes desirable in crawlers because it is impossible for any crawler to index the entire Web. Because the indegree of a Web page is often closely related to its *PageRank*, a measure of a Web page's quality, this bias is not necessarily undesirable. Crawlers use a variety of other selection strategies defined by the algorithm $\mathcal{A}$.

Because most universal crawlers are incremental crawlers that are intended to refresh previous crawls, it is desirable to crawl frequently changing pages. The explicit detection of whether a Web page has been changed can be done at a relatively low cost using the HEAD request of the HTTP protocol. The HEAD request receives only the header information from a Web page at a lower cost than crawling the Web page. The header information also contains the last date at which the Web document was modified. This date is compared with that obtained from the previous fetch of the Web page (using a GET request). If the date has changed, then the Web page needs to be crawled again.

The use of the HEAD request reduces the cost of crawling a Web page, although it still imposes some burden on the Web server. Therefore, the crawler needs to implement some internal mechanisms in order to estimate the frequency at which a Web page changes (without actually issuing any requests). This type of internal estimation helps the crawler in minimizing the number of fruitless requests to Web servers. Specific types of Web pages such as news sites, blogs, and portals might change frequently, whereas other types of pages may change slowly. The change frequency can be estimated from repeated previous crawls of the same page or by using learning algorithms that factor in specific characteristics of the Web page. Some resources such as news portals are updated frequently. Therefore, frequently updated pages may be selected by the algorithm $\mathcal{A}$. Other than the change frequency, another factor is the *popularity and usefulness* of Web pages to the general public. Clearly, it is desirable to crawl popular and useful pages more frequently. Therefore, the selection algorithm $\mathcal{A}$ may specifically choose Web pages with high *PageRank* from frontier list. The computation of *PageRank* is discussed in Section 9.6.1. The use of *PageRank* as a criterion for selecting Web pages to be crawled is closely related to that of preferential crawlers.

9.4.2 Preferential Crawlers

In the preferential crawler, only pages satisfying a user-defined criterion need to be crawled. This criterion may be specified in the form of keyword presence in the page, a topical criterion defined by a machine learning algorithm, a geographical criterion about page location, or a combination of the different criteria. In general, an arbitrary predicate may be specified by the user, which forms the basis of the crawling. In these cases, the major change is to the approach used for updating the frontier list during crawling, and also the order of selecting the URLs from the frontier list.

1. The Web page needs to meet the user-specified criterion in order for its extracted URLs to be added to the frontier list.

2. In some cases, the anchor text may be examined to determine the relevance of the Web page to the user-specified query.

3. In context-focused crawlers, the crawler is trained to learn the likelihood that relevant pages are within a short distance of the page, even if the Web page is itself not directly relevant to the user-specified criterion. For example, a Web page on *"data mining"* is more likely to point to a Web page on *"information retrieval,"* even though the data mining page may not be relevant to the query on *"information retrieval."* URLs from such pages may be added to the frontier list. Therefore, heuristics need to be designed to learn such context-specific relevance.

Changes may also be made to the algorithm $\mathcal{A}$. For example, URLs with more relevant anchor text, or with relevant tokens in the Web address, may be selected first by algorithm $\mathcal{A}$. A URL such as http://www.golf.com, with the word *"golf"* in the Web address may be more relevant to the topic of *"golf,"* than a URL without the word in it. The bibliographic notes contain pointers to a number of heuristics that are commonly used for preferential resource discovery.

A number of simple techniques, such as the use of *PageRank* to preferential crawl Web pages, greatly enhance the popularity of the Web pages crawled. This type of approach ensures that only "hot" Web pages are crawled, which are of interest to many users. Therefore, these types of preferential crawlers are desirable when one has limited resources to crawl pages, although the goal is often similar to that of a universal crawler algorithm.

Other types of preferential crawlers include *focused crawling* or *topical crawling*. In focused crawling, the crawler is biased using a particular type of classifier. The first step is to build a classification model based on an open resource of Web pages such as the *Open Directory[9] Project (ODP)*. When a crawled Web page belongs to a desired category, then the URLs inside it are added to frontier list. The URLs can be scored in various ways for ordering on frontier list, such as the level of classifier relevance of their parent Web pages as well as their recency.

In topical crawling, labeled examples are not available to the crawler. Only a set of seed pages and a description of the topic of interest is available. In many cases, a description of the topic of interest is provided by the user with a short query. A common approach in this case is to use the *best-first* algorithm in which the (crawled or seed) Web pages containing the most number of user-specified keywords are selected, and the URLs inside them are preferentially added to the frontier list. As in the case of focused crawling, the topical relevance (e.g., matching between query and Web page) of the parent Web pages of the URLs on frontier list as well as their recency can be used to order them.

9.4.3 Multiple Threads

Crawlers typically use multiple threads to improve efficiency. You might have noticed that it can sometimes take a few seconds for your Web browser to fulfill your URL request. This situation is also encountered in a Web crawler, which idles while the server at the other end satisfies the GET request. It makes sense for Web crawlers to use this idle time in order to fetch more Web pages. A natural way to speed up the crawling is by leveraging concurrency. The idea is to use multiple threads of the crawler that update a shared data structure for visited URLs and the page repository. In such cases, it is important to implement concurrency control mechanisms for locking or unlocking the relevant data structures during updates. The concurrent design can significantly speed up a crawler with more efficient use of resources. In practical implementations of large search engines, the crawler is distributed geographically with each "sub-crawler" collecting pages in its geographical proximity.

One problem with this approach is that if hundreds of requests are made to the Web server at a single site, it would result in an unreasonable load on the server, which also has to serve URL requests from other clients. Therefore, Web crawlers often use politeness policies, in which a page is not crawled from a Web server, if one has been crawled recently. This is achieved by creating per-server queues, and disallowing a fetch from a particular queue, if a page has been crawled from it within a particular time window.

9.4.4 Combatting Spider Traps

The main reason that the crawling algorithm always visits distinct Web pages is that it maintains a list of previously visited URLs for comparison purposes. However, some shopping sites create dynamic URLs in which the last page visited is appended at the end of the user sequence to enable the server to log the user action sequences within the URL for future analysis. For example, when a user clicks on the link for *page2* from http://www.examplesite.com/page1, the new dynamically created URL will be http://www.examplesite.com/page1/page2. Pages that are visited further will continue to be appended to the end of the URL, even if these pages were visited before. A natural way to combat this is to limit

---

[9]http://www.dmoz.org.

the maximum size of the URL. Furthermore, a maximum limit may also be placed on the number of URLs crawled from a particular site.

9.4.5 Shingling for Near Duplicate Detection

One of the major problems with the Web pages collected by a crawler is that many duplicates of the same page may be crawled. This is because the same Web page may be mirrored at multiple sites. Therefore, it is crucial to have the ability to detect near duplicates. An approach known as *shingling* is commonly used for this purpose.

A k-shingle from a document is simply a string of k consecutively occurring words in the document. A shingle can also be viewed as a k-gram. For example, consider the document comprising the following sentence:

Mary had a little lamb, its fleece was white as snow.

The set of 2-shingles extracted from this sentence is *"Mary had"*, *"had a"*, *"a little"*, *"little lamb"*, *"lamb its"*, *"its fleece"*, *"fleece was"*, *"was white"*, *"white as"*, and *"as snow"*. Note that the number of k-shingles extracted from a document is no longer than the length of the document, and 1-shingles are simply the set of words in the document. Let S_1 and S_2 be the k-shingles extracted from two documents D_1 and D_2. Then, the shingle-based similarity between D_1 and D_2 is simply the Jaccard coefficient between S_1 and S_2.

$$J(S_1, S_2) = \frac{|S_1 \cap S_2|}{|S_1 \cup S_2|} \tag{9.19}$$

The advantage of using k-shingles instead of the individual words (1-shingles) for Jaccard coefficient computation is that shingles are less likely than words to repeat in different documents. There are r^k distinct shingles for a lexicon of size r. For $k \geq 5$, the chances of many shingles recurring in two documents becomes very small. Therefore, if two documents have many k-shingles in common, they are very likely to be near duplicates. To save space, the individual shingles are hashed into 4-byte (32-bit) numbers that are used for comparison purposes. Such a representation also enables better efficiency.

9.5 Query Processing in Search Engines

The broad framework for query processing in search engines is inherited from traditional information retrieval, as discussed in Sections 9.2 and 9.3. All the data structures introduced in earlier sections, such as dictionaries (cf. Section 9.2.1) and inverted indexes (cf. Section 9.2.2) are used in search engines, albeit with some modifications to account for the large size of the Web and the immense burden placed on Web servers by the large numbers of queries.

After the documents have been crawled, they are leveraged for query processing. The following are the primary stages in search index construction:

1. *Preprocessing:* This is the stage in which the search engine preprocesses the crawled documents to extract the tokens. Web pages require specialized preprocessing methods, which are discussed in Chapter 2. A substantial amount of meta-information about the Web page is also collected, which is often useful in enabling query processing. This meta-information might include information like the date of the document, its geographical location, or even its *PageRank* based on the methodology in

Section 9.6.1. Note that measures such as *PageRank* need to be computed up front because they are expensive to compute, and cannot be reasonably computed during query processing time.

2. *Index construction:* Most of the data structures discussed in Section 9.2 carry over to the search engine setting. In particular, the inverted indexes are required to respond to queries, and dictionary data structure is required to map terms to offsets in inverted files and conveniently access the relevant inverted lists for each term. However, there are many issues of scalability when using such data structures. For example, the size of the Google index is of the order of a hundred trillion documents as of 2018, and it continues to grow over time. In most cases, it becomes cost effective and economical to build distributed indexes that are highly fault tolerant. Fault tolerance is achieved by replicating the index over multiple machines. Therefore, distributed *MapReduce* methods [145] are often used in the process of index construction.

3. *Query processing:* This preprocessed collection is utilized for online query processing. The relevant documents are accessed and then ranked using both their relevance to the query and their quality. The basic technique of query processing with an inverted index is discussed in Section 9.2.4, and a number of additional information retrieval models for scoring are discussed in Section 9.3. User feedback on search results is often used to construct machine learning methods on extracted meta-features such as zones, positional data, document meta-information, and linkage features such as the *PageRank*. As long as all these characteristics can be combined in an additive way to create a score, accumulators can be used in conjunction with inverted indexes for efficient query processing (cf. page 266). However, query processing in search engines is more challenging than in a traditional information retrieval system because of the high volume of the search and the distributed nature of the indexes. In such cases, the use of techniques like *tiered* inverted lists and skip pointers becomes essential. These types of optimizations can lead to large speedups because huge portions of the inverted lists are not accessed at all.

Some of these issues are discussed in the following subsections.

9.5.1 Distributed Index Construction

In most large search engines, the indexes are distributed over multiple nodes. One can partition the inverted indexes by storing different inverted lists at different nodes (term-wise partitioning) or by partitioning the different documents over multiple nodes. Although partitioning by terms might seem to be natural in the inverted list setting, it creates some challenges for query processing. Consider a situation, where the user enters the keywords "*text mining*" and the inverted lists for "*text*" and "*mining*" are located in different nodes. This means that one now has to send one of the two inverted lists from one node to the other to perform the intersection operation. This can create inefficiencies during query processing.

On the other hand, document-wise partitioning distributes the documents across different nodes and maintains all the inverted lists for that subset of documents in a single node. This means that all the intersection operations and intermediate computations can occur within the individual nodes. However, the global statistics such as the frequencies of terms in documents (for computing idf) need to be computed in a global way across all documents using distributed processes and then stored at individual nodes along with the terms. A hash function is used on each URL in order to distribute the different Web pages to nodes in a uniform way. For query processing, a given query is first broadcast to all the nodes, and computations/mergings are performed within the nodes. The top results from each node are returned and then combined in order to yield the top results for the query.

9.5.2 Dynamic Index Updates

Another important issue with search engines is that new documents are continually created and deleted, which creates problems when multiple inverted lists are stored in a single file. This is because inserting postings in a single file containing multiple inverted lists will affect the offset locations of all the postings in the dictionary. Although some techniques like *logarithmic merging* are available in the literature, the drawback is that such methods increase the complexity of query processing, and it is crucially required to very efficient at query processing time. Therefore, many large search engines periodically reconstruct the indexes from scratch as new crawled pages are discovered by spiders, and old pages are removed. The trade-offs associated with different ways of dynamic index maintenance are discussed in [322].

9.5.3 Query Processing

Web pages have different types of text in them, which can be used for various types of ad hoc query optimizations. An example is that of treating the anchor and title text differently from the body of a document during query processing. For a given set of terms in a query, all the relevant inverted lists are accessed, and the intersection of these inverted lists is determined. One can use accumulators to score and rank the different documents. Typically, to speed up the process, two indexes are constructed. A smaller index is constructed on only the titles of the Web page, or anchor text of pages *pointing to the page*. If enough documents are found in the smaller index, then the larger index is not referenced. Otherwise, the larger index is accessed. The logic for using the smaller index is that the title of a Web page and the anchor text of Web pages pointing to it, are usually highly representative of the content in the page.

If only the textual content of a Web page is used, the number of pages returned for common queries may be of the order of millions or more. Obviously, such a large number of query results will not be easy for a human user to assimilate. A typical browser interface will present only the first few (say 10) results to the human user in a single view of the search results, with the option of browsing other less relevant results. Therefore, the methodology for ranking needs to be very robust to ensure that the top results are highly relevant, As discussed in Section 9.3, information retrieval techniques always extract meta-features beyond the tf-idf scores in order to enable high-quality retrieval. While the exact scoring methodology used by commercial engines is proprietary, a number of factors are known to influence the content-based score:

1. A word is given different weights, depending upon whether it occurs in the title, body, URL token, or the anchor text of a pointing Web page. The occurrence of the term in the title or the anchor text of a Web page pointing to that page is generally given higher weight. This is similar to the notion of zoning discussed in Section 9.2.4.4. The weights of various zones can be learned using the machine learning techniques discussed in Section 9.2.4.5.

2. The prominence of a term in font size and color may be leveraged for scoring. For example, larger font sizes will be given a larger score.

3. When multiple keywords are specified, their relative positions in the documents are used as well. For example, if two keywords occur close together in a Web page, then this increases the score.

4. The most important meta-feature that is used by search engines is a *reputation-based* score of the Web page, which is also referred to as the *PageRank*.

The weighting of many of the aforementioned features during query processing requires the use of machine learning techniques in information retrieval. When a user chooses a Web page from among the responses to a search result in preference to earlier ranked results, this is clear evidence of the relevance of that page to the user. As discussed in Section 9.2.4.5 such data can be used by search engines to learn the importance of the various features. Furthermore, as long as the scoring computation is additive over the various meta-features, one can use the inverted index in combination with accumulator variables in order to enable efficient query processing (cf. page 266).

9.5.4   The Importance of Reputation

One of the most important meta-features used in Web search is the *reputation*, or the *quality*, of the page. This meta-feature corresponds to the *PageRank*. It is important to use such mechanisms because of the uncoordinated and open nature of Web development. After all, the Web allows anyone to publish almost anything, and therefore there is little control on the quality of the results. A user may publish incorrect material either because of poor knowledge on the subject, economic incentives, or with a deliberately malicious intent of publishing misleading information.

Another problem arises from the impact of *Web spam*, in which Website owners intentionally serve misleading content to rank their results higher. Commercial Website owners have significant economic incentives to ensure that their sites are ranked higher. For example, an owner of a business on golf equipment, would want to ensure that a search on the word *"golf"* ranks his or her site as high as possible. There are several strategies used by Website owners to rank their results higher.

1. *Content-spamming:* In this case, the Web host owner fills up repeated keywords in the hosted Web page, even though these keywords are not actually visible to the user. This is achieved by controlling the color of the text and the background of the page. Thus, the idea is to maximize the content relevance of the Web page to the search engine, without a corresponding increase in the *visible* level of relevance.

2. *Cloaking:* This is a more sophisticated approach, in which the Website serves different content to crawlers than it does to users. Thus, the Web site first determines whether the incoming request is from a crawler or from a user. If the incoming request is from a user, then the actual content (e.g., advertising content) is served. If the request is from a crawler, then the content that is most relevant to specific keywords is served. As a result, the search engine will use different content to respond to user search requests from what a Web user will actually see.

It is obvious that such spamming will significantly reduce the quality of the search results. Search engines also have significant incentives to improve the quality of their results to support their paid advertising model, in which the *explicitly marked* sponsored links appearing on the side bar of the search results are truly paid advertisements. Search engines do not want advertisements (disguised by spamming) to be served as *bona fide* results to the query, especially when such results reduce the quality of the user experience. This has led to an adversarial relationship between search engines and spammers, in which the former use reputation-based algorithms to reduce the impact of spam. At the other end of Website owners, a *Search Engine Optimization (SEO)* industry attempts to optimize search results by using their knowledge of the algorithms used by search engines, either through the general principles used by engines or through reverse engineering of search results.

For a given search, it is almost always the case that a small subset of the results is more informative or provides more accurate information. How can such pages be determined? Fortunately, the Web provides several natural voting mechanisms to determine the reputation of pages. The citation structure of Web pages is the most common mechanism used to determine the quality of Web pages. When a page is of high quality, many other Web pages point to it. A citation can be logically viewed as a vote for the Web page. While the number of in-linking pages can be used as a rough indicator of the quality, it does not provide a complete view because it does not account for the quality of the pages pointing to it. To provide a more holistic citation-based vote, an algorithm referred to as *PageRank* is used.

It should be pointed out, that citation-based reputation scores are not completely immune to other types of spamming that involve coordinated creation of a large number of links to a Web page. Furthermore, the use of anchor text of *pointing* Web pages in the content portion of the rank score can sometimes lead to amusingly irrelevant search results. For example, a few years back, a search on the keyword *"miserable failure"* in the Google search engine, returned as its top result, the official biography of a previous president of the United States. This is because many Web pages were constructed in a coordinated way to use the anchor text *"miserable failure"* to point to this biography. This practice of influencing search results by coordinated linkage construction to a particular site is referred to as *Googlewashing*. Such practices are less often economically motivated, but are more often used for comical or satirical purposes.

Therefore, the ranking algorithms used by search engines are not perfect but have, nevertheless, improved significantly over the years. The algorithms used to compute the reputation-based ranking score will be discussed in the next section.

9.6 Link-Based Ranking Algorithms

The *PageRank* algorithm uses the linkage structure of the Web for reputation-based ranking. The *PageRank* method is independent of the user-query, because it only precomputes the reputation portion of the score. Eventually, the reputation portion of the score is combined with other content-scoring methods like BM25, and the weights of the different components are regulated by learning-to-rank methods (e.g., ranking SVM of Section 9.2.4.6). The *HITS* algorithm is query-specific. It uses a number of intuitions about how authoritative sources on various topics are linked to one another in a hyperlinked environment.

9.6.1 PageRank

The *PageRank* algorithm models the importance of Web pages with the use of the citation (or linkage) structure in the Web. The basic idea is that highly reputable documents are more likely to be cited (or in-linked) by other reputable Web pages.

A random surfer model on the Web graph is used to achieve this goal. Consider a random surfer who visits random pages on the Web by selecting random links on a page. The long-term relative frequency of visits to any particular page is clearly influenced by the number of in-linking pages to it. Furthermore, the long-term frequency of visits to any page will be higher if it is linked to by other frequently visited (or *reputable*) pages. In other words, the *PageRank* algorithm models the reputation of a Web page in terms of its long-term frequency of visits by a random surfer. This long-term frequency is also referred to as the *steady-state probability*. This model is also referred to as the *random walk model*.

The basic random surfer model does not work well for all possible graph topologies. A critical issue is that some Web pages may have no outgoing links, which may result in the random surfer getting trapped at specific nodes. In fact, a probabilistic transition is not even meaningfully defined at such a node. Such nodes are referred to as *dead ends*. An example of a dead-end node is illustrated in Figure 9.7(a). Clearly, dead ends are undesirable because the transition process for *PageRank* computation cannot be defined at that node. To address this issue, two modifications are incorporated in the random surfer model. The first modification is to add links from the dead-end node (Web page) to all nodes (Web pages), including a self-loop to itself. Each such edge has a transition probability of $1/n$. This does not fully solve the problem, because the dead ends can also be defined on *groups of nodes*. In these cases, there are no outgoing links from *a group of nodes* to the remaining nodes in the graph. This is referred to as a *dead-end component*, or *absorbing component*. An example of a dead-end component is illustrated in Figure 9.7(b).

Dead-end components are common in the Web graph because the Web is not strongly connected. In such cases, the transitions at individual nodes can be meaningfully defined, but the steady-state transitions will stay trapped in these dead-end components. All the steady-state probabilities will be concentrated in dead-end components because there can be no transition out of a dead-end component after a transition occurs into it. Therefore, as long as even a minuscule probability of transition into a dead-end component[10] exists, *all* the steady-state probability becomes concentrated in such components. This situation is not desirable from the perspective of *PageRank* computation in a large Web graph, where dead-end components are not necessarily an indicator of popularity. Furthermore, in such cases, the final probability distribution of nodes in various dead-end components is not unique and it is dependent on the starting state. This is easy to verify by observing that random walks starting in different dead-end components will have their respective steady-state distributions concentrated within the corresponding components.

While the addition of edges solves the problem for dead-end nodes, an additional step is required to address the more complex issue of dead-end components. Therefore, aside from the addition of these edges, a *teleportation*, or *restart step* is used within the random surfer model. This step is defined as follows. At each transition, the random surfer may either jump to an arbitrary page with probability α, or it may follow one of the links on the page with probability $(1 - \alpha)$. A typical value of α used is 0.1. Because of the use of teleportation, the steady state probability becomes unique and independent of the starting state. The value of α may also be viewed as a *smoothing* or *damping probability*. Large values of α typically result in the steady-state probability of different pages to become more even. For example, if the value of α is chosen to be 1, then all pages will have the same steady-state probability of visits.

How are the steady-state probabilities determined? Let $G = (N, A)$ be the directed Web graph, in which nodes correspond to pages, and edges correspond to hyperlinks. The total number of nodes is denoted by n. It is assumed that A also includes the added edges from

---

[10]A formal mathematical treatment characterizes this in terms of the *ergodicity* of the underlying Markov chains. In ergodic Markov chains, a necessary requirement is that it is possible to reach any state from any other state using a sequence of one or more transitions. This condition is referred to as *strong connectivity*. An informal description is provided here to facilitate understanding.

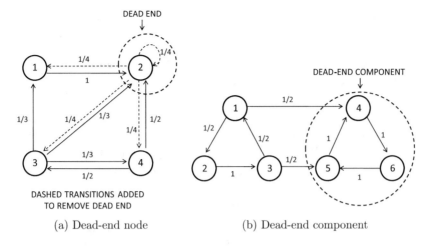

(a) Dead-end node   (b) Dead-end component

Figure 9.7: Transition probabilities for *PageRank* computation with different types of dead ends

dead-end nodes to all other nodes. The set of nodes incident on i is denoted by $In(i)$, and the set of end points of the outgoing links of node i is denoted by $Out(i)$. The steady-state probability at a node i is denoted by $\pi(i)$. In general, the transitions of a Web surfer can be visualized as a *Markov chain*, in which an $n \times n$ transition matrix P is defined for a Web graph with n nodes. The *PageRank* of a node i is equal to the steady-state probability $\pi(i)$ for node i, in the Markov chain model. The probability[11] p_{ij} of transitioning from node i to node j, is defined as $1/|Out(i)|$. Examples of transition probabilities are illustrated in Figure 9.7. These transition probabilities do not, however, account for teleportation which will be addressed[12] separately below.

Let us examine the transitions into a given node i. The steady-state probability $\pi(i)$ of node i is the sum of the probability of a teleportation into it and the probability that one of the in-linking nodes directly transitions into it. The probability of a teleportation into the node is exactly α/n because a teleportation occurs in a step with probability α, and all nodes are equally likely to be the beneficiary of the teleportation. The probability of a transition into node i is given by $(1-\alpha) \cdot \sum_{j \in In(i)} \pi(j) \cdot p_{ji}$, as the sum of the probabilities of transitions from different in-linking nodes. Therefore, at steady-state, the probability of a transition into node i is defined by the sum of the probabilities of the teleportation and transition events are as follows:

$$\pi(i) = \alpha/n + (1-\alpha) \cdot \sum_{j \in In(i)} \pi(j) \cdot p_{ji} \tag{9.20}$$

For example, the equation for node 2 in Figure 9.7(a) can be written as follows:

$$\pi(2) = \alpha/4 + (1-\alpha) \cdot (\pi(1) + \pi(2)/4 + \pi(3)/3 + \pi(4)/2)$$

---

[11]In some applications such as bibliographic networks, the edge (i,j) may have a weight denoted by w_{ij}. The transition probability p_{ij} is defined in such cases by $\frac{w_{ij}}{\sum_{j \in Out(i)} w_{ij}}$.

[12]An alternative way to achieve this goal is to modify G by multiplying existing edge transition probabilities by the factor $(1-\alpha)$ and then adding α/n to the transition probability between each pair of nodes in G. As a result G will become a directed clique with bidirectional edges between each pair of nodes. Such strongly connected Markov chains have unique steady-state probabilities. The resulting graph can then be treated as a Markov chain without having to separately account for the teleportation component. This model is equivalent to that discussed in the chapter.

There will be one such equation for each node, and therefore it is convenient to write the entire system of equations in matrix form. Let $\overline{\pi} = (\pi(1) \ldots \pi(n))^T$ be the n-dimensional column vector representing the steady-state probabilities of all the nodes, and let $\overline{e}$ be an n-dimensional column vector of all 1 values. The system of equations can be rewritten in matrix form as follows:

$$\overline{\pi} = \alpha \overline{e}/n + (1 - \alpha)P^T \overline{\pi} \qquad (9.21)$$

The first term on the right-hand side corresponds to a teleportation, and the second term corresponds to a direct transition from an incoming node. In addition, because the vector $\overline{\pi}$ represents a probability, the sum of its components $\sum_{i=1}^{n} \pi(i)$ must be equal to 1.

$$\sum_{i=1}^{n} \pi(i) = 1 \qquad (9.22)$$

Note that this is a linear system of equations that can be easily solved using an iterative method. The algorithm starts off by initializing $\overline{\pi}^{(0)} = \overline{e}/n$, and it derives $\overline{\pi}^{(t+1)}$ from $\overline{\pi}^{(t)}$ by repeating the following iterative step:

$$\overline{\pi}^{(t+1)} \Leftarrow \alpha \overline{e}/n + (1 - \alpha)P^T \overline{\pi}^{(t)} \qquad (9.23)$$

After each iteration, the entries of $\overline{\pi}^{(t+1)}$ are normalized by scaling them to sum to 1. These steps are repeated until the difference between $\overline{\pi}^{(t+1)}$ and $\overline{\pi}^{(t)}$ is a vector with magnitude less than a user-defined threshold. This approach is also referred to as the *power-iteration method*. It is important to understand that *PageRank* computation is expensive, and it cannot be computed on the fly for a user query during Web search. Rather, the *PageRank* values for *all* the known Web pages are pre-computed and stored away. The stored *PageRank* value for a page is accessed only when the page is included in the search results for a particular query for use in the final ranking. Typically, this stored value is used as one of the features in a learning-to-rank procedure (cf. Section 9.2.4.6).

The *PageRank* values can be shown to be the n components of the largest left eigenvector[13] of the stochastic transition matrix P, for which the eigenvalue is 1. However, the stochastic transition matrix P needs to be adjusted to incorporate the restart within the transition probabilities. This approach is described in Section 12.3.3 of Chapter 12.

9.6.1.1 Topic-Sensitive PageRank

Topic-sensitive PageRank is designed for cases in which it is desired to provide greater importance to some topics than others in the ranking process. While personalization is less common in large-scale commercial search engines, it is more common is smaller scale site-specific search applications. Typically, users may be more interested in certain combinations of topics than others. The knowledge of such interests may be available to a personalized search engine because of user registration. For example, a particular user may be more interested in the topic of automobiles. Therefore, it is desirable to rank pages related to automobiles higher when responding to queries by this user. This can also be viewed as the *personalization* of ranking values. How can this be achieved?

The first step is to fix a list of base topics, and determine a high-quality sample of pages from each of these topics. This can be achieved with the use of a resource such as the *Open*

---

[13]The left eigenvector $\overline{X}$ of P is a row vector satisfying $\overline{X}P = \lambda \overline{X}$. The right eigenvector $\overline{Y}$ is a column vector satisfying $P\overline{Y} = \lambda \overline{Y}$. For asymmetric matrices, the left and right eigenvectors are not the same. However, the eigenvalues are always the same. The unqualified term "eigenvector" refers to the right eigenvector by default.

Directory Project (ODP),[14] which can provide a base list of topics and sample Web pages for each topic. The *PageRank* equations are now modified, so that the teleportation is only performed on this sample set of Web documents, rather than on the entire space of Web documents. Let $\overline{e_p}$ be an n-dimensional personalization (column) vector with one entry for each page. An entry in $\overline{e_p}$ takes on the value of 1, if that page is included in the sample set, and 0 otherwise. Let the number of nonzero entries in $\overline{e_p}$ be denoted by n_p. Then, the *PageRank* Equation 9.21 can be modified as follows:

$$\overline{\pi} = \alpha \overline{e_p}/n_p + (1 - \alpha)P^T\overline{\pi} \tag{9.24}$$

The same power-iteration method can be used to solve the personalized *PageRank* problem. The selective teleportations bias the random walk, so that pages in the structural locality of the sampled pages will be ranked higher. As long as the sample of pages is a good representative of different (structural) localities of the Web graph, in which pages of specific topics exist, such an approach will work well. Therefore, for each of the different topics, a separate *PageRank* vector can be precomputed and stored for use during query time.

In some cases, the user is interested in specific *combinations of* topics such as sports and automobiles. Clearly, the number of possible combinations of interests can be very large, and it is not reasonably possible or necessary to prestore every personalized *PageRank* vector. In such cases, only the *PageRank* vectors for the base topics are computed. The final result for a user is defined as a weighted linear combination of the topic-specific *PageRank* vectors, where the weights are defined by the user-specified interest in the different topics.

9.6.1.2 SimRank

The notion of *SimRank* was defined to compute the structural similarity between nodes. *SimRank* determines *symmetric* similarities between nodes. In other words, the similarity between nodes i and j, is the same as that between j and i. Before discussing *SimRank*, we define a related but slightly different asymmetric ranking problem:

Given a target node i_q and a subset of nodes $S \subseteq N$ from graph $G = (N, A)$, rank the nodes in S in their order of similarity to i_q.

Such a query is very useful in recommender systems in which users and items are arranged in the form of a bipartite graph of preferences, in which nodes corresponds to users and items, and edges correspond to preferences. The node i_q may correspond to an item node, and the set S may correspond to user nodes. Alternatively, the node i_q may correspond to a user node, and the set S may correspond to item nodes. Refer to [3] for a discussion on recommender systems. Recommender systems are closely related to search, in that they also perform ranking of target objects, but while taking user preferences into account.

This problem can be viewed as a limiting case of topic-sensitive *PageRank*, in which the teleportation is performed to the *single node* i_q. Therefore, the personalized *PageRank* Equation 9.24 can be directly adapted by using the teleportation vector $\overline{e_p} = \overline{e_q}$, that is, a vector of all 0s, except for a single 1, corresponding to the node i_q. Furthermore, the value of n_p in this case is set to 1.

$$\overline{\pi} = \alpha \overline{e_q} + (1 - \alpha)P^T\overline{\pi} \tag{9.25}$$

The solution to the aforementioned equation will provide high ranking values to nodes in the structural locality of i_q. This definition of similarity is *asymmetric* because the similarity value assigned to node j starting from query node i is different from the similarity

---
[14]http://www.dmoz.org.

value assigned to node i starting from query node j. Such an *asymmetric* similarity measure is suitable for *query-centered* applications such as search engines and recommender systems, but not necessarily for arbitrary network-based data mining applications. In some applications, symmetric pairwise similarity between nodes is required. While it is possible to average the two topic-sensitive *PageRank* values in opposite directions to create a symmetric measure, the *SimRank* method provides an elegant and intuitive solution.

The *SimRank* approach is as follows. Let $In(i)$ represent the in-linking nodes of i. The *SimRank* equation is naturally defined in a recursive way, as follows:

$$SimRank(i,j) = \frac{C}{|In(i)| \cdot |In(j)|} \sum_{p \in In(i)} \sum_{q \in In(j)} SimRank(p,q) \qquad (9.26)$$

Here C is a constant in $(0,1)$ that can viewed as a kind of decay rate of the recursion. As the boundary condition, the value of $SimRank(i,j)$ is set to 1 when $i = j$. When either i or j do not have in-linking nodes, the value of $SimRank(i,j)$ is set to 0. To compute *SimRank*, an iterative approach is used. The value of $SimRank(i,j)$ is initialized to 1 if $i = j$, and 0 otherwise. The algorithm subsequently updates the *SimRank* values between all node pairs iteratively using Equation 9.26 until convergence is reached.

The notion of *SimRank* has an interesting intuitive interpretation in terms of random walks. Consider two random surfers walking *in lockstep* backwards from node i and node j till they meet. Then the number of steps taken by each of them is a random variable $L(i,j)$. Then, $SimRank(i,j)$ can be shown to be equal to the expected value of $C^{L(i,j)}$. The decay constant C is used to map random walks of length l to a similarity value of C^l. Note that because $C < 1$, smaller distances will lead to higher similarity and vice versa.

Random walk-based methods are generally more robust than the shortest path distance to measure similarity between nodes. This is because random walks measures implicitly account for the *number* of paths between nodes, whereas shortest paths do not.

9.6.2   HITS

The <u>H</u>ypertext <u>I</u>nduced <u>T</u>opic <u>S</u>earch *(HITS)* algorithm is a *query-dependent* algorithm for ranking pages. The intuition behind the approach lies in an understanding of the typical structure of the Web that is organized into hubs and authorities.

An *authority* is a page with many in-links. Typically, it contains authoritative content on a particular subject, and, therefore, many Web users may trust that page as a resource of knowledge on that subject. This will result in many pages linking to the authority page. A *hub* is a page with many out-links to authorities. These represent a compilation of the links on a particular topic. Thus, a hub page provides guidance to Web users about where they can find the resources on a particular topic. Examples of the typical node-centric topology of hubs and authorities in the Web graph are illustrated in Figure 9.8(a).

The main insight used by the *HITS* algorithm is that good hubs point to many good authorities. Conversely, good authority pages are pointed to by many hubs. An example of the typical organization of hubs and authorities is illustrated in Figure 9.8(b). This mutually reinforcing relationship is leveraged by the *HITS* algorithm. For any query issued by the user, the *HITS* algorithm starts with the list of relevant pages and expands them with a *hub ranking* and an *authority ranking*.

The *HITS* algorithm starts by collecting the top-r most relevant results to the search query at hand. A typical value of r is 200. This defines the *root set* R. Typically, a query to a commercial search engine or content-based evaluation is used to determine the root

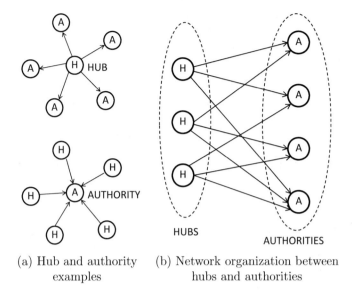

(a) Hub and authority   (b) Network organization between
        examples   hubs and authorities

Figure 9.8: Illustrating hubs and authorities

set. For each node in R, the algorithm determines all nodes immediately connected (either in-linking or out-linking) to R. This provides a larger *base set* S. Because the base set S can be rather large, the maximum number of in-linking nodes to any node in R that are added to S is restricted to k. A typical value of k used is around 50. Note that this still results in a rather large base set because *each* of the possibly 200 root nodes might bring 50 in-linking nodes, along with out-linking nodes.

Let $G = (S, A)$ be the subgraph of the Web graph defined on the (expanded) base set S, where A is the set of edges between nodes in the root set S. The entire analysis of the *HITS* algorithm is restricted to this subgraph. Each page (node) $i \in S$ is assigned both a *hub score* $h(i)$ and *authority score* $a(i)$. It is assumed that the hub and authority scores are normalized, so that the sum of the squares of the hub scores and the sum of the squares of the authority scores are each equal to 1. Higher values of the score indicate better quality. The hub scores and authority scores are related to one another in the following way:

$$h(i) = \sum_{j:(i,j)\in A} a(j) \quad \forall i \in S \tag{9.27}$$

$$a(i) = \sum_{j:(j,i)\in A} h(j) \quad \forall i \in S \tag{9.28}$$

The basic idea is to reward hubs for pointing to good authorities and reward authorities for being pointed to by good hubs. It is easy to see that the aforementioned system of equations reinforces this mutually enhancing relationship. This is a linear system of equations that can be solved using an iterative method. The algorithm starts by initializing $h^0(i) = a^0(i) = 1/\sqrt{|S|}$. Let $h^t(i)$ and $a^t(i)$ denote the hub and authority scores of the ith node, respectively, at the end of the tth iteration. For each $t \geq 0$, the algorithm executes the following iterative steps in the $(t+1)$th iteration:

for each $i \in S$ set $a^{t+1}(i) \Leftarrow \sum_{j:(j,i) \in A} h^t(j)$;
for each $i \in S$ set $h^{t+1}(i) \Leftarrow \sum_{j:(i,j) \in A} a^{t+1}(j)$;
Normalize L_2-norm of each of hub and authority vectors to 1;

For hub-vector $\overline{h} = [h(1) \ldots h(n)]^T$ and authority-vector $\overline{a} = [a(1) \ldots a(n)]^T$, the updates can be expressed as $\overline{a} = A^T \overline{h}$ and $\overline{h} = A\overline{a}$, respectively, when the edge set A is treated as an $|S| \times |S|$ adjacency matrix. The iteration is repeated to convergence. It can be shown that the hub vector $\overline{h}$ and the authority vector $\overline{a}$ converge in directions proportional to the dominant eigenvectors of AA^T and $A^T A$, respectively. This is because the relevant pair of updates can be shown to be equivalent to power-iteration updates of AA^T and $A^T A$, respectively.

9.7 Summary

This chapter discusses the data structures and query processing methods involved in information retrieval, and their generalizations to search engines. A proper design of the inverted index is crucial in obtaining efficient responses to queries. Many types of additive functions over terms can be computed with an inverted index and accumulator variables. Many vector space and probabilistic models of retrieval are used by search engines. Some of these models use relevance feedback, whereas others are able to score documents with respect to queries without using relevance feedback. An important aspect of search engine construction is the discovery of relevant resources with the use of crawling techniques. In search engines, the linkages can be used to create a measure Web page quality with *PageRank* measures. These quality measures are combined with match-based measures to provide responses to queries.

9.8 Bibliographic Notes

A discussion of several data structures, such as hash tables, binary trees and B-Trees, may be found in [493]. All these data structures are useful for dictionary construction. The use of k-gram dictionaries for spelling corrections and error-tolerant retrieval may be found in [636, 637].

A detailed discussion of the inverted file in the context of search engines may be found in [589, 639], including various optimizations like skip pointers. Skip pointers were introduced in [404]. The inverted file is the dominant data structure for indexing documents, although some alternatives like the signature file [185] have also been proposed in the literature. A comparison of the inverted files with the signature file is provided in [638], and it is shown that the signature file is inefficient as compared to the inverted file. Several methods for constructing inverted indexes are discussed in [245, 589]. The construction of distributed indexes is discussed in [35, 84, 222, 380, 469]. Most recent techniques are based on the *MapReduce* framework [145]. Dynamic index construction methods like logarithmic merging are discussed in [82]. The trade-offs of different ways of index maintenance are discussed in [322].

The technique of using accumulators with early stopping is discussed in [404]. Other efficient methods for query processing with early termination and pruning are discussed in [25, 27]. Methods for using inverted indexes in phrase queries are discussed in [585]. Machine learning approaches for search engine optimization with the use of the ranking SVM were pioneered in [276]. However, the view of information retrieval as a classification problem was recognized much earlier and also appears in van Rijsbergen's classical book [556], which was written in 1979. The earliest methods for learning to rank with pairwise training data points were proposed in [122]. The work in [547] optimized the parameters of the BM25 function on the basis of the NDCG measure. Refer to Chapter 7 for a discussion of the NDCG measure. The ranking SVM idea has been explored in [87] in the context of information retrieval. A structured SVM idea that optimizes average precision is discussed in [610]. The work in [81] discusses ways of ranking using gradient-descent techniques with the *RankNet* algorithm. It is stated in [349] that the *RankNet* algorithm was the initial approach used in several commercial search engines. The listwise approach for learning to rank is discussed in [88]. The extraction of document-specific features for improving ranking with machine learning approach is discussed in [470]. An excellent overview of different learning-to-rank algorithms with a focus on search engines is provided in [349].

The use of champion lists, pruning, and tiered indexes for large-scale search is discussed in [90, 354, 367, 422]. Dictionary compression is discussed in [589], and variable byte codes were proposed in [501]. Word aligned codes have also been proposed that improve over variable byte codes [26, 29]. The delta coding scheme was proposed in [175]. The use of caching for improving retrieval performance is studied in [35, 318, 495]. Many of these techniques show how one can use caches of multiple levels to improve performance. The combination of inverted list compression and caching for improved performance is explored in [622]. In general, compression improves caching performance as well. A good overview of caching and compression may also be found in Zobel and Moffat's survey on indexing [639].

The vector space model for information retrieval was introduced in [492], and term weighting methods were studied in [489]. Over the years, numerous term weighting and document length normalization methods have been proposed. Pivoted length document normalization is a notable approach that is often used [519]. The use of idf normalization was first conceived in [522]. The binary independence model was proposed in [476, 556], and the final form of the retrieval status value is a confirmation of the importance of idf in retrieval applications. A number of theoretical arguments for idf normalization may also be found in [475]. A number of experiments about the probabilistic model of information retrieval are provided in [525]. This paper also adapts the binary independence model into the BM25 model. The BM25 model has had a significant impact on the matching function used in the search engines. A variant of the BM25 model that uses the different fields in the document is referred to as BM25F [477]. The use of language models in information retrieval were pioneered by Ponte and Croft with the Bernoulli approach [443]. A language model by Hiemstra [243] also appeared at approximately the same time, which used a multinomial approach. The use of Hidden Markov Models for language modeling were proposed by Miller et al. [395]. The role of smoothing in language modeling approaches is studied by Zhai and Lafferty [618]. An overview of language models for information retrieval may be found in [617].

Detailed discussions on crawling and search engines may be found in several books [35, 82, 92, 137, 345, 367, 589]. Focused crawling was proposed in [96]. The work in [110] discusses the importance of proper URL ordering in being able to efficiently crawl useful pages. The *PageRank* algorithm is described in [74, 426]. The *HITS* algorithm was described in [299]. A detailed description of different variations of the *PageRank* and *HITS* algorithms may be found in [92, 321, 345, 367]. The topic-sensitive *PageRank* algorithm is described in [234].

9.8.1 Software Resources

Numerous open source search engines are available such as Apache Lucene and Solr [683, 684], Datapark search engine [682], and Sphinx [685]. Some of the search engines also provide crawling capabilities. The Lemur project [678] provide an open source framework for language modeling approaches in information retrieval. Numerous open source crawlers are available such as Heritrix [681] (Java), Apache Nutch [679] (Java), Datapark search engine [682] (C++), and Python Scrapy [680]. The package *scikit-learn* [646] has an implementation of the computation of the principal eigenvector, which is useful for *PageRank* evaluation and HITS. The Snap repository at Stanford University also includes a *PageRank* implementation [686]. The **gensim** software package [464] has an implementation of some ranking functions like BM25.

9.9   Exercises

1. Show that the space required by the inverted index is exactly proportional to that required by a sparse representation of the document-term matrix.

2. The index construction of Section 9.2.3 assumes that document identifiers are processed in sorted order. Discuss the modifications required when the document identifiers are not processed in sorted order. How much does this modification increase the time complexity?

3. Discuss an efficient algorithm to implement the OR operator in Boolean retrieval with two inverted lists that are available in sorted form.

4. Show that a dictionary, which is implemented as a hash table with linear probing, requires constant time for insertions and lookups. Derive the expected number of lookups in terms of the fraction of the table that is full.

5. Write a computer program to implement a hash-based dictionary and an inverted index from a document-term matrix.

6. Suppose that the inverted index also contains positional information. Show that the size of the inverted index is proportional to the number of tokens in the corpus.

7. Consider the string *ababcdef*. List all 2-shingles and 3-shingles, using each alphabet as a token.

8. Show that the *PageRank* computation with teleportation is an eigenvector computation on an appropriately constructed probability transition matrix.

9. Show that the hub and authority scores in *HITS* can be computed by dominant eigenvector computations on AA^T and $A^T A$ respectively. Here, A is the adjacency matrix of the graph $G = (S, A)$, as defined in the chapter.

10. Propose an alternative to the ranking SVM based on logistic regression. Discuss how you would formulate the optimization problem and how the stochastic gradient-descent steps are related to traditional logistic regression.

11. The ranking SVM is a special case of the classical SVM in which each class variable is +1 in the training data (but not necessarily at the time of prediction) and the bias variable is 0. Show that any classical SVM in which the bias variable is 0 but the class variables are drawn from $\{-1, +1\}$ can be transformed to the case in which each class variable is +1. Why do we need the bias variable to be 0 for this transformation?

Chapter 10

Language Modeling and Deep Learning

"A sequence works in a way a collection never can."–George Murray

10.1 Introduction

Much of the discussion in the previous chapters has focused on a bag-of-words representation of text. While the bag-of-words representation is sufficient in many practical applications, there are cases in which the sequential aspects of text become more important. There are two primary reasons for the sequence representation to become particularly useful:

1. *Data-centric reasons:* In some settings, the lengths of the text units are small. For example, text segments corresponding to micro-blogs and tweets are relatively short. In such cases, there is simply not sufficient information within the bag-of-words representation to make meaningful inferences. When the underlying document is large, the bag of words contains sufficient evidence in the form of word frequencies. On the other hand, it is more important to enrich short text snippets with sequencing information to extract the most out of limited data.

2. *Application-centric reasons:* Many applications like text summarization, information extraction, opinion mining, and question answering require semantic insights. Semantic understanding can be gained only by treating sentences as sequences.

Word ordering conveys semantics that cannot be inferred from the bag-of-words representation. For example, consider the following pair of sentences:

The cat chased the mouse.
The mouse chased the cat.

Clearly, the two sentences are very different (and the second one is unusual), but they are identical from the point of view of the bag-of-words representation. For longer segments of text, term frequency usually conveys sufficient evidence to robustly handle simple machine learning decisions like binary classification. This is one of the reasons that sequence

© Springer Nature Switzerland AG 2022
C. C. Aggarwal, *Machine Learning for Text*,
https://doi.org/10.1007/978-3-030-96623-2_10

information is rarely used in simpler settings like classification. On the other hand, more sophisticated applications with fine-grained nuances require a greater degree of linguistic intelligence.

A common approach is to convert text sequences to multidimensional embeddings because of the wide availability of machine learning solutions for multidimensional data. However, unlike the matrix factorization techniques of Chapter 3, the goal is to incorporate the sequential structure of the data within the embedding. For example, consider the word analogy "king is to queen as man is to woman." We would like a multidimensional embedding $f(\cdot)$ of the terms *"king,"* *"queen,"* *"man,"* and *"woman,"* which satisfies the following [436]:

$$f(king) - f(queen) \approx f(man) - f(woman) \qquad (10.1)$$

Such embeddings can only be created with the use of sequencing information because of its semantic nature.

There are two broader approaches for incorporating linguistic structure from text sentences. These methods are as follows:

1. *Language-specific methods:* These techniques use the parts of speech and other linguistic features in the learning process, which requires input from linguists of the specific language at hand. An example of a language-specific method is the use of a *probabilistic context free grammar*, which uses grammar rules to define the *syntax* of a language. Although these rules are supplemented with statistical analysis, the basic approach is inherently language specific because it starts with a set of grammar rules that are specific to the language at hand. A key point is that the incorporation of linguistic domain knowledge with the use of rules is essential to the workings of such a system. Therefore, such learning systems are not purely inductive, because a certain amount of deductive learning with rules is involved.

2. *Language-independent methods:* These techniques create a language model purely using statistical analysis of the sequential ordering of words. A *statistical language model* is a probability distribution over sequences of words, and it learns the statistical likelihood of a word following a sequence of words in a sentence. Examples of such models include the *unigram*, *bigram*, *trigram*, and *n-gram* models. Neural language models use neural networks to encode the grammatical structure of a language from text examples. These models can be used with arbitrary languages and applications, because the underlying representations are learned in a data-driven manner without significant domain knowledge.

Even though language-independent methods start with a handicap (i.e., no specified set of rules encoding a grammar), they generally work surprisingly well when sufficient data is available. The surprising effectiveness of language-illiterate methods over linguistically-literate methods is a result of the inherently intuitive way in which many ideas are conveyed in spoken language. Human language is sufficiently inexact and complex in terms of the variations in usage that it is hard to decode the semantic interpretation of a sentence purely based on grammatical rules. From this point of view, it can be an advantage to *not* use language-specific input, because it avoids the incorrect biases arising from our inherent limitation to encode complex semantic ideas with grammatical rules. Very often, our understanding of sentences is based on our intuitive life experiences of learning usages and semantic meanings of sentences from examples, which cannot be encoded in a literal way. The same insights apply to machine learning, where *given sufficient training data*, the algorithm should learn much of the grammar *and usage* on its own in a *goal-directed*

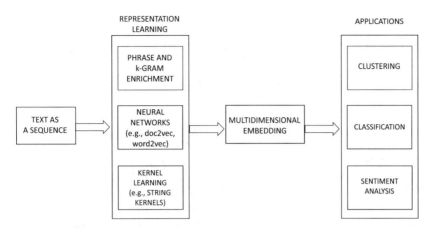

Figure 10.1: Representation learning for converting sequences into semantically knowledge-able embeddings

way for the specific application at hand (e.g., machine translation). This also elucidates another advantage of such methods in which the engineered features are often learned in an application-dependent manner. For example, the optimal feature representation of a sentence for an image captioning application may not be the same as that for a machine translation application, and the typical objective functions in inductive learning automatically account for these differences. In general, inductive learning methods have several advantages over deductive learning methods, although the debate between the two is not yet settled. Linguistic input is indeed helpful in many cases, because part-of-speech tags and other linguistically derived features are used as additional input to improve the learning. In this point of view, linguistic features primarily serve as guide rails to deal with the challenges associated with limited data in the same way as a regularizer improves performance. However, linguistically-oriented methods are often used only on the margins, and rarely serve as the primary basis of sequence-centric applications. Since this book is primarily focused on language-independent methods, the second category of inductive methods will be considered in this chapter.

A closely related problem to that of statistical language modeling is that of encoding all the rich structural information in text sequences in a multidimensional format that is friendly to the use of text mining algorithms. Indeed, feature engineering is the holy grail of machine learning and text mining methods. The common techniques for feature engineering use statistical language models as well as some amount of syntactic information in order to encode the sequencing information into a multidimensional format. The statistical language models are used indirectly within feature engineering methods. For example, one might use phrase and k-gram enrichment, neural networks, and kernel methods. The use of statistical language models in the design of kernel similarity functions like the string kernel (cf. Section 3.6.1.3 of Chapter 3) is more subtle, and it is implicit. Other methods like the convolution tree kernel (cf. Section 13.3.3.3) use language-dependent input in the form of context-free grammars. All these methods typically return a high-dimensional representation of each word and/or document. Once the multidimensional representation of words and/or documents has been constructed, it can be used in conjunction with any off-the-shelf mining algorithm. This process is illustrated in Figure 10.1.

Chapter Organization

This chapter will study both statistical language models and representation learning within a unified framework. Since the process of representation learning and feature engineering depends on the underlying language models, these models will be discussed first in Section 10.2. Some embedding methods are good for creating word embeddings, whereas others are good for creating document embeddings. Kernel methods are discussed in several chapters of this book, and Section 10.3 summarizes these methods from the point of view of feature engineering. In word-context factorization models (cf. Section 10.4), the occurrence frequencies of different contextual elements in a given window are used to create word embeddings. The use of graph-based methods for representation learning is discussed in Section 10.5. As discussed in Section 10.6, similar goals to word-context factorization can also be achieved using neural networks. Recurrent neural networks for language modeling are discussed in Section 10.7. Applications of recurrent neural networks are introduced in Section 10.8. Convolutional neural networks are text are discussed in Section 10.9. The conclusions are discussed in Section 10.10.

10.2 Statistical Language Models

A statistical language model assigns a probability to a sequence of words. Given a sequence of words $w_1, w_2, \ldots, w_m$, the language model estimates its probability $P(w_1, w_2, \ldots, w_m)$. Sequences of words that are grammatically correct and occur with high frequency in the collection will typically be assigned high probabilities. Common applications of statistical language models include speech recognition, machine translation, part-of-speech tagging, and information retrieval. The use of statistical language models for information retrieval is discussed in Section 9.3.4 of Chapter 9. Statistical language models also provide the intuition needed for feature engineering of text sequences into multidimensional data.

The simplest way to compute $P(w_1, w_2, \ldots, w_m)$ is to use the chain rule in sequences:

$$P(w_1, \ldots w_m) = P(w_1) \cdot P(w_2|w_1) \cdot P(w_3|w_1, w_2) \cdot \ldots \cdot P(w_m|w_1, w_2, \ldots, w_{m-1})$$
$$= \prod_{i=1}^{m} P(w_i|w_1, \ldots, w_{i-1})$$

Each of the terms $P(w_i|w_1 \ldots w_{i-1})$ needs to be estimated in a data-driven manner. This is achieved by expressing the conditional with the Bayes rule as follows:

$$P(w_i|w_1, \ldots, w_{i-1}) = \frac{P(w_1, \ldots w_i)}{P(w_1, \ldots, w_{i-1})} = \frac{\text{Count}(w_1, \ldots w_i)}{\text{Count}(w_1, \ldots, w_{i-1})}$$

The counts in the numerator and denominator are estimated directly from the data. Unfortunately, the count of the group $(w_1, \ldots, w_i)$ is hard to robustly estimate for large values of the group-size i. In such cases, the numerator and the denominator in the aforementioned estimation can be close to 0. Even with Laplacian smoothing, such an estimation is unlikely to be robust.

In order to address this problem, the *Markovian* assumption is used, which is also referred to as the *short-memory* assumption. According to this assumption, only the last $(n-1)$ tokens are used in order to estimate the conditional probability of a token, which results in an *n-gram model*. Mathematically, the short-memory assumption for the n-gram model can be written as follows:

$$P(w_i|w_1, \ldots, w_{i-1}) \approx P(w_i|w_{i-n+1}, \ldots, w_{i-1}) \tag{10.2}$$

With this simplified assumption, the conditional probabilities can be estimated as follows:

$$P(w_i|w_1,\ldots,w_{i-1}) \approx P(w_i|w_{i-n+1},\ldots w_{i-1}) = \frac{\text{Count}(w_{i-n+1}\ldots w_i)}{\text{Count}(w_{i-n+1},\ldots,w_{i-1})} \qquad (10.3)$$

Larger values of n provide better theoretical discrimination, but the amount of data is usually not sufficient to obtain reliable estimates. Therefore, one can view the choice as a *reliability* versus *discrimination* trade-off, which is a form of the bias-variance trade-off that is popularly used in machine learning. Setting $n = 2$ is referred to as the *bigram model*, whereas setting $n = 3$ is referred to as the *trigram model*. Setting $n = 1$ results in the unigram language model, which does not use any sequence information at all. This model is equivalent to using independent rolls of a die in order to decide the choice of tokens in a sentence. The unigram language model is a probabilistic avatar of the conventional bag-of-words model, and it results in a multinomial distribution of term frequencies. As discussed in Section 9.3.4 of Chapter 9, this model is used frequently for query-likelihood estimation in search. Furthermore, this model has also been (implicitly) discussed for probabilistic clustering and classification in Chapters 4 and 5, respectively. Since this chapter is focused in the sequential aspects of text, the unigram language model is of little interest to us, and we will focus on the n-gram language model only for $n \geq 2$.

In the case of the bigram model at $n = 2$, the aforementioned estimation can be written as follows:

$$P(w_i|w_1\ldots w_{i-1}) \approx P(w_i|w_{i-1}) = \frac{\text{Count}(w_{i-1}, w_i)}{\text{Count}(w_{i-1})} \qquad (10.4)$$

An n-gram essentially incorporates the effect of sequential context. For example, the value of $P(\text{"}sky\text{"}|\text{"}blue\text{"})$ will usually be greater than that of $P(\text{"}tree\text{"}|\text{"}blue\text{"})$ in a bigram model, because "*sky*" is more likely to follow "*blue*" than the word "*tree*". The most common values of n range between 2 and 5. However, larger values can also be used with increasing data.

Smoothing is particularly important in the case of language models because of the paucity of data. In such cases, the counts in both the numerator and denominator of Equation 10.4 can be zero, as a result of which the estimation becomes difficult. Furthermore, the presence of very rare words can have a confounding effect on the estimation process. Words that are present with very low frequency in the training data are often replaced with a special token denoted by ⟨UNK⟩. This token is then treated as any other term in the estimation process. Furthermore, for a lexicon of size d, Equation 10.4 can be modified in order to incorporate smoothing as follows:

$$P(w_i|w_{i-1}) = \frac{\text{Count}(w_{i-1}, w_i) + \beta}{\text{Count}(w_{i-1}) + d \cdot \beta} \qquad (10.5)$$

Here, $\beta > 0$ is the smoothing parameter that controls the degree of regularization. Large values of β lead to a greater degree of smoothing.

The n-gram models can be expressed as generative models in the form of *finite state automatons*. This type of Markovian model contains a set of states that transition to one another while generating a word at each transition. The sum of the probabilities of the transitions out of a state is always equal to 1. Each state is labeled with a string of length $(n - 1)$. At any given point in the transition, the value of the state provides the previous string of length $(n - 1)$. For a lexicon of size d, there are d possible transitions to states, which are obtained by appending the generated word to the last $(n - 2)$ words. The probabilities of these transitions are precisely the conditional probabilities that are estimated using Equation 10.4.

The number of states in an n-gram model is d^{n-1}. Consider a toy lexicon of size 4 with elements $\{the, mouse, chased, cat\}$. The data set contains two sentences:

> The mouse chased the cat.
> The cat chased the mouse.

The bigram model has 4 possible states, corresponding to the individual words, and the trigram model has 4^2 possible states. However, most of these states are not present even once in the training data. Therefore, in practice, one drops such states from the model. Furthermore, assume that the n-gram models do not cross sentence boundaries, and therefore each of the two sentences above is treated as a unit. The corresponding bigram and trigram models are illustrated in Figure 10.2(a) and (b), respectively. A state is defined by a single word in a bigram model and by two words in a trigram model. Note that only a small subset of the transitions are valid in the trigram model Figure 10.2(b), because two successive states must have an overlapping word. Therefore, such transitions are invalid (with zero theoretical probability) and are not included at all. The probabilities of the various transitions are estimated by using the fractional counts discussed earlier with Laplacian smoothing. Some of the (valid) transitions shown in Figure 10.2(a) and (b) are not reflected in the two training sentences, but will nevertheless have nonzero probability because of Laplacian smoothing. For example, the words "*cat*" and "*mouse*" do not occur consecutively in any of the two sentences above, but the transition in Figure 10.2(a) from the state "*cat*" to the state "*mouse*" will have nonzero probability because of Laplacian smoothing.

In the case of the trigram model, infrequent states (with zero frequency) have been pruned, and therefore the number of states in quite modest. It is noteworthy that any text segment of length m can create at most $(m - n + 1)$ n-grams, which provides a practical limitation on the size of the Markovian model that is created with pruning. One consequence of state pruning is that it can cause generalization problems when using the Markovian model in an application-specific setting on a text segment where that state was never seen before. In such cases, it becomes essential to use *back-off* models to be able to handle such states. For example, one might use the trigram model of Figure 10.2(b) when a sufficient amount of data is available. However, if one encountered a word pair such as "*cat cat*" in a (grammatically incorrect or rare) test segment, one might back-off to the bigram model of Figure 10.2(a). In general, one tries the n-gram model first, then the $(n - 1)$-gram model, and so on in order to estimate the conditional probability of Equation 10.4.

Such models are considered *visible* Markov models, as opposed to *hidden* Markov models. The reader should take a minute to examine Figure 10.2 and verify the fact that the precise sequence of states for a given training sentence can be inferred in a deterministic way. This is always true for n-gram models in which the precise sequence of states is always visible to the analyst. In a hidden Markov model, the states are defined with a hidden semantic notion, and multiple paths through the model could generate the same training sentence. Hidden Markov models will be discussed in Chapter 13 on information extraction. It is noteworthy that visible Markov models have a much simpler parameter estimation procedure because of the ability to deterministically infer the sequence of transitions corresponding to the training data.

10.2.1 Skip-Gram Models

The problem of data sparsity can never be fully solved with methods like smoothing or back-off. After all, smoothing and back-off are forms of regularization that do not provide

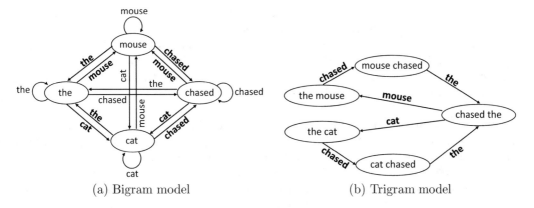

(a) Bigram model   (b) Trigram model

Figure 10.2: The Markovian models for bigrams and trigrams. The transitions are labeled with the symbols (words) that they generate

very refined models, and they only encode prior beliefs into the model. Therefore, if the model is dominated by these regularization aspects, the quality of prediction is often low. Extensive studies have shown that the vast majority of valid trigrams are not available even in very large training data sets.

One problem is that the required value of n encodes the length of the context that may vary from sentence to sentence. Small variations of a given sentence can have large effects on the estimation in an n-gram model. For example, consider the following pair of sentences:

Adam and Eve ate an apple.
Adam ate the red apple.

Although each of the two sentences contain information that is not available in the other, the key idea of Adam eating an apple is available in both sentences. Unfortunately, the two sentences do not share even a *single* bigram or trigram. Part of the problem is caused by the noise in words like *"red,"* which are not important to the core idea, but they nevertheless have an outsized effect on the n-gram representation.

In order to handle this problem, several recent models have been proposed that allow one to skip tokens while creating the model. One such model is the skip-gram model. The skip-gram model predicts a context word of the target word from the given word. In the skip-gram model, one is allowed to skip at most k tokens while predicting the context of a word. Then, the set $S(k, n)$ of all k-skip-n-grams of a sentence sequence $w_1 w_2 w_3 \ldots w_m$ is defined as follows:

$$S(k, n) = \{w_{i_1} w_{i_2} \ldots w_{i_n} : \sum_{j=2}^{n} (i_j - i_{j-1} - 1) \leq k, i_j > i_{j-1} \ \forall j\} \tag{10.6}$$

Note that all n-grams can be considered 0-skip-n-grams. Therefore, skip-grams represent a natural generalization of n-grams. As we will see later, there is also a difference in terms of how skip-grams are used for predictive modeling.

For example, the 2-skip-2-grams of the sentence *"Adam and Eve ate an apple."* are as follows:

Adam and, Adam Eve, Adam ate, and Eve, and ate, and an, Eve ate, Eve an, Eve apple, ate an, ate apple, an apple

We can already see that the skip-grams contain key phrases like "*Adam ate*", "*Eve ate*", and "*ate apple*", which are very useful for understanding the semantics of a sentence. In fact, both "*Adam ate*" and "*ate apple*" can be shown to be 2-skip-2-grams of the second sentence above in which the apple is qualified to be red. This qualifier can be skipped by the modeling process. The skip-grams provide the context that can be used to extract the semantic information that are hidden within the word occurrences.

Unfortunately, these desirable characteristics of skip-grams come at the expense of expanding the number of word sets, many of which are nonsensical. Nevertheless, the additional contextual information obtained from such skip-grams usually outweighs the noise effects of the low-quality skip-grams. In some cases, however, the storage and computational costs of this representation can be large, when it is used for creating word embeddings and feature engineering. A closely related model to the skip-gram is the continuous bag of words, with the main difference being in terms of the direction of the prediction. In the skip-gram model, one predicts the context C from the target word w, whereas in the continuous bag of words, one predicts the target word w, given the context. These types of predictive models are useful for creating embeddings.

10.2.2   Relationship with Embeddings

There are several ways in which skip-grams can be directly or indirectly used for embeddings. In many cases, the relationship with skip-grams is only an indirect one. Most of these methods respect sentence boundaries because the individual sentences are assumed to be sufficiently independent, so that it does not make sense for the skip-gram to cross sentence boundaries. Furthermore, some embeddings are performed at the word level, whereas others are performed at the document or sentence level.

The key methods for using language models for creating embeddings are as follows:

1. *Kernel-based embeddings:* Kernel-based embeddings compute similarities between pairs of sentences/documents in order to create multidimensional representations of the data. Note that these embeddings are typically performed at the level of sentences or documents rather than individual words. Various methods for kernel-based embeddings are discussed in Chapters 3 and 13. Therefore, this chapter will primarily summarize these methods and discuss their relationships with other methods. Some kernel-based methods directly enrich the document with n-gram representation to compute high-quality similarities. Other methods like string kernels (cf. Section 3.6.1.3) indirectly use the principles in skip-grams for similarity computation with the use of decaying context.

2. *Distributional semantic models:* Distributional semantic models are also referred to as count models, and they use the counts of the words occurring in the contexts of other words in order to create an embedding of words. Typically, a word-context co-occurrence matrix C is constructed, in which each row contains the frequencies of various contextual elements occurring in the vicinity of a row-specific target word. This matrix is factorized $C = UV^T$ in order to extract the matrix U, in which each row contains the embedding of a target word. Although the contextual elements can also be words, they can be generalized beyond words to features such as the parts-of-speech and so on.

3. *Contextual neural-network models:* These models are similar to distributional semantic models under the covers, although a neural network architecture is used to achieve

the same goals. In these models, a word and its surrounding text (i.e., context) is used to create a supervised learning problem. The use of the neural network is particularly popular, although any supervised method can be used in principle. These methods have also been shown to be closely related to count-based matrix factorization and kernel methods [323, 420]. These methods (e.g., *word2vec*) are designed to create word embeddings, although some modifications like *doc2vec* can also create document embeddings.

4. *Recurrent neural-network models:* These methods use recurrent neural networks in order to create a *neural* language model. Unlike contextual models, which are primarily designed for word embeddings, recurrent neural network models are designed for sentence-level embeddings. These are among the most powerful methods that can be used for applications like image captioning and sequence-to-sequence learning.

Finally, multidimensional embeddings are not the only way to represent text [16]. Rather, it is possible to use *distance graphs* in order to represent text. Using a graphical representation has the advantage that one can use many combinatorial graph-mining algorithms (rather than multidimensional algorithms) for learning purposes. Graph mining is a rich area of machine learning in its own right, which provides many off-the-shelf options. Interestingly, the factorization of distance graphs provides similar representations to count-based models.

10.2.3 Evaluating Language Models with Perplexity

Numerous language models in the literature generate segments of text with machine learning. As we will see later in this chapter and the next chapter, deep learning methods can generate text segments that are often grammatically and semantically correct. How realistic are these generated text snippets? The evaluation of language models is based on statistically characterizing the likelihood of the generation of a piece of text with respect to a base collection. Note that a language model tries to maximize the probability $P(w_i|w_1 \ldots w_{i-1})$ that a word w_i is generated in a sentence, given all the words $w_1 \ldots w_{i-1}$ preceding it. The perplexity $PL(s)$ of a sentence with t words $s = w_1 \ldots w_t$ is defined by the geometric mean of the inverse probability of generation of each of the words in it:

$$PL(s) = \left(\prod_{i=1}^{t} \frac{1}{P(w_i|w_1 \ldots w_{i-1})} \right)^{(1/t)}$$

Note that the logarithm of this value is the cross-entropy per word in the sentence. Both the perplexity and the cross-entropy are low when the language model is of high quality. The best-case value of the perplexity is 1. The worst-case value of the perplexity is the vocabulary size, when one simply places a random token at each position with equal probability. The perplexity is, therefore, highly sensitive to the vocabulary size, with larger vocabulary sizes causing greater perplexity. One way of understanding perplexity is that subtracting 1 from it (very roughly) yields the number of times that one would obtain a different token at a randomly sampled position by generating tokens from the language model. Clearly, a high-quality language model will tend to give similar tokens each time (and therefore lower perplexity). One can also write the perplexity of a sentence of length t in terms of probability $p(s)$ of generating the sentence as follows:

$$PL(s) = \left(\frac{1}{p(s)} \right)^{(1/t)}$$

Now consider a corpus C with m sentences $s_1 \ldots s_m$, so that the sum of the number of words over all sentences is given by n. Then, the perplexity of the corpus is defined as follows:

$$PL(C) = \left(\prod_{i=1}^{t} \frac{1}{P(s_i)} \right)^{(1/n)}$$

The perplexity of a corpus is often used as a surrogate for evaluating the quality of fit of the language model that was constructed by training on it.

10.3 Kernel Methods for Sequence-Centric Learning

Kernel methods typically extract an embedding at the sentence or the document level. Consider a data set containing N different sentences/documents, and similarity between the ith and jth sentences is given by s_{ij}. Therefore, an $N \times N$ matrix $S = [s_{ij}]$ is computed. In cases where the similarities between documents need to be computed, the pairwise similarities between sentences can be aggregated into pairwise similarities between documents. For example, the 1-nearest neighbor similarities of each sentence in one document to the sentences in the other document are computed. These values are first averaged over the sentences in each document, and then averaged between the two documents.

Then, the matrix S can be symmetrically factorized using the $N \times K$ matrix U as follows:

$$S \approx UU^T \tag{10.7}$$

Here, the N rows of U contain the K-dimensional representations of the various data instances. Note that the matrix U is not unique as one can rotate the axis system with the orthogonal transformation $U' = UP$ with a $K \times K$ matrix P with orthogonal columns, such that the product of the symmetric factors is not changed. In kernel methods, the convention is to choose an embedding matrix U whose columns are mutually orthogonal. This is achieved by diagonalizing the positive semi-definite and symmetric matrix S into an $N \times k$ matrix Q with orthonormal columns and a $k \times k$ diagonal matrix Σ as follows:

$$S \approx Q\Sigma^2 Q^T = \underbrace{(Q\Sigma)}_{U} \underbrace{(Q\Sigma)^T}_{U^T} \tag{10.8}$$

The sequential information is incorporated within the process of similarity computation. Since many of these methods are discussed in Chapters 3 and 13, we do not revisit them here. We summarize pointers to these methods below:

1. *Kernels based directly on n-grams and skip-grams:* These kernels are discussed in detail in Section 3.6.1.2 of Chapter 3. The simplest possible approach is to enrich the vector space representation with n-grams and skip-grams and then compute the vector-space similarity between document pairs. The symmetric factorization $S = UU^T$ of the resulting similarity matrix yields the document embedding U.

2. *String subsequence kernels:* String subsequence kernels are discussed in Section 3.6.1.3 of Chapter 3. Like skip-grams, these kernels can allow the skipping of words by using decay factors to weight a skip-gram. Furthermore, string subsequence kernels can be extended to allow features to be associated with the tokens in the string. Such kernels are very useful for incorporating linguistic knowledge in the embedded representation (cf. Section 13.3.3.2 of Chapter 13).

3. *Kernels based on language-specific grammars:* It is also possible to incorporate some amount of linguistic knowledge directly into the kernel by creating *constituency-based parse trees* of sentences, and then computing *convolution tree kernels* (cf. Section 13.3.3.3 of Chapter 13). Furthermore, this approach can be extended to handle features associated with tokens, which is useful for information extraction.

It is noteworthy that all these methods use *symmetric factorization*, and many feature engineering methods use matrix factorization in one form or the other.

10.4   Word-Context Matrix Factorization Models

Matrix factorization models compute a word-context matrix based on the counts of contextual elements for each target word. The context of a target word is generally defined in terms of windows of equal size on either side of the target word. For the purpose of this section, assume that the size of the windows on either side of the target is denoted by t. The window size varies between 2 and 10 depending on training data size. Larger training data sets enable larger window sizes.

10.4.1   Matrix Factorization with Counts

In the simplest case, we factorize the matrix of co-occurrence frequencies in order to create embeddings of *words*. This was approach used in <u>H</u>yperspace <u>A</u>nalogue to <u>L</u>anguage (HAL) model. HAL [358] was an early and relatively unoptimized model that was not followed up for several years. In this section, we describe HAL along with several recent learning, optimization, and postprocessing techniques that are general-purpose techniques for all types of embeddings.

For each word i, let c_{ij} represent the number of times that the word j occurs in the context (i.e., separated from it by a distance of at most t in a sentence) of word i on either side. For example, consider the sentence:

Adam and Eve lived in Eden and frequently ate apples.

Consider the case in which we use a window of length 2 on either side of the word. Then, the context of the word "*Eden*" is any of the four words from {*lived, in, and, frequently*}. Therefore, the full context window has size $m = 2 \cdot t$. Then, for any given word, we can extract the number of times each context occurs for that word.

Then, the $d \times d$ contextual co-occurrence matrix is denoted by matrix $C = [c_{ij}]$. Note that each row of C can *directly* be used as a new high-dimensional feature representation of the word. However, it is possible to create a more compact feature representation by dimensionality reduction of C.

The matrix C can be factorized into rank-p matrices U and V (of size $d \times p$) as follows:

$$C \approx UV^T \qquad (10.9)$$

The rows of the matrix U yield the embedding that we are interested in. A popular choice is the singular value decomposition (SVD) method of Section 3.2. However, most of the entries in C are 0s, and these entries are not as informative about context as the non-zero entries. For example, for a context window of $m = 10$, a word that occurs only 20 times in the corpus will have at most 200 non-zero entries out of a vocabulary of possible size 10^6. As a result the factorization is dominated by the zero terms. Direct use of stochastic gradient descent has the advantage that one can indirectly change the objective function of factorization to de-emphasize the zero entries by sampling zero entries at a lower rate. Such an approach results in *weighted* matrix factorization. Furthermore, the computational complexity of this approach is dependent only on the number of non-zero entries.

One can write the relationship of Equation 10.9 in element-wise form by expressing $C = [c_{ij}]$ in terms of the dot product of the ith row $\overline{u}_i$ of matrix U and the jth row $\overline{v}_j$ of matrix V.

$$c_{ij} \approx \hat{c}_{ij} = \overline{u}_i \cdot \overline{v}_j \tag{10.10}$$

Note the circumflex on top of $\hat{c}_{ij}$ differentiating this *predicted* value from the observed value of c_{ij}. The rows of the $d \times p$ matrix U provide the p-dimensional embeddings of the individual words. The rows of the matrix V provide the embeddings of the contextual elements, which (in this case) are words as well. One can further improve this matrix factorization by using *biased* matrix factorization with row biases b_i^r and column biases b_j^c as follows:

$$\hat{c}_{ij} = b_i^r + b_j^c + \overline{u}_i \cdot \overline{v}_j \tag{10.11}$$

These bias variables also need to be learned using the optimization model of the factorization.

The straightforward optimization model of matrix factorization is as follows:

$$\text{Minimize } J = \sum_i \sum_j (c_{ij} - b_i^r - b_j^c - \overline{u}_i \cdot \overline{v}_j)^2 + \lambda \cdot \underbrace{\left(\sum_i ||\overline{u}_i||^2 + \sum_i (b_i^r)^2 + \sum_j ||\overline{v}_j||^2 + \sum_j (b_j^c)^2 \right)}_{\text{Regularizer}}$$

The regularizer can be dropped when the amount of data is large.

Stochastic gradient descent initializes U and V randomly. In each iteration, it updates U and V based on the error $e_{ij} = c_{ij} - \hat{c}_{ij}$ of a randomly selected entry (i, j) in C according to the following updates:

$$\overline{u}_i \Leftarrow \overline{u}_i (1 - \alpha\lambda) + \alpha e_{ij} \overline{v}_j$$
$$\overline{v}_j \Leftarrow \overline{v}_j (1 - \alpha\lambda) + \alpha e_{ij} \overline{u}_i$$
$$b_i^r \leftarrow b_i^r (1 - \alpha\lambda) + \alpha e_{ij}$$
$$b_j^c \leftarrow b_j^c (1 - \alpha\lambda) + \alpha e_{ij}$$

Here, $\alpha > 0$ is the learning rate. One can cycle through all the entries of C in random order, and make these updates. Such cycles are repeated until convergence is reached. It is possible to improve the factorization by using the notion of *negative sampling* in which a single cycle of stochastic gradient descent samples through all the non-zero entries of C but only a random sample of zero entries. The random sample may be different in each cycle, but its size is always k times the number of non-zero entries. The value of $k > 1$ is a user-driven parameter, and it implicitly controls the weight of positive and negative samples in the factorization. In such a case, each cycle of stochastic gradient descent no longer requires time proportional to $O(d^2)$, but it is linearly proportional to the number of non-zero entries in the matrix. It is noteworthy that this type of negative sampling is ubiquitous in contextual embeddings. There are other structured ways of negative sampling in which each positive sample (i, j) is matched with k negative samples $(i, j_1) \ldots (i, j_k)$. The indices $j_1 \ldots j_k$ are sampled with probability proportional to their frequencies in the underlying corpus. One can also use damped values of c_{ij} (e.g., populate C with $\sqrt{c_{ij}}$ to reduce the impact of frequent words.

Postprocessing Issues

Several postprocessing issues are common to all types of word embedding methods discussed in this chapter (including neural methods). For example, why does one use the rows of U as embeddings? Why not the rows of V? Also how should one normalize the rows of U and V? This section will provide an integrated discussion of these issues.

It is common to use the same context vocabulary as the lexicon. In such cases, one can either *concatenate* or *add* the word embedding $\overline{u}_i$ and the contextual embedding $\overline{v}_i$ to create the embedding of word i. This first of these ideas was proposed in the original HAL paper, and the second was proposed by the GloVe approach (see next section). However, other definitions of contexts are possible in which the number of contexts d' may not be the same as d. In such cases, a matrix C of size $d \times d'$ may be factorized (see Section 10.4.5), and one must follow the conventional practice of working only with $\overline{u}_i$. In general, adding $\overline{u}_i$ and $\overline{v}_i$ could either help or hurt the embedding. It is worthwhile to try.

The factorization $C \approx UV^T$ is not unique. For example, multiplying the first column of U by 2 and dividing the first column of V by 2 will not change UV^T. How should U and V be normalized to create a unique embedding. There are several ways to perform the normalization. If the rows of U and V are added, then both matrices must be normalized in the same way. One can either normalize the rows of U to unit norm, or the columns to unit norm. Furthermore, the row and column normalization can be applied in succession. If SVD is used for factorizing $C = Q\Sigma P^T$, then a viable alternative is to use $U = Q\sqrt{\Sigma}$ and $V = P\sqrt{\Sigma}$, respectively [324]. This trick can, in fact, be used for any type of factorization by converting it into a standardized three-way factorization like SVD (cf. Section 3.1.1).

10.4.2 The GloVe Embedding

One problem with the factorization approach is that the wide variations of the frequencies of the different words can cause the embedding to be dominated by the effect of frequent words. The problem is particularly severe because the raw co-occurrence counts can span over eight or nine orders of magnitude. Therefore, the GloVe method (<u>Glo</u>bal <u>Ve</u>ctors for Word Representation) [436] makes two modifications to the basic matrix factorization approach, which are as follows:

1. The (i, j)th entry of matrix C is defined as $\log(1+c_{ij})$ rather than as c_{ij}. As discussed in Section 2.4 of Chapter 2, this type of frequency *damping function* is used in all types of text mining applications to reduce the impact of frequent words.

2. The term in the optimization objective function of matrix factorization (see Section 10.4.1), corresponding to the error of the (i, j)th entry, is weighted as a function of c_{ij} with the use of a maximum threshold M and parameter α as follows:

$$\text{Weight}(i, j) = \min\left\{1, \frac{c_{ij}}{M}\right\}^{\alpha} \tag{10.12}$$

The values of M and α are recommended to be 100 and 3/4, respectively, based on empirical considerations.

One can now write the modified objective function as follows:

$$\text{Minimize } J = \sum_i \sum_j \text{Weight}(i,j) \cdot [\log(1 + c_{ij}) - b_i^r - b_j^c - \overline{u}_i \cdot \overline{v}_j]^2$$

$$= \sum_i \sum_j \min\left\{1, \frac{c_{ij}}{M}\right\}^\alpha \cdot [\log(1 + c_{ij}) - b_i^r - b_j^c - \overline{u}_i \cdot \overline{v}_j]^2$$

The sum of $\overline{u}_i$ and $\overline{v}_i$ is used as the embedding of word i.

The original GloVe paper does not use any regularization term. The optimization model can be solved using either stochastic gradient-descent or coordinate-descent. The stochastic gradient descent steps of GloVe are similar to those in the previous section, except that each entry is sampled with probability proportional to Weight(i,j), and that log-normalized frequencies are used. It is noteworthy that zero entries are not sampled at all because the value of Weight(i,j) is 0 for such entries. In other words, *zero values of c_{ij} are disregarded* by GloVe. Therefore, the complexity of stochastic gradient-descent only depends on the number of non-zero entries. The implications of dropping the zero entries are potentially intriguing. For example, if we have a data set in which each non-zero c_{ij} does not vary much (but there are many zero entries), the approach discussed in the previous section can still discover reasonable embeddings because of the contrast between zero and non-zero entries. However, GloVe could discover trivial matrices U and V in which all entries are the same (see Exercise 8). Of course, in real settings, such pathological situations do not occur. GloVe is known to work quite well in practice.

10.4.3   PPMI Matrix Factorization

Glove uses logarithmic normalization for damping very frequent words. Damping can also be accomplished by using various types of correlation measures to construct the matrix, such as positive pointwise mutual information (PPMI) [77]. For each word, we extract the contexts of a particular length on either side that surround the word. The length of the window size on either side is denoted by $t = m/2$ for some even number m.

Let c_{ij} be the number of times that a word j appears in the context of target word i. Let $f_i = \sum_j c_{ij}$, and $q_j = \sum_i c_{ij}$. Furthermore, we denote $N = \sum_{i,j} c_{ij}$. Then, the positive pointwise mutual information, PPMI, is defined as follows:

$$PPMI(i,j) = \max\left\{\log\left(\frac{N \cdot c_{ij}}{f_i \cdot q_j}\right), 0\right\} \tag{10.13}$$

These PPMI values are stored in a $d \times d$ matrix M. The matrix M can be factorized into rank-p matrices U and V, respectively:

$$M \approx UV^T \tag{10.14}$$

The matrices U and V are both of size $d \times p$. The rows of U will contain the p-dimensional embeddings of the various words, whereas the rows of V contain the embeddings of contexts. The optimization approach is similar to that discussed for matrix factorization in Section 10.4.1. An alternative is to factorize the matrix $M \approx Q\Sigma P^T$ using (truncated) singular value decomposition in p dimensions. The matrices U and V are set to $Q\sqrt{\Sigma}$ and $P\sqrt{\Sigma}$, respectively [324].

10.4.4 Shifted PPMI Matrix Factorization

A shifted PPMI matrix (SPPMI) uses a parameter k to shift the matrix and sparsify it:

$$SPPMI(i,j) = \max \left\{ \log \left(\frac{N \cdot c_{ij}}{f_i \cdot q_j} \right) - \log(k), 0 \right\} \tag{10.15}$$

Using $k = 1$ is equivalent to PPMI. Better results are obtained for values of $k > 1$, and the results can be shown to be closely related to one of the popular negative sampling-based variants of *word2vec* [323]. Let $M^{(s)}$ be the $d \times d$ matrix, whose entries correspond to the SPPMI values computed above. As in the case of PPMI matrix factorization, the singular value decomposition of $M^{(s)}$ can be used to extract the word embeddings. An alternative is to simply use the rows of the SPPMI matrix *directly* without any factorization. Since the SPPMI matrix is sparse, it is possible to work efficiently with these types of vectors.

10.4.5 Incorporating Syntactic and Other Features

A key point in the design of count-based methods is that it is relatively easy to incorporate syntactic and other features in the factorization process. For example, the case of word-word embedding, the matrix C is a word-word co-occurrence matrix. Strictly speaking, the columns need not correspond to words, but they may correspond to any type of contextual elements. For example, one might have counts of various parts-of-speech, entity types and so on within the context window. In such a case, the matrix C might not even be a square matrix. Alternatively, the context words might be tagged with their parts of speech. For example, the word "*bear*" might have the variants "*bear*-V" and "*bear*-N" corresponding to the verb and noun form, respectively. Clearly, building such matrices requires a certain amount of natural language processing. The choice of features is limited only by the type of linguistic preprocessing one is willing to perform in these systems.

10.5 Graphical Representations of Word Distances

Feature engineering generally refers to the creation of multidimensional embeddings, because they can be paired with most learning algorithms. However, in recent years, many algorithms have been designed to work with more powerful data types like graphs. Therefore, it also makes sense to create graph representations. Furthermore, such methods also provide an indirect route to the creation of multidimensional embeddings by the factorization of the adjacency matrices of such graphical representations. The resulting representations are similar to those created by factorizations of word-context matrices. Graphical representations have some advantages over using context-based windows because they encode more information about proximity and ordering of words. The following discussion will present the *distance graph* [16].

The distance graph representation captures the sequential relationships between words, and therefore it makes sense to create distance graphs from individual sentences. In other words, each distance graph corresponds to an individual sentence. Each distinct word in the sentence corresponds to a node, and a directed edge occurs between nodes when the corresponding words occur in sufficient proximity in a sentence. This proximity level is defined by the *order* $k \geq 1$ of the distance graph. One can view an order-k distance graph as a graphical representation of the skip-gram model in which each edge represents a $(k-1)$-skip-2-gram. A value of $k = 1$ corresponds to a bigram. The value of k is analogous to the context window used in the factorization and context-based neural network models of this chapter. Furthermore, the weight of the edge is defined by the number of times[1] that the r-skip-2-gram occurs in the sentence for $r = k - 1$. It is noteworthy that the weight from edge i to edge j is not the same as that between edge j to edge i. Therefore, the distance graph captures a greater amount of precedence information between words than most of the factorization models.

For more refined analysis, it is also possible to weight the edges based on the number of skips between words. For example, consider a pair of nodes i and j, in which word i precedes word j for a total of t times. Furthermore, assume that the number of skips in each of these cases is $r_1, r_2, \ldots r_t$, where each $r_i \leq k - 1$. Then, the weight c_{ij} of the edge (i, j) is given by the following sum of decay weights for a decay parameter $\lambda \in (0, 1)$:

$$c_{ij} = \sum_{s=1}^{t} \lambda^{r_s + 1} \tag{10.16}$$

Note that choosing $\lambda = 1$ results in a non-decayed version of the distance graph. Adding this type of decayed weight also results in a more refined representation of context in which the distances between the individual words matter to a greater degree.

Although the distance graph is naturally created at the sentence level, it can be converted into a document-level distance graph by aggregating all the sentence-level distance graphs. The sentence-level distance graphs are aggregated by defining a new distance graph over the union of all the nodes in the individual (sentence-level) distance graphs, and adding the weights of the corresponding edges. It is also possible to create a *corpus*-level distance graph, factorizing which yields similar results to many models of this chapter.

As in all other text representation methods, frequent words and stop words have a confounding effect on the accuracy of the representation. Therefore, one solution that is proposed in [16] is to remove the stop-words before creating the distance graph. However, it is also possible to down-sample the frequent words to create the distance graph. Four examples of distance graphs created from a well known nursery rhyme are shown in Figure 10.3. The self-loops in this distance graph are slightly different from those presented in [16] because of a slightly different criterion for their inclusion. The distance graph can be leveraged in two distinct ways:

1. It is possible to apply graph mining algorithms directly on the distance graph representations of individual sentences. Graph mining algorithms are much better developed in the literature as compared to sequence mining algorithms [2], and off-the-shelf options are often available [16].

2. The distance graphs of different sentences can be aggregated into a single distance graph. It is possible to factorize the (aggregated) distance graphs to create similar embeddings to word-context factorization methods (see Exercise 6).

It is possible to create undirected variants of distance graphs in the event that the precedence information is not important [16].

---

[1]The original definition of distance-graph [16] differs from the skip-gram definition here in a very minor way. The original definition always assumes that a word is connected to itself. Such a definition also allows a distance graph of order $k = 0$, which corresponds to a traditional bag-of-words with only self-loops. For example, for the sentence *"Adam ate an apple"* there would always be a self-loop at each of the four nodes even though there are no adjacent repetitions of words. The slightly modified definition here would include a self-loop only when words actually occur in their own context. For many traditional applications, however, this distinction does not seem to affect the results.

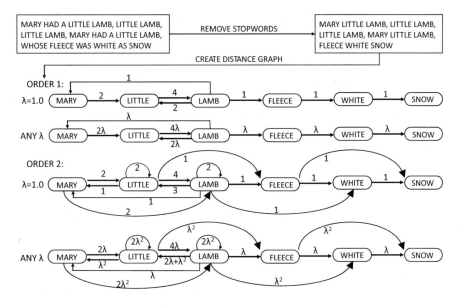

Figure 10.3: Distance graphs of orders $k = \{1, 2\}$ for different λ

10.6 Neural Networks and Word Embeddings

Neural embeddings like *word2vec* [389, 390] create embeddings from a context window like GLoVe and other matrix factorization methods. First, we will start with a brief introduction to neural networks.

10.6.1 Neural Networks: A Gentle Introduction

Neural networks are a model of simulation of the human nervous system. The human nervous system is composed of cells, referred to as neurons. Biological neurons are connected to one another at contact points, which are referred to as synapses. Learning is performed in living organisms by changing the strength of synaptic connections between neurons. Typically, the strength of these connections change in response to external stimuli. Neural networks can be considered a simulation of this biological process.

As in the case of biological networks, the individual nodes in artificial neural networks are referred to as *neurons*. These neurons are units of computation that receive input from other neurons, make computations on these inputs, and feed them into yet other neurons. The computation at a neuron is affected by the weights on the input connections to that neuron because the input to the neuron is scaled by the weight. This weight can be viewed as analogous to the strength of a synaptic connection. By changing these weights appropriately, the overall computation function of the artificial neural network can be learned, which is analogous to the learning of the synaptic strength in biological neural networks. The "external stimulus" in artificial neural networks for learning these weights is provided by the training data. The idea is to incrementally modify the weights whenever incorrect predictions are made by the current set of weights.

A wide variety of architectures exist, starting from a simple *perceptron* to complex multilayer networks. The use of a large number of layers is referred to as *deep learning*. In particular, the recurrent neural networks in this chapter can be considered deep models.

10.6.1.1   Single Computational Layer: The Perceptron

The most basic architecture of a neural network is referred to as the perceptron. An example of the perceptron architecture is illustrated in Figure 10.5(a). The perceptron contains two layers of nodes, which correspond to the input nodes and a single output node. The number of input nodes is exactly equal to the dimensionality d of the underlying data. Each input node receives and transmits a single numerical attribute to the output node. Therefore, the input nodes only *transmit* input values and do not perform any *computation* on these values. In the basic perceptron model, the output node is the only node that performs a mathematical function on its inputs. The individual features in the training data are assumed to be numerical. Categorical attributes are handled by creating a separate binary input for each value of the categorical attribute, and only one of these inputs take on the value of 1 and the other inputs take on the value of 0. This is logically equivalent to binarizing the categorical attribute into multiple attributes. This type of encoding is referred to as *one-hot encoding* and is useful in the text setting where there are multiple inputs corresponding to different words. In the binary classification problem, there is a single output node with two possible values drawn from $\{-1, +1\}$.

Consider the simplest possible setting with numerical inputs and a single binary output. In this case, each input node is connected with a weighted connection to the output node. These weights define a function from the values transmitted by the input nodes to a binary value drawn from $\{-1, +1\}$. This value can be interpreted as the perceptron's prediction of the class variable of the data instance fed to the input nodes. Just as learning is performed in biological systems by modifying synaptic strengths, the learning in a perceptron is performed by modifying the weights of the links connecting the input nodes to the output node whenever the predicted label does not match the true label.

The function learned by the perceptron is referred to as the *activation function*, which is a signed linear function. This function is very similar to that learned in support vector machines for mapping training instances to binary class labels. Let $\overline{W} = (w_1 \ldots w_d)$ be the weights for the connections of d different inputs to the output neuron for a data record of dimensionality d. In addition, the $(d+1)$th input to the neuron has a constant value of 1, whose coefficient is the bias b. The output $\hat{y} \in \{-1, +1\}$ for the feature set $(x_1 \ldots x_d)$ of the data point $\overline{X}$, is as follows:

$$\hat{y} = \text{sign}\{\sum_{j=1}^{d} w_j x_j + b\} \tag{10.17}$$

The value $\hat{y}$ (with the circumflex on top) represents the *prediction* of the perceptron for the class variable of $\overline{X}$. The sign function is used in order to transform the output to a form that is suitable for binary classification. The sign function can be dropped if the target to be learned is real valued.

It is noteworthy that the use of a dummy input with a constant value of 1 allows us to drop an explicit use of the bias from the above function, because the bias can be incorporated within the coefficient vector $\overline{W}$:

$$\hat{y} = \text{sign}\{\sum_{j=1}^{d} w_j x_j\} \tag{10.18}$$

This type of feature engineering trick is similar to what is used in the case of linear models (see Section 6.1.2 of Chapter 6). An equivalent approach is to create a *bias neuron* that always emits an output value of 1. This neuron is not connected to any of the nodes in its previous layer, but it has an input into each of the nodes of the layer where a bias is required. The coefficient of each such connection provides the node-specific bias.

It is desired to learn the weights, so that the value of $\hat{y}$ is equal to the true value y of the class variable for as many training instances as possible. The goal in neural network algorithms is to learn the vector of weights $\overline{W} = (w_1 \ldots w_d)$, so that $\hat{y}$ approximates the true class variable y as closely as possible. Therefore, one implicitly tries to minimize the sum of squared errors $(\hat{y} - y)^2$ over various training instances. This type of objective function is, however, well suited to numerical data, and binary outputs can be viewed as special cases of numerical data. In cases where the ground-truth is a multi-way categorical value (e.g., identity of word in text data), multiple output nodes are used with a similar form of probabilistic prediction as multinomial logistic regression (cf. Section 6.4.4). The objective function therefore takes the form of log-likelihood maximization. This type of output is referred to as *softmax*, and we will discuss a specific example in the context of the *word2vec* model. The key point to understand is that neural networks enable a variety of architectures to handle different types of learning problems and data types.

The basic perceptron algorithm starts with a random vector of weights. The algorithm then feeds each input data instance $\overline{X}$ into the neural network one by one to create the prediction $\hat{y}$. The weights are then updated, based on the error value $E(\overline{X}) = (y - \hat{y})$. Specifically, when the data point $\overline{X}$ is fed into the network, the weight vector $\overline{W}$ is updated as follows:

$$\overline{W} \Leftarrow \overline{W} + \alpha(y - \hat{y})\overline{X} \tag{10.19}$$

The parameter α regulates the learning rate of the neural network. The perceptron algorithm repeatedly cycles through all the training examples in random order and iteratively adjusts the weights until convergence is reached. Note that a single training data point may be cycled through many times. Each such cycle is referred to as an *epoch*. One can also write the gradient-descent update in terms of the error $E(\overline{X}) = (y - \hat{y})$ as follows:

$$\overline{W} \Leftarrow \overline{W} + \alpha E(\overline{X})\overline{X} \tag{10.20}$$

The perceptron implicitly uses a loss function, which is not quite the same as that used in least-squares classification (cf. Equation 6.21). Let us examine the incremental term $(y - \hat{y})\overline{X}$ in the update of Equation 10.19, without the multiplicative factor α. The computed error $(y - \hat{y})$ is always an integer value in the perceptron but not in least-squares classification. However, the two methods use their respective versions of the error in an identical way. In other words, the perceptron algorithm is unique with respect to other related least-squares methods.

The perceptron updates are performed on a tuple-by-tuple basis, rather than globally, over the entire data set, as one would expect in a global least-squares optimization. The basic perceptron algorithm can be considered a *stochastic gradient-descent method*, which implicitly minimizes the squared error of prediction by performing *local* gradient-descent updates with respect to randomly chosen training points. The assumption is that the neural network cycles through the points in random order during training with the goal of reducing prediction error. It is easy to see that non-zero updates are made to the weights only when $y \neq \hat{y}$, and errors are made in categorization. In *mini-batch* stochastic gradient descent, the aforementioned updates of Equation 10.20 are implemented over a randomly chosen subset S of training points:

$$\overline{W} \Leftarrow \overline{W} + \alpha \sum_{\overline{X} \in S} E(\overline{X})\overline{X} \tag{10.21}$$

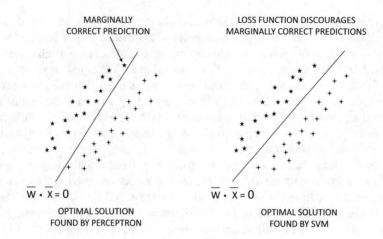

Figure 10.4: Comparing the SVM and the perceptron

As in all linear models of Chapter 6, it is possible to penalize the weights in order to prevent overfitting. The most common form of penalization is L_2-regularization, in which case the updates are as follows:

$$\overline{W} \Leftarrow \overline{W}(1 - \alpha\lambda) + \alpha \sum_{\overline{X} \in S} E(\overline{X})\overline{X} \qquad (10.22)$$

Here, $\lambda > 0$ is the regularization parameter. One can view this type of penalization as a kind of weight decay during the updates. Note that this type of penalization is applied in all types of neural networks, and not just the perceptron. Regularization is particularly important when the amount of available data is limited.

Relationship to Support Vector Machines

The perceptron turns out to very closely related to the support vector machine (SVM). This similarity becomes particularly evident, when the perceptron updates are rewritten as follows:

$$\overline{W} \Leftarrow \overline{W}(1 - \alpha\lambda) + \alpha \sum_{(\overline{X},y) \in S^+} y\overline{X} \qquad (10.23)$$

Here, S^+ is defined as the set of $\overline{X} \in S$ satisfying $y(\overline{W} \cdot \overline{X}) < 0$. Although the above update looks different from the error-based update of the perceptron, the (integer) error value $E(\overline{X}) = (y - \text{sign}\{\overline{W} \cdot \overline{X}\}) \in \{-2, +2\}$ is equal to $2y$ *for misclassified points*, and one can absorb the factor of 2 within the learning rate. The resulting perceptron update (on replacing $E(\overline{X})$ with y) is similar to that used by the primal SVM algorithm, except that the updates are performed only for the misclassified points, and do not include the marginally correct predictions near the decision boundary. Note that the SVM uses the condition $y(\overline{W} \cdot \overline{X}) < 1$ to define S^+, which causes the marginal predictions to be included as well. This is the *only* difference between the perceptron and the primal SVM algorithm. Refer to Section 6.3.3 of Chapter 6 for a discussion of the primal SVM algorithm.

The perceptron implicitly optimizes the *perceptron criterion* [58] defined by the following:

$$L = \max\{-y(\overline{W} \cdot \overline{X}), 0\} \qquad (10.24)$$

The reader is encouraged to verify that the gradient of this smoothed objective function leads to the perceptron update. A remarkable observation is that the perceptron criterion is a shifted version of the hinge-loss used in the SVM:

$$L_{svm} = \max\{1 - y(\overline{W} \cdot \overline{X}), 0\} \tag{10.25}$$

This shifting by one unit also explains the minor difference in the condition under which the training point $(\overline{X}, y)$ contributes to the update. Illustrative examples of optimal solutions found by the perceptron and the SVM are shown in Figure 10.4. The SVM discourages marginally correct predictions, as a result of which it will tend to generalize better to unseen test data near the decision boundary.

Choice of Activation Function

In the case of the perceptron, the sign function is used as the activation function. This particular type of activation is useful for creating discrete outputs, although it may not be well suited to every type of output. For example, if it is desired to predict the probability of the positive class, one might choose to use a sigmoid function. In such cases, the optimization of the squared error is no longer appropriate (as in the perceptron), and it makes sense to maximize the likelihood of the observed data. Therefore, the choice of the activation and loss functions may also vary with the desired goal of the learning algorithm. Many of these loss functions are similar to those used in the linear and nonlinear models of Chapter 6. The reason is that these models are highly dependent on optimization methods like gradient-descent, which are easy to generalize to neural network architectures. Mastery of these models is highly recommended for readers interested in neural networks and deep learning.

Common choices of activation functions are the sign, sigmoid, or hyperbolic tangent functions. We use the notation Φ in order to denote the activation function:

$$\hat{y} = \Phi(\overline{W} \cdot \overline{X}) \tag{10.26}$$

Some examples of the function $\Phi(\cdot)$ are as follows:

$\Phi(v) = v$ (identity function) $\qquad \Phi(v) = \text{sign}(v)$ (sign function)
$\Phi(v) = \frac{1}{1+e^{-v}}$ (sigmoid function) $\qquad \Phi(v) = \max\{v, 0\}$ (ReLU)
$\Phi(v) = \frac{e^{2v}-1}{e^{2v}+1}$ (tanh function) $\qquad \Phi(v) = \max\{\min[v, 1], -1\}$ (hard tanh)

The specific choice of the activation function is determined by the analyst depending on her experience and insight about different types of problems. For example, the sigmoid function always maps the output in $(0, 1)$, which is interpreted as a probability value. The tanh function is similar to the sigmoid function, except that it maps the output in the range $(-1, +1)$ rather than $(0, 1)$. It is desirable when the output needs to be mapped to both positive and negative values. The Rectified Linear Unit (ReLU) and hard tanh functions are piecewise linear approximations of the sigmoid and tanh functions, respectively.

As we will see later, such nonlinear activation functions are also very useful in multilayer networks, because they help in creating more powerful compositions of different types of functions. Many of these functions are referred to as *squashing* functions, as they map the outputs from an arbitrary range to bounded outputs.

Choice of Output Nodes

In many basic variations of the perceptron architecture, it is possible to use multiple output nodes when there are multiple classes. Multiple output nodes are useful in other types of architectures where one is trying to reconstruct more than one attribute of the data from a subset of other attributes. A classical example of such a setting is an *autoencoder* that tries to reconstruct all the features with a multi-layer architecture. Multiple output nodes are also of interest to us, because the output may be a set of t contextual words that are predicted from a single word. The discrete nature of the target word(s) also requires us to choose a form of the output that is different from the sign activation function. In the case where the output is a target word with d possible values, the output is typically computed using a *softmax* layer with d outputs, and each output is a probability value indicating the likelihood that the particular word is selected. Therefore, if $\overline{v} = [v_1, \ldots v_d]$ be the d output values using a linear model for each word, then the ith softmax output is computed as follows:

$$\Phi(\overline{v})_i = \frac{\exp(v_i)}{\sum_{j=1}^{d} \exp(v_j)} \quad \forall i \in \{1 \ldots d\} \tag{10.27}$$

One can view a softmax output as a form of multinomial logistic prediction (cf. Section 6.4.4 of Chapter 6), and it is particularly useful for discrete variable prediction in the multi-way setting. In such cases, the loss function of the neural network is computed using the negative log loss over the output probabilities rather than the squared error used in the perceptron.

Choice of Loss Function

It is evident from the aforementioned discussion that the choice of activation function, the nature of the output nodes, and the goal of the specific application have a role to play in deciding the loss function. For example, least-squares regression with numeric outputs requires a simple squared loss of the form $(y - \hat{y})^2$ for a single training instance with real-valued target y and prediction $\hat{y}$ (with identity activation). One can also use other types of loss like *hinge loss*, which is defined for $y \in \{-1, +1\}$ and real-valued prediction $\hat{y}$ (with identity activation):

$$L = \max\{0, 1 - y \cdot \hat{y}\} \tag{10.28}$$

The reader should satisfy herself that this type of loss function can be used to implicitly define a support vector machine with single-layer neural networks (cf. Section 6.3 of Chapter 6). The resulting algorithm is different from the perceptron in only a very small way (see discussion surrounding Equation 10.23).

For multiway predictions (like predicting word identifiers or one of multiple classes), the softmax output is particularly useful. However, a softmax output is probabilistic, and therefore it requires a different type of loss function. In fact, for probabilistic predictions, two different types of loss functions are used, depending on whether the prediction is binary or whether it is multiway:

1. **Binary targets (logistic regression):** In this case, it is assumed that the observed value y is drawn from $\{-1, +1\}$, and the prediction $\hat{y}$ is a an arbitrary numerical value on using the identity activation function. In such a case, the loss function for a single instance with observed value y and prediction $\hat{y}$ is defined as follows:

$$L = \log(1 + \exp(-y \cdot \hat{y})) \tag{10.29}$$

The reader should satisfy herself that this loss function is identical to that used in logistic regression (cf. Section 6.4 of Chapter 6). Alternatively, one can use a sigmoid activation function to output $\hat{y} \in (0, 1)$, which indicates the probability that the observed value y is 1. Then, the negative logarithm of $|y/2 - 0.5 + \hat{y}|$ provides the loss,

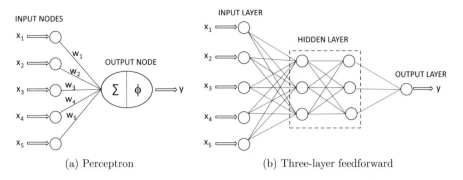

(a) Perceptron · (b) Three-layer feedforward

Figure 10.5: Single and multilayer neural networks

assuming that y is coded from $\{-1, 1\}$. This is because $|y/2 - 0.5 + \hat{y}|$ indicates the probability that the prediction is correct, which creates a negative log-loss.

2. **Categorical targets:** In this case, if $\hat{y}_1 \ldots \hat{y}_k$ be the probabilities of the k classes (using the softmax activation), and the rth class is the ground-truth class, then the loss function for a single instance is defined as follows:

$$L = -\log(\hat{y}_r) \tag{10.30}$$

This function is referred to as *cross-entropy loss*, and is the same as the one used in multinomial logistic regression (see Equation 6.35 of Section 6.4.4). Furthermore, this loss function is a direct generalization of the loss function discussed above for the binary case.

Predicting the probability of an output word (e.g., next word in language model) is a classical example of the use of such an approach, because one needs to predict a single discrete possibility (word) out of a lexicon of d terms. The *word2vec* method provides a specific example of the use of such an approach. The key point to remember is that the nature of the output nodes, the activation function, and the loss function that is optimized depend on the application at hand. Even though the perceptron is often presented as the quintessential representative of single-layer networks, it is only a single representative out of a very large universe of possibilities.

10.6.1.2 Multilayer Neural Networks

The perceptron model is the most basic form of a neural network, containing only a single input layer and an output layer. Because the input layers only transmit the attribute values without actually applying any mathematical function on the inputs, the function learned by the perceptron model is only a simple linear model based on a single output node. In practice, more complex models may need to be learned with multilayer neural networks.

Multilayer neural networks have a *hidden layer*, in addition to the input and output layers. The nodes in the hidden layer can, in principle, be connected with different types of topologies. For example, the hidden layer can itself consist of multiple layers, and nodes in one layer might feed into nodes of the next layer. This is referred to as the *multilayer feedforward network*. The nodes in one layer are also assumed to be fully connected to the nodes in the next layer. Therefore, the topology of the multilayer feedforward network is automatically determined, after the number of layers and the number/type of nodes in

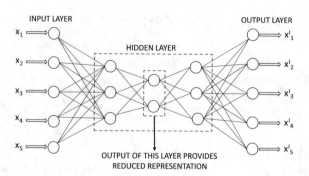

Figure 10.6: An example of an autoencoder with multiple outputs

each layer have been specified by the analyst, although the choice of the loss function is also critical. The basic perceptron may be viewed as a single-layer feedforward network. A popularly used model is one in which a multilayer feedforward network contains only a single hidden layer. Such a network may be considered a two-layer feedforward network. An example of a three-layer feedforward network is illustrated in Figure 10.5(b). Note that the number of layers refers to the number of *computational* layers, and does not include the input layer (which only transmits the data to the next layer).

The example of Figure 10.5(b) illustrates a multilayer network, which is relevant for the classification problem. However, it is possible to use different types of outputs, depending on the goal at hand. A classical example of this setting is the autoencoder, which recreates the outputs from the inputs. Therefore, the number of outputs is equal to the inputs, as shown in Figure 10.6. The constricted hidden layer in the middle outputs the reduced representation of each instance, just as principal component analysis or matrix factorization produces lower dimensional representations of data points. In fact, a shallow variant of this scheme can be shown to be mathematically equivalent to principal component analysis. Another common example of multiple outputs is a simulation of multinomial logistic regression. This approach is particularly useful in the context of language models, where one predicts the probabilities of various words with the use of a context window as the input. This chapter will discuss one such method.

What does a multi-layer network mean in terms of the functions computed at individual nodes? A path of length 2 in the neural network in which the function $f(\cdot)$ follows $g(\cdot)$ can be considered a composition function $f(g(\cdot))$. Furthermore, if $g_1(\cdot)$, $g_2(\cdot) \ldots g_k(\cdot)$ are the functions computed in layer m, and a particular layer-$(m+1)$ node computes $f(\cdot)$, then the composition function computed by the layer$(m+1)$ node in terms of the layer-m inputs is $f(g_1(\cdot), \ldots g_k(\cdot))$. The use of nonlinear activation functions increases the power of multiple layers. If all layers use an identity activation function, then a multilayer network simplifies to linear regression. It has been shown [258] that a network with a single hidden layer in which the outputs of multiple sigmoidal units (or any reasonable squashing function) are combined linearly can compute any function *in theory* (given sufficient data). As a result, neural networks are often referred to as *universal function approximators*. In practice, however, since the number of units required in the hidden layer may be very large (causing over-fitting), the practical usefulness of this paradigm has its own limitations.

It is helpful to view a neural network as a *computational graph*, which is constructed by piecing together many of the basic parametric models discussed in earlier chapters. Neural networks are fundamentally more powerful than their building blocks because the

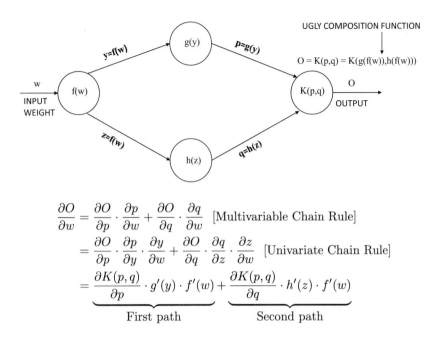

$$\frac{\partial O}{\partial w} = \frac{\partial O}{\partial p} \cdot \frac{\partial p}{\partial w} + \frac{\partial O}{\partial q} \cdot \frac{\partial q}{\partial w} \quad \text{[Multivariable Chain Rule]}$$

$$= \frac{\partial O}{\partial p} \cdot \frac{\partial p}{\partial y} \cdot \frac{\partial y}{\partial w} + \frac{\partial O}{\partial q} \cdot \frac{\partial q}{\partial z} \cdot \frac{\partial z}{\partial w} \quad \text{[Univariate Chain Rule]}$$

$$= \underbrace{\frac{\partial K(p,q)}{\partial p} \cdot g'(y) \cdot f'(w)}_{\text{First path}} + \underbrace{\frac{\partial K(p,q)}{\partial q} \cdot h'(z) \cdot f'(w)}_{\text{Second path}}$$

Figure 10.7: **Illustration of chain rule in computational graphs:** The products of node-specific partial derivatives along paths from weight w to output O are aggregated. The resulting value yields the derivative of output O with respect to weight w. Only two paths between input and output exist in this simplified example

parameters of these models are learned *jointly* to create a highly optimized composition function of these models. The common use of the term "perceptron" to refer to the basic unit of a neural network is somewhat inexact, because the multilayer network puts together models of different types (which frequently have nonlinear activations) in order to gain its power.

In the single-layer neural network, the training process is relatively straightforward because the error (or loss function) can be computed as a direct function of the weights, which allows easy gradient computation. In the case of multi-layer networks, the problem is that the loss is a complicated composition function of the weights in earlier layers. The gradient of a composition function is computed using the backpropagation algorithm. The backpropagation algorithm leverages the chain rule of differential calculus, which computes the error gradients in terms of summations of local-gradient products over the various paths from a node to the output. Although this summation has an exponential number of components (paths), one can compute it efficiently using *dynamic programming*. The backpropagation algorithm is a direct application of dynamic programming. It contains two main phases, referred to as the *forward* and *backward* phases, respectively. The forward phase is required to compute the output values and the local derivatives at various nodes, and the backward phase is required to accumulate the products of these local values over all paths from the node to the output:

1. *Forward phase:* In this phase, the inputs for a training instance are fed into the neural network. This results in a forward cascade of computations across the layers, using the current set of weights. The final predicted output can be compared to that of the training instance and the derivative of the loss function with respect to the output is

computed. The derivative of this loss now needs to be computed with respect to the weights in all layers in the backwards phase.

2. *Backward phase:* The main goal of the backward phase is to learn the gradient of the loss function with respect to the different weights by using the chain rule of differential calculus. These gradients are used to update the weights. Since these gradients are learned in the backward direction, starting from the output node, this learning process is referred to as the backward phase. Consider a sequence of hidden units $h_1, h_2, \ldots, h_k$ followed by output o, with respect to which the loss function L is computed. Furthermore, assume that the weight of the connection from hidden unit h_r to h_{r+1} is $w_{(h_r, h_{r+1})}$. Therefore, the input into the rth unit, which is contributed by the $(r-1)$th hidden unit, is $w_{(h_{r-1}, h_r)} \cdot h_{r-1}$. Then, in the case that a single path exists from h_1 to o, one can derive the gradient of the loss function with respect to any of these edge weights using the chain rule:

$$\frac{\partial L}{\partial w_{(h_{r-1}, h_r)}} = \frac{\partial L}{\partial o} \cdot \left[\frac{\partial o}{\partial h_k} \prod_{i=r}^{k-1} \frac{\partial h_{i+1}}{\partial h_i} \right] \frac{\partial h_r}{\partial w_{(h_{r-1}, h_r)}} \quad \forall r \in 1 \ldots k \qquad (10.31)$$

The aforementioned expression assumes that only a *single path* from h_1 to o exists in the network, whereas an exponential number of paths might exist in reality. A generalized variant of the chain rule, referred to as the *multivariable chain rule*, computes the gradient in a computational graph, where more than one path might exist. This is achieved by adding the composition along each of the paths from h_r to o. An example of the chain rule in a computational graph with two paths is shown in Figure 10.7. Therefore, one generalizes the above expression to the case where a set $\mathcal{P}$ of paths exist from h_r to o:

$$\frac{\partial L}{\partial w_{(h_{r-1}, h_r)}} = \frac{\partial L}{\partial o} \cdot \underbrace{\left[\sum_{[h_r, h_{r+1}, \ldots h_k, o] \in \mathcal{P}} \frac{\partial o}{\partial h_k} \prod_{i=r}^{k-1} \frac{\partial h_{i+1}}{\partial h_i} \right]}_{\text{Backpropagation computes } \Delta(h_r, o) = \frac{\partial L}{\partial h_r}} \frac{\partial h_r}{\partial w_{(h_{r-1}, h_r)}} \qquad (10.32)$$

The computation of $\frac{\partial h_r}{\partial w_{(h_{r-1}, h_r)}}$ on the right-hand side is straightforward, and is discussed later in Equation 10.35. However, the path-aggregated term above [annotated by $\Delta(h_r, o) = \frac{\partial L}{\partial h_r}$] is aggregated over an exponentially increasing number of paths (with respect to path length), which seems to be intractable at first sight. A key point is that the computational graph of a neural network does not have cycles, and it is possible to compute such an aggregation in a principled way in the backwards direction by first computing $\Delta(h_k, o)$ for nodes h_k closest to o, and then recursively computing these values for nodes in earlier layers in terms of nodes in later layers. This type of dynamic programming technique is used frequently to efficiently compute all types of path-centric functions in directed acyclic graphs, which would otherwise require an exponential number of operations. The recursion for $\Delta(h_r, o)$ can be expressed as follows:

$$\Delta(h_r, o) = \frac{\partial L}{\partial h_r} = \sum_{h:h_r \Rightarrow h} \frac{\partial L}{\partial h} \frac{\partial h}{\partial h_r} = \sum_{h:h_r \Rightarrow h} \frac{\partial h}{\partial h_r} \cdot \Delta(h, o) \qquad (10.33)$$

The value of $\Delta(o, o)$ is initialized to $\frac{\partial L}{\partial o}$ at the beginning of the backwards pass. Since each h is in a later layer than h_r, $\Delta(h, o)$ has already been computed while evaluating

$\Delta(h_r, o)$. However, we still need to evaluate $\frac{\partial h}{\partial h_r}$ in order to compute Equation 10.33. Consider a situation in which the edge joining h_r to h has weight $w_{(h_r, h)}$, and let a_h be the value computed in hidden unit h just *before* applying the activation function $\Phi(\cdot)$. In other words, we have $h = \Phi(a_h)$, where a_h is a linear combination of its inputs from earlier-layer units incident on h. Then, by the univariate chain rule, the following expression for $\frac{\partial h}{\partial h_r}$ can be derived:

$$\frac{\partial h}{\partial h_r} = \frac{\partial h}{\partial a_h} \cdot \frac{\partial a_h}{\partial h_r} = \frac{\partial \Phi(a_h)}{\partial a_h} \cdot w_{(h_r, h)} = \Phi'(a_h) \cdot w_{(h_r, h)}$$

By substituting $\frac{\partial h}{\partial h_r}$ in Equation 10.33, one obtains the following:

$$\Delta(h_r, o) = \sum_{h: h_r \Rightarrow h} \Phi'(a_h) \cdot w_{(h_r, h)} \cdot \Delta(h, o) \tag{10.34}$$

These gradients are successively accumulated in the backwards pass. Finally, the Equation 10.32 requires the computation of $\frac{\partial h_r}{\partial w_{(h_{r-1}, h_r)}}$, which is easily computed as follows:

$$\frac{\partial h_r}{\partial w_{(h_{r-1}, h_r)}} = h_{r-1} \cdot \Phi'(a_{h_r}) \tag{10.35}$$

The approach requires linear time in the number of network connections.

The recurrence condition of Equation 10.34 is derived above by applying the chain rule to the variables representing hidden values *after* applying the activation function. A slightly different recurrence condition may be obtained [58, 210] by replacing $\Delta(h_r, o)$ with $\delta(h_r, o)$, which is defined in terms of variables representing pre-activation values $a_{h_1} \ldots a_{h_k}$ and a_o in the units $h_1 \ldots h_k$ and o. Therefore, Equation 10.32 is adjusted as follows:

$$\frac{\partial L}{\partial w_{(h_{r-1}, h_r)}} = \frac{\partial L}{\partial o} \Phi'(a_o) \cdot \underbrace{\left[\sum_{[h_r, h_{r+1}, \ldots h_k, o] \in \mathcal{P}} \frac{\partial a_o}{\partial a_{h_k}} \prod_{i=r}^{k-1} \frac{\partial a_{h_{i+1}}}{\partial a_{h_i}} \right]}_{\text{Backpropagation computes } \delta(h_r, o) = \frac{\partial L}{\partial a_{h_r}}} \cdot \underbrace{\frac{\partial a_{h_r}}{\partial w_{(h_{r-1}, h_r)}}}_{h_{r-1}} \tag{10.36}$$

The recurrence of $\delta(h_r, o) = \frac{\partial L}{\partial a_{h_r}}$ is defined as $\delta(h_r, o) = \Phi'(a_{h_r}) \sum_{h: h_r \Rightarrow h} w_{(h_r, h)} \delta(h, o)$ and each $\delta(o, o)$ is initialized to $\frac{\partial L}{\partial o} \Phi'(a_o)$. The recurrence using pre-activation values is more commonly used, and is the one found in most textbooks. It is easy to generalize the recurrence to multiple outputs by adding all contributions.

10.6.2 Neural Embedding with Word2vec

The two variants of *word2vec* are as follows:

1. *Predicting target words from contexts:* This model tries to predict the ith word, w_i, in a sentence using a window of width t around the word. Therefore, the words $w_{i-t} w_{i-t+1} \ldots w_{i-1} w_{i+1} \ldots w_{i+t-1} w_{i+t}$ are used to predict the target word w_i. This model is also referred to as the *continuous bag-of-words (CBOW) model*.

2. *Predicting contexts from target words:* This model tries to predict the context $w_{i-t} w_{i-t+1} \ldots w_{i-1} w_{i+1} \ldots w_{i+t-1} w_{i+t}$ around word w_i, given the ith word in the

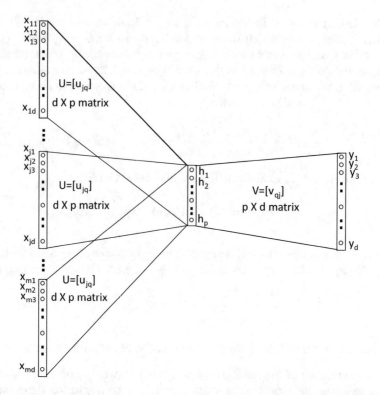

Figure 10.8: Word2vec: The CBOW model. One could equivalently choose to collapse the m sets of d one-hot encoded input nodes into a single set of d real-valued input nodes, which are fed the averages of the m one-hot encoded inputs to achieve the same effect

sentence, denoted by w_i. This model is referred to as the *skip-gram model*. There are, however, two ways in which one can perform this prediction. The first technique is a *multinomial* model which predicts one word out of d outcomes. The second model is a Bernoulli model, which models whether or not each context is present for a particular word. The second approach uses *negative sampling* for efficiency.

Each of these methods will be discussed in this section.

10.6.2.1   Neural Embedding with Continuous Bag of Words

In the continuous bag-of-words (CBOW) model, the training pairs are all context-word pairs in which a window of context words is input, and a single target word is predicted. The context contains $2 \cdot t$ words, corresponding to t words both before and after the target word. For notational ease, we will use the length $m = 2 \cdot t$ to define the length of the context. Therefore, the input to the system is a set of m words. Without loss of generality, let the subscripts of these words be numbered so that they are denoted by $w_1 \dots w_m$, and let the target (output) word in the middle of the context window be denoted by w. Note that w can be viewed as a categorical variable with d possible values, where d is the size of the lexicon. The goal of the neural embedding is to compute the probability $P(w|w_1 w_2 \dots w_d)$ and maximize the product of these probabilities over all training samples.

The overall architecture of this model is illustrated in Figure 10.8. In the architecture, we have a single input layer with $m \times d$ nodes, a hidden layer with p nodes, and an output layer with d nodes. The nodes in the input layer are clustered into m different groups, each of which has d units. Each group with d input units is the one-hot encoded input vector of one of the m context words being modeled by CBOW. Only one of these d inputs will be 1 and the remaining inputs will be 0. Therefore, one can represent an input x_{ij} with two indices corresponding to contextual position and word identifier. Specifically, the input $x_{ij} \in \{0, 1\}$ contains two indices i and j in the subscript, where $i \in \{1 \ldots m\}$ is the position of the context, and $j \in \{1 \ldots d\}$ is the identifier of the word.

The hidden layer contains p units, where p is the dimensionality of the hidden layer in *word2vec*. Let $h_1, h_2, \ldots h_p$ be the outputs of the hidden layer nodes. Note that each of the d words in the lexicon has m different representatives in the input layer corresponding to the m different context words, but the weight of each of these m connections is the same. Such weights are referred to as shared. Sharing weights is a common trick used for regularization in neural networks, when one has specific insight about the domain at hand. Let the shared weight of each connection from the jth word in the lexicon to the qth hidden layer node be denoted by u_{jq}. Note that each of the m groups in the input layer has connections to the hidden layer that are defined by the same $d \times p$ weight matrix U. This situation is shown in Figure 10.8.

It is noteworthy that $\overline{u}_j = [u_{j1}, u_{j2}, \ldots u_{jp}]$ (i.e., jth row of U) can be viewed as the p-dimensional embedding of the jth input word over the entire corpus, and $\overline{h} = [h_1 \ldots h_p]$ provides the embedding of a specific instantiation of an input context. Then, the output of the hidden layer is obtained by averaging[2] the embeddings of the words present in the context. One can write this relationship in vector form:

$$\overline{h} = \frac{1}{m} \sum_{i=1}^{m} \sum_{j=1}^{d} \overline{u}_j x_{ij} \tag{10.37}$$

In essence, the one-hot encodings of the input words are aggregated, which implies that the ordering of the words within the window of size m does not affect the output of the model. This is the reason that the model is referred to as the continuous bag-of-words model.

The embedding $[h_1 \ldots h_p]$ is used to predict the probability that the target word is one of each of the d outputs with the use of the softmax function. The weights in the output layer are parameterized with a $d \times p$ matrix $V = [v_{jq}]$. The jth row of V is denoted by $\overline{v}_j$. Since the hidden-to-output matrix is a $p \times d$ matrix, it is annotated by V^T instead of V in Figure 10.8. Note that the multiplication of the hidden vector $\overline{h}$ with V^T to create $\overline{h} V^T$ will create a real-valued vector $[\overline{h} \cdot \overline{v}_1, \ldots, \overline{h} \cdot \overline{v}_d]$. The softmax function is applied to this vector to convert it to d probabilities $\hat{y}_1 \ldots \hat{y}_d$ summing to 1. These outputs are the d probabilities for the target word, given the context:

$$\hat{y}_j = P(y_j = 1 | w_1 \ldots w_m) = \frac{\exp(\overline{h} \cdot \overline{v}_j)}{\sum_{k=1}^{d} \exp(\overline{h} \cdot \overline{v}_k)} \tag{10.38}$$

The *ground-truth* value of only one of the outputs $y_1 \ldots y_d$ is 1 and the remaining values are 0 for a given training instance: follows:

$$y_j = \begin{cases} 1 & \text{if the target word } w \text{ is the } j\text{th word} \\ 0 & \text{otherwise} \end{cases} \tag{10.39}$$

---

[2] When m is fixed over all sampled windows, it is possible to omit the factor of m in the denominator on the right-hand side of Equation 10.37. This would make the transformation identical to that of a traditional linear layer.

For a particular target word $w = r \in \{1 \ldots d\}$, the loss function is given by $L = -\log[P(y_r = 1|w_1 \ldots w_m)] = -\log(\hat{y}_r)$. The use of the negative logarithm turns the multiplicative like-lihoods over different training instances into an additive loss function using the negative log-likelihoods.

The updates on the rows of U and V are defined by computing the gradients using the backpropagation algorithm, as training instances are passed through the neural network one by one. The update equations with learning rate α are as follows:

$$\overline{u}_i \Leftarrow \overline{u}_i - \alpha \frac{\partial L}{\partial \overline{u}_i} \quad \forall i$$

$$\overline{v}_j \Leftarrow \overline{v}_j - \alpha \frac{\partial L}{\partial \overline{v}_j} \quad \forall j$$

Next, we provide expressions for the aforementioned partial derivatives of this loss function [389, 390, 480]. The probability of making a mistake in prediction on the jth word in the lexicon is defined by $|y_j - \hat{y}_j|$. However, we use *signed* mistakes ϵ_j, in which only the correct word with $y_j = 1$ is given a positive mistake value, while all the other words in the lexicon receive negative mistake values. This is achieved by dropping the modulus:

$$\epsilon_j = y_j - \hat{y}_j \tag{10.40}$$

Note that ϵ_j is also the negative derivative of the cross-entropy loss with respect to jth input into the softmax layer. Further backpropagation results in the following updates:

$$\overline{u}_i \Leftarrow \overline{u}_i + \frac{\alpha}{m} \sum_{j=1}^{d} \epsilon_j \overline{v}_j \quad [\forall \text{ words } i \text{ present in context window}]$$

$$\overline{v}_j \Leftarrow \overline{v}_j + \alpha \epsilon_j \overline{h} \quad [\forall j \text{ in lexicon}]$$

Here, $\alpha > 0$ is the learning rate. Repetitions of the same word i in the context window trigger multiple updates of $\overline{u}_i$. The use of an aggregate context window in the input has a smoothing effect on the CBOW model, which is particularly helpful with smaller data sets.

The training examples of context-target pairs are presented one by one, and the weights are trained to convergence. It is noteworthy that the *word2vec* model provides not one but two different embeddings that correspond to the p-dimensional rows of the matrix U and the p-dimensional columns of the matrix V. The former type of embedding of words is referred to as the *input* embedding, whereas the latter is referred to as the *output* embedding. In the CBOW model the input embedding represents context, and therefore it makes sense to use the output embedding. However, the input embedding (or the sum/concatenation of input and output embeddings) can also be helpful for many tasks.

10.6.2.2 Neural Embedding with Skip-Gram Model

In the skip-gram model, the target words are used to predict the m context words. There-fore, we have one input word and m outputs. One issue with the CBOW model is that the averaging effect of the input words in the context window (which creates the hidden repre-sentation) has a (helpful) smoothing effect with less data, but fails to take full advantage of a larger amount of data. The skip-gram model is the technique of choice when a large amount of data is available.

The skip-gram model uses a single target word w as the input and outputs the m context words denoted by $w_1 \ldots w_m$. Therefore, the goal is to estimate $P(w_1, w_2 \ldots w_m|w)$, which

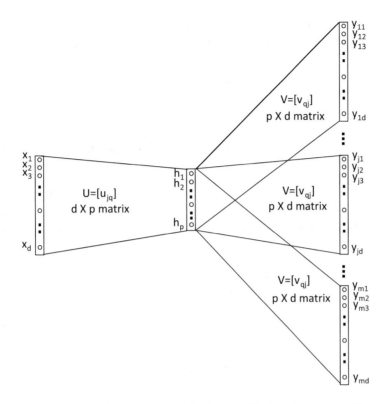

Figure 10.9: Word2vec: The skip-gram model. One could also choose to collapse the m sets of d output nodes into a single set of d outputs, and mini-batch the m instances in a single context window during stochastic gradient descent to achieve the same effect

is different from the quantity $P(w|w_1 \ldots w_m)$ estimated in the CBOW model. As in the case of the continuous bag-of-words model, we can use one-hot encoding of the (categorical) input and outputs in the skip-gram model. After such an encoding, the skip-gram model will have d binary inputs denoted by $x_1 \ldots x_d$ corresponding to the d possible values of the single input word. Similarly, the output of each training instance is encoded as $m \times d$ values $y_{ij} \in \{0, 1\}$, where i ranges from 1 to m (size of context window), and j ranges from 1 to d (lexicon size). Each $y_{ij} \in \{0, 1\}$ indicates whether the ith contextual word takes on the jth possible value for that training instance. However, the (i, j)th output node only computes a soft probability value $\hat{y}_{ij} = P(y_{ij} = 1|w)$. Therefore, the probabilities $\hat{y}_{ij}$ in the output layer for fixed i and varying j sum to 1, since the ith contextual position takes on exactly one of the d words. The hidden layer contains p units, the outputs are denoted by $h_1 \ldots h_p$. Each input x_j is connected to all the hidden nodes with a $d \times p$ matrix U. Furthermore, the p hidden nodes are connected to each of the m groups of d output nodes with the same set of shared weights. This set of shared weights between the p hidden nodes and the d output nodes of each of the context words is defined by the $p \times d$ matrix V. Note that the input-output structure of the skip-gram model is an inverted version of the input-output structure of the CBOW model. The neural architecture of the skip-gram model is illustrated in Figure 10.9.

The output of the hidden layer can be computed from the input layer using the $d \times p$ matrix of weights $U = [u_{jq}]$ between the input and hidden layer as follows:

$$\overline{h} = \sum_{j=1}^{d} \overline{u}_j x_j \qquad (10.41)$$

The above equation has a simple interpretation because of the one-hot encoding of the input word w in terms of $x_1 \ldots x_d$. If the input word w is the rth word, then one simply copies $\overline{u}_r$ to the hidden layer. In other words, the rth row $\overline{u}_r$ of U is copied to the hidden layer. As discussed above, the hidden layer is connected to m groups of d output nodes, each of which is connected to the hidden layer with a $d \times p$ matrix $V = [v_{jq}]$. Since the hidden-to-output matrix is of size $p \times d$, the matrix V^T is used instead of V. Each of these m groups of d output nodes computes the probabilities of the various words for a particular context word. The jth column of V is denoted by $\overline{v}_j$ and represents the output embedding of the jth word. The output $\hat{y}_{ij}$ is the probability that the word in the ith context position takes on the jth word of the lexicon. However, since the same matrix V is shared by all groups, the neural network predicts the same multinomial distribution for each of the context words. Therefore, we have the following:

$$\hat{y}_{ij} = P(y_{ij} = 1|w) = \underbrace{\frac{\exp(\overline{h} \cdot \overline{v}_j)}{\sum_{k=1}^{d} \exp(\overline{h} \cdot \overline{v}_k)}}_{\text{Independent of context position } i} \qquad \forall i \in \{1 \ldots m\} \qquad (10.42)$$

Note that the probability $\hat{y}_{ij}$ is the same for varying i and fixed j, since the right-hand side of the above equation does not depend on the exact location i in the context window.

The loss function for the backpropagation algorithm is the negative of the log-likelihood values of the ground truth $y_{ij} \in \{0, 1\}$ of a training instance. This loss function L is given by the following:

$$L = -\sum_{i=1}^{m} \sum_{j=1}^{d} y_{ij} \log(\hat{y}_{ij}) \qquad (10.43)$$

Note that the value outside the logarithm is a ground-truth binary value, whereas the value inside the logarithm is a predicted (probability) value. Since y_{ij} is one-hot encoded for fixed i and varying j, the objective function has only m non-zero terms. For each training instance, this loss function is used in combination with backpropagation to update the weights of the connections between the nodes. The update equations with learning rate α are as follows:

$$\overline{u}_i \Leftarrow \overline{u}_i - \alpha \frac{\partial L}{\partial \overline{u}_i} \quad \forall i$$

$$\overline{v}_j \Leftarrow \overline{v}_j - \alpha \frac{\partial L}{\partial \overline{v}_j} \quad \forall j$$

Next, we derive the gradients above. The probability of making a mistake in predicting the jth word in the lexicon for the ith context is defined by $|y_{ij} - \hat{y}_{ij}|$. However, we use *signed* mistakes ϵ_{ij} in which only the predicted words (positive examples) have a positive probability. This is achieved by dropping the modulus:

$$\epsilon_{ij} = y_{ij} - \hat{y}_{ij} \qquad (10.44)$$

Then, the updates for a particular input word r and its output context are as follows:

$$\overline{u}_r \Leftarrow \overline{u}_r + \alpha \sum_{j=1}^{d} \left[\sum_{i=1}^{m} \epsilon_{ij} \right] \overline{v}_j \qquad [\text{Only for input word } r]$$

$$\overline{v}_j \Leftarrow \overline{v}_j + \alpha \left[\sum_{i=1}^{m} \epsilon_{ij} \right] \overline{h} \qquad [\text{For all words } j \text{ in lexicon}]$$

Here, $\alpha > 0$ is the learning rate. The p-dimensional rows of the matrix U are used as the embeddings of the words. In other words, the convention is to use the input embeddings in the rows of U rather than the output embeddings in the rows of V. It is stated in [324] that adding the input and output embeddings can help in some tasks (but hurt in others). The concatenation of the two can also be useful.

Practical Issues

Several practical issues are associated with the accuracy and efficiency of the *word2vec* framework. Increasing the embedding dimensionality improves discrimination, but it requires a greater amount of data. In general, the typical embedding dimensionality is of the order of several hundred, although it is possible to choose dimensionalities in the thousands for very large collections. The size of the context window typically varies between 5 and 10, with larger window sizes being used for the skip-gram model as compared to the CBOW model. Using a random window size is a variant that has the implicit effect of giving greater weight to words that are placed close together. The skip-gram model is slower but it works better for infrequent words and for larger data sets.

Another issue is that the effect of frequent and less discriminative words (e.g., *"the"*) can dominate the results. Therefore, a common approach is to downsample the frequent words, which improves both accuracy and efficiency. Note that downsampling frequent words has the implicit effect of increasing the context window size because dropping a word in the middle of two words brings the latter pair closer. The words that are very rare are misspellings, and it is hard to create a meaningful embedding for them without overfitting. Therefore, such words are ignored.

From a computational point of view, the updates of output embeddings are expensive. This is caused by applying the softmax over a lexicon of d words, which requires an update of each $\overline{v}_j$. One possibility is to use a *hierarchical softmax model* [389, 390, 480]. A more popular option changes the loss function (and neural architecture) slightly, which is referred to as *skipgram with negative sampling*. This approach is discussed in the next section.

10.6.2.3 Skip-Gram with Negative Sampling

An efficient alternative to the hierarchical softmax technique is a method known as *skip-gram with negative sampling (SGNS)* [390], in which both presence or absence of word-context pairs are used for training. As the name implies, the negative contexts are artificially generated by sampling words in proportion to their frequencies in the corpus (i.e., unigram distribution). This approach optimizes a different objective function from the skip-gram model, which is based on ideas from *noise contrastive estimation* [226, 402, 403].

The basic idea is that instead of directly predicting each of the m words in the context window, we try to predict whether or not each of the d words in the lexicon is present in the window. In other words, the final layer of Figure 10.9 is not a softmax prediction, but a Bernoulli layer of sigmoids. The output unit for each word at each context position in Figure 10.9 is a sigmoid providing a probability value that the position takes on that word. As the ground-truth values are also available, it is possible to use the logistic loss function over all the words. Therefore, in this point of view, even the prediction problem is defined differently. Of course, it is computationally inefficient to try to make binary predictions for all d words. Therefore, the SGNS approach uses all the positive words in a context window and a *sample* of negative words. The number of negative samples is k times the number of positive samples. Here, k is a parameter controlling the sampling rate. Negative sampling becomes essential in this modified prediction problem to avoid learning trivial weights that predict all examples to 1. In other words, we cannot choose to avoid negative samples entirely (i.e., we cannot set $k = 0$).

How does one generate the negative samples? The vanilla unigram distribution samples words in proportion to their relative frequencies $f_1 \ldots f_d$ in the corpus. Better results are obtained [390] by sampling words in proportion to $f_j^{3/4}$ rather than f_j. As in all *word2vec* models, let U be a $d \times p$ matrix representing the input embedding, and V be a $p \times d$ matrix representing the output embedding. Let $\overline{u}_i$ be the p-dimensional row of U (input embedding of ith word) and $\overline{v}_j$ be the p-dimensional column of V (output embedding of jth word). Let $\mathcal{P}$ be the set of positive target-context word pairs in a context window, and $\mathcal{N}$ be the set of negative target-context word pairs which are created by sampling. Therefore, the size of $\mathcal{P}$ is equal to the context window m, and that of $\mathcal{N}$ is $m \cdot k$. Then, the (minimization) objective function for each context window is obtained by summing up the logistic loss over the m positive samples and $m \cdot k$ negative samples:

$$O = -\sum_{(i,j) \in \mathcal{P}} \log(P[\text{Predict } (i,j) \text{ to } 1]) - \sum_{(i,j) \in \mathcal{N}} \log(P[\text{Predict } (i,j) \text{ to } 0]) \qquad (10.45)$$

$$= -\sum_{(i,j) \in \mathcal{P}} \log\left(\frac{1}{1 + \exp(-\overline{u}_i \cdot \overline{v}_j)}\right) - \sum_{(i,j) \in \mathcal{N}} \log\left(\frac{1}{1 + \exp(\overline{u}_i \cdot \overline{v}_j)}\right) \qquad (10.46)$$

This modified objective function is used in the skip-gram with negative sampling (SGNS) model in order to update the weights of U and V. SGNS is mathematically different from the basic skip-gram model discussed earlier. SGNS is not only efficient, but it also provides the best results among the different variants of skip-gram models.

10.6.2.4 What Is the Actual Neural Architecture of SGNS?

Even though the original *word2vec* paper seems to treat SGNS as an efficiency optimization of the skip-gram model, it is using a fundamentally different architecture in terms of the activation function used in the final layer. Unfortunately, the original *word2vec* paper does not explicitly point this out (and only provides the changed objective function), which causes confusion.

The modified neural architecture of SGNS is as follows. The softmax layer is no longer used in the SGNS implementation. Rather, each observed value y_{ij} in Figure 10.9 is *in-dependently* treated as a *binary* outcome, rather than as a multinomial outcome in which the probabilistic predictions of different outcomes at a contextual position depend on one another. Instead of using softmax to create the prediction $\hat{y}_{ij}$, it uses the sigmoid activation to create $\hat{y}_{ij}$, which is the predicted probability whether or not the jth word occurs in each contextual position i. One then uses the logarithmic loss with respect to *both* positive and negative occurrences of each word in each contextual position. Then, one can add up the loss of $\hat{y}_{ij}$ over all $m \cdot d$ possible values of (i, j) to create the full loss function of a context window. However, this is impractical because the number of zero values of y_{ij} is too large and zero values are noisy anyway. Therefore, SGNS uses negative sampling to approximate this *modified* objective function. This means that for each context window, we are back-propagating from only a subset of the $m \cdot d$ outputs in Figure 10.9. The size of this subset is $m + m \cdot k$. This is where efficiency is achieved. However, since the final layer uses binary predictions (with sigmoids), it makes the SGNS architecture fundamentally different from the vanilla skip-gram model even in terms of the basic neural network it uses (i.e., logistic instead of softmax activation). The difference between the SGNS model and the vanilla skip-gram model is analogous to the difference between the Bernoulli and multinomial models in naïve Bayes classification (with negative sampling applied only to the Bernoulli model). Obviously, one cannot be considered a direct efficiency optimization of the other.

10.6.3 Word2vec (SGNS) Is Logistic Matrix Factorization

The architecture of the skip-gram models look suspiciously similar to an autoencoder (except that words decode to their context). Autoencoders are often indirect ways of performing matrix factorization. The SGNS model of *word2vec* can be simulated with logistic matrix factorization. The SGNS embedding can be shown to be roughly equivalent to shifted PPMI matrix factorization [323] of Section 10.4.4. However, this equivalence is only *implicit* in terms of a derived PPMI matrix. This section discusses a more direct relationship in terms of a binary matrix of actual outcomes.

Let $B = [b_{ij}]$ be a binary matrix in which the (i, j)th value is 1 if word j occurs at least once in the context of word i in the data set, and 0 otherwise. The weight c_{ij} for any word (i, j) that occurs in the corpus is defined by the number of times word j occurs in the context of word i. The weights of the zero entries in B are defined as follows. For each row i in B we sample $k \sum_j b_{ij}$ different entries from row i, among the entries for which $b_{ij} = 0$, and the frequency with which the jth word is sampled is proportional to $f_j^{3/4}$. These are the negative samples, and one sets the weights c_{ij} for the negative samples (i.e., those for which $b_{ij} = 0$) to the number of times that each entry is sampled. As in *word2vec*, the p-dimensional embeddings of the ith word and jth context are denoted by $\overline{u}_i$ and $\overline{v}_j$, respectively. The simplest way of factorizing is to use *weighted* matrix factorization of B with the Frobenius norm:

$$\text{Minimize}_{U,V} \sum_{i,j} c_{ij}(b_{ij} - \overline{u}_i \cdot \overline{v}_j)^2 \qquad (10.47)$$

Even though the matrix B is of size $O(d^2)$, this matrix factorization only has a limited number of nonzero terms in the objective function, which have $c_{ij} > 0$. Like GloVe, these weights are dependent on co-occurrence counts, but (unlike GloVe) some zero entries also have positive weight. Therefore, the stochastic gradient-descent steps only have to focus on entries with $c_{ij} > 0$, as in GloVe. Each cycle of stochastic gradient-descent is *linear* in the number of non-zero entries, as in the SGNS implementation of *word2vec*. This objective function goes beyond Glove's logarithmic damping by damping counts all the way to binary values. Therefore, contrast between the entries is achieved by negative sampling, whereas GloVe ignores negative samples and only contrasts the variation in non-zero entries.

However, this objective function also looks somewhat different from *word2vec*, which has a logistic form. Just as it is advisable to replace linear regression with logistic regression in supervised learning of binary targets, one can use the same trick in matrix factorization of binary matrices [277]. We can change the squared error term to the familiar likelihood term

L_{ij}, which is used in logistic regression:

$$L_{ij} = \left| b_{ij} - \frac{1}{1 + \exp(\overline{u}_i \cdot \overline{v}_j)} \right| \tag{10.48}$$

The value of L_{ij} always lies in the range $(0, 1)$, and higher values indicate greater likelihood (which results in a maximization objective). The modulus in the above expression flips the sign only for the negative samples in which $b_{ij} = 0$. Now, one can optimize the following objective function in minimization form:

$$\text{Minimize}_{U,V} \; J = -\sum_{i,j} c_{ij} \log(L_{ij}) \tag{10.49}$$

The main difference from the objective function (cf. Equation 10.46) of *word2vec* is that this is a global objective function over all matrix entries, rather than a local objective function over a particular context window. Using mini-batch stochastic gradient-descent in matrix factorization (with an appropriately chosen mini-batch) makes the approach almost identical to *word2vec*'s backpropagation updates.

How can one interpret this type of factorization? Instead of $B \approx UV$, we have $B \approx f(UV)$, where $f(\cdot)$ is the sigmoid function. More precisely, this is a *probabilistic* factorization in which one computes the product of matrices U and V, and then applies the sigmoid function to obtain the parameters of the Bernoulli distribution from which B is generated:

$$P(b_{ij} = 1) = \frac{1}{1 + \exp(-\overline{u}_i \cdot \overline{v}_j)} \qquad \text{[Matrix factorization analog of logistic regression]}$$

It is also easy to verify from Equation 10.48 that L_{ij} is $P(b_{ij} = 1)$ for positive samples and $P(b_{ij} = 0)$ for negative samples. Therefore, the objective function of the factorization is in the form of log-likelihood maximization. This type of logistic matrix factorization is commonly used [277] in recommender systems with binary data (e.g., user click-streams).

It is suggested [42] that models like SGNS, which make binary predictions, are fundamentally better than count-based factorization models (cf. Section 10.4). However, as the above argument shows, SGNS is also equivalent to a factorization model, where the counts are used as weights in the factorization rather than as matrix entries. Therefore, any difference in performance is only the result of the specific choice of *how* the counts are used in the objective function. It is also shown in [436] that count-based factorization can outperform *word2vec*, especially when the counts are properly damped and also used to weight the objective function. Therefore, the claim that predictive models are fundamentally better than count-based models should be taken with a grain of salt. It is likely that the real issue is one of properly handling the order-of-magnitude variation in the word counts within the objective function in a graceful way. Leveraging the zero entries of the count co-occurrence matrix in a careful way is also helpful, depending on the problem setting (see Exercise 8). If the SGNS variant of the skip-gram model is logistic matrix factorization, then what about the vanilla skip-gram model? It turns out that it is possible to show that the vanilla skip-gram model is equivalent to *multinomial matrix factorization* (see Exercise 9).

Gradient Descent

It is also helpful to examine the gradient-descent steps of the factorization. One can take the derivative of J with respect to the input and output embeddings:

$$\frac{\partial J}{\partial \overline{u}_i} = - \sum_{j:b_{ij}=1} \frac{c_{ij}\overline{v}_j}{1 + \exp(\overline{u}_i \cdot \overline{v}_j)} + \sum_{j:b_{ij}=0} \frac{c_{ij}\overline{v}_j}{1 + \exp(-\overline{u}_i \cdot \overline{v}_j)}$$

$$= - \underbrace{\sum_{j:b_{ij}=1} c_{ij}P(b_{ij}=0)\overline{v}_j}_{\text{Positive Mistakes}} + \underbrace{\sum_{j:b_{ij}=0} c_{ij}P(b_{ij}=1)\overline{v}_j}_{\text{Negative Mistakes}}$$

$$\frac{\partial J}{\partial \overline{v}_j} = - \sum_{i:b_{ij}=1} \frac{c_{ij}\overline{u}_i}{1 + \exp(\overline{u}_i \cdot \overline{v}_j)} + \sum_{i:b_{ij}=0} \frac{c_{ij}\overline{u}_i}{1 + \exp(-\overline{u}_i \cdot \overline{v}_j)}$$

$$= - \underbrace{\sum_{i:b_{ij}=1} c_{ij}P(b_{ij}=0)\overline{u}_i}_{\text{Positive Mistakes}} + \underbrace{\sum_{i:b_{ij}=0} c_{ij}P(b_{ij}=1)\overline{u}_i}_{\text{Negative Mistakes}}$$

Since this is a minimization problem, the optimization procedure uses gradient descent to convergence:

$$\overline{u}_i \Leftarrow \overline{u}_i - \alpha \frac{\partial J}{\partial \overline{u}_i} \quad \forall i$$

$$\overline{v}_j \Leftarrow \overline{v}_j - \alpha \frac{\partial J}{\partial \overline{v}_j} \quad \forall j$$

It is noteworthy that the derivatives can be expressed in terms of the probabilities of making mistakes in predicting b_{ij}. This is common in gradient descent with log-likelihood optimization. It is also noteworthy that the derivative of the SGNS objective in Equation 10.46 yields a similar form of the gradient. The only difference is that the derivative of the SGNS objective is expressed over a smaller batch of instances, defined by a context window. We can also solve the probabilistic matrix factorization with mini-batch stochastic gradient descent. With an appropriate choice of the mini-batch, the stochastic gradient descent of matrix factorization becomes identical to the backpropagation update of SGNS. The only difference is that SGNS samples negative entries *for each set of updates* on the fly, whereas matrix factorization fixes the negative samples up front. Of course, on-the-fly sampling can also be used with matrix factorization updates.

10.6.4 Beyond Words: Embedding Paragraphs with Doc2vec

The broader principle of *word2vec* can be generalized to embedding paragraphs as well, by treating the paragraph identifier as another word in the context. For example, consider the CBOW model of *word2vec*, which has m input words with an one-hot encoding of d possibilities per input word. We can add an $(m+1)$th context "word" corresponding to a paragraph identifier; however, this paragraph identifier is drawn from a different vocabulary of d' paragraph identifiers than the other context words. Therefore, the one-hot encoding of this paragraph identifier requires d' binary input units. Another difference is that we now have an additional $d' \times p$ matrix U' containing the weights of the connections from the

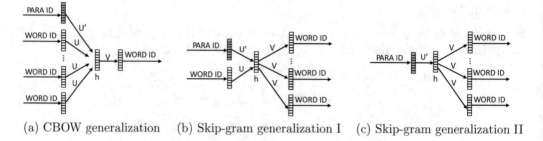

(a) CBOW generalization   (b) Skip-gram generalization I   (c) Skip-gram generalization II

Figure 10.10: Different generalizations of *word2vec* to *doc2vec*

paragraph identifier input nodes to the hidden layer, in addition to the $d \times p$ matrix U that connect the word input nodes to the hidden layer. This is because the weights of the "words" corresponding to paragraph identifiers are not shared with the regular vocabulary of words. The rows of the matrix U' provide the embeddings of the various paragraphs. The training process then samples contexts from the various paragraphs and uses the same gradient-descent method in combination with the backpropagation algorithm. The representations of paragraph vectors (in the training data) and word vectors are jointly learned because the entries of U', U, and V are updated simultaneously during training. It is also possible to quickly learn the embedding of out-of-sample paragraphs by applying the same process of gradient-descent to update only U' while keeping U and V fixed. The basic idea here is that word embeddings are considered stable beyond a certain amount of training, and paragraph embeddings can be efficiently learned directly from them while holding the word embeddings fixed.

There are several variations of the architecture that one might use:

1. One can add an additional input for paragraph vectors to the CBOW model of *word2vec*. This generalization is shown in Figure 10.10(a).

2. One can add an additional input for paragraph vectors to the skip-gram model of *word2vec*. This generalization is shown in Figure 10.10(b). However, this generalization is not explored in the original *doc2vec* paper.

3. One can *replace* the one-hot encoded word identifier input in the skip-gram model of *word2vec* with a one-hot encoded target *paragraph* identifier input. Not that in this case, we are dropping the matrix U entirely, and learning only U' and V. This generalization is shown in Figure 10.10(c).

There are some additional differences proposed in the *doc2vec* work over the original *word2vec* framework:

1. Rather than using a word at the center of the context window, the *doc2vec* framework uses the historical context window to predict the next word. There is, however, considerable flexibility in what one might choose as context for a particular application.

2. The *word2vec* framework averages the embeddings of the words in the context in order to create the output of the hidden layer for a particular context. The *doc2vec* framework recommends an additional choice, corresponding to the concatenation of the vectors. Note that concatenating the vectors increases the dimensionality of the hidden layer but it does have the benefit of preserving more ordering information.

The incorporation of the paragraph identifiers within the context helps in creating memory about memberships of text segments in various paragraphs. This memory can help in the creation of a more refined model, although it does increase the number of parameters. Therefore, the *doc2vec* model will generally require a larger corpus than *word2vec* in order to work effectively.

10.7   Recurrent Neural Networks

Recurrent neural networks (RNNs) allow the use of loops within the neural network architecture to model language dependencies. Recurrent networks take as input a *sequence* of inputs, and produce a sequence of outputs. In other words, such models are particularly useful for sequence-to-sequence learning. Some examples of applications include the following:

1. The input might be a sequence of words, and the output might be the same sequence shifted by 1. This is a classical language model in which we are trying the predict the next word based on the sequential history of words.

2. The input might be a sentence in one language, and the output might be a sentence in another language.

3. The input might be a sequence (e.g., sentence), and the output might be a vector of class probabilities, which is triggered by the end of the sentence.

The simplest recurrent neural network is shown in Figure 10.11(a). A key point here is the presence of the self-loop in Figure 10.11(a), which will cause the hidden state of the neural network to change after the input of each word in the sequence. In practice, one only works with sequences of finite length, and it makes sense to unfurl the loop into a "time-layered" network that looks more like a feedforward network. This network is shown in Figure 10.11(b). Note that in this case, we have a different node for the hidden state at each time-stamp and the self-loop has been unfurled into a feedforward network. This representation is mathematically equivalent to Figure 10.11(a), but it is much easier to comprehend because of its similarity to a traditional network. The copies of the weight matrices in the different temporal layers, which are created by unfurling, *are shared* to ensure that the same mathematical transformation is applied at each time-stamp. The annotations W_{xh}, W_{hh}, and W_{hy} of the weight matrices in Figure 10.11(b) make the sharing evident.

Given a sequence of words, their one-hot encoding is fed to the network one at a time to the network in Figure 10.11(a), which is equivalent to feeding the words to the adjacent inputs in Figure 10.11(b). In the setting of language modeling, the output is a vector of probabilities predicted for the next word in the sequence. For example, consider the sentence:

> The cat chased the mouse.

When the word "*The*" is input, the output will be a vector of probabilities of the entire lexicon that includes the word "*cat*," and when the word "*cat*" is input, we will again get a vector of probabilities predicting the next word. This is, of course, the classical way in which language models are defined in which the probability of a word is estimated based on the immediate history of previous words. For the purpose of discussion, let us refer to each step of processing the sequence as a *time-stamp*. In general, the input vector at time t (e.g., one-hot encoded vector of tth word) is $\overline{x}_t$, the hidden state at time t is $\overline{h}_t$, and the output vector at time t (e.g., predicted probabilities of $(t+1)$th word) is $\overline{y}_t$. Both $\overline{x}_t$ and $\overline{y}_t$ are d-dimensional for a lexicon of size d. The hidden vector $\overline{h}_t$ is p-dimensional, where p regulates the complexity of the embedding. For the purpose of discussion, we will assume that all these vectors are column vectors. In many applications like classification, the output is not produced at each time unit but is only triggered at the last time-stamp in the end of the sentence. Although output and input units may be present only at a subset of the time-stamps, we examine the simple case in which they are present in all time-stamps. Then, the hidden state at time t is given by a function of the input vector at time t and the hidden vector at time $(t-1)$:

$$\overline{h}_t = f(\overline{h}_{t-1}, \overline{x}_t) \tag{10.50}$$

This function is defined with the use of weight matrices and activation functions (as used by all neural networks for learning), and *the same weights are used at each time-stamp.* Therefore, even though the hidden state evolves over time, the weights and the underlying function $f(\cdot, \cdot)$ remain fixed over all time-stamps (i.e., sequential elements) after the neural network has been trained. A separate function $\overline{y}_t = g(\overline{h}_t)$ is used to learn the output probabilities from the hidden states.

Next, we describe the functions $f(\cdot, \cdot)$ and $g(\cdot)$ more concretely. We define $p \times d$ input-hidden matrix W_{xh}, a $p \times p$ hidden-hidden matrix W_{hh}, and a $d \times p$ hidden-output matrix W_{hy}. Then, one can expand Equation 10.50 and also write the condition for the outputs as follows:

$$\overline{h}_t = \tanh(W_{xh}\overline{x}_t + W_{hh}\overline{h}_{t-1})$$
$$\overline{y}_t = W_{hy}\overline{h}_t$$

Here, the "tanh" notation (cf. Section 10.6.1.2) is used in a relaxed way, in the sense that the function is applied to the p-dimensional column vector in an element-wise fashion to create a p-dimensional vector with each element in $[-1, 1]$. Throughout this section, this relaxed notation will be used for several activation functions such as tanh and sigmoid. At the very first time-stamp, $\overline{h}_{t-1}$ is assumed to be some default constant vector, because there is no input from the hidden layer at the beginning of a sentence. Although the hidden states change at each time-stamp, the weight matrices stay fixed over the various time-stamps. Note that the output vector $\overline{y}_t$ is a set of continuous values with the same dimensionality as the lexicon. A softmax layer is applied on top of $\overline{y}_t$ so that the results can be interpreted as probabilities. *The p-dimensional output $\overline{h}_t$ of the hidden layer at the end of a text segment of t words yields its embedding, and the p-dimensional columns of W_{xh} yield the embeddings of individual words.* The latter provides an alternative to *word2vec* embeddings. In fact, if the training data set is not very large, it is recommended to fix the weights of this layer to pre-trained *word2vec* encodings (in order to avoid overfitting).

How can one train such networks? The negative logarithms of the softmax probabilities of the correct words at various time-stamps are aggregated to create the loss function. The backpropagation algorithm will need to account for the temporal weight sharing during updates. This special type of backpropagation algorithm is referred to as *backpropagation through time (BPTT)*. It works just like the backpropagation algorithm on the unfurled network by (i) running the input sequentially in the forward direction through time and computing the error/loss at each time-stamp, (ii) computing the changes in edge weights in the backwards direction on the unfurled network without any regard for the fact that weights in different time layers are shared, and (iii) adding all the changes in the (shared) weights corresponding to different instantiations of an edge in time. The last of these steps is unique to BPTT. Readers are referred to [210] for details.

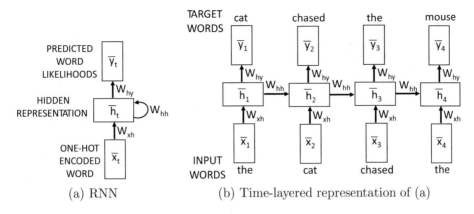

Figure 10.11: A recurrent neural network and its time-layered representation

Practical Issues

The entries of each weight matrix are initialized to small values in $[-1/\sqrt{r}, 1/\sqrt{r}]$, where r is the number of columns in that matrix. One can also initialize each of the d columns of W_{xh} to the *word2vec* embedding of the corresponding word. Another detail is that the training data often contains a special <START> and an <END> token at the beginning and end of each training segment. These types of tokens help the model to recognize specific text units such as sentences, paragraphs, or the beginning of a particular module of text. It is also noteworthy that multiple hidden layers (with long short-term memory enhancements) are used in all practical applications, which will be discussed in Section 10.7.4. However, the following application-centric exposition will use the simpler single-layer model for clarity. The generalization of each of these applications to enhanced architectures is straightforward.

10.7.1 Language Modeling Example of RNN

In order to illustrate the workings of the RNN, we will use a toy example of a single sequence defined on a vocabulary of four words. Consider the sentence:

The cat chased the mouse.

In this case, we have a lexicon of four words, which are { *"the," "cat," "chased," "mouse"* }. In Figure 10.12, we have shown the probabilistic prediction of the next word at each of time-stamps from 1 to 4. Ideally, we would like the probability of the next word to be predicted correctly from the probabilities of the previous words. Each one-hot encoded input vector $\overline{x}_t$ has length four, in which only one bit is 1 and the remaining bits are 0s. The main flexibility here is in the dimensionality p of the hidden representation, which we set to 2 in this case. As a result, the matrix W_{xh} will be a 2×4 matrix, so that it maps a one-hot encoded input vector into a hidden vector $\overline{h}_t$ vector of size 2. As a practical matter, each column of W_{xh} corresponds to one of the four words, and one of these columns is copied by the expression $W_{xh}\overline{x}_t$. Note that this expression is added to $W_{hh}\overline{h}_t$ and then transformed with the tanh function to produce the final expression. The final output $\overline{y}_t$ is defined by $W_{hy}\overline{h}_t$. Note that the matrices W_{hh} and W_{hy} are of sizes 2×2 and 4×2, respectively.

In this case, the outputs are continuous values (not probabilities) in which larger values indicate greater likelihood of presence. Therefore, the word *"cat"* is predicted in the first time-stamp with a value of 1.3, although this value seems to be (incorrectly) outstripped

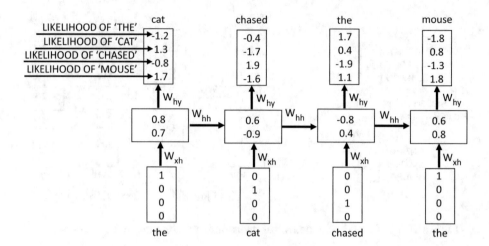

Figure 10.12: Example of language modeling with a recurrent neural network

by "*mouse*" for which the corresponding value is 1.7. However, the word "*chased*" seems to be predicted correctly. As in all learning algorithms, one cannot hope to predict any value exactly, and such errors are more likely to be made in the early iterations of the backpropagation algorithm. However, as the network is repeatedly trained over multiple iterations, it makes fewer errors over the training data.

10.7.1.1 Generating a Language Sample

Such an approach can also be used to generate an arbitrary sample of a language, once the training has been completed. How does one use such a language model at testing time, since each state requires an input word, and none is available during language generation? The likelihoods of the tokens at the first time-stamp can be generated using the <START> token as input. Since the <START> token is also available in the training data, the model will typically select a word that often starts text segments. Subsequently, the idea is to sample one of the tokens generated at each time-stamp (based on the predicted likelihood), and then use it as an input to the next time-stamp. To improve the accuracy of the sequentially predicted token, one might use beam search to expand on the most likely possibilities by always keeping track of the b best sequence prefixes of any particular length. The value of b is a user-driven parameter. By recursively applying this operation, one can generate an arbitrary sequence of text that reflects the particular language at hand. If the <END> token is predicted, it indicates the end of that particular segment of text. Although such an approach often results in syntactically correct text, it might be nonsensical in meaning. For example, a character-level RNN[3] (available/described in [290, 715]) was trained on William Shakespeare's plays. A character-level RNN requires the neural network to learn both syntax *and* spelling. After only five iterations of learning, the following was a sample of the output:

KING RICHARD II:
Do cantant,-'for neight here be with hand her,-

---

[3]A variation on the vanilla RNN, referred to as the LSTM, was used. This model is discussed later in this chapter.

Eptar the home that Valy is thee.

NORONCES:
Most ma-wrow, let himself my hispeasures;
An exmorbackion, gault, do we to do you comforr,
Laughter's leave: mire sucintracce shall have theref-Helt.

Note that there are a large number of misspellings in this case, and a lot of the words are gibberish. However, when the training was continued to 50 iterations, the following was generated as a part of the sample:

KING RICHARD II:
Though they good extremit if you damed;
Made it all their fripts and look of love;
Prince of forces to uncertained in conserve
To thou his power kindless. A brives my knees
In penitence and till away with redoom.

GLOUCESTER:
Between I must abide.

This generated piece of text is largely consistent with the syntax and spelling of the archaic English in William Shakespeare's plays, although there are still some errors. Furthermore, the approach also indents and formats the text in a manner similar to the plays by placing newlines at reasonable locations. Continuing to train for more iterations makes the output almost error-free, and some impressive samples are also available at [291].

Of course, the semantic meaning of the text is limited, and one might wonder about the usefulness of generating such nonsensical pieces of text from the perspective of machine learning applications. The key point here is that by providing an additional *contextual* input, such as the neural representation of an image, the neural network can be made to give intelligent outputs such as a gramatically correct description (i.e., caption) of the image.

The primary goal of the language-modeling RNN is not to create arbitrary sequences of the language, but to provide an architectural base that can be modified in various ways to incorporate the effect of the specific context. For example, applications like machine translation and image captioning learn a language model that is *conditioned* on another input such as a sentence in the source language or an image to be captioned. Therefore, the precise design of the application-dependent RNN will use the same principles as the language-modeling RNN, but will make small changes to this basic architecture in order to incorporate the specific context. In all these cases, the key is in choosing the input and output values of the recurrent units in a judicious way, so that one can backpropagate the output errors and learn the weights of the neural network in an application-dependent way. Examples of such applications will be discussed in Section 10.8.

10.7.2 Backpropagation Through Time

The negative logarithms of the softmax probability of the correct words at the various time-stamps are aggregated to create the loss function. If the output vector $\overline{y}_t$ can be written as $[\hat{y}_t^1 \ldots \hat{y}_t^d]$, it is first converted into a vector of d probabilities using the softmax function:

$$[\hat{p}_t^1 \ldots \hat{p}_t^d] = \text{Softmax}([\hat{y}_t^1 \ldots \hat{y}_t^d])$$

If j_t is the index of the ground-truth word at time t in the training data, then the loss function L for all T time-stamps is computed as follows:

$$L = -\sum_{t=1}^{T} \log(\hat{p}_t^{j_t}) \tag{10.51}$$

The derivative of the loss function with respect to the raw outputs may be computed using the same approach as used in multinomial logistic regression:

$$\frac{\partial L}{\partial \hat{y}_t^k} = \hat{p}_t^k - I(k, j_t) \tag{10.52}$$

Here, $I(k, j_t)$ is an indicator function that is 1 when k and j_t are the same, and 0, otherwise. Starting with this partial derivative, one can use the straightforward backpropagation update (on the unfurled temporal network) to compute the gradients with respect to the weights in different layers. However, one also has to account for the sharing of weight matrices in various temporal layers; it is not difficult to modify the backpropagation algorithm to handle shared weights.

The main trick for handling shared weights is to first "pretend" that the parameters in the different temporal layers are independent of one another. For this purpose, we introduce the temporal variables $W_{xh}^{(t)}$, $W_{hh}^{(t)}$ and $W_{hy}^{(t)}$ for time-stamp t. Conventional backpropagation is first performed by working under the pretense that these variables are distinct from one another. Then, the contributions of the different temporal avatars of the weight parameters to the gradient are added to create a unified update for each weight parameter. This special type of backpropagation algorithm is referred to as *backpropagation through time (BPTT)*. We summarize the BPTT algorithm as follows:

(i) Run the input sequentially in the forward direction through time and compute the errors (and the negative-log loss of softmax layer) at each time-stamp.

(ii) Compute the gradients of the edge weights in the backwards direction on the unfurled network without any regard for the fact that weights in different time layers are shared. In other words, it is assumed that the weights $W_{xh}^{(t)}$, $W_{hh}^{(t)}$ and $W_{hy}^{(t)}$ in time-stamp t are distinct from other time-stamps. As a result, one can use conventional backpropagation to compute $\frac{\partial L}{\partial W_{xh}^{(t)}}$, $\frac{\partial L}{\partial W_{hh}^{(t)}}$, and $\frac{\partial L}{\partial W_{hy}^{(t)}}$. Note that we have used matrix calculus notations where the derivative with respect to a matrix is defined by a corresponding matrix of element-wise derivatives.

(iii) Add all the derivatives with respect to different instantiations of an edge in time as follows:

$$\frac{\partial L}{\partial W_{xh}} = \sum_{t=1}^{T} \frac{\partial L}{\partial W_{xh}^{(t)}}$$

$$\frac{\partial L}{\partial W_{hh}} = \sum_{t=1}^{T} \frac{\partial L}{\partial W_{hh}^{(t)}}$$

$$\frac{\partial L}{\partial W_{hy}} = \sum_{t=1}^{T} \frac{\partial L}{\partial W_{hy}^{(t)}}$$

In backpropagation methods with shared weights, we are using the fact that the partial derivative of the loss with respect to each parameter can be obtained by adding the derivative with respect to each copy of the parameter. Furthermore, the computation of the partial derivatives with respect to the temporal copies of each parameter is not different from traditional backpropagation at all. Therefore, one only needs to wrap the temporal aggregation around conventional backpropagation in order to compute the update equations. The original algorithm for backpropagation through time can be credited to Werbos's seminal work in 1990 [576], long before the use of recurrent neural networks became popular.

Truncated Backpropagation Through Time

One of the computational problems in training recurrent networks is that the underlying sequences may be very long, as a result of which the number of layers in the network may also be very large. This can result in computational, convergence, and memory-usage problems. This problem is solved by using *truncated backpropagation through time*. This technique may be viewed as the analog of stochastic gradient descent for recurrent neural networks. In the approach, the state values are computed correctly during forward propagation, but the backpropagation updates are done only over segments of the sequence of modest length (such as 100). In other words, only the portion of the loss over the relevant segment is used to compute the gradients and update the weights. The segments are processed in the same order as they occur in the input sequence. The forward propagation does not need to be performed in a single shot, but it can also be done over the relevant segment of the sequence as long as the values in the final time-layer of the segment are used for computing the state values in the next segment of layers. The values in the final layer in the current segment are used to compute the values in the first layer of the next segment. Therefore, forward propagation is always able to accurately maintain state values, although the backpropagation uses only a small portion of the loss.

Practical Issues

The entries of each weight matrix are initialized to small values in $[-1/\sqrt{r}, 1/\sqrt{r}]$, where r is the number of columns in that matrix. One can also initialize each of the d columns of the input weight matrix W_{xh} to the *word2vec* embedding of the corresponding word. This approach is a form of pretraining, which is helpful when the amount of training data is small.

Another detail is that the training data often contains a special <START> and an <END> token at the beginning and end of each training segment. These types of tokens help the model to recognize specific text units such as sentences, paragraphs, or the beginning of a particular module of text. The distribution of the words at the beginning of a segment of text is often very different than how it is distributed over the whole training data. Therefore, after the occurrence of <START>, the model is more likely to pick words that begin a particular segment of text. There are other approaches that are used for deciding whether to end a segment at a particular point. A specific example is the use of a binary output that decides whether or not the sequence should continue at a particular point. Note that the binary output is in addition to other application-specific outputs. Such an approach is useful with real-valued sequences.

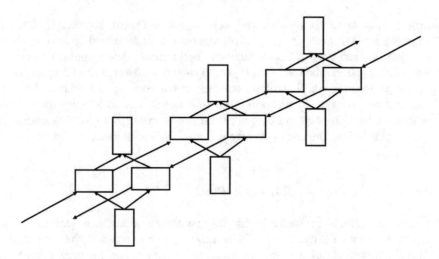

Figure 10.13: Showing three time-layers of a bidirectional recurrent network

10.7.3 Bidirectional Recurrent Networks

One disadvantage of recurrent networks is that the state at a particular time unit only has knowledge about the past inputs up to a certain point in a sentence, but it has no knowledge about future states. In certain applications like inference of word semantics, the results are vastly improved with knowledge about both past and future states. For example, the specific semantics of the polysemous word "*right*" in "*right choice*" can be inferred only after processing the word "*choice*." Therefore, the correct semantics of a word may be hard to infer without processing positions occurring after the word. Traditional recurrent networks are unable to achieve this goal at least to some extent, because any predictions made by an intermediate state occurring before such informative positions are likely to be lacking in a sufficient amount of context.

Bidirectional recurrent neural networks are appropriate for applications in which the predictions are not causal based on a historical window. A classical example of a causal setting is a stream of symbols in which an event is predicted on the basis of the history of previous symbols. Even though language-modeling applications are formally considered causal applications (i.e., based on immediate history of *previous* words), the reality is that a given word can be predicted with much greater accuracy through the use of the contextual words on each side of it. In general, bidirectional RNNs work well in applications where the predictions are based on bidirectional context. Examples of such applications include handwriting recognition and speech recognition, in which the properties of individual elements in the sequence depend on those on either side of it. For example, if a handwriting is expressed in terms of the strokes, the strokes on either side of a particular position are helpful in recognizing the particular character being synthesized. It has increasingly become common to use bidirectional recurrent networks in various language-centric applications as opposed to unidirectional networks.

In the bidirectional recurrent network, we have separate hidden states $\overline{h}_t^{(f)}$ and $\overline{h}_t^{(b)}$ for the forward and backward directions. The forward hidden states interact only with each other and the same is true for the backward hidden states. The main difference is that the forward states interact in the forwards direction, while the backwards states interact in the backwards direction. Both $\overline{h}_t^{(f)}$ and $\overline{h}_t^{(b)}$, however, receive input from the same vector $\overline{x}_t$ (e.g., one-hot encoding of word) and they interact with the same output vector $\hat{y}_t$. An example of three time-layers of the bidirectional RNN is shown in Figure 10.13.

In the case of the bidirectional network, we have separate forward and backward parameter matrices. The forward matrices for the input-hidden, hidden-hidden, and hidden-output interactions are denoted by $W_{xh}^{(f)}$, $W_{hh}^{(f)}$, and $W_{hy}^{(f)}$, respectively. The backward matrices for the input-hidden, hidden-hidden, and hidden-output interactions are denoted by $W_{xh}^{(b)}$, $W_{hh}^{(b)}$, and $W_{hy}^{(b)}$, respectively.

The recurrence conditions can be written as follows:

$$\overline{h}_t^{(f)} = \tanh(W_{xh}^{(f)}\overline{x}_t + W_{hh}^{(f)}\overline{h}_{t-1}^{(f)})$$
$$\overline{h}_t^{(b)} = \tanh(W_{xh}^{(b)}\overline{x}_t + W_{hh}^{(b)}\overline{h}_{t+1}^{(b)})$$
$$= \overline{y}_t = W_{hy}^{(f)}\overline{h}_t^{(f)} + W_{hy}^{(b)}\overline{h}_t^{(b)}$$

It is easy to see that the bidirectional equations are simple generalizations of the conditions used in a single direction. It is assumed that there are a total of T time-stamps in the neural network shown above, where T is the length of the sequence. One question is about the forward input at the boundary conditions corresponding to $t = 1$ and the backward input at $t = T$, which are not defined. In such cases, one can use a default constant value of 0.5 in each case, although one can also make the determination of these values as a part of the learning process.

An immediate observation about the hidden states in the forward and backwards direction is that they do not interact with one another at all. Therefore, one could first run the sequence in the forward direction to compute the hidden states in the forward direction, and then run the sequence in the backwards direction to compute the hidden states in the backwards direction. At this point, the output states are computed from the hidden states in the two directions. After the outputs have been computed, the backpropagation algorithm is applied to compute the partial derivatives with respect to various parameters. First, the partial derivatives are computed with respect to the output states because both forward and backwards states point to the output nodes. Then, the backpropagation pass is computed only for the forward hidden states starting from $t = T$ down to $t = 1$. The backpropagation pass is finally computed for the backwards hidden states from $t = 1$ to $t = T$. Finally, the partial derivatives with respect to the shared parameters are aggregated. Therefore, the BPTT algorithm can be generalized to bidirectional networks as follows:

1. Compute forward and backwards hidden states independently in separate passes.

2. Compute output states from backwards and forward hidden states.

3. Compute partial derivatives of loss with respect to output states and each copy of the output parameters.

4. Compute partial derivatives of loss with respect to forward states and backwards states independently using backpropagation. Use these computations to evaluate partial derivatives with respect to each copy of the forwards and backwards parameters.

5. Aggregate partial derivatives over shared parameters.

A bidirectional recurrent network shares some resemblance to an ensemble of two independent recurrent networks—in one, the input is presented in original form, and in the other, the input is reversed. The main difference is that the parameters of the forwards and backwards states are interdependent and therefore trained jointly in bidirectional networks.

10.7.4 Multilayer Recurrent Networks

In all of the aforementioned applications, a single-layer RNN architecture is used for ease in understanding. However, in real applications, a multilayer architecture is used in order to build models of greater complexity. Furthermore, this multilayer architecture can be used in combination with advanced variations of the RNN, such as the LSTM architecture.

An example of a deep network containing three layers is shown in Figure 10.14. Note that nodes in higher-level layers receive input from those in lower-level layers. The relationships among the hidden states can be generalized directly from the single-layer network. First, we rewrite the recurrence equation of the hidden layers (for single-layer networks) in a form that can be adapted easily to multilayer networks:

$$\overline{h}_t = \tanh(W_{xh}\overline{x}_t + W_{hh}\overline{h}_{t-1})$$
$$= \tanh W \left[\begin{array}{c} \overline{x}_t \\ \overline{h}_{t-1} \end{array} \right]$$

Here, we have put together a larger matrix $W = [W_{xh}, W_{hh}]$ that includes the columns of W_{xh} and W_{hh}. Similarly, we have created a larger column vector that stacks up the state vector in the first hidden layer at time $t-1$ and the input vector at time t. In order to distinguish between the hidden nodes for the upper-level layers, let us add an additional superscript to the hidden state and denote the vector for the hidden states at time-stamp t and layer k by $\overline{h}_t^{(k)}$. Similarly, let the weight matrix for the kth hidden layer be denoted by $W^{(k)}$. It is noteworthy that the weights are shared across different time-stamps (as in the single-layer recurrent network), but they are not shared across different layers. Therefore, the weights are superscripted by the layer index k in $W^{(k)}$. The first hidden layer is special because it receives inputs both from the input layer at the current time-stamp and the adjacent hidden state at the previous time-stamp. Therefore, the matrices $W^{(k)}$ will have a size of $p \times (d+p)$ only for the first layer (i.e., $k = 1$), where d is the size of the input vector $\overline{x}_t$ and p is the size of the hidden vector $\overline{h}_t$. Note that d will typically not be the same as p. The recurrence condition for the first layer is already shown above by setting $W^{(1)} = W$. Therefore, let us focus on all the hidden layers k for $k \geq 2$. It turns out that the recurrence condition for the layers with $k \geq 2$ is also in a very similar form as the equation shown above:

$$\overline{h}_t^{(k)} = \tanh W^{(k)} \left[\begin{array}{c} \overline{h}_t^{(k-1)} \\ \overline{h}_{t-1}^{(k)} \end{array} \right]$$

In this case, the size of the matrix $W^{(k)}$ is $p \times (p+p) = p \times 2p$. The transformation from hidden to output layer remains the same as in single-layer networks. It is easy to see that this approach is a straightforward multilayer generalization of the case of single-layer networks. It is common to use two or three layers in practical applications.

10.7.5 Long Short-Term Memory (LSTM)

Recurrent neural networks have problems associated with *vanishing and exploding gradients* [254, 433]. This is a common problem in neural network updates where successive multiplication by the matrix $W^{(k)}$ is inherently unstable; it either results in the gradient

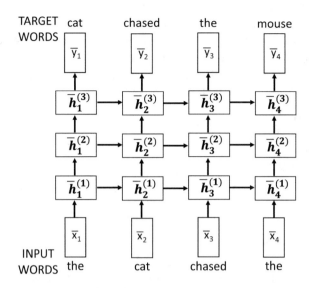

Figure 10.14: Multi-layer recurrent neural networks

disappearing during backpropagation or in blowing up to large values in an unstable way. This type of instability is the direct result of successive multiplication with the (recurrent) weight matrix at various time-stamps, which either continually increases or decreases the prediction. Another problem is that this model erases some of the information learned in previous time stamps at each update. One way of viewing this problem is that a neural network that uses only multiplicative updates is good only at learning over short sequences, because it scrambles all the hidden states. Such an approach is therefore inherently endowed with good short-term memory but poor long-term memory [254]. To address this problem, a solution is to change the recurrence equation for the hidden vector with the use of the LSTM.

The LSTM is an enhancement of the recurrent neural network architecture of Figure 10.14 in which we change the recurrence conditions of how the hidden states $\overline{h}_t^{(k)}$ are propagated. In order to achieve this goal, we have an additional hidden vector of p dimensions, which is denoted by $\overline{c}_t^{(k)}$, and it is referred to as the *cell state*. One can view the cell state as a kind of long-term memory that retains at least a part of the information in earlier hidden states by using a combination of partial "forgetting" and "increment" operations on previous cell states. It has been shown in [290] that the nature of the memory in $\overline{c}_t^{(k)}$ is occasionally interpretable when it is applied to text data such as literary pieces. For example, one of the p values in $\overline{c}_t^{(k)}$ might change in sign after an opening quotation and then revert back only when that quotation is closed. The upshot of this phenomenon is that the resulting neural network is able to model long-range dependencies in the language or even a specific pattern (like a quotation) extended over a large number of tokens. This is achieved by using a gentle approach to update these cell states over time, so that there is greater persistence in information storage.

As with the multilayer recurrent network, the update matrix is denoted by $W^{(k)}$ and is used to premultiply the column vector $[\overline{h}_t^{(k-1)}, \overline{h}_{t-1}^{(k)}]^T$. However, this matrix is of size $4p \times 2p$, and therefore pre-multiplying a vector of size $2p$ with $W^{(k)}$ results in a vector of size $4p$. In this case, the updates use four intermediate, p-dimensional vector variables $\overline{i}, \overline{f}$,

$\bar{o}$, and $\bar{c}$. The intermediate variables $\bar{i}$, $\bar{f}$, and $\bar{o}$ are respectively referred to as *input*, *forget*, and *output* variables, because of the roles they play in updating the cell states and hidden states. The determination of the hidden state vector $\overline{h}_t^{(k)}$ and the cell state vector $\overline{c}_t^{(k)}$ uses a multi-step process of first computing these intermediate variables and then computing the hidden variables from these intermediate variables:

$$\begin{bmatrix} \bar{i} \\ \bar{f} \\ \bar{o} \\ \bar{c} \end{bmatrix} = \begin{pmatrix} \text{sigm} \\ \text{sigm} \\ \text{sigm} \\ \tanh \end{pmatrix} W^{(k)} \begin{bmatrix} \overline{h}_t^{(k-1)} \\ \overline{h}_{t-1}^{(k)} \end{bmatrix} \quad \textbf{[Setting up intermediate variables]}$$

$$\overline{c}_t^{(k)} = \bar{f} \odot \overline{c}_{t-1}^{(k)} + \bar{i} \odot \bar{c} \quad \textbf{[Selectively forget and/or add to long-term memory]}$$

$$\overline{h}_t^{(k)} = \bar{o} \odot \tanh(\overline{c}_t^{(k)}) \quad \textbf{[Selectively leak long-term memory to hidden state]}$$

Note the difference between $\bar{c}$ and $\overline{c}_t^{(k)}$. Here, the element-wise product of vectors is denoted by "$\odot$," and the notation "sigm" denotes a sigmoid operation. In practical implementations, biases are also used in the above updates, although they are omitted here for simplicity. The aforementioned update seems rather cryptic, and therefore it requires further explanation.

The first step in the above sequence of equations is to set up the intermediate variable vectors $\bar{i}$, $\bar{f}$, $\bar{o}$, and $\bar{c}$, of which the first three should *conceptually* be considered binary values, although they are continuous values in $(0, 1)$. Multiplying a pair of binary values is like using an AND gate on a pair of boolean values. We will henceforth refer to this operation as gating. The vectors $\bar{i}$, $\bar{f}$, and $\bar{o}$ are referred to as input, forget, and output gates, and $\bar{c}$ is the newly proposed content of the cell state. In particular, these vectors are conceptually used as boolean gates for deciding (i) whether to add to a cell-state, (ii) whether to forget a cell state, and (iii) whether to allow leakage into a hidden state from a cell state. The use of the binary abstraction for the input, forget, and output variables helps in understanding the types of decisions being made by the updates. In practice, a continuous value in $(0, 1)$ is contained in these variables, which can enforce the effect of the binary gate in a probabilistic way if the output is seen as a probability. In the neural network setting, it is essential to work with continuous functions in order to ensure the differentiability required for gradient updates.

The four intermediate variables $\bar{i}$, $\bar{f}$, $\bar{o}$, and $\bar{c}$, are set up using the weight matrices $W^{(k)}$ for the kth layer using the first equation above. Let us now examine the second equation that updates the cell state with the use of some of these intermediate variables:

$$\overline{c}_t^{(k)} = \underbrace{\bar{f} \odot \overline{c}_{t-1}^{(k)}}_{\text{Reset?}} + \underbrace{\bar{i} \odot \bar{c}}_{\text{Increment?}}$$

This equation has two parts. The first part uses the p forget bits in $\bar{f}$ to decide which of the p cell states from the previous time-stamp to reset[4] to 0, and it uses the p input bits in $\bar{i}$ to decide whether to add the corresponding components from $\bar{c}$ to each of the cell states. Note that such updates of the cell states are in additive form, which is helpful in avoiding the vanishing gradient problem caused by multiplicative updates. One can view the cell-state vector as a continuously updated long-term memory, where the forget and input bits respectively decide (i) whether to reset the cell states from the previous time-stamp and

---

[4]Here, we are treating the forget bits as a vector of binary bits, although it contains continuous values in $(0, 1)$, which can be viewed as probabilities. As discussed earlier, the binary abstraction helps us understand the conceptual nature of the operations.

forget the past, and (ii) whether to increment the cell states from the previous time-stamp to incorporate new information into long-term memory from the current word. The vector $\overline{c}$ contains the p amounts with which to increment the cell states, and these are values in $[-1, +1]$ because they are all outputs of the tanh function.

Finally, the hidden states $\overline{h}_t^{(k)}$ are updated using leakages from the cell state. The hidden state is updated as follows:

$$\overline{h}_t^{(k)} = \underbrace{\overline{o} \odot \tanh(\overline{c}_t^{(k)})}_{\text{Leak } \overline{c}_t^{(k)} \text{ to } \overline{h}_t^{(k)}}$$

Here, we are copying a functional form of each of the p cell states into each of the p hidden states, depending on whether the output gate (defined by $\overline{o}$) is 0 or 1. Of course, in the continuous setting of neural networks, partial gating occurs and only a fraction of the signal is copied from each cell state to the corresponding hidden state. It is noteworthy that the use of the tanh in the last equation is not always essential to ensure good performance. The update can be even as simple as a (partial) copy from each cell state to the corresponding hidden state. This alternative update is as follows, which can be used to replace the above equation:

$$\overline{h}_t^{(k)} = \overline{o} \odot \overline{c}_t^{(k)}$$

As in the case of all neural networks, the backpropagation algorithm is used for training purposes. It is noteworthy that all the applications discussed in previous sections (for single-layer RNNs) are always implemented using multi-layer LSTMs for best performance. Another similar model is the *Gated Recurrent Unit (GRU)*, which also provides excellent performance. This model is discussed in the next section.

10.7.6 Gated Recurrent Units (GRUs)

The Gated Recurrent Unit (GRU) can be viewed as a simplification of the LSTM, which does not use explicit cell states. Another difference is that the LSTM directly controls the amount of information changed in the hidden state using separate forget and output gates. On the other hand, a GRU uses a single reset gate to achieve the same goal. However, the basic idea in the GRU is quite similar to that of an LSTM, in terms of how it partially resets the hidden states. As in the previous sections, we assume a multilayer architecture (in addition to time layering); therefore, the notation $\overline{h}_t^{(k)}$ represents the hidden states of the kth layer at time-stamp t for $k \geq 1$. For notational convenience, we also assume that the input layer $\overline{x}_t$ can be denoted by $\overline{h}_t^{(0)}$ (although this layer is obviously not hidden). As in the case of LSTM, we assume that the input vector $\overline{x}_t$ is d-dimensional, whereas the hidden states are p-dimensional. The sizes of the transformation matrices in the first layer are accordingly adjusted to account for this fact.

In the case of the GRU, we use two matrices $W^{(k)}$ and $V^{(k)}$ of sizes[5] $2p \times 2p$ and $p \times 2p$, respectively. Pre-multiplying a vector of size $2p$ with $W^{(k)}$ results in a vector of size $2p$, which will be passed through the sigmoid activation to create two intermediate, p-dimensional vector variables $\overline{z}_t$ and $\overline{r}_t$, respectively. The intermediate variables $\overline{z}_t$ and $\overline{r}_t$ are respectively referred to as update and reset gates. The determination of the hidden state vector $\overline{h}_t^{(k)}$ uses a two-step process of first computing these gates, then using them to

---

[5]In the first layer ($k = 1$), these matrices are of sizes $2p \times (p + d)$ and $p \times (p + d)$.

decide how much to change the hidden vector with the weight matrix $V^{(k)}$:

$$\begin{array}{c} \text{Update Gate:} \\ \text{Reset Gate:} \end{array} \left[\begin{array}{c} \overline{z} \\ \overline{r} \end{array} \right] = \left(\begin{array}{c} \text{sigm} \\ \text{sigm} \end{array} \right) W^{(k)} \left[\begin{array}{c} \overline{h}_t^{(k-1)} \\ \overline{h}_{t-1}^{(k)} \end{array} \right] \quad \textbf{[Set up gates]}$$

$$\overline{h}_t^{(k)} = \overline{z} \odot \overline{h}_{t-1}^{(k)} + (1 - \overline{z}) \odot \tanh V^{(k)} \left[\begin{array}{c} \overline{h}_t^{(k-1)} \\ \overline{r} \odot \overline{h}_{t-1}^{(k)} \end{array} \right] \quad \textbf{[Update hidden state]}$$

Here, the element-wise product of vectors is denoted by "$\odot$," and the notation "sigm" denotes a sigmoid operation. For the very first layer (i.e., $k = 1$), the notation $\overline{h}_t^{(k-1)}$ in the above equation should be replaced with $\overline{x}_t$. Furthermore, the matrices $W^{(1)}$ and $V^{(1)}$ are of sizes $2p \times (p+d)$ and $p \times (p+d)$, respectively. One can write the entirety of the function represented in the equations above by a single GRU function:

$$\overline{h}_t^{(k)} = \text{GRU}\left(\overline{h}_t^{(k-1)}, \overline{h}_{t-1}^{(k)}, W^{(k)}, V^{(k)} \right) \tag{10.53}$$

Here, $W^{(k)}$ and $V^{(k)}$ are the matrices used in the kth layer for setting up the gates and updating the hidden states respectively. We have also omitted the mention of biases here, but they are usually included in practical implementations. In the following, we provide a further explanation of these updates and contrast them with those of the LSTM. Just as the LSTM uses input, output, and forget gates to decide how much of the information from the previous time-stamp to carry over to the next step, the GRU uses the update and the reset gates. The GRU does not have a separate internal memory and also requires fewer gates to perform the update from one hidden state to another. Therefore, a natural question arises about the precise role of the update and reset gates. The reset gate $\overline{r}$ decides how much of the hidden state to carry over from the previous time-stamp for a matrix-based update (like a recurrent neural network). The update gate $\overline{z}$ decides the *relative* strength of the contributions of this matrix-based update and a more direct contribution from the hidden vector $\overline{h}_{t-1}^{(k)}$ at the previous time-stamp. By allowing a direct (partial) copy of the hidden states from the previous layer, the gradient flow becomes more stable during backpropagation. The update gate of the GRU simultaneously performs the role of the input and forget gates in the LSTM in the form of $\overline{z}$ and $1 - \overline{z}$, respectively. However, the mapping between the GRU and the LSTM is not precise, because it performs these updates directly on the hidden state (and there is no cell state). Although the GRU is a closely related simplification of the LSTM, it should not be seen as a special case of the LSTM.

In order to understand why GRUs provide better performance than vanilla RNNs, let us examine a GRU with a single layer and single state dimensionality $p = 1$. In such a case, the update equation of the GRU can be written as follows:

$$h_t = z \cdot h_{t-1} + (1 - z) \cdot \tanh[v_1 \cdot x_t + v_2 \cdot r \cdot h_{t-1}] \tag{10.54}$$

Note that layer superscripts are missing in this single-layer case. Here, v_1 and v_2 are the two elements of the 2×1 matrix V. Then, it is easy to see the following:

$$\frac{\partial h_t}{\partial h_{t-1}} = z + (\text{Additive Terms}) \tag{10.55}$$

Backward gradient flow is multiplied with this factor. Here, the term $z \in (0, 1)$ helps in passing *unimpeded* gradient flow and makes computations more stable. Furthermore, since the additive terms heavily depend on $(1 - z)$, the overall multiplicative factor tends to be closer to 1 even when z is small.

A comparison of the LSTM and the GRU is provided in [115, 282]. The two models are shown to be roughly similar in performance, and the relative performance seems to depend on the task at hand. The GRU is simpler and enjoys the advantage of greater ease of implementation and efficiency. It might generalize slightly better with less data because of a smaller parameter footprint [115], although the LSTM would be preferable with an increased amount of data. The work in [282] also discusses several practical implementation issues associated with the LSTM. The LSTM has been more extensively tested than the GRU, simply because it is an older architecture and enjoys widespread popularity. As a result, it is generally seen as a safer option, particularly when working with longer sequences and larger data sets. The work in [220] also showed that none of the variants of the LSTM can reliably outperform it in a consistent way. This is because of the explicit internal memory and the greater gate-centric control in updating the LSTM.

10.7.7 Layer Normalization

Layer normalization is a technique, which is intended to address the inherent instability in training recurrent neural networks. In layer normalization, the normalization is performed only over a single training instance, and the normalization factor is obtained by using all the current activations in that layer of only the current instance. In order to understand how layer-wise normalization works, we repeat the hidden-to-hidden recursion of page 342:

$$\overline{h}_t = \tanh(W_{xh}\overline{x}_t + W_{hh}\overline{h}_{t-1})$$

This recursion is prone to unstable behavior because of the multiplicative effect across time-layers. We will show how to modify this recurrence with layer-wise normalization. The normalization is applied to *pre-activation* values before applying the tanh activation function. Therefore, the pre-activation value at the tth time-stamp is computed as follows:

$$\overline{a}_t = W_{xh}\overline{x}_t + W_{hh}\overline{h}_{t-1}$$

Note that $\overline{a}_t$ is a vector with as many components as the number of units in the hidden layer (which we have consistently denoted as p in this chapter). We compute the mean μ_t and standard σ_t of the pre-activation values in $\overline{a}_t$:

$$\mu_t = \frac{\sum_{i=1}^{p} a_{ti}}{p}, \qquad \sigma_t = \sqrt{\frac{\sum_{i=1}^{p} a_{ti}^2}{p} - \mu_t^2}$$

Here, a_{ti} denotes the ith component of the vector $\overline{a}_t$.

Note that we have additional learning parameters, associated with each unit. Specifically, for the p units in the tth layer, we have a p-dimensional vector of *gain parameters* $\overline{\gamma}_t$, and a p-dimensional vector of *bias parameters* denoted by $\overline{\beta}_t$. The purpose of these parameters is to re-scale the normalized values and add bias in a learnable way. The hidden activations $\overline{h}_t$ of the next layer are therefore computed as follows:

$$\overline{h}_t = \tanh\left(\frac{\overline{\gamma}_t}{\sigma_t} \odot (\overline{a}_t - \overline{\mu}_t) + \overline{\beta}_t\right) \tag{10.56}$$

Here, the notation $\odot$ indicates elementwise multiplication, and the notation $\overline{\mu}_t$ refers to a vector containing p copies of the scalar μ_t. The effect of layer normalization is to ensure that the magnitudes of the activations do not continuously increase or decrease with time-stamp (causing vanishing and exploding gradients), and the learnable parameters add flexibility.

10.8 Applications of Recurrent Neural Networks

Recurrent neural networks have numerous applications in the text domain because of their ability to construct language models. The hidden states in the neural network also encode the representations of full sentences, which are used in various sentence-wise classification applications. Because of the versatility of recurrent neural networks in terms of generating representations of both sentences and specific positions in sentences, they can be used for sentence-wise applications and token-wise applications (e.g., classifying specific words). Furthermore, by conditioning the recurrent neural network on a specific input, they can be used to generate sentences (e.g., image caption) that are relevant to the context (e.g., image) at hand. However, the vanilla application of building a language model is itself quite powerful for transfer learning. It is noteworthy that many of these applications are based on variations of recurrent neural networks like LSTMs or GRUs rather than the vanilla forms of recurrent networks. Nevertheless, we will use the generic architecture of recurrent neural networks in order to retain simplicity in presentation.

10.8.1 Contextual Word Embeddings with ELMo

Recurrent neural networks are extremely powerful at constructing pre-trained language models, which are then used for a variety of downstream applications. An important observation about language models is that *they are unsupervised multitask learners.* For example, most of the applications discussed in this section can be accomplished more efficiently, if one starts with a language model, and then performs task-specific fine-tuning. The amount of effort involved in performing the fine tuning is typically much smaller than what is required than training the model from scratch. This is essentially an example of *transfer learning,* wherein the training work that is done up front is reused for task-specific training. The main caveat for building a useful pre-trained language model is that it needs to be large and built on a sufficient training data. This type of model is also useful in building *contextualized* word representations that have an advantage over models like GloVe or *word2vec* in terms of creating *sentence-specific* embeddings. For example, consider the following pair of sentences:

> He used a stick to pry open the gate.
> The postage stamp did not stick properly to the envelope.

The GloVe (or *word2vec*) embedding is not sentence-specific, and it will therefore give the same representation to the word "stick" in both cases, which can be viewed as a weighted average of the embeddings of the two contexts, based on relative presence in the *training* corpus. However, in some applications, one might want to derive a word embedding that is *specific to the particular sentence at hand.* This type of approach is useful in many applications like *word sense disambiguation* in which we wish to determine the correct sense in which a particular word is used.

The ELMo language model [438] is constructed by using a bidirectional recurrent neural network (cf. Section 10.7.3), and more specifically, a bidirectional LSTM. After it has been trained, contextualized embeddings for the words in a given sentence can be generated by feeding the sentence to the trained network. The contextualized representation of a word is given by the concatenation of the forward and backward (hidden) states at that position of the sentence. If a multilayer recurrent neural network is used, then the forward and backward states of all the layers in a particular position are concatenated. This type of word representation can be very useful for token-wise applications like identifying the part of speech in a particular sentence. These applications are discussed later in this section. Furthermore, the forward hidden state at the end of the sentence and the backward hidden state at the beginning of the sentence can be concatenated in order to obtain a representation of the full sentence. Such embeddings are very useful for applications like question answering (cf. Chapter 14). *Language models can also be used for multitask training, because the training sequences can start with a task-specific prefix (e.g., "Translate from English to German:"), which is followed by the source and target sequences separated by special tokens.* The pretrained language model of ELMo has been used for many of the applications discussed in later chapters, such as text summarization (cf. Chapter 12), entity extraction (cf. Chapter 13), sentiment analysis (cf. Chapter 15), *co-reference resolution, semantic role labeling, word-sense disambiguation, textual entailment,* and *grammar checking.* Some of these applications are discussed briefly as applications of *transformer models* in Section 11.5 of Chapter 11. It is also noteworthy that the use of pre-trained language models based on transformers has now become more common as compared to the use of recurrent models like ELMo.

10.8.2 Application to Automatic Image Captioning

In image captioning, the training data consists of image-caption pairs. For example, the image[6] in the left-hand side of Figure 10.15 is obtained from the National Aeronautics and Space Administration Web site. This image is captioned *"cosmic winter wonderland."* One might have hundreds of thousands of such image-caption pairs. These pairs are used to train the weights in the neural network. Once the training has been completed, the captions are predicted for unknown test instances. Therefore, one can view this approach as an instance of image-to-sequence learning.

One issue in the automatic captioning of images is that a separate neural network is required to learn the representation of the images. A common architecture to learn the representation of images is the *convolutional neural network.* A detailed discussion of convolutional neural networks is beyond the scope of this book, and readers are referred to [210]. Consider a setting in which the convolutional neural network produces the q-dimensional vector $\overline{v}$ as the output representation. This vector is then used as an input to the neural network, but only[7] at the first time-stamp. To account for this additional input, we need another $p \times q$ matrix W_{ih}, which maps the image representation to the hidden layer. Therefore, the update equations for the various layers now need to be modified as follows:

$$\overline{h}_1 = \tanh(W_{xh}\overline{x}_1 + W_{ih}\overline{v})$$
$$\overline{h}_t = \tanh(W_{xh}\overline{x}_t + W_{hh}\overline{h}_{t-1}) \quad \forall t \geq 2$$
$$\overline{y}_t = W_{hy}\overline{h}_t$$

An important point here is that the convolutional neural network and the recurrent neural network are not trained in isolation. Although one might train them in isolation in order to

---

[6]https://www.nasa.gov/mission_pages/chandra/cosmic-winter-wonderland.html.

[7]In principle, one can also allow it to be input at all time-stamps, but it only seems to worsen performance.

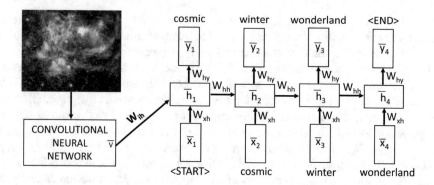

Figure 10.15: Example of image captioning with a recurrent neural network. An additional convolutional neural network is required for representational learning of the images. The image is represented by the vector $\overline{v}$, which is the output of the convolutional neural network. The inset image is by courtesy of the National Aeronautics and Space Administration (NASA)

create an initialization, the final weights are always trained jointly by running each image through the network and matching up the predicted caption with the true caption. In other words, for each image-caption pair, the weights in both networks are updated when errors are made in predicting the caption. Such an approach ensures that the learned representation $\overline{v}$ of the images is sensitive to the specific application of predicting captions. After all the weights have been trained, a test image is input to the entire system and passed through both the convolutional and recurrent neural network. For the recurrent network, the input at the first time-stamp is the <START> token and the representation of the image. At later time-stamps, the input is the most likely token predicted at the previous time-stamp. One can also use beam search to keep track of the b most likely sequence prefixes to expand on at each point. This approach is not very different from the language generation approach discussed in Section 10.7.1.1, except that it is conditioned on the image representation that is input to the model in the first time-stamp of the recurrent network. This results in the prediction of a relevant caption for the image.

10.8.3   Sequence-to-Sequence Learning and Machine Translation

Just as one can put together a convolutional neural network and a recurrent neural network to perform image captioning, one can put together two recurrent networks to translate one language into another. Such methods are also referred to as *sequence-to-sequence* learning because a sequence in one language is mapped to a sequence in another language. In principle, sequence-to-sequence learning can have applications beyond machine translation. For example, even text summarization and question-answering (QA) systems can be viewed as sequence-to-sequence learning applications.

In the following, we provide a simple solution to machine translation with recurrent neural networks, although such applications are rarely addressed directly with the simple forms of recurrent neural networks. Rather, a variation of the recurrent neural network, such as the long short-term memory (LSTM) or Gated Recurrent Unit (GRU) model is used. More recently, *transformers* (cf. Section 11.3 of Chapter 11) have been used widely.

In the machine translation application, two different RNNs are hooked end-to-end, just as a convolutional neural network and a recurrent neural network are hooked together for

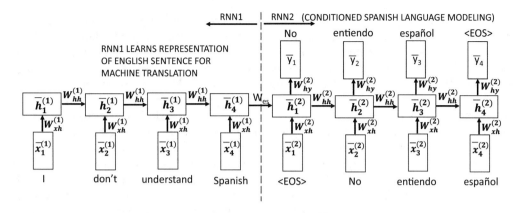

Figure 10.16: Machine translation with recurrent neural networks. Note that there are two separate recurrent networks with their own sets of shared weights. The output of $\overline{h}_4^{(1)}$ is a fixed length encoding of the 4-word English sentence

image captioning. The first recurrent network uses the words from the source language as input. No outputs are produced at these time-stamps and the successive time-stamps accumulate knowledge about the source sentence in the hidden state. Subsequently, the end-of-sentence symbol is encountered, and the second recurrent network starts by outputting the first word of the target language. The next set of states in the second recurrent network output the words of the sentence in the target language one by one. These states also use the words of the target language as input, which is available for the case of the training instances but not for test instances (where predicted values are used instead). This architecture is shown in Figure 10.16.

The architecture of Figure 10.16 is similar to that of an autoencoder, and can even be used with pairs of identical sentences in the same language to create fixed-length representations of sentences. The two recurrent networks are denoted by RNN1 and RNN2, and their weights are not the same. For example, the weight matrix between two hidden nodes at successive time-stamps in RNN1 is denoted by $W_{hh}^{(1)}$, whereas the corresponding weight matrix in RNN2 is denoted by $W_{hh}^{(2)}$. The weight matrix W_{es} of the link joining the two neural networks is special, and can be independent of either of the two networks. This is necessary if the sizes of the hidden vectors in the two RNNs are different because the dimensions of the matrix W_{es} will be different from those of both $W_{hh}^{(1)}$ and $W_{hh}^{(2)}$. As a simplification, one can use[8] the same size of the hidden vector in both networks, and set $W_{es} = W_{hh}^{(1)}$. The weights in RNN1 are devoted to learning an encoding of the input in the source language, and the weights in RNN2 are devoted to using this encoding in order to create an output sentence in the target language. One can view this architecture in a similar way to the image captioning application, except that we are using two recurrent networks instead of a convolutional-recurrent pair. The output of the final hidden node of RNN1 is a fixed-length encoding of the source sentence. Therefore, irrespective of the length of the sentence, the encoding of the source sentence depends on the dimensionality of the hidden representation.

---

[8]The original work in [537] seems to use this option [313]. In the Google Neural Machine Translation system [716], this weight is removed. This system is now used in Google Translate.

The grammar and length of the sentence in the source and target languages may not be the same. In order to provide a grammatically correct output in the target language, RNN2 needs to learn its language model. It is noteworthy that the units in RNN2 associated with the target language have both inputs and outputs arranged in the same way as a language-modeling RNN. At the same time, the output of RNN2 is conditioned on the input it receives from RNN1, which effectively causes language translation. In order to achieve this goal, training pairs in the source and target languages are used. The approach passes the source-target pairs through the architecture of Figure 10.16 and learns the model parameters with the use of the backpropagation algorithm. Since only the nodes in RNN2 have outputs, only the errors made in predicting the target language words are backpropagated to train the weights in both neural networks. The two networks are jointly trained, and therefore the weights in both networks are optimized to the errors in the translated outputs of RNN2. As a practical matter, this means that the internal representation of the source language learned by RNN1 is highly optimized to the machine translation application, and is very different from one that would be learned if one had used RNN1 to perform language modeling of the source sentence. After the parameters have been learned, a sentence in the source language is translated by first running it through RNN1 to provide the necessary input to RNN2. Aside from this contextual input, another input to the first unit of RNN2 is the <EOS> tag, which causes RNN2 to output the likelihoods of the first token in the target language. The most likely token using beam search (cf. Section 10.7.1.1) is selected and used as the input to the recurrent network unit in the next time-stamp. This process is recursively applied until the output of a unit in RNN2 is also <EOS>. As in Section 10.7.1.1, we are generating a sentence from the target language using a language-modeling approach, except that the specific output is conditioned on the internal representation of the source sentence.

The use of neural networks for machine translation is relatively recent. Recurrent neural network models have a sophistication that greatly exceeds that of traditional machine translation models. The latter class of methods uses phrase-centric machine learning, which is often not sophisticated enough to learn the subtle differences between the grammars of the two languages. In practice, deep models with multiple layers are used to improve the performance. A discussion of deep variations of recurrent neural networks is provided in Section 10.7.4.

One weakness of such translation models is that they tend to work poorly when the sentences are long. Numerous solutions have been proposed to solve the problem. A recent solution is that the sentence in the source language is input in the *opposite order* [537]. This approach brings the first few words of the sentences in the two languages closer in terms of their time-stamps within the recurrent neural network architecture. As a result, the first few words in the target language are more likely to be predicted correctly. The correctness in predicting the first few words is also helpful in predicting the subsequent words, which are also dependent on a neural language model in the target language.

Sequence-to-sequence encoders are also used in a host of other natural language tasks. As discussed in Chapter 12, the approach can be used for text summarization by treating the original text as the source passage and the summary as the target passage. Similarly, sequence-to-sequence learners are used for a variety of subtasks in question answering. One such subtask is to convert an unstructured natural language query to a structured query that can work on *knowledge graphs*. Such methods are discussed in Chapter 14. Furthermore, *almost any natural-language processing task that has a pre-defined text input and text output can treated as a sequence-to-sequence task.* For example, all the tasks discussed later in this section (with various other architectures) can also be addressed with the sequence-to-sequence architecture. For example, a sentence-level classification task (see next section) can be treated as a source sentence followed by a target "sequence" containing the text class label. A token-level classification task (see Section 10.8.5) can be treated as a sequence-to-sequence task by treating the output as a sequence of token labels. This general principle is used by the T5 transformer (cf. Section 11.4.3 of Chapter 11) to create a multi-task learner from sequence-to-sequence models.

10.8.3.1 BLEU Score for Evaluating Machine Translation

The BLEU score is used to evaluate machine translation, and it stands for *Bilingual Evaluation Understudy*. BLEU tries to measure how well the translation of a machine corresponds to the ideal translation that one would obtain from a human. The BLEU score is a number between 0 and 1, with higher values indicating better quality. A score of 1 is obtained when the translated sentence matches a single reference (ground-truth) translation exactly. It is assumed that a set of good-quality reference translations are available in order to compute the BLEU score. In general, since the reference translations may be created in multiple ways, one would rarely obtain a BLEU score of 1. However, better translations will almost always yield higher BLEU scores, as long as the reference translations are of good quality. For each sentence, multiple reference translations might be available, which are used to provide greater robustness of the BLEU score for a given sentence.

The corpus is segmented at the sentence level and the BLEU score for each sentence is first computed. The BLEU score for the full corpus is equal to the geometric mean of the BLEU score over all sentences, which is then adjusted with a *length-wise penalty factor* (described later). For each translated sentence s, let $s_1, s_2, \ldots, s_k$ be the reference translations (i.e., alternative translations) provided as the ground truth. Interestingly, the BLEU score focuses on computing the precision of the distinct words present in the translated sentence, rather than the recall. Let $w_1 \ldots w_r$ be the distinct words in the sentence s, and let $|s|$ be the number of words in translated sentence s. Let $n(w_i)$ be the *minimum* number of times that the word w_i is present in any of the reference translations $s_1 \ldots s_k$ and also the predicated translation s. Note that since $n(w_i)$ is at most equal to the number of times w_i is present in s, the sum over all $n(w_i)$ is at most equal to $|s|$. Then, the sentence-wise BLEU score is the following precision value:

$$BLEU(s) = \frac{\sum_{i=1}^{r} n(w_i)}{|s|}$$

Since, the BLEU score focuses on the precision rather than the recall, it will tend to favor translations with shorter sentences. For example, single-word translations of long sentences will obtain BLEU scores of 1, as long as the word in them is contained in each of the reference translations. In order to avoid this type of bias, a penalty value on shorter translations is applied at the corpus level to encourage longer translations. First, the geometric mean of the BLEU scores of all the sentence-wise translations in the corpus is computed. Let m_r be the average number of tokens at the corpus level over all reference translations, and let m_p be the number of tokens at the corpus level in the predicted translations. Then, if the predicted translation has shorter length than the average over all reference translations ($m_p < m_r$), the corpus-level BLEU score is multiplied with the penalty factor $\exp(1 - m_r/m_p)$. Note that since $m_p < m_r$ for the penalty factor to apply, the corpus-level BLEU score gets multiplied with a factor less than 1 for shorter translations. Another important point about the BLEU score is that it is rarely applied at the word level. Rather, it is computed with respect to k-grams in the text collection over small values of k. The optimal value of k, which provides the best fluency was found to be 4. Machine translation can also be evaluated using the

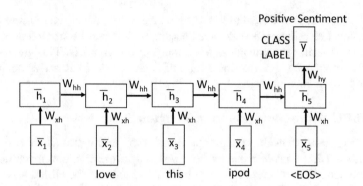

Figure 10.17: Example of sentence-level classification in a sentiment analysis application with the two classes *"positive sentiment"* and *"negative sentiment"*

Rouge score, which doubles as an evaluation metric for text summarization. The Rouge score is discussed on page 409 in the summarization chapter (cf. Chapter 12).

10.8.4   Application to Sentence-Level Classification

In this problem, each sentence is treated as a training (or test) instance for classification purposes. Sentence-level classification is generally a more difficult problem than document-level classification because sentences are short, and there is often not enough evidence in the vector space representation to perform the classification accurately. However, the sequence-centric view is more powerful and can often be used to perform more accurate classification. The RNN architecture for sentence-level classification is shown in Figure 10.17. Note that the only difference from Figure 10.18 is that we no longer care about the outputs at each node but defer the class output to the end of the sentence. In other words, a single class label is predicted at the very last time-stamp of the sentence, and it is used to backpropagate the class prediction errors.

Sentence-level classification is often leveraged in sentiment analysis (cf. Section 15.3 of Chapter 15). For example, one can use sentence-level classification to determine whether or not a sentence expresses a positive sentiment by treating the sentiment polarity as the class label. In the example shown in Figure 10.17, the sentence clearly indicates a positive sentiment. Note, however, that one cannot simply use a vector space representation containing the word *"love"* to infer the positive sentiment. For example, if words such as *"don't"* or *"hardly"* occur before *"love"*, the sentiment would change from positive to negative. Such words are referred to as *contextual valence shifters* [442], and their effect can be modeled only in a sequence-centric setting. Recurrent neural networks can handle such settings because they use the accumulated evidence over the specific sequence of words in order to predict the class label. One can also combine this approach with linguistic features. In the next section, we show how to use linguistic features for token-level classification; similar ideas also apply to the case of sentence-level classification.

10.8.5   Token-Level Classification with Linguistic Features

The numerous applications of token-level classification include information extraction and text segmentation (cf. Chapters 13 and 16). In information extraction, specific words or combinations of words are identified that correspond to persons, places, or organizations.

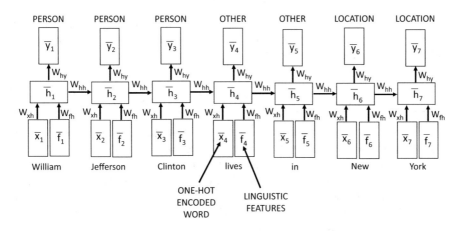

Figure 10.18: Token-wise classification with linguistic features

The linguistic features of the word (capitalization, part-of-speech, orthography) are more important in these applications than in typical language modeling or machine translation applications. Nevertheless, the methods discussed in this section for incorporating linguistic features can be used for any of the applications discussed in earlier sections. For the purpose of discussion, consider a *named-entity recognition application* in which every entity is to be classified as one of the categories corresponding to person (P), location (L), and other (O). In such cases, each token in the training data has one of these labels. An example of a possible training sentence is as follows:

$$\underbrace{\text{William}}_{P}\ \underbrace{\text{Jefferson}}_{P}\ \underbrace{\text{Clinton}}_{P}\ \underbrace{\text{lives}}_{O}\ \underbrace{\text{in}}_{O}\ \underbrace{\text{New}}_{L}\ \underbrace{\text{York}}_{L}.$$

In practice, the tagging scheme is often more complex because it encodes information about the beginning and end of a set of contiguous tokens with the same label (cf. Section 13.2.2). For test instances, the tagging information about the tokens is not available.

The recurrent neural network can be defined in a similar way as in the case of language modeling applications, except that the outputs are defined by the tags rather than the next set of words. The input at each time-stamp t is the one-hot encoding $\overline{x}_t$ of the token, and the output $\overline{y}_t$ is the tag. Furthermore, we have an additional set of q-dimensional linguistic features $\overline{f}_t$ associated with the tokens at time-stamp t. These linguistic features might encode information about the capitalization, orthography, capitalization, and so on. The hidden layer, therefore, receives two separate inputs from the tokens and from the linguistic features. The corresponding architecture is illustrated in Figure 10.18. We have an additional $p \times q$ matrix W_{fh} that maps the features $\overline{f}_t$ to the hidden layer. Then, the recurrence condition at each time-stamp t is as follows:

$$\overline{h}_t = \tanh(W_{xh}\overline{x}_t + W_{fh}\overline{f}_t + W_{hh}\overline{h}_{t-1})$$
$$\overline{y}_t = W_{hy}\overline{h}_t$$

The main innovation here is in the use of an additional weight matrix for the linguistic features. The change in the type of output tag does not affect the overall model significantly. The overall learning process is also not significantly different. In token-level classification applications, it is sometimes helpful to use *bidirectional recurrent networks* in which recurrence occurs in both temporal directions [503]. The main advantage of bidirectional recurrent neural networks is that it can incorporate context from both sides of a position in a sentence while making a prediction at that position. This model is discussed in Section 10.7.3.

10.9   Convolutional Neural Networks for Text

While convolutional neural networks have been used frequently for image data, it has become increasingly popular to also use them for text processing. The input to the convolutional neural network contains a sequence of words, corresponding to a sentence or other sequential unit. However, the words are not represented in one-hot form but are represented by their *word2vec* encodings. Therefore, we assume that the text segment is represented by an $L \times d$ matrix, where L corresponds to the maximum number of words in an input sequence and d corresponds to the dimensionality of the *word2vec* encodings. For inputs that are shorter than L, it is possible to use *padding* with additional dummy words. The value of d is also referred to as the *depth* of the input layer, and it corresponds to the number of *channels* in the input.

The convolutional operation is achieved with the use of *filters* or *kernels*, which is also a 2-dimensional matrix of size $F \times d$ containing trainable weights, where $F \leq L$. It is noteworthy that the depth of each filter is equal to that of the input to which it is applied. Multiple filters are used to produce the activations of the next layer, and the depth of the next layer is equal to the number of filters. Let x_{ij} be the (i,j)th element of the input, where i refers to the word position and j refers to channel number. Similarly, let f_{ij}^k be the (i,j)th element of the kth filter. Then, the activations of the kth channel are created by aligning the filter with each possible starting position in the input sequence, performing an element-wise multiplication of filter with input, and adding then. Note that there are a total of $L - F + 1$ possible alignments and therefore the output for the next layer is of size $(L - F + 1) \times k$, where k is the number of filters used in the current layer. Therefore, the convolution operation from the input layer to the first hidden layer may be expressed as follows:

$$h_{sr}^1 = \sum_{i=1}^{F}\sum_{j=1}^{d} f_{ij}^r x_{s+i,j} \quad \forall s \in \{1 \ldots L - F + 1\}, \forall r \in \{1 \ldots k\}$$

Note that one can recursively use this convolution operation to produce the $(m+1)$th hidden layer h_{ij}^{m+1} from the activations h_{ij}^m in the mth hidden layer. It is typical to apply an activation function immediately after the convolutional layer. Common activation functions include the use of the ReLU and tanh activation functions. Since the value of s in the above equation varies from 1 to $L - F + 1$, one undesirable effect of this is an uncontrolled reduction in the size of the sentence from L to $(L - F + 1)$ after each layer—this is the direct result of the *edge effects* of a convolution, wherein the activations corresponding to the words at the beginning or end of a sentence are involved in fewer convolutions. One way of avoiding such edge effects is to pad both temporal sides of the input sentences with $(F-1)/2$ dummy values containing 0s in each channel. As a result, the output of a convolutional network will continue to have length L. If one wishes to reduce the length of the temporal footprint by a particular factor, it is possible to perform *strided convolutions*, wherein the beginning position of the filter is always aligned with a multiple of the stride $m > 1$, where m is an integer. As a result, the temporal footprint of the hidden layer in the next hidden layer reduces by a factor of m (assuming that padding is used to get rid of the edge-effect reductions).

Finally, an important operation that is common to convolutional neural networks is that of *max-pooling*, which is also referred to as *max-pooling over time*. A max-pooling operation is applied to each channel of a layer separately. Let l be the length (i.e., number of time steps) of an input channel after the convolution operation has already been applied (along with the activation function). For example, in the first layer, the value of l might be $L - F + 1$ (if padding is not used), and the new depth might be k, resulting in an input map of size $l \times k$. The max-pooling operation divides the input map of size $l \times k$ into p segments of length $l/p \times k$ each. For each of the k channels, the maximum value out of the l/p inputs is computed and reported, and mapped to the corresponding value for that channel. Therefore, the output from the max-pooling operation has a reduced size of $p \times k$. Note that the number of channels is not affected by the max-pooling operation, because max-pooling over the ith channel input results in the ith channel output of max-pooling. Therefore, there is a one-to-one correspondence between input and output channels in max-pooling. The max-pooling is often applied to the final (convolutions) layer, and the value of p in such cases is chosen to be the entire temporal length of the input, so that the resulting output of a max-pool operation is a single value. Each channel will produce one such output, and there is no sequential dependence among units any more, which is similar to a conventional neural network. These layer may be subjected to further operations similar to those in a traditional multilayer neural network.

After applying the convolution, activation, and max-pooling for a certain number of layers, a number of fully connected layers are used. Consider the case where the hidden layer size at a particular point is $q \times r$, where q is the sequential length and r is the depth, and one wishes to follow up with fully connected layers. In such a case, one can apply a weight matrix to each q-dimensional input channel to produce an unordered r-dimensional output (since there are r channels). This r-dimensional output then feeds into the fully connected layer. Instead of using a weight matrix, it is more common to apply a max-pooling over time to each channel, which results in an r-dimensional conventional neural network layer (without sequential ordering among the units). What is the effect of a max-pooling over time, when applied to the entire sentence? One can view the process of max-pooling as that of finding the dominant concepts along each *word2vec* dimension (channel), and then viewing the input sentence as an amalgamation of these concepts. For example, each of the channels of a max-pooling operation will be able to capture when a sentence is about a particular concept (e.g., restaurants, reviews) and if it conveys a positive sentiment.

After an unordered r-dimensional output has been produced, a number of fully connected layers feed off this r-dimensional layer like a multilayer perceptron. The fully connected layers feed into an output layer, which is relevant to the application at hand. Text convolutional neural networks have become increasingly popular as compared to recurrent neural networks. One advantage of convolutional neural networks is that they can be parallelized easily, which is not possible with recurrent neural networks because of their inherently sequential computations. Parallelization always leads to the ability to use larger training data sets, which in turn lead to higher accuracy. The TextCNN model has been successfully used in *sentiment analysis* applications.

10.10 Summary

Feature engineering is a useful process for encoding the sequential structure of text into a multidimensional representation. Multidimensional representations are particularly convenient because they can be used in conjunction with many off-the-shelf tools. This chapter discusses a number of matrix factorization, graph-based, and neural network models for feature engineering. Feature engineering methods find their genesis in statistical language models, which provide the mathematical basis for converting the data into multidimensional representations in a systematic way. The kernel methods discussed in earlier chapters are also defined using language modeling principles. Recurrent neural networks are powerful methods for end-to-end sequential analysis of text, which have applications to language modeling, sequence classification, image captioning and machine translation. Some classes of recurrent networks, such as LSTMs and GRUs, are more robust than straightforward implementations of recurrent networks, especially when it pertains to the representation of longer sentences.

10.11   Bibliographic Notes

An early overview of vector-space models of semantics may be found in [555]. A comparison of count-based and prediction-based models for word embeddings is provided in [42]. String subsequence and n-gram kernels are discussed in [353]. The incorporation of different types of token-specific features in kernels is discussed in [78, 79, 124, 139, 452, 616, 624, 625]. The use of Latent Semantic Analysis and matrix factorization methods for document and word embedding is discussed in Chapter 3. These methods do not use the sequential structure or the context at all for embedding purposes. An early method that uses context is the Hyperspace Analog to Language (HAL) [358], which creates term-term matrices, in which the co-occurrence is defined based on proximity within a context window. One problem with the technique in [358] is that it is dominated by words with very high frequencies. A general framework for distributional methods for creating embeddings is discussed in [43]. The value of using word embeddings that leverage sequential information for various natural language tasks is discussed in [125].

The GloVe system for word representation was proposed in [436], which uses logarithmic normalization to reduce the effect of very high-frequency words. The use of pointwise mutual information has a rich history in natural language processing [116]. The earliest methods espousing PPMI for semantic representation may be found in [77], and the notion of noise contrastive estimation was proposed in [226]. The vLBL and ivLBL models, as well as methods based on noise contrastive estimation, are discussed in [402, 403]. Related log bilinear models were proposed in early work [401] on statistical language modeling. An embedding based on a different pointwise mutual information metric is proposed in [325]. Numerous practical ideas for the best implementation of contextual models like *word2vec* are provided in [324]. The use of graphical models for text representation and processing is discussed in [16], although its relationship to word-context factorization models is not shown in this work. In essence, the factorization of aggregated distance graphs yields intuitively similar results to the factorization of word-context matrices (see Exercise 6).

General discussions on neural networks are available in [58, 210]. Neural language models have become increasingly popular in recent years [53, 208]. Various methods for using skip-grams and continuous bag-of-word predictions with neural networks are discussed in the *word2vec* and *doc2vec* methods [314, 389]. A related model with negative sampling is discussed in [390], although this model uses a different objective function than the skip-gram model. An empirical evaluation of *doc2vec* may be found in [312]. Some good explanations of the *word2vec* model are provided in [209, 480]. The relationship of neural word embeddings to matrix factorization and kernel methods is shown in [323, 420]. The types of embeddings found by methods like *word2vec*, *doc2vec*, GloVe, and distance graphs have been shown to be useful in many applications. These methods have been used in traditional clustering/classification [16, 323], word analogy tasks [392], word-to-word machine translation [393], named entity recognition [516], opinion mining, and sentiment analysis [314].

The LSTM model was proposed in [253] and its use for language modeling is discussed in [536]. The problem associated with the poor ability of recurrent networks to store long-term dependencies is discussed in [254, 433]. Several variations of recurrent neural networks and LSTMs for language modeling are discussed in [111, 115, 224, 225, 370, 391]. The particular discussion of LSTMs in this chapter is based on [224], and an alternative gated recurrent unit (GRU) is presented in [111, 115]. A guide to understanding recurrent neural networks is available in [290]. Further discussions on the sequence-centric applications of recurrent neural networks are available in [341]. LSTM networks are also used for sequence labeling [223], which is useful in sentiment analysis [712]. The use of a combination of convolutional neural networks and recurrent neural networks for image captioning is discussed in [562]. Sequence-to-sequence learning methods for machine translation are discussed in [111, 284, 537]. The use of convolutional neural networks for different types of deep learning tasks in text is explored in [125, 126, 298].

10.11.1 Software Resources

The DISSECT (Distributional Semantics Composition Toolkit) [705] is a toolkit that uses word co-occurrence counts in order to create embeddings. The GloVe method is available from Stanford NLP [706], and it is also available in the **gensim** library [464]. The *word2vec* tool is available [661] under the terms of the Apache license. The **TensorFlow** version of the software is available at [662]. The **gensim** library has Python implementations of *word2vec* and *doc2vec* [464]. Java versions of *doc2vec*, *word2vec*, and GloVe may be found in the **DeepLearning4j** repository [707]. In several cases, one can simply download pre-trained versions of the representations (on a large corpus that is considered generally representative of text) and use them directly, as a convenient alternative to training for the specific corpus at hand. Many of these repositories also provide implementations of recurrent neural networks and LSTMs. For example, the LSTM component of the deep learning software provided by **DeepLearning4j** is available at [714]. The software for using LSTM networks in the context of sentiment analysis is available at [712]. This approach is based on the sequence labeling technique presented in [223]. Blogs with pointers to further resources in the topic of LSTMs may be found in [711, 713]. A notable piece of code [715] is a character-level RNN, and it is particularly instructive for learning purposes. The conceptual description of this code is provided in [290, 711]. The implementation of the ELMo is available at [738].

10.12 Exercises

1. Consider the following sentence, *"The sly fox jumped over the lazy dog."* Enumerate all the 1-skip-2-grams and 2-skip-2-grams.

2. Implement an algorithm to discover all 2-skip-2-grams from a given sentence.

3. Suppose you are given an embedding found by *word2vec* of each of the d terms in the lexicon. You are also given an $n \times d$ document-term matrix D containing the term frequencies of each document in its rows with the same lexicon of size d. Propose a heuristic to find the coordinates of the documents in terms of this word embedding.

4. Suppose you have additional syntactic features of words such as the part-of-speech, orthography, and so on. Show how you would incorporate such features in *word2vec*.

5. How can you use an RNN to predict grammatical errors in a sentence?

6. Suppose that you have the n distance graphs $G_1, \ldots G_n$ in a document corpus. You create the union of all these distance graphs by taking the union of their nodes/edges and aggregating the weights of any parallel edges.

   (a) Discuss the relationship of the factorization of the adjacency matrix of this graph with word-context factorization models.

   (b) How would you change the factorization objective function to address the effect of wide variation in counts.

7. For each of the CBOW and skip-gram models, show the following:

   (a) Express the loss function only as a function of the inputs and weights, after eliminating the hidden layer variables.

   (b) Compute the gradients of the loss function with respect to the weights in the input and output layers.

8. Suppose that you use GloVe on a count matrix $C = [c_{ij}]$ in which each count c_{ij} is either 0 or 10000. A sizeable number of counts are 0s.

   (a) Show that GloVe can discover a trivial factorization with zero error in which each word has the same embedded representation.

   (b) Suppose we generate a sequence of 10^7 tokens in which the first half are randomly selected from word identifiers in $\{1 \ldots 100\}$, and the second half are randomly selected from $\{101 \ldots 200\}$. Discuss why the ignoring of negative samples is a bad idea for creating meaningful factorization-based word embeddings in this case.

9. **Multinomial matrix factorization:** Consider a $d \times d$ word-word context matrix $C = [c_{ij}]$ in which c_{ij} is the frequency of word j in the context of word i. The goal is to learn $d \times p$ and $p \times d$ matrices U and V, respectively, so that applying softmax to each *row* of UV matches the relative frequencies in the corresponding row of C. Create a loss function for this probabilistic factorization of C into U and V. Discuss the relationship with the skip-gram model.

10. **Multiclass Perceptron:** Consider a multiclass setting in which each training instance is of the form $(\overline{X_i}, c(i))$, where $c(i) \in \{1 \ldots k\}$ is the class label. We want to find k linear separators $\overline{W_1} \ldots \overline{W_k}$ so that $\overline{W}_{c(i)} \cdot \overline{X_i} > \overline{W_r} \cdot \overline{X_i}$ for any $r \neq c(i)$. Consider the following loss function for the ith training instance:

$$L_i = \max_{r:r \neq c(i)} \max(\overline{W_r} \cdot \overline{X_i} - \overline{W}_{c(i)} \cdot \overline{X_i}, 0)$$

Propose a neural architecture for this loss function and the corresponding stochastic gradient-descent steps. Discuss the relationship with the perceptron at $k = 2$.

11. Use the representer theorem of Chapter 6 to propose an objective function and gradient-descent steps for the kernelized version of a regularized perceptron.

Chapter 11

Attention Mechanisms and Transformers

"The true art of memory is the art of attention." – Samuel Johnson

11.1 Introduction

Attention mechanisms have revolutionized the field of natural language processing. Attention helps in focusing the learning process on important parts of the data, so that portions of the data that are most relevant for prediction are emphasized. A classical example of an application of attention occurs in machine translation, where the translation of a specific part of a target sentence often focuses on important parts of the source sentence. For example, while translating the sentence *"I don't understand Spanish"* to *"No entiendo espanol,"* it is important to pay attention to the word *"understand"* while generating the target word *"entiendo."*

The concept of attention in machine learning has roots in biological notions of attention. Humans do not actively use all the information available to them from the environment at any given time. Rather, they focus on specific portions of the data that are relevant to the task at hand. This type of attention applies to almost all sensory perceptions, such as sight, smell, sound and so on. Correspondingly, the notion of attention is used in various aspects of machine learning, such as computer vision, natural language processing, and graph processing.

In order to understand the notion of attention, let us consider how a human important component of the taskn focuses their visual attention when trying to find an address defined by a specific house number on a street. A key aspect of this process is to identify the specific place on the house where the number is written. Although the retina often has an image of large portions of the frontal regions of the house, one rarely focuses on the full image. The retina has a small portion, referred to as the *macula* with a central *fovea*, which has an extremely high resolution compared to the remainder of the eye. This region has a high concentration of color-sensitive cones, whereas most of the non-central portions of the eye have relatively low resolution with a predominance of color-insensitive rods. The different regions of the eye are shown in Figure 11.1. When reading a street number, the fovea *fixates* on the number, and its image falls on a portion of the retina that corresponds to the macula

© Springer Nature Switzerland AG 2022
C. C. Aggarwal, *Machine Learning for Text*,
https://doi.org/10.1007/978-3-030-96623-2_11

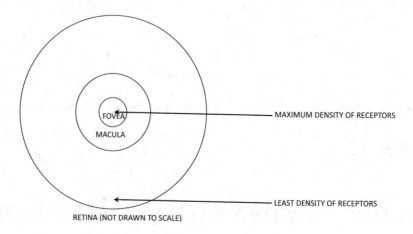

Figure 11.1: Resolutions in different retinal regions: the fovea is the center of visual attention

(and especially the fovea). Although one is aware of the other objects outside this central field of vision, it is virtually impossible to use images in the peripheral region to perform detail-oriented tasks. For example, it is very difficult to read letters projected on peripheral portions of the retina. The foveal region is a tiny fraction of the full retina, and it has a diameter of only 1.5 millimeters. The eye effectively transmits a high-resolution version of less than 0.5% of the surface area of the image that falls on the full retina. This approach is biologically advantageous, because only a carefully selected part of the image is transmitted in high resolution, and it reduces the internal processing required *for the specific task at hand.* Similarly, while hearing a question, one often focuses on the most informative words (e.g., entities and their relations) in order to provide a useful answer. One also naturally uses attention in order to derive important contexts of words, which helps one differentiate between the various usages of the word *"right"* in *"right turn"* and *"right choice."* By paying attention to the connection between *"turn"* and *"right"* it becomes easy to identify the specific usage of the word *"right."* As we will see in this chapter, this precise approach to identifying context is very useful in natural language processing.

There are two types of attention that are commonly used in the literature, namely *hard attention* and *soft attention.* In hard attention, the irrelevant parts of the data are dropped with techniques that are drawn from concepts of *reinforcement learning* in artificial intelligence. However, this type of approach is not the dominant one now, especially as far as natural language processing is concerned. The dominant techniques today use soft weighting instead of selecting specific parts of the data in a hard way. The idea of soft attention is closely related to that of supervised nearest neighbor classification in which the weights of best matching data objects are learned using a combination of distances and learned parameters [446]. The close relationship between nearest neighbor methods and attention will be evident throughout this chapter, and many of the mechanisms for attention even adopt the terminology used in database query processing.

Chapter Organization

This chapter is organized as follows. The next section will discuss the use of attention mechanisms for neural machine translation, as they apply to the standard translation mechanism of a recurrent neural network. The use of a transformer for machine translation instead of a recurrent neural network is discussed in Section 11.3. The use of transformers for building language models is discussed in Section 11.4. Applications of language models are introduced in Section 11.5. A summary is given in Section 11.6.

11.2 Attention Mechanisms for Machine Translation

As discussed in Section 10.8.3 of Chapter 10, recurrent neural networks (and especially their variants like LSTMs/GRUs) are used frequently for machine translation. In the following, we use a vanilla recurrent neural network with a single layer for simplicity in exposition (and ease in illustration of neural architectures), although more sophisticated architectures like LSTMs and GRUUs are used in practice. When multiple hidden layers are used, the attention mechanism is almost always applied to the top hidden layer. The two most common forms of RNN-based attention in neural machine translation are *Luong attention* [359] and *Bahdanau attention* [37]. Here, we present Luong attention.

11.2.1 The Luong Attention Model

We start with the architecture discussed in Section 10.8.3 of Chapter 10. For ease in discussion, we replicate the neural architecture of that section in Figure 11.2(a). Note that there are two recurrent neural networks, of which one is tasked with the encoding of the source sentence into a fixed length representation, and the other is tasked with decoding this representation into a target sentence. The hidden states of the source language (encoder) and target language (decoder) networks are denoted by $h_t^{(1)}$ and $h_t^{(2)}$, respectively, where $h_t^{(1)}$ corresponds to the hidden state of the tth word in the source sentence, and $h_t^{(2)}$ corresponds to the hidden state of the tth word in the target sentence. These notations are borrowed from Section 10.8.3 of Chapter 10.

In attention-based methods, the hidden states $h_t^{(2)}$ are transformed to enhanced states $H_t^{(2)}$ with some additional processing from an *attention layer*. The goal of the attention layer is to incorporate context from the source hidden states into the target hidden states to create a new and enhanced set of target hidden states.

In order to perform attention-based processing, the goal is to find a source representation that is close to the current target hidden state $h_t^{(2)}$ being processed. This is achieved by using the similarity-weighted average of the source vectors to create a context vector $\bar{c}_t$:

$$\bar{c}_t = \frac{\sum_{j=1}^{T_s} \exp(\overline{h}_j^{(1)} \cdot \overline{h}_t^{(2)})\overline{h}_j^{(1)}}{\sum_{j=1}^{T_s} \exp(\overline{h}_j^{(1)} \cdot \overline{h}_t^{(2)})} \tag{11.1}$$

Here, T_s is the length of the source sentence. Attention mechanisms couch this type of similarity weighting in terms of the language of database query processing. At any given time stamp t (i.e., for the tth word), the state $\overline{h}_t^{(2)}$ being processed in the target language is treated as a query over the "database" of all states (words) in the source language to identify the best equivalent word (context). Like any database, the source language "database" has a set of *values-key pairs*, where the jth value and key in the source language "database" are both equal to $\overline{h}_j^{(1)}$. A *query* word representation $\overline{h}_t^{(2)}$ in the target sentence creates a context vector, $\bar{c}_t$, by "soft-searching" for the best matching *key* $\overline{h}_j^{(1)}$ in the source sentence and presenting its *value*. By *soft-searching* we refer to the fact that rather than presenting a

Table 11.1: Examples of alignments between source and target sentences

	I	don't	understand	Spanish
No		X		
entiendo	X		X	
espanol				X

single best matching source-language state, the approach weights the corresponding values $\overline{h}_j^{(1)}$ of these keys using the query-key similarities (which can be interpreted as probabilities of best matching). This type of approach is a *cross-attention mechanism* as the similarity values are computed *across* two different sets of objects (hidden representations at different positions of source and target sentences). The similarity value used for weighting is the *attention weight* $a(t,s)$, which indicates the importance of (key) source vector at position s to (query) target vector at position t:

$$a(t,s) = \frac{\exp(\overline{h}_s^{(1)} \cdot \overline{h}_t^{(2)})}{\sum_{j=1}^{T_s} \exp(\overline{h}_j^{(1)} \cdot \overline{h}_t^{(2)})} \tag{11.2}$$

These similarity weights sum to 1 like probabilities, and the entire vector of weights $[a(t,1), a(t,2), \ldots a(t,T_s)]$ for target $\overline{h}_t^{(2)}$ is denoted by the *attention vector* $\overline{a}_t$. The attended representation of the target position t is the expectation over *all* source hidden vectors $\overline{h}_j^{(1)}$ with probabilistic attention weights $a(t,j)$ for fixed t and varying j:

$$\overline{c}_t = \sum_{j=1}^{T_s} a(t,j)\overline{h}_j^{(1)} \tag{11.3}$$

Here, $\overline{c}_t$ provides the matching source-centric context for $\overline{h}_t^{(2)}$, which can be added to the information already contained in $\overline{h}_t^{(2)}$. Therefore, one can apply a single-layer feed-forward network to the concatenation of $\overline{c}_t$ and $\overline{h}_t^{(2)}$ to create a new target hidden state $H_t^{(2)}$:

$$\overline{H}_t^{(2)} = \tanh\left(W_c \begin{bmatrix} \overline{c}_t \\ \overline{h}_t^2 \end{bmatrix}\right) \tag{11.4}$$

This new hidden representation $\overline{H}_t^{(2)}$ is used in lieu of the original hidden representation $\overline{h}_t^{(2)}$ for the final prediction. The overall architecture of the attention-sensitive system is given in Figure 11.2(b). Note the enhancements from Figure 11.2(a) with the addition of an attention layer. When translating "*I don't understand Spanish*," the attention mechanism is able to align relevant positions (e.g., *understand* and *entiendo*) in the English and Spanish versions of the sentence by returning high attention values for the corresponding state pairs. An example of the typical alignment between the aforementioned source and target sentence is shown in Table 11.1. The source-target pairs with high levels of cross-attention are marked with 'X.' It is noteworthy that a one-to-one alignment of words typically does not exist between a pair of languages. For example, a word such as "*entiendo*" implies both the use of the "I" pronoun as well as the concept of "*understand*."

This model is referred to as the *global attention model* in [359], because it uses all the words in the source sentence. The work in [359] also proposes a *local* attention model, which restricts the weighting to a subset of the most relevant words and can therefore be considered a combination of soft and hard attention. This approach focuses on a small window of context around the best aligned source state by using the importance weighting generated by the attention mechanism. Such an approach is able to implement a limited level of hard attention without incurring the training challenges of reinforcement learning. The reader is referred to [359] for details.

11.2.2 Variations and Comparison with Bahdanau Attention

Several refinements can improve the basic attention model. First, the attention vector $\overline{a}_t$ is computed by exponentiating the raw dot products between $\overline{h}_t^{(2)}$ and $\overline{h}_s^{(1)}$, as shown in Equation 11.2. It is possible to increase the capacity of the model further by adding learnable parameters within the attention computation. The parameterized alternatives for computing similarity between source and target states were as follows [359]:

$$
\text{Score}(t,s) = \begin{cases} \overline{h}_s^{(1)} \cdot \overline{h}_t^{(2)} & \text{[Dot product]} \\ (\overline{h}_t^{(2)})^T W_a \overline{h}_s^{(1)} & \text{[Parameterized dot product with } W_a] \\ \overline{v}_a^T \tanh\left(W_a \begin{bmatrix} \overline{h}_s^{(1)} \\ \overline{h}_t^2 \end{bmatrix} \right) & \text{[Parameterized FFN with } W_a \text{ and } \overline{v}_a] \end{cases} \tag{11.5}
$$

The first of these options is identical to the dot product attention of Equation 11.2 and the second option (referred to as *general* in [359]) is its parameterized version. The third option applies a two-layer feed-forward network (FFN) to the concatenated source-target pair $[\overline{h}_s^{(1)}, \overline{h}_t^{(2)}]$ with the parameters of the two layers respectively set to matrix W_a and vector $\overline{v}_a$. This model was referred to as *concat* in [359] and as *additive attention* in the original Bahdanau paper where it was proposed [37] — it is, therefore, popularly referred to as *Bahdanau attention*. The parametrization of the last two models provides additional flexibility by enabling learning during training. The attention values can be computed from the vector of scores using softmax activation (as in the previous section):

$$
a(t,s) = \frac{\exp(\text{Score}(t,s))}{\sum_{j=1}^{T_s} \exp(\text{Score}(t,j))} \tag{11.6}
$$

These attention probabilities are used to compute the expected context vector $\overline{c}_t$ from the source sentence. Surprisingly, it was shown in [359] that the unparameterized dot-product was more accurate than parameterized alternatives in global attention models, although the parameterized alternatives did better in local attention models. It is possible that the good performance of the unparameterized dot product was a result of its regularizing effect on the model. The parameterized dot product model generally seemed to perform better than the *concat* (additive attention) model, which was inherited from the original version of Bahdanau attention.

Finally, we comment on the differences of Luong attention from the original Bahdanau attention model (aside from its use of additive attention). First, the encoder in Bahdanau attention is bidirectional and the decoder is uni-directional, whereas both are uni-directional in Luong attention. Therefore, each forward-backward state pair of the Bahdanau encoder was concatenated before additive attention computations. Second, for each decoder time-stamp t, the Bahdanau attention computation is applied to the target hidden state $\overline{h}_{t-1}^{(2)}$ from

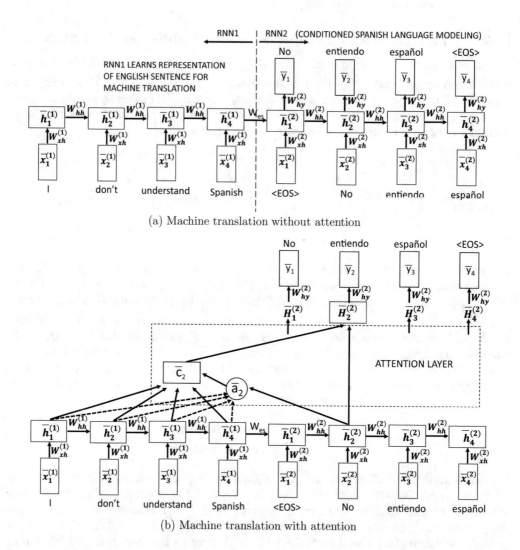

(a) Machine translation without attention

(b) Machine translation with attention

Figure 11.2: The neural architecture in (a) is the same as the one illustrated in Figure 10.16 of Chapter 10. An extra attention layer has been added to (b)

the *previous* time stamp to create context vector $\bar{c}_t$. This context vector is concatenated with $\bar{h}_{t-1}^{(2)}$ and *then* passed through one step of the recurrent neural network to create the current hidden state $\bar{h}_t^{(2)}$. The word prediction at t is made by passing this hidden state $\bar{h}_t^{(2)}$ through a deep output layer. Unlike Luong attention, Bahdanau attention does not have a separate attention layer to create a *post-attentive* hidden state such as $\overline{H}_t^{(2)}$, because its computation flow is $\bar{h}_{t-1}^{(2)} \to (\bar{a}_t, \bar{c}_t) \to \bar{h}_t^{(2)}$, whereas that of Luong is $\bar{h}_t^{(2)} \to (\bar{a}_t, \bar{c}_t) \to \overline{H}_t^{(2)}$. Therefore, Luong attention is more modular and easily generalizable to an off-the-shelf architecture with the addition of an attention layer.

11.3 Transformer Networks

As evident from the machine translation example of the previous section, attention can be added to recurrent neural networks to improve accuracy. It turns out that can process sequences *without* using recurrent neural networks by combining attention mechanisms with *positional encodings*. This leads to a much simpler architecture with better performance. This approach drops the use of the recurrent neural network, and moves back to traditional multilayer neural network with important roles for attention and positional encodings. Such architectures are referred to as *transformer networks* [560]. The use of transformer networks has now become a worthy alternative to the recurrent neural network for natural language processing. One advantage of transformer networks over recurrent neural networks is that the underlying computations can be parallelized easily, which enables more efficient use of hardware advancements such as GPUs. The greatest weakness of the recurrent neural network architecture has been its resistance to parallelizeability as a result of its dependence on sequentially computed states; by doing away with this type of sequential computation, the transformer has a clear computational advantage. Gains in computational efficiency often translate to improved accuracy because of the ability to build larger models with more robust training.

11.3.1 How Self Attention Helps

The transformer learns embeddings of words (like *word2vec*) but it adjusts the embeddings for each specific usage of the word based both on its position and context within the sequence (or sentence). Consider a sequence of n words, each of which has codes $\bar{h}_1 \ldots \bar{h}_n$. These codes can either be derived from another learning mechanism like *word2vec* or learned/fine-tuned via backpropagation within the neural architecture of the transformer. While the representation $\bar{h}_j$ of the jth word in the sentence depends on its usage and context in the entire training data (from which it is derived), it is also desirable to adjust these representations based on the usage of the term in the specific sentence being processed by the transformer. This adjustment is achieved with the use of a *self-attention mechanism*, wherein the words in a sentence are used to provide additional context. In particular, the representations $\bar{h}_1 \ldots \bar{h}_n$ can be mapped to contextual representations $\bar{c}_1 \ldots \bar{c}_n$ by using dot-product self attention as follows:

$$\bar{c}_t = \frac{\sum_{j=1}^{n} \exp(\bar{h}_j \cdot \bar{h}_t)\bar{h}_j}{\sum_{j=1}^{n} \exp(\bar{h}_j \cdot \bar{h}_t)} \tag{11.7}$$

After this adjustment, the code for "*right*" in the sentence fragment "*turn right*" will be different from that in "*do right*." While the recurrent neural network is also able to achieve contextual encoding for entire *sentences*, it seems to have greater problems with longer sentences such as the following: "*Selecting the right branch at the fork was a mistake, and I should have understood that it made better sense to choose the left one.*" This is because of the sequential updates of the states in the recurrent neural network in which each update scrambles some of the information in previous states. On the other hand, the transformer will be able to adjust the coding for "*right*" using attention from related words like "*left*" without regard to the distance between these two words.

One can also express this operation in terms of the relative similarity values a_{tj} between vectors $\overline{h}_j$ and $\overline{h}_t$:

$$a_{tj} = \frac{\exp(\overline{h}_j \cdot \overline{h}_t)}{\sum_{j=1}^{n} \exp(\overline{h}_j \cdot \overline{h}_t)} \tag{11.8}$$

The relative similarity values a_{tj} are positive and add to 1 over fixed t and varying j — these are used to create re-weighted codes as follows:

$$\overline{c}_t = \sum_{j=1}^{n} a_{tj}\overline{h}_j \tag{11.9}$$

The magnitude of the vanilla dot product tends to grow with the square-root of the dimensionality — therefore, it is common to use *scaled* dot product attention in which each $\overline{h}_j \cdot \overline{h}_t$ is divided by the square-root of the dimensionality of $\overline{h}_j$ before exponentiation.

11.3.2   The Self-Attention Module

The attention operation can be expressed using the notions of queries, keys, and values, which derives its motivation from the fact that attention adjusts word representations based on "matching" query words with related words in the sentence. For each *query* vector $\overline{h}_t$, it is converted into a corresponding attention-modified vector $\overline{c}_t$, which is a weighted average of *values* (value vectors) $\overline{h}_j$ contained in the sentence (see Equation 11.9). Each value vector is associated with a *key vector* with which queries are "matched" in a soft way with dot product attention. In this simplified case, the value and key vectors are the same. The query-specific "matching" weight of a value vector (for attention re-adjustment) is obtained by applying softmax to the dot product similarity between the query vector and the key vector of that value vector. Therefore, the notation $\overline{h}_t$ in Equation 11.8 is the query, and the notation $\overline{h}_j$ for $j \in \{1 \ldots n\}$ in the same equation refers to the key. By using this approach, the representation for a query word on a particular topic (e.g., sports) is more likely to be affected by the representations of other sports-related words in a sentence. Therefore, the new representation is more context-sensitive.

In this simplified exposition of attention, the queries, keys, and the values associated with word j have the same p-dimensional column vector representation $\overline{h}_j$, which begs the question as to why we need separate terminologies for different portions of the attention equation. First, it provides an analogy to query-matching weights for the attention mechanism. Second, in many applications of *cross-attention* like machine translation, the queries in a target language may be different from the key/value pairs in a source language. Finally, even in self-attention, it is common to transform queries, keys and values to *different* q-dimensional vectors $W^Q\overline{h}_j$, $W^K\overline{h}_j$, and $W^V\overline{h}_j$ using learned $q \times p$ matrices W^Q, W^K, and W^V, which results in greater model capacity. Consider the $p \times n$ matrix H obtained by stacking the column vectors $[\overline{h}_1 \ldots \overline{h}_n]$ adjacent to one another. The queries, keys, and values for a particular input sentence of length n can be respectively represented by $n \times q$ matrices $Q = (W^Q H)^T$, $K = (W^K H)^T$, and $V = (W^V H)^T$. It is also possible for the value matrix W^V to have a different size $q_1 \times p$, although the original transformer paper

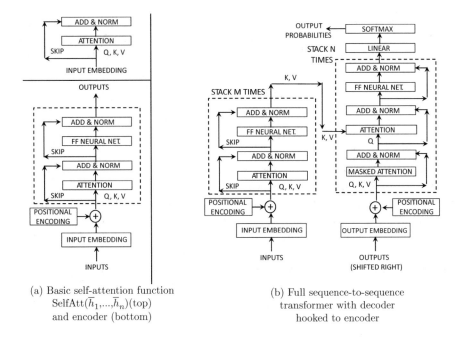

(a) Basic self-attention function
SelfAtt$(\overline{h}_1,...,\overline{h}_n)$(top)
and encoder (bottom)

(b) Full sequence-to-sequence
transformer with decoder
hooked to encoder

Figure 11.3: The Transformer Architecture

uses $q_1 = q$. The matrices W^Q, W^K, and W^V are considered a part of the attention subunit, and they are learned during backpropagation. This additional step is important to prevent the attention model from under-fitting and improve accuracy.

The attention operation (with scaling) can be implemented by applying the softmax operation to each row of $(QK^T)/\sqrt{q}$ to yield a matrix of the same size and then multiplying with V to create the $n \times q$ re-weighted value matrix Softmax$(QK^T/\sqrt{q})V$. These operations can be implemented with efficient and parallelizeable matrix operations on GPUs. In order to ensure that the attention layer outputs a vector of the same p-dimensional size as the input $\overline{h}_j$ (and also to enable additional learning), a linear dimensionality-fixing transformation $W^{fix}\overline{x}$ using $p \times q$ matrix W^{fix} is applied to each transposed row $\overline{x}$ of the $n \times q$ matrix Softmax$(QK^T/\sqrt{q})V$. This entire sequence of attention-centric transformations from p-dimensional word vector embeddings $\overline{h}_1 \ldots \overline{h}_n$ to another set of p-dimensional embeddings is referred to as the Att-Sublayer.

In order to avoid losing any important information in the original representation $\overline{x}$ of a particular word when it is converted to Att-Sublayer$(\overline{x})$ with the use of the attention sublayer, a skip-level connection is used in which the attention-modified vector is added to the original vector and then layer-normalized (cf. Section 10.7.7) to create the normalized output $\overline{x}' = $ LayerNorm$(\overline{x} + $ Att-Sublayer$(\overline{x}))$. This principle is similar to that of residual learning popularly used in the gating mechanisms of recurrent neural networks (via LSTMs) and convolutional neural networks (via ResNets) to avoid losing features that have already been learned. The architecture for this basic attention-based mechanism is shown in the upper portion of Figure 11.3(a). Note that the queries, keys, and values (denoted by Q, K, and V) are shown as inputs to this attention block for simplification, although they are technically created by this block using the learned matrices W^Q, W^K, and W^V on the input matrix H into the block, and therefore these matrices belong to the attention block as well. We refer to the overall function computed by this block as $[\overline{h}_1' \ldots \overline{h}_n'] = $ SelfAtt$(\overline{h}_1, \ldots, \overline{h}_n)$.

Finally, the p-dimensional attended vectors are subjected to a two-layer feed-forward network, with hidden layer dimensionality of $4p$ and output dimensionality of p (i.e., same as that of input). Therefore, this feed-forward network FFN($\cdot$) will contain $p \times 4p$ and $4p \times p$ matrices W_1 and W_2, which are also learned during backpropagation:

$$\text{FFN}(\overline{x}) = W_1(\text{ReLU}(W_2 \overline{x}))$$

As in the case of the attention sublayer, the output FFN($\overline{x}$) of the feed-forward sublayer is subjected to skip-connections and normalization to create the output LayerNorm($\overline{x} +$ FFN($\overline{x}$)). The combination of a self-attention sublayer and the feed-forward s sublayer creates a single "layer" of the transformer (demarcated by a dotted line in Figure 11.3(a)), and its output dimensionality is unchanged from its input. A total of M such layers are stacked in order to create a *deep encoder* with more expressive learning abilities. The p-dimensional output of a layer (denoted above by LayerNorm($\overline{x} +$ FFN($\overline{x}$))) for each of the n words in the sequence can be used to create another $p \times n$ matrix H', which can be input to the next layer and transformed with the matrices W^Q, W^K, W^V, to create new query, key, and value matrices of the next layer. The parameters of the different layers are not shared. Therefore, while we use shared notations such as W^Q, W^K, W^V, W^{fix}, W_1, and W_2 repeatedly across layers to reduce notational complexity, these matrices are different in each layer. The original transformer [560] used $M = 6$, although recent language models use much deeper architectures. The encoder processes each word in the sentence independently and will therefore create an output code for each word in the input sequence. This is different from a traditional recurrent neural network that creates fixed-length representations of full sentences (which can be suboptimal). This entire process represents the *encoding* portion of a sequence-to-sequence learner (see bottom portion of Figure 11.3(a)), although we have yet to discuss the creation of the positional encodings depicted in this diagram.

11.3.3   Incorporating Positional Information

The attention mechanism does not use the relative positions of the words while computing contextual representations. Note that the vectors $\overline{h}_1 \ldots \overline{h}_n$ are processed independently, and changing the order of words in a sentence does not change the learning in any way. Therefore, unlike the encoding of a recurrent neural network, the encodings of the transformer will be the same even after sequence permutation. This can be a problem because the ordering of the words in phrases like "*the hot dog fell on the pavement*" and "*the dog fell on the hot pavement*" can be very significant. Therefore, the transformer adds *positional encodings* to the word codes *before* performing any other operation. Like the word embedding of dimensionality p, the positional encoding is a p-dimensional vector $\overline{z}_t$ for the word in the tth position. Then, the jth component z_t^j of vector $\overline{z}_t$ is defined as follows:

$$z_t^j = \begin{cases} \sin(t \cdot \omega_j) & \text{if } j \text{ is even} \\ \cos(t \cdot \omega_{j-1}) & \text{if } j \text{ is odd} \end{cases}$$

The value of the frequency ω_j is defined as $\omega_j = 10000^{-(j/p)}$. The basic idea here is to capture the positional information in sinusoidal waves with varying periods (corresponding to varying levels of granularity). Positional vector components that are close to one another at any particular level of granularity will have similar values for the positional encoding. The positional encodings are then added to the word encodings in order to create encodings that incorporate positional information:

$$\overline{h}_t \Leftarrow \overline{h}_t + \overline{z}_t$$

The initial word embeddings $\bar{h}_t$ (to which positional encodings are added) are obtained by multiplying a parameter matrix containing encodings for each word with one-hot encoded input vectors for each word in the sequence. This weight matrix (at the bottom of Figure 11.3(a)) can be initialized to *word2vec* encodings and fine-tuned during backpropagation.

11.3.4 The Sequence-to-Sequence Transformer

As in any sequence-to-sequence learning method, one needs a decoder along with the encoder. Consider the case of the machine-translation application, in which we are trying to translate the sentence "*I don't understand Spanish*" into "*No entiendo espanol*". In Figure 11.3(b), we have shown the decoder on the right. It is immediately evident that the structure of the decoder is very similar to that of the encoder. The decoder layers stacked N times, just as the encoder layers are stacked M times. However, both self-attention and cross-attention are used in different parts of the decoder in order to enable translation.

The *entire source sentence* is input to the encoder, whereas only the leftmost portion of the target sentence is input into the decoder in order to output the probability of the word occurring after this segment. This process is repeated for each word, and therefore the decoder requires a training step for each word in the sentence (which makes it slower). As an example, the encoder is always taking as input the full English sentence "*I don't understand Spanish*," but it takes as input only the leftmost portion such as' "*<START> No*" or '*<START> No entiendo*" When "*<START> No*" is input, the goal is to predict the probability of the next word "*entiendo*" without any awareness of the occurrence of this word and "*espanol*" in the sentence. This is achieved with a masking mechanism, so that words occurring after the current prediction word are forced to have an attention weight a_{tj} of 0 during the first self-attention round. A "*<START>*" tag is added to the beginning of each target sentence to shift it right by one unit. Therefore, when "*<START>*" is input, the goal is to predict "*No.*"

After passing the target sentence (fragment) through the self-attention layer, one applies a second cross-attention mechanism. In this case, the queries are defined by the words of the input target sentence (segment), and the keys/values are defined from the representation of the source sentence produced by the encoder. This point is highlighted in Figure 11.3(b) by the fact that the pair (K, V) is created by multiplying the transformations W^K and W^V with the encoder input matrices, whereas the query matrix Q (representation matrix of words in incompletely translated sentence) is obtained by multiplying W^Q to the decoder input matrices. This type of cross-attention mechanism gives greater weights to the most relevant words in the source sentence while translating the next word, which is similar in principle to the RNN-based cross-attention computation in Section 11.2. The difference is that the translation model no longer uses a recurrent network — attention is all that is needed along with positional encodings. The decoder blocks are similar to encoder blocks in that they contains feedforward networks, skip connections, and normalizations. The decoder also contains multiple layers like the encoder. After passing each word in the partially translated sentence through the decoder, their vectors are passed through a linear layer (or averaging layer) to create a single code, and the softmax function is applied on the similarity between this code and the language dictionary codes in order to predict the probability of the next target word in the translated language. The predicted word is compared to the ground-truth word with cross-entropy loss for training.

11.3.5 Multihead Attention

As discussed earlier, the attention mechanisms need not be applied to the original codes corresponding to queries keys and values, but to transformed vectors $W^Q \overline{h}_j$, $W^K \overline{h}_j$, and $W^V \overline{h}_j$, where W^Q, W^K, and W^V are $q \times p$ matrices. It is typical for q to be less than p, which results in *projections* to smaller vectors. These *projection* matrices are learned parameters within the neural architecture. The projection matrices focus on specific dimensions (meta-concepts) of the words, and the effect is to focus attention on these meta-concepts.

A single word may contain multiple meta-concepts of interest. The point of the multi-head mechanism is that we can use k different sets of such projection matrices (denoted by W_i^Q, W_i^K and W_i^V) in parallel in order to learn different output vectors for the words in the same input sequence. Therefore, the attention mechanism can focus on different sets of meta-concepts. Each of these parallel pipelines is an *attention head*. The q-dimensional output vectors of these different attention heads are then concatenated into one long vector of length $k \cdot q$ that contains the information about all the different meta-concepts. However, since $k \cdot q$ may be different from p, one needs to multiply this vector with the $p \times (k \cdot q)$ matrix W^{fix}, which is also a part of the attention block as a linear feed-forward unit. This modification with multiple attention heads is performed to the basic attention block at the top of Figure 11.3(a), and so all matrices such as W_i^Q, W_i^K, W_i^V, and W^{fix} belong to that attention block (without sharing across layers). The matrices W_i^Q, W_i^K, W_i^V, and W^{fix} can be learned during the backpropagation of the end-to-end model of sequence-to-sequence learning. Multihead attention is more robust and represents the practical implementation of the transformer network. It is also highly parallelizeable, since different heads can be implemented on different GPUs or machines. Although multiple attention heads do improve performance, an interesting observation [385] is that most layers require only a single head for good performance, and the redundant heads can be pruned either in a validation phase or (more efficiently) by estimating the sensitivity of the loss function to each head via the backpropagation algorithm.

11.4 Transformer-Based Pre-trained Language Models

Transformers have fundamentally changed the landscape of natural language processing by gradually replacing recurrent neural networks across the board. One mechanism to achieve this goal is to create a large-scale, pre-trained language model (that works off an internet-scale corpus); two examples include OpenAI's *Generative Pretrained Transformer (GPT-n)* [75], and Google's *Bidirectional Encoder with Transformer (BERT)* [149]. However, an important difference of these models from the transformer model of the previous section is that GPT-n and BERT are more generic models for unsupervised learning and language modeling rather than machine translation — such unsupervised models can be fine-tuned for a large number of natural language processing tasks and they use either the encoder or the decoder of the transformer, or they can use the original transformer architecture with both the encoder and decoder. Therefore, the basic architecture of these models can be somewhat different from that of the vanilla transformer architecture, which always has a connected encoder-decoder pair. Aside from the fact that the encoder feeds into the decoder in the vanilla transformer, the main difference between the encoder and decoder is that the former is bidirectional, whereas the latter is unidirectional (with the use of masked inputs only in one direction). GPT-n uses only the decoder portion of the architecture (thereby inheriting its unidirectional nature), whereas BERT uses the encoder (with its bidirectional characteristics). On the other hand, T5 [460] uses the encoder-decoder structure of the original transformer, and is therefore a text-to-text model. In the case of decoder use (GPT-n), cross-attention units are replaced with self-attention units, whereas in the case of encoder use (BERT), the outputs to the decoder now feed into an additional output layer predicting words at some positions. All these models are massive and use large training data sets with hundreds of billions of parameters. The effect of training data and model size has been truly remarkable in terms of learning the ability to perform a wide variety of language processing tasks.

11.4.1 GPT-n

The GPT-n series of architectures has had several editions, which are GPT, GPT-2, and GPT-3, each of which has been an improvement over the previous one (typically by increasing the scale of the training data and by making small architectural modifications). GPT-n uses the decoder portion of the sequence-to-sequence architecture of Figure 11.3(b) both for training and for prediction. The keys and values are no longer obtained from an encoder, but from the decoder input itself. Like all practical transformer architectures, a multihead version is used.

Unlike the original transformer architecture, GPT-n learns a language model from the data by predicting the next word, given all the words to the left of the current word. Therefore, the input to GPT-n is a sequence in which all words to the right of the target word are masked. The output corresponds to the softmax probabilities of predicting the next word. The latest model (GPT-3) uses a maximum sequence input length of 2048, and pads empty positions for shorter sequences. Using all distinct words in the vocabulary as tokens can blow up vocabulary size. Therefore, GPT-3 uses *byte-pair encodings* [508] where subwords of two characters are used as tokens instead of entire words. For example, instead of "love," the tokens "lo" and "ve" are used. Since a byte reflects $2^8 = 256$ possible characters, a byte-pair encoding has a maximum vocabulary size of $256^2 = 65536$. However, *merge rules* are used to select about $50,000$ byte-pairs by merging the most frequently co-occurring character pairs. The original 256 character tokens and special end-of-text tokens are also included as GPT-3 tokens. These tokens are then converted into real-valued embedding vectors of dimensionality 12288 in GPT-3, much as *word2vec* converts words to vectors. GPT-n creates these vectors by learning the one-hot encoded token to embedding transformation during backpropagation. Therefore, the overall dimensionality of an input sequence to GPT-3 is a matrix of size 2048×12288. Throughout this section, we will use "word" and "token" interchangeably, even though a GPT-3 (or BERT) token is not really a word from the semantic perspective.

As in the decoder of Figure 11.3(b), positional encodings are added to the sequence and then self-attention modules are used along with feed-forward layers in order to create context-sensitive word embeddings. Unlike the decoder in the original transformer architecture, cross-attention is not used (since there is no encoder input to decode). The latest (GPT-3) model adopts the *sparse transformers* [109] variant among its layers. The feed-forward network of GPT-3 contains a single hidden layer with $4 * 12888$ units. The largest GPT-3 model contained 96 self-attention layers with 96 attention heads of dimensionality 128. The softmax output is trained with cross-entropy loss in order to learn the weights of the model. Aside from Wikipedia, Books1, Books2, and WebText2 datasets, GPT-3 uses a very large common crawl of the Web (via the Common Crawl dataset), which is filtered for quality to reduce the effect of contamination. The Common Crawl data set contains 410 billion tokens out of the total of 499 billion tokens in the full training data used by GPT-3.

The details of these datasets are discussed in [75].

A notable characteristic of GPT-n is that it uses a single model for all downstream tasks. Part of the reason is that next-word prediction is a fundamental task in language modeling, which can be used for other tasks. A few examples can be optionally added to the model to improve performance and fine-tune the results. Note that one can easily use a language model for the original transformer application of machine translation as long as the concatenated pair of sentences in the two languages is treated as a single sequence. However, a few examples of sentences in the two languages need to be provided, in order for GPT-3 to achieve state-of-the-art performance.

The most interesting application of GPT-3 is a conditional generative model that creates near-human-quality text by providing it contextual input such as the prompt "Article:", a title, and a subtitle. GPT-3 is able to write short articles of about 200 words that cannot be easily distinguished from human-written articles. In addition GPT-3 was able to achieve state-of-the-art performance on several closed-book question-answering tasks. GPT-3 performed well on some common-sense reasoning on tasks and poorly on others. Overall, GPT-3 was able to able to match or outperform a majority out of twelve natural language processing tasks. This is particularly impressive, considering the generic nature of the model. The performance on arithmetic tasks was particularly poor, as it achieved only 9%-accuracy on five-digit operations. This is not particularly surprising, as it learns a language model from text (rather than arithmetic computation), and rarely encounters all pairs of five digit numbers in sequences. Yet, the 9% accuracy is much higher than the type of learning in a language model would indicate. This suggests that GPT-3 is learning at least some common patterns in arithmetic operations. However, the poor performance in reasoning and arithmetic strongly suggests that the level of "understanding" in the model is limited to semantic patterns inherent in languages.

11.4.2   BERT

GPT-n uses a uni-directional language model, which can be harmful for some tasks like question-answering, where context from both directions in available. For example, the embedding of the word "*right*" is different in the two phrases "*right turn*" and "*right choice*," even though the useful context for disambiguation is available only after the occurrence of "*right*." Unlike GPT-n, BERT trains the encoder of the transformer of Figure 11.3 (rather than the decoder). The tokenization approach in BERT uses *WordPiece* [592], which is similar to byte-pair encodings, except that it maximizes the likelihood of the training data in order to perform merges (rather than using frequent co-occurrence). The positional embeddings are learned in BERT [567] rather than fixed sinusoidal functions. The training strategy is also different. BERT uses a loss function that is a combination of missing word prediction and sentence contiguity prediction. Therefore, each training example is a pair of sentences. For the missing word component, about 15% of the words in the input are randomly chosen for prediction, of which 80% are replaced with [MASK] in the input, and the remaining are equally likely to be replaced by the true word or a randomly chosen word (like a de-noising autoencoder). The entire masked sequence is input into the encoder. BERT starts with the encoder on the left-hand side of Figure 11.3(b), and uses a stack of either 12 or 24 layers (instead of the 6 layers used in the original transformer paper). BERT places a word-prediction layer at the end of the encoder to associate a loss with the 15% positions associated with prediction. For each word in the input sequence, the encoder provides a contextual embedding of that word. Subsequently, the dot product of each of these output word representations with the original embedding of each possible word in the dictionary is used to compute a real-valued similarity. The softmax function is applied to this real-valued similarity in order to predict a probability distribution of lexicon words for each word in the sequence. A cross-entropy loss with respect to the true value of the word (before masking) was used in order to train the model. Note that the training process is inherently bidirectional, since the words on both sides of the masked words were used by BERT as input. This is different from the decoder architecture of GPT-3, which does not look ahead of the word to be predicted.

The second component of the loss is based on *next-sentence prediction*. The model is fed pairs of sentences, and the task is to predict whether these pairs occur consecutively in a document. Therefore, 50% of the training pairs occur consecutively in the document, and the other 50% are randomly selected. The output is 0 or 1, depending on whether or not the sentence pairs occur consecutively. A [SEP] token is added at the end of each sentence, and a special [CLS] token is added at the beginning of the first sentence. The [CLS] token derives its name from "classification" since it is often viewed as a representation of the sentence following it, and is therefore used for sentence-level classification tasks. An embedding for either of the tokens "Sentence A" and "Sentence B" is added to each word token. The embedding for "Sentence A" is added to each token in the first sentence, and the embedding for "Sentence B" is added to each token in the second sentence. Adding this embedding helps the model distinguish between the two sentences beyond just the use of the [SEP] token. The encoder output of the [CLS] token is passed through the classification layer to output a probability of whether or not the two sentences are consecutive. The two components of the loss function (missing word and sentence contiguity prediction) are then consolidated into a single loss function. The idea is that the first loss component learns semantic relationships across words, whereas the second loss component learns semantic relationships across sentences. BERT uses all of the text portion of English Wikipedia and BooksCorpus [634].

Another difference between BERT and GPT is that BERT needs an additional task-specific neural layer or changed loss function and corresponding fine tuning to use it for specific applications. This difference is necessitated by the fact that BERT does not use a wide-spectrum language model for the training process (which can be easily adapted to tasks like machine translation without architectural changes). For token-wise tasks like tagging, labels are associated with output word embeddings, whereas for sentence-wise tasks, labels are associated with sentence-wise tokens like [CLS]. For sentence-level classification tasks, a single sentence can be input to BERT, with a [CLS] at the beginning and [SEP] at the end, which is followed by a *degenerate null sentence*. This is because BERT really expects to see sentence pairs rather than a single sentence, and the second sentence of the pair is simply a null sentence. The representation of [CLS] is then mapped to a class probability with an additional output layer. In general, the nature of the output layer varies with the task at hand, since token-level tasks will need to associated weights with the output representations of tokens. The output-layer weights need to be fine-tuned with the changed architecture and loss function (specific to the task at hand). The amount of fine tuning required is typically modest, and BERT seems to perform quite well on a variety of tasks such as sentiment analysis, question answering, and named entity recognition. Overall, BERT achieved state-of-the-art performance on eleven natural language processing tasks. Google search queries are now processed by BERT.

11.4.3 T5

The T5 model [460] uses a text-to-text model for pretraining, and it therefore follows the original transformer architecture more closely than the other two models. The basic idea behind using text-to-text models is that *sequence-to-sequence learners can be treated as multi-task learners by appending a task-specific prefix*. For example, if one wanted to translate from English to German, one can use the following input and output sequences:

> **Input Sequence:** translate English to German: That is good.
> **Output Sequence:** Das ist gut.

Note that it is possible to translate to a different language by using a different prefix. Similarly, for a summarization task (cf. Chapter 12), the input and output sequences would be as follows:

> **Input Sequence:** summarize: state authorities dispatched emergency crews tuesday to survey the damage after an onslaught of severe weather in mississippi. ⟨ Paragraph continues specifying number of hospitalized people and identity of county ⟩
> **Output Sequence:** six people hospitalized after a storm in attala county.

The grammar checking task, which is often applied to the *CoLA* benchmark (i.e., *Corpus of Linguistic Acceptability*), is achieved by adding the prefix "cola sentence" to the input, and the output can be "acceptable" or "not acceptable."

> **Input Sequence:** cola sentence: the course is jumping well.
> **Output Sequence:** not acceptable

The aforementioned example provides an instance of sentence-wise classification. For a sentence classification task, the input sequence be the prefix would contain a cue corresponding to the dataset-specific classification instance (e.g., sentiment analysis or cola), followed by the sentence — the output sequence would be the text label. Another example is *co-reference resolution*, which resolves ambiguous mentions of a word. For example, in one of the tasks supported by the *Winograd Natural Language Inference (WNLI) data set*, a passage is presented with an ambiguous pronoun, and one wishes to determine which noun in the passage that pronoun refers to. In such a case, the ambiguous pronoun can be highlighted in the input sequence as follows: "The city councilmen refused the demonstrators a permit because *they* feared violence." Note the highlighting of the word "they," which could either refer to the demonstrators or the city councilmen. The output sequence in this case is "the city councilmen".

In order to support these types of diverse tasks, the T5 transformer uses *a single unsupervised pretraining* phase followed by task-specific fine-tuning. The pre-training approach is motivated by BERT and it uses masked language modeling on a large corpus (described at the end of this section). The basic idea is to drop out about 15% of the tokens and replace *consecutive* spans of dropped tokens with *sentinel tokens*. For example, consider the following sentence in which the crossed-out portions need to be masked:

> Thank you ~~for inviting~~ me to your party ~~last~~ week.

Then, the input to the encoder of T5 removes the crossed-out segments, and replaces them with sentinel tokens that are unique to the specific sequence at hand. In this case, the two tokens ⟨X⟩ and ⟨Y⟩ are used for the two cross-out spans:

Thank you $\langle X \rangle$ me to your party $\langle Y \rangle$ week.

Note that since "for inviting" is a consecutive set of masked out tokens, it is replaced with a single sentinel token. The outputs then contain these sentinel tokens followed by the corresponding sequences that they represent in the original (unmasked) input. An additional sentinel token $\langle Z \rangle$ is reserved as the final token delimiting the end of the output. Note that all the sentinel tokens are different from one another in order to distinguish them. Therefore, the ground-truth sequence provided to the decoder of T5 during training is as follows:

$\langle X \rangle$ for inviting $\langle Y \rangle$ last $\langle Z \rangle$

The training of encoder-decoder architecture used 12 layers in both the encoder and the decoder. All sublayers and embeddings have a dimensionality of 768. The feed-forward networks in each block contained an intermediate layer with a dimensionality of 3072 followed by a ReLU nonlinearity. All attention mechanisms used 12 heads, each with an inner dimensionality of 64. For task-specific tuning, the (best) model uses a *gradual unfreezing approach*, wherein only the top block of the encoder and decoder were tuned first (with the task-specific data) while freezing the parameters in lower blocks. Then, the next lower blocks of the encoder-decoder stack were tuned together with the top blocks (while keeping the lower 10 blocks frozen). This approach was continued all the way down the 12 blocks in the encoder-decoder stack.

Note that this approach uses pretraining followed by fine-tuning for a single task. It is also possible to perform true multi-task learning by mixing examples together from different tasks (during the fine-tuning phase). Since each task has its own prefix, the model learns to recognize the specific tasks based on the prefix in the input at prediction time. The main challenge with multi-task learning is that it is possible for the model to overfit to specific tasks, and the training data for some tasks may interfere with others. This problem is referred to as *negative transfer*. Therefore, the T5 transformer mixes examples roughly in proportional to their task-specific data set sizes (i.e., concatenating the data sets), while reducing the proportion of those data sets that are too large (to avoid crowding out the tasks with smaller data sets). This is achieved by placing a threshold on the maximum size of each data set. A variant of this approach uses *temperature-based mixing*, wherein the probability of sampling from a data set is proportional to $s^{1/t}$, where s is the data set size and t is the temperature parameter. At $t = 1$, this approach reduces to concatenating the data sets, and larger values of t resulted in a damping effect with respect to data set size (to avoid crowding out). The best results were obtained at $t = 2$, although the multi-task learning approach could not match the performance of the model that used single-task-specific fine tuning after pretraining. This difference is natural, because there is some level of cross-interference between the models of different tasks in multi-task learning.

The pre-training data that was used for building the masked model was *The Colossal Clean Crawled Corpus*, which leveraged the *Common Crawl* as its original source after cleaning it. The cleaning process removed text containing menus, error messages, gibberish, duplicate text, code, or offensive words. Lines that did not contain punctuation or contained fewer than three words were dropped. Any page with fewer than five sentences was removed. A number of other heuristics that were used to clean the data set are described in [460].

11.5 Natural Language Processing Applications

Pretrained language models have revolutionized the landscape of natural language processing. These models are used for a variety of applications, many of which are discussed in later chapters of this book. This section will primarily focus on how BERT, GPT, and T5 are used for various types of natural language applications. It is also noteworthy that there are significant differences between how GPT-2 [459] and GPT-3 [75] in that the former is focused on fine tuning, whereas the latter is focused on zero-shot learning. This section will focus more the fine-tuning rather than the zero-shot learning use case in which natural language descriptions of tasks are provided to the model. Note that GPT-3 can also be used in conjunction with fine-tuning, and it would typically be more accurate, given that it is constructed on more data.

Some of the applications, such as text summarization, information extraction, question answering and sentiment analysis are discussed in later chapters of this book, and are therefore not addressed in detail here. This section will therefore focus on some of the applications that are not discussed elsewhere in this book. In this context, a well-known benchmark is the *General Language Understanding Benchmark* also referred to as *GLUE*. This benchmark can be viewed as a decathlon of natural language processing tasks, and have special utility in evaluating models with wide generalizability across tasks. Pre-trained language models represent a particularly important use case for this setting. This section will discuss the implementation of several of these tasks with pre-trained language models (along with a host of other tasks that are not included in the GLUE or SuperGLUE benchmarks).

11.5.1   The GLUE and SuperGLUE Benchmarks

The GLUE Benchmark [739] contains nine sentence- or sentence-pair language understanding tasks, which are based on existing data sets, and they are selected to cover a wide range of language understanding settings. A public leaderboard is also provided to track the performance on the benchmark together with visualization capabilities. the following exposition will provide an overview of some of these tasks (not discussed elsewhere in the book), together with how they are addressed by various pretrained language models. The SuperGLUE Benchmark [740] is an enhancement of the GLUE Benchmark containing some of the more difficult tasks of the GLUE benchmark and also contains additional challenging tasks (like co-reference resolution and question answering). Note that tasks like co-reference resolution cannot be expressed in terms of individual sentences or sentence pairs (like the GLUE tasks). Therefore, the SuperGLUE benchmark tests a wider range of problem settings. Aside from the GLUE benchmark, pointers to recent progress in several Natural Language Processing Tasks, together with corresponding data sets may be found at [742].

11.5.2   The Corpus of Linguistic Acceptability (CoLA)

The *Corpus of Linguistic Acceptability* [572] contains over 10,657 sentences in the English language that are grammatically correct or incorrect. The CoLA task is to flag these sentences as "acceptable" or "unacceptable." At its core, *this task is that of sentence-level classification*, since the sentence needs to be tagged with one of two labels based on its sequence of tokens. The original paper [572] describes the use of recurrent neural networks for this task, which are similar in principle to the model discussed in Section 10.8.4 of Chapter 10.

Pre-trained language models can be easily adapted to the CoLA task. For example, the T5 transformer uses a sequence-to-sequence model in which in which the input is the source sentence and the output sequence is the text class label containing the acceptability judgement. An example of how T5 is used for sentence-level classification is provided in Section 11.4.3. It is also easy to adapt GPT-n to the CoLA problem, because the concatenation of the sentence together with its label can be used as a new set of sentences for fine-tuning. The labels can be represented by tokens not present elsewhere in the vocabulary, and a special token can be added to the beginning of the sentence indicating that it is a CoLA task. This is a generic approach that can be used to fine-tune any pretrained language model to the classification. The original paper for GPT-2 [459], however, suggests to train an additional classification layer in order to output the class label based on the output of special token at the end of the sentence. In the case of BERT, an additional layer needs to be trained during fine tuning that maps to the [CLS] token to either "acceptable" or "unacceptable." Note that the BERT family always uses the representation of the [CLS] token to model the sentence at hand. Furthermore, since BERT works with sentence pairs as inputs during pre-training, the second sentence of the pair is assumed to be a null sentence during fine tuning.

11.5.3 Sentiment Analysis

The sentiment analysis task classifies sentences as "positive sentiment" or "negative sentiment" depending on the tone of the sentence. Therefore, like CoLA, sentiment analysis is a classification task. The use of recurrent neural networks to perform sentiment analysis is discussed in Section 10.8.4 of Chapter 10. Since sentiment analysis is a sentence-wise classification task, the methods used are similar to those in CoLA. An important data set that is used for sentiment analysis is the *Stanford Sentiment Treebank* [741]. Further details on the use of the deep learning for sentiment analysis are discussed in Section 15.3.4.2 of Chapter 15.

11.5.4 Token-Level Classification

The aforementioned applications, such as linguistic acceptability checking and sentiment analysis use the model corresponding to sentence-level classification. However, several applications, such as *part-of-speech tagging*, *named entity extraction*, and *semantic role labeling* use token-level classification, in which labels are associated with individual tokens. In semantic role labeling, the tokens in sentences are associated with roles, such as agent, subject, and so on.

There are many variations of the problem of token-wise classification problem, depending on whether or not each token is associated with a label. Furthermore, in some cases, the labels may be associated with a consecutive sequence of tokens (i.e., a *span* of tokens), as in the case of named entity recognition where named entities like "William Jefferson Clinton" have multiple tokens. It is noteworthy that when only a subset of the tokens are associated with labels, it is possible to use a catch-all label like "other" in order to associate labels with all tokens. Furthermore, in order to address cases in which the labels are associated with a group of tokens, it is possible to create different labels corresponding to the beginning, continuation, and end, of the span of that label sequence in order to identify the fact that the entire span should be treated as an indivisible group of tokens. For example, even though the label for *person entity* might be "P," one might have three labels PB, PC, and PE, corresponding to the beginning continuation and end of the person entity. Similarly, labels like LB can be used to designate the beginning of a location entity. The "Other" label does not need "beginning" and "end" designations, because it is not an indivisible group of tokens (like the name of a person or city). Therefore, the "O" label is used for such tokens.

The following is an example of the tagged sentence:

$$\underbrace{\text{William}}_{PB} \ \underbrace{\text{Jefferson}}_{PC} \ \underbrace{\text{Clinton}}_{PE} \ \underbrace{\text{lives}}_{O} \ \underbrace{\text{in}}_{O} \ \underbrace{\text{New}}_{LB} \ \underbrace{\text{York}}_{LE}.$$

It is relatively easy to train (i.e., fine-tune) a layer of weights in GPT-3 to output the token at each step of the autoregressive output of the decoder. In the case of BERT, the embeddings of the individual tokens are reported, and an additional layer of weights can be fine-tuned to report their labels. Since BERT assumes sentence pairs as inputs, the second sentence of the pair is assumed to be a null sentence in token-wise classification. In the case of T5, the fine-tuning can use a set of training instances in which the output sequence corresponds to the sequence of labels over the input sentence. A slightly more detailed discussion of token-wise classification in the context of named entity recognition is provided in Section 13.2.6.2 of Chapter 13.

11.5.5   Machine Translation and Summarization

Transformers were originally proposed in the context of machine translation, and therefore it should come as no surprise that the pre-trained language models based on transformers can be used for machine translation. For some models like T5, which are already text-to-text models, the application to machine translation is trivial. In fact, the original paper [460] provides some pretrained presents for machine translation over languages like French and German. In the case of GPT, only the decoder is used, and therefore some adaptations are required for machine translation. For fine-tuning, the sentences in the two languages can be concatenated into a single sequence, and the sentences in the two languages can be demarcated with a special separator. Subsequently, the fine-tuning process will coax GPT-3 to treat the concatenated sequence as an example of a valid language, whenever it encounters the special token corresponding to the separator between the two languages. Of all the pretrained language models presented in this chapter, BERT is the hardest to adapt to machine translation. This is because of its use of the encoder portion of the transformer, which does not naturally output a sequence of variable length. A discussion of the challenges of using BERT as well as a (somewhat limited) solution for using BERT in machine translation is provided in [632].

A closely related problem to machine translation is *abstractive summarization* (cf. Section 12.7.5 of Chapter 12), in which one is given a document, and a summary needs to be created from the document. Abstractive summarization is most general case of summarization in which the summaries are not necessarily created by using sentences from the original document, but the goal is to mimic the human style of summarization. When examples of human-provided summaries are available, the problems of abstractive summarization and machine translation are quite similar — after all, both machine translation and abstractive summarization are both sequence-to-sequence learning tasks. However, translation tasks are done at the sentence level, which makes them much simpler for length-sensitive learners like recurrent neural network (or LSTMs). On the other hand, transformers are natural solutions in such scenarios, since they work much more effectively in settings where the sequences are long. As discussed with the help of an example in Section 11.4.3, it is relatively easy to adapt a text-to-text pretrained model to the summarization problem. GPT-3 can also be adapted to summarization, as it can generate fluent pieces of text that are conditioned on specific inputs (including a source document). However, since BERT is not naturally suited to sequence-to-sequence tasks, the challenges associated with machine translation with BERT are also faced in abstractive summarization. The work in [351] shows how BERT may be used for abstractive summarization by adding a decoder to it. This approach can also be used for machine translation. We refer the reader to Section 12.7.5 for a brief description of this approach along with other deep learning methods for abstractive summarization.

11.5.6 Textual Entailment

Textual entailment is also referred to as *natural language inference.* The task of textual entailment is that of reading two sentences and then determining the relationship between them. The two sentences are the premise and the hypothesis. The relationship could be that the premise entails (i.e., implies) the hypothesis, that they are contradictory, or that the relationship is neutral. Labels are available indicating the relationships between pairs of sentences. In simpler variations of this task, the output is simply True or False, wherein the label is True only when the premise implies the hypothesis. The *Winograd Natural Language Inference* (WNLI) and *Winograd Schema Challenge* (WSC) are part of the GLUE and SuperGLUE benchmarks, and are commonly used for evaluating these tasks. A number of other data sets for natural language inference are available at [743].

The GPT-3 model concatenates the premise and hypothesis with a $ token separating them, which directly converts it into a single sequence classification task. This representation can then be classified with any of the models discussed above. In the case of BERT, the [CLS] token is fed to the output layer for classification. On the other hand, a T5 transfomer would treat the class label as the second sequence to be inferred.

11.5.7 Semantic Textual Similarity

The goal of sentence similarity is to identify whether or not two sentences are semantically similar. In some data sets, one has to predict a real-valued score between 1 and 5. This case is similar to textual entailment, except that the task is similar to regression rather than classification and there is no ordering between the two sentences. As in the case of textual entailment, the two sentences are concatenated, but they are concatenated in both orders, thereby creating two new sequences. Subsequently, both sequences are fed into the transformer in order to create two representations. In the case of GPT-3, the two representations are concatenated and fed into a fine-tubed output layer that can perform regression. In the case of BERT, the [CLS] representations of the two cases are concatenated and fed into a fine-tuned output layer. For T5, the problem of regression is more difficult since the output is supposed to be a text string. Therefore, T5 discretizes the data into 21 groups using an increment of 0.2, and converts real-valued scores to strings like "2.2". This results in 21 classes, since the first string is "1.0" and the final string is "5.0." Subsequently, T5 treats this problem in a similar manner to any other classification problem. The semantic textual similarity (STS) data set is a part of the GLUE benchmark, and it is commonly used for evaluating this task.

11.5.8 Word Sense Disambiguation

In word-sense disambiguation, one wishes to determine the specific mode of use of a polysemous word. For example, the word "stick" may be used in two different ways, and the specific use depends on the context of the sentence at hand. It is assumed that training data are available that provide examples of the usage of each possible sense of the word. Recurrent language models like ELMo (cf. Section 10.8.1 of Chapter 10) have been used in the past for this type of application by leveraging the hidden states of the recurrent network as representations of the word. However, attention-mechanisms are even better suited to word-sense disambiguation becuase of their use of context. All transformer-based models output context-sensitive representations of words from sequence. First, training data containing example uses of each word is required. This training data is used in order to create an average representation of each sense of the word. Subsequently, for a given test use of the word, the nearest representation from the each of the senses of the word is identified, and that word sense is reported.

For some data sets from the GLUE benchmark, two sentences are provided together with a specific word, and it is desired to determine whether the two sentences correspond to the same usage of that word. In this case, the two sentences can be concatenated in any order, together with the identified word, and the problem boils down to a binary classification task of True or False, depending on whether the usage is the same. Such a task can be addressed using the sentence-wise classification methods discussed earlier in this section in the context of the CoLA task.

11.5.9   Co-Reference Resolution

The co-reference resolution problem uses the Winograd Natural Language Processing Inference (WNLI) data set. This problem was discussed briefly in Section 11.4.3 in the context of resolving ambiguous use of a pronoun. We repeat the example of an input to the T5 transformer:"The city councilmen refused the demonstrators a permit because *they* feared violence." The word "they" could either refer to the demonstrators or the city councilmen. The output sequence in this case is "the city councilmen". It is fairly straightforward to use T5 for this task. The BERT and GPT-3 papers do not explicitly describe a co-reference resolution approach. However, it is fairly straightforward to generalize any task that is achieved with a text-to-text transformer (like T5) to an autoregressive model like GPT-3. After all, any sequence-to-sequence prediction task can be recast as a sentence completion task by concatenating the two sequences. Language models like GPT-3 are masters at sentence completion and the language model will simply complete the second "sentence" containing the chunk referred to by the pronoun (such as "the city councilmen"). The Winograd Natural Language Inference (WNLI) data set is commonly used for evaluating this task.

11.5.10   Question Answering

While question answering is addressed in a more general way in Chapter 14, we discuss a simpler use case in which multiple choices for the different answers to a question are available. In addition, a context paragraph is provided, which contains the answer to the question in some form. GPT-3 concatenates the question, the context, each of the different choices for the answers, where the answer is separated by a delimiter from the question and context. For each choice of answer, a different sequence is produced. The different sequences are processed independently by GPT-3, and then normalized with a softmax distribution to obtain the probabilities of different answers. Both BERT and T5 are trained on the SQuAD data set, which is introduced in detail in Chapter 14. Therefore, a further discussion on the use of pre-trained language models for question-answering is deferred to Chapter 14.

11.6   Summary

Attention mechanisms have been found to be very useful in improving the performance of a wide variety of natural language models, such as machine translation. The initial approach to attention was to use it in the context of improving the performance of recurrent neural networks. Subsequently, with the development of the transformer, it was shown that one can use these attention mechanism independent of the recurrent neural network. This is particularly useful because of its parallelizeability. The transformer has gradually begun to replace the recurrent neural network in a wide variety of settings, and has been used to construct pretrained language models. Pre-trained language models have revolutionized the field of natural language processing and are used for a variety of downstream tasks. This chapter discusses an overview of some of these tasks, and additional tasks are discussed in later chapters.

11.7 Bibliographic Notes

Early techniques for using attention in neural network training were proposed in [83, 311]. The two most well known models are neural machine translation with attention are discussed in [37, 359]. The ideas of attention have also been extended to image captioning [596]. The use of attention models for text summarization is discussed in [483]. The notion of attention is also useful for focusing on specific parts of the image to enable visual question-answering [465, 594, 608]. Transformer networks that process text without RNNs were proposed in [560]. Transformers have been used for unsupervised natural language generation models and meta-learning in the form of Google BERT [149] and Open AI's GPT-*n* [75, 459]. An optimized version of BERT is referred to as *Roberta* [352], and it is often used as a pretrained language model for off-the-shelf uses. The T5 model was proposed in [460]. Other notable language models include T5 [460], Microsoft Turing Natural Language Generator (NLG) [717], XLNet [611], and the Switch Transformer [192]. A repository containing a discussion of the progress in various NLP tasks is available at [742]. This repository contains a discussion of the diverse uses of various pretrained language models.

11.7.1 Software Resources

The MATLAB code for the attention mechanism for neural machine translation discussed in this chapter (from the original authors) may be found in [642]. Implementations of transformer networks may be found from HuggingFace and Google Brain Trax among others [718, 727, 728]. The implementations of numerous language models are available, such as those of ELMo [723], BERT [719], T5 [720], XLNet [722], and Switch Transformer [721]. The GLUE and SuperGLUE benchmarks are available at [739, 740].

11.8 Exercises

1. What is the computational complexity of a self-attention mechanism on a sentence of length n?

2. Discuss how the Luong attention model can be used for self attention in language modeling. Provide an example of when attention might help predictions.

3. Discuss the behavior of the softmax function on a vector whose components have large magnitudes. What are the goals of scaling in scaled dot-product attention? Can you think of ways of optimizing the scaling for a specific training data set?

4. What changes would you make to the transformer architecture to make it work for sets instead of sequences?

Chapter 12

Text Summarization

"Less is more."– Ludwig Mies van der Rohe

12.1 Introduction

Text summarization creates a short summary of a document, which can be easily assimilated by the user. The most basic form of text summarization creates a summary from a single document, although it is also possible to do so from multiple documents. The key applications of text summarization are as follows:

1. *News articles:* A short summary of a news article enables quick perusal. It may also be useful to summarize the titles of a large number of related news articles in order to understand the common theme.

2. *Search engine results:* A query on a search engine may return multiple results that need to be presented on a single page. Typically, the title is followed by short summaries on that page.

3. *Review summarization:* Reviewers at sites such as Amazon produce large numbers of short documents describing their assessment of a particular product. It may be desirable to condense these reviews into a shorter summary.

4. *Scientific articles:* Impact summarization is a way of extracting the most influential sentences in a particular article. This type of summarization provides a broad understanding of what the article is about.

5. *Emails:* A thread of email corresponds to a discourse between two participants. In such cases, it is important to take the interactive nature of the dialog into account during the summarization process.

6. *Improving other automated tasks:* An unexpected benefit of text summarization is that it sometimes improves the performance of other tasks in text analytics. For example,

C. C. Aggarwal, *Machine Learning for Text*,
https://doi.org/10.1007/978-3-030-96623-2_12

it has been reported [485] that the precision of information retrieval applications improves when terms from the summary are used to expand the query.

The applications associated with single-document and multi-document summarization are quite different. Multi-document summarization often arises in settings in which the set of documents is closely related, such as the articles related to a particular news event, the tweets in response to an event [91], or the search-engine results of a particular query.

In many cases, the specific *context* or *application domain* plays an important role in deciding the choice of the summarization technique. For example, in *query-focused summarization*, the documents returned by a query processing system are often summarized to show short snippets of the documents to the user for ease in browsing. In such cases, the summaries are tailored to be more inclusive of the specific query words entered by the user. The use of context provides additional hints, and this is often helpful in tailoring the results to the application domain at hand. In the multi-document setting, the presence of context is particularly common, because the documents are interrelated by their context. It is sometimes argued [523] that context is so important that one should not attempt to summarize at all without the presence of some context. Nevertheless, the broader literature on text summarization proposes a number of generic methods that can be used for summarization without the presence of context. This chapter will primarily focus on generic methods.

12.1.1   Extractive and Abstractive Summarization

The two main types of summarization are either *extractive* or *abstractive*, which are defined below:

1. *Extractive summarization:* In extractive summarization, a short summary is created by extracting sentences from the original document without modifying the individual sentences in any way. In such cases, an important step is often that of scoring the importance of different sentences. Subsequently, a subset of the top-scored sentences are retained to maximize the topical coverage and minimize redundancy.

2. *Abstractive summarization:* Abstractive summarization creates a summary that contains new sentences not available in the original document. In some cases, such methods may use phrases and clauses from the original representation although the overall text is still considered new. Of course, generating new text is often challenging because it requires the use of a language model to create a meaningful sequence of words. Abstractive summarization shares a number of similarities with machine translation, when done in a supervised setting. However, machine translation tends to occur at the sentence-to-sentence level, whereas the source text in abstractive summarization is much larger. Because of the increased lengths of the source (and even target) instances, one is not guaranteed that the generated summary is always of high quality. In general, abstractive summarization is much harder, although recent advances in transformers and pretrained language models have brought abstractive summarization closer to reality. This is because transformers are much better than the traditional approach (recurrent neural networks) at handing long sequences in sequence-to-sequence learning.

It is noteworthy that abstractive summarization requires coherence and fluency. This requires a high level of semantic understanding of the underlying text, which is beyond the capabilities of modern systems. Completely fluent abstractive summarization represents an unsolved problem in artificial intelligence, and most summarization systems are extractive.

However, recent advancements in deep learning have made abstractive summarization a realistic possibility. Because of the preponderance of extractive summarization in the text mining literature, most of the discussed methods belong to the extractive summarization category. However, recent advancements for abstractive summarization will also be discussed.

12.1.2   Key Steps in Extractive Summarization

Most extractive text summarization methods use two stages, and the specific choices used at each stage regulates the overall design of the method at hand:

1. *Sentence scoring:* The first step in many techniques is to score sentences based on their importance towards the creation of a coherent summary. Some methods use only the content of the sentences whereas others use various types of meta-information such as its length or positioning of the sentence. In many cases, an intermediate representation is created in order to perform the modeling. For example, one might create a table of important words for the summary or a graph representation of sentence-sentence similarities. Sentences are scored based on their ability to represent key themes in the document at hand, which are eventually useful in creating the summary.

2. *Sentence selection:* Based on the scores, the sentences are selected in order to represent the summary. During this process, it is important to not only account for the score of a document but also its redundancy with respect to the other selected sentences. Reducing overlap is the key mechanism in controlling summary size.

In many cases, the processes of sentence scoring and selection are independent of one another, but in other cases (such as sentence-sentence similarity methods) the scoring and selection process are tightly integrated. In cases where the scoring and selection are independent, it is possible to reuse a particular sentence selection technique across multiple scoring methods.

12.1.3   The Segmentation Phase in Extractive Summarization

Text segmentation [244] is an important step in extractive summarization. The basic idea in text segmentation is to break up a long document into shorter or more coherent segments, each of which is contiguous within the document. These shorter segments might be based on grammatical rules (e.g., sentences/paragraphs), or they might be based on topical contiguity. In spite of the widespread use of sentences as the units of summarization, it has sometimes been argued that the use of longer segments like paragraphs is often more useful [491]. Although this chapter will consistently use a sentence as the unit segment (because of its preponderance in the literature), we point out that most of the techniques in the chapter can be generalized to any type of segment without changing the underlying algorithms in a significant way. For example, one can segment the text based on topical continuity. The problem of text segmentation is discussed in Section 16.2 of Chapter 16.

Chapter Organization

This chapter is organized as follows. The next section will discuss methods based on topic words. Latent methods for summarization are discussed in Section 12.3. The use of traditional machine learning for extractive summarization is discussed in Section 12.4. The use of deep learning methods for extractive summarization is discussed in Section 12.5. Methods for multi-document summarization are presented in Section 12.6. Abstractive summarization methods are discussed in Section 12.7. The summary is presented in Section 12.8.

12.2 Topic Word Methods for Extractive Summarization

Topic word methods create a table of words and their weights, where a larger weight is more indicative of the topic at hand. The earliest work on extractive summarization with topic words was done by Luhn [357]. The basic idea in his work was to find the most topical words based on frequencies. Words that are too frequent or too infrequent are not helpful for identifying the topical content of a document. Very frequent words are often stopwords, whereas very infrequent words are misspellings or obscure words. By identifying lower and upper thresholds on the frequencies, the remaining words are identified as topical words and used to score sentences. Luhn's original work introduced the notion that topical words that are placed close to one another should have more impact on the score of a sentence than scattered words. Therefore, Luhn proposed to place a bracket around the segment of a sentence in which the topically significant words (i.e., words satisfying upper and lower thresholds) are separated by a gap of no larger than g. This type of bracketing enables subsequent scoring. Note that the gap is measured only between consecutive pairs of topically significant words. The value of g was set to around 4 or 5. Then, the square of the number of words in each segment divided by the length of the segment provides a significance score of the *bracketed segment*. The score of a sentence is the maximum score over all its bracketed segments.

These basic ideas of Luhn provided a starting point for much of the eventual research on topic-word methods. The following will discuss the key ideas along this line, such as the use of word probabilities, tf-if, and log-likelihoods. The last of these is considered state-of-the-art among topic-word techniques.

12.2.1 Word Probabilities

Consider a document $\overline{X} = (x_1 \ldots x_d)$, in which x_j is the raw frequency of the jth word. Then, the word probability p_j may be computed as the fractional presence of that word:

$$p_j = \frac{x_j}{\sum_{j=1}^{d} x_j} \tag{12.1}$$

Now consider a summary with M tokens, in which the jth term occurs m_j times, and therefore we have:

$$\sum_{j=1}^{d} m_j = M \tag{12.2}$$

The likelihood $\mathcal{L}$ of this summary can be computed using the multinomial distribution:

$$\mathcal{L} = P(m_1, m_2, \ldots m_d) = \frac{M!}{\prod_{j=1}^{d} m_j!} \prod_{j=1}^{d} p_j^{m_j} \tag{12.3}$$

Why is it desirable to maximize likelihood? The core basis of this assumption is the hypothesis is that summaries reflecting the frequency distribution of the terms in the original document are more likely to be informative.

One heuristic way of selecting summaries with high likelihood is the *SumBasic* method. Let t_j denote the jth word (term). Then, *SumBasic* computes the average probability over all the words in each sentence S_r as follows:

$$\mu(S_r) = \frac{\sum_{t_j \in S_r} p_j}{|\{t_j : t_j \in S_r\}|} \tag{12.4}$$

Then, the sentence S^* with the largest value of $\mu(S_r)$ is selected and included in the summary. At this point, the probability p_j of each term in S^* is reduced by setting it to the square of its original value. The idea here is that users are unlikely to select a summary with too many repeated terms because it would cause redundancy. The entire process of computing $\mu(S_r)$ is repeated with these adjusted probabilities, and the next sentence with the largest value of $\mu(S_r)$ is selected. This process is repeated until the summary is of the desired length. Therefore, the algorithm can be described as follows:

1. Compute the probability p_j of each word according to Equation 12.1.

2. Compute the average word-probability $\mu(S_r)$ of each sentence S_r according to Equation 12.4.

3. Select the sentence $S^* = \text{argmax}_r \, \mu(S_r)$ with the largest value of $\mu(S_r)$ and add it to the summary.

4. Reduce the probability of each word included in the added sentence, S^*, by squaring it.

5. If the desired summary length has not been reached, then go to step 2.

The bag of sentences at the end of the process provides the summary. This approach tightly integrates sentence scoring with selection, while accounting for redundancy. Redundancy is avoided by reducing the probabilities of already included words by squaring them. An alternative is to multiply the probability with a factor less than 1.

12.2.2 Normalized Frequency Weights

This approach distinguishes the frequent words in a particular document from those that are present in a generic corpus. Frequent words in a generic corpus are often caused by the fact that they are stopwords like articles, prepositions, or conjunctions. However, in a specific corpus, some of the frequent words may be germane to the topics in the collection, and are worthy of using at least a few times in the summary. For example, the word "*election*" may be very common in a specific document that needs to be summarized, but it may not be so common in a generic *background* corpus. Therefore, it is desirable to always use background information when performing stopword removal, rather than using frequency thresholding with respect to the specific document being summarized. A common approach is to use stopword lists in order to remove the irrelevant words. Furthermore, terms that occur an extremely small number of times (e.g., once or twice) are also removed because they might be misspellings or too unusual to be germane to a summary.

A key step in this approach is to use inverse document frequency normalization before deciding on the significant words. Document length normalization is used by dividing the tf-idf weight with the maximum frequency of any word in the document. For any document

$\overline{X} = (x_1 \ldots x_d)$ in which the jth term has inverse document frequency of idf_j, the weight w_j can be computed as follows:

$$w_j = \frac{x_j \cdot idf_j}{\max\{x_1, \ldots x_d\}} \tag{12.5}$$

All words with weight w_j below a particular threshold are reset to a weight of 0, because such words are presumed to be noisy words. These word weights can then be used to score sentences. The simplest approach is to use the average weight of a word in a sentence S_r in order to compute its significance. The average weight $\mu^w(S_r)$ of a word in a sentence can be computed as follows:

$$\mu^w(S_r) = \frac{\sum_{t_j \in S_r} w_j}{|\{t_j : t_j \in S_r\}|} \tag{12.6}$$

Note that this type of averaging is almost identical to that used in *SumBasic*. The sentences are then ordered in decreasing order by weight and the top sentences are selected. However, unlike *SumBasic*, it does not seem to account for the redundancy between different sentences in the selection process. Nevertheless, it is relatively easy to incorporate such modifications by multiplying the frequency of each selected word by a small factor less than 1. It is noteworthy that the process of sentence selection is largely independent of the scoring process, and a wide variety of selection methods can be paired with different scoring methods. Therefore, the key methods for sentence selection (which are often reusable over different ways of scoring) will be discussed in Section 12.2.4.

12.2.3   Topic Signatures

Topic signatures are important words for summarization that are identified with a log-likelihood ratio test. In order to implement the log-likelihood ratio test, the term frequency in a particular document is compared with that in a background corpus. The basic idea is to identify words that occur frequently in the document to be summarized, but are rare with respect to the background collection. Statistical hypothesis testing is a well-defined methodology with a probabilistic interpretation, which also provides appropriate thresholds for word selection.

The probability of the term t_j in the document $\overline{X}$ with term frequencies $(x_1 \ldots x_d)$ is denoted by p_j:

$$p_j = \frac{x_j}{\sum_{j=1}^{d} x_j} \tag{12.7}$$

Note that this way of defining p_j is similar to Equation 12.1. The value of p_j can also be defined for a set of documents instead of a single document by using the corresponding frequencies across n documents as follows:

$$p_j = \frac{\sum_{i=1}^{n} x_j^{(i)}}{\sum_{i=1}^{n} \sum_{j=1}^{d} x_j^{(i)}} \tag{12.8}$$

Here, $x_j^{(i)}$ is the frequency of the jth term in the ith document. A similar approach can also be used to define the probabilities in the background collection. Let $b_j \in (0,1)$ be the corresponding probability of the jth term in the background collection and p_j be the probability only in the document(s) being summarized. Similarly, we compute the probability $a_j \in (0,1)$ of the jth term belonging to the union of the background collection and the specific document(s) being summarized. The main idea in the likelihood-ratio test is to assume that the number of occurrences of each token in both the background and the document are each generated by repeatedly flipping a biased coin for each token. The goal is to find out whether the same coin is used for both document and background. Therefore, the two hypotheses are as follows:

H_1: **[For jth token]** The number of occurrences of the jth token in the document for summarization and background are both generated by repeatedly flipping a biased coin with probability a_j.

H_2: **[For jth token]** The number of occurrences of the jth token in the document for summarization is generated from a biased coin with probability p_j, and that in the background with a biased coin with probability b_j.

Furthermore, we are only interested in testing terms t_j that satisfy $p_j > b_j$, because other terms cannot be topic signatures. Now consider a situation in which the document to be summarized contains n tokens of which n_j correspond to term t_j. For the background corpus, the corresponding numbers are $n(b)$ and $n_j(b)$, respectively. Then, under the hypothesis H_1, the probabilities of n_j and $n_j(b)$ are defined by binomial distributions with different numbers of trials, but the same sampling parameter a_j. Therefore, the probability distribution of the number of occurrences of the jth token can be computed as follows:

$$\text{Document to be summarized: } P(n_j|H_1) = \binom{n}{n_j} a_j^{n_j}(1-a_j)^{n-n_j}$$

$$\text{Background collection: } P(n_j(b)|H_1) = \binom{n(b)}{n_j(b)} a_j^{n_j(b)}(1-a_j)^{n(b)-n_j(b)}$$

The joint probability of n_j and $n_j(b)$ under hypothesis H_1 is computed as the product of the above two quantities:

$$P(n_j, n_j(b)|H_1) = P(n_j|H_1) \cdot P(n_j(b)|H_1) \qquad (12.9)$$

In the case of hypothesis H_2, the main difference is that we are using different coins with face probabilities p_j and b_j, respectively, to model the distribution of term t_j.

$$\text{Document to be summarized: } P(n_j|H_2) = \binom{n}{n_j} p_j^{n_j}(1-p_j)^{n-n_j}$$

$$\text{Background collection: } P(n_j(b)|H_2) = \binom{n(b)}{n_j(b)} b_j^{n_j(b)}(1-b_j)^{n(b)-n_j(b)}$$

The joint probability of n_j and $n_j(b)$ based on hypothesis H_2 is given the product of the above two quantities:

$$P(n_j, n_j(b)|H_2) = P(n_j|H_2) \cdot P(n_j(b)|H_2) \qquad (12.10)$$

Then, the likelihood ratio λ is defined as the ratio of the quantities estimated in Equations 12.9 and 12.10:

$$\lambda = \frac{P(n_j, n_j(b)|H_1)}{P(n_j, n_j(b)|H_2)} \qquad (12.11)$$

The value $-2\log(\lambda)$ has a χ^2-distribution, which enables the use of a threshold on $-2\log(\lambda)$ at a specific level of probabilistic significance. For example, one can use a 99.9% level of confidence in order to select a threshold value from the χ^2 distribution tables. Therefore, unlike some of the methods discussed earlier in this section, one is able to choose thresholds that are statistically better justified.

Given the topic signatures, the score of a sentence is equal to the number of tokens that are topic signatures. An alternative is to set the score of a sentence to the fraction of the tokens in it that are topic signatures. The first method tends to favor longer sentences for inclusion in the summary, whereas the second method normalizes for the length of the sentence in the summary. As discussed in the next section, it is also possible to account for redundancy in sentence selection.

12.2.4   Sentence Selection Methods

Sentence selection is a key step in summarization that follows the scoring process. The presence of redundancy defeats one of the key goals of summarization. Therefore, various ad hoc techniques are sometimes used to reduce redundancy when the sentence scoring is tightly integrated with sentence selection. For example, the *SumBasic* method adjusts the probabilities of the words after selecting each sentence, so that the probabilities of words that are included in previously selected sentences are reduced. However, it is desirable to sometimes decouple sentence scoring from sentence selection. In such cases, generic methods for redundancy removal are required. Such methods can be used in combination with arbitrary scoring methods.

A method in [89] proposed techniques for sentence selection in query-focused summarization. However, such techniques have also been adapted to generic summarization settings [219, 338]. The method proposed in [89] is a greedy technique for selecting sentences with the largest maximum marginal relevance (MMR). The basic idea is to add sentences to the summary one by one, while ensuring that the score of the added sentences is as favorable as possible, but the overlap with the previously selected sentences is as little as possible. There are several ways of operationalizing this technique. The original idea, which is proposed in [89], is to use a convex combination of the relevance score and a novelty score. The relevance score can be computed in a variety of ways, including any of the scoring methods discussed in this section.

Let $\mathcal{S} = \{S_1 \ldots S_r\}$ be the sentences that have been added to the summary so far from document $\overline{X}$. The novelty score $N(S)$ of the sentence S from document $\overline{X}$ is high when the document is dissimilar to other documents that have been included in the summary. Therefore, the novelty score has been quantified by using the negative[1] of the cosine similarity of the sentence S_j with the other sentences in document $\mathcal{S}$.

$$N(S) = 1 - \max_{S_j \in \mathcal{S}} \text{cosine}(S, S_j) \tag{12.12}$$

The cosine is computed using the vector space representation of the sentences. Note that the novelty score always lies in $(0, 1)$ and larger values of the novelty are desirable. Let $T(S)$ be the score of a sentence S using any of the scoring methods such as tf-idf or topic signature method with the same convention that larger values are more desirable. Then, the overall score $F(S)$ of a sentence is a linear combination of the scores using the two criteria and combination parameter $\lambda \in (0, 1)$.

$$F(S) = \lambda T(S) + (1 - \lambda)N(S) \tag{12.13}$$

---

[1]In the original paper [89], the novelty component is set to $N(S) = -\max_{S_j \in \mathcal{S}} \text{cosine}(S, S_j)$. We have added the additional value of 1 in order to create a score in the range $(0, 1)$, which is easier to interpret. The addition of 1 does not change the final results of the computation.

Here, $\lambda \in (0, 1)$ regulates the trade-off between diversity and sentence relevance. The MMR algorithm always adds a sentence S with the largest value of $F(S)$ to the summary $\mathcal{S}$ from the remaining sentences in the document $\overline{X}$. One can vary on the similarity function that is used to compute the novelty. For example, some methods [338] use the percentage overlap instead of the cosine in order to compute novelty.

Another simplified approach, which is discussed in [338], is to add a sentence S with the highest value of $T(S)$ to the summary, while constraining the novelty score between this sentence and previously selected sentences to be above a particular threshold. This approach boils down to setting a minimum threshold on the novelty score $N(S)$ in order to ensure that every sentence that is added has a minimum level of novelty. Among all such sentences, the sentence with the largest score $T(S)$ is added to the summary. Although the greedy approach does not necessarily find the optimal solution, it usually finds a high-quality solution in most practical settings.

12.3 Latent Methods for Extractive Summarization

Latent methods borrow ideas from latent semantic analysis, matrix factorization, and chains of co-occurring words in order to identify summary sentences.

12.3.1 Latent Semantic Analysis

An important property of latent semantic analysis is that it exposes the independent latent concepts in the data. Therefore, by selecting sentences with large components along these latent directions, one is able to create a summary of sentences that express dominant concepts in the document.

One of the earliest methods using latent semantic analysis was proposed in [219]. Consider a document $\overline{X}$, which contains m sentences denoted by $S_1 \ldots S_m$. The sentence S_i is denoted by the d-dimensional vector $\overline{Y_i}$. The $m \times d$ matrix for which the ith row contains the vector $\overline{Y_i}$ is denoted by D_y. Then, one can use latent semantic analysis in order to create a rank-k latent decomposition of the matrix D_y as follows:

$$D_y \approx Q \Sigma P^T \tag{12.14}$$

Here, Q is an $m \times k$ matrix, Σ is a $k \times k$ diagonal matrix, and P is a $d \times k$ matrix. The value of k can be chosen to be $\min\{m, d\}$ to ensure that the above approximation is satisfied as an exact equality. Here, the k columns of P provide the k orthogonal basis vectors along which the sentences are represented. Therefore, sentences with large projections along these independent concepts are likely to be relatively independent of one another and represent good choices for summarization. The rows of the matrix $Q\Sigma$ contain the k-dimensional reduced representations of each of the sentences. The matrix $Q = [q_{ij}]$ (rather than $Q\Sigma$) contains the normalized coordinates of the sentences after adjusting for the relative frequency of each concept. The m-dimensional column vectors of the $m \times k$ matrix Q are useful for extracting the sentences that correspond to each of the k independent concepts. Large absolute values of q_{ij} in a particular column of Q indicate that the corresponding sentence has a strong projection along that concept. Therefore, the approach for summary extraction processes the columns of Q one by one in decreasing order of singular value to add sentences. While processing the jth column of Q, we pick the entry q_{ij}, whose absolute value is larger than all values of q_{rj} for $r \neq i$. The index i provides the sentence S_i that should be added next to the summary. The sentences are added one by one to the summary using this approach until the desired summary length is reached. By using different eigenvectors to generate different sentences, the impact of redundancy is minimized.

One problem with this approach is that it uses only one representative sentence for each latent concept (singular vector of D_y). In practice, it may be possible that a single sentence might not be sufficient to represent each concept. In particular, latent concepts with large singular values might require more than one sentence to represent them because of their preponderance in the collection in terms of frequency. Furthermore, it is often the case that good summary sentences do not just discuss one concept but they may discuss several concepts.

Therefore, several modifications [527, 528] were proposed in the later literature to address this issue. One approach[2] is to use the normalized matrix $U = Q\Sigma^2$. Note that $U = [u_{ij}]$ is an $m \times k$ matrix like Q except that the dominant columns are scaled up with the square of the singular values. Then, the score s_i of the ith sentence, S_i (i.e., ith row of D_y) is computed as follows:

$$s_i = \sqrt{\sum_{p=1}^{k} u_{ip}^2} \tag{12.15}$$

Then, a large value of the score s_i for sentence S_i is indicative of the fact that it should be included in the summary. Unlike the original LSA approach [219], this type of scoring does not provide a natural way to check for redundancy. Therefore, one can combine this type of scoring with the MMR approach (cf. Section 12.2.4) to create the summary. It is noteworthy that one can use a wide variety of matrix factorization techniques discussed in Chapter 3 in order to generate such summaries, and this avenue is explored to some extent in multi-document summarization.

12.3.2 Lexical Chains

Methods like latent semantic analysis drive their power in large part because of their ability to capture semantic similarity in a *data-driven manner*, which is able to adjust for natural linguistic effects such as synonymy and polysemy. In contrast, the lexical chain methods use a *manually constructed thesaurus* in order to find groups of closely related words. For this purpose, *WordNet* [396] is used, which is an automated thesaurus.

12.3.2.1 Short Description of WordNet

WordNet is a lexical database of English nouns, verbs, adjectives, and adverbs. These words are grouped into sets of cognitive synonyms, which are also referred to as *synsets*. Although WordNet serves some of the same functions as a thesaurus, it captures richer relationships in terms of the complexity of the relationships it encodes. WordNet can be expressed as a network of relationships between words, which go beyond straightforward notions of similarity. The main relationship among the words in WordNet is synonymy, which naturally creates a total of about $117,000$ synsets. Polysemous words occur in multiple synsets, which provides useful information for the mining process. An important point is that a lexical chain is a sequence of words extracted from the same text, which often serves the purpose of

---

[2]There is some difference between the presentations in [527] and [528]. The former suggests to use $U = Q\Sigma$, whereas the latter uses $U = Q\Sigma^2$. Using $U = Q\Sigma$ is almost equivalent to setting the sentence-wise scores to the L_2-norms of the original vector space representations (rows of D_y) except that the truncation of LSA removes some noise. The equality becomes exact when k is set to $\min\{m, d\}$. In such a case, the use of LSA is not even necessary.

disambiguating the word. For example, the word *"jaguar"* is polysemous because it might be either a car of a cat. The sense of this word in a chain such as *"jaguar-safari-forest,"* would be different from that in *"jaguar-race-miles."* This is a similar type of disambiguation to what is achieved by many latent and matrix factorization methods. Synsets also have encoded relationships between them, such as between the general and the specific. For example, a specific form of *"furniture"* is *"bed."* WordNet distinguishes between types and instances. For example, a *"bunkbed"* is a type of bed, whereas *"Bill Clinton"* is an instance of a president. The specific types of relationships depend deeply on the parts of speech of the constituent words. For example, verbs can have relationships corresponding to intensity (e.g., *"like"* and *"love"*) whereas adjectives can have relationships corresponding to antonymy (e.g., *"good"* and *"bad"*). There are also a few relationships across different parts of speech, such as words arising from the same stem. For example, *"paint"* and *"painting"* arise from the same stem but are different parts of speech. In general, one can view WordNet as a kind of graph in which groups of synonyms (synsets) are nodes and edges are relations.

12.3.2.2 Leveraging WordNet for Lexical Chains

The relationships between words in the document are first categorized by using the graph structure of WordNet. Relationships between words are classified as *extra strong, strong,* or *moderate.* An extra strong relationship is a word and its repetition, and a strong relationship is the presence of a WordNet relation. A moderate relationship is the presence of a path of length greater than one in the WordNet graph. Some restrictions [249] are also placed on the patterns of paths between two words in the graph.

The earliest lexical chain generation algorithms appeared in [249, 526]. The process of chain generation consists of successive insertion of words in the current set of chains. This insertion step is discussed in [44] as follows:

1. Find a set of candidate words. Generally, nouns are used as candidate words.

2. For each candidate word, find a chain that satisfies one of the above relatedness criteria (based on the strength of the WordNet relation) of the candidate word to a member of the chain. As discussed in detail later, the insertion decision also depends on the physical distance of the candidate word to the chain-specific words in the segment of the text being summarized. For example, insertion is allowed if very highly related words are far apart in the text being summarized.

3. If a chain is found that satisfies the relatedness criterion, then insert the word in the chain and update it accordingly. When a related chain cannot be found, a new chain is started containing the word together with links to all its synsets.

In order to insert words in a lexical chain, the inserted words need to be related to a member of the lexical chain. For selecting the lexical chain, extra-strong relations are preferred to strong relations, and strong relations are preferred to moderately strong relations. In order to insert a word in a chain based on a relatedness criterion to another word in the chain, the two words need to be no more than a certain distance apart from one another in the text being summarized. For extra-strong relations, there is no limit on the distance between words, for strong relations the maximum window length of the text segment is 7, and for moderate relations the window length is 3.

At this point, it is useful to understand how lexical chains disambiguate between polysemous uses of the same word, because it relates to the update of a lexical chain when a word is inserted. A polysemous word has more than one synset corresponding to its multiple senses. When a chain is started with a single word, the links to all its synsets are retained. However, when new words are inserted, all unconnected synsets of the word are removed. This type of removal leads to automatic disambiguation of different senses of the word as the chain grows over time and unconnected synsets are removed.

In order to use the chains for text summarization, the first step is to score the chains based on their relevance to the main topic at hand. In order to score a chain, the number of occurrences of the members of the chains (including repetition) in the text segment is computed. Furthermore, the number of *distinct* occurrences of the members of the chain is computed. The difference between the two is quantified as the score of the lexical chain. A chain is considered strong if its score is more than two standard deviations above the average score of all the chains that were identified in the first step.

Once the strong chains have been identified, they are used to extract significant sentences from the base text in order to create the summary. A key point is that not all words in a chain are equally good indicators (i.e., representatives) of the subject matter in the text being summarized. A word in a chain is considered representative if its frequency is no less than the frequency of other words in the chain with respect to the text segment being summarized. For *each* strong chain that is identified in the previous step, we choose a *single* sentence that contains the *first* appearance of a representative word from the chain in the text. Note that this step is similar to latent semantic analysis in which a single sentence is extracted for each latent concept in the collection. In this sense, lexical chains serve the same purpose as latent concepts except that they are mined with the help of significant linguistic input such as the WordNet database.

12.3.3   Graph-Based Methods

Graph-based methods use *PageRank* on the sentence-sentence similarity graph in order to determine the significant sentences. The *PageRank* method is described in Section 9.6.1 of Chapter 9 in the context of Web ranking. At first sight, it might seem that such graph-based methods have nothing in common with latent techniques. However, like latent methods, they use the overall similarity structure between sentences. This relationship will be explained in greater detail at the end of this section. The steps underlying the *PageRank* method are as follows:

1. Create a node for each sentence in the document to create m nodes. For any pair of sentences between which the cosine similarity exceeds a pre-defined threshold, add an undirected edge. Use the cosine similarity as the weight of the edge.

2. Compute the transition probabilities for the edges using this weighted adjacency matrix. Note that the transition probability from node i to node j is equal to its fractional weight among all edges incident on node i. Furthermore, the transition probability from node i to node j may not be the same as that from node j to node i. The resulting matrix $m \times m$ is denote by A_P.

3. As discussed in Section 9.6.1, the *PageRank* method requires restart with probability α. Let A_R be an $m \times m$ matrix in which every entry is $1/m$. Then, update A_P to a new stochastic transition matrix A with restart:

$$A = A_P(1 - \alpha) + \alpha A_R \qquad (12.16)$$

The matrix A incorporates the restart within the transition probabilities.

4. The dominant left eigenvector of the matrix A provides the *PageRank* values in its m entries. Each of these values can be shown to be the steady-state probability of a random walk on the stochastic transition graph represented by A. For greater details on the *PageRank* method, the reader is referred to Section 9.6.1 of Chapter 9.

The *PageRank* values provide scores that can be used for ranking sentences. Although one can simply select the top-scoring sentences, it sometimes helps to use the redundancy elimination methods of Section 12.2.4 in order to create a more informative summary. The earliest methods that used *PageRank* for summarization were the *TextRank* [387] and *LexRank* methods [176]. These methods have also been extended to multi-document summarization [176]. One advantage of the *PageRank* approach is that it is relatively easy to incorporate linguistic and semantic information within the similarity graph [97].

How is the *PageRank* method related to the latent semantic methods of Section 12.3.1? The latent semantic methods of Section 12.3.1 also use eigenvectors of similarity matrices, just as the *PageRank* method uses the dominant left eigenvector of a modified similarity matrix (which is a stochastic transition matrix). Note that the LSA method of Section 12.3.1 first constructs the sentence-term matrix D_y, and then performs SVD of the matrix $D_y = Q\Sigma P^T$. One can also extract the matrices Q and Σ by using the top eigenvectors of the sentence-sentence similarity matrix $D_y D_y^T$.

The *PageRank* technique can be viewed as a closely related method except that it constructs the similarity matrix in the form of a transition matrix, and uses a single dominant eigenvector of this matrix. The approach for selecting sentences is also different from the latent semantic analysis method because one no longer uses multiple eigenvectors of the similarity matrix, but the elements of a single eigenvector of the transition matrix. Therefore, one has to be careful about not selecting redundant sentences (with the use of MMR-like methods) when using the *PageRank* technique. In the case of the latent semantic analysis method, a single sentence is selected with respect to each eigenvector.

12.3.4 Centroid Summarization

Although centroid summarization is naturally designed for multi-document collections [457, 458], one can also adapt it in a simple way for single document summarization by treating each sentence as a document. In fact, such an adaptation is very similar to the latent semantic analysis method, if one uses a single representative from each cluster. The overall approach for centroid summarization with single documents proceeds as follows:

1. Treat each sentence in the document as a document. Cluster the sentences into groups of k clusters. It is often difficult to cluster short segments of text like sentences. In such cases, one can use nonnegative matrix factorization to create a latent representation of the documents before clustering. Alternatively, some of the feature engineering techniques discussed in Chapter 10 can be helpful for short text clustering.

2. For each cluster, use the tf-idf frequency of words in the cluster in order to determine the importance of topic words and also score sentences. One can use any of the topic-word methods discussed in Section 12.2, except that all the sentences in each cluster are aggregated into a single document during the computation of term frequencies in the scoring process. In other words, the scoring of the sentences in a particular cluster is independent of other clusters by defining the term-frequency of a word in a cluster only with respect to sentences in the cluster (rather than all sentences in the original document).

3. Create a summary using the top-scoring sentence of each cluster. Therefore, the summary will contain exactly k sentences as in latent semantic analysis. The sentences in the summary are in the same order as they occur in the original document.

This approach is very similar to the latent semantic analysis method, except that each cluster is treated as a latent component. Furthermore, the approach discussed here is an adaption of its (more common) use case in the multi-document setting. The original centroid-summarization method [457, 458], which was proposed for multi-document summarization, is presented in Section 12.6.

12.4   Traditional Machine Learning for Extractive Summarization

Most of the methods discussed so far use only content in the summarization process. However, there are important characteristics about the positioning of various sentences and other meta-information that provide useful information for the importance of various sentences in summarization. This point of view leads to the broader perspective that one should extract indicator features that reflect the importance of a sentence belonging to a summary. Such an approach also paves the way towards a machine learning view of text summarization. The earliest work by Edmundson [172] was an unsupervised technique, which noticed that multiple characteristics of a sentence in a document such as the length and the location of the sentence within the text played an important role in deciding whether it should be part of the summary. In fact, it is quite common in many systems to include the first sentence in the text segment as a part of the summary. These observations led to the natural conclusion that it makes sense to extract different types of *features* about sentences based on both content-centric as well as non-content-centric criteria, and then use machine learning techniques [307] in order to score the importance of sentences for summarization. The key point here is that machine learning methods require *training examples* in order to perform the learning. The training examples take the form of a text segment with binary annotations indicating whether or not the sentence should be a part of the summary. One can then use a binary classifier in which features are associated with individual sentences and the label indicates whether or not it should be a part of the summary. The need for human annotation is the main bottleneck in the use of machine learning systems for text summarization.

12.4.1   Feature Extraction

The first work on machine learning for text summarization [307] proposed using a set of features that were motivated by the work of Edmundson [172], Luhn [357], and Paice [427]. The initial work of Paice proposed the use of specific types of features associated with sentences such as frequency-based features (i.e., number and frequency of topic words), presence of title words, and location features (e.g., beginning or end of paragraph). In addition, indicator phrases often accompany summary material. For example, the phrase such as "This report..." often occurs at the beginning of a summary sentence. A related notion is that of *cue words*, containing *bonus* and *stigma* words, which are positively or negatively correlated with summary sentences.

The work in [307], which was motivated by Paice's initial feature set, proposed to use several related and additional features, all of which were discrete. It was proposed to use a *sentence length cut-off feature* that is set to 1 when the sentence length is greater than 5.

The basic idea is that summaries generally do not contain very small sentences. In addition, a *fixed-phrase feature* was used that was set to 1, when the sentence contained phrases like "This report..." or keywords like "conclusion." Sentences that contained one of a set of 26 indicator phrases or which contained keywords from the section heading had this feature set to 1. A *paragraph feature* indicated whether a sentence occurred in the beginning, middle, or ending of a paragraph. A *thematic feature* was used, which can be viewed as setting the binary feature to 1, if a frequency-based topic-word score of a sentence (cf. Section 12.2) is larger than a particular threshold. The occurrence of a proper noun several times or the explanatory expansion of an acronym sets the *uppercase word feature* to 1.

An immediate observation is that some of the features, such as thematic features, are actually used to score sentences on a standalone basis in some of the techniques discussed in previous sections. The supervised approach is therefore potentially more powerful because it extracts many of the features used for scoring sentences (along with other indicator features) and then learns the importance of a specific combination of features from the training data. This broader approach of using different scoring methods to create features was exploited in other ways. For example, the *PageRank* feature of Section 12.3.3 is also used. The work in [320] also proposed to use various structural features (including *PageRank*) from a graph in which nodes correspond to words and phrases rather than sentences.

12.4.2 Which Classifiers to Use?

The original work in [307] used a naïve Bayes classifier on the training data. However, almost any machine learning algorithm can be used in practice. Refer to the bibliographic notes for pointers to supervised methods. A recent trend has been on the use of *hidden Markov models*, which treat sentences as sequential entities rather than as independent entities [127]. The basic idea here is that the likelihood of a sentence belonging to a summary is not independent of whether its preceding sentence has been included in the summary. In general, the machine learning approach has often improved the performance of summarization methods in many domains, although the unsupervised methods often perform competitively in generic cases. The main constraint on the use of the technique is the presence of labeled training data. In recent years, deep learning methods have become increasingly popular because of their ability to use the sequential ordering of words in the modeling process.

12.5 Deep Learning for Extractive Summarization

The traditional machine learning methods discussed in the previous section use the sequential ordering of the sentences in only a limited way when creating a summary. However, deep learning models allow the use of sequencing information in order to create summaries with the use of either recurrent neural networks or with the use of pre-trained language models (i.e., pre-trained transformers like GPT and BERT). This section will discuss both these methods.

12.5.1 Recurrent Neural Networks

There are many variations of recurrent neural networks that are used for extractive summarization. In most cases, bidirectional forms of long-short term memory networks and GRUs are favored. In bidirectional recurrent neural network, a hidden state feeds both into the state at the next time stamp, as well as the state at the previous time stamp. The model is

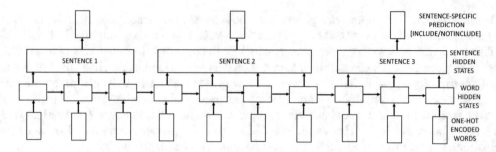

Figure 12.1: General framework of recurrent neural network for extractive summarization

otherwise similar to a unidirectional neural network; the main advantage of a bidirectional architecture is that it can incorporate context from both directions. However, in this exposition, we will use a straightforward unidirectional recurrent neural network for simplicity and then discuss the specific modifications made in [410] to improve the model.

The primary idea is to treat a piece of text as a nested sequence-of-sequences. Each sentence can be treated as a sequence of words, whereas the entire article can be treated as a sequence of sentences. Correspondingly, we create a two-layer recurrent neural network in the first layer contains hidden states corresponding to words (like a normal recurrent network) and the second layer contains hidden states corresponding to sentences. The hidden states for the words feed into the corresponding hidden states for the sentences (as long as that word is contained in that sentence). Finally, the sentence-level hidden states feed into a sigmoid output layer, which emits the probability of whether or not a sentence should be included in the summary. An example of such a recurrent neural network is shown in Figure 12.1. It is noteworthy that the output layer requires a ground-truth, the construction of which will be discussed slightly later. Furthermore, depending on how the ground truth is constructed, it is also possible for the output layer to be real-valued.

The model presented in Figure 12.1 is a gross simplification of a bidirectional recurrent neural network presented in [410], referred to as *SummaRuNNer*. The goal of the simplification in this section is to present the essential elements and concepts of the model, although the refinements of this basic idea are essential for good performance. In particular, the model in [410] uses a bidirectional GRU model. Furthermore, it has a state corresponding to the representation of the entire document. This state is obtained as a nonlinear transformation of the concatenation of all the sentence-wise hidden states (where the nonlinearity is in the form of the tanh activation function). Let $\overline{h}$ be the column vector corresponding to the concatenation of the hidden states for all the sentences. Then, the global state $\overline{g}$ representing the entire document is as follows:

$$\overline{g} = \tanh\left(W_{global}\overline{h}\right)$$

This representation of the entire document feeds into each of the outputs and improves performance by proving a global view of the document to each of the outputs.

In order to extract the summaries, ground truth is needed in the form of labels. This is of course a challenge, because most text corpora do not have such ground truth available, although they may have abstractive summaries written by humans. Therefore, the approach discussed in [410] proposes a method that maximizes the *Rouge Score* with respect to the abstractive (human gold-standard) summaries. The Rouge score is defined in terms of a precision component and a recall component, which can then be aggregated into a single score (by using the harmonic, geometric, or arithmetic means of the precision and recall scores). Let $\bar{c}$ and $\bar{c}^*$ be the word0count vectors in the candidate and target (abstractive) summaries, respectively, over the lexicon of words. Then, the number of words in common between the two summaries is given by $\|\min\{\bar{c}, \bar{c}^*\}\|_1$, respectively. where the minimum function is applied in element-wise fashion to the two vectors. Then, the precision- and recall-based Rouge Scores $Rouge_p$ and $Rouge_r$ are as follows:

$$Rouge_p = \frac{\|\min\{\bar{c}, \bar{c}^*\}\|_1}{\|\bar{c}\|_1}$$

$$Rouge_r = \frac{\|\min\{\bar{c}, \bar{c}^*\}\|_1}{\|\bar{c}^*\|_1}$$

The two scores can be consolidated into a single score by using various types of means. One can define different types of Rouge scores by varying on the types of lexicon used. For example, one can use bigrams, trigrams, or even n-grams as the base lexicon.

Since it is computationally difficult to find the optimal subset of documents maximizing the Rouge score, the approach in [410] greedily adds sentences one by one to the summary so that the overall Rouge score of the current summary is maximized with respect to the gold-standard abstractive summary created by a human. This greedily constructed summary is used as the ground truth for training the recurrent neural network (in terms of the binary labels for each sentence in the original article).

The combination of word-level and sentence-level representations is notable in this model. It is generally impractical to represent an entire article with the use of a word-level recurrent neural network. Using hidden states corresponding to sentences helps in preserving some of the long-range dependencies that would not be otherwise possible with a word-level recurrent network. The model can be further improved with the use of attention mechanisms. However, a more direct approach to using attention is obtained with the use of transformers.

12.5.2 Using Pre-Trained Language Models with Transformers

Chapter 11 discusses how transformers can be used to build pretrained language models, such as BERT and GPT-n. The use of pre-trained models like BERT as general-purpose vehicles for various natural language processing tasks has become increasingly popular (cf. Section 11.4.2 of Chapter 11). This is because these language models often deliver close to state-of-the-art performance on a wide variety of natural language processing tasks with only minor fine tuning over the base model. One such example, referred to as BERTSUM, was proposed in [350], where it is shown how BERT can be fine-tuned in order to deliver state-of-the-art performance on extractive summarization.

The vanilla BERT model, discussed in Section 11.4.2, needs to be modified in a number of ways in order to make it work for the summarization problem. The original BERT model uses a [CLS] token before each pair of sentences (representing a unit of input), whereas the summarization variant contains a [CLS] token before each sentence (and the input might contain multiple sentences). The vector representation of the [CLS] token in the top BERT layer provides the representation for the sentence following the [CLS] token. As in the original version of BERT, a [SEP] token occurs after each sentence. Similarly, odd or even sentences in the original article have an embedding added to them "Sentence A" and "Sentence B." This is similar to the original BERT, except that the original BERT adds the embedding for "Sentence A" to the first sentence and the embedding for "Sentence B" to the second sentence, whereas the summarization application does not contain pairs of sentences. Therefore, the odd/even indicator is used to add the embeddings for "Sentence A" (odd sentences) and "Sentence B" (even sentences) appropriately. This difference is because the summarization application treats the input as a sequence of multiple sentences, whereas the original BERT treats inputs as pairs of sentences.

The vector representations of the sentences (i.e., the vector representations of their corresponding [CLS] tokens) can then be fed into an output layer with a sigmoid function indicating whether or not a sentence should be included in the summary. These outputs are then trained with a gold-standard summary to determine whether or not a sentence should be included in the summary. In addition two other variations were tested:

- Instead of feeding the sentence representations directly into the output layer, an additional summarization layer is used between the sentence representations of BERT and the output layer. In this summarization layer, a transformer is applied to the *sequence of sentences* in order to extract the global document-level features and incorporate them into the sentence embeddings. Therefore, the sequence of sentence representations that are input into the summarization layer are used to create a new set of sentence representations, which incorporate the context with respect to other sentences (because of the use of the attention-centric transformer). The outputs of the sentence representations after the contextual transformations are used in order to make predictions of whether or not a sentence should be included in the summary. The model is trained using the gold-standard summaries.

- One can apply an LSTM to the sequence of sentences instead of the transformer in order to create the summarization layer.

It was found in [350] that the use of a transformer for refining intersentential representations provided the best results, even though even the simplest model of directly using BERT sentence representations for feeding into the output layer seemed to work very well.

12.6   Multi-Document Summarization

In a multi-document summarization, the summary is not just germane to *one* article but to *multiple* articles that are closely related. Most of these techniques use either a clustering method or a topic model on the document collection in order to identify sentences that are locally relevant to each cluster.

12.6.1   Centroid-Based Summarization

In centroid-based summarization [457, 458], a clustering algorithm is used to partition the corpus into groups of related documents. The centroid of each cluster is defined by averaging the tf-idf representations of its documents, although words with low tf-idf scores are truncated. Note that the inverse document frequency (idf) is computed with the help of a background corpus rather than the corpus being summarized. Therefore, the steps for centroid-based summarization are as follows:

1. Cluster the documents using any off-the-shelf method. In early works [457, 458], clustering methods from the topic detection and tracking (TDT) effort [20] were used. The use of the k-means technique is particularly desirable because of its natural tendency to create centroids of topical words from the cluster.

2. For each cluster, create a pseudo-document corresponding to the centroid of all documents in the cluster. The centroid is created by adding the tf-idf frequencies of each term across different documents in the cluster and truncating the terms with aggregate frequency below a particular threshold. Select the terms with the largest tf-idf frequencies as the topical words of the cluster.

3. Score each sentence in the cluster based on the tf-idf frequencies of its words in the cluster centroid. This type of scoring is similar to that of single document summarization in Section 12.2.2. The main difference is that the scoring of the sentence is done with respect to the documents in its cluster rather than the frequency in the document itself. This notion is referred to as *cluster-based sentence utility* [457, 458]. In addition, documents are scored higher based on positional factors. Early sentences in documents are given higher credit, which reduces linearly with the order of the sentence. The first sentence in a document gets an additional credit C_{max} which is equal to the score of the highest-ranking sentence based only on centroid scoring. Subsequent sentences have a linearly reducing credit which is equal to $C_{max}(m-i+1)/m$, where m is equal to the number of sentences in the document, and i is the positional index of the sentence. An additional credit is given based on the dot product similarity with the first sentence in the collection. The idea is that the first sentence is indicative of topical salience and therefore overlap with the first sentence is desirable. Therefore, if s_c is the centroid score, s_p is the positional score, and s_f is the first-sentence overlap score, then the overall score (without accounting for redundancy) is as follows:

$$s_{all} = w_c s_c + w_p s_p + w_f s_f \qquad (12.17)$$

Here, w_c, w_p, and w_f are user-driven parameters, which can be tuned but are often set to 1.

4. Since multiple documents in a cluster might contain similar sentences, such redundant sentences need to be removed. Although it is possible to use the maximum marginal relevance (MMR) principles for sentence selection, the work in [457, 458] uses the notion of *cross-sentence informational subsumption (CSIS)*. If W_1 and W_2 are the respective sets of words in two sentences, then the value of CSIS between the two sentences is as follows:

$$CSIS(W_1, W_2) = w_R \frac{2|W_1 \cap W_2|}{|W_1| + |W_2|} \qquad (12.18)$$

Here, w_R is the weight of the redundancy penalty, which is set at the maximum value of s_{all} over all sentences according to Equation 12.17. The score value s_{all} for each sentence is then adjusted with the redundancy penalty defined by Equation 12.18. For a sentence, we subtract the redundancy penalties only with respect to sentences that have higher scores. Of course, since the ranking of sentences is itself influenced by the redundancy penalty, this approach for adjusting the scores with the redundancy penalty is circular. Therefore, the approach starts with an initial ranking without any redundancy penalty, computes the penalties based on this fixed ranking, and then re-ranks documents. The approach is repeated iteratively until the ranking does not change. At the end of the process, the top-ranked documents across all clusters are included in the summary.

How does one order the sentences in a summary drawn from multiple documents? It is assumed that the original document collection has some pre-defined order (e.g., chronological), which also provides an ordering for the sentences from different documents. The sentences from a particular document appear contiguously in the summary.

12.6.2   Graph-Based Methods

It is evident from the discussion in the previous section that issues associated with redundancy are slightly more complex in the case of multi-document models as compared to single-document models. An approach is proposed in [339] that uses the MMR method directly in the graph context. As discussed in Section 12.6, a graph is constructed by treating individual sentences as nodes. However, in this case, a sentence may be drawn from any of the documents[3] in the collection. An edge is added between each pair of nodes for which the cosine similarity exceeds a pre-defined threshold. The weight w_{ij} between nodes i and j is equal to the cosine similarity between the corresponding pair of sentences.

Let $\mathcal{U}$ represent the universe of all the sentences in a multi-document collection. Let $\mathcal{S}$ be the set of sentences to be included in a summary. Ideally, a summary should select sentences that are as representative of the entire collection as possible. This goal is achieved by ensuring that the sentences in $\mathcal{S}$ are as similar as possible to those in $\mathcal{U} - \mathcal{S}$. At the same time, one should ensure that the sentences within $\mathcal{S}$ are as dissimilar as possible to one another. Therefore, one attempts to identify the set of sentences $\mathcal{S}$ that maximize the following *submodular* function:

$$f(\mathcal{S}) = \sum_{i \in \mathcal{S}} \sum_{j \in \mathcal{U}-\mathcal{S}} w_{ij} - \lambda \sum_{i \in \mathcal{S}} \sum_{j \in \mathcal{S}} w_{ij} \qquad (12.19)$$

The balancing parameter $\lambda > 0$ regulates the relative importance of content coverage and redundancy. The goal is to maximize the value of $f(\mathcal{S})$ while imposing a budget on the maximum *cost* of the summary. For example, the cost associated with the ith sentence might be the number of bytes in it, although one might use other types of costs as well.

A submodular function like $f(\mathcal{S})$ satisfies the law of diminishing returns as applied to set-wise functions. In other words, adding a sentence to a larger superset $\mathcal{S}_1$ is not incrementally as rewarding as adding it to the subset $\mathcal{S}_2 \subseteq \mathcal{S}_1$. Such set functions are known to work well with greedy algorithms, with provable approximation bounds in special cases [412]. Therefore, the approach in [339] adds sentences to the summary greedily (i.e., largest incremental increase in $f(\mathcal{S})$ per unit cost), until the budget limit is reached.

12.7   Abstractive Summarization

The goal of extractive summarization is to generate summaries that reuse sentences (or segments) from the original document(s). However, such a summary is often not fluent from a human perspective. Humans often rewrite portions of the document completely for greater clarity and fluency. Abstractive summarization is designed to create summaries that do not necessarily reuse portions of the document in a verbatim way.

In general, the construction of completely fluent summaries that mimic human performance is quite difficult, even though recent advancements in deep learning have shown encouraging results. The high level of difficulty of this setting also explains why most modern systems are extractive.

There are two broad types of methods for abstractive summarization. The first type of method, which can be considered "abstractive" only in a rather limited way, first creates

---

[3]The approach can also be used for a single document summary.

extractive summaries from the input and then modifies portions of these summaries in order to improve the presentation and output. However, these methods have met only limited success. In fact, some works [613] have explicitly demonstrated that attempts to incorporate abstractive methods on extractive summaries can sometimes worsen concrete evaluation measures for summarization. In spite of the mixed results achieved by such methods, the modifications made by such systems do sometimes agree with typical changes made by human participants. The second type of abstractive summarization method typically constructs summaries directly with the use of deep learning methods. These techniques have shown significantly greater promise in generating summaries of high quality. The following discussion introduces some of these techniques.

12.7.1 Sentence Compression

When humans summarize documents, they often shorten longer sentences in terms of linguistic and writing style. Inessential phrases are removed. Automated methods for sentence compression are motivated by similar goals, and try to mimic human performance as an ideal. The following techniques are often used for sentence compression:

1. *Rule-based methods:* Rule-based methods often use linguistic knowledge in order to identify the phrases that are removed from a summary. The different grammatical parts of a sentence can be identified with the use of a linguistic parser. Those parts of the sentence that are not essential to the grammatical integrity of the sentence, and are not closely related to the subject of the overall article are removed [270]. These types of conditions are encoded in terms of rules that are used to modify an extracted summary. Many other methods use syntactic heuristics [128, 518] to simplify sentences, and these methods have been shown to improve the quality of the summarization. The syntactic heuristics are often based on linguistic rules in a direct or indirect way.

2. *Statistical methods:* In this case, the portions of the sentences to be removed are learned by the model. The approach discussed in [300] constructs different *parse trees* with the use of *probabilistic context free grammars (PCFG)*. This method uses the *Ziff-Davis corpus* [687] as training data for the learning process. The goodness of a sentence is computed with the use of PCFG scores and a bigram language model. Several later works [206, 552] improved on these methods by avoiding undesirable deletions and acquiring training data in innovative ways.

   These methods have led to increasing interest in the field of *sentence compression* [118], which has taken a life of its own (beyond text summarization applications). Several recent methods use integer linear programming [118] in order to identify which words to remove from a sentence. These methods do not use the parse trees, and therefore they cannot ensure that the grammar of the output is accurate.

Both types of methods have had mixed success in terms of their performance with respect to human annotators. Sentence compression methods are also used for *headline generation* [161, 588], in which the length of the compressed summary is small enough to be considered a headline. Headline generation has often been seen as a related but independent problem from text summarization tasks.

12.7.2 Information Fusion

The aforementioned methods for sentence compression almost have a one-to-one correspondence of the created summary from an extractive summarization approach. The main difference is that the generated sentences are compressed using either rule-based or machine-learning techniques. A more generic form of summarization is one in which the information from multiple sentences is integrated into a single sentence. Such an approach is also referred to as "cut-and-paste" approach [271].

A approach, referred to as *MultiGen* [46], proposed techniques for fusing more than two sentences in the context of multi-document summarization. The basic idea is that there are many documents that are similar to one another in a cluster and they can often be fused into a grammatically correct sentence by identifying phrases that occur in common across different sentences. This general line of work has led to the standardized problem of finding the best union of two sentences that conveys all the information in the two sentences as well as possible [196, 369].

12.7.3   Information Ordering

The order in which the extracted sentences are presented in a summary need not be the same as that in the original document. In the multi-document setting, this issue is particularly important because the ordering of the sentences is often based on the chronological order of the underlying articles. The chronological order of the underlying articles may not correctly reflect the order in which they appear in the summary. A more natural approach [239] is to cluster sentences with similar topical content contiguously in the summary. Furthermore, the ordering of the different topics may not be the same across different documents. Therefore, a graph is constructed in which each vertex represents a topic cluster. An edge is placed between two vertices if one topic precedes another in a document. The *majority ordering* across the various documents is used to decide which topics should precede one another in the final summary. Another idea that combines topical locality with chronological ordering is discussed in [45].

12.7.4   Recurrent Neural Networks for Summarization

Deep learning methods generally perform much better than other machine learning techniques for abstractive summarization. This is because deep learning methods can build relatively fluent summaries, although they are not always very effective at content selection.

One of the earliest methods in this respect was proposed with the use of a recurrent neural network [411]. The basic idea is to use sequence-to-sequence recurrent neural networks in order to perform abstractive summarization. The input sequence corresponds to the original text and the output sequence corresponds to the summarized text. A bidirectional recurrent neural network with attention is used for performing the sequence-to-sequence transformation. Note that examples of original text representations and abstractive summaries must be available in order to perform the training. In this context, abstractive summarization is treated in a manner similar to machine translation. However, abstractive summarization is much more challenging than machine translation. The reason is that machine translation works only at the sentence-to-sentence level with relatively short inputs. On the other hand, the inputs and outputs in the case of abstractive summarization are much longer, which makes the task more challenging.

A *large vocabulary trick* is used in order to further improve performance. One bottleneck during the training occurs in the output layer of the decoder when the lexicon size is large. However, an important observation is that the effects of words not present in the current minibatch on the updates to the weights are relatively small. Therefore, words that are not present in the minibatch can be dropped from the output layer of the decoder while making the updates in that specific mini-batch. Making this change helps in creating a more efficient training model.

12.7.5 Abstractive Summarization with Transformers

Since the original transformer [560] was constructed using the machine translation application in mind, it naturally maps easily to the text summarization problem. The advantage of using transformers is that it can handle longer sequences much more effectively than recurrent neural networks. The summarization application is much more likely to be presented with longer sequences, as a single training pair contains a full article and its summary. One such example of the adaptation of transformers for abstractive summarization is presented in [252]. Up until a decade back abstractive summarization was considered to be too challenging for traditional machine learning techniques. However, the recent implementations of abstractive summarization with transformers turn out to be quite fluent.

It is also possible to adapt pre-trained models such as T5, GPT-n, BERT in order to perform summarization. It is relatively simple to generalize T5 to summarization tasks, as it is already a text-to-text model. In the case of T5, a prefix such as as "summarize" is placed in front of the document as a cue that it needs to be summarized. After pretraining (cf. Section 11.4.3 of Chapter 11), the model is fine-tuned on training pairs of documents and their abstractive summaries. An example of a training pair for summarization is as follows:

> **Input Sequence:** summarize: state authorities dispatched emergency crews tuesday to survey the damage after an onslaught of severe weather in mississippi. ⟨ Paragraph continues specifying number of hospitalized people and identity of county ⟩
> **Output Sequence:** six people hospitalized after a storm in attala county.

After fine-tuning, input paragraphs are provided to T5, which provides a summary based on the fine tuning. The quality of the summaries generated by T5 are improved greatly because of the massive pretraining that precedes the fine-tuning.

The adaptation of GPT-n to summarization is also straightforward, as each input can be a concatenation of the original document and its summary (with a separator token indicating the segmentation between them). This special separator token is essentially a context clue that is learned during fine tuning [295]. In the current version of GPT-3, the document is followed by the token "TL;DR", which is a common English abbreviation for "too long didn't read." Since GPT-n learns language models, it treats the entire sequence containing both document and summary (including the TL;DR token) as a single sequence — therefore, the original document is simply the initial part of the sequence followed by the cue TL;DR, which prompts the GPT language model to generate the remaining part of the full "sequence."

It is noteworthy that the adaptation of GPT-n to abstractive summarization is straightforward because of its decoder architecture. On the other hand, it is not quite as simple to adapt BERT to abstractive summarization because of its encoder architecture. One issue with BERT is that it uses a masked form of token-wise training, which does not adapt as naturally to sequence-to-sequence tasks. Therefore, additional modifications are required to the BERT architecture in order to make it suitable for abstractive summarization. One such approach is discussed in [351], This approach converts BERT into a standard encoder-decoder framework by using the pre-trained BERT encoder, but adding a decoder to it, which must be trained from scratch. One challenge with using this approach is that it can cause some instability — the decoder might overfit to the additional training data, whereas the encoder might underfit. Therefore, the approach in [351] addresses this problem by separating the optimizers for the training of the encoder and the decoder. We refer the readers to [351] for more specific details. It is noteworthy that BERT is inherently not suited to either summarization or machine translation, which can be considered sister problems.

Self-Supervised Training

One challenge with many of these abstractive summarization tasks is that they require human-annotated examples of ground-truth summaries. While such ground-truth summaries are essential at the end of the day, a question arises as to whether one can generate more easily obtained pre-training pairs in order to build an initial model. This initial model can then be fine-tuned with human-annotated abstractive summaries. Such an approach is presented in the PEGASUS summarizer [623, 726].

It has been proposed in [623] that it is possible to use a self-supervised form of *pre-training*, wherein one starts with a large corpora of unannotated articles and removes the important sentences from them in an automated way. These important sentences can be removed by either using the Rouge score or by using an unsupervised extractive summarizer. These sentences are removed from the articles, and they are used as the "summaries" of the articles with the missing sentences. Note that this is a rather difficult task because some of the removed sentences may have highly pertinent information that is often missing from the article (after these sentences have been removed). It turns out that this type of approach is excellent for pre-training because it forces the learner to infer key semantic connections between the sources (with missing sentences) and the targets (with the key sentences). Although the training data is obviously flawed because the target summaries will often contain source information not present in the original article, it does serve its purpose of pretraining. This approach is also attractive because it van generate a huge amount of data for pre-training; after all, the generation of the source-target pairs from the original articles is fully automated. An encoder-decoder transformer can be pre-trained with these pairs. Subsequently, human-annotated pairs can be used for fine tuning. It had been shown in [623] that this type of approach provided state-of-the-art performance at the time, although the large language models that were built recently exceed its performance.

12.8   Summary

Summarization methods are either extractive or abstractive. The former type represents the majority of the available techniques in the literature, and is defined as a methodology in which sentences from the original document(s) are put together to create the summary. Various methods have been proposed for document summarization such as topic-word methods, latent methods and machine learning techniques. These methods have also been extended to the multi-document scenario. In recent years, some progress has also been made on abstractive summarization methods. Most of these techniques start with an extracted summary and then modify it in order to remove redundant sentences, fuse related sentences, and reorder the underlying segments more appropriately. Encouraging results have also been obtained in recent years with the use of deep learning methods and sequence-to-sequence learning. Summarization methods can also be improved significantly by incorporating domain-specific knowledge.

12.9 Bibliographic Notes

Numerous surveys have been written in the literature on text summarization [143, 413, 414, 524]. Among these, the survey by Nenkova and McKeown [413] is excellent and comprehensive; furthermore, it is very systematic in providing an overview of the different topics. A lot of the work in text summarization derives its motivation and ideas from the early work of Luhn [357] and Edmundson [172]. Luhn's work led to the popularization of topic representation methods, whereas Edmundson's work popularized indicator methods in general and machine learning methods in particular. The *SumBasic* method is discussed in [559]. The use of topic signatures is proposed in [337]. Methods for selecting non-redundant methods based on various scoring methods in combination with the greedy MMR approach are discussed in [89, 338]. Methods that go beyond the greedy approach and use global optimization are discussed in [215].

The latent semantic method for summarization was first proposed in [219] and subsequent improvements were proposed in [527, 528]. The method of lexical chains was introduced in [249, 526], and its use for summarization was first discussed in [44]. The use of *PageRank* for summarization was first explored in the *LexRank* [176] and *TextRank* [387] methods. The *TextRank* method also provides techniques for keyword extraction. The earliest work on graph-based summarization was proposed in [491]. The first centroid-based summarization method was proposed in [457, 458], although it was proposed for multi-document summarization. The single-document summarization method presented in Section 12.3.4 is a simplification of this approach.

The works by Edmundson [172] and Paice [427] set the stage for feature extraction for machine learning techniques, although these methods were themselves not machine learning methods. The first machine learning method in text summarization [307] used many of these features in conjunction with a naïve Bayes classifier. A number of supervised methods for text summarization are discussed in [203, 230, 259, 320, 423, 590]. The use of hidden Markov models for text summarization was first introduced in [127]. The extractive summarization method discussed in this chapter is adapted from [410]. Other notable RNN-based neural models for extractive summarization are presented in [107, 626]. The Rouge score is introduced in [336]. The use of BERT for extractive summarization is discussed in [350].

The centroid-based technique for multi-document summarization is based on [457, 458]. Topic models provide an attractive alternative [229, 569] to clustering, and they use the latent semantic structure of the relationships between the sentences, documents, and corpus in order to create a summary. Graph-based methods for multi-document summarization are discussed in [176, 339].

Numerous methods for abstractive summarization are presented in [252, 295, 351, 411]. The PEGASUS model for abstractive summarization is presented in [623]. Although abstractive summarization has been historically considered a hard problem, recent results with the use of deep learning have been quite encouraging. For more discussion on these topics, the interested reader is referred to [413].

12.9.1 Software Resources

A number of open source libraries such as Apache OpenNLP [644], NLTK [652], and Stanford NLP [650] support preprocessing tasks (including sentence-wise segmentation) that are crucial for text summarization. Numerous software packages are available for text summarization such as a component of **gensim** [464] and ROUGE [688]. A summarizer that is based on Latent Semantic Analysis is presented in [690]. A summarization functionality is also available from **TensorFlow** [691]. The CNN and Dailymail data sets, which are frequently used for evaluating text summarization are available at [729].

The code for SummaRuNNer is available in [725], whereas that for BERTSUM is available in [724]. A multi-document summarizer, referred to as MEAD, is also available in the public domain [692]. The ICSI document multi-document summarizer [689] uses integer linear programming techniques, and is known to one of the best performing systems in various evaluations. The code for the PEGASUS model for abstractive summarization is presented in [726].

12.10 Exercises

1. Suppose that you are only given pairwise similarities between text sentences in a document, but you are not given the sentences themselves. Show how you can use these pairwise similarities in order to create a summary of the document.

2. Consider a background corpus with 100,000 tokens in which the word *"politics"* occurs 250 times. Furthermore, a document with 70 tokens contains this word twice. Calculate the likelihood ratio that this word is a topic signature.

3. Suppose that you use nonnegative matrix factorization instead of LSA for the latent method discussed in Section 12.3.1. Discuss the intuitive relationship of such a technique with the clustering method discussed in Section 12.3.4.

4. Implement the topic-signature method for single-document summarization.

5. Implement the latent semantic analysis method for single-document summarization.

Chapter 13

Information Extraction and Knowledge Graphs

"We are drowning in information, while starving for wisdom."–E. O. Wilson

13.1 Introduction

In its most basic form, text is a sequence of tokens, which is not annotated with the properties of these tokens. The goal of information extraction is to discover specific types of useful properties of these tokens and their interrelationships relationships. The umbrella term *"information extraction"* refers to a family of the following closely related tasks:

1. *Named entity recognition:* The tokens in the text may refer to *named entities*, such as locations, people, and organizations. For example, consider the following sentences:

   > Bill Clinton lives in New York at a location that is a few miles away from an IBM building. Bill Clinton and his wife, Hillary Clinton, relocated to New York after his presidency.

   For this text segment, it needs to be determined which tokens correspond to which type of entity. In this case, the system needs to recognize that *"New York"* is a location, *"Bill Clinton"* is a person, and *"IBM"* is an organization.

2. *Relationship extraction:* Relationship extraction generally follows named entity recognition, and it attempts to find the relationships among different named entities in a sentence. Examples of relationships may be as follows:

   > **LocatedIn**(*Bill Clinton, New York*)
   > **WifeOf**(*Bill Clinton, Hillary Clinton*)

   In general, the types of relationships to be mined will be specified in the application at hand.

© Springer Nature Switzerland AG 2022
C. C. Aggarwal, *Machine Learning for Text*,
https://doi.org/10.1007/978-3-030-96623-2_13

A related problem, which is not discussed in this chapter, is one in which different terms may refer to the same entity. This problem is referred to as *co-reference resolution*. For example, both *"International Business Machines"* and *"IBM,"* refer to the same named entity, and the co-reference resolution system needs to recognize this in an automated way. As another example, consider the following pair of sentences:

> Bill Clinton lives in New York at a location that is a few miles away from an IBM building. He and his wife, Hillary Clinton, relocated to New York after his presidency.

Note that the word *"He"* at the beginning of the second sentence is not a named entity, but it refers to the same entity as Bill Clinton. These types of references to the same entity also need to be captured by co-reference resolution.

It is worth noting that there are many different settings in which information extraction systems are used. An *open* information extraction task is unsupervised and has no idea about the types of entities to be mined up front. Furthermore, weakly supervised methods either expand a small set of initial relations, or they uses other knowledge bases from external sources in order to learn the relations in a corpus. Although such methods have recently been proposed[1] in the literature, this chapter will focus on a more traditional view of information extraction that is fully supervised. In this view, it is assumed that the types of entities and the relationships among them to be learned are *pre-defined*, and tagged training data (i.e., text segments) are available containing examples of such entities and/or relationships. Therefore, in named entity extraction, tagged examples of persons, locations, and organizations may be provided in the training data. In relationship extraction, examples of specific relationships to be mined may be provided along with the free text. Subsequently, the entities and relations are extracted from untagged text with the use of models learned on the training data. As a result, many of the important information extraction methods are supervised in nature, since they learn about specific types of entities and relationships from previous examples. In this chapter, we will focus on such supervised settings. In each case, the types of entities to be mined and the relationships among them depend on the specific application at hand. In many cases, the extracted information is organized into *knowledge graphs*, where the entities correspond to nodes and the relationships correspond to edges between nodes. These types of knowledge graphs play an important role in a variety of applications such as search and question=answering. The common applications in which the need for information extraction arises are as follows:

1. *News tracking:* This is one of the oldest applications in information extraction, which involves the tracking of different events from news sources and the various interactions/relations between different entities. Several early competitions were also organized around this application, which has also facilitated the availability of research prototypes in this context. The relationship between information extraction and event detection is also explored in Section 16.4.3 of Chapter 16.

2. *Counter-terrorism:* In these applications, law enforcement agencies may need to go through large numbers of articles in order to identify different types of individual entities, organizations, events, and relationships among them.

3. *Business and financial intelligence:* There are numerous events that occur between corporations such as mergers, takeovers, business agreements, and so on. It is often

---

[1]The bibliographic notes contain pointers to such methods.

useful to extract the different types of named entities and the relationships among them in order to obtain an idea of key trends.

4. *Biomedical data:* Biomedical data may have different entity types such as gene, protein, drug, and disease names. It may be useful to determine when different terms refer to the same thing, and also find the relationships between different entities.

5. *Entity-oriented search:* Some information retrieval systems provide the ability to search for specific types of entities in documents such as persons, locations, and organizations. For example, entering the query *"restaurants in Manhattan"* in the Google search engine yields results containing several popular restaurants, which are named entities. The first step in building searchable entity indexes is to identify different types of named entities within the documents. These entities are often organized into knowledge graphs in order to facilitate search.

6. *Question-answering systems:* Question-answering systems are often built on top of information extraction systems and entity-oriented search. This is because the process of understanding a question requires one to extract the different types of entities in the question and the relationships among them. These extracted entities are used to construct *knowledge graphs* on top of which question answering systems are built. Knowledge graphs are introduced in Section 13.4 of this chapter together with a brief description of how they might support applications like search—more details of building question-answering systems with knowledge graphs provided in Section 14.5 of Chapter 14.

7. *Scientific libraries:* The automated indexing of scientific libraries requires the extraction of specific fields of the document such as the author, title, and so on. Each of these fields is a named entity.

8. *Text segmentation:* Although text segmentation is often studied as a separate problem from named entity recognition, many existing algorithms for named entity recognition can be used for text segmentation. The problem of text segmentation is studied in detail in Chapter 16.

Over the years, it has been shown that these diverse applications can be addressed by solving a pair of *application-independent* tasks, which are named entity recognition and relationship extraction. Therefore, these problems have become the primary focus of work in the field.

13.1.1 Historical Evolution

The earliest problems in information extraction converted unstructured to structured data using *slot-filling* tasks over a pre-defined *template*. For example, imagine a case where we have training data containing Wikipedia pages of US politicians and some associated tables containing the relevant fields. In Wikipedia, this information is available in tables, which are referred[2] to as *infoboxes*. For example, consider the following statistics for *John F. Kennedy*, most of which is available both in the text of the Wikipedia page[3] as well as in the tables on the right-hand side of the page:

---

[2]https://en.wikipedia.org/wiki/Help:Infobox.
[3]https://en.wikipedia.org/wiki/John_F._Kennedy.

Slot/Field	Value
Born	May 29, 1917
Political Party	Democratic
Spouse(s)	Jacqueline Bouvier
Parents	Joseph Kennedy Sr. Rose Kennedy
Alma mater	Harvard University
Positions	US House of Representatives US Senate US President
Military Service	Yes

Here, it is evident that some fields (such as political party and military service) are obtained from a predefined set, whereas others like the names of spouses and parents can be arbitrary values. Early variants of the information extraction task were designed to extract such tables from unstructured text.

One problem with the slot-filling task is that it is highly application dependent, and such systems are not generalizable across different application settings. For example, a slot-filling system for US politicians on Wikipedia might not work for terrorism-centric slot filling on news articles. The main problem is that the various settings either require significant customization, or they require significantly increased complexity in terms of how the input to the problem is defined. This situation makes it difficult to create off-the-shelf software for such tasks. An important evolution of the slot-filling task was defined in MUC-6 [221], which was one of a series[4] of the early *Message Understanding Conferences (MUC)*. It was recognized that more crisply defined tasks like named entity recognition and relationship extraction were often used as *subtasks* of slot filling, and they also had the advantage of being template independent. As a result, these subtasks eventually became the predominant forms of information extraction. This chapter will, therefore, primarily focus on these subtasks.

13.1.2   The Role of Natural Language Processing

Information extraction draws techniques from natural language processing, information retrieval, machine learning, and text mining. The roots of the field lie within the natural language processing community, and many algorithms for information extraction are motivated by natural language processing. For example, various types of Markovian models are used in information extraction, and also for natural-language processing tasks like recognizing parts of speech. Information extraction also requires a pipeline of preprocessing tasks related to natural language processing, which is as follows:

1. *Tokenization and preprocessing:* Tokenization represents the first task in any text-mining application, and this task is discussed in detail in Chapter 2. For many applications, additional preprocessing steps such as stemming may be required.

2. *Parts-of-speech tagging:* Each token is assigned to a part of speech such as noun, verb, adjective, adverb, pronoun, conjunction, preposition, and article. However, there are many refined categorization of these basic types, which can lead to as many as 179 tags. There are standardized tags corresponding to the Brown [693] or the Penn Treebank

---

[4]This early series of conferences played an important role in the evolution of the field of information extraction, which had been largely sporadic till then.

tags [694]. For example, one might differentiate among different types of nouns with tags such as NN, NNP, and NNS, to represent singular, proper, and plural nouns, respectively. Similarly, one might distinguish between the base form of a verb and its past tense as VB and VBD, respectively. Common articles, such as "*a*," "*an*," and "*the*," are referred to as *determiners* with a tag of DT. An example of a sentence with parts-of-speech tags is as follows:

The/DT rabbit/NN ate/VBD the/DT carrot/NN.

Recognizing the parts of speech is crucial for information extraction, because entities comprise nouns or groups of nouns, whereas the expression of relationships often requires the usage of verbs.

3. *Parser:* A parser extracts a hierarchical structure from each sentence in the form of a *parse tree*, in which the lower-level subtrees group the parts of speech into syntactically coherent phrases like noun phrases and verb phrases. The former is useful for named entity recognition, whereas the latter is useful for relationship extraction. For example, one can recognize the noun phrases (NP) and verb phrases (VP) for the previous sentence as follows:

One can already see the tree-like nesting structure of the phrases, which results in a *constituency-based parse tree* that includes both the parts-of-speech tags and the phrase tags. The leaf nodes of the tree are the actual tokens, and the pre-terminal nodes are the parts-of-speech tags, each of which has a single child with the corresponding token. An example of the constituency-based parse tree is shown in Figure 13.1. Named entities are usually noun phrases, whereas relationships are often inferred from verb phrases.

4. *Dependency analyzer:* Some of the words depend on others, which are useful for relationship extraction. For example, both "*rabbit*" and "*carrot*" depend on "*ate*." One can encode these dependencies in the form of a dependency graph with nodes representing words and directed edges representing dependencies. This type of graph is useful for relationship extraction, and is referred to as a *dependency graph*. An example of a dependency graph is shown in Figure 13.5.

The aforementioned preprocessing tasks are available in many off-the-shelf natural language processing libraries, such as Stanford CoreNLP [650], NLTK toolkit [652], and the Apache OpenNLP effort [644]. It is noteworthy that some parts of speech, such as adjectives, are useful for *opinion mining*, which is closely related to information extraction (cf. Chapter 15).

Chapter Organization

This chapter is organized as follows. The next section discusses the problem of named entity recognition. The problem of relation extraction is discussed in Section 13.3. Knowledge graphs are introduced in Section 13.4. A summary is given in Section 13.5.

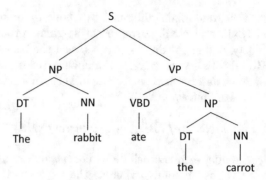

Figure 13.1: The tree-structure obtained by parsing

13.2   Named Entity Recognition

A named entity typically refers to a sequence of words that correspond to a specific entity in the real world (i.e., an entity with a *name*). Examples of such entities include "*Bill Clinton*," "*New York*," and "*IBM*." Most named entity recognition methods focus on three types of entities corresponding to *person, location*, and *organization*. The original definition of named entity recognition, as provided in MUC-6 [221], also allowed for the detection of *dates, times, monetary values*, and *percentages*. Although these types of entities are not really *named* entities, they can still be useful in some applications. In other domains such as biological data and online advertising, entities can be biological or they can be names of products. Although these types of entities lie outside the person/location/organization categorization, the broader principles of extracting any particular type of entity remain roughly the same across all domains. Most of the principles and methods discussed in this chapter apply to these diverse settings.

It is assumed that the training data consists of a set of unstructured texts together with all occurrences of the relevant entities marked at the appropriate places. For example, the entities in the training data might be tagged as follows:

⟨*Person*⟩ Bill Clinton ⟨*/Person*⟩ lives in ⟨*Location*⟩ New York ⟨*/Location*⟩ in a neighborhood that is a few miles away from an ⟨*Organization*⟩ IBM ⟨*/Organization*⟩ building. He and his wife, ⟨*Person*⟩ Hillary Clinton ⟨*/Person*⟩, relocated to ⟨*Location*⟩ New York ⟨*/Location*⟩ after his presidency.

Given this training data, the goal is to tag the test segments in which the entities are not marked a priori. Note that the entities might comprise multiple tokens, and therefore one needs to mark the starting point, the ending point, and the name of the entity.

Named entity recognition is the most fundamental problem in information extraction because it provides the basic building block on top of which many other information extraction methods are built. For example, it would be impossible to perform relationship extraction, if one did not have the named entities between which to extract relationships. In fact, the entire pipeline of information extraction can be combined with the linguistic preprocessing pipeline as follows:

Tokenization ⇒ POS Tagging ⇒ Named Entity Recognition ⇒ Relationship Extraction

Linguistic Preprocessing

In the above, the abbreviation POS tagging refers of parts-of-speech tagging. We have not shown the (optional) preprocessing steps of parsing and dependency extraction, although these steps are also used in many information extraction settings. Furthermore, the linguistic analysis is always carried out at the sentence level, and therefore a sentence segmentation module is required.

At first sight, it might seem that one can use a dictionary of all known entities on the planet and simply extract all occurrences of these entities from the text. Such dictionaries of different types of entities do exist, and are referred to as *gazetteers*. However, such a simplistic solution is an incomplete one at best. First, the set of known entities is not constant, but it evolves over time. For some types of entities like locations, the names of entities evolve slowly, whereas for others like people and organizations, the incompleteness problem is very significant. As a result, any particular list of entities used for matching with a document would always be incomplete. The second problem is that there is significant ambiguity in defining a named entity by using a sequence of words. In particular, the use of abbreviations in a contextual setting often leads to the same entity name referring to multiple instances. For example, when a news article uses the term *"Texas quarterback James Street,"* the word *"Texas"* refers to the University of Texas at Austin, and this point cannot be inferred without considering the context of the word that corresponds to its surrounding tokens. A different piece of text might use the term *"Texas"* to refer to a US State, whereas another might refer to a British pop band. To provide an idea of the magnitude of this problem, the surface form *"Texas"* is used to refer to twenty different entities in Wikipedia [138]. Clearly, the *context* in which a particular term is used is crucial in making a clear judgement about the type of entity at hand. Although gazetteers are frequently used as one of the inputs to named entity recognition systems, they are not sufficient on a standalone basis.

There are two primary classes of methods for information extraction in text. The first class of methods uses rule-based methods on extracted features in order to perform information extraction. The second class of methods, referred to as statistical learning methods, uses hidden Markov models, maximum entropy Markov models, and conditional random fields. We will discuss each of these different types of models in this section.

13.2.1 Rule-Based Methods

Rule-based methods work as follows. Each token in the text is converted into a set of features. These features are typically helpful properties of the token or their context for entity extraction. For example, one obvious feature could be the information about whether or not that token starts with a capital letter. These features, therefore, help in defining various patterns on the left-hand side of the rule. The process of feature extraction is an important aspect of feature engineering in rule-based methods, which will be discussed later in this section. Subsequently, a set of rules are mined from the data, which are of the following form:

$$\text{Contextual Pattern} \Rightarrow \text{Action}$$

The contextual pattern on the left-hand side of the rule is a combination of conditions corresponding to the features associated with a sequence of tokens. Therefore, a rule is fired if a sequence of tokens in the text matches this pattern. The action on the right-hand side could correspond to labeling that sequence as a named entity. More generally, it could correspond to inserting the start of an entity tag at a particular position, the end of an entity tag, or multiple tags. The simplest and most general case is one in which the right-hand side of the rule is an entity tag.

The nature of the left-hand side of the rule can vary with the specific rule system being constructed. In general, the pattern will always contain a *regular expression* matching the tokens in the entity. Note that the "matching" may be based on the extracted features of the tokens rather than the tokens themselves. The typical features associated with each token are as follows:

1. The most basic feature of a token is the string representation of the token itself, which is also referred to as its *surface* value. In some cases, the surface value may be sufficiently informative for entity extraction.

2. The *orthography* type of the token can capture characteristics of the token such as its capitalization, punctuation, or specific choice of spelling.

3. Linguistic preprocessing provides the part of speech of the token. Some features (e.g., noun phrases) might correspond to a sequence of *multiple* tokens.

4. A number of dictionaries are used to identify whether a token belongs to a specific type such as titles, locations, organizations, and so on. Furthermore, dictionaries can even identify whether a token occurs as a part of a specific name. An example of a title is "*Mr.*" and an example of a company ending is "*Inc*" or "*LLC*".

5. In some rule-based methods, the text is sequentially tagged in phases. In such cases, the tags in earlier phases are used as features in rule conditions of later phases.

In addition to the structured patterns matching the token, the left-hand side might optionally contain patterns corresponding to the context preceding or following an entity. Examples of two possible rules are as follows:

$$(\text{Token} = \text{``}Ms.\text{''}, \text{Orthography} = FirstCap) \Rightarrow \text{Person Name}$$
$$(\text{Orthography} = FirstCap, \text{Token} = \text{``}Inc\text{''}) \Rightarrow \text{Organization Name}$$

The first rule matches a sequence of two tokens beginning with "*Ms.*" and a capitalized letter. The second rule matches a sequence of two tokens starting with a capitalized letter and ending with the "*Inc*" abbreviation. Many useful dictionaries of titles and company endings are available for constructing such rules. Therefore, one might have a feature such as *Dictionary-Class* to describe such tokens. An alternative set of rules is as follows:

$$(\text{Dictionary-Class} = \text{Titles}, \text{Orthography} = FirstCap) \Rightarrow \text{Person Name}$$
$$(\text{Orthography} = FirstCap, \text{Dictionary-Class} = \text{Company-End}) \Rightarrow \text{Organization Name}$$

It is noteworthy that the regular expression on the left-hand side of the rule can be quite complex, and many alternatives might exist. Furthermore, dictionaries of all types exist, corresponding to person names, location names, and so on. Given the large number of ways in which one might create a matching rule for the same expression, there are significant efficiency challenges in creating rule-based systems.

For a given text segment, the fired rules are used to recognize the entities in the text. As in all rule-based systems (cf. Section 5.6 of Chapter 5), there will always be conflicts in the fired rules. Two rules might have different actions on the right-hand side, while matching overlapping spans of text on the left-hand side. Therefore, such systems always have conflict resolution mechanisms like the following:

1. There is no ordering among the rules, but one can have specific policies about which type of rule is favored over the other. For example, if the left-hand side of one rule matches a larger span of text than the other, then the former rule is given priority.

2. An ordering is imposed among the rules that indicates which rule has priority over the other. Rules with greater precision and coverage (with respect to the training data) are given priority. This approach is not different from the use of support and confidence measures in traditional rule-based systems (cf. Section 5.6).

It is easy to see that there are similarities in rule-based systems as used in information extraction to those used in the classification. The main difference is that the structure of the rules is often more complex in information extraction.

13.2.1.1 Training Algorithms for Rule-Based Systems

The simplest training algorithms for rule-based systems use manual and hand-crafted rules. This approach encodes natural domain knowledge about the structure of person, location, and organization names, and represents a form of *deductive* learning. However, such rules are tedious to construct, and there are limitations on what humans can achieve with arbitrary corpora. Therefore, most of the rule-based systems are automated, and they learn from labeled training data with the use of *inductive* learning.

As discussed earlier, the training data contains unstructured texts together with occurrences of tagged entities. Starting with this training data, the learning algorithm iteratively adds rules that have good precision and coverage with respect to these tagged entities. An example of a primitive "master-algorithm" for rule generation is illustrated below:

$\mathcal{R} = \{\}$;
repeat;
  Select a tagged entity E in the training data that is uncovered;
  Create a rule R that covers E;
  $\mathcal{R} = \mathcal{R} \cup \{R\}$;
until no more uncovered entities;

At the end of this algorithm, a post-processing approach may be applied in order to create the policies required to avoid conflicts, remove the redundant or weak rules, create default rules for situations not covered by the training data, and so on. Within this basic framework, there is significant flexibility in how a specific rule R is discovered from the training data. Existing algorithms for rule generation are either top-down or they are bottom-up. In top-down rule-generation methods, we start with more general conditions in the antecedent (i.e., covering lots of positive examples) and then add constraints in order to make them more specific. For example, adding a conjunct to the antecedent (as in Learn-One-Rule of Section 5.6) is a way of making a rule more specific. In bottom-up rule generation, one starts with very specific rules and then generalizing it so as to allow the rule to cover more positive examples. Top-down systems often have low precision on the training data, but they generalize well to the test data. Bottom-up systems often have higher precision on the training data, but do not generalize as well to the test data. A lot of the successful systems for named entity recognition in recent years have been bottom-up systems, although a combination of the two has also been successfully used. Since the branching factor for a top-down rule (i.e., number of ways to specialize the rule) is often quite high, as a result of which such systems are computationally expensive. In the following, we give a brief overview of these two methods.

Top-Down Rule Generation

One of the earliest methods, referred to as WHISK [521], was implemented as top-down rule specialization method. Strictly speaking, WHISK is an *active learning* method because it interleaves the user tagging activities (i.e., creating new training instances) with the process of rule creation. Nevertheless, it is possible to also use WHISK without user interaction. The approach starts with a seed instance of the tagged data and creates the most general instance that covers the rule. Subsequently, terms are added to the antecedent of the rule one at a time, so as to minimize the expected error of the rule. This type of rule-growth is similar to the *Learn-One-Rule* algorithm of Section 5.6, although the growth of an instance is somewhat different in this case.

First, since a seed instance is used to grow a rule, only terms from this instance need to be used in the growth. Furthermore, in information extraction, one often does not add the term to the antecedent of the rule, but a semantic class matching the rule (e.g., when the dictionary-class is *"title"*). Therefore, the WHISK approach checks for not only the term itself but all its matching semantic classes while adding to the rule. The addition with the lowest error is selected for extension. Laplacian smoothing is used during the computation of the error rate in order to minimize the impact of overfitting, when the amount of training data is limited. Other than the error rate, WHISK tries to use the least restrictive rule, when making growth choices between rules with similar error. It is possible to continue growing the rules, until zero error is achieved on the training data. However, such a choice would lead to overfitting. Therefore, WHISK uses a pre-pruning approach, in which rule growth is prematurely stopped when the error falls below a particular threshold. In addition, a post-pruning step is also used in which rules with low coverage and high error on the training data are discarded. An alternative is to use the error rate on the validation set in order to prune the rules. The top-down rule generation methods historically preceded the bottom-up methods, which might partially be a result of the fact that the traditional rule learning methods in machine learning tend to resemble top-down methods.

Bottom-Up Rule Generation

In bottom-up rule generation, one always starts with a specific instance, and uses it to construct an antecedent matching this instance exactly. This will result in a rule with 100% accuracy on the training data. Unfortunately, however, such a rule will perform poorly on unseen test data and will also have low coverage. Such a rule needs to be generalized. It is noteworthy that the generalization process will reduce the precision on the training data, but it will typically improve the performance on the test data at least in the initial phase. This is the reverse of what happens in the top-down approach. Two well-known methods for bottom-up rule generation are Rapier [85] and (LP)^2 [117].

The broad approach in these methods is summarized by the pseudo-code below:

$\mathcal{R} = \{\}$;
repeat;
  Select a tagged entity E in the training data that is uncovered;
  Create a rule R that covers E by:
      starting with the most specific rule covering the entity
      and successively generalizing this specific rule;
  $\mathcal{R} = \mathcal{R} \cup \{R\}$;
  Remove instances covered by R;
until no more uncovered entities;

A seed rule might be the most specific rule that covers an instance. For example, a series of generalizations may be as follows:

$$(\text{Token}=\text{``}Ms.\text{''}, \text{Token}=\text{``}Smith\text{''}) \Rightarrow \text{Person Name}$$

$$(\text{Token}=\text{``}Ms.\text{''}, \text{Orthography}=FirstCap) \Rightarrow \text{Person Name}$$

$$(\text{Dictionary-Class}=\text{Titles}, \text{Orthography}=FirstCap) \Rightarrow \text{Person Name}$$

In order to include the effect of context, the set of tokens preceding or following a particular pattern is also included in the left-hand side of the rule. In such a case, the action on the right-hand side must also indicate where the tag begins or ends. For example, consider the following form, which includes the verb after the person occurrence:

$$((\text{Token}=\text{``}Ms.\text{''}, \text{Token}=\text{``}Smith\text{''}):\mathbf{p}, \text{Token}=\text{``}studies\text{''}) \Rightarrow \text{Person Name before } \mathbf{p}$$

$$((\text{Dictionary-Class}=\text{Titles}, \text{Orthography}=FirstCap):\mathbf{p}, (\text{POS}=\text{VB})) \Rightarrow \text{Person Name bef. } \mathbf{p}$$

Note that the right-hand side of the rule is now an action about the positioning of the end point of tag just before the verb, rather than the placement of the entire tag. One can also have rules about the placement of the beginning of the tag.

It is evident from the above examples that the number of ways in which the rules may be generalized increases exponentially with successive branching. Therefore, one always greedily performs the best generalization, while maintaining a limited number of best generalizations that have been seen so far. The quality of the generalization is quantified using a number of factors such as a combination of precision and coverage.

13.2.2 Transformation to Token-Level Classification

In token-level classification, an entity with multiple consecutive tokens can be tagged by using four types of tokens $\{B, C, E, O\}$, which stand for Begin, Continue, End, and Other, respectively [496]. An example of a possible token-level classification is as follows:

$$\underbrace{\text{William}}_{B} \ \underbrace{\text{Jefferson}}_{C} \ \underbrace{\text{Clinton}}_{E} \ \underbrace{\text{lives}}_{O} \ \underbrace{\text{in}}_{O} \ \underbrace{\text{New}}_{B} \ \underbrace{\text{York}}_{E}.$$

Consecutive occurrences of B, C, and E, correspond to an entity. The above labeling scheme does not include the entity type. For example, the person-level and location-level tokens have been given the same classification. Nevertheless, it is easy to integrate the specific entity type within this framework with additional labels like L, P, and O, for location, person, and organization, respectively.

$$\underbrace{\text{William}}_{PB} \ \underbrace{\text{Jefferson}}_{PC} \ \underbrace{\text{Clinton}}_{PE} \ \underbrace{\text{lives}}_{O} \ \underbrace{\text{in}}_{O} \ \underbrace{\text{New}}_{LB} \ \underbrace{\text{York}}_{LE}.$$

One can use the features associated with tokens (such as orthography and dictionary membership) to classify the tokens with a traditional classifier like a decision tree. However, such an approach does not use sequential ordering and context, which is undesirable. For example, it is hard to guess that the token *"New"* is the beginning of a place (in spite of capitalization), because it could be placed at the beginning of a sentence and it can also correspond to the beginning of an organization. Therefore, classifiers that treat each token independently are not useful in such settings, because the tokens preceding or following a token are informative. The first step of the process is to extract the features associated with the tokens, which are typically not very different from those used in extracting rules. In particular, the features such as the token itself, the orthography, the parts of speech tags, and the dictionary lookup features are commonly used in these settings. The following sections will discuss several other ways of performing token-wise classification with methods such as hidden Markov models, maximum entropy Markov models, conditional random fields, and deep learning.

13.2.3 Hidden Markov Models

Hidden Markov models transition through a sequence of hidden *states*, and each state produces a token. One can view the state of a hidden Markov model in the same way as one views a mixture component (i.e., hidden latent component), which generates a data point from a given distribution. Just as the hidden latent component is used to generate an instance within a particular cluster of closely related points (cf. Chapter 4), a state in a hidden Markov model is used to generate a token (i.e., word) in a given sequence of text words. However, the key differences between the generative process used in a hidden Markov model and the mixture modeling approach discussed earlier is the fact that the states in a hidden Markov model are dependent on one another. In other words, the hidden components at the sequential generations are not independent of one another, and this fact needs to be accounted for in the estimation process. Each transition from one state to another generates some type of data, which is a symbol in its most basic form. However, it can also generate some multidimensional combination of features from a multivariate probability distribution. For each transition, a categorical distribution (i.e., loaded die roll with faces showing symbols) is used to generate the tokens. The outcome of the die roll decides the token at this position. This is, of course, the simplest form of the model, which does not include the effect of extracted features. For now, we will discuss this simplified version.

Visible Versus Hidden Markov Models

In visible forms of Markov models, the states of the models are often directly associated with the generated symbols. As a result, the observed sequence of states can be used to trivially infer the state of the model. For example, a *bigram language model* is a visible model in which the state is defined by the last word that was generated. As a result, the number of states is equal to the number of possible tokens (i.e., words), and each token is generated based on the one before it. On the other hand, hidden Markov models allow a more generalized definition of a hidden state with some semantic interpretation. For example, a hidden state could correspond to the fact whether or not a token lies inside an entity of a particular type in a text document. Although the location of entities is known in training documents, it is not known in test documents. Therefore, a hidden Markov model that uses the occurrences of tokens within specific entity types as hidden states needs to learn the transition probabilities from the training data, and use it to make predictions for the test data. One such system is referred to as *Nymble*.

The Nymble System

Since there is considerable variations in different systems, we describe an early representative called *Nymble* [56], which provides a broader idea about how such systems work. The Nymble approach associates a state with each of the different types of entities in the data. However, within each state corresponding to an entity, it creates a set of word states, in which one transition corresponds to the generation of a word. This is a standard bigram model in

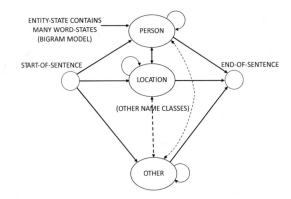

Figure 13.2: The hidden Markov model used by Nymble

which each state generates the next word in the sequence (cf. Section 10.2 of Chapter 10). The basic idea here is that each entity state generates the tokens within a multi-token entity with a bigram model that is specific to that entity. In addition, we have a special *"end"* token[5] after each entity, which provides the cue that one is moving from one entity-state to another within the model. Therefore, the different entity states inside Figure 13.2 are connected through transitions using the special *"end"* token. Although Figure 13.2 shows only the entity states and not the word states inside them, the total number of combinations of word-entity states is equal to the product of the lexicon size with the number of different types of entities. This type of architecture ensures that the generation of an observation is dependent on the preceding label and/or observation. The transition probabilities in this model are estimated from a training data set in which the entities are marked. In such cases, the states of the Markovian model become visible because of the entity tagging information. This fact simplifies the parameter estimation process. Furthermore, labels of tokens in a test document can be inferred by estimating the states of the most likely path through the Markov model for an unlabeled sequence of tokens. The determination of the most likely path through the Markov model is somewhat more complex, and is discussed in Section 13.2.3.2.

Formally, let $\overline{x} = (x_1 \ldots x_m)$ be the sequence of tokens in a text document, and let $\overline{y} = (y_1 \ldots y_m)$ be the corresponding sequence of labels. Note that each y_i is a tag like *"Person,"* *"Location,"* *"Other,"* and so on. These tags are referred to as *name classes* by Nymble, although many other Markovian models use more complex name classes like $\{PB, PC, PE, LB, LC, LE, O\}$ and so on. Another simplification is that we are ignoring the features associated with the tokens for simplicity, although we will come back to this point later. For tagged documents in the training data, the sequence $\overline{y}$ is known, whereas for untagged documents in the test data, the sequence $\overline{y}$ is unknown. One must therefore, determine the sequence $\overline{y}$ for test data that maximizes $P(\overline{y}|\overline{x})$.

It is noteworthy that Nymble is considered a hidden Markov model in the broader literature, but there are a number of subtle differences from the traditional way in which a hidden Markov model is used. In particular, the number of states (including the word states inside the name-class states) is equal to the lexicon size, and each word state produces a word deterministically. This is different from the traditional usage of a hidden Markov

---

[5]Note that this approach is slightly different from directly tagging the last token of an entity that a state that indicates its end point. In the approach of the previous section, a person entity would be ended by a token with a PE tag, whereas Nymble simply uses a separate *"end"* token.

model in which a small number of states with semantic significance is used, and the token generation from each state is inherently probabilistic. As a result of this modeling approach, the word-states become at least partially visible if the entity-states are known. As we will see later, this fact helps in the modeling process. The Nymble model uses the following generative process to create each token x_i:

1. Select the current name-class y_i based on the previous label y_{i-1} and the previous word x_{i-1} with probability $P(y_i|y_{i-1}, x_{i-1})$. Note that the value of $P(y_i|y_{i-1}, x_{i-1})$ will eventually be estimated in a data-driven manner.

2. Based on the selected y_i, generate x_i using one of the following two rules:

   - If x_i is the first word of a named entity, then generate it based on the previous state y_{i-1} and current state y_i with probability $P(x_i|y_i, y_{i-1})$. Note that y_{i-1} and y_i are different in this case. This way of generating the first token of an entity (including "*Other*") is helpful in accounting for the effect of context after an entity.

   - If x_i is inside a named entity, then generate it based on the current name-class state y_i and previous token x_{i-1} with probability $P(x_i|y_i, x_{i-1})$.

It is assumed that the training and test data are generated using this repetitive process. The main difference between the two is that the entity tags and special "*end*" token are available in the training data, but they have been removed from the test data.

13.2.3.1   Training

For such generative processes, the model parameters need to be *estimated* using a maximum-likelihood approach. The model parameters correspond to the probabilities in the generative process shown above. In traditional hidden Markov models, the parameters are estimated using a process known as the *Baum-Welch* algorithm, which is an adaptation of the expectation-maximization approach to a setting in which the latent states are dependent on one another. However, the specific structure of the model by Nymble is simplified by the fact that the word-states inside the name-class states generate the terms deterministically, if the tokens and their tags are known (as is the case for training data). As a result, for a given training data set, the exact sequence of visited states are fully visible and deterministic. In such cases, the parameter estimation process simplifies to a counting problem that is a slight generalization of bigram probability estimation. The three steps in the generative process above require the estimation of three probabilistic parameters, which are $P(y_i|y_{i-1}, x_{i-1})$, $P(x_i|y_i, y_{i-1})$, and $P(x_i|y_i, x_{i-1})$. To estimate these parameters, one simply needs to run the training data through the model, and (i) count the number/fraction of times y_i follows (x_{i-1}, y_{i-1}); (ii) count the number/fraction of times x_i occurs in combination with (y_{i-1}, y_i) (when y_i is different from y_{i-1}); (iii) count the number/fraction of times that x_i occurs in combination with (x_{i-1}, y_i) (with y_i is same as y_{i-1}). In other words, we have:

$$P(y_i|y_{i-1}, x_{i-1}) = \frac{\text{Count}(x_{i-1}, y_{i-1}, y_i)}{\text{Count}(x_{i-1}, y_{i-1})}$$

$$P(x_i|y_i, y_{i-1}) = \frac{\text{Count}(y_{i-1}, y_i, x_i)}{\text{Count}(y_{i-1}, y_i)} \quad [y_i \text{ and } y_{i-1} \text{ are different}]$$

$$P(x_i|y_i, x_{i-1}) = \frac{\text{Count}(x_{i-1}, y_i, x_i)}{\text{Count}(x_{i-1}, y_i)} \quad [y_i \text{ and } y_{i-1} \text{ are the same}]$$

Laplacian smoothing can be used in order to make the estimation more robust in the presence of sparsity. We emphasize that this greatly simplified training process is largely a consequence of the simplified type of hidden Markov model used in Nymble; the states are not really hidden for the training data, but they are visible as in bigram models. However, for the test data, the states are hidden and they need to be inferred probabilistically.

13.2.3.2 Prediction for Test Segment

After the estimations have been performed, a test segment can be classified with the use of the estimated parameters. For a test segment $\overline{x} = x_1 \ldots x_t$ that is *not* annotated with entities, the path through the Markov model is no longer deterministic. In other words, there are multiple possible label sequences $\overline{y} = y_1 \ldots y_t$ that can result in the sequence of transitions $x_1 \ldots x_t$, and each has its own probability $P(\overline{y}|\overline{x})$. Therefore, one must determine the optimal sequence $\overline{y}$ so that $P(\overline{y}|\overline{x})$ is maximized. Note that maximizing $P(\overline{y}|\overline{x})$ is equivalent to maximizing $P(\overline{y}, \overline{x})$ as long as we are comparing different sequences of states $\overline{y}$, and the test segment $\overline{x}$ is fixed:

$$P(\overline{y}|\overline{x}) = P(\overline{y}, \overline{x})/P(\overline{x}) \propto P(\overline{y}, \overline{x}) \tag{13.1}$$

Therefore, once the parameters of the model have been estimated, it suffices to find the tag sequence $\overline{y}$ that maximizes $P(\overline{y}, \overline{x})$. This probability is the product of the probabilities of the transitions in each of the paths through the Markov model. A naïve approach would enumerate all possible paths through the model that match $\overline{x}$ and select the one that maximizes the probability. However, it turns out that this maximization can be done in a much more efficient way with dynamic programming, when using the *Viterbi algorithm*. Refer to [2, 456] for details of the Viterbi algorithm.

13.2.3.3 Incorporating Extracted Features

The approach discussed so far uses the surface values of the tokens and does not incorporate the extracted features such as the orthography, dictionary features, and so on. Nymble uses the features in a simplified way. Specifically, the different feature values are prioritized (such as capitalization and first word of a sentence). The lowest priority feature is a single catch-all keyword, referred to as "*other*." The highest priority feature value is associated with each token as an additional keyword, and this keyword is selected from one of fourteen different possibilities. Therefore, in addition to the token sequence $x_1 \ldots x_m$, we have an additional feature sequence $f_1 \ldots f_m$, so that each f_i is one of fourteen keywords, like "*AllCaps*," "*FirstWord*," and "*other*." The problem changes only in a minor way with the incorporation of features, because we can now pretend that $\langle x_i, f_i \rangle$ is the generated token at the ith position rather than x_i. All other steps remain the same.

13.2.3.4 Variations and Enhancements

There are several natural variations and enhancements to the basic model that was proposed in [56]. Many of these enhancements are related to the fact that hidden Markov models face severe challenges related to sparsity and overfitting. These problems are addressed as follows:

1. One approach that was discussed in the original work [56] was the use of *back-off* models. The idea of back-off models is to use a generalized model to handle cases in which sufficient data does not exist to estimate the parameters of a complex model. An example is the use of a unigram to estimate the probability of a token in cases

where a bigram model does not work because of sparsity of data. For example, the back-off estimate of $P(x_i|y_i, x_{i-1})$ is simply $P(x_i|y_i)$, when the word-pairs $x_{i-1}x_i$ do not occur frequently enough in the training data. The use of such back-off models is widespread in language modeling.

2. The Nymble model uses one word-state for each term in the lexicon within an entity-state. However, this results in a large number of states. It was suggested in [510] that one could merge related states with the same entity label in order to create a model that generalizes better. This type of merging does increase the training complexity of the model.

3. One can use unlabeled data [510] in order to improve generalizability. The unlabeled data can be used in an iterative way in order to improve the parameter estimations in cases where the amount of labeled data is limited. This is a natural extension of the semi-supervised approach discussed in Chapter 5 to the problem of information extraction.

Hidden Markov models have been recently outperformed by related models like maximum entropy Markov models, conditional random fields, and recurrent neural networks. One problem with hidden Markov models is that they use the features associated with tokens in a rather rudimentary way compared to many other models.

13.2.4   Maximum Entropy Markov Models

Hidden Markov models *generate* sequences using transitions between states. On the other hand, *maximum entropy Markov models* directly model the probability of labeling based on the states. Such models are referred to as *discriminative* models, like the logistic regression model discussed in Chapter 6. The multinomial logistic regression model discussed in Section 6.4.4 is a maximum entropy model without the Markovian assumption of sequential dependencies among data items. This section discusses the generalization of this model with the Markovian assumption, which is required for sequence data.

It is noteworthy that the way in which features are used in the case of the Nymble system (with hidden Markov models) is rather limited in scope. In particular, features are extracted only from individual positions in the sequence. Furthermore, Nymble prioritizes the features, and it uses only the highest prioritized feature associated with each token in order to ease the estimation process with limited data. We would like to be able to extract many different types of features from overlapping portions of the text segment and use them simultaneously for the modeling process.

Let $\overline{x} = (x_1 \ldots x_m)$ be the sequence of tokens in a text document, and let $\overline{y} = (y_1 \ldots y_m)$ be the corresponding sequence of label tags. Furthermore, let $\overline{x}_{i-q}^{i+q}$ denote the segment $(x_{i-q}, x_{i-q+1}, \ldots x_{i+q})$ of $\overline{x}$ from the $(i-q)$th position to the $(i+q)$th position. Similarly, let $\overline{y}_{i-1}^{i-p}$ denote the segment $(y_{i-p}, y_{i-p+1}, \ldots y_{i-1})$ of $\overline{y}$ from the $(i-p)$th position to the $(i-1)$th position.

A key point here is that one can now extract features from the neighborhood of the tokens in the ith position and the history of labels *including and before* the ith position. In other words, it is assumed that the labels up to the $(i-1)$th position have been inferred, but the labels including and after position i are not known. In other words, features are extracted from the contiguous sequence $\overline{x}_{i-q}^{i+q}$ of tokens and the contiguous sequence $\overline{y}_{i-p}^{i-1}$ of labels. For example, consider the case where $p = q = 1$, and the token x_i follows the token

"*Ms.*" at x_{i-1}. In such a case, binary feature $f_1(y_i, y_{i-1}, \overline{x}_{i-1}^{i+1})$ is defined as follows:

$$f_1(y_i, y_{i-1}, \overline{x}_{i-1}^{i+1}) = \begin{cases} 1 & \text{if } [y_{i-1} == PB] \text{ AND } [x_{i-1} == \text{"Ms."}] \text{ AND } [y_i == PC] \\ & \text{AND } [\text{Dictionary-Class}(x_{i+1}) == \text{Person-End}] \\ 0 & \text{otherwise} \end{cases}$$

Note that PB and PC correspond to the beginning and continuation tags for the person entity type. This is a binary feature, which is extracted in the form of a boolean expression, although it is possible to extract numerical features as well. For example, a contiguous window of three tokens such as "*Thomas Watson Jr.*" and x_i set to "*Watson*" could be represented as follows:

$$f_2(y_i, y_{i-1}, \overline{x}_{i-1}^{i+1}) = \begin{cases} 1 & \text{if } [\text{FirstCap}(x_{i-1}) == 1] \text{ AND } [\text{FirstCap}(x_i) == 1] \\ & \text{AND } [\text{Dictionary-Class}(x_{i+1}) == \text{Person-End}] \\ 0 & \text{otherwise} \end{cases}$$

Note that the first feature uses both tokens and labels, whereas the second uses only tokens. In general, one can use any subset of the arguments to define the feature. These features combine the effects of multiple token and label properties over a contiguous sequence of tokens, and can often capture excellent semantics when sufficient effort and thought is put into feature engineering. Such features are quite powerful because they naturally encode the context required for making inferences. Furthermore, they are naturally more expressive than the simplistic features extracted from individual tokens in hidden Markov models. Now imagine a setting in which d such features are extracted. Then, one can model the label y_i directly using the same form as logistic regression. However, since the entity tags have multiple possible values, we need to use *multinomial* logistic regression (cf. Section 6.4.4) over the entire set of possible entity tags denoted by $\mathcal{Y}$:

$$P(y_i | \overline{y}_{i-p}^{i-1}, \overline{x}_{i-q}^{i+q}) = \frac{\exp(\sum_{j=1}^{d} w_j f_j(y_i, \overline{y}_{i-p}^{i-1}, \overline{x}_{i-q}^{i+q}))}{\sum_{y' \in \mathcal{Y}} \exp(\sum_{j=1}^{d} w_j f_j(y', \overline{y}_{i-p}^{i-1}, \overline{x}_{i-q}^{i+q}))} \tag{13.2}$$

The values $w_1 \ldots w_d$ are the regression coefficients that needed to be *learned* in a data-driven manner. Learning these parameters is the essence of the modeling process.

The parameters can be learned so as maximize the conditional probabilities of the labels in the training data. The conditional probability of the sequence of labels $\overline{y} = (y_1 \ldots y_m)$ in the training data is defined as the product of the conditional probabilities of the constituent labels y_i, given the observed tokens in the neighborhood of the ith position and the labels that occur prior to the ith position:

$$P(\overline{y} | \overline{x}) = \prod_{i=1}^{m} P(y_i | \overline{y}_{i-p}^{i-1}, \overline{x}_{i-q}^{i+q}) \tag{13.3}$$

One can then compute the likelihood $\mathcal{L}(\mathcal{D})$ of all pairs $(\overline{x}, \overline{y})$ in the training data $\mathcal{D}$:

$$\mathcal{L}(\mathcal{D}) = \prod_{(\overline{x}, \overline{y})} P(\overline{y} | \overline{x}) \tag{13.4}$$

The log-likelihood needs to be maximized in order to learn the coefficients, which is achieved with the *generalized iterative scaling* [364, 372]. Furthermore, the optimal sequence of labels for an unlabeled test sequence can also be determined using dynamic programming.

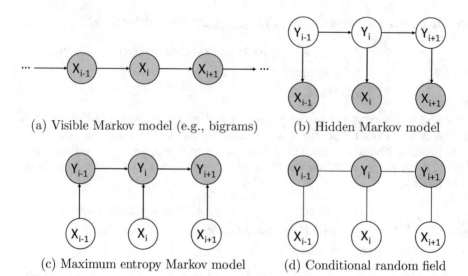

(a) Visible Markov model (e.g., bigrams)   (b) Hidden Markov model

(c) Maximum entropy Markov model   (d) Conditional random field

Figure 13.3: A comparison of the probabilistic graphical models associated with Markovian models (visible states), hidden Markov models, maximum entropy Markov models, and conditional random fields

Note that the Viterbi algorithm for hidden Markov models is also a dynamic programming algorithm. A detailed discussion of these algorithms is beyond the scope of this book, and the reader is referred to [364, 372, 456].

13.2.5   Conditional Random Fields

Conditional random fields are among the best performing models for information extraction, and they are closely related to maximum entropy Markov models. In the maximum entropy Markov model, a restriction is placed on the probabilistic modeling of y_i, which depends only on the *labels* $y_{i-p} \ldots y_{i-1}$ occurring before it but not after it (although one can use the *tokens* occurring both before and after it). This restriction is removed in a conditional random field, where the inference of y_i depends on both the labels occurring before it and those occurring after it. The removal of this restriction increases the training complexity of the model. Therefore, one simplification is to use only the *immediately* neighboring labels y_{i-1} and y_{i+1} to predict y_i, although one can use a larger window of tokens. Such models are referred to as *linear chain* conditional random fields. The use of long-range dependencies among labels can make conditional random fields too expensive to train. One can illustrate these differences among the linear-chain versions of hidden Markov models, maximum entropy Markov models, and conditional random fields with the use of probabilistic graphical models, which are depicted in Figure 13.3. A shaded state represents an value *generated* by the model, whereas the unshaded value is hidden, and therefore not generated by the model. Therefore, hidden Markov models generate token sequences, whereas maximum entropy Markov models and conditional random fields generate label sequences. Furthermore, conditional random fields are *undirected* graphical models because the dependency can occur in any direction.

As in the case of the maximum entropy Markov model, it is assumed that the label sequence is denoted by $\overline{y} = (y_1 \ldots y_m)$ and the corresponding token sequence is denoted by $\overline{x} = (x_1 \ldots x_m)$. However, the features are extracted somewhat differently with respect to the *edge* joining y_i and y_{i-1}, and also all the tokens in $\overline{x}$ with a specific focus on their contextual relationship to position i. Let us represent the jth feature by $f_j(y_i, y_{i-1}, \overline{x}, i)$. Note that the index i has been added to the argument of the feature function so that the tokens of $\overline{x}$ may be used for feature extraction with a specific focus on the context with respect to position i. Assume that a total of d features denoted by $f_1() \ldots f_d(\cdot)$ are extracted for the position i.

How does one model the predictive process in a conditional random field? As in the case of maximum entropy Markov models, the goal is to maximize $P(\overline{y}|\overline{x})$ for the label sequence $\overline{y} = (y_1 \ldots y_m)$ given $\overline{x} = (x_1 \ldots x_m)$. However, in this case, the probability is expressed by multiplying the likelihood of various *edges* between y_{i-1} and y_i:

$$P(\overline{y}|\overline{x}) \propto \prod_{i=2}^{m} \exp\left(\sum_{j=1}^{d} w_j f_j(y_i, y_{i-1}, \overline{x}, i)\right) = \exp\left(\sum_{i=2}^{m} \sum_{j=1}^{d} w_j f_j(y_i, y_{i-1}, \overline{x}, i)\right)$$

As before, the parameters $w_1 \ldots w_d$ represent the coefficients of various features and learning them is the key to the prediction process. The constant of proportionality in the aforementioned equation can be removed by rewriting it as follows:

$$P(\overline{y}|\overline{x}) = \frac{\exp\left(\sum_{i=2}^{m} \sum_{j=1}^{d} w_j f_j(y_i, y_{i-1}, \overline{x}, i)\right)}{\sum_{\overline{y}'} \exp\left(\sum_{i=2}^{m} \sum_{j=1}^{d} w_j f_j(y_i', y_{i-1}', \overline{x}, i)\right)} \tag{13.5}$$

Note that the normalization factor is defined using all possible combinations of label sequences, which can be rather expensive. As in the case of maximum entropy models, the likelihood function can be computed, and the optimization process yields the model parameters. A number of quasi-Newton methods, such as L-BFGS, are used in order to determine the optimal parameters.

13.2.6 Deep Learning for Entity Extraction

Since entity extraction is a token-level classification task, it is relatively easy to adapt various sequence-centric methods to this scenario. Both recurrent neural networks and transformer-based pretrained language models can be adapted to this setting.

13.2.6.1 Recurrent Neural Networks for Named Entity Recognition

Recurrent neural networks represent a natural choice for entity extraction in these settings, because the token-level labels can be represented as the outputs of the various temporal states of a neural networks, and the words of the sentence can correspond to the input tokens. Since recurrent neural networks provide a high level of flexibility in designing the input layer, the linguistic features of the word (capitalization, part-of-speech, orthography) can be represented as features to the input layer. These features can be extracted separately in a preprocessing stage and can be incorporated into the input of the recurrent neural network. Note that the linguistic features, such as parts of speech, can be extracted in an automated way with off-the-shelf tools. Since modern deep learners are end-to-end systems, such features are often not used in the learning process. Nevertheless, we have assumed the availability of such features for greater generality in presentation.

The training data contains sentences that are tagged with their entities. Each tag contains a concatenation of entity type corresponding to person (P), location (L), and other

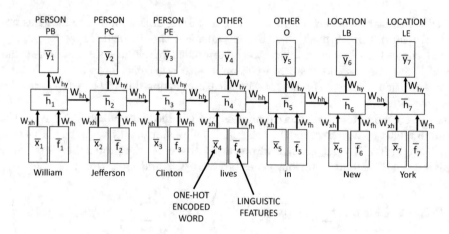

Figure 13.4: Token-wise classification with linguistic features

(O) along with its position corresponding to beginning (B), continuation (C)+ and end (E) within the entity. Therefore, if a token is tagged with PB, it corresponds to the fact that the token begins an entity type. It is not necessary to pair "Other" tokens with a beginning, continuation, and end designation, because such tokens are not closely related like entities, which should be treated as indivisible groups of tokens. An example of a possible training sentence is as follows:

$$\underbrace{\text{William}}_{PB} \ \underbrace{\text{Jefferson}}_{PC} \ \underbrace{\text{Clinton}}_{PE} \ \underbrace{\text{lives}}_{O} \ \underbrace{\text{in}}_{O} \ \underbrace{\text{New}}_{LB} \ \underbrace{\text{York}}_{LE}.$$

A variety of different tagging schemes are available, beyond the specific example used here. In some tagging schemes, the end of the entity also gets a continuation (i.e., C) tag rather than its own E tag. Tokens that do not belong to any named entity, get their own O tag and nothing else. This is, in fact, the most widely used format, corresponding to the *Inside-Outside-Beginning (IOB)* format. For test instances, the tagging information about the tokens is not available.

The outputs of the recurrent neural network are the tags of the various tokens. The input at each time-stamp t is the one-hot encoding $\overline{x}_t$ of the token, and the output $\overline{y}_t$ is the tag. Furthermore, we have an additional set of q-dimensional linguistic features $\overline{f}_t$ associated with the tokens at time-stamp t. These linguistic features might encode information about the capitalization, orthography, capitalization, and so on. The hidden layer, therefore, receives two separate inputs from the tokens and from the linguistic features. The corresponding architecture is illustrated in Figure 13.4. We have an additional $p \times q$ matrix W_{fh} that maps the features $\overline{f}_t$ to the hidden layer. Then, the recurrence condition at each time-stamp t is as follows:

$$\overline{h}_t = \tanh(W_{xh}\overline{x}_t + W_{fh}\overline{f}_t + W_{hh}\overline{h}_{t-1})$$
$$\overline{y}_t = W_{hy}\overline{h}_t$$

Most recurrent architectures omit the use of linguistic features, and train directly with the tokens. In token-level classification applications, it is sometimes helpful to use bidirectional recurrent networks in which recurrence occurs in both temporal directions [503].

13.2.6.2   Use of Pretrained Language Models with Transformers

It is relatively easy to adapt pretrained language models such as GPT and BERT in order to perform named entity recognition. GPT-n is particularly easy to adapt for named entity recognition, since it is implemented as a transformer decoder, which already outputs a token-wise output corresponding to the next word (cf. Section 11.4.1 of Chapter 11). The next word is typically predicted by applying a softmax layer to a numerical representation $\overline{r}_i$ of a sentence immediately preceding the word $(i+1)$. In addition to outputting the next word at position $(i+1)$, an additional set of weights can be trained (during fine-tuning), whose job it is to convert the numerical output $\overline{r}_i$ of the decoder into a prediction of the tag at the current position i. This tag could be drawn from the IOB convention.

In the case of BERT, it already has an additional token-level-classification functionality built into it for additional fine-tuning. Note that BERT creates enhanced representations of word vectors after passing through the encoder of a transformer. These word vectors incorporate bidirectional context, as most of the words in the entire sentence are made available for prediction of missing words within the BERT model (cf. Section 11.4.2 of Chapter 11). The token-level-classification of BERT trains an additional weight layer, which takes as input the enhanced word representations and outputs the token label (which, in this case, happens to be the entity tag in IOB format). Since BERT works with sentence pairs during pretraining (whereas token-level classification requires only a single sentence as input), it is assumed that the second sentence of the pair is a null sentence. An example of an implementation of token-wise classification for Named Entity Recognition with BERT is provided in [730].

One can also use T5 in order to perform token-wise classification. In this case, the input sequence to T5 is simply the prefix containing the type of task (e.g., named entity recognition) along with the sentence that needs to be labeled. The output sequence is a sequence of tags corresponding to the IOB convention. The transformer can be fine-tuned with this type of training data in order to learn to predict sequence of labels for the individual tokens, when the sentence is provided as input.

13.3   Relationship Extraction

The task of relationship extraction is built on top of entity extraction. In other words, once the entities in the text have been extracted, the relationships among them can be mined. Some examples of relations between various entities are as follows:

LocatedIn(*Bill Clinton, New York*)
WifeOf(*Bill Clinton, Hillary Clinton*)
EmployeeOf(*ABC Corporation, John Smith*)

The most common relationship types include *physical location* relations, *social relations*, and *organizational affiliation*, which are shown above. The above examples illustrate binary relationships. In general, it is also possible to have relationships between more than two entities, which are referred to as *multi-way* relationships. This chapter will focus on binary relationships, which are considered fundamental. One can see that each relationship is a triple, containing a *head entity*, a *tail entity*, and a *relationship type*. Depending on how the relationship type is expressed, the head and tail entities may ne interchangeable. For example, the relationships *EmployeeOf* and *Employs* will typically swap the head and tail entity types.

The problem of relationship extraction is defined as follows. Given a fixed set $\mathcal{R}$ of relations, the goal is to identify all occurrences of these relations in a test document where the *entities* have already been tagged but the relations among them are missing. In supervised settings, one also has a training corpus in which *both* the entities and the relations between specific occurrences of the entity have been identified. Since the entities are tagged in both the training and test corpus, one can view the entity extraction task as a more fundamental task that precedes relationship extraction. Therefore, if one is given a training corpus with entities and their relations and a test document with no annotations at all, then one will first extract the entities in the test document and then the relationships among them.

One issue is that the same entity like *"Bill"* might have multiple *mentions* in the same document, which can cause a large number of candidate pairs. However, a commonly used assumption is that the task of relationships between sentence mentions do not cross sentence boundaries. This chapter will adhere to this assumption. For any pair of entity mentions in a sentence, the task is to determine whether or not a relationship exists between them from the set $\mathcal{R}$. It is assumed that the set $\mathcal{R}$ contains a special relationship type referred to as **"Null"**, which applies in cases where the pair of entities occur in the same sentence but no relationship between them has been specified.

13.3.1   Transformation to Classification

The relationship extraction problem can be naturally posed as a classification problem. Since the relationships are extracted only between entity mentions with sentence boundaries, one can extract the pairs of entity mentions in the same sentence in both the training and test data. Therefore, the key is to create one data instance for each pair of entities within a sentence. The instance is also labeled with the relationship type for sentences in training documents, and is not labeled for test documents. For pairs of entity mentions in the same sentence that are not labeled in the training data, one creates a negative training instance, and uses the label **"Null"**. For example, consider the following sentence in the training data in which three person entities are marked:

$$\underset{\text{Person}}{\underline{\text{Bill}}}\text{ , who is a brother of }\underset{\text{Person}}{\underline{\text{Roger}}}\text{, is married to }\underset{\text{Person}}{\underline{\text{Hillary}}}.$$

From this single sentence, as many as three training instances can be extracted for each pair of persons. In this particular case, the entities belong to the same type, but this may not be the case in general. Furthermore, one would have to predefine relations like **brother**, and **wife** in the training data in order to label the training instances properly. For example, it may be possible that the analyst might not spend the time to pre-define the **brother-in-law** relation between *"Hillary"* and *"Roger."* In such a case, the training instance between this pair might be labeled **"Null."** Such a training instance might be useful as a negative example with respect to the types of relations one is interested in. In general, for any sentence containing q entities, one can extract as many as $\binom{q}{2}$ training instances.

The information required for inferring the relationships between a pair of entity mentions is hidden in the vocabulary and grammatical structure of the sentence in which the pair occurs. For example, consider the following test sentence in which the entities have already been marked but the relationships have not been marked:

$$\underset{\text{Person}}{\underline{\text{Bill Clinton}}}\text{ lives in }\underset{\text{Location}}{\underline{\text{New York}}}.$$

Here "*Bill Clinton*" and "*New York*" are two named entities, and one can infer the fact that the person entity lives in the location entity by using the training data to learn the fact that the phrase "*lives in*" provides useful clues for learning the following relation:

LocatedIn(*Bill Clinton, New York*)

In other words, one needs to extract features from various regions of the sentence (e.g., tokens between entity pair) to make inferences about relationships. Such a learning process in relationship extraction can be implemented by extracting the appropriate features from the sentence containing a pair of entity mentions. An alternative approach is to use kernel similarity functions defining similarities between pairs of instances (i.e., pairs of marked relationships). At the end of the day, kernel methods are also indirect ways of performing feature engineering. In the case where explicit feature engineering is used, it is common to use a linear support vector machine (cf. Chapter 6), especially if many features are extracted. In the case where kernel similarity functions are used, the natural approach is to use kernel support vector machines. However, in the case of explicit feature engineering, one advantage is that a wider variety of classification methods can be used. In fact, the earliest techniques were rule-based methods, which are specialized types of classifiers. This section will focus on these two different ways of performing the feature engineering.

13.3.2 Relationship Prediction with Explicit Feature Engineering

The features extracted from the words in a sentence use various properties of the words, such as the surface tokens, the parts-of-speech tags, and the features extracted from the syntactic parse-tree structure. The features may be extracted both from within and outside the entity. Features that are extracted from within the entity are referred to as *entity features*. Furthermore, the features that are extracted from the regions of the sentence surrounding the argument entities, or those located between the two entities are useful for making inferences. Such features are referred to as *contextual features*.

The entity features and contextual features are used somewhat differently during the feature engineering process. However, in both cases, similar features are extracted from the individual tokens, which are not very different from those used in entity extraction. These features (associated with individual tokens) are as follows:

1. *Surface tokens:* Consider the following sentence:

   Bill Clinton lives in New York .
      Person   Location

   In this sentence, the word "*lives*" as well as the phrase "*lives in*" provides useful information about the relationship between the person and location entity. In many cases, the training data might contain sufficient number of such occurrences of these tokens that tell us a lot about relationships among entities. In other cases, the morphological roots of the sentence can also be extracted.

2. *Parts-of-speech tags:* The word "*lives*" is often used both as a noun and as a verb. In the sentence above, the fact that the word "*lives*" is used as a verb is useful for making meaningful inferences about the relationship between the person and the location entity.

3. *Constituency-based parse-tree structure:* In many cases, the sentence structure may be complicated, which will cause challenges in making inferences. For example, a

sentence may contain more than two named entities and there will be some ambiguity in deciding which pairs of entities are more closely related or in how the clues extracted from the sentence should be used. In such cases, the constituency-based parse tree structure, such as the one shown in Figure 13.1, is very useful. For example, consider the following sentence:

$$\underbrace{\text{Bill}}_{\text{Person}} \text{, who is a brother of } \underbrace{\text{Roger}}_{\text{Person}} \text{, is married to } \underbrace{\text{Hillary}}_{\text{Person}}.$$

In the sentence above, the word *"Roger"* is located closer to *"Hillary"* and it is easy for an automated learning algorithm to use the surface token *"married"* to make the wrong assumption that the two are married. However, a parse-tree will place the entire phrase *"who is a brother of Roger"* in a completely different subtree. This is very helpful in knowing that the entities *"Bill"* and *"Hillary"* are respectively the subject and the object of the verb *"married."* Parse trees are, however, quite expensive to construct. Therefore, simplified structural representations, referred to as *dependency graphs*, are often used. The features extracted from the parse tree are not related to individual tokens, but they may correspond to groups of tokens that comprise subtrees of the sentence. As we will see later, this type of feature extraction falls within the general approach of graph-based methods [269].

Individual word features, however, are somewhat limited in their ability to extract relations between entities. It is often more informative to extract combinations of features from the sentence. Most of the features are extracted using either the sequential structure of the sentence or the parse tree/dependency graph of the sentence.

Feature Extraction from Sentence Sequences

Consider two named entities E_1 and E_2 (in either the training or the test data) between which we wish to predict the relations. Let $S = x_1 x_2 \ldots x_m$ be the sentence containing these two entities, where x_i is the ith token of S. We note that the tokens that are inside the entities are treated slightly differently from those outside the entity during feature engineering because they have somewhat different significance for relationship extraction.

We assume that each token x_i is associated with a fixed set of p properties corresponding to its surface token, orthography, part-of-speech, or even its entity label (in the event that x_i is inside an entity). As a practical matter, one can assume that a set of p keywords is associated with each x_i, and this set of features is denoted by F_i. For example, the keywords associated with *"Bill"* and *"Hillary"* might correspond to the fact that they have orthography of *FirstCaps*, they are both person entities, and the actual values of the tokens. In practice, the number of such values will be much larger, but assume for now that only the following $p = 3$ keywords are extracted:

$$\text{Features for } \text{``Bill''}: F_1 = \{\text{``Bill''}, \text{person}, \text{FirstCaps}\}$$
$$\text{Features for } \text{``Hillary''}: F_2 = \{\text{``Hillary''}, \text{person}, \text{FirstCaps}\}$$

An important set of features is created by using all possible combination of the features inside F_i, F_j, where x_i and x_j are inside the two different entities between which the relationships need to be extracted. Therefore, in the example above, there are $3 \times 3 = 9$ possible combinations of features that can be extracted with respect to *"Bill"* and *"Hillary"*. Then, we can create a composite feature corresponding to the concatenation of each pair of these features. For example, the feature *person-person* is very useful because it will often correspond to various types of social relations in the training data. Although not all combinations of features are as discriminative, machine learning algorithms will have various mechanisms for deciding which combination pairs can be extracted.

Another set of features is extracted using the broader structure of the sentence, which could include tokens from outside the entity. These types of features are more challenging to extract because they depend on the broader syntactic structure of the sentence at hand. In this respect, the work in [269] is notable, because it provides a generic graph-based approach to extract features from the sentence. As a starting point, consider the parse tree associated with a particular sentence. We can treat the parse tree as a graph $G = (N, A)$, where N is the set of nodes in the tree, and E is the set of edges indicating the relationships. Each node $i \in N$ is associated with the feature set F_i, which is similar to the feature set discussed earlier for individual tokens (with the same notation). However, in this case, the node i can not only be a token (leaf node) but it can also be a portion of a sentence such as a noun phrase (internal node). Therefore, the extracted features can include the phrase-type tag at a given node such as a noun phrase or a verb phrase. Furthermore, since nodes correspond to phrases and segments of sentences, it is useful to include information about how these segments are related to the two entities E_1 and E_2 for which the training or test instance is being constructed. Therefore, an additional *flag* feature is included within F_i for each node i. The *flag* feature is defined as follows:

$$flag(i) = \begin{cases} 0 & \text{Node } i \text{ does not subsume either } E_1 \text{ or } E_2 \\ 1 & \text{Node } i \text{ subsumes } E_1 \text{ but not } E_2 \\ 2 & \text{Node } i \text{ subsumes } E_2 \text{ but not } E_1 \\ 3 & \text{Node } i \text{ subsumes both } E_1 \text{ and } E_2 \end{cases} \tag{13.6}$$

Note that the *flag* feature is categorical rather than numeric, because no ordering is assumed among the different values.

For example, let us revisit an earlier sentence that includes mentions of three entities "*Hillary,*" "*Bill,*" and "*Roger*":

$$\underbrace{\text{Bill}}_{\text{Person}} \text{ , who is a brother of } \underbrace{\text{Roger}}_{\text{Person}}, \text{ is married to } \underbrace{\text{Hillary}}_{\text{Person}}.$$

The sentence fragment "*who is a brother of Roger*" corresponds to a node of the parse-tree. A particular instance in the training or test data always corresponds to the relationship between two entities, and the value of the *flag* feature for this node will depend on which two entities are being considered. The value of the *flag* feature for this node will be either 1 or 2, when one of the two entities in the instance is "*Roger*". However, if we are trying to create an instance that includes only "*Hillary*" and "*Bill*" (but not "*Roger*") from the same sentence, then the value of the *flag* feature will be 0.

At the most basic level, one can create a feature out of any subgraph of the parse tree containing a particular number of nodes. In general, however, one tends to use only two nodes or three nodes of the tree in order to create the features. Typically, the sets of nodes selected to create the features are either all adjacent leaf nodes or they are sets of two or three nodes that are directly connected to one another by parent-child relationships in the parse-tree. In the former case, the adjacent leaf nodes correspond to adjacent words in the sentence, whereas in the latter case, at least one internal node is included in the node set. For any pair of nodes i and j, the bigram features correspond to all the feature combinations

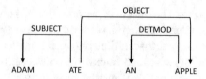

Figure 13.5: A dependency graph

in $F_i \times F_j$. Similarly, for any three nodes i, j, and k, the trigram features correspond to all the feature combinations in $F_i \times F_j \times F_k$. Note that this approach is not too different from how the entity-based features are created. The only difference is in terms of how pairs of nodes are selected for the creation of the features, and in terms of the additional *flag* features.

Simplifying Parse Trees with Dependency Graphs

One challenge in using parse trees is that they are rather expensive to construct. An alternative and complementary way of representing sentence structure is with the use of a dependency graph. A dependency graph is a directed acyclic graph constructed on the tokens in the sentence that tells us about the dependencies between tokens. For example, the subject and object of a verb depend on it. An example of a dependency graph for a sentence is shown in Figure 13.5.

Here, we omit the details of how dependency graphs are constructed, and refer the reader to [652] for both a description and open-source software. The key point is that dependency graphs can be constructed much more efficiently than parse trees, and bigram/trigram features can be extracted using the edges of the dependency graph. In addition, bigram/trigram features can be extracted from adjacent words as discussed earlier.

13.3.3 Relationship Prediction with Implicit Feature Engineering: Kernel Methods

Kernel methods are also feature engineering methods, which represent the underlying feature representation *indirectly* with the use of kernel functions. For example, instead of extracting features from dependency graphs, one might directly try to compute the similarities between pairs of training instances (using linguistic knowledge) to create a kernel similarity matrix. As discussed in Section 3.6 of Chapter 3, such matrices are very useful in extracting an engineered representation, or they can be directly used with kernel support vector machines. Such kernel methods are very powerful because they encode language-specific knowledge within the similarity function. Although we discuss the kernel methods in the context of information extraction, it is important to emphasize that many of these methods can be easily adapted to general-purpose settings with minor modifications.

Since the training instances are usually extracted from sentences containing the two entities, one can compute the similarities between the structured representations of the two sentences, rather than "flattening" the structural text into a multidimensional representation. For example, one can extract kernel similarity matrices from dependency graphs, sequence representations, parse trees, or any other structured representation of the sentences. For any training or test set of data instances (i.e., sentences with entity pairs), the kernel-based approach is as follows:

1. Create a structured representation of each data instance, which can be a sequence, dependency graph, or a parse tree. Each element or node of this structured representation is typically a token, but it can also be a phrase (in the case of a parse tree). The ith element is associated with the set of features F_i, which is defined in a similar way as in the case of explicit feature engineering. For example, each element could be associated with its part-of-speech tag, entity type, and so on.

2. Using these structured representations of sentences, one defines a concrete way of computing similarities between them. Therefore, if we have n_1 training instances, then one can create an $n_1 \times n_1$ kernel similarity matrix among them. The *explicitly* engineered multidimensional representation is given by symmetrically factorizing the similarity matrix K into the form UU^T to yield the $n_1 \times k$ embedding U. One can then use a linear support vector machine on U. Of course, in the kernel methods, the whole point is to not do this explicitly, but use the similarity matrix K directly with nonlinear support vector machines via the kernel trick. The results are mathematically equivalent to those obtained by using a linear support vector machine on U, but the kernel trick provides better computational and space efficiency.

The key point in the above exercise is in creating an appropriate definition of the similarity function between a pair of sentences and the pairs of entities inside them. In the specific case of information extraction, the entity arguments inside the sentences must be used in the computation of similarity in order to ensure that the similarity function is sufficiently discriminative with respect to the relation being mined. However, for other natural language applications, which do not use entities, minor modifications of these similarity functions can be constructed. The following discussion will focus on various structured representations from which the similarities can be extracted.

13.3.3.1 Kernels from Dependency Graphs

One of the earliest methods [78] proposed to extract kernels from dependency graphs (cf. Figure 13.5). Consider two sentences S_1 and S_2, each of which contains a pair of marked entities. The shortest paths between entity arguments in the in the undirected versions of the dependency graphs of these two sentences are used to compute the similarity. The intuition of this approach is that most of the information about the relationships between two entities is often concentrated in the segment between the two sentences.

The first step is to compute the shortest path P_1 between the two entity arguments of S_1 and the corresponding shortest path P_2 between the two entity arguments of S_2. If the paths P_1 and P_2 are of different lengths, then the kernel similarity between S_1 and S_2 is set to 0. However, if the two paths $P_1 = i(1), i(2), \ldots, i(m)$ and $P_2 = j(1), j(2), \ldots j(m)$ are of the same length, then the similarity between the sentences S_1 and S_2 is computed by using the similarities between the features sets $F_{i(r)}$ and $F_{j(r)}$ of the corresponding tokens (which are corresponding nodes in the paths P_1 and P_2). In other words, we have:

$$\text{Similarity}(S_1, S_2) = \prod_{r=1}^{m} |F_{i(r)} \cap F_{j(r)}| \tag{13.7}$$

Since the features contain many characteristics of the words such as the parts of speech and the surface tokens, one can capture a high level of semantic similarity with this approach. For example, consider the following two sentences containing named persons as entities:

Romeo loved Juliet
Harry met Sally

Even though the tokens of the two sentences are completely disjoint, there is still nonzero similarity between the two sentences. This is because the features corresponding to the parts-of-speech tags in the dependency path are the same. In particular, the feature sequence in the shortest path in the dependency graph in both cases is NNP←VBD→NNP. Note that if we change the word "*met*" in the second sentence to "*loved,*" the similarity would increase further since the engineered feature set $F_{i(2)}$ and $F_{j(2)}$ will usually contain the surface token as well.

13.3.3.2   Subsequence-Based Kernels

The subsequence-based kernel treats sentences as sequences and it considers two sentences similar, if many subsequences of the two sentences can be found that have a large amount of similarity between them. The approach described in this section is a trivial modification of the subsequence-based kernel described in Section 3.6.1.3. The key modification required over the approach described in Section 3.6.1.3 is to account for the fact that the each token in the sentence is associated with a feature set. The feature set might contain the surface value of the token, its part-of-speech tag, entity type, and so on. The choice of the feature set depends on the application at hand. For example, for applications beyond information extraction, the entity type might not be available, but the part-of-speech tag and surface token might be available.

In order to understand the generalization of subsequence kernels of Section 3.6.1.3 to the relationship extraction problem, we advise the reader to revisit Sections 3.6.1.3 and 3.6.1.4 before reading further. A dynamic programming approach is described for the computation of the string kernel. The notations used below are also borrowed from that section, and are not redefined here for brevity. The recursive steps from that section are replicated below:

$$K_0'(\overline{x}, \overline{y}) = 1 \ \ \forall \overline{x}, \overline{y}$$
$$K_h'(\overline{x}, \overline{y}) = K_h(\overline{x}, \overline{y}) = 0 \quad \text{if either } \overline{x} \text{ or } \overline{y} \text{ has less than } h \text{ tokens}$$
$$K_h''(\overline{x} \oplus w, \overline{y} \oplus v) = \lambda K_h''(\overline{x} \oplus w, \overline{y}) + \lambda^2 K_{h-1}'(\overline{x}, \overline{y}) \cdot M(w, v) \ \ \forall h = 1, 2 \ldots k-1$$
$$K_h'(\overline{x} \oplus w, \overline{y}) = \lambda K_h'(\overline{x}, \overline{y}) + K_h''(\overline{x} \oplus w, \overline{y}) \ \ \forall h = 1, 2 \ldots k-1$$
$$K_k(\overline{x} \oplus w, \overline{y}) = K_k(\overline{x}, \overline{y}) + \sum_{j=2}^{l(\overline{y})} K_{k-1}'(\overline{x}, \overline{y}_1^{j-1}) \lambda^2 \cdot M(w, y_j)$$

Here, $l(\overline{y})$ denotes the length of $\overline{y}$. The description of Section 3.6.1.4 assumes that only the surface tokens in the two strings $\overline{x} = x_1 x_2 \ldots x_m$ and $\overline{y} = y_1 y_2 \ldots y_p$ are used, and therefore a binary match function $M(w, v)$ is assumed. The match function is defined to be 1 when w and v are the same; otherwise, it is 0. Here, the match function is modified, because the tokens w and v are associated with feature sets $F_x(w)$ and $F_y(v)$, respectively. Correspondingly, the *only difference* to the recursive approach of Section 3.6.1.4 is to use a feature-centric match function:

$$M(w, v) = |F_x(w) \cap F_y(v)| \tag{13.8}$$

Note that this match function specializes to the approach in Section 3.6.1.4 when a token is associated with only a single feature corresponding to its surface value.

13.3.3.3 Convolution Tree-Based Kernels

Tree-based kernels are able to encode complex grammatical relationships by comparing the parse trees of two relation instances. The main idea of this approach is that similar substructures in two parse trees are indicative of the fact that the concepts in the underlying sentences are similar. In the following discussion, it will be assumed that the parse-trees of individual sentences are already available using off-the-shelf linguistic preprocessing [368].

All kernel methods implicitly create a feature space under the covers. In the case of parse-trees, this feature space is defined by the subtrees of the parse-tree. However, only specific types of subtrees of the parse tree are used, which correspond to *all-or-none* subtrees.

Definition 13.3.1 (All-or-None Subtree) *An all-or-none subtree T_s of a given tree T is such that if a node i and its child in T are included in T_s, then all children of i in T must be included in T_s. In other words, either all children of i in T are included in T_s or no children are included.*

The reason that all-or-none subtrees are useful because of their semantic significance in the field of linguistics. Each node corresponds to a *grammar-production rule*, which is defined by a node and all its children. For example, one of the production rules of the noun phrase (NP) node of Figure 13.6(a) is as follows:

$$NP \rightarrow DT\ NN$$

While picking subtrees to define the feature space, it is important to preserve the underlying grammar production rule that creates the sentence for better semantic similarity. Therefore, if any children of a node are selected, then all of them are always selected in the all-or-none subtrees so that the grammar production rules of selected nodes (with children) remain undisturbed. An example of a parse tree and its valid subtrees are shown in Figure 13.6(a). Examples of some invalid subtrees are shown in Figure 13.6(b). Note that one of these subtrees corresponds to the following grammar production rule:

$$NP \rightarrow DT$$

Clearly, this grammar production rule is invalid, and it would be counter-productive to use it for kernel similarity computation. Therefore, it is not used.

The engineered representation uses the subtrees in order to create features. If the ith subtree occurs q times in the parse tree of a given sentence S, then the value of the ith dimension of the engineered representation is set to q. Note that this engineered representation will have a very large number of dimensions because of the exponential number of possibilities for subtree creation. It turns out to be far more efficient to directly compute the kernel similarity between pairs of trees rather than explicitly compute an engineered representation. As in the case of string kernels, one can use a recursive approach for similarity computation. For two trees T_1 and T_2, the kernel similarity is defined as follows:

$$K(T_1, T_2) = \Phi(T_1) \cdot \Phi(T_2) \tag{13.9}$$

Let $I_i(n, T)$ be an indicator function, that takes on the value of 1 when the ith subtree of the engineered representation occurs in T and is rooted at node n. Then, the ith dimension of the engineered representation (i.e., number of occurrences of the ith subtree) in T is given by summing this indicator function over all nodes in the tree:

$$\Phi_i(T) = \sum_{n \in T} I_i(n, T) \tag{13.10}$$

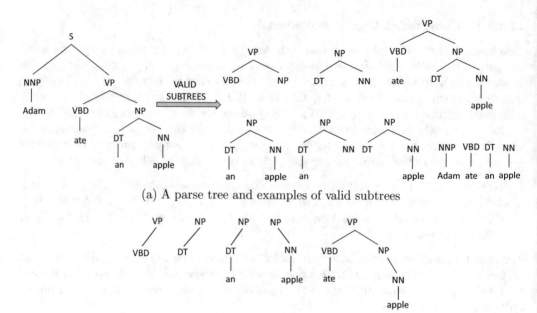

(a) A parse tree and examples of valid subtrees

(b) Examples of invalid subtrees

Figure 13.6: The parse tree and examples of valid and invalid subtrees

One can compute the dot product of Equation 13.9 by summing over the individual dimensions of the engineered representation $\Phi(\cdot)$ as follows:

$$K(T_1, T_2) = \sum_i \Phi_i(T_1) \cdot \Phi_i(T_2)$$

$$= \sum_i [\sum_{n_1 \in T_1} I_i(n_1, T_1)][\sum_{n_2 \in T_2} I_i(n_2, T_2)]$$

$$= \sum_i \sum_{n_1 \in T_1} \sum_{n_2 \in T_2} I_i(n_1, T_1) I_i(n_2, T_2)$$

$$= \sum_{n_1 \in T_1} \sum_{n_2 \in T_2} [\sum_i I_i(n_1, T_1) I_i(n_2, T_2)]$$

The key here is to be able to compute the expression in the square brackets above:

$$C(n_1, n_2) = \sum_i I_i(n_1, T_1) I_i(n_2, T_2) = \# \text{ Common subtrees rooted at both } n_1 \text{ and } n_2$$

One can view each node in a parse tree as a production rule of the grammar in which the parent is on the left-hand side of the rule and the children are on the right-hand side of the rule. Matching subtrees will have matching production rules as well because the corresponding parents and children have the same labels. Therefore, the value of $C(n_1, n_2)$ can be computed recursively as follows:

1. If the grammar productions at n_1 and n_2 are different, then set $C(n_1, n_2) = 0$.

2. If the grammar productions at n_1 and n_2 are the same, and n_1, n_2 are pre-terminals (i.e., nodes just above tokens), then set $C(n_1, n_2) = 1$.

3. If the grammar productions at n_1 and n_2 are the same, and n_1, n_2 are not pre-terminals, then we have:

$$C(n_1, n_2) = \prod_{j=1}^{nc(n_1)} (1 + C(ch(n_1, j), ch(n_2, j))) \tag{13.11}$$

Here, $nc(n_1)$ denotes the number of children of node n_1. This value is the same as $nc(n_2)$ because the production rules at n_1 and n_2 are the same. The notation $ch(n_1, j)$ denotes the jth child of node n_1.

This recursion can be shown to be correct by verifying the fact that each case of the recursion amounts to counting the number of common subtrees rooted at nodes n_1 and n_2. Furthermore, the running time is modest because it depends on the product of the number of nodes in the two trees T_1 and T_2. It is possible to enrich the nodes of the parse tree structure with additional features and further enrich the similarity function. Refer to the bibliographic notes for several variants of tree kernels, which have been used for entity extraction.

13.3.4 Relationship Extraction with Pretrained Language Models

In recent years, pretrained language models such as GPT and BERT have been increasingly used for relationship extraction. For example, a method discussed in [22] shows how one might use pretrained language models like GPT for relationship extraction. In order to perform the relationship extraction, it is assumed that the inputs contain a set of sentences, each of which is tagged with a position indicating the head of the entity, the position containing the tail of the entity, and the relationship type. First, the GPT model is pretrained in a manner that is not specific to entity extraction.

Subsequently, fine-tuning is performed by training additional layers for entity classification. It is assumed that the inputs correspond to entity pairs, their relationship type label, and the sentence in which the relation occurs. The representation of each sentence is extracted by passing it through the decoder used in GPT. Note that the representation $\bar{s}_i$ of the sentence corresponds to the hidden representation of the last state in the final layer of the transformer. All sentences that correspond to a particular pair of entities and a particular relation (which obviously have the same relation label l) are then collected in set S averaged in a weighted way:

$$\bar{s} = \sum_{i \in S} \alpha_i \bar{s}_i$$

The weight α_i of each sentence is proportional to the exponentiated similarity between between the sentence representation and a learned representation $\bar{r}$ of the relation:

$$\alpha_i \propto \exp(\bar{s}_i \cdot \bar{r})$$

The constant of proportionality is chosen so that the weights α_i sum to 1 (as is common in all attention mechanisms).

This representation $\bar{s}$ is then input into a feedforward network with a softmax output predicting the probabilities of labels:

$$P(\bar{s}) = \text{softmax}(W\bar{s})$$

The weight matrix W is trained based on the cross-entropy loss between the predicted relation label and the true label l for that set. At prediction time, one can also input representations of sentences in order to predict the relevant relation labels. An alternative approach that uses BERT for relationship extraction is discussed in [514].

13.4   Knowledge Graphs

The entities and relationships that are extracted from unstructured data are often organized into knowledge graphs, which organize the entities and the relationships among them into a graph structure. Knowledge graphs are closely related to the field of information networks studied by the data mining community. Information networks correspond to graphs of entities with various types of relationships among them. In many cases, the nodes in knowledge graphs are associated with a hierarchical taxonomy or a schema (ontology), although this requirement is not essential. A knowledge graph is defined as *any network structure of entities and different types of relations among them.* The nodes are often either instances of objects, or object types, which define the ontologies in the knowledge graph. Therefore, nodes are either *instance-nodes*, or they are *concept nodes*. The entity types of different objects in the graph (instance-type nodes) are usually specified with the use of hierarchical ontologies, when they are present.

We first motivate knowledge graphs with an example of entity-based search, which is where the use of knowledge graphs is most common. Search applications are particularly instructive, because many queries are associated with finding entities that are related to other known entities or for finding general information about specific entities. Consider the case where a user types in the search term "chicago bulls" in a purely content-centric search engine (without the use of knowledge graphs). This term corresponds to the well-known basketball team in Chicago, rather than a specific species of animal located in Chicago. If one did not use knowledge graphs, the search results would often not contain sufficiently relevant information to the basketball team. Even when relevant Web pages are returned, the context of the search results will fail to account for the fact that the results correspond to a specific type of entity. In other words, it is important to account for the specific context and relationships associated with entity-based search. On the other hand, if the reader tries this query on Google, it becomes evident that the proper entity is returned (together with useful information about entities related to the team), even when the search is not capitalized as a proper noun. Similarly, a search for "chinese restaurants near me" yields a list of Chinese restaurants near the GPS or network location of a user. This would often be hard to achieve with a content-centric search process. In all these cases, the search engine is able to identify entities with a specific name, or entities obeying specific relationships to other entities. In some cases, such as in the case of the search "Chicago Bulls," *infoboxes* are returned together with the results of the search. Infoboxes correspond to a tabular representation of the various properties of an entity. Infoboxes are returned with many types of named entities, such as people, places, and organizations. An example of an infobox for a search of the president "*John F. Kennedy*" was shown earlier in this chapter. The data contained in infoboxes is quite rich, and it includes data about different types of family relations, affiliations, dates, and so on. This information is encoded in Google as a knowledge graph. The nature of the presented information depends on the type of search, and the returned infoboxes may also vary, depending on the type of search at hand. Google creates these knowledge graphs by harvesting open sources of unstructured data; this is the task of information extraction because the nodes often corresponds to entities and the edges often correspond to relationships.

In the context of search, one often combines machine learning techniques with the structure of the graph in order to discover what other types of entities a user *might* be interested in. For example, if a user searches for the movie *Star Wars*, it often implies that they may be interested in particular types of actors starring in the movie (e.g., Ewan McGregor), or they may be interested in particular genres of movies (e.g., science fiction). In many cases, this type of additional information is returned together with the search results. The knowledge graph is, therefore, a rich representation that can be used in order to extract useful connections and relationships between the entities. This makes it particularly useful for answering search queries that are posed as relationship-centric questions, such as the following: "*Who played Obi Wan Kenobi in Star Wars?*" Then, by searching for the relations of the *Obi Wan Kenobi* and *Star Wars* entities, one can often infer that the entity *Ewan McGregor* is connected to these two entities with the appropriate types of edges. The key point in responding to such queries is to able to translate natural language queries into graph-structured queries. Therefore, a separate process is required to translate the unstructured language of free-form queries to the structured language of knowledge graphs. This is an issue that will be discussed later in this chapter. Such forms of search are also referred to as *question answering* systems.

A knowledge graph may be viewed as an advanced form of a knowledge base in which a network of relations among entities is represented, together with the hierarchical taxonomy of the entities, referred to as an ontology. An ontology is built on top of basic objects, which serve as the leaf nodes of the taxonomy and belong to specific *classes*. The classes correspond to types of objects. From the perspective of a knowledge graph, each object or class may be represented as a node. The attributes corresponds to the types of properties that objects and classes might have. The upper levels of the taxonomy correspond to concept nodes, whereas the lower levels of the taxonomy correspond to instance (entity) nodes. The *relations* correspond to the ways in which the classes or instances of classes are related to one another. The relations are represented by the edges in the knowledge graph and they may correspond to edges between nodes at any level of the taxonomy. However, it is most common for the relationship edges to occur among instance nodes, and a few may exist among concept nodes.

In knowledge graphs, the relations associated with concept nodes of the ontology are often hierarchical in nature, and form a tree or (more commonly) a directed acyclic graph. This is because the concept nodes in a hierarchy are often organized from the general to the specific. For example, a movie is a type of object, and an action movie is a type of movie. On the other hand, the instance nodes of the knowledge graph (i.e., leaf nodes of the ontology) might be connected together with an arbitrary topology, because they correspond to the arbitrary way in which *instances* of objects are related to one another. The concept nodes (or classes) in a knowledge graph often naturally correspond to the classes in the object-oriented programming paradigm. Therefore, they may inherit various types of properties from their parent nodes.

In order to understand the nature of knowledge graphs, we provide an example of a sample knowledge graph in Figure 13.7. The illustration of Figure 13.7(a) shows the hierarchy of the different object types in a movie database. Note that only a snapshot of the different object types is shown in Figure 13.7(a). Although Figure 13.7(a) shows a hierarchy of object types in the form of a tree structure, this hierarchy often exists as a directed acyclic graph (rather than a tree). Furthermore, different types of objects (e.g., movies and actors) would have different classifications with their own hierarchical structure, which suits their specific domain of discourse. The lowest level of leaf nodes in this hierarchy contains the instances of the objects. The instances of the objects are also connected by relations,

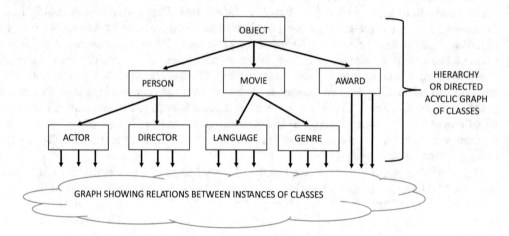

(a) The hierarchical class relations of a knowledge graph (concept nodes)

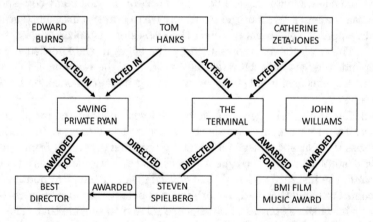

(b) A small snapshot of the knowledge graph between class instances (instance nodes)

Figure 13.7: A knowledge graph contains hierarchical class relations and the relationships among instances

Table 13.1: Examples of RDF triplets based on the knowledge graph of Figure 13.7

Subject	Predicate	Object
Edward Burns	Acted In	Saving Private Ryan
Tom Hanks	Acted In	Saving Private Ryan
Tom Hanks	Acted In	The Terminal
Catherine Zeta-Jones	Acted in	The Terminal
Steven Spielberg	Directed	Saving Private Ryan
Steven Spielberg	Directed	Saving Private Ryan
Steven Spielberg	Awarded	Best Director
Best Director	Awarded For	Saving Private Ryan
John Williams	Awarded	BMI Music Award
BMI Music Award	Awarded For	The Terminal

depending on how one node is related to another for that specific pair of instances. For example, person instances may be connected to movie instances using the *"directed in"* relation. The instance-level graph may have a completely arbitrary structure as compared to the concept-level graph, which is a directed acyclic graph. The instance-level relations are shown by the cloud at the bottom of Figure 13.7(a). The nodes shown in Figure 13.7(a) correspond to the concept nodes, and they correspond to a tree structure in this particular example. However, it is more common for such nodes to be arranged in the form of a directed acyclic graph (where a node may not have a unique parent). An expansion of this cloud of Figure 13.7(a) is shown in Figure 13.7(b). In this case, we focus on a small snapshot of two movies, corresponding to *Saving Private Ryan* and *The Terminal*. Both movies were directed by Steven Spielberg, and the actor Tom Hanks stars in both movies. Steven Spielberg won the award for best director in *Saving Private Ryan*. Furthermore, John Williams received the BMI music award for *The Terminal*. These relationships are shown in Figure 13.7(b), and they correspond to edges of different types, such as *"acted in," "directed,"* and so on. In a sense, each of these edges can be viewed as an assertion in a knowledge base, when one considers a knowledge graph as a specific example of a knowledge base. For example, the edge between a director and a movie could be interpreted as the assertion *"Steven Spielberg directed Saving Private Ryan."* In many knowledge graphs, the class portion of the ontology is captured by a *type* attribute. Each object type has its own set of instances, methods, and relations. In fact, each object instance node of Figure 13.7(b) can be connected with an edge labeled "type" to a class node of Figure 13.7(b). Furthermore, one can associate additional attributes with entities depending on their type. This additional information creates a rich representation that can be useful in a wide variety of applications.

In many cases, knowledge graphs are represented formally with the use of the *Resource Description Framework*, also referred to as *RDF*. Edges are often represented as triplets, containing the source (entity) of the edge, the destination (entity) of the edge, and the type of relation between the two entities. This triplet is sometimes referred to as *subject, predicate,* and *object*. Examples of triplets based on Figure 13.7 are shown in Table 13.1. Note that the subject is the source and the object is the destination of the directed link in the knowledge graph. It is occasionally possible for the edges in the knowledge graph to be associated with additional attributes. For example, if a relationship type between a person and place corresponds to an event, it is possible for the edge to be associated with the date of the event. Therefore, the edges can be associated with additional attributes, which can be represented by expanding the RDF triplets associated with relationships among entities. In general, knowledge graphs are quite rich, and they may contain a lot of information beyond what is represented in the form of RDF triplets.

It is possible to view a knowledge graph from the relational database perspective of RDF triplets, and borrow many database concepts to apply to knowledge graphs. A key point is that there are many different types of triplets corresponding to different relationship types and entity types. When using a knowledge graph, it is sometimes difficult for the end user to know what types of entities are connected with particular types of relationships. Relational databases are often described using *schema*, which provide information on which attributes are related to one another in various tables and how the tables are linked to one another via shared columns. In a knowledge graph, "tables" correspond to relationships (RDF triplets) of a particular type between particular types of entities, and one column may be shared by multiple tables. These "shared columns" correspond to nodes of a particular type that are incident on other nodes using various types of relationships. The relationship type is also one of the columns in the resulting table of triplets. For example, only a person-type can be a director (relationship type) of a movie type in the aforementioned knowledge graph. This information is important for understanding the structure of the knowledge graph. Therefore, knowledge graphs are often provided together with schema, which describe which types of entities link to one another with particular types of relationships. Not all knowledge graphs are provided together with schema. A schema is a desirable but not essential component of a knowledge graph. The main challenge in incorporating schemas is that databases require schemas to be set in stone up front, whereas incremental and collaboratively edited knowledge bases require a greater degree of flexibility in modifying and appending to existing schemas. This challenge was addressed in one of the earliest knowledge bases, *Freebase*, with the use of a novel graph-centric database design, referred to as *Graphd*. The idea was to allow the community contributor to modify existing schema based on the data that was added.

In some knowledge graphs, location and time are added for richness of representation. This is particularly important when the facts in the knowledge graph can change over time. Some facts in a knowledge graph do not change with time, whereas others do change with time. For example, even though the director of a movie is decided up front and remains fixed after the movie's release, the heads of all countries change over time, albeit at different time scales. Similarly, the locations of annual events (e.g., scientific conferences) might change from year to year. As a result, it is critical to allow knowledge graphs to be updateable with time, as new data comes in and entities/relationships are updated. Most graph-centric database design methods support this type of functionality. The actual information about the place and time stamp can be stored along with the corresponding RDF triplet.

In general, there is no single representation that is used to consistently represent all types of knowledge graphs. In most cases, they have a number of common characteristics:

- They always contain nodes representing entities.

- They always contain edges corresponding to relationships. These edges are represented by RDF triplets, which are *instances* of the objects. This portion of the knowledge graph is also referred to as the *ABox*.

- In most cases, hierarchical taxonomies and ontologies are associated with nodes. The ontological portion is sometimes referred to as the *TBox*. In such cases, schema may also be associated with the knowledge graph that set up the plan for the database tables of the RDF triplets.

Knowledge graphs can cover either a broad variety of domains, or they can be domain-specific. The former types of knowledge graphs are referred to as open-domain knowledge graphs, and often cover a wide variety of entities searchable over the Web (and can therefore be used in Web search). Such knowledge graphs can also be used in open-domain question answering systems like Watson. Examples of open-domain knowledge graphs include Freebase, Wikidata, and YAGO. On the other hand, a domain-specific knowledge graph like WordNet or Gene Ontology will cover the entities specific to a particular domain (like English words or gene information). Domain-specific knowledge graphs have also been constructed in various commercial settings, such as the Netflix knowledge graph, the Amazon product graph, as well as various types of tourism-centric knowledge graphs. Such knowledge graphs tend to be useful in somewhat narrower applications like protein search or movie search. In each case, the entities are chosen based on the application domain at hand. For example, an Amazon product graph might contain entities corresponding to products, manufacturers, brand names, book authors, and so on. The relationships among these different aspects of products can be very useful for product search as well as for performing customer recommendations. In fact, a product graph can be viewed as an enriched representation of the content of products, and can be used for designing content-based algorithms (which are inherently inductive learning algorithms). Similarly, a Netflix product graph will contain entities corresponding to movies, actors, directors, and so on. A tourism-centric knowledge graph will contain entities corresponding to cities, historical sites, museums, and so on. Another example is the Gene Ontology knowledge graph, which contains information on the functions of genes [28]. In each case, the rich connections in the knowledge graph can be leveraged to perform domain-specific inferences. In the following, we will provide some examples of real-world knowledge graphs in various types of settings.

Perhaps the most widely used knowledge graph is Wikidata, which is closely related to the online encyclopedia, referred to as Wikipedia. Wikidata is a collaboratively hosted knowledge base by the Wikimedia foundation, and its data is used by other Wikimedia projects such as Wikipedia and Wiki Commons. Wikipedia represents knowledge in unstructured format, whereas Wiki Commons is a repository of media objects, such as images, sounds, and video. Wikidata provides the facility to users to query the knowledge base with the use of the SPARQL query language (which it has popularized). Wikidata is collaboratively edited, and it provides users with the ability to add new facts to the knowledge base.

The SPARQL query language (pronounced as *sparkle*) is a database query language that is well suited to graph databases. SPARQL is a recursive acronym for $\underline{S}$PARQL $\underline{P}$rotocol $\underline{a}$nd $\underline{Q}$uery $\underline{L}$anguage. The language has a structure and syntax that is very similar to the *structured query language* that is used in traditional multidimensional databases. However, it also has the functionality to support the querying and editing of graph databases. SPARQL has now become the de facto standard in working with knowledge graphs in general and RDF representations in particular.

Wikidata is one of the largest and most comprehensive sources of openly available knowledge bases today. Wikidata, Wikipedia, and Wiki Commons are closely related, and the data in one of these repositories is often used to augment data in the other repositories. Among these, only Wikidata can be considered a comprehensive knowledge graph containing all types of objects. The data from Wikidata is used in more than half of all Wikipedia's articles. Since many knowledge bases rely on data crawled from Wikipedia, it is clear that much of the data in a variety of knowledge bases is inherited by Wikidata. In this sense, the source of most of the existing knowledge bases comes in one form or another through collaboratively edited information, even though a knowledge base might be constructed through the use of semi-structured and unstructured data processing. This is not particularly surprising, given that the Web is itself a collaboratively edited endeavor at the most basic level.

Table 13.2: Methodologies for knowledge graph construction, as presented in [417]

Construction Method	Schema	Examples
Curated	Yes	Cyc/OpenCyc [319], WordNet [396], UMLS [65]
Collaborative	Yes	Wikidata [564], Freebase [197]
Auto. Semi-Structured	Yes	YAGO [532], DBPedia [702], Freebase [197]
Auto. Unstructured	Yes	Knowledge Vault [160], NELL [86], PATTY [408], PROSPERA [409], DeepDive/Elementary [421]
Auto. Unstructured	No	ReVerb [698], OLLIE [699], PRISMATIC [187]

13.4.1   Constructing a Knowledge Graph

A knowledge graph is a very rich and structured representation of the entities in the real world, and it needs to be explicitly curated or constructed from semi-structured to highly unstructured data. The way in which the knowledge graph is constructed depends on the sources from which the data is collected in order to construct the graph. The methodology of construction depends on the source from which the raw data is obtained. For open-source knowledge graphs like Wikidata, the effort is largely collaborative. On the other hand, knowledge graphs like WordNet are created via the process of curation by experts. A table of the different ways in which knowledge graphs are constructed is provided in [417]. We provide this information in Table 13.2.

Based on Table 13.2, we list the four main ways in which knowledge graphs are created:

1. In *curated methods*, the knowledge graph is created by a small group of experts. In other words, the group of people contributing is closed, and it is restricted to a small set of people. Presumably, the restriction to a small set of people, ensures that the recruited people are experts, and the resulting knowledge graph is of high quality. The main problem with this approach is that it does not scale particularly well to large knowledge bases. However, such an approach is particularly effective in specialized domains, where a standardized knowledge graph of high quality is required.

2. In *collaborative methods*, the methodology for knowledge graph construction is similar to that of curated methods, except that the construction of the knowledge graph is done by an open group of volunteers. Even though the knowledge graph is open to contribution from the "wisdom of the crowds," there may still be some partial controls on who might contribute. This is done in order to avoid the effects of spam or other undesirable characteristics associated with open platforms. The collaborative approach does scale better than the curated approach, but the graph may sometimes contain errors or inconsistencies. As a result, confidence or trust values are often associated with the relationships in the constructed knowledge graphs.

3. In automated semi-structured approaches, the edges of the knowledge graph are extracted automatically from semi-structured text. This extraction can take on a variety of forms, such as domain-specific rules and machine learning methods. An example of this type of semi-structured data are the infoboxes on Wikipedia (which are themselves crowdsourced, albeit not in knowledge graph form).

4. In automated unstructured approaches, the edges are extracted automatically from unstructured text via machine learning and natural language processing techniques. Examples of such machine learning techniques include entity and relation extraction.

This broader area is referred to as *information extraction* in natural language processing. Note that there are two separate entries in Table 13.2, depending on whether or not a database schema exists in conjunction with the knowledge graph.

The aforementioned list of methods is not exhaustive. In many cases, the data from diverse sources need to be combined in order to create the knowledge graph, or some of the above methods may be used in combination. In some cases, the knowledge graph may be constructed via a combination of curation and collaborative effort. Similarly, even though a knowledge graph of movies can be constructed from the data corresponding to the movies, the data for various movies may come from diverse sources, such as relational data or unstructured data. This is because content from large producers may be available as relational data, whereas data about small home productions may need to be scraped from unstructured sources. In such cases, the knowledge graph needs to be meticulously constructed from the data obtained from diverse sources.

There are numerous special challenges when knowledge graphs need to be created from products for recommendation and related applications. This is because such graphs often cannot be created from open-source information, but one needs to rely on multiple retailers who might provide this information in a variety of different formats. Similarly, maintaining the freshness of the knowledge graph may be a challenge, as the information continually evolves with time. As a result, data integration and dynamic updates are critical in knowledge graphs. Therefore, most knowledge graphs are supported with graph databases that have the capabilities to perform these types of dynamic updates.

First-Order Logic to Knowledge Graphs

The above different ways of constructing knowledge graphs provide an understanding of the role that domain-specific rules may play in knowledge graph construction. These rules could have been extracted from a traditional knowledge base. Traditional knowledge bases contain rules in the form of *first-order logic*. An example of such a rule could be the following [529]:

$$\forall\, x, y\ [Married(x, y) \Rightarrow SameLocation(x, y)]$$

One could use this type of rule to rapidly populate edges in the knowledge base by repeatedly identifying pairs of nodes with *"married to"* relations between them, and then inserting the edge *"lives in same location"* between them. The reverse process of extracting rules with the use of machine learning methods on the graph is also possible. For example, if one identifies that a knowledge graph contains a *"lives in same location"* edge a vast majority of the times that a *"married to"* relation exists, the above rule can be extracted using association mining methods [2]. This can be achieved by creating a list of relations for each pair of entities, and then finding frequent patterns from these sets of entities. These patterns can be used to create rules using association mining methods discussed in [2]. Although the rules may not be absolute truths, a domain expert may often be used to decide which rules make sense from a semantic point of view. Subsequently, the extracted rules can be used to populate additional edges in the graph.

Extraction from Unstructured Data

Among the aforementioned methods, the extraction from unstructured data is the most interesting case, because the construction of the knowledge graph is itself a machine learning task. In the case of unstructured data extraction, even the entities may not be directly available, and they may need to be identified from unstructured data. Both the methods of named entity recognition and relationship extraction turn out to be very useful for building the knowledge graph as follows:

1. *Named entity recognition:* This process helps identify the persons, places, or other entities used to populate the nodes of the knowledge graph. For example, consider the following:

   > Bill Clinton lives in New York at a location that is a few miles away from an IBM building. Bill Clinton and his wife, Hillary Clinton, relocated to New York after his presidency.

   After performing the named entity recognition, one can create nodes for entities like *"New York"*, *"Bill Clinton"*, and *"IBM"*.

2. *Relation extraction:* Once the entities have been extracted, the relations among them need to be extracted in order to create the edges in the knowledge graph. These relations are used in order to create the edges in the knowledge graph. Examples of relationships may be as follows:

   > **LocatedIn**(*Bill Clinton, New York*)
   > **WifeOf**(*Bill Clinton, Hillary Clinton*)

   The nature of the types of relations to be extracted will depend on the type of knowledge graph that one is trying to construct.

There are many different settings in which information extraction systems are used. An *open* information extraction task is unsupervised and has no idea about the types of entities to be mined up front. Furthermore, weakly supervised methods either expand a small set of initial relations, or they uses other knowledge bases from external sources in order to learn the relations in a corpus. Although such methods have recently been proposed in the literature, it is more common to use supervised methods. In this view, it is assumed that the types of entities and the relationships among them to be learned are *pre-defined*, and tagged training data (i.e., text segments) are available containing examples of such entities and/or relationships. Therefore, in named entity extraction, tagged examples of persons, locations, and organizations may be provided in the training data. In relationship extraction, examples of specific relationships to be mined may be provided along with the free text. Subsequently, the entities and relations are extracted from untagged text with the use of models learned on the training data. As a result, many of the important information extraction methods are supervised in nature, since they learn about specific types of entities and relationships from previous examples. Most of this chapter has discussed supervised methods for information extraction.

13.4.2   Knowledge Graphs in Search

Knowledge graphs are used in a wide variety of search applications. Indeed, the term "knowledge graph" was coined by Google in the context of search, although the broader idea was explored in many fields such as the Internet of Things, the Semantic Web, and artificial intelligence. Like the Google knowledge graph, Microsoft uses a knowledge base called Satori with its search engine. Furthermore, many domain-specific search applications use product graphs, which are specialized forms of knowledge graphs. In these cases, the search applications target other products or entities that are related to known products.

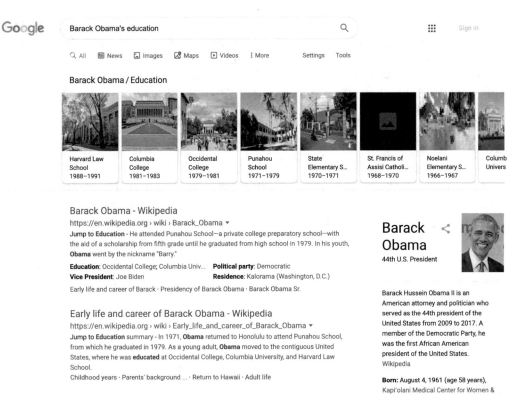

Figure 13.8: An example of the results yielded by the Google search query *"Barack Obama's education"*

There are several ways in which search applications can be used in the context of knowledge graphs. For example, the search query *"Barack Obama's education"* yields a chronological list of the educational institutions attended by Obama, together with their images (cf. Figure 13.8). This type of response is hard to achieve with purely content-based search. Presumably, a knowledge graph was used to return the institutional entities to which Obama is connected via a link indicating affiliation.

The real issue in resolving such queries is in understanding the semantics of the query at hand. For example, in the case of the query on Obama's education, one needs to be able to infer that a portion of the query string, "Obama" refers to an entity, and the remaining portion refers to a relationship of the entity. This is often the most difficult part of resolving such queries. In many cases, complex natural language queries corresponding to complete sentences are used in order to query knowledge graphs. For example, consider the following search query: *"Find all actor-movie pairs have received an award for a movie directed by Steven Spielberg."* In such a case, the challenge is even greater in being able to provide responses to such queries. This problem can often be posed into a learning problem, where natural language queries are transformed into more structured queries that can be mechanically applied to the knowledge graph. One can view this problem to be somewhat similar to machine translation, in which a sentence in one language is transformed into a sentence of another. Note that machine translation methods are discussed in Section 10.8.3 of Chapter 10.

What type of structured query language is appropriate for a knowledge graph? The key language for querying RDF-based knowledge graphs is SPARQL. The SPARQL language is similar in syntax to the SQL querying language, except that it is designed for RDF databases rather than relational databases. Like SQL, it contains commands like SELECT and WHERE in order to create a clear syntax for what needs to be returned. Like any programming language, it can be easily understood by a parser and compiler in a non-ambiguous and unique way (unlike the case of natural language, which is always more challenging to understand). Many publicly available ontologies such as YAGO are tightly integrated with SPARQL-based query systems, and therefore it is relatively easy to build search functionalities into such knowledge bases.

In practice, one often wants to use natural language queries rather than SPARQL queries. Therefore, a natural step is to transform a natural language query into a SPARQL query. This can be performed by constructing a machine translation model between natural language queries and SPARQL queries. The training data for such a model can first be constructed manually (or via ad hoc translation methods) in order to handle the cold start problem in learning. For example, manually constructed rules by experts can be used to create candidate queries, which can be further curated (manually) by the human experts. The resulting pairs of natural language queries and SPARQL queries can be used to train a machine learning model, such as a sequence-to-sequence autoencoder (cf. Section 10.8.3). Subsequently, implicit feedback from user clicks on search engine queries can be used to construct further training data from the outputs of this machine learning model. For example, for a natural language query (and correspondingly translated SPARQL query), if the user clicks on a particular search result, it is positive feedback for the SPARQL query generated to create that search result. This positive feedback can be used to generate further training data for the sequence-to-sequence learning algorithm. Some discussions on translating between natural language queries and SPARQL are provided in [164, 294, 568, 612]. Note that some of these techniques [294, 568] use traditional machine translation methods such as parse trees. However, if sufficient data are available, it makes more sense to construct a trained machine translation system that can provide more accurate results.

Training between pairs of query representations is, however, not the only approach used for responding to search queries in knowledge graphs. In many cases, one can directly train between pairs of questions and the subgraphs of the knowledge graph that represent answers to these questions. However, such an approach requires the learning system to have access to the knowledge graph in the first place, via a machine learning representation, such as a *memory network*. Examples of such methods are discussed in [67, 68].

13.5 Summary

The problem of entity and relationship extraction lies on the interface of information retrieval, text mining, and natural language processing. In named entity recognition, one is trying to extract names, places, and organizations, although other types of entities are also possible. The earliest techniques used rule-based methods for named entity recognition. More recently, machine learning techniques like hidden Markov models, maximum entropy Markov models, and conditional random fields have become increasingly popular. A second task is that of relationship extraction in which machine learning techniques are very popular. Machine learning techniques either use explicit feature engineering or use kernel methods. Many of the kernel techniques used for relationship extraction have dual use in natural language processing.

13.6 Bibliographic Notes

The earliest methods for structured information extraction from text date back to the FRUMP program [146]. However, the research during the seventies and eighties on information extraction was sporadic and limited. An important source of advancement in the field was the group of *Message Understanding Conferences (MUC)* that started in the nineties. Among these, MUC-6 [221] is particularly notable for introducing the key information extraction subtasks as they are known today. Many challenges have since been held in the problems of named entity recognition and relationship extraction. For named entity recognition, the key challenges include the Automatic Content Extraction (ACE) program [695], several tasks defined by the Conference on Natural Language Learning (CoNLL) [697], and the BioCreAtIvE challenge evaluation [696]. The last of these is specifically focused on the biological domain. Excellent surveys on information extraction may be found in [268, 496]. It is noteworthy that this chapter focuses on information extraction from free text, although there exist various information extraction methods for semi-structured data, such as HTML and XML. Such procedures are referred to as *wrappers*. This chapter is focused on information extraction methods from free text and does not discuss such specialized methods for the Web domain. Interested readers may refer to [345, 496] for excellent discussions of these methods.

Rule-based methods for named entity recognition are either top-down methods like WHISK [521], or they are bottom-up [85, 117]. Among machine learning methods, the early methods used only a limited amount of sequential information [153, 541]. There is a significant similarity between the models used for parts-of-speech tagging and named entity recognition. For example, hidden Markov models [304, 456] were used for parts-of-speech tagging in [306]. Subsequently, hidden Markov models were used [56, 198, 510] for entity extraction. Maximum entropy Markov models were first introduced for part-of-speech tagging in [463] and then used for named entity recognition in [372]. Other popular models in this family are discussed in [51, 140]. Various learning algorithms and their comparisons for maximum entropy Markov models are discussed in [364].

The use of conditional random fields for information extraction was first discussed in [308]. A segment-wise model for information extraction with conditional random fields was proposed in [497]. Both the maximum entropy models and the conditional random field models were originally proposed with dual applications in entity extraction and text segmentation. This is not particularly surprising, given the similarity in the underlying problem formulation. A detailed discussion of conditional random fields may be found in [538].

The use of feature-based classification for relation learning is explored in [99, 269, 285]. The work in [269] is particularly notable because it explores systematic ways of feature engineering for relationship extraction. A second method for relationship extraction is with the use of kernel methods. Shortest-path and subsequence kernels are discussed in [78, 79]. Tree-based kernels are discussed in [124, 139, 452, 616]. In the context, the works in [452, 624] are notable because they specifically explore the domain of relationship extraction while constructing convolution tree kernels. Methods for combining structured and sequence kernels are referred to as composite kernels, and are discussed in [625, 627].

13.6.1 Weakly Supervised Learning Methods

The main challenge in problems like relationship extraction is that the amount of (labeled) training data required is very large. In such cases, one can reduce the requirement on the amount of training data by using methods like *bootstrapping* and *distant supervision*. In bootstrapping [18, 73], one starts with a small set of seed relations, and iteratively labels nearby entity pairs using this training data. This newly labeled data is used to learn relevant aspects of the context and label other records. This type of training data is usually noisy, and therefore it is important to have the ability to filter out the noisy patterns and learn the relevant features.

A second method is that of *distant supervision*, in which large knowledge bases are available in which known target relations are available. Such knowledge bases are created using crowd-sourced efforts from users. Such type of data can be used to training in settings where the data is drawn from different sources. Such methods are discussed in [398, 416].

13.6.2   Unsupervised and Open Information Extraction

Most of this chapter focuses on supervised information extraction in which a significant amount of effort needs to be spent in both defining the appropriate entity and relation structures, as well as in labeling training data. In unsupervised information extraction, the goal is to group similar entries and entity pairs based on their syntactic, lexical, or contextual similarities. In open extraction, the goals are even more fundamental, in which one tries to extract various types of relations from a large corpus like the Web.

The problem of unsupervised entity extraction from the Web is discussed in [180]. A clustering-based approach for unrestricted relationship discovery is discussed in [515], in which the important relations from a corpus are discovered. A later work explicitly discussed the problem of grouping entity pairs into clusters [481].

In *open information extraction*, relations are extracted from a large and open corpus like the Web [40, 182]. In such cases, target relations are not pre-specified, and one learns new relations along with a phrase to describe them. This type of approach is useful when one does not want to have the restriction of a pre-defined set of relations. Open relationship extraction is closely related to *distant supervision*, in which a corpus from a different domain is used [466].

13.6.3   Software Resources

Information extraction is a field that lies on the interface of machine learning and natural language processing. As a result, many applications in information extraction require the use of natural language processing tools. Apache OpenNLP [644] supports many information extraction tasks, such as sentence segmentation, named entity extraction, chunking, and coreference resolution. Many of the underlying preprocessing tasks and natural language processing tasks are also supported by Apache OpenNLP. Other open source repositories that support these tasks include NLTK [652] and Stanford NLP [650]. Stanford NLP also specifically supports open information extraction [700]. The **MALLET** toolkit supports many of the Markov models for sequence tagging and entity recognition [701]. The **Re-Verb/Ollie** packages provide open information extraction capabilities for extracting binary relations from English sentences [698, 699]. The **DBpedia Spotlight** [702] provides tools for named entity recognition and annotation. A conditional random field implementation from Sunita Sarawagi's group is available at [703]. Information extraction capabilities are also provided by **OpenCalais**, which is a software provided by the company *ClearForest* [704].

13.7 Exercises

1. Although this chapter does not discuss deep learning methods in detail, these techniques can be used for entity extraction. Which deep learning model discussed in Chapter 10 can be used for entity extraction? Pick out the specific model discussed in Chapter 10, and how the inputs and outputs of the neural network would be defined.

2. Show how you can extend the methodology discussed in Exercise 1 to relationship extraction.

3. Suppose you have a corpus in which words and phrases indicating positive/negative sentiment are tagged. You want to tag corresponding words and phrases with various sentiments in an unmarked corpus. Discuss the relationship of this problem to entity extraction.

4. The Nymble approach uses the features of the tokens in a relatively rudimentary way. Discuss an approach based on HMMs in which each state of the model can output a multidimensional feature vector.

5. Show that the running time of the convolutional kernel is at most proportional to the product of the number of nodes in the two trees between which the kernel similarity is being computed.

6. Implement the string kernel discussed in this chapter.

Chapter 14

Question Answering

"Every answer begins with a question." – T.A. Uner

14.1 Introduction

Question answering has become one of the most popular applications in natural language processing in recent years because of an inherent need to forage for information from massive sources of unstructured data. Interestingly, a large percentage of searches on the Web are now formulated as natural language questions, and the responses from search engine often go beyond just returning relevant documents. Rather, the relevant portions of the best-matching document may often be highlighted to emphasize the answer to that specific query. An example of the Google search query, *"When did Einstein win the Nobel Prize?"* is shown in Figure 14.1. It is evident that the response is not just a document containing the data, but it provides the specific year, as well as the context distinguishing the fact that the official date of the Nobel Prize was different from the date on which it was received.

One of the reasons why question answering is considered a fundamental problem in natural language processing is that it is one of the key ways of evaluating whether a machine

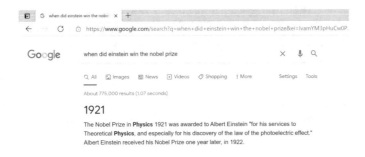

Figure 14.1: An example of a Google search query

© Springer Nature Switzerland AG 2022
C. C. Aggarwal, *Machine Learning for Text*,
https://doi.org/10.1007/978-3-030-96623-2_14

truly understands a piece of natural language. This is also true of humans, wherein reading comprehension questions are used to test language skills in all types of standardized tests. Reading comprehension is a simpler version of question answering, in which a small passage (or set of passages) is initially provided, and the answer is extracted from the provided text. The main characteristic of the reading comprehension task is that text used to extract the answer is usually quite limited. In fact, the question answering task can be viewed as an extension of the reading comprehension task, in which the text from which the answer is extracted is a large document collection (or, in some cases, not even provided, so that it becomes the responsibility of the analyst to collect the source documents up front).

Question answering may also be viewed as an enhanced form of search in which a certain level of semantic comprehension of the source documents is required — in other words, it integrates the reading comprehension task with that of information retrieval. In fact, a vast number of Google searches are in the form of questions, and the trend towards question-like searches has increased in recent years because of the ability of search engines to respond intelligently to such queries. Such intelligent responses often show a high level of semantic understanding of the question being asked, and the ideas behind providing such responses are rooted in question-answering techniques. Unlike traditional information retrieval in which the search is based on keyword matching and page importance (based on metrics like *PageRank*), the process of answering questions requires a certain level of semantic *understanding* of the question in terms of how it relates to the *evidence blocks* retrieved by the query. Therefore, question answering often boils down to a combination of accurate retrieval of evidence blocks and reading comprehension with respect to these blocks. The different ways in which search is distinguished from question answering are as follows:

1. In search, it suffices to return the documents that are relevant to a specific query. However, in question answering, one is looking for specific responses to queries rather than documents, and the nature of the queries also make the specificity explicit. For example, when one is searching for the query "*Olympics,*" returning documents corresponding to various types of Olympic games may be useful. On the other hand, consider the following question: "*Who won the 100 meters Olympics gold medal in 2016?*" In this case, the intent of the person formulating the query is clear, and it would be a waste of their time to force them to sift through the returned documents for the actual answer — indeed Google searches that are formulated as questions often highlight relevant portions of documents. In fact, the discovery of relevant facts from passages (i.e., the reading comprehension problem) is an important part of question answering. The reading comprehension problem is itself posed as an important subproblem in the field of question answering, where the specific passage of text in which the answer exists is provided up front, and the specific goal of question answering is to ferret out the nugget of information relevant to the question at hand. More general forms of question answering do not assume that such passages are available up front, but they first extract such passages from large collections (via information retrieval) and then apply reading comprehension techniques on them. The challenge is, therefore, to integrate these two complementary aspects of question answering.

2. Traditional keyword search requires only a limited amount of understanding of the intent of the question, whereas question answering requires a high level of understanding of intent.

3. Traditional keyword search places little attention to the entities and the relationships among them. On the other hand, question answering is often about entities and their

relationships, especially when dealing with questions like *"who"*, *"what,"* and *"when."* Many questions are really about filling of one or more of the three pieces of information in an edge (or RDF triplet) in a knowledge graph. As a consequence, knowledge graph representations have emerged as an important source of data for question answering.

It is noteworthy that this chapter is primarily focused on *factoid question answering*, in which the answer is typically a fact, which can be identified as a short span of text within the base collection. For example, asking for the capital of a country is a *factoid* setting, whereas asking for the details of how one might use a particular product like a smartphone is not a factoid setting. This type of question is referred to as a *long-form question*. Another example of a long-form question is as follows:

What are the similarities and differences between whales and fishes?

This type of question is much more challenging for automated machine learning because one must extract the relevant pieces of text from different documents and then summarize them in a coherent way. Therefore, long-form question answering integrates aspects of text retrieval, reading comprehension, and abstractive text summarization, all of which are challenging tasks. As a result, the progress in long-form question answering has been quite limited. Methods used for factoid question answering have the potential to be generalized to more complex settings (which largely remain unexplored in the research literature on question answering). This chapter is, therefore, primarily devoted to factoid question answering, with some limited discussions on long-form question answering.

There are several variations of the (factoid) question-answering task, depending on the trade-off it explores between document search and reading comprehension. The most difficult tasks, referred to as *open-domain question answering tasks* mingle document search and reading comprehension in a seamless way. The different variations of the question-answering tasks are as follows:

1. *Reading comprehension with a single document:* In this case, a single document is provided containing one or more paragraphs of text, and a question is posed, whose answer is available in contained in this paragraph. The task of the question-answering system is, therefore, to *comprehend* both the provided question and the paragraph in order to ferret out the information relevant to the question at hand. The reading comprehension problem is an important subproblem in the domain of question answering, which is used in more complex settings.

2. *Question answering with pre-defined document collections or subjects:* In this case, a pre-defined database of documents is provided up front, and the answers are extracted from this database. Therefore, this scenario is significantly more difficult than the first case, because one has to search both for the relevant document(s) and the information within these document(s) that is relevant to the question at hand. In most cases, such systems perform question answering based on a restricted subject. For example, a question-answering system on biomedical literature would be based only on documents related to this subject. Such systems are referred to as closed-domain systems.

3. *Open-domain question answering:* In this case, the questions may be on any subject, and the system often answers questions based on a large crawl of the Web, online encyclopedias, or publicly available knowledge graphs like *Wikidata*. The IBM Watson system for play Jeopardy! can be considered an open-domain question answering system. The main difference is that open-domain systems put a larger burden for

document collection on the analyst, whereas closed-domain systems often have standardized collections. However, the methodology for question answering does not vary too much between closed-domain and open-domain systems, once the base document collection has been defined. This is because both systems assume availability of document collections during query time. This availability makes both systems *open book* systems.

4. *Closed-book question answering:* In this case, the base document collection is stored up front within the parameters of a neural network model in much the same way as a human would store knowledge in the parameters of their neural network. In other words, access to an externally stored document collection is disallowed during question answering. Pre-trained language models are commonly used for closed-book question answering.

The technical difference between open-domain and closed-domain question answering is quite small as long as a base document collection is used for answering a question. In both cases, one must retrieve segments of text, referred to as *evidence blocks*, from the base collection and then zero in on the answer using these evidence blocks. However, the difference between open book and closed book systems is much greater.

It is noteworthy that other than the reading comprehension problem, all other question answering systems not only require one to comprehend the question, but also to search for the relevant document containing the answer. This tight integration of information retrieval and reading comprehension is what makes the question-answering task challenging. Many question-answering systems either integrate the information in multiple documents with knowledge graph representations or decouple the search and reading comprehension processes to some extent. In general, methods that use tight integration of search and comprehension always work better than those that do not use tight integration. This is because the errors in multiple components often feed on one another in unexpected ways, leading to noisy results. This type of deterioration of error caused by pipelined components was one of the key problems faced by early question-answering systems, which were very large and had multiple components making sequential predictions.

In order to address the problem of rapidly multiplying errors of pipelined systems, the early systems heavily relied on increasing robustness by combining the predictions of multiple systems, since each individual system was often quite inaccurate. This inaccuracy was often the result of the inherent limitations of machine comprehension and natural language understanding, which were not quite as well developed at the time as they are today. Therefore, these systems heavily relied in information redundancy within multiple sources and querying methods, multiple occurrences of candidate answers were used as evidence of accuracy and subsequently used to construct the answer using hand-crafted methods. In recent years, advances in natural language processing (based on deep learning methods) have brought us much closer to designing end-to-end systems without too many pipelined parts. Many of these systems extend a basic reading comprehension system to perform simultaneous retrieval and reading comprehension.

Although the basic reading comprehension task is overly simplified from an application-centric perspective, it is a critical one because it is focused on the issue of semantic understanding of natural language. Developing this basic ability has the potential to generalize to more complex scenarios.

Chapter Organization

This chapter is organized as follows. The next section introduces the problem of reading comprehension along with some sequence-centric models for accomplishing this task. The design of retrieval methods for open-domain question answering is discussed in Section 14.3. The direct use of pretrained language models for question answering is discussed in Section 14.4. The use of knowledge graphs for question answering is introduced in Section 14.5. A brief discussion of long-form question answering is provided in Section 14.6. A summary is given in Section 14.7.

14.2　The Reading Comprehension Task

The most basic form of question answering is the reading comprehension task, in which we have a single document or paragraph, and we wish to answer questions related to this single document. Therefore, this task is purely one of reading comprehension in which one needs to extract information related to the specific question from the passage at hand. The reading comprehension task is considered fundamental in the field of question answering, because this module is often reused in more general forms of question answering. Another reason that reading comprehension is considered so fundamental for machine understanding is that it is also the ideal way to test human understanding of reading skills — almost all standardized tests for humans in languages use some form of reading comprehension. However, since machine understanding is not as well developed as human understanding, the types of questions (e.g., factoid questions) used in machine reading comprehension are much simpler than what one would encounter in a human reading comprehension task.

The reading comprehension task was explored more extensively in recent years, as an increased number of data sets became available for the task. A discussion of several data sets that have been used in recent years for reading comprehension tasks is given in Section 14.8.1 of the bibliographic notes. A famous data set in this regard is the *Stanford Question Answering Data Set (SQuAD)* [461, 462], which contains a list of reading passages together with associated questions. In order to illustrate this point, we provide an example passage from the SQuAD data set:

> The Normans (Norman: Nourmands; French: Normands; Latin: Normanni) were the people who in the 10th and 11th centuries gave their name to Normandy, a region in France. They were descended from Norse ("Norman" comes from "Norseman") raiders and pirates from Denmark, Iceland and Norway who, under their leader Rollo, agreed to swear fealty to King Charles III of West Francia. Through generations of assimilation and mixing with the native Frankish and Roman-Gaulish populations, their descendants would gradually merge with the Carolingian-based cultures of West Francia. The distinct cultural and ethnic identity of the Normans emerged initially in the first half of the 10th century, and it continued to evolve over the succeeding centuries.

Based on this paragraph, a number of questions are asked, and the answers to all of them are subsequences of the original paragraph. This is a specific property of this data set in which the answers are constrained to be a *span* in the paragraph, and the corresponding system is also referred to as *extractive* question answering. A sample of the questions (and one of the three possible ground-truth answers) provided by the SQuAD data set are as follows:

- In what country is Normandy located? [Answer: France]

- When were the Normans in Normandy? [Answer: In the 10th and 11th centuries]

- From which countries did the Norse originate? [Answer: Denmark, Iceland and Norway]

- Who gave their name to Normandy in the 1000's and 1100's? [Answer: ⟨No Answer⟩]

The SQuAD data set allows for three candidate answers, which were usually minor rephrasings of the same answer, but extracted from different positions in the passage. It is noteworthy that the last question in the above list does not have an answer within the passage. Although the question does have a valid answer (when using information sources outside the passage), a reading comprehension system is not expected to provide an answer since it is not contained within the passage at hand. Therefore, the correct answer is to state that no answer to this question exists. Since the reading comprehension system is expected to provide answers based on only the selected passage, the correct answer to the question is the *"No Answer"* response. The inclusion of questions without answers was a feature of SQuAD 2.0 [462], which included questions without answers within the reading comprehension passage. The original version of SQuAD (SQuAD 1.1 [461]) only contained questions with answers. Including questions without answers tests the robustness of the reading comprehension system and challenges it to demonstrate that it knows what it does not know. The ability to detect the situation when the answer is *not* contained in the data set is an important part of question answering, because it forces the system to have a robust measure evaluating the accuracy of its response to a particular question. When the best answer is not deemed to be accurate enough, it is assumed that the answer to the question is not contained in the passage.

The original articles in the data set were derived from Wikipedia, and human participants on Amazon Mechanical Turk were used both to construct the questions and provide ground-truth answers. Three candidate answers were available for each question, and providing a prediction that matched the correct answer was scored as correct. A second approach for evaluation was that the ground-truth answers were used as a bag of words, with respect to which the precision and recall of the provided answer could be calculated. The harmonic mean of the precision and recall provided the F1-score, which was reported as the evaluation measure.

It is noteworthy that there are several criticisms of the SQuAD data set, in terms of how it has been constructed. The fact that the answer is always extracted by using a contiguous span from the passage can be problematic because the answers in reading comprehension could be phrased in a slightly different way, as a result of which they may not be drawn from the original data set. Secondly, the questions in SQuAD were constructed after reading the article, which often caused biases towards question wordings that were directly drawn from the data set. Finally, the SQuAD data set could not test reasoning abilities, as the answers were always constrained to be in the span of the data set. For example, all human comprehension tasks include the ability to make indirect inferences based on reasoning, and the answer may not necessarily be included in the data set In general, factoid question answering is not designed to make indirect or subtle inferences. In spite of these shortcomings, this data set has been used as a gold standard for the reading comprehension task, and it is fairly standard to use this for all types of evaluation.

The SQuAD data sets were released with a simple baseline classifier using logistic regression, which did not seem to provide anything close to human-level performance. Since then, several systems have been proposed for question-answering that drastically improve this baseline performance. Most of these systems use sequence-centric models. The early systems used RNN-style models, whereas later systems worked with transformer-based language models. The recent transformer-based systems seem to have exceeded human-level performance. In the following, we will introduce some of these systems. Other data sets, such as the CNN/Daily Mail data set [246], were released even earlier than the SQuAD data set for question answering — however, the SQuAD data set is the de facto standard for the reading comprehension task and will be the focus of our discussion. Nevertheless, the CNN/Daily Mail data set continues to be very popular, and it also has been adapted to the text summarization task. In the following, we will focus on some of the key reading comprehension systems that were developed in recent years.

14.2.1 Using Recurrent Neural Networks with Attention

A common class of methods for reading style comprehension is that of using recurrent neural networks (or rather, a suitable variant such as a bidirectional LSTM) with attention. Numerous systems such as the Stanford Attentive Reader and DrQA [105], are based on the notion of combining recurrent neural networks with question answering. The basic principle in all these methods is very similar:

1. The question is embedded into a fixed-length neural representation with the use of a recurrent neural network (or suitable variant, which is typically a bidirectional LSTM or GRU).

2. Another recurrent neural network is used to create representations for each position in the passage, where the hidden state for a position is used as its representation. This is easy to achieve with a recurrent neural network, which naturally defines a hidden state for each position in a sequence.

3. Attention is used to match the embedded representation of the question with the suitable position in the passage. This helps identify the relevant span for that question.

In the following, we will discuss DrQA [105], which is one of the simpler models designed for question answering. Note that DrQA can also work with a *collection of documents* with the help of an initial filtering stage of narrowing down the number of documents (with a classical retrieval system), although we currently discuss its specific adaptation for reading comprehension, where it works reasonably well (when compared to other recurrent neural network-based models). It is noteworthy, however, that recent advances based on pretrained language models (like BERT) have greatly exceeded the performance of recurrent neural networks, and are therefore preferred.

Note that even though each passage may contain multiple questions, a passage and a single question are treated as input pairs together with a corresponding prediction, which indicates the span of the passage where the question is answered. Both the question and the passage are embedded into a latent representation using a bidirectional LSTM. This overall architecture is shown in Figure 14.2(a). In order to embed a question, a bidirectional LSTM is used for training. The weights of the input layer of this lSTM are fixed to *word2vec* or *GloVe* vectors in order to reduce the training data requirements. The states in the LSTM were then converted into a fixed-length representation of the entire question by concatenating all the forward and backward states in the bidirectional LSTM. Attention can be further used to enhance the embedded representation, by setting setting the weight of a state $\overline{h}$ to be proportional to $\exp(\overline{w} \cdot \overline{h})$, where $\overline{w}$ is a learned weight vector. An

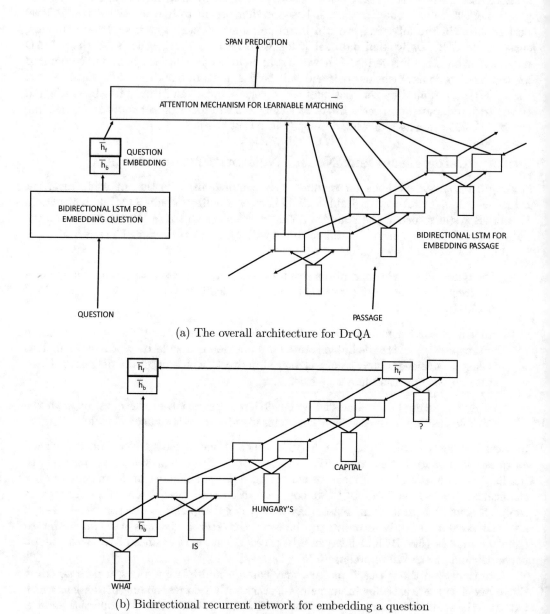

(a) The overall architecture for DrQA

(b) Bidirectional recurrent network for embedding a question

Figure 14.2: Recurrent model for DrQA. The figure in (b) shows details of the question embedding block in (a)

alternative approach and simpler approach is to assume that the only the last forward and backward states are the most important ones, since they encode the entirety of the sentence. Therefore, one can simply concatenate the columns vector representations of the last forward state defined by column vector $\overline{h}_f$ and the last backward state defined by column vector $\overline{h}_b$ into the longer column vector $\overline{h}_q$ as follows:

$$\overline{h}_q = \left[\begin{array}{c} \overline{h}_f \\ \overline{h}_b \end{array} \right]$$

This form of concatenation is depicted pictorially in Figure 14.2. The column vector $\overline{h}_q$ is the representation for the full question. The architecture for question embedding is illustrated in Figure 14.2(b).

The passage on which the questions need to be answered is embedded using a separate bidirectional LSTM from the one used for question embedding. As in the case of the LSTM embedding of the question, GloVe or *word2vec* representations are used in the embedding layer to input the words sequentially into the LSTM used for passage embedding. However, additional features are used as input to the LSTM for passage embedding beyond the words in the sentence. These features are as follows:

- **Hard word relevance:** A binary indicator is input at each word position to the passage LSTM, corresponding to whether or not that word is included in the question at hand. Note that the inputs to the system correspond to question-passage training pairs as shown in Figure 14.2(a). This input turns out to be quite useful as it emphasizes important words in the passage.

- *Soft word relevance:* While the binary indicator discussed above captures word relevance based on exact match, a soft word relevance is also used. Specifically, for each word position in the paragraph, the weighted average of all the question words is input at that position, where the weight of a (question) word is based on the dot-product attention similarity between the current passage word and that question word. For example, consider the case where the *word2vec* or GloVe embedding of the various question words are $\overline{z}_1 \ldots \overline{z}_r$, and the current paragraph word being input at that position of the LSTM has an embedding of $\overline{x}$. Then, the input (corresponding to soft word relevance) at that position is defined by $\sum_i \alpha_i \overline{z}_i$, where the nonnegative scalar α_i is defined as follows:

$$\alpha_i = \frac{\exp(F[\overline{x}] \cdot F[\overline{z}_i])}{\sum_{k=1}^{r} \exp(F[\overline{x}] \cdot F[\overline{z}_k])}$$

Here, $F[\cdot]$ is a vector-to-vector function defined by single-layer neural network with a ReLU activation function.

- *Miscellaneous features:* Additional features about the word at each position, such as its named entity recognition tag, its part-of-speech tag, and normalized term frequency within the paragraph are included. Named entity tags are useful, as the answers to factoid questions are often entities of various types.

After the embeddings of the question and the recurrent representation of the passage have been created, an attention-based matching approach is applied in order to find the span in the passage that most closely matches the question representation in the next layer (which is the attention layer). However, this approach is a somewhat different application of attention, in which the attention-centric matching is itself the prediction (as opposed to using it to focus prediction to more informative parts of the representation).

Let $\overline{p}_i$ represent the concatenation of the forward and backward hidden states at the ith position. Then, two learned matrices W_s and W_e are used to predict the probability $P_s(i)$ and $P_e(i)$ of each position i to be the (starting) starting and ending positions of the relevant span. Specifically, these probabilities are computed using the following bilinear computations:

$$P_s(i) = \overline{p}_i^T W_s \overline{h}_q$$
$$P_e(i) = \overline{p}_i^T W_e \overline{h}_q$$

Note that the two predictions have exactly the same mathematical form, except that they use different weight matrices, and the training process will automatically force these matrices to take on values that forces the learned span to relevant starting and ending positions. All possible spans of a maximum fixed length are tried and the one with the maximum value of $P_s(i) \times P_e(i)$ is reported as the prediction. The loss with respect to the accuracy of span prediction is used for training the neural network. This simple model provided an F1-accuracy of 79.4%, which was considered state-of-the-art till late 2017. However, this accuracy is nowhere close to the state-of-the-art methods today, many of which are based on pretrained language models.

Although this exposition has described the use of DrQA in the context of reading comprehension, it can also be used for answering questions over document collections like Wikipedia. Achieving this goal requires one to decompose the problem into two separate subproblems, one of which corresponds to narrowing down the number of documents being considered (i.e., information retrieval), and the second corresponds to the identification of a specific answer from this restricted set of documents (i.e., reading comprehension).

The work in [105] based the document filtering approach on a traditional information retrieval system. An inverted index lookup was used in conjunction with term-vector model scoring. These methods are discussed in Chapter 9. The system was further improved by accounting for local word order by incorporating n-gram features. Several data structure optimizations like hashing were used to improve retrieval efficiency, and then the best five documents were passed through the reading comprehension system in order to identify the exact answer. The span with the highest score was reported as the correct one. It is noteworthy that this broad approach is referred to as the *retriever-reader* approach, and it is one of the many important methods in open-domain question-answering. The accuracy of such systems can stem from either the retriever or the reader; therefore, a lot of research in question answering is focused on improving the quality of the retriever. This will be the topic of discussion in Section 14.3.

14.2.2   Leveraging Pretrained Language Models

Most pretrained language models are excellent at question answering, which is considered a litmus test for natural language understanding. In fact, Google regularly uses BERT to resolve natural language queries, and many Google searches are posed in natural language as factoid questions. Such searches represent a much more complex scenario as compared to the reading comprehension task, since they need to address the additional burden of large-scale information retrieval.

In the following, we describe the use of BERT for the reading comprehension task. Before reading further, the reader is advised to revisit Section 11.4.2 of Chapter 11 in which the basics of the BERT language model are described. As discussed in that section, BERT uses a fine-tuning phase that is task-specific — the task in this case happens to be question answering. The fine-tuning phase of BERT needs to use a different output layer from the pre-training phase; the purpose of this output layer is to learn the span of the answer within a particular passage, given a question. Recall that BERT trains pairs of sentences for making the prediction of whether or not a particular pair of sentences occur successively in a passage. These two sentences are referred to as sentence A and sentence B (see Section 11.4.2). During pretraining, an embedding for sentence A is added to to each token of sentence A, and an embedding for sentence B is added to each token of sentence B.

For the question-answering task, the question is treated is sentence A, whereas the passage is treated as "sentence" B. The output layer contains a representation of each token in both the sentences A and B. A start vector $\bar{v}_s$ and an end vector $\bar{v}_e$ are used in order to identify which tokens in sentence B of the output layer correspond to the starting and ending positions. The output layer of BERT reproduces all the words in the input sequence — however, the fine tuning ensures that the vector representations of the corresponding output tokens are affected by their likelihood of occurring at either the start or end of the span containing the answer. In particular, a token in the output in sentence B (i.e., passage) that is at the start of the answer span will tend to be more similar to $\bar{v}_s$, whereas the token that is at the end of the span will be similar to the vector $\bar{v}_e$. Both the vectors $\bar{v}_s$ and $\bar{v}_e$ have an output dimensionality that matches the dimensionalities of the reproduced sentences in the output layer. The similarity between a token in the output layer and the start vector can be measured with the dot product. Therefore, for any candidate pair of output vectors $\bar{o}_i$ and $\bar{o}_j$, corresponding to the word vectors at positions i and j in the passage (so that $j \geq i$), one can define the score of this candidate span as $\bar{o}_i \cdot \bar{v}_s + \bar{o}_j \cdot \bar{v}_e$. The candidate span with the largest score is reported as the prediction.

In order to perform the training, one needs to define a crisp loss function with respect to the prediction. First, the likelihood of each position $\bar{o}_i$ to be the start of the span is defined by applying the softmax function to the various values of $\bar{o}_i \cdot \bar{v}_s$ obtained by varying i. Similarly, the probabilistic likelihood of each position $\bar{o}_j$ to be the end of the span is defined by applying the softmax function to the different values of $\bar{o}_j \cdot \bar{v}_e$ obtained by varying j. The loss is defined by the sum of the log likelihoods of the correct start and end positions; this is equivalent to the negative of the sum of the logarithms of the computed likelihood probabilities at these positions.

Anatural question arises as to how one might handle cases in which the answer does not exist in the data set. For example, the recent version of the SQuAD data set also contained questions for which an answer did not exist in the data set. BERT was able to answer such questions by using the [CLS] token as a valid start and position. Therefore, a question was considered not to have a valid answer in the passage, when the start and end position was predicted to be the [CLS] token. Specifically, the score of the best non-null position was compared with that of $\bar{v}_s \cdot \bar{c} + \bar{v}_e \cdot \bar{c}$, where $\bar{c}$ is the output vector representation of the [CLS] token. The question was assumed not to have a valid answer within the data set, when the score of the span defined by the [CLS] position was higher than any non-null span. One interpretation of this situation is that the optimal starting and ending position both occur at the end of the sentence ([CLS] token), which corresponds to the null span for a question that does not have an answer within the passage.

The best BERT model [149] exhibited a test accuracy of 93.2%, which exceeded human performance, and was significantly better than methods based on recurrent neural networks. Almost all reading comprehension systems today are based on pre-trained language models because of their superior performance to recurrent neural networks.

14.3   Retrieval for Open-Domain Question Answering

Although Section 14.2.1 is focused on the reading comprehension task in DrQA, it provides a foundation for one of the important methods in open-domain question answering, which is the *two-stage retriever-reader approach.* All these methods use a two-stage approach, wherein the first stage uses traditional information retrieval to filter a small subset of relevant documents, and the second stage uses reading comprehension methods in order to identify the answers within these documents. The idea of using retrieval to augment document sources for open-domain question answering is also referred to as *distant supervision.* One challenge associated with such distant supervision methods is that incorrect selection of relevant documents in the first stage can lead to poor performance of the reader. Therefore, high-quality retrieval approaches are critical. This can only be achieved by some level of merging between the retrieval and reading tasks.

Most of the known retrieval methods (cf. Section 9) rely on the original sparse representations of the documents. In such sparse representations, the lexicon of text may have size that is on the order of 10^5 to 10^6 words, but a single document may contain only a few hundred words. This a relevant point because it means that the passages obtained during the retrieval phase do not have a very large dimensionality. On the other hand, latent representations are dense, in which the overall dimensionality is only about 500, but almost all dimensions have nonzero values. These different representations have different trade-offs both in terms of search efficiency and search quality. Natural language understanding is somewhat different from search engine queries in that the natural language queries have greater semantics embedding in them, and it makes sense to use latent representations of the documents to retrieve documents of high quality. As discussed in Chapter 3, latent representations can identify semantic similarities between phrases such as *"played the bad guy"* and *"acted in the role of villain,"* which are not possible with the use of sparse retrieval methods relying on original word representations. Note that the two phrases have the same meaning, but use completely different language. In contrast to traditional information retrieval with keyword queries, such situations are even more common with natural language queries. However, latent representations are typically dense, which do not allow the use of efficient indexing techniques (like the inverted index). So how should one navigate this trade-off?

It turns out that the trade-off is less important than one might imagine at first glance. While it is natural to hypothesize that latent representations are important for semantic matching, sparse matching techniques also have certain advantages, which turn out to be more important to the overall accuracy. For example, consider the following question:

> *"When was George Washington born?"*

In this case, the entity *"George Washington"* is absolutely critical to the question, and an exact matching of the type that is common in tf-idf and BM25 is more suitable. A dense retrieval method might dilute the all-important name in a latent space, which can result in poor performance. In fact, *latent methods had never actually been shown to provide higher quality retrieval for open-domain question answering in practice (until 2019).* Aside from the fact that dense methods fail to pay sufficient importance to situations in which exact matching is important, it is critical to use latent methods that can retrieve useful evidence blocks from the large collection. A poor retriever will eventually lead to an inaccurate question-answering system, especially because the training data for the reader is obtained from the retriever. If few answers are contained in the retrieved evidence blocks, it will be impossible for the reader to provide useful results.

Simpler forms of latent representations (like latent semantic analysis) do not encode sufficient semantic or syntactic concepts to be of use for such systems. It is critical for such latent methods to be *supervised.* but sufficient supervision is often hard to get. Therefore, question-answering systems need to be trained with limited supervision, and this is often achieved with high-quality unsupervised pretraining (which can then be fine-tuned in a supervised manner). In recent years, advances in latent embedding methods for natural language processing with the use of transformers and deep learning has provided an opportunity for achieving this goal. These pretrained models provide a base of latent representations on top of which further supervision and fine-tuning can be performed.

14.3.1 Dense Retrieval in Open Retriever Question Answering

The first success of a dense retrieval method for open-domain question answering was achieved in 2018 with the Open Retriever Question Answering (ORQA) system. In this system, the retriever learns to identify relevant documents from an open corpus, and the supervision of the retriever is achieved only with question-answer pairs without the use of passages. This is a natural property of an retrieval-centric system in which the passages are identified during the overall answering process rather than provided as additional information along with the input questions (like SQuAD). In this sense, data sets like SQuAD are not considered completely suitable for open-domain question answering, even when all passages are collated as a single corpus for question answering. Since the questions are often lexically related to the passages by virtue of how the questions were constructed after looking at the passages – this approach may not be completely realistic and the danger is that the problem setting may be simpler than real-world settings. Therefore, more data sets have been created for these settings (see bibliographic notes), even though a collated version of the SQuAD data set is also used.

In the ORQA system, the relevant passages are directly identified by the retriever, and also referred to as *evidence blocks.* The supervised latent representations in ORQA are built on top of multiple pre-trained BERT models (cf. Section 11.4.2 of Chapter 11), which form critical parts of the architecture along with some additional parameters that ensure that these different models work together effectively for question answering. Three different BERT models are used, corresponding to (i) the embedding of the questions for the retriever, (ii) the embedding of the retrieved passages (required by retriever), and (iii) the embedding of the question-block pair required by the reader for answering the questions from the retrieved passages (in terms of the relevant span of the answer in the retrieved block). These different BERT models and corresponding parameters are then fine tuned together along with additional parameters that connect them into a single question-answering system. The training is based on an objective function that includes *both* the retrieval and reader performance so that the retrieval and reading comprehension models are well integrated. This performance is quantified with a *scoring function* for the question-answering system.

For the purpose of scoring, the retrieved documents are treated as a set of evidence blocks, each with a finite set of spans corresponding to the starting and ending position of an answer. The score of a specific span s within a block b of retrieved text is the sum of the goodness measures of the retrieval relevance of that block for the question q and the accuracy of the span s within the block with respect to the answer:

$$S(b, s, q) = S_{retr}(b, q) + S_{read}(b, s, q)$$

As we will see later, this type of scoring ensures that the retriever and the reader are jointly trained. Note that the retrieval score is useful for identifying documents with high relevance to the question at hand, and the reader then zeros into the relevant span from the restricted set. The scoring of the reader is more computationally intensive on a per-document basis, but it only has to process a small number of documents pre-selected by the retriever, whereas the retriever also has to decide which evidence blocks are good enough to include in the scoring of the reading comprehension component. One can even convert this score into a a probability with the use of the softmax function:

$$P(b, s|q) \propto \exp(S(b, s, q))$$

The best span and block pair is then obtained as the one that maximizes this score (or corresponding probability):

$$[s^*, b^*] = \text{argmax}_{s,b}\{S(b, s, q)\}$$

Finding the best matching span-block pairs is accomplished with the help of locality sensitive hashing, as discussed in [317]. Traditionally, information retrieval of evidence blocks is achieved with the use of measures like tf-idf and BM25 (cf. Chapter 9). However, in this case, the scores are parameterized within neural network layers, and therefore the retrieval process itself becomes trainable. Each of the retriever scores $S_{retr}(b, q)$ and the reader scores $S_{read}(b, s, q)$ is computed with the help of the BERT pretrained language model.

Note that the retriever itself uses the portion $S_{retr}(b, q)$ of the score in order to rank and identify relevant documents from the corpus. For the case of the retriever, the question q and the evidence block b are each embedded in latent space with the use of a pretrained BERT language model:

$$\overline{h}_q = W_q BERT_Q(q)[CLS]$$
$$\overline{h}_b = W_b BERT_B(b)[CLS]$$

The question and the evidence block are embedded with the use of different BERT models, and therefore the notations $BERT_B$ and $BERT_Q$ are used to indicate this distinction. Here, W_q and W_b are weight matrices that are designed to embed the questions and the blocks into 128-dimensional space, and are learned in order maximize the overall score. Therefore, $\overline{h}_q$ and $\overline{h}_b$ are 128-dimensional representations of the questions and the blocks. As is common in BERT-based models, the embedding of the question or block is the same as that of the [CLS] token in the BERT model (see Section 11.4.2). The score of the retrieval component is defined by the dot product between the query and the block:

$$S_{retr}(b, q) = \overline{h}_b^T \overline{h}_q$$

The scoring of the reading component is a span-based variant of the reading comprehension model in the original BERT paper [149], which is also discussed in Section 14.2.2. The hidden representations of the start and end positions of a particular span s span are then defined as follows:

$$\overline{h}_{start} = BERT_R(q, b)[START(s)]$$
$$\overline{h}_{end} = BERT_R(q, b)[END(s)]$$

Here, $START(s)$ and $END(s)$ represent the starting and ending positions of the span s. The reading score is obtained by applying a multilayer perceptron on the concatenation $\overline{h}_s$ of $\overline{h}_{start}$ and $\overline{h}_{end}$:

$$S_{read}(b, s, q) = MLP(\overline{h}_s)$$

Note that $\bar{h}_s$ represents a hidden representation of the candidate span.

In order to perform the learning, the probability for each span is computed by applying a softmax on the combined retriever-reader score $S(b, s, q) = S_{retr}(b, q) + S_{read}(b, s, q)$ as follows:

$$P(b, s|q) \propto \exp(S(b, s, q))$$

The normalization constant of the proportionality relationship is chosen to ensure that the probabilities over all the candidate blocks and spans within these blocks sum to 1. The candidate blocks are the top-k blocks returned by the retrieval, and the value of k used is typically around 5. The loss function is the negative of the logarithm of this probability over the correct span-block pair corresponding to the gold-standard answer. Note that this loss function uses both the retriever component and the reader component, since the overall score of a span-block pair is obtained by adding the retriever and reader scores (which are exponentiated to obtain the probabilities).

One challenge in this training is that if sufficient numbers of correct answer spans are not included in the retrieved blocks, the training of the reader component will be poor. In distant supervision, the quality of the training data is poor, and one can often end up with training examples that are not very informative. It is also noteworthy that reader training is more computationally expensive on a per-block basis. This problem is also referred to as the *cold start problem*. Therefore, the technique in [317] proposes an optimization in which an early update is first performed using only retriever scores, but over larger numbers of retrieved blocks (i.e., larger values of k in top-k retrieval), and the full update is done over a smaller number of blocks. In other words, the early loss function defines the likelihood that the answer is contained within a block (without worrying about the span of the reading comprehension component):

$$P(b|q) \propto \exp(S_{retr}(b, q))$$

The early loss is defined as the negative logarithm of this probability value on ground-truth blocks that do contain the answer. However, a larger value of k is used in the top-k retrieval process in order to ensure more aggressive learning. Note that the early update ensures that numerous updates occur only because of the retriever component, which can be especially helpful when the weights are suboptimal and fewer relevant blocks are being accessed.

Good performance of the retriever is critical in ensuring that the reader can be trained properly, given that the retriever obtains the training data for the reader (which was referred to earlier as the cold-start problem). To achieve the goal of addressing cold starts, the retriever is initialized with unsupervised pretraining with respect to a task that is closely related to good retrieval performance in the question-answering task. This task is referred to as the *inverse cloze task*. The pretraining task uses an unsupervised analog of question-evidence pairs, which corresponds to *sentence-context pairs*. A context of a sentence is semantically relevant and it can be used to infer information that is missing from the sentence. This is similar to how the evidence in response to a question discusses entities, events, and relations from the question, but the answer contains extra information that is missing from the question. One of the classical procedures in this regard is the *Cloze* procedure, which tries to predict a missing sentence, given the surrounding context. In contrast, the *inverse* Cloze procedure does the inverse job of predicting the surrounding context, given a randomly chosen sentence. For example, consider the random sentence (italicized) together with the context surrounding it:

> "...Zebras have four gaits: walk, trot, canter and gallop. *They are generally slower than horses, but their great stamina helps them outrun predators.* When chased, a zebra will zigzag from side to side...."

Note that in this case, the italicized sentence (selected randomly) is the "question" q, and the remaining part of the text is the context b. The objective of the inverse Cloze procedure is to select the true context b_t from a candidate set of possibilities $b_1 \ldots b_t$. For this purpose, the inverse Cloze procedure applies the softmax function to the vector whose ith element is $S_{retr}(b_i, q)$ in order to generate probability values for each b_i. In order to generate the pretraining data, a random sentence was selected as the source and the text in the block around it was chosen as the true target. This sentence is removed from the predicted context in 90% of the cases, and therefore some degree of exact matching was encouraged by the remaining 10% of the cases. Note that removing the sentence in 100% of the cases leads to poor performance, because the system is unable to properly handle cases in which specific keywords in the question are critical for finding the correct answer.

14.3.2 Salient Span Masking

An additional approach for improving performance is to pretrain the language models by leveraging *salient span masking*. This idea is proposed in REALM, which stands for *Retrieval Augmented Language Models* [227]. Note that the vanilla BERT model is trained with masking of tokens within sentences, which are randomly selected and not specifically selected to emphasize the importance of tokens containing the typical answers to *factoid* questions. In the context of factoid question answering, not all tokens are equally important. The answers to factoid questions are more commonly named entities such as persons and places. For example, consider the following pair of masked sentences:

> Nelson Mandela was born in the village [MASK].
> Nelson Mandela [MASK] born in the village Mvezo.

In this case, the first masking helps learn a fact, whereas the second masking helps learn a syntactic aspect of English grammar. Both are arguably important, although the first masking is more helpful for learning the facts needed for question answering. Therefore, the work in [227] performs additional pretraining of BERT with masked entities, which is referred to as salient span masking. In this case, only named entities and dates are used for span masking. In addition, the inverse Cloze task is used in a manner similar to the ORQA paper. It has been shown in [227] that salient span masking outperforms other forms of masking by emphasizing the learning of relevant factoids from the data.

14.4 Closed Book Systems with Pretrained Language Models

The use of the retriever for question answering is done in order to access the information blocks that are required to answer a specific question. Therefore, the corpus of knowledge is held in external memory, which is accessed as needed during question answering. The logic of holding the corpus in external memory is its very large size. However, in recent years, main memory sizes have increased rapidly, which has also contributed to the unprecedented success and flexibility of pretrained language models. A natural question arises as to whether one can directly query a pretrained language model for answers. It is noteworthy that such language models are typically pretrained on large document collections like Wikipedia, which are the same collections used by the retriever. If the pre-trained language model can store (i.e., memorize) most of the relevant knowledge from such large corpora within its parameters, there is no need to use a retriever at all. In this context, it has been asserted [439] that *large-scale language models can be viewed as knowledge bases*, because they store not just the syntactic aspects of the language but are also large enough to memorize specific facts for question answering.

A retrieval-based language model can be viewed as an *open book* question answering system [474], whereas a model that does not have access to external retrieval can be viewed as a *closed book* system. The external corpus of an open book system is similar to that of external sources of knowledge available to a student in an open-book test. On the other hand, the pretrained parameters of a language model in a closed-book system are analogous to the knowledge stored in the biological neural network of a student taking a closed book test. A closed-book system has the advantage of effectively using the pretrained model when accessing the information needed for answering a particular question. The seamless integration of retrieval and reading comprehension is clearly an advantage of closed-book systems, because any looseness in the integration of the retrieval and reading comprehension components of an open-book system almost always leads to a reduction in accuracy.

It is noteworthy that language models are already indirectly trained on a special type of question answering task. A unidirectional language model like GPT-n is trained on predicting the next word, which can be considered as a task of answering the next word. Similarly, a bidirectional language model like BERT is trained on predicting a masked word. The prediction of a masked word is a very special type of question-answering task, referred to as *slot filling*, although the slot filling in BERT is done in order to build a knowledge of the language rather than building a knowledge of specific facts. Nevertheless, with increasing model sizes and training data sets, such massive language models also indirectly function as memorized knowledge bases storing large amounts of factual data. This is because large-scale language models allow the memorization of facts that have only a small number of mentions in the data.

Language models have many attractive qualities in comparison with knowledge graphs. First, note that they are schema-free, which makes them easier to use. Furthermore, if the language model is held entirely in main memory and there is no search involved, it can be extremely fast in comparison with systems requiring access to external storage. Language models are particularly useful in handling *Cloze* queries in which a particular portion of the sentence (e.g., named entity) can be blanked out. For example, consider the following query:

> Which village was Nelson Mandela born in?

This type of query is a Cloze query, because it is equivalent to the following masking query:

> Nelson Mandela was born in the village [MASK].

Language models like BERT are trained on precisely these types of masked sentences, and therefore are likely to work well at providing accurate responses as long as the model can comprehend the query sentence being asked. Note that if the query is already provided in the form of a masked sentence, then one can use existing language models directly *with no changes*. On the other hand, if the query is provided in the form of a natural language question, one can either fine-tune a pretrained language model like BERT/GPT (suited to slot filling) to answer such questions or one can use a sequence-to-sequence learner (e.g., transformer or T5 language model) in order to map the question sequence into the answer sequence. The latter approach is usually more effective.

The early models [265, 439] directly used masked sentences as queries to evaluate these models. The work in [474] used fine tuning on an existing language model to support pre-

Table 14.1: Performance of different open-book systems with respect to T5 closed-book system

Method	Natural Questions	Web Questions	TriviaQA (dev)	TriviaQA (test)
DrQA	-	20.7%	-	-
ORQA	33.3%	36.4%	47.1%	-
REALM	40.4%	40.7%	-	-
Karpukhin *et al.* [292]	41.5%	42.4%	57.9%	-
T5 (large version)	35.2%	42.8%	51.9%	61.6%

diction. In particular, it chose the T5 language model [460], which pretrains a sequence-to-sequence encoder (like a transformer). This type of model is particularly amenable to fine tuning because the model is already designed for text-to-text learning. The T5 language model is fine tuned in a manner similar to its other sequence-to-sequence applications like machine translation and text summarization (see Section 11.4.3 of Chapter 11). In other words, the T5 model is fine-tuned with examples of sample questions and answers, just as it would be fine tuned in a machine translation task with examples of sentences in the two languages. The work in [474] adopted salient span masking for pretraining the language model, which was adopted from the ideas presented in [227]. The results are reported for the Natural Questions, Web Questions, and TriviaQA data sets. For the last data set, both the validation (dev) and test accuracy is reported. The reader is asked to refer to the bibliographic notes in order to obtain pointers to these data sets.

The results in [474] show that closed-book systems can perform surprisingly well in comparison with with open-book systems, even though they are constrained not to have access to external storage. In this context, we present the results from [474] with respect to some competitive open-book systems on well-known question answering data sets in Table 14.1. The T5 model was used in this case. In the case of TriviaQA, the validation and test accuracy were separately reported. It is noteworthy that some of the entries in the table have been left blank because certain types of data sets do not naturally work with particular types of systems. The winning entries are underlined in the table — it is evident that even though open-book systems have an inherent advantage over closed-book systems by virtue of having access to greater information during execution time, the closed-book T5 system outperformed all open-book systems in two of the cases, and also provided competitive performance in other cases. Furthermore, the strength of closed-book systems will increase more dramatically than that of open-book systems with increases in model sizes stemming from greater computational power and memory capacity. On the other hand, open-book systems are hampered more significantly by scalability challenges during retrieval.

14.5　Question Answering with Knowledge Graphs

Knowledge graphs encode relationships between entities with the use of a graph structure. Knowledge graphs are described in detail in Section 13.4 of Chapter 13. The knowledge graph setting is particularly suitable for question answering systems, especially since a lot of work of converting unstructured data into structured form has already been done up front. As long as a question can be converted into structural form, it can often be answered relatively easily with the information encoded in a knowledge graph. Knowledge graphs are also useful for Web search, especially when the answers to queries occur in the form of named entities. There are a number of broad approaches that use knowledge graphs in order to respond to queries. The first of these methods uses a query translation approach to convert a natural language query into a structured language such as SPARQL. A second approach fuses text with knowledge graphs for query responses. The third approach translates a knowledge graph into a corpus to enable question answering.

14.5.1 Leveraging Query Translation

Many queries are relatively simple in terms of requiring entity-centric responses that are related to other entities. In such cases, one can convert natural language questions into queries that are properly posed in terms of entity-oriented search. A key problem is that the questions are posed in natural language, as a result of which understanding what is being asked (in terms of a properly represented query) is sometimes more difficult than answering the query itself. Therefore, it is critical to be able to convert the query into a structured format that the knowledge base can easily work with. In such cases, the training pairs will correspond to the informal and structured representations of questions. For example, one might have a pair as follows:

What is the capital of China? <EOQ1>   **CapitalOf(** *China*, ?) <EOQ2>
　　　　Natural language question　　　　　　　Formal Representation

The expression on the right-hand side is a structured question, which queries for entities of the type discussed in Chapter 13. The first step would be to convert the question into an internal representation like the one above, which is more prone to query answering. This conversion can be done using training pairs of questions and their internal representations in conjunction with an recurrent network. In practice, the structured representation of the question is in the form of a SPARQL query (see Section 13.4 of Chapter 13):

What is the capital of China? <EOQ1>
　　Natural language question
　　　　　　SPARQL representation of CapitalOf(*China*, ?) <EOQ2>
　　　　　　　　　Formal Representation

Once the question is understood as an entity-oriented search query (by virtue of conversion to the SPARQL query), it can be posed to the indexed corpus, from which relevant relationships might already have been extracted up front. Therefore, the knowledge base is also preprocessed in such cases, and the question resolution boils down to matching the query with the extracted relations. It is noteworthy that this approach is limited by the complexity of the syntax in which questions are expressed, and the answers might also be simple one-word responses. Therefore, this type of approach is often used for more restricted domains. An important challenge in using this type of approach is that knowledge bases are often incomplete, and certain types of relations are often not encoded in knowledge bases. In such cases, it is critical to design responses that fuse the knowledge from text domains with structured data.

14.5.2 Fusing Text and Structured Data

In recent years, a number of question answering methods have been proposed that fuse text and structured data for question answering. Therefore, both a corpus and a knowledge graph is provided for answering the question. In this section, we provide a brief overview of *GRAFTNet* [533] and *PullNet* [534], which are methods fusing graph and natural language processing in order to respond to queries. The approach requires the use of graph convolutional networks, an understanding of which is beyond the scope of this book. Therefore, this section will only provide a brief conceptual overview of this approach. For details, we refer the reader to the bibliographic notes, where pointers to important methods in this area are provided.

The *GRAFT-NET* [533] and *PullNet* [534] are closely related methods, and the latter can be viewed as an enhancement of the former. *GRAFT-Net* [533] which stands for *Graph of Relations Among Facts and Text Networks*, and it first builds a *question subgraph* based on the entities and text in the natural language question. This step is analogous to the retrieval step in retriever-reader approaches in the sense that the question subgraph is (ideally) a modest-sized representation of the corpus and knowledge graph, which is supposed to contain the answer to the question. The question subgraph is built using both the knowledge graph and the corpus and therefore this part of the approach results in *early fusion* of the information contained in the corpus and knowledge graph in a manner that is specific to the question at hand. The nodes in the question subgraph contain text from the corpus, and it contains entities and relations from the knowledge base. A graph convolutional network is then applied to this question subgraph in order to produce the final answer. Therefore, like the retriever-reader approaches, this technique also combines a retriever and reader except that it retrieves information from both the natural language corpus and the knowledge graph, and it uses a graph convolutional network as the reader. In order to build the question subgraph, a *PageRank*-style traversal approach is used on the knowledge graphs, starting with *seed nodes* (i.e., entities) contained in the question. Because of the ability of this approach to extract chains of relations in the knowledge graph, the approach has the ability to integrate relational information, which might originally have been incorporated in the knowledge graph from disparate sources (corresponding to *multi-hop* question answering). This is a distinct advantage over many of the approaches discussed earlier in this chapter, which try to extract the answer from a single evidence block in the corpus — in many cases, the complete answer may be contained in multiple evidence blocks. The subgraph is constructed in such a a way that one or more nodes in it correspond to the answer. Therefore, instead of selecting a span in an evidence block (as in retriever-reader approaches), this technique boils down to selecting and returning *nodes* in the question subgraph. Therefore, the problem boils down to the binary classification of nodes using a graph convolutional network over the question graph. The original question is also provided as one of the inputs to the graph convolutional network in order to enable the use of context. To achieve this goal, the question is embedded into a latent representation using an LSTM, and provided as one of the inputs to the graph convolutional network. In subsequent layers, the question representation is updated based on an aggregation of the hidden representations of the content contained in the seed entities of the knowledge subgraph. A number of changes are made to the basic graph convolutional network architecture, which are discussed in [533]. For example, attention is used to perform convolutions in the graph neural network favoring relations (edges) relevant to the question at hand. Similarly, the nature of the update rules for entities extracted from the knowledge graph and the text extracted from the corpus were different. The reader is referred to [533] for a discussion of these enhancements.

As in the case of retriever-reader approaches, the success of the approach depends heavily on the ability to extract a question subgraph of high quality. *GRAFT-Net* uses an ad hoc approach for extracting the knowledge graph, which results in question subgraphs of low quality. This problem is addressed with the use of *PullNet*, which uses successive "*pull*" operations on the knowledge graph and corpus in order to iteratively construct the question subgraph. *PullNet* learns how to construct the subgraph by using a succession of retrieval operations either on the knowledge graph or on the corpus in order to retrieve new information. These operations are referred to as *pull* operations, and the decision of where and when to apply them is achieved with another graph convolutional network. This classifier, referred to as the *pull classifier*, uses question-answer pairs for supervision. Therefore, the overall approach is a learned iterative approach for subgraph extraction, starting with only the question text and its entities, and expands to incorporate information sequentially. This type of intelligent subgraph extraction approach tends to make the extracted subgraphs smaller and more likely to contain the eventual answer. The specific details of the pull operation are presented in [534].

14.5.3   Knowledge Graph to Corpus Translation

One of the challenges associated with knowledge graph processing for question answering is that one is trying to process two different types of data simultaneously and therefore the models need to account for the differences in data format. For example, the methods discussed in the previous section need to integrate convolutional neural networks tightly with sequential models such as LSTMs. These types of integrations are sometimes challenging. In this sense, the structured format of knowledge graphs does not always help, when the information from the knowledge graphs need to be fused with natural language. Nevertheless, the quality of information in knowledge graphs is often higher than that in unstructured text, especially because many knowledge graphs are created through the process of successive community curation and correction.

An approach that has been proposed recently is the conversion of knowledge graphs to synthetic natural language representations of high quality. A recent approach, referred to as *knowledge-enhanced language model (KeLM)*, uses precisely this approach. The basic ideas in KeLM comprise the following steps:

- Convert the knowledge graph into a synthetic natural language corpus.

- Leverage the existing natural language corpora in combination with the synthetic natural language corpus for question answering.

It is noteworthy that there is a wide variation in the ways in which the second step could be potentially implemented. For example, one could leverage one of the retrieval-based language models (like ORQA), or one could leverage a pre-trained language model for question answering.

The key step in the process is the generation of a synthetic natural language corpus from a knowledge graph. In order to achieve this goal, the *TEKGEN* model is constructed, which is a sequence-to-sequence model for converting the knowledge graph to text. In order to use this model, the TEKGEN training corpus is created. This training corpus aligns knowledge graphs with text by encoding a wide variety of entities and the relations among them along with natural language text describing these entities and relations. For this purpose, Wikipedia is an ideal training data set, as the text in Wikipedia matches with known entities and relations in Wikidata (see page 455 of Chapter 13). Therefore, the Wikipedia-Wikidata pair provides a natural approach for aligning knowledge graphs to natural language. It is noteworthy that simply creating pseudo-sentences using raw RDF triples does not seem to work well, as the resulting pseudo-sentence is not in natural language format, and any pretrained natural language-based approach (like BERT and T5) will not integrate well with such a corpus at fine-tuning time.

For each entity in Wikidata, KeLM restricts the relations to its root Wikipedia page, because the natural language descriptions of how the subject entity are related to other entities are obtained at the root of this page. All triples that have this entity as the subject in Wikidata are then matched to sentences in Wikipedia. Subsequently, all sentences on this page that have the object entity in the sentence (or its alias) are matched to the RDF triplet. This approach allows an RDF triplet of Wikidata to be aligned to multiple sentences in Wikipedia; at the same time, it allows a sentence in Wikipedia to be aligned with multiple triplets in Wikidata. Different approaches are used for processing the triplets, depending on whether or not the object entity has a Wikipedia page, or it is a quantity or date.

When multiple triplets for a given subject map to the same Wikipedia sentence, a single training example is generated by collating the triplets together as follows:

> Subject, Relation(1), Object(1), Relation(2), Object(2), ... Relation(n), Object(n)

This provides the source sequence for building a translator. The target sequence is simply the sentence in Wikipedia. Subsequently, the T5 model (see Section 11.4.3 of Chapter 11) was fine-tuned on this training data. This type of model is referred to as a *triplet-to-text model*. The triplet-to-text model was used to generate sentences from triplets that were not a part of the training data. Note that this leads to the natural language representation of structured information that may not be present in other unstructured sources (such as the original Wikipedia), since many of the triplets are curated via community contributions.

In order to improve the quality of the generated text, a semantic quality-based filtering was used on the se sentences generated by the triplet-to-text module. The filtering was performed with the help of a quality score, which was generated using a BERT model, which was trained on a WebNLG human assessment data, which contains human assessments for semantics and fluency on a given corpus. The KELM corpus was then created by grouping sentences into documents. Specifically, the sentences on the sa,e subject entities were grouped to create 5.7 million documents.

Once this corpus has been generated, it can be integrated with any existing open-domain (or even closed-book) question answering system in order to perform enhanced training. The key point is that the synthetically generated corpus is simply an example of *data augmentation* that can be used to enhance existing methods. The work in [19] specifically integrates the generated corpus with the REALM model for question answering, and it shows that there are significant improvements to the quality of the generated answers by using this type of approach.

14.6   Challenges of Long-Form Question Answering

Long-form question answering is distinct from factoid question answering, because it gives detailed and descriptive answers to questions. To a large extent, long-form question answering remains a challenge as of the writing of this book. The reason is that long-form question answering requires the extraction of pieces of information from disparate sources and then their integration into a coherent answer. This problem is, in fact, more difficult than text summarization, where one tries to achieve the same goal, except that it is done for the purpose of summarization rather than for answering a question.

Since long-form question answering is, as yet, not a fully developed field, the following discussion enumerates some of the promising directions in this area, without a specific discussion of the underlying approaches. Most of the underlying approaches do acknowledge that there is significant scope for improvement of the underlying techniques. The following are some of the key directions used for long-form question answering:

- *Extractive methods:* In extractive methods, two steps are used for constructing the answer to a question. In [186], a bidirectional RNN was used to identify relevant sentences from Web search information, and these sentences were pasted together to create the answer. Even though the work in [186] uses the BidAF model, one can use any existing method including DrQA. These models are used to extract multiple spans of text, which are then put together to create the final answer.

- *Abstractive methods:* These techniques directly synthesize information from multiple sources to create an integrated answer in the form of a paragraph. The model is improved by training it on multiple tasks related to question answering. For example, one also trains the model on the standard question-answering task of reading the question and documents, and then providing the answer.

- *Unified model-based approach:* Most methods for question-answering use a retrieval-comprehension paradigm, wherein the retrieved documents are then searched for to find the relevant span of the answer. A notable exception is that of closed-book language models, which use pre-trained language models to achieve the same goal. Unified model-based approaches follow the same path in terms of using a model-based approach to provide responses. The work in [384] provides a broad framework for achieving this goal, although the paper does not spell out specific details (or experimental results) that accomplish this task.

Note that the extractive and abstractive methods for question answering are analogous to the corresponding methods used in text summarization. Overall, long-form question answering is a developing area, which is very much an aspirational goal, as of the writing of this book.

14.7 Summary

This chapter introduces question answering systems that operate on natural language questions and text collections. Question answering is different from search in that a greater amount of semantic understanding is required for responding to queries. Therefore, a fundamental problem in question answering is that of reading comprehension, in which questions are answered on a pre-defined passage. Open-domain question answering is more challenging because the questions are not guaranteed to be present in a particular passage, but over a vast collection of documents. Therefore, these systems have to combine a retriever with the reading comprehension component and train them jointly. In recent years, pre-trained language models also been used for question answering, as the internal representation of a language model often memorizes sufficient information in order to respond to queries. The most advanced methods for querying are based on a combination of the use of knowledge graphs and natural language text. An open challenge in question-answering is that of long-form question answering, in which the goal is to give descriptive answers to questions that do not have short answers. The existing models for this task generally do not perform well, and there is significant scope for further research in the area.

14.8   Bibliographic Notes

Before 2015, most of the methods designed for question answering were large systems with many pipelined components. Examples include Watson's Jeopardy! [195] and the Yoda QA system [47]. Such systems have inherently low accuracy because of the tendency of pipelined system errors to multiply rapidly. In spite of this fact, there were some notable successes with such systems, such as IBM's Question-Answering system for *Jeopardy!* [195]. Other early methods used paraphrasing [183, 184]. There were also some early attempts to use recurrent neural networks for question answering [263], although the limited amount of training data made it difficult for these methods to achieve high accuracy. Some methods also used memory networks [577, 578] for question answering, although the accuracy using these methods is quite limited from modern standards.

Building question-answering as end-to-end deep learning systems had always been a goal that remained elusive because of the difficulty in having access to large data sets of knowledge. The recent advances in question answering are largely a result of huge data sets that are now available for this task. The main advantage of the large data sets was the fact that they brought in the ability to construct supervised systems, which are inherently more accurate than the unsupervised systems that the models had been trained on before. Given that the data sets had changed fundamentally, it also led to a change in the types of algorithms used for resolving questions on these data sets. Section 14.8.1 provides a brief overview of some of these data sets.

In the early years, there were only a relatively limited number of question answering data sets, such as the TREC question-answering data set [731]. The question-answering task largely contains two tasks, one of which is the identification of a relevant block from a corpus containing the answer, and the other is the identification of the span of the answer within an appropriate evidence block. In this sense, the data set delineates the two most important subtasks in question answering. Later models integrated these two tasks for open-domain question answering. The two earliest reading comprehension data sets included the CNN/Daily Mail [246] and the SQuAD [461, 462] data sets. The DrQA system was introduced in [105], and it provided performance that was considered state of the art in 2017. There are several notable systems that use recurrent neural networks, such as BiDAF [509] and R-Net [571].

An early (and popular) model that integrated retrieval with reading comprehension was the ORQA model [317]. The REALM model [227] introduced the concept of salient spans, which are useful for pretraining language models in a manner that is more sensitive to the problem of question answering (by masking named entities and dates instead of arbitrary words). It was later proposed [292] that it is not necessary to perform pretraining with the inverse Cloze task as discussed in [317]. Rather, choosing appropriate positives and negatives while training can provide sufficient fine-tuning of the (already pretrained) BERT model, so that high-quality retrieval is achieved.

The exploration of direct use of language models for closed-book question answering is done in [265, 439, 474, 542]. Among these the earlier models such as [265, 439] directly used masked sentences, whereas later models such as [474] work with natural language queries, and fine-tine the pre-trained language model in order to respond to such queries. The work in [542] evaluates reasoning capabilities. Various methods that use fusion between knowledge graphs and text corpora include *GRAFT-Net* [533], *PullNet* [534], *Knowledge-aware Reader* [595] and *Knowledge-Guided Text Retrieval* [397]. The KeLM model that is discussed in this book is based on the presentation in [19]. A number of early proposals for long-form question answering are given in [186, 384]. An excellent tutorial on the state of the art in question answering may be found in [734].

14.8.1 Data Sets for Evaluation

A revolution in question answering occurred after 2016 because of the explosion in the large number of training data sets, such as SQuAD [461, 462] and CNN/Daily Mail [246, 729]. The CNN/Daily Mail was one of the earliest data sets released for the reading comprehension task, and it has subsequently adapted to the problem of text summarization. The original version of the SQuAD data set was introduced in [461], and this version only contains questions that have answers within the passages. Subsequently, a variant that contains questions without answers in the data set was introduced in [462].

The SQuAD data set is designed for question answering in the context of the reading comprehension task. A number of data sets have since been used for the open-domain question answering task. Some of these data sets are also used for reading comprehension tasks, since they have passages associated with individual questions. The most general assumption in an open=domain question-answering task is that one or more fixed set of sources are used for answering the question, and specific paragraphs that are relevant to a particular question are not available up front. Therefore, the open-domain setting requires both retrieval of the relevant *evidence blocks* from the fixed source of text as well as a reading comprehension task on top of these collected evidence blocks. However, many of the data sets such as *TriviaQA* [281], *SearchQA* [169], *Quasar-T* [152], and *Natural Questions* [303] do include passages with the questions. Nevertheless, when the testing is done for open-domain question answering, the passages are not used in a manner that is specific to the question at hand. Rather, a retriever is used on an appropriate source (either built from the provided passages or other external source) in order to identify relevant passages. In addition, a number of data sources use the Wikipedia as a fixed corpus. Examples include *Curated-TREC* [47], *WebSearch* [54], and *WikiMovies* [394]. Another area in which numerous data sets are available in question answering includes *multihop question answering*, in which pieces of information from multiple sources need to be integrated in order to yield a result. In this context, the *HotSpotQA* [609] and the *HybridQA* [106] benchmarks are commonly used. A data set for long-form question answering is ELI5 [186].

14.8.2 Software Resources

Implementations of DrQA are available at [732, 733]. The code and implementation of ORQA is available at [735]. The implementation of the closed-book question answering system discussed in [474] is available at [736]. The *GRAFT-Net* implementation may be found in [737].

14.9 Exercises

1. Implement a model that uses the same framework as DrQA, except that it uses a trsnsformer instead of a recurrent neural network for the comprehension portion of the task.

2. Implement a closed-book question answering system with the use of T5.

Chapter 15

Opinion Mining and Sentiment Analysis

"Opinion is the medium between knowledge and ignorance."–Plato

15.1 Introduction

The recent proliferation of social media has enabled users to post views about entities, individuals, events, and topics in a variety of formal and informal settings. Examples of such settings include reviews, forums, social media posts, blogs, and discussion boards. The problem of opinion mining and sentiment analysis is defined as the computational analytics associated with such text. In some settings, such as review text, the problem of opinion mining can be viewed as the natural language analog/complement to recommender systems. Whereas recommender systems analyze quantitative ratings to make predictions about user likes and dislikes, opinion mining analyzes review text to infer user likes, dislikes, and sentiments. Opinion mining therefore provides a more subjective and detailed point of view, which is complementary to the predictions of a recommender system. Furthermore, opinion mining is not restricted to product reviews, but it may pertain to user attitudes, political opinions, and so on. For example, the sentiment from Twitter users has been used to predict election results [131]. In contrast, recommender systems are almost always focused on maximizing the sales of products. Opinion mining refers to the discovery of positive and negative sentiments about objects (e.g., a computer) and their attributes (e.g., computer battery) with the use of text processing. An example[1] of an opinion is as follows:

> "The Logitech X300 is a compact wireless Bluetooth speaker that offers decent sound for its size and features an attractive, sturdy design at a modest price point. It has a built-in microphone for speakerphone calls and can be laid down horizontally or stood up vertically. Battery life could be better. At around $60 online, the decently performing and well-designed Logitech X300 is a relative bargain in the mini Bluetooth speaker category."

---

[1] https://www.cnet.com/products/logitech-x300-mobile-wireless-stereo-speaker/review/

C. C. Aggarwal, *Machine Learning for Text*,
https://doi.org/10.1007/978-3-030-96623-2_15

Note that this opinion expresses points of view not only about the main product (which is the Logitech X300), but also about several of its attributes such as the battery, design, and price point. Each of these sentiments might be positive or negative, which provides an overall point of view of the product. The task of opinion mining and sentiment analysis is not only about finding the opinions about the whole entity but also the opinions about individual attributes of the entity and summarizing them. These individual attributes are also referred to as *aspects*. The person making the opinion is referred to as the *opinion holder*, and the nature of the sentiment expressed (e.g., positive or negative) is referred to as its *orientation* or *polarity*. The entity or aspect that the opinion is expressed about is referred to as the *opinion target*.

Within the broader umbrella of topics associated with opinion mining, there is a significant variation in the types of questions that one tries to answer using opinion mining. Some examples are as follows. Does a piece of text represent a positive or a negative sentiment? What are the entities being discussed about, and are they being discussed about in a positive or negative way? What attributes of the entity are discussed, and what are the sentiments expressed about them? Are entities being compared to one another? Is a particular opinion a spam?

It is evident that the discovery of useful opinions from raw text requires a nontrivial amount of natural language processing. Furthermore, the problem of opinion mining uses various methodologies discussed in earlier chapters as building blocks for its analysis. Examples of such methodologies include feature engineering, entity extraction, and classification. In this sense, opinion mining may be considered an application-centric topic, which builds on many of the recent advancements in text mining.

The process of opinion mining can be performed at several levels. Opinions can be discovered at the document level, sentence level, or at the entity level. We briefly summarize these different ways of processing:

1. *Document-level sentiment analysis:* In this case, the implicit assumption is that a single document expresses opinions about a particular target (which is known). The goal of the task is to discover whether the document expresses positive or negative sentiments. The problem of document-level sentiment analysis can be viewed as a special case of classification with natural language data.

2. *Sentence- and phrase-level sentiment analysis:* Document-level sentiment analysis is often broken up into smaller units corresponding to individual words, phrases, or sentences as an intermediate step. These finer grained classifications are then aggregated into a higher-level prediction at the document level. In this context, phrase- and sentence-level sentiment analysis are important subproblems in their own right because they enable document-level classification. Furthermore, the output is sometimes used for opinion summarization.

   In sentence-level sentiment analysis, each sentence is analyzed one by one, and sentences are classified one by one as positive, negative, or neutral. In some cases, it is tricky to determine how a statement of fact should be treated. For example, consider the following sentence from the aforementioned review of the Logitech X300:

   > "It has a built-in microphone for speakerphone calls and can be laid down horizontally or stood up vertically."

   Although this sentence is a statement of fact about the features of the Logitech X300, it can also be considered positive because it conveys the flexibility associated with the product. However, from a practical point of view, such sentences could have a confounding effect on the classification, and they are therefore prevented from influencing the document classification process by removing them [429]. In order to achieve this goal, sentences need to be first classified as either *subjective* or *objective*. Subjective sentences often contain many adjectives and emotional phrases, whereas objective sentences contain statements of fact. It is generally much easier to classify the polarity of subjective sentences with the use of an *opinion lexicon*, whereas objective sentiments pose significant challenges from the point of view of sentiment polarity classification. Therefore, an important problem in sentence-level sentiment analysis is that of *subjectivity classification*.

3. *Entity and aspect-level opinions:* Many sentences in an opinion text may not refer to the entity itself. For example, consider the following sentence picked out of a larger review of a computer security product:

"Hackers have made our lives miserable in this day and age."

Even though this statement expresses a negative sentiment, it does not state anything about the product and in fact (implicitly) emphasizes the necessity of the broader class of computer security products. This implies that it is extremely important to specify the opinion *targets* in order to make opinion mining truly useful. Similarly, the previous opinion about the Logitech X300 states a positive point of view about the overall product, its flexibility in features, and price, but it states a negative point of view about the battery life. The "battery life" is an aspect of the broader entity "Logitech X300" and, therefore, an opinion about it can be viewed as fine-grained analysis. Entity- and aspect-level sentiment analysis expresses opinions about the fine-grained characteristics of an entity.

Entity extraction is closely related to opinion mining because entities need not always be persons, places, and organizations, but they could be products or other types of entities. Furthermore, the extraction of an *opinion lexicon* is similar to entity extraction, except that one is trying to identify adjectives/phrases associated with sentiments rather than noun-phrases associated with named entities.

15.1.1 The Opinion Lexicon

Certain types of words, referred to as *opinion words* or *sentiment words* are particularly important from the point of view of opinion mining and sentiment analysis. Typically, the opinion lexicon contains words like *"good" "bad,"* *"excellent,"* *"wonderful,"* and so on. In many cases, the opinion lexicon might contain phrases like *"blows away,"* *"gets under my skin,"* or *"silver lining."* Opinion words are often adjectives and adverbs, although nouns (e.g., *"trash"*) or verbs (e.g., *"annoy"*) can be considered opinion words. The topic of finding an opinion lexicon is a problem in its own right, and pre-compiled lists of opinion words are often used as simple solutions for off-the-shelf applications. Examples of such lists are available at [34, 708]. One can isolate an opinion lexicon with either a *dictionary-based approach* or a *corpus-based approach*. Both methods start with seed sets of words, which are expanded with the use of either a dictionary or a corpus. This process is referred to as *opinion lexicon expansion*.

Dictionary-Based Approaches

This approach combines a seed set of opinion words and a dictionary like WordNet in order to expand the seed set. A brief description of WordNet is provided in Section 12.3.2.1 of Chapter 12. The main approach is to first select a modest set of positive and negative seed words, and then grow them by leveraging the online dictionary by using their synonyms and antonyms. This approach is applied recursively because the words that are found are added to the seed set, and their synonyms/antonyms in the dictionary are again explored. Furthermore, it is possible to use machine learning to improve the quality of the lexicon found. Numerous such methods for opinion expansion are discussed in [23, 178, 179, 286]. The main shortcoming of this approach is that it does not take into account the context of a word in making judgements. For example, the word "*hot*" might be slang for a desirable product, or it might simply refer to an overheated computer with a poorly working fan. Corpus-based methods are more effective at handling problems associated with context.

Corpus-Based Approaches

Corpus-based methods also start with a seed set of positive and negative words, and then leverage their usage in a corpus to infer whether other co-occurring words in the vicinity of one of the known words are positive or negative. The notion of vicinity is often defined with the use of connectives like "*or*," "*and*," "*but*," and so on. For example, two adjectives that are joined together with the connective "*and*" are often likely to be of the same orientation. Therefore, if one of them is already known to be a positive member of the opinion lexicon, then the other one can also be added to this set. In general, linguistic rules can be defined using various types of connectives.

These types of rules provides natural ways to recursively expand the seed set of positive and negative words [240]. In general, it is possible that the same pair of adjectives may be connected in both a positive and a negative way in different parts of the same corpus. For example, even though the words "*good*" and "*bad*" are of different orientations, it is possible for them to be occasionally connected with an "*and*." A natural way of resolving such conflicts is to create a graph in which adjectives correspond to nodes, and links correspond to relationships of different orientations. Clustering can be applied on this graph in order to identify the words of the same orientation.

Another idea is to assume that the same sentence or neighboring sentences are of similar orientation, which provides useful hints about the orientations of the adjectives and opinion words inside those sentences [158, 288]. This is because opinions do not abruptly change in tone and orientation within a continuous piece of text. This concept is referred to as *inter-sentential* or *intra-sentential consistency*. Of course, this type of consistency is only an empirical phenomenon, and the tone may indeed change in many settings. For example, the use of words like "*but*" might precede changes in opinion orientation. It is also natural to integrate the concept of context within inter-sentential and intra-sentential consistency. In the work in [158, 355], the opinions are mined as aspect-opinion word pairs to distinguish between different uses of the word. For example, a *warm soda* and a *warm blanket* have completely different orientations of the word "*warm*" because they are applied to different entities. Instead of using the inter-sentential and intra-sentential consistency, an alternative idea is to use syntactic consistency like the parts-of-speech patterns for learning the orientation of opinion words [553]. This approach is described in Section 15.2, and therefore we do not discuss this approach in detail here.

Finally, many of the ideas discussed in Chapter 13 for entity extraction are also useful for finding opinion words. For example, the conditional random field method [308] (cf. Section 13.2.5 of Chapter 13) is often used for extracting opinion words with the use of supervised learning. These methods have the natural ability to incorporate context within the extraction process like all entity extraction methods. The use of such methods for joint entity and opinion word extraction is discussed in [453]. In general, opinion mining can be considered a variation on the information extraction task (cf. Section 15.1.1) in which instead of trying to find person/location/organization entities and their relations, one is trying to find product/person/location/organization entities and their related opinion "entities" (even though opinions are usually adjectives rather than noun phrases). In both cases, similar types of models can be used by tagging the appropriate portions of the text and applying supervised learning. This point is discussed in detail in the following section.

Opinion Mining as a Slot Filling and Information Extraction Task

One can view the opinion mining task as a closely related task to information extraction and slot filling. Refer to Chapter 13 for a definition of these terms. In general, it is not only important to know the polarity of the opinion but also the target. In some cases, it may also be useful to know the opinion holder, although this might not be specified, or it might be implicit in some cases. At the most general level, one can view the problem of opinion mining as that of a *slot filling* task of finding the opinion holder, entity, aspect, opinion polarity, and the time at which the opinion was made. These slots are shown in the table below:

Slot	Value
Entity Holder	CNET Editor
Entity	Logitech X300
Aspect	Battery
Orientation/polarity	Negative
Time	October 20, 2014

For the *Aspect* slot, the value "*General*" is reserved for the case when one of talking about the entity itself (e.g., Logitech X300) rather than a specific attribute (e.g., battery). Even though the five slots above provide a comprehensive view of the opinion, the values of many of these slots are often implicitly assumed (or not considered important) in many opinion mining tasks. In other words, the individual opinion mining tasks are often much simpler than slot filling in the same way as the slot filling task in information extraction has been simplified to the tasks of entity extraction and relation extraction (see Chapter 13). For example, the extraction of the opinion holder and the time of the opinion is not very different from off-the-shelf tasks in information extraction, which provide the ability to discover named entities as well as dates from unstructured text. In some simplified tasks like document-level sentiment classification, it is assumed that the targets of the opinion are known, and one only predicts the polarity of the opinion. This problem is very similar to binary classification, although the approach is often customized to the domain of opinion mining. In entity- and aspect-level opinion mining, one predicts a subset of the aforementioned slots corresponding to the entity, aspect, and orientation. Many recent techniques for opinion mining discover these different slot values simultaneously along with the corresponding opinion words. This type of simultaneous discovery of entities, opinion words, and polarities, shares many methodological and conceptual similarities with information extraction.

Finally, it should be pointed out that there are some variations of opinion mining that cannot be fully captured by the above slot-filling task. For example, sentence-level slot filling requires the creation of slots at the sentence-level, and the information about whether a sentence is subjective or objective can be represented by additional slots. Opinion mining can therefore be considered a simplification of the slot-filling task, just as entity extraction and relation mining are considered simplifications of slot-filling tasks. However, this is a very general view of opinion mining. Many document-level opinion mining methods assume that the document is about a single entity, and other slots are either implicitly known or they are not considered important. In such cases, there is little difference between the classification problem and the sentiment analysis task.

Chapter Organization

This chapter is organized as follows. Document-level sentiment classification is discussed in Section 15.2. The determination of sentence subjectivity and classification is explored in Section 15.3. Section 15.4 discusses a more generic view of opinion mining, in which it is viewed as information extraction problem. The problem of opinion spam detection is discussed in Section 15.5. Methods for opinion summarization are discussed in Section 15.6. The conclusions and summary of the chapter are presented in Section 15.7.

15.2 Document-Level Sentiment Classification

Document-level sentiment classification is the simplest setting of opinion mining in which the sentiment classification (e.g., positive, negative, or neutral polarity) is done at the document level. Furthermore, the classification is about the "*General*" aspect of the entity. Certain types of documents, such as Amazon reviews, are usually about a single entity or product, and therefore document-level methods are particularly relevant for these settings. A useful characteristic of product-centric settings is that ratings are usually available with the text of the reviews, which can be leveraged for supervised learning. For example, Amazon product reviews are associated with a rating on a five-point scale, which can be transformed to positive, negative, or neutral ratings.

This problem can be considered an off-the-shelf classification problem, and any of the existing methods for supervised learning in Chapters 5 and 6 can be used on the bag-of-words representation of text. However, a pure bag-of-words approach to text classification does not work well in this specialized setting because of the importance of linguistic subtleties and opinion words. In order to use richer information about the underlying natural language, one can also use the sequence-centric feature engineering tricks (e.g., *doc2vec* method) discussed in Chapter 10. For example, the work in [314] shows how one can use the *doc2vec* approach for sentiment analysis. Although the technique in [314] is tested in a sentence-level setting, the ideas can be easily adapted to document-level sentiment analysis.

If the bag-of-words representation is used, the tf-idf representation is usually not sufficient, and additional features are required. Sentiment analysis is a specific domain in which some features are particularly important for learning. For example, adjectives should be weighted differently than other parts of speech. However, the work by [432] showed that the use of only adjectives provided worse results than using frequent unigrams. When parts of speech are used, their primary goal is word-sense disambiguation by concatenating the specific part of speech with the word. For example "*bear-V*" is considered a different word than "*bear-N*" corresponding to the verb and noun forms of "*bear*," respectively. Furthermore, the parts of speech are more helpful in identifying useful phrases containing adjectives or adverbs.

Rather than using specific parts of speech, it is more important to distinguish words in terms of whether or not they belong to the opinion lexicon. Note that the opinion lexicon is already somewhat biased towards specific parts of speech such as adjectives or adverbs. The following are the common features that are used for document-level classification:

1. *Opinion lexicon:* Words that belong to the opinion lexicon have greater significance than those that do not. Methods for finding words that belong to the opinion lexicon are discussed in Section 15.1.1. When a word belongs to the opinion lexicon, information about the orientation of that word (such as positivity/negativity) is also incorporated among the features. A specific study of the effect of this type of lexical knowledge on classification accuracy is provided in [381].

2. *Specialized phrase extraction with adjectives or adverbs:* A specialized type of phrase extraction has been shown to be specially helpful to classification of opinions. In particular, the work in [553] showed that phrases with adjectives and adverbs have significant discriminatory power in terms of classification of opinions. Although this work was designed for unsupervised classification of reviews, it can also be incorporated for supervised classification.

3. *Term presence versus frequency:* While the frequency of a term plays an important role in traditional information retrieval, a finding in sentiment analysis is that is that the presence or absence of a term is often sufficiently significant, and the repeated use of a term does not add to significance, and it may detract from it in some cases [432].

4. *Term positioning:* The position of a token within the document plays an important role in its effect on the sentiment polarity. For example, the last sentence of a review often has a special significance in terms of summing up the feeling of the reviewer, and the presence of an opinion word towards the beginning and the middle also has specific significance. The feature position is therefore sometimes encoded into the documents. Examples of such methods may be found in [297, 432]. The basic idea is to append the feature positioning of a token (e.g., *first, middle, end*) to create a separate feature. Therefore, a feature such as "*excellent-middle*" would be treated differently from "*excellent-end.*"

5. *Negation:* Negation has an unusually important role to play in sentiment analysis, which does not have a corresponding parallel in traditional information retrieval. For example, in topic-oriented classification, the presence of the word "*not*" often tells us little about whether or not a document belongs to a specific category (e.g., "*politics*"). However, the presence of a negation while indicating whether or not one *likes* politics is often a strong indicator. The basic idea is to first create a representation that is independent of the presence of a negation, and then convert to a representation that is negation-aware. For example, the technique in [144] suggests to add the word "*not*" to words that occur near the negation. Therefore, the word "*like*" becomes "*like-not.*" Other methods [407] search for phrases that are considered *negation phrases*, which might possibly be different for different negation words.

6. *Valence shifters:* A generalization of the idea of notion of negation is that of *valence shifters* [442]. A valence shifter is any word (such as negation) that changes the value of a base word. For example, the word "*very*" can be considered a valence shifter. Valence shifters include negations, intensifiers, downtoners, and irrealis markers.

7. *Context and topic features:* The context and the topic of a document might play an important role in how the document is interpreted. For example, consider the following sentence: *"Hillary Clinton is polling really well in the final stretch."*

This sentence can be considered to be indicative of either a positive or a negative sentiment, depending on whether the opinion holder is a Democrat or a Republican. Similarly, consider the following statement: *"It was over so quickly!"*

This sentiment can be either positive or negative, depending on whether one is referring to a surgery or to a vacation. Some discussions on the use of context and topical features for sentiment classification may be found in [228, 297, 406, 586].

8. *Syntactic features:* It has been shown [586] that the use of syntactic features like parse trees can be helpful in determining the orientation of specific mentions of opinion words. Determining the orientation of specific mentions of opinion words is a first step towards determining the polarity of the sentence or the document.

In many of the supervised settings, such as product reviews, labels (in the form of ratings) are available for learning. Therefore, techniques like ordinal regression can be used. There are several works that use regression modeling rather than binary classification in order to determine the *degree* of polarity. An example of such a work may be found in [430].

15.2.1 Unsupervised Approaches to Classification

Although labeled data is often easily available when documents correspond to product reviews, this is not the case when the documents correspond to posts on a social network site, blogs, or discussion boards. In other words, paucity of labeled data becomes a major problem. In such cases, *active learning techniques* can be employed. The basic idea in active learning is to provide the user with good candidates for labeling documents, so that robust models can be learned with a small amount of training data. A discussion on active learning may be found in [1]. Documents that contain a lot of opinion words are often good candidates for active learning, because they provide evidence of the importance of specific opinion words to documents of various polarities.

Another approach is to use unsupervised learning. In fact, unsupervised learning methods were among the earliest techniques used for opinion mining [553]. The work in [553] mines phrases according to the rules shown in the table below. Here, the tags[2] starting with NN correspond to noun variations, the tags starting with VB are verb variations, the tags starting with RB are adverb variations, and the tag JJ is an adjective. The idea is to extract two consecutive words based on their parts-of-speech and then also check a third word immediately following these two words. The specific rules for the two consecutive words and the third word following it are as follows:

First Word	Second Word	Third Word (Not Extracted)
JJ	NN or NNS	anything
RB, RBR, or RBS	JJ	not NN or NNS
JJ	JJ	not NN or NNS
NN or NNS	JJ	not NN or NNS
RB, RBR, or RBS	VB, VBD, VBN, or VBG	anything

---

[2]See [694] for the complete list of tags according to the Penn Treebank project.

The extraction of parts-of-speech tags is a well-known problem in natural language processing [283, 368], and many off-the-shelf tools are available[3] for this purpose. The parts-of-speech tags are then used to extract phrases according to the rules above. Once the phrases have been extracted, their semantic orientation in terms of positivity or negativity is determined. In order to achieve this goal, the notion of *pointwise mutual information* is used. Let $\text{Count}(p, w)$ be the number of times that the phrase p and word w co-occur in a document. The notion of co-occurrence is defined in terms of closeness in occurrence of the phrase and the word. Then, the pointwise mutual information $PMI(p, w)$ is defined as follows:

$$PMI(p, w) = \log_2 \left(\frac{\text{Count}(p, w)}{\text{Count}(p) \cdot \text{Count}(w)} \right) \tag{15.1}$$

Then, the semantic orientation $SO(p)$ of phrase p is defined using two special words corresponding to "*excellent*" and "*poor*," respectively.

$$SO(p) = PMI(p, \text{"}excellent\text{"}) - PMI(p, \text{"}poor\text{"}) \tag{15.2}$$

These two specific words are chosen because they often correspond to high and low ratings in reviews, respectively. In some sense, these can be viewed as two (extremely strong) opinion words with special significance.

Which corpus is used to perform the aforementioned computations? The work in [553] issues queries to a search engine in order to discover documents in which the discovered phrases (according to the rules of the earlier table) occur in proximity of these two opinion words. Note that the phrases are extracted from the corpus that one is trying to classify whereas the computation of the semantic orientation of the phrases is done using the results from a search engine. The approach in [553] used the *AltaVista* search engine in order to discover documents containing each of the two special opinion words, as well as documents that contain phrases in proximity of either of the two special opinion words "*excellent*" and "*poor*." The notion of proximity is defined in terms of a distance of at most ten tokens. Then, the semantic orientation is defined in terms of the search hits as follows:

$$SO(p) = \log_2 \left(\frac{\text{Hits}(p \text{ NEAR } \text{"}excellent\text{"}) \cdot \text{Hits}(\text{"}poor\text{"})}{\text{Hits}(p \text{ NEAR } \text{"}poor\text{"}) \cdot \text{Hits}(\text{"}excellent\text{"})} \right) \tag{15.3}$$

How is the computed semantic orientation used to classify reviews? Given a review, the approach computes the semantic orientation of all the phrases in them and averages them. If the semantic orientation is positive, then the review is classified as positive. Otherwise, the review is classified as negative.

A later follow-up work by Turney and Littman [554] discusses how one can use latent semantic analysis in order to find the semantic orientation of words. Another interesting lexicon-based approach for finding semantic orientation is discussed in [540]. This technique combines parts-of-speech analysis with valence shifters in order to discover the semantic orientation of words.

15.3 Phrase- and Sentence-Level Sentiment Classification

Phrase- and sentence-level sentiment classification problems are not standalone problems in their own right, but they are often used as approaches to enable methods for document

---

[3]See the section on software resources at the end of this chapter.

classification. In general, a document can be broken up into smaller units corresponding to paragraphs, sentences, or phrases, and at least some of these units can be meaningfully classified. These classifications can then be aggregated into document-level classification. It is also noteworthy that phrase-level classification is performed at the granularity of individual words, which is similar to the problem of finding an opinion lexicon. However, the main difference is that in phrase-level classification, one is attempting to classify the polarity of individual *mentions* of words and phrases, rather than the typical polarity of a particular word over all mentions.

As discussed in the introduction of this chapter, individual sentences can be either subjective or objective. Objective sentences, which are statements of fact rather than expressions of polarity, often dilute the effectiveness of a document-level (sentiment) classifier. This is because the content and tone of such sentences is often neutral, and the use of such sentences in the sentiment classifier worsens the accuracy. Therefore, there are two separate problems at the sentence level, because it is often hard to classify the sentiment of objective sentences. These two problems are as follows:

1. **Subjectivity classification:** Given a sentence, is it subjective or objective?

2. **Sentiment classification:** If the sentence is subjective, then is its polarity positive or negative?

Although both of the above two problems are binary classification problems, they have very different domain-specific characteristics. In both cases, the individual units are short text segments, and therefore it can sometimes be useful to leverage techniques for classification of short text (e.g., the embedding tricks in Chapter 10). Furthermore, the enrichment of text with additional features can also be useful in these settings.

In subjectivity classification, we have labeled sentences available with the binary class *subjective/objective*, one can use off-the-shelf classifiers for determining whether unlabeled sentences are subjective or objective. However, such an approach ignores a lot of information about the proximity of the different sentences to one another. Just as the notion of inter-sentential consistency applies to polarity classification, it applies to subjectivity classification as well.

15.3.1 Applications of Sentence- and Phrase-Level Analysis

Sentence- and phrase-level analysis is often used as an intermediate step in opinion mining applications. There are two key applications of this type of analysis:

1. *Preprocessing step for document-level classification:* The sentence- and phrase-level analysis of documents helps in providing intermediate results that are often used for document-level classification. For example, objective sentences might be removed in order to improve classification accuracy. Furthermore, some document-level classification methods use an approach in which fine-grained classification at the word and phrase level is aggregated in order to perform document-level classification.

2. *Summarization in terms of polar phrases and sentences:* A closely related task to opinion classification is that of opinion summarization in which an explanation is provided about why a particular document is classified in a certain way. In many cases, the sentences and phrases with the greatest polarity are extracted from a review and presented to the user along with the overall classification.

A discussion of the related task of opinion summarization is provided in Section 15.6.

15.3.2 Reduction of Subjectivity Classification to Minimum Cut Problem

The work in [429] transforms the problem to a minimum cut problem. In this transformation, each sentence is treated as a node. Furthermore, a source and a sink node is added, and goal is to create a partition of nodes in which the source and sink are on opposite sides. All nodes on the same side of the source are deemed to be subjective sentences, whereas all nodes on the side of the sink are deemed to be objective sentences. Therefore, the key is to define the weights of the edges in the network in such a way that the effect of the output of a *subjective/objective* classifier and the inter-sentential consistency are reflected in these weights. The weights of the edges from the source to the various nodes (sentences) are defined to be the (standalone classifier) output probabilities that these sentences are subjective. Similarly, the weights of the edges from the sink to the various nodes (sentences) are defined to be the (standalone classifier) output probabilities that these sentences are objective. Although some classifiers like the naïve Bayes classifier do produce probabilities, it is also possible to use classifiers that output some type of numerical weight by normalizing the numerical weights for the two classes to sum to 1. For example, in a support vector machine, one can convert the distance from the hyperplane into a probability with a logistic function. Other types of heuristic functions were tested in [429]. Note that if these are the only edges that are used, then a minimum *s-t* cut in this network will simply assign each sentence independently to the class to which it has the best affinity. Therefore, inter-sentential consistency is enforced by adding edges between each pair of nodes with an *association weight*.

Consider a setting in which the sentences are ordered $s_1, s_2, \ldots s_r$ as they appear in the text. The distance between two sentences s_i and s_j is therefore given by $|j - i|$. The association weight between a pair of sentences s_i and s_j is defined as follows:

$$\text{Assoc}(s_i, s_j) = \begin{cases} C \cdot f(|j - i|) & \text{if } |j - i| \leq T \\ 0 & \text{otherwise} \end{cases} \tag{15.4}$$

Here, C is a scaling parameter and T is a threshold parameter. The function $f(x)$ is a non-increasing function of x. The values of $f(x)$ used in [429] were 1, $1/x^2$ and $\exp(1 - x)$. The edges between the sentences s_i and s_j are then weighted with the association weight defined above. The minimum *s-t* cut was found in the graph, and all nodes (sentences) on the same side of the source were deemed to be subjective. It was found in [429] that such an approach significantly improved the effectiveness of subjectivity classification.

15.3.3 Context in Sentence- and Phrase-Level Polarity Analysis

Once the subjective and objective sentences have been determined, a common approach is to classify the polarity of subjective sentences. These can be used in order to classify the polarity at the document level. At the fine-grained level of sentences, the context becomes particularly important in identifying polarity. For example, consider the sentence:

"I hardly consider my experience with this product satisfying."

Even though the word "*satisfying*" is a positive member of the opinion lexicon, the presence of the valance shifter "*hardly*", changes the semantic orientation of this word. Therefore, the approach in [586] distinguishes between the prior orientation of a word, and a specific orientation of the word mention depending on its usage in the corpus. This is achieved

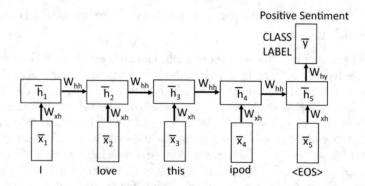

Figure 15.1: Re-visiting Figure 10.17: Example of sentence-level classification in a sentiment analysis application with the two classes *"positive sentiment"* and *"negative sentiment."*

by mining features like the presence of negations or other valence shifters that change the tone of the sentence. Furthermore, structural features extracted from the parse tree and the topical content of the document are used as features. These features are extracted in order to perform off-the-shelf classification of words and phrases at the level of specific mentions. In particular, the *AdaBoost* classifier was used in [586]. Such word- and phrase-level classification methods can also be used to classify sentences more accurately because the orientations of the words and phrases have been properly corrected by their context.

15.3.4   Sentiment Analysis with Deep Learning

Various types of recurrent neural network architectures (cf. Section 10.8.4 of Chapter 10) are also used for sentiment analysis. In particular, the use of long short term memory (LSTM) variants of the recurrent architecture is very common. One advantage of using LSTM is that the effect of valence shifters can be automatically learned in a data-driven manner, because the long-term and short-term memory captured by this architecture is able to learn the effect of valence shifters directly from the training data. In recent years, transformer networks have also gained increasing attention in terms of performing sentiment analysis tasks. In the following, we provide a brief overview of the common deep learning methods that are used.

15.3.4.1   Recurrent Neural Networks

As discussed in Section 10.8.4 of Chapter 10, recurrent neural networks are used for sentence-level classification tasks. Since sentiment analysis is sentence-level classification task, one can leverage any deep learning model that works for sentence-level classification. The architecture of a recurrent neural network for sentiment analysis is illustrated in Figure 15.1. In this case, the recurrent neural network outputs a class label at the very end of the sentence corresponding to whether the sentence has a positive sentiment or a negative sentiment. One can use a similar approach in order to classify sentences as subjective or as objective.

Although the architecture shown in Figure 15.1 uses a vanilla recurrent network, a more sophisticated variant, such as an LSTM or a GRU is used in order to deal with longer sentences. Furthermore, it is common to use bidirectional networks in order to incorporate bidirectional context. In particular, the advantages of using gated networks for sentiment analysis are discussed in [544].

15.3.4.2 Leveraging Pretrained Language Models with Transformers

The pre-trained transformer-based language models in Chapter 11 can be naturally used for sentence-level classification tasks. In particular, GPT-n, T5, and BERT can be adapted to sentence-level classification as follows:

- *GPT-n:* In this case, an additional token can be added to the vocabulary, which separates the end of the sentence from the sentiment label. When the model encounters this token, it is a cue for the model to output the sentiment label. During the fine-tuning phase, the next token after the special token will be the class label, since this is what the fine-tuning phase will encounter in the training data. Another possibility is to train a special layer of weights, which takes as input the representation of this special token and outputs the class label. Earlier versions of GPT used the approach of training an additional weight layer for classification purposes.

- *T5:* The T5 pre-trained model is introduced in Section 11.4.3. T5 is a text-to-text transformer, and therefore the input is a sentence, and the output is a "sequence" corresponding to the raw text label. The T5 model uses as prefix, such as "sentiment" to specify the task, followed by the input sequence, which is trained to map to the output sequence. A training example for the sentiment analysis task in T5 might be as follows:

  Input Sequence: sentiment: the movie was so boring that I slept through it.
  Output Sequence: negative

  Note that the input format is quite similar to CoLA (see Section 11.4.3), except that the prefix is different, corresponding to the specification of the sentiment analysis task. In principle, it is possible for to train T5 for multiple tasks together, and the prefix tells the model which task to perform.

- *BERT:* BERT already uses a [CLS] token to represent the input sentence. Therefore, a weight layer is fine-tuned with respect to the representation of [CLS]. Another important point is that BERT uses sentence pairs for pretraining, whereas sentence-level classification requires only a single sentence as input. Therefore, during fine-tuning, it is assumed that the second sentence of the pair is a null sentence. Note that a similar approach is used in other sentence-level classification applications, such a linguistic grammar checking (cf. Section 11.5.2 of Chapter 11).

Pre-trained language models often provide a high-degree of accuracy, and can even be used in cases where one is trying to perform paragraph- or document-level sentiment analysis. This is because transformer-based language models are much more effective at handling long sentences as compared to recurrent neural networks.

15.4 Aspect-Based Opinion Mining as Information Extraction

As discussed in the introduction section, one can view the opinion mining task in a similar way to information extraction at the most general level. This is because each sentence in an opinion mining task can state an opinion about a target, which might not be the base entity but it might only be an aspect or attribute of the entity. As in document-based opinion mining, there are both unsupervised and supervised variants to the aspect-based opinion mining problem. Information extraction is traditionally defined in the supervised setting, although many recent methods for information extraction (like open-domain information extraction) are defined in the unsupervised setting. It is not surprising that supervised methods have better accuracy than unsupervised methods; the main challenge in using supervised methods is that a sufficient amount of training data is not available. However, with the availability of an increasing number of crowd-sourcing platforms like *Amazon Mechanical Turk*, this problem is likely to become less severe over time.

Interestingly, some of the unsupervised systems for information extraction (like OPINE [445]) are built on top of unsupervised information extraction systems (like *Know-ItAll* [180]). Furthermore, many supervised methods for opinion mining (like the one in [264]) adapt methods like conditional random fields from information extraction to opinion mining. As in document-based sentiment classification, significant customizations are needed to adapt the information extraction approaches to the domain of opinion mining. In fact, some of the early unsupervised methods for aspect-based opinion mining (such as Hu and Liu's seminal approach [260]) developed in parallel with the field of unsupervised information extraction. Therefore, many aspect-based opinion mining methods are not formally recognized as information extraction tasks, although it is easy to see the similarities between these methods and information extraction tasks.

15.4.1 Hu and Liu's Unsupervised Approach

Hu and Liu's work [260] was one of the earliest unsupervised methods for aspect-based opinion mining. Central to the problem of aspect-based opinion mining is the problem of discovering product features. This is achieved by identifying sets of words that co-occur in many texts and are nouns. Nouns can be identified using parts-of-speech tagging methods from natural language processing [283, 368]. Subsequently, a flat file of "transactions" is created in which each line contains the nouns and noun phrases in each sentence. All other words are discarded under the assumption that they do not reflect product features. The approach first applies the *Apriori* algorithm from *frequent pattern mining* [2] in order to discover groups of three or less words that co-occur in many opinion texts. A minimum threshold on the co-occurrence frequency, which is referred to as the *minimum support*, is used to identify the relevant sets of words.

Subsequently, a *compactness criterion* is applied to these sets of words. Let F be a feature set containing the words in the specific order $w_1 \ldots w_n$ in a particular sentence S. This specific instantiation of the feature set is said to be compact, if the distance between adjacent words w_i and w_{i+1} is no greater than three. A feature set F is said to be compact, if at least two instantiations of it in specific sentences are compact. For example, consider the following two sentences [260]:

> "This is the best digital camera on the market."
> "This camera does not have a digital zoom."

The first sentence is compact for the feature set {*"digital"*, *"camera"*}, but the second is not.

After the compact features have been identified, the redundant features with only single words are removed. For example, if the phrase *"battery life"* is both compact and frequent, it is often possible for the single word *"life"* to be redundant with respect to this phrase. In order to find redundant words, the p-support (i.e., pure support) of a word is found, which is the number of sentences containing the word as a noun, in which the frequent superset phrases are not present. For example, if the support of *"life"* is 10, that of *"battery life"* is 4, and that of *"life guarantee"* is 5 (with the last two in disjoint sentences), then the p-support of *"life"* is $10 - 5 - 4 = 1$. The p-support of single words must be greater than a separate threshold on that quantity in order for it to not be considered redundant. This threshold was set to 3 in [260].

After product feature (aspect) extraction, the next phase is that of opinion word extraction. An important observation is that opinion words are often adjectives that occur near the product feature words within the sentences. Therefore, the following rule is used [260]:

> For each sentence in the review database, if it contains any frequent feature, extract the nearby adjective. If such an adjective is found, it is considered an opinion word. A nearby adjective refers to the adjacent adjective that modifies the noun/noun phrase that is a frequent feature.

The adjacency rule also provides a way of identifying product features that were missed by the initial attempt. This is because some of the features are inherently *infrequent*, which cannot be found by frequent pattern mining algorithms. For example, consider the phrase:

> "Red eye is easy to correct."

The phrase *"red eye"* might be a useful product feature, but with insufficient frequency. It is noteworthy that this product feature occurs close to the word *"easy."* Therefore, in order to identify the elusive features, which are infrequent, one uses the opinion words to determine the features as follows:

> For each sentence in the review database, if it contains no frequent feature but one or more opinion words, find the nearest noun/noun phrase of the opinion word. The noun/noun phrase is then stored in the feature set as an infrequent feature.

The *nearest* noun/noun phrase of an opinion word is the noun or noun phrase that the opinion word modifies.

After extraction of the opinion features, the semantic orientation (i.e., polarity) of each sentence is identified by using WordNet to determine the orientation of the opinion words in it. The opinion orientation is identified by using the dominant orientation of the opinion words in the sentence. Note that this approach can also be considered sentence-level classification, although it can also be used for aspect-level classification because of the fact that it identifies the product features in individual sentences.

15.4.2 OPINE: An Unsupervised Approach

One of the earliest uses of information extraction for opinion mining was the OPINE system [445], which is directly based on an information extraction system [180] referred to as *KnowItAll*. The basic idea in this approach is to use the steps of (i) identifying products and their attributes, (ii) Mining opinions about product features, and (iii) determining opinion polarity. As a final step, one can also rank the opinions if needed.

The OPINE system distinguishes between *explicit* features and *implicit* features. The former can be mined more easily using certain syntactic characteristics. For example, the word *"scan quality"* refers to an explicit product feature, although it might sometimes be implicitly referred to as *"scans."* In order to find the explicit product features, one needs to find the related concepts such as parts and properties of the product. The first step is to find noun phrases from reviews that have frequency greater than an experimentally set threshold. Each such noun phrase is assessed by computing the pointwise mutual information (PMI) scores between the phrase and *meronymy discriminators* associated with the product class. Examples of such discriminators include *"of scanner," "scanner has,"* and so on. Note that this basic idea is similar to that of [553], as discussed in Section 15.2.1. However, the goal here is to find product features rather than opinion words.

Once the product features have been extracted, the opinion words are assumed to the terms that occur in the vicinity of these features. This assumption is similar to that used in [260]. However, rather than using simple proximity conditions, ten specific extraction rules are defined in order to identify opinion phrases. At this point, one is able to generate opinion-feature-sentence triplets, each of which is assigned a semantic orientation. This achieved by first starting with a semantic orientation of the (opinion) word, generalizing it to a semantic orientation of the word-feature pair, and then generalizing it to the word-feature-sentence triplet (i.e., opinion-feature-sentence triplet). The basic idea is to successively incorporate context into the process. For example, *"hot coffee"* might be positive whereas, *"hot room"* might be negative. This is achieved with the use of *relaxation labeling* in which the neighborhoods of word-feature pairs are defined. For example, if *"hot room"* appears in the neighborhood of many words with negative orientation like *"stifling kitchen"* then the relaxation labeling approach will eventually assign it a negative label in an iterative update process. The initial semantic orientation of a word is defined using the same PMI-based approach [553] discussed in Section 15.2.1. The neighborhood relationships in relaxation labels are inferred from conjunctions and disjunctions in review text, syntactic dependency rules, relationships between words (e.g., grammatic/tense variations of the same word), and WordNet-specified dependencies. Note that the final result of the approach assigns polarities to specific mentions of opinion words and features in individual sentences.

15.4.3　Supervised Opinion Extraction as Token-Level Classification

An interesting method for supervised opinion extraction with hidden Markov models was the *OpinionMiner* system [266]. This approach is particularly notable because it transforms the (supervised) opinion mining problem to almost the same form that is used in information extraction. The approach is a holistic technique that answers several useful questions that arise in the aspect-based opinion mining. These questions pertain to (i) the extraction of potential product entities and opinion entities from the reviews, (ii) the identification of opinion sentences that describe each extracted product entity, and (iii) the determination of opinion orientation (positive or negative) given each recognized product entity.

To achieve these goals, one needs to define *entity tags*, as is used in information extraction (cf. Chapter 13). The work in [266] defined two types of entities, which are referred to as *aspect entities* and the *opinion entities*. Examples of aspect entities and opinion entities in the context of opinions provided on a camera are shown below.

Tag Set	Corresponding Entities
⟨PROD_F⟩	Feature entity (e.g., camera color, speed, size, weight, clarity)
⟨PROD_P⟩	Component (part) entity (e.g., LCD, battery)
⟨PROD_U⟩	Function of entity (e.g., move playback, zoom)
⟨OPIN_P⟩	Positive opinion entity (e.g., "*love*")
⟨OPIN_N⟩	Negative opinion entity (e.g., "*hate*")
⟨BG⟩	Background words (e.g., "*the*")

The original work [266] also defines notions of *implicit* and *explicit* opinions with a more refined tag set, although we omit this distinction for simplicity. As in traditional information extraction, entities can be represented by either individual words or phrases. However, it is much easier to build Markovian models by tagging at the word level. Therefore, we append one of the three symbols corresponding to $\{B, C, E\}$ corresponding to beginning, continuation, and end word of the tag. For example, the beginning word of a product feature becomes tagged by ⟨PROD_FB⟩, a continuation word by ⟨PROD_FC⟩, and an end word by ⟨PROD_FE⟩. This type of *hybrid* tagging approach is used commonly in information extraction (cf. Section 13.2.2 of Chapter 13) in order to transform the problem to token-level classification. Token-level classification can be handled much more easily with a variety of sequence-centric models. Consider the sentence, "*I love the ease of transferring the pictures to my computer.*" Then, this sentence is tagged as follows:

I   love   the   ease   of   transferring   the   pictures to my computer .
BG OPIN_P BG PROD_FB PROD_FM   PROD_FM   PROD_FM PROD_FE BG   BG   BG

In the above example (based on [266]), opinion entities and background tokens are not encoded in hybrid format, because they are treated as independent entities. However, it is also possible to encode these types of entities in hybrid format, especially in cases where the presence of phrases is common. *This transformation creates the same token-wise classification problem, as is used in the information extraction problem* (see Section 13.2.2). As a result, almost all the information extraction methods such as hidden Markov models, maximum entropy Markov models, and conditional random fields (cf. Sections 13.2.3, 13.2.4, and 13.2.5) can be used for this problem. Although the approach in [266] uses a hidden Markov model for solving the problem, the specific details of the approach are less important than the fact that the transformation in this paper opens the door to use of a well-known family of techniques from information extraction. Indeed, a later work [329] used a similar kind of token-level classification in combination with conditional random fields for aspect-based opinion mining. Furthermore, it is also possible to focus exclusively on extracting other types of properties of opinions (e.g., opinion sources or only opinion targets) rather than mining opinion phrases. The main difference among all these cases lies in the preprocessing step of deciding how the tokens in the sentence are labeled. The work in [264] proposed a method based on conditional random fields in order to identify the opinion targets. The sources of opinions are identified in [114] with conditional random fields.

It is also noteworthy that this family of Markovian models for information extraction allows the use of a wide variety of features associated with each token, such as the orthography, part-of-speech and so on. One can, therefore, engineer useful features in the opinion mining problem as in the case of the information extraction problem. The work on feature engineering for opinion mining is limited compared to that in information extraction, and there is significant scope of using such methods in opinion mining. Furthermore, the recurrent neural networks and long short-term memory networks discussed in Chapter 10 can also be used for token-level classification in opinion mining (cf. Section 10.8.5).

15.5 Opinion Spam

Opinion mining is often performed on reviews of products, in which good reviews are rewarded with customer interest and better sales. This fact provides a significant incentive to the sellers and manufacturers of items to cheat on the reviews. For example, the author of a book on Amazon might post fake reviews about his or her item. The problem of opinion mining is closely related to that of *shilling recommender systems* [3] in which users post fake ratings about items. Such users are referred to as "shills" in the parlance used in the field of recommender systems.

The main difference between the field of shilling recommender systems and opinion spam is that the former is exclusively devoted to the analysis of numerical ratings, whereas the latter is (almost) exclusively focused on the textual component of the review. In the case of opinion mining, numerical ratings are not always available, and therefore it makes sense to focus on the textual component. Nevertheless, in cases where textual reviews are posted with numerical ratings, it makes sense to use both types of methods to detect fake reviewers. There are two primary types of spam discovery. The first uses supervised learning in order to discover fake reviews. The second uses unsupervised methods to detect fake *reviewers* who have certain types of atypical behaviors.

15.5.1 Supervised Methods for Spam Detection

In supervised methods, it is assumed that training data is available indicating which reviews are fake, and which ones are not. The main problem with this approach is that it is very hard to obtain labeled training data. Furthermore, unlike many other types of topical classification, manual annotation is extremely difficult. After all, spammers have an inherently deceptive intent, and it is difficult to manually label a review as a spam just by reading it. However, some types of reviews are easier to classify using manual labeling. The work in [267] defined three types of reviews:

- *Type I:* The first type of spam contains untruthful opinions on products. These types of reviews might either promote a product, or they might maliciously try to harm the reputation of a product. This is the most common type of spam and it is often difficult to identify such types of spam manually because a user might be careful while crafting it.

- *Type II:* The second type of spam contains opinions about product *brands* but not about the manufacturers. This type of spam might be caused by the employee of a specific brand trying to market all the products for that brand. In many cases, such type of spam tries to praise or criticize the brand rather than the product itself, when the spammer has not spent the time needed to craft careful product-specific opinions.

- *Type III:* The third type of spam contains advertisements and other non-informative content about individual products.

The second and third types of spam are relatively easy to manually identify and label. Therefore, the work in [267] treats the prediction of a spam of the first type differently from the spams of the second two types. The first type of spam is referred to as *deceptive spam*, whereas the second two types of spam are referred to a *disruptive spam* [425].

15.5.1.1 Labeling Deceptive Spam

The creating of labeled data is the key in using supervised methods for spam detection. Since disruptive spam is relatively easy to label manually, the main challenges arise in terms of how deceptive spam may be labeled. Therefore, this section will focus on the problem of labeling deceptive spam. Several techniques have been proposed in the literature for labeling deceptive spam.

1. *Leveraging duplicates:* One approach to labeling is by identifying reviews that were duplicates or near duplicates. Although duplicates might sometimes be legitimately caused by a user clicking twice on the "submit" button by accident, there are other types of duplicates that are often spam. In particular, near duplicates form the same userid on different products, or duplicates from different userids on either the same or different products are often spam. The work in [267] marks these reviews as spam, and the remaining reviews are marked as non-spam. Duplicates were detected by extracting the bigrams from each review, and then marking a pair of reviews as similar if the Jaccard coefficients between their bags-of-bigrams was greater than 0.9.

2. *Constructing a spam review training data set using Amazon Mechanical Turk:* A second approach is the use of Amazon Mechanical Turk to have users hand-craft fake reviews [425] (of a positive nature). This approach uses reviews of hotels as a test case. Truthful reviews were obtained by using 5-star reviews from *TripAdvisor* after removing some of the reviews that were more likely to be spam. In particular, reviews that were unreasonably short or were submitted from first-time authors were removed. The idea here is to ensure that truthful reviews are contaminated with as little spam as possible. The fake reviews were obtained from 400 Amazon Turk participants, who were asked to comment on a particular hotel (from the same set used in the truthful reviews). The Turker was asked to assume that they work for the hotel's marketing department, and to pretend that their boss wants them to write a fake review. The review needed to sound realistic and portray the hotel in a positive light. Finally, the (larger number of) truthful reviews were subsampled down to 400 in such a way that the length distribution of the reviews in the two sets was the same.

In general, the approach of manual labeling with crowd-sourced system like *Amazon Mechanical Turk* does come at a cost, although it provides the benefit of more realistic training data. There are still some residual biases in the data, because a simulated user on *Amazon Mechanical Turk* has a different motivation for submitting a fake review than a fake reviewer in the real world.

15.5.1.2 Feature Extraction

The features that are used for identifying review spam are extracted from the review content, the reviewer who posted the review, and the product being reviewed. Some commonly used features [267] are as follows:

1. Many reviewer sites like Amazon have feedback mechanisms. In such cases, the number of feedbacks, the number of positive feedbacks, and the percentage of positive feedbacks are used as features.

2. Longer reviews and titles tend to get more positive feedbacks (which is recognized by spammers). Therefore, the lengths of the review title and/or body could be used as features.

3. Earlier reviews tend to get more attention. Therefore, the temporal rank of the review of a product is used as a feature.

4. Spammers tend to be unbalanced in their use of opinion words. Therefore, the percentage of positive and negative opinion words are used as features.

5. Spammers often provide excessive detail and over-emphasize with capitalization. Therefore, the percentages of numerals, capitals, and all-capital words in the review were extracted.

6. Some features are also extracted based on the similarity between the review and the description of the product, or the number of times that the brand name is mentioned. These types of spam occur in cases where an advertisement is posted or the opinion is about a specific brand.

7. For product-oriented systems, ratings are available. Therefore, the deviation of the rating from the average, the rating value, and the orientation of the rating are extracted as features.

8. A feature can be extracted when a good review is posted just after a bad review and vice versa. The former is often common when a spammer tries to do damage control.

9. The percentage of time that a reviewer was the first to post a review can be extracted as a feature. Furthermore, the percentage of time that the reviewer posted the only review can be extracted.

10. Spammers often tend to give similar ratings to various products to save time. Therefore, the standard deviation of ratings posted by the reviewer can be extracted. The average rating and the distribution of good and bad ratings can be extracted.

With these extracted features, the work in [267] used logistic regression as the classifier of choice. Other work has also used support vector machines. The optimal choice of the classifier might be sensitive to the particular data set at hand.

15.5.2 Unsupervised Methods for Spammer Detection

Unsupervised methods are generally focused on detecting *spammers* (i.e., users posting spam) rather than spam reviews. The basic idea here is that spamming users often have certain types of undesirable behaviors that are easy to detect with unsupervised methods. Spammers are sometimes also referred to as *trolls* in some online settings. Some sites, such as Slashdot, Wikipedia, and Epinions, allow users to specify trust and distrust links about each other, which can be used to create signed network representations of trust and distrust [545]. Such signed networks can be used in combination with *PageRank*-like algorithms in order to find trustworthy and untrustworthy users [593]. Such networks can be used in order to identify the reliability of users and reviews jointly. In settings where users are not allowed to specify trust and distrust links about one another, it is still possible to *predict* distrust links between users by using their feedbacks about each other's reviews [546]. Finding pairs of users that distrust one another is helpful in constructing a signed network, which is eventually helpful in finding spammers. These methods do not use text analysis in any way, but are, nevertheless, useful for the discovery of spammer behavior.

The work in [334] proposes to find spammers with the use of rating data and content associated with reviews. This approach uses four different models referred to as (i) the targeting product, (ii) the targeting group, (iii) general rating deviation, and (iv) early rating deviation. The scores of these models can be combined to create an integrated score. For example, to identify the targeted product, multiple reviews on the same product by the same user with very high/low ratings and similar review text is suspicious. In some cases, a reviewer may give spam ratings to many different products of the same brand, which corresponds to the targeting group model. When a reviewer assigns a rating that is very different from other reviews for the product, then the review is more likely to be spam. Finally, the early rating deviation model identifies reviewers that rate products soon after it becomes available for review. Most of these methods do not use deep text analysis, and therefore we omit a detailed discussion of these methods.

15.6 Opinion Summarization

Opinion summarization is closely related to the problem of text summarization discussed in Chapter 12. However, there several unique characteristics of the problem in the opinion mining domain. Like traditional document summarization, one can summarize either a single document or multiple documents. However, the nature of the summaries can be either textual or non-textual. Textual summaries are often constructed using considerations beyond topical characteristics of the document, because the subjectivity of sentences also plays an important role in deciding whether or not that sentence should be included in the summary. Another way of classifying summarization methods is in terms of whether or not they are aspect-based. In the following, we provide an overview of the different types of summaries that are created from documents. Many of these techniques can be enabled by using variations of methods discussed both in this chapter and in the book. In many cases, opinion summarization is used as the final step after sentiment classification [261], and therefore the intermediate steps and outputs of sentiment classification provide the raw statistics needed for summarization.

Rating Summary

This is the simplest type of summary associated with rating-based review systems. For example, Amazon always provides a summary rating of each item, which is typically the arithmetic mean of the user provided ratings. Some rating systems also allow users to rate different aspects of a product, such as the battery. In such cases, one can also provide a summary of the ratings of each individual aspect. Note that this type of approach does not use any natural language processing, and it will work only when the opinions are associated with *explicitly specified* ratings.

Sentiment Summary

The notion of sentiment summary is similar to that of a rating summary, except that the techniques of this chapter are used to perform sentiment-based classification of either the individual documents or the aspects in the documents. Most classifiers will return a numerical score along with each such classification, which can be viewed as the analog of a user-specified rating. Subsequently, a summary can be presented at either the overall (general) level or at the aspect level, which shows the means of the scores. More commonly, the percentage of positive and negative orientations ate presented in order to provide a more interpretable summary.

Sentiment Summary with Phrases and Sentences

In this case, the sentiment summary is provided together with key phrases that provide an explanatory point of view of the summary. For example, if the battery aspect receives a 70% positive response, one can add the phrases *"long lasting," "sturdy,"* and *"too noisy"* to the sentiment summary in order to provide better insights. Such methods can be implemented by extracting the key phrases, sentences, and opinion words that contribute to the classification process. The sentence- and phrase-level classification methods of Section 15.3 are useful in the process of generating such summaries. It makes sense to use only subjective sentences that are oriented in one direction or the other for the summarization process. A review may often contain both strong points and weak points about a product. A contrastive summary presents the summaries of the strong points and the weak points separately.

Extractive and Abstractive Summaries

These types of summaries are similar to those presented in Chapter 12. In the setting of opinion mining, opinion words often have greater importance in the creation of summaries. Furthermore, multi-document summaries are particularly important, since each product may be associated with multiple reviews. Abstractive summaries try to create more coherent summaries from a large number of opinions. However, the creation of abstractive summaries is currently considered too difficult, and is an emerging area of research.

15.7   Summary

Opinion mining and sentiment analysis has gained increasing attention in recent years because of the vast amount of reviews and other textual data being contributed on social platforms. Sentiment analysis can be done at the document-level, at the sentence-level, or at the aspect level. Document-level sentiment mining is an instantiation of the classification problem, although unsupervised methods also exist for such cases. Aspect-level opinion mining is a special case of information extraction, and many of the existing information extraction techniques can be generalized to this case. The quality of opinion mining can be improved by using various techniques for spam detection. The final process in opinion mining is that of opinion summarization, in which the intermediate steps of sentiment classification are used to create summaries.

15.8   Bibliographic Notes

An excellent book on sentiment analysis may be found in [347]. Shorter reviews of sentiment analysis from the same author may be found in individual chapters of [15, 345]. An earlier review of opinion mining and sentiment analysis from a different set of authors may be found in [431].

Dictionary-based approaches for opinion lexicon expansion are discussed in [23, 178, 179, 286]. Corpus-based methods with the use of connectives are discussed in [240]. The use of sentential consistency for corpus-based lexicon expansion is discussed in [158, 288, 355]. Syntactic methods for opinion-based lexicon expansion are discussed in [553]. The use of double propagation with entity-extraction methods for lexicon extraction are discussed in [453]. In this approach, entities/aspects and opinion words are extracted together. The discovery of multi-word opinion expressions with the use of context is discussed in [69]. The work in [586] distinguishes between the notion of prior polarity of words and that of specific *instantiations* of words in a particular context.

Numerous techniques have been proposed for document-based sentiment classification. In this context, the work in [432] is particularly notable in making numerous advances to machine-learning methods. Feature engineering methods are particularly important for document-centric sentiment classification. In this context, methods for using term positioning [297, 432], negation [144], valence shifters [442, 540], and topical context [228, 297, 406] for engineering features have been proposed. The supervised learning of a vector-space representation of text with the use of ratings is discussed in [361]. Methods for using regression in combination with sentiment analysis in order to infer the precise degree of polarity may be found in [430, 587]. The integration of opinions from multiple sources with the use of semi-supervised topic modeling is discussed in [356]. The works in [540, 553, 554] proposed the use of unsupervised and lexicon-based methods for sentiment classification. The work in [554] is notable because it shows that latent semantic analysis can be useful in finding the semantic orientation of words. The use of joint topic modeling and sentiment classification is proposed in [344, 379].

The earliest definition of sentence subjectivity seems to be attributed to the work in [582]. An overview of subjectivity in opinion mining may be found in [346]. The effect of the semantic orientation of adjectives on sentence subjectivity was studied in [241]. The work in [586] showed how various types of contextual features such as negations and valance shifters can change the default polarity of a word. The work in [429] is among the earliest works on emphasizing the importance of document subjectivity on sentiment classification. A rule-based approach for discovering subjective and objective sentences from unannotated texts may be found in [583]. The use of conditional random fields for converting sentence classification to document classification is discussed in [375].

The earliest methods for aspect-based information extraction were unsupervised techniques [260]. The OPINE system [445] was built on top of an unsupervised information extraction system [180]. A discussion of open information extraction and its relevance to unsupervised aspect-based opinion mining is provided in [181]. One of the earliest techniques for supervised information extraction was proposed in [635]. The work in [266] formally recognized supervised aspect-based opinion mining as an information extraction task by transforming the problem to token-wise classification. This work used hidden Markov models in order to perform joint opinion mining and aspect extraction. The works in [264, 329] used conditional random fields to provide a solution to the opinion mining problem with a similar token-wise classification. The work in [114] focuses on the problem of opinion *source* mining with the use of conditional random fields.

Methods for finding review spam were proposed in [267, 425]. This work is complementary to the techniques for finding shills in recommender systems [3]. Signed network methods for finding trustworthy users in social media are discussed in [545, 593]. Techniques for finding opinion spammers with rating analysis and text mining are discussed in [52, 334]. The works on shilling recommender systems, trust analysis, and opinion spam have largely been conducted independently by multiple communities of researchers. It is likely that many of these methods are complementary to one another. A future direction of work could compare/contrast these methods and possibly combine them where needed. Overviews of methods for opinion summarization may be found in [296, 431].

15.8.1   Software Resources

The seminal survey by Pang and Lee [431] contains a dedicated section on publicly available resources. A number of data sets for benchmarking opinion mining and sentiment analysis algorithms are available at [708]. The *Stanford Sentiment Treebank* [741] is an important data set that is used by researchers for evaluation of sentiment analysis algorithms. A key part of opinion mining is that of linguistic processing, such as part-of-speech tagging. The Stanford NLP [650] and NLTK site [652] contain several natural language processing tools that can be used for tokenizing, term extraction operations, and part-of-speech tagging. Furthermore, opinion lexicons are also available at [34, 708]. A methodical approach for creating a sentiment-centric corpus from Twitter is discussed in [428]. For the classification and entity extraction tasks, which are often used in opinion mining, refer to software sections of Chapters 5, 6, and 13. Furthermore, recent methods for opinion mining use modern feature engineering tricks like *word2vec*, which is discussed in Chapter 10. The software for using LSTM networks in the context of sentiment analysis is available at [712], and this approach is based on the sequence labeling technique proposed in [223]. Some of the text summarization methods are also useful for opinion summarization, and an overview of the software resources for text summarization may be found in Chapter 12.

15.9   Exercises

1. Write a computer program that uses the package available at [712] for sentence-level sentiment classification.

2. Design a framework and architecture for aspect-based sentiment classification with the use of recurrent neural networks.

Chapter 16

Text Segmentation and Event Detection

"To improve is to change; to be perfect is to change often."– Winston Churchill

16.1 Introduction

Although text segmentation and event detection might seem like different problems, they are closely related. In both cases, the text in one or more documents is scanned sequentially in order to detect key changes. Therefore, the concept of *change detection in a sequential context* is the overarching theme of this chapter. This chapter covers the following topics:

1. *Text segmentation:* The goal of text segmentation is to divide a single document into coherent linguistic or topical units. Linguistic segmentation corresponds to the segmentation into words, sentences, or paragraphs, and is often based on punctuation and language-specific issues. On the other hand, topical segmentation is based on semantic content. In all cases, the unit of segmentation is a smaller portion of a document, such as a line, sentence, or a fixed number of tokens.

2. *Mining text streams:* In the text stream setting, the text is analyzed sequentially within the context of multiple documents. Therefore, the unit of analysis is an individual document. In this setting, it is particularly important to process the streams in a single pass. A foundational problem in the stream setting is that of clustering text streams, because it provides a summary representation of the stream and serves as a starting point for solutions to other problems like event detection.

3. *Event Detection:* Event detection refers to the occurrence of unusual and sudden changes in a text stream structure, which are often caused by external shocks. A first step in event detection is that of creating a summary representation of the data stream. This summary representation is then leveraged for finding unusual deviations that correspond to informative events.

This chapter will provide an overview of the key topics and ideas underlying each of these subject areas.

© Springer Nature Switzerland AG 2022
C. C. Aggarwal, *Machine Learning for Text*,
https://doi.org/10.1007/978-3-030-96623-2_16

16.1.1   Relationship with Topic Detection and Tracking

It is noteworthy that many of these topics are closely related to the *topic detection and tracking effort* (TDT), which was a DARPA-sponsored[1] initiative in finding events in a stream of broadcast news stories. The original effort contained three major tasks: (i) segmenting a stream of data, especially recognized speech, into distinct stories; (ii) identifying those news stories that are the first to discuss a new event occurring in the news; and (iii) given a small number of sample news stories about an event, finding all following stories in the stream. A more detailed description may be found in the final report of the pilot study [20]. The last of these tasks is an online classification task, and almost all the methods in Chapters 5 and 6 can be used for this problem. Therefore, this chapter will primarily focus on the first two tasks. Furthermore, some related problems that are not discussed in the original TDT effort will also be discussed in this chapter.

The current state-of-the-art supports many applications beyond the ones discussed in the original TDT effort. For example, the ideas in the segmentation task have now been generalized to the information extraction domain. Similarly, from a chronological point of view, the streaming setting was investigated several years later than the original TDT effort, but it has more stringent computational and memory constraints. There are, however, close similarities in the TDT and streaming setting, which are both temporal in nature. This chapter will, therefore, provide an integrated overview of stream mining, segmentation, and event detection.

Chapter Organization

This chapter is organized as follows. Section 16.2 discusses the problem of text segmentation. The problem of mining text streams is discussed in Section 16.3. Section 16.4 studies the problem of event detection. A summary of the chapter is provided in Section 16.5.

16.2   Text Segmentation

Text segmentation can be either linguistic or topical. In linguistic segmentation, the linguistic characteristics of a text document, such as words, sentences, or paragraphs, are used to divide it into smaller units. In topical segmentation, a long document containing multiple topics is divided into contiguous units that are topically coherent. This chapter will be primarily focused on topical segmentation in the unsupervised case, although supervised settings can handle both cases.

Topical segmentation has a number of useful applications in information retrieval and word-sense disambiguation. In many search applications, it may be helpful to highlight the coherent segments of a long document that are most relevant to the search. Early experiments by Salton *et al.* [488] showed that comparing a query against coherent units of the text is more effective than against the full document. In a different work, Salton *et al.* [491] showed the utility of topical segmentation for text summarization. It is noteworthy that the extractive summarization approaches discussed in Chapter 12 extract individual *sentences* from the document to create summaries. An alternative approach is to use topical segments, which was recommended by Salton *et al.* [491]. Finally, an important application of text segmentation is word-sense disambiguation. It is much easier to disambiguate between the

---

[1]DARPA stands for *Defense Advanced Research Projects Agency*, which is an agency of the United States Department of Defense. It is responsible for the development of emerging technologies for use by the military, and often funds academic research efforts.

multiple senses of a word, when a coherent segment of the document in which the word occurs is available.

A natural question arises as to why one cannot simply use paragraphs as natural topical segments. Paragraphs are often used to demarcate topical changes, although these expectations are often not met in real-world applications. In many newspapers, the text is segmented just to break up its physical appearance for improved readability [242]. Nevertheless, the known boundaries of paragraphs do provide useful hints about topical changes. Therefore, several topical segmentation algorithms use the boundaries of paragraphs as the *candidate* segmentation points that are used in conjunction with other types of topical analysis for demarcation of topical segments.

Text segmentation can be unsupervised or supervised. In unsupervised methods, the text is segmented based on significant changes in the underlying topics. In supervised methods, examples of valid segmentations are provided to guide the results. Although supervised methods have the disadvantage of requiring labeled data, they are more useful when one is looking for a particular type of segmentation. For example, a common application of supervised segmentation is that of segmenting a list of frequently asked questions (FAQ) into question and answer segments. In such cases, the specific punctuation, orthography, and capitalized tokens like "*What*" can be used to learn the break between questions and answers, even though such linguistic features are usually ignored in unsupervised topic segmentation. Therefore, the supervised approach to segmentation is more powerful, and it extends beyond topical segmentation applications. Supervised segmentation can be transformed into the problem of token-level classification, which is identical to the setting used in information extraction (cf. Chapter 13). Consequently, the problems of supervised information extraction and segmentation are sometimes discussed in the same research papers [308, 372].

16.2.1 TextTiling

TextTiling is an unsupervised approach to topic segmentation [242]. The first step in Text-Tiling is to divide the text into *token sequences* of fixed length, which are of relatively small size compared to the overall length of the document. It is suggested in [242] to use about 20–40 tokens as the length of each such sequence. The position in the text sequence between two adjacent token sequences is referred to as a *gap*. The TextTiling algorithm has three primary components, which correspond to the *cohesion scorer*, the *depth scorer*, and the *boundary selector*.

Since a gap is the dividing point between two token sequences, a natural step is to measure the amount of *cohesion* (i.e., similarity in topic) between the sequences lying on either side of the gap. The cohesion is measured in terms of the cosine similarity between the token sequences on either side of the gap. Typically, a *block* of k token sequences on either side of the gap is treated as a bag of words for the similarity computation. A high amount of cohesion indicates that a similar topic is discussed on both sides of the gap. Therefore, a lower level of cohesion tends to be indicative of a segmental change point, especially if the surrounding gaps have higher cohesion. This type of comparison is performed with a depth scorer. For example, if the cohesion is low at a particular gap, as compared to surrounding gaps, then the depth is considered high. Otherwise, the depth is low. Then, if c_i is the cohesion at the ith gap, then a simplistic definition for the depth at the ith gap would be $(c_{i-1} - c_i) + (c_{i+1} - c_i)$. Large (positive) values of this depth would imply that the cohesion is low at the ith gap compared to surrounding gaps. However, such a definition does not account for the gradual change across multiple token sequences. Intuitively, one wants to find a deep "valley" of cohesion values when plotting the cohesion against the gap index as a kind of time-series. Therefore, starting at gap i, one chooses the first gap index $l < i$ to the left of gap i such that $c_{l-1} < c_l$. Similarly, starting at gap i, one chooses the first gap index $r > i$ to right of gap i, such that $c_{r+1} < c_r$. Then, the depth is defined as $(c_l - c_i) + (c_r - c_i)$. In order to address the problems of noisy variations between adjacent values, one can smooth this "time-series" of cohesion values using a variety of time-series smoothing methods before performing the computation.

The final segmentation is done using the boundary detector. The first step is to sort the depth scores by decreasing value. The gaps are added one by one to the list of valid segmentation points in this order. However, a gap is not added to the list of segmentation points, when at least three token sequences do not occur between it and at least one of the gaps already added to the list of segmentation points. In order to terminate the process of adding the segmentation points, a termination criterion based on the mean μ and standard deviation σ of the depth scores is used. The process of adding segmentation points is terminated when the depth score falls below $\mu - t\sigma$, where $t \in [0.5, 1]$ is a user-driven parameter. A token-sequence length of 20, and block size of $k = 10$ token sequences was used in [242]. However, the proper length of the token sequence and the value of k are likely to be sensitive to the corpus at hand.

16.2.2 The C99 Approach

The C99 approach [112] is an unsupervised method that maximizes the average pairwise similarity between sentences in the same segment with the use of a divisive clustering approach. The first step is to compute the pairwise cosine similarity between each pair of sentences in the entire document using the vector-space representations of the sentences. Therefore, for a document containing m sentences, one can create an $m \times m$ matrix $S = [s_{ij}]$. The rows (and columns) of this matrix are ordered so that adjacent rows (and columns) correspond to adjacent sentences. Note that pairwise similarity on short text segments (like sentences) using the vector-space model can be notoriously unreliable. Furthermore, there is significant variation in the intra-sentence similarity depending on location in the document. For example, the sentences in the abstract might vary quite a lot, but the sentences in a later portion of the document might be tightly related. This implies that it makes sense to perform some kind of local normalization of the similarity values.

The C99 technique converts these similarity values into ranks based on *relative locality analysis*. To achieve this goal, we use a locality threshold $t \ll m$, which defines the local region with respect to which a similarity value s_{ij} is compared. Each similarity value s_{ij} in S is replaced with a rank value r_{ij}, which is equal to the number of neighboring entries of (i, j) in S, which are less than s_{ij}. How is the neighborhood defined? Formally, the neighborhood of (i, j) is the square region centered at (i, j) such that each side has length $2 \cdot t + 1$. Thus, the use of the value of $t = 1$ results in a 3×3 square matrix centered at (i, j). Then, the value of r_{ij} is equal to the number of entries among these $3 \times 3 = 9$ cells that are strictly less than s_{ij}. Formally, we can define r_{ij} as follows:

$$r_{ij} = |\{(p, q) : |p - i| \le t, |q - j| \le t, s_{pq} < s_{ij}\}| \qquad (16.1)$$

Note that a different region of size $(2t + 1) \times (2t + 1)$ needs to be examined for each cell in order to compute its rank value. Examples of four steps of the conversion of similarity values to rank values are shown in Figure 16.1. This example is directly adapted from [112], and it uses integer similarity values for simplicity, although the cosine values always lie in the range $(0, 1)$ in practice. Furthermore, this example uses $t = 1$ for simplicity, but a value of $t = 5$ was used in [112]. Let the resulting matrix be denoted by R.

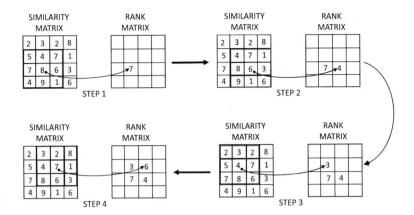

Figure 16.1: Converting absolute similarity into localized rank-centric similarity

The final step is to use divisive hierarchical clustering in order to find the segmentation points. It is assumed that the segmentation points are always at sentence boundaries, and therefore there are only $(m - 1)$ segmentation points. Note that picking any of these segmentation points will result in two segments. For each such candidate segmentation, its quality is computed as follows. The similarity between each pair of sentences (according to matrix R) in the same segment is aggregated. For example, if one segment contains 10 sentences and another contains 5 sentences, then a total of $10 \times 10 + 5 \times 5 = 125$ similarity values is retrieved from matrix R and aggregated. This value is divided by 125 to provide the quality of the candidate segmentation. The candidate segmentation with the best quality is selected out of all the $(m - 1)$ possible segmentations. If needed, constraints can be imposed on the minimum length of a segment. This process is repeated recursively until the use of a division no longer improves the average segmentation quality significantly. The work in [112] also suggests the use of a number of efficiency optimizations to improve the speed of the approach.

Divisive clustering is not the only way one might perform segmentation. Other methods like [491] uses pairwise similarity analysis for segmentation. In particular, the method in [491] creates a weighted graph of pairwise similarity values, drops links with low similarity weight, and then performs graph partitioning for segmentation.

16.2.3 Supervised Segmentation with Off-the-Shelf Classifiers

The basic idea in this approach [50] is to associate a "*Yes*" or "*No*" tag with the end of each sentence, which is a *potential* segmentation point. This model is supervised because it is assumed that each such point is like an instance in either the training or test data. Each potential segmentation point is converted into a multidimensional feature vector by extracting *contextual features* from it. Let the set of features at a potential segmentation point be denoted by $\overline{X}$ with the use of (an as yet unspecified) feature engineering procedure. Note that one can extract labeled pairs $(\overline{X}, y)$ from each labeled segmentation point in the training data. For the test segmentation points, the binary labels are not available. This is a classical supervised learning setting. Then, the problem can be decomposed into the following two subtasks:

1. The first step is to extract a set of words $\overline{X}$ at a potential training or test segmentation point. This set $\overline{X}$ is defined by 500 words to either side of the segmentation point.

2. Engineer a new set of features $\overline{Z} = (z_1, \ldots, z_d) = (f_1(\overline{X}), \ldots f_d(\overline{X}))$, which are more sensitive to the label of the potential segmentation point. This step is described in more detail later. Note that these features are extracted for all the (potential) training and test segmentation points.

3. Build a binary classification model using the engineered features. For any potential segmentation point in the test data (with features $\overline{Z}$), compute the probabilities $P(Yes|\overline{Z})$ and $P(No|\overline{Z})$. Although a logistic regression classifier (cf. Chapter 6) naturally provides probabilistic predictions, one can use any off-the-shelf classifier (e.g., decision tree) that outputs a numerical score instead of a probability value.

4. Given the aforementioned probabilities at each potential segmentation point, identify those segmentation points for which $P(Yes|\overline{Z}) > \alpha$ and no other (selected) segmentation point lies within ϵ units to the left or right of that position. Note that this step can be implemented in a similar way as the final step of the TextTiling algorithm.

It remains to be described how the features of each potential segmentation point are extracted. Blindly using the context of 500 words directly is not likely to be very discriminative for segmentation point detection. In the following, we provide a very simplified variation of the feature engineering process in [50]. Our goal here is to convey the basic principles of [50] in a compact way without introducing a large amount of mathematical formalism. The work in [50] engineers two types of features:

- *Topicality feature:* Imagine a sequential word predictor that predicts the next word from the previous word using a language model. When the topic changes very suddenly at a segmentation point, the predicted probability of a word based on a bag of larger context (long-range prediction) would be different from that predicted by a bag of shorter context. At a potential segmentation point with the next word as w, one can use the tri-gram probability $P(w|w_{-1}w_{-2})$ as the short-range prediction. Furthermore, one can use logistic regression on the subset of words in $\overline{X}$ corresponding to the *history* of last 500 words to predict w. In other words, the probability $P(w|\overline{X})$ is the long-range prediction. The ratio of these two quantities provides the topicality feature. Our description of the topicality feature is a highly simplified variation of the "trigger pair" approach discussed in [50], but it captures the basic principle of what is achieved with the use of such a feature. Note that one can use histories of different lengths to construct multiple such features if needed.

- *Cue-word features:* The cue-word features are defined for each word in the lexicon at a potential segmentation point corresponding to (i) whether each word occurs in the next few words after a potential segmentation point; (ii) whether each word occurs in the next few sentences; (iii) whether each word occurs in the previous few words/sentence; (iv) whether each word occurs in the previous few sentences but not in the next few sentences; and (v) whether a word begins the preceding/next sentence.

  The above description uses the word "few" several times. The word "few" is not interpreted as a single value, but multiple features are extracted for different sizes of the history or future context. Typically, this value ranges between 1 and 10. Clearly, such an approach would extract a large number of features because these questions are asked for each word in the vocabulary.

For such high-dimensional settings, linear or generalized linear models (like logistic regression) are effective. Furthermore, it is extremely important to use feature selection mechanisms or other classifiers with built-in feature selection mechanisms like regularization.

It is noteworthy that this type of approach is not restricted to *Yes/No* types of topical segmentations. One can even recognize *specific* types of segments, such as the segmentation of a frequently-asked-questions (FAQ) list into *question, answer,* and *other* segments. One can view this problem as that of multi-label classification as opposed to binary *Yes/No* classification. This type of *domain-specific* segmentation goes beyond topical segmentation, and it is possible using only supervised models. For such multi-way labeling settings, it is more common to use another class of sophisticated models known as Markovian models. These include methods like *maximum entropy Markovian models* and *conditional random fields.* Many of these models are also used for information extraction, and their use is discussed in the next section.

16.2.4 Supervised Segmentation with Markovian Models

The feature engineering approach discussed in the previous section predicts the labels of different sequence points with the use of off-the-shelf classifiers by treating them as *independent* instances with engineered features. It is only in the final postprocessing phase that the predictions at various segmentation points are integrated with the use of ϵ-length constraints on the segments. These types of oversimplified ways of integrating independent predictions become particularly inaccurate in domain-specific settings, where the segments might have specific interpretations (like *"question"* and *"answer"* labels in a FAQ application). This is somewhat of an over-simplification because the class label prediction at a particular segmentation point should influence the one at the next segmentation point. Markovian models are able to learn the labels at different segmentation points *jointly* so as to maximize the overall prediction probability over the whole text segment.

The problem of supervised segmentation is very similar to that in information extraction. The main difference is that information extraction is *token-level* classification, whereas segmentation is *sentence-level* classification. Just as each entity in information extraction might contain multiple tokens, a text segment might contain multiple sentences. (see Section 13.2.2 of Chapter 13). Therefore, the unit of analysis is different, although the broader principles are similar. As a result, almost all the information extraction methods such as hidden Markov models, maximum entropy Markov models, and conditional random fields (cf. Sections 13.2.3, 13.2.4, and 13.2.5) can be used for this problem. It is noteworthy that the many of the papers on Markovian models for information extraction [308, 372] explicitly consider text segmentation as an additional (or main) application of these sequence classification methods. Refer to Chapter 13 for a discussion of these methods. Furthermore, a natural approach for performing token-level classification with recurrent neural networks is discussed in Section 10.8.5 of Chapter 10.

Since these models are already described in Chapter 13, we restrict our discussion in this chapter to *how* these models are applied in the context of text segmentation. For example, consider the FAQ segmentation application in which the training data is labeled into four classes corresponding to *"question," "answer," "head,"* or *"tail."* A snapshot of the training data, based on [372], is shown in Figure 16.2.

This data contains four different labels (with three of them shown in the figure), and the nature of the segments are highly domain-specific. The typical cue-words or the topicality features that are extracted in the previous section are unlikely to work very well in this type of application. In such settings, a considerable amount of *domain-specific* effort needs to be invested in the feature engineering process associated with each unit of the sequence. Each unit in the sequence might correspond to the end of a sentence or a line, which is a potential segmentation point. For example, the work in [372] used the end of each line as a

<head> X-NNTP-Poster: NewsHound v1.33
<head>
<head> Archive-name:acorn/faq/part2
<head> Frequency: monthly
<head>
<question> 2.6) What configuration of serial cable should I use
<answer>
<answer> Here follows a diagram of the necessary connections
<answer> programs to work properly. They are as far as I know t
<answer> agreed upon by commercial comms software developers fo
<answer>
<answer> Pins 1, 4, and 8 must be connected together inside
<answer> is to avoid the well known serial port chip bugs. The

Figure 16.2: Example of training data labeling for FAQ segmentation

Table 16.1: Features extracted from each line for FAQ segmentation application

begins-with-number	contains-question-mark
begins-with-ordinal	contains-question-word
begins-with-punctuation	ends-with-question-mark
begins-with-question-word	first-alpha-is-capitalized
begins-with-subject	indented
blank	indented-1-to-4
contains-alphanum	indented-5-to-10
contains-bracketed-number	more-than-one-third-space
contains-http	only-punctuation
contains-non-space	prev-is-blank
contains-number	prev-begins-with-ordinal
contains-pipe	shorter-than-30

potential segmentation point. Furthermore, a total of 24 features were extracted from each line, which are shown in Table 16.1.

The label of the previous state can also be used as a feature, which results in a model requiring *joint* prediction of labels rather than independent prediction. Similar to information extraction, one combines the features extracted from the lines as well as the potential labels of the previous tokens in order to create new features. This creates a powerful sequence-driven model, although it requires a more expensive learning procedure like *generalized iterative scaling*, as compared to off-the-shelf classification models. Although we have not provided a detailed description of the features in Table 16.1, their names are self explanatory. It is easy to see that these features are highly optimized to the FAQ segmentation application, and they require some thought on the part of the analyst. In most of these applications, the feature engineering phase is easily the most important part of the learning process.

16.3 Mining Text Streams

The problem of mining text streams is closely related to that of topic segmentation. Just as topic segmentation looks for changes within a text document, a streaming setting often looks for sudden changes in the stream. The latter problem can also be used to detect new events in the stream. In fact, the topic detection and tracking project [20] defined the problem of *first story detection*, in which the first story on a specific topic is identified in the temporal setting. In the streaming setting, one can detect deviations by creating a summary model of the data stream. Such deviations may correspond to new events (or their representative stories). The natural way to create a summary model from a text stream is to use clustering. In the following, we describe a simple approach for streaming text clustering, which can be used for various down-stream tasks like event detection or first-story detection.

16.3.1 Streaming Text Clustering

A fundamental assumption in the streaming setting is that it is not possible to maintain the entire history of the data stream in detail. Therefore, a summary model of the stream is often enabled with the use of clustering. In the following, we describe a simplification of an online streaming approach that is discussed in [14].

The basic idea is to always maintain a summary representation of each cluster. This summary representation can be viewed as the time-decayed centroid of each cluster. To create a time-decayed centroid, we have a decay parameter λ that regulates the weight of the document. At current time t_c, the weight of the ith document that arrived at time t_i is denoted by $2^{-\lambda(t_c-t_i)}$. Therefore, if a cluster $\mathcal{C}$ contains the documents $\overline{X_1} \ldots \overline{X_r}$, which have arrived at times $t_1 \ldots t_r$, then the centroid $\overline{Y}(\mathcal{C}, t_c)$ of the cluster $\mathcal{C}$ is given by the following:

$$\overline{Y}(\mathcal{C}, t_c) = \frac{\sum_{i=1}^{r} \overline{X_i} \cdot 2^{-\lambda(t_c-t_i)}}{r} \tag{16.2}$$

Note that this type of decay function changes each centroid at every time stamp by a multiplicative factor of $2^{-\lambda}$, and the value of $1/\lambda$ is the *half-life* of the decay. Therefore, it would seem that a clustering approach that adapts k-means would need to update the centroids at every time-stamp. However, it turns out that such updates can be performed effectively with the use of a lazy approach. The key point is that the decay-based approach needs to be applied to a centroid only when a document is added to the cluster, because all documents in the cluster decay at the same rate of $2^{-\lambda}$, and this is also the decay rate of the term frequencies in the centroid. Because of the fact that cosine similarity computations are normalized, the value of the similarity between a centroid and a document is not affected by the decay. In other words, the cosine similarity between a document and a centroid at time t_c will be the same as that at t_c+1, even though the terms in the centroid have decayed with time. In this sense, the use of exponential decay is particularly convenient. Furthermore, exponential decay is theoretically well accepted because of its memoryless property.

Therefore, the technique in [14] uses a k-means approach in which we continuously maintain the centroids of the k clusters. The value of k is an input parameter to the algorithm. For the jth cluster we maintain the time stamp l_j, which is the last time that a document was added to the cluster. At time t_c, when a new document $\overline{X}$ arrives, the cosine similarity is computed between $\overline{X}$ and each centroid, and the index m of the closest centroid is determined. At this point, the multiplicative decay-based update is applied only to cluster m. Since the cluster m was last updated at time-stamp l_m, the term frequencies in its centroid are multiplied with the decay factor $2^{-\lambda(t_c-l_m)}$, and then the term frequencies of $\overline{X}$ are added to that centroid. At this point, the last update time l_m of the mth cluster is updated to t_c. This approach can be continuously applied for each incoming document in order to maintain the clusters in real time. The value of λ regulates the rate of decay. In fast-evolving environments, large values of λ should be used. From an intuitive point of view, one should set the half-life $1/\lambda$ to a time-span over which the patterns are expected to change significantly. This approach does not yet discuss how outliers or sudden changes in the patterns are handled. Such changes in patterns are often useful in discovering the first story on specific events. The work in [14] makes some additional changes to this basic approach in order to handle outliers. These changes are discussed in the next section.

16.3.2   Application to First Story Detection

The clustering model of [14] can be used in order to identify novelties from the underlying data stream. Instead of using a fixed number of clusters k, the value of k reflects the maximum number of clusters. The approach starts with a single cluster and incrementally adds new clusters when an incoming point is not sufficiently close to the existing clusters (because it is a novelty). Such an approach identifies the novelties and clusters simultaneously. The basic idea is to maintain the mean μ and standard deviation σ of the similarity values of the assignments of documents to clusters in the history of the stream. Note that these quantities can be maintained incrementally by maintaining the number t of assignments, first-moment (sum) F of assignment similarities, and second-moment (squared sum) S of assignment similarities. The value of the mean μ is simply F/t, and the value of σ is computed as follows:

$$\sigma = \sqrt{\frac{S}{t} - \frac{F^2}{t^2}} \tag{16.3}$$

Then, a data point is an outlier when its similarity to closest centroid is less than $\mu - r \cdot \sigma$, where r is a user-driven parameter. The work in [14] uses $r = 3$. When a data point is deemed to be an outlier, a new cluster is created containing that singleton data point. In order to make room for this new cluster, one of the older clusters need to be ejected in a fixed-memory setting. To decide which cluster is ejected, the last update time of a cluster is used. The cluster j with the least-recent update time (i.e., smallest value of l_j) is removed. If the amount of memory is large enough, one can afford to maintain a large number of clusters, and one may not even need to use the entire amount of available memory. In such cases, the "cluster" ejected will often be another singleton cluster containing an outlier, which was created as a new cluster but was never updated. In such cases, one can identify which novelties lead to the eventual creation of new clusters (i.e., true events containing multiple documents) and which novelties are simply isolated outliers. The former type of outlier can be considered a first story of a larger event. However, if the amount of memory is relatively small, one will often eject the singleton clusters too early to identify whether it results in the creation of a new event.

Both probabilistic and deterministic clustering methods have been used in various unsupervised event detection applications. It is noteworthy that the use of one-pass clustering for event detection was also proposed in the context of the TDT project [607]. However, the approach in [14] is more sophisticated in the sense that it can allow for different levels of evolution with a decay parameter. It also provides the ability to perform analysis of the events and evolving clusters over different time horizons.

16.4 Event Detection

An event is something that happens in a particular place at a particular time (e.g., meeting, conference, terrorist attack, earthquake, tsunami), which has an effect on online streams of documents such as emails, publications, news, blogs, or tweets. For large events like earthquakes, the occurrence of an event can often be detected from extrinsic sources like news outlets. However, for smaller events, the effects are often far more subtle, and can be detected from pattern changes in the underlying stream of articles. Furthermore, some sources of document streams like Twitter might sometimes show event indications earlier than traditional news streams. At the same time, such sources are noisy, and they might contain all types of noise and other spurious events like rumors. Early (or even predictive) event detection is often more useful than late event detection, because the occurrence of larger events is often known in retrospect through external sources. However, for smaller and inconspicuous events, even retrospective event detection can be useful.

The problem of first-story detection, which is discussed in the previous section, is directly related to event detection, because a novel story is often the first indicator of a newsworthy event. The clustering-related approach discussed in the previous section is only one out of a broader class of methods for event detection. This section will provide a holistic view of different event detection tasks. In general, there are two settings in event detection:

1. *Unsupervised event detection:* In unsupervised event detection, examples of relevant events are not available. Therefore, a significant change in the textual patterns over time is interpreted as an event.

2. *Supervised event detection:* In supervised event detection, examples of relevant events are available for learning purposes.

Both these tasks have different applications. It is also noteworthy that there are many similarities in the methods used in text segmentation and event detection. For example, in text segmentation, one looks for natural change boundaries within a single document. On the other hand, in event detection, one looks for change boundaries across multiple documents. However, since a single document is more likely to be coherent than a set of multiple documents, there are some differences in how the change analysis is performed. These differences are more significant in the unsupervised case, because supervised methods more adaptable to different settings.

16.4.1 Unsupervised Event Detection

In unsupervised event detection, no prior knowledge is assumed about the types of events that one is trying to detect. As a result, the problem is very similar to that of change detection in data streams. The first-story detection technique discussed in Section 16.3.2 is one example of an unsupervised method. The following discusses other popular methods.

16.4.1.1 Window-Based Nearest-Neighbor Method

An early approach that was proposed in [607] used a window-based nearest neighbor method. In this technique the current document $\overline{Z}$ was compared to the window of the previous m documents. Let the vector-space representations of these m documents received in the stream of documents be denoted by $\overline{X_1} \ldots \overline{X_m}$ where the documents are received in this temporal order as well. Then, the score of the document $\overline{Z}$ is denoted by the inverse of the maximum similarity of $\overline{Z}$ with all documents in the window. Let the similarity between document $\overline{Z}$ and $\overline{X_i}$ be denoted by $S(\overline{Z}, \overline{X_i}) \in (0, 1)$. Then, the event detection score of the document $\overline{Z}$ is as follows:

$$\text{Score}(\overline{Z}) = 1 - \text{MAX}_{i \in \{1 \ldots m\}} \left\{ S(\overline{Z}, \overline{X_i}) \right\} \qquad (16.4)$$

The value $S(\overline{Z}, \overline{X_i})$ is typically computed with the use of the cosine similarity function. It is noteworthy that this quantification shows some similarity with that of Section 16.3.2 for first-story detection in which the centroid of a cluster is used for score computation rather than the actual documents in the window. Furthermore, an exponential decay-based approach is used in Section 16.3.2, rather than the window-based approach discussed here. The work in [607] also provides a way of scoring that incorporates linear decay within the window itself. In other words, if the previous window of documents (in temporal order) is denoted by $\overline{X_i} \ldots \overline{X_m}$, then the decay-based score of the incoming document $\overline{Z}$ is computed as follows:

$$\text{Score}(\overline{Z}) = 1 - \text{MAX}_{i \in \{1 \ldots m\}} \left\{ \frac{i}{m} S(\overline{Z}, \overline{X_i}) \right\} \qquad (16.5)$$

Note that the only difference of this scoring method from Equation 16.4 is the presence of the linear decay factor (i/m) within the window. As a result, earlier documents in the window are less likely to have an impact, because they may not have the maximum (adjusted) similarity with respect to the test document $\overline{Z}$.

16.4.1.2   Leveraging Generative Models

Another approach is to use a generative model and assume that every document in the corpus was generated by this model. Documents that have low fit with respect to this model are deemed as outliers. There are several choices about how one might design the generative model:

- One can assume that each document was generated from one of k mixture components $\mathcal{G}_1 \ldots \mathcal{G}_k$ with prior probabilities $\alpha_1 \ldots \alpha_k$, and each document is generated using a multinomial distribution $P(\overline{X} | \mathcal{G}_r)$ specific to the rth component. The prior probabilities and the parameters of the distribution are learned using exactly the same process as discussed in Section 4.4.2 of Chapter 4. The main difference is that parameter estimation process (i.e., E- and M-step) is done incrementally using r passes over the last window of m points after the arrival of a batch of m points. This ensures that the effect of previous documents decays with time because points before that window are not used for iterative updates. The values of r and m affect the rate of decay. The novelty score of each data point $\overline{Z}$ in the window of m points is computed with respect to the model *before* updating it iteratively. The score is computed using the inverse of the maximum likelihood fit of the document to the model. This is achieved using the prior probabilities and multinomial model parameters as follows:

$$\text{Score}(\overline{Z}) = 1 - \sum_{r=1}^{k} \alpha_r P(\overline{Z} | \mathcal{G}_r) \qquad (16.6)$$

  Points with high novelty scores are deemed as the starting points of novel events. Note that this approach can be viewed as a probabilistic variant of the clustering method. After all, the EM-based clustering approach is closely related to methods based on k-means (cf. Section 4.5.3 of Chapter 4).

- *Use of PLSA and LDA:* The generative process described above uses a single-membership mixture model. However, it is also possible to use a mixed membership model like PLSA or LDA in cases where individual documents are likely to discuss more than one subject. Such an approach is discussed in [621].

It is noteworthy that this approach can be used in conjunction with almost any type of data and almost any type of generative model. In each case, the maximum likelihood fit for a document provides it novelty score. For example, one could impose a sequential dependency among the mixture components and use a hidden Markov model to compute the probabilities of transitions from one state to another. Such methods have been used in text segmentation [597], and they can also be adapted to event detection.

16.4.1.3 Event Detection in Social Streams

In recent years, *social streams* have become increasingly popular, which contain continuous posts from social network users. A specific example is the case of tweet streams from Twitter. Twitter provides a dedicated application programming interface (API) that allows the collection of a certain percentage of the stream both on a free and on a subscription basis. Such streams are often very valuable because they can be used to make predictions about key events.

The work in [11] uses clustering of the tweets in the Twitter social stream in order to detect events. However, social streams are very noisy, and each tweet contains only about 140 characters. Therefore, the known list of followers of each user are added to a tweet in order to facilitate the clustering process. The keyword and the follower tokens in a tweet can be weighted differently in order to provide varying importance to content and structure in the clustering process. Another problem in the social stream setting is that simply using a tweet which is very different from other tweets (as in the first-story detection method of Section 16.3.2) will only detect noise. Therefore, the approach in [11] defines events as time instants in which the *relative fraction* of points in the cluster has increased over a fixed time-horizon by a minimum factor α. The intuition is that cascading discussions on a specific and focused topic in the social network will lead to the sudden growth of particular clusters. This approach is also generalized to the supervised setting in [11] by treating the ratio of the change in cluster size over a specific horizon as a feature. The training data is used to learn the importance of these features to the occurrence of specific types of events.

16.4.2 Supervised Event Detection as Supervised Segmentation

In supervised event detection, one often has examples of key event points in the training data. These training examples are used to identify the event points in the test data. As a practical matter, one can assume that a *Yes/No* label exists after each document that indicates whether or not a new event has occurred. It turns out that supervised event detection is closely related to supervised segmentation in text streams.

It is easy to see that supervised event detection is very similar to the supervised segmentation task. An event point is similar to a segmentation point. The main difference is in terms of how the potential event boundaries are defined. In the segmentation task, the potential segmentation points are defined at the end of sentences in a single document. In the event detection task, one can pretend that the entire sequence of text documents is one long document and the potential "segmentation" (i.e., event) points correspond to the document boundaries. Just like the supervised segmentation task, we have a *Yes/No* label at each potential event point in the training data. It is also possible to have refined labels corresponding to multiple types of events. For the test data, the potential event points are the document boundaries in the test sequence. Note that one can use all the supervised methods discussed in Section 16.2 for this problem. The main difference is in terms of how the features are extracted. Clearly, longer context windows need to be used in order to extract the features compared to a text segmentation task. Furthermore, for specific types of event detection tasks (e.g., tsunami event or terrorist attack) some amount of domain-specific feature engineering is required. For example, specific cue words or features may be extracted for these tasks. This is not particularly surprising because such feature engineering is also required in the case of the segmentation task. The use of off-the-shelf classifiers is generally more efficient with these models rather than the use of Markovian models. This is because Markovian models often require complex dynamic programming or generalized iterative scaling methods at prediction time.

16.4.3   Event Detection as an Information Extraction Problem

The event detection methods discussed so far depend only on broad topical changes between documents. A different view of event detection is at the *mention-level* in current documents like news articles. In such cases, a single document might contain mentions of two different events such as the *"tsunami"* and *"earthquake."* Such detections at the mention level are handled by the creation of *event entities*, which are phrases corresponding to the occurrence of events. The mention-level event detection problem is generally supervised, and it requires heavily annotated training data. This is a potential drawback of the approach, although the existence of crowd-sourcing resources like *Amazon Mechanical Turk* has simplified this problem to a large extent.

There are two settings for event extraction, which correspond to the supervised and unsupervised scenarios. In the supervised setting, the individual phrases corresponding to event occurrences are already tagged with specific labels, and the goal is to mark similar events in other documents. In the unsupervised settings, such a tagged training data set may not be available, and even the definition of the events might depend on the problem setting.

16.4.3.1   Transformation to Token-Level Classification

The guidelines for annotating training data for event extraction are available in a seminal paper introducing the *Timebank corpus* [448]. Events are tagged the same way as other types of named entities. Furthermore, events are often tagged along with other types of entities such as dates. The basic idea is that such types of extraction can also facilitate sophisticated question-answering systems that can answer who/what/when/why questions about events. For example, consider a setting in which one is mining events of type *"death"* being caused by events of type *"accident,"* and we are also interested in the dates at which the accidents occurred. Now consider the sentence:

Fifty passengers <u>lost their lives</u> in the <u>May 5</u> <u>train collision.</u>

There are three entities of interest in this sentence, which are underlined. There are many ways to annotate this type of event extraction task at the token level. Consider, a case in which we use four types of token tags $\{A, D, T, O\}$ corresponding to *accident, death, time,* and *other* types of tokens, respectively. One can annotate this sentence *at the token level* as follows:

$$\underbrace{\text{Fifty}}_{O} \; \underbrace{\text{passengers}}_{O} \; \underbrace{\text{lost}}_{D} \; \underbrace{\text{their}}_{D} \; \underbrace{\text{lives}}_{D} \; \underbrace{\text{in}}_{O} \; \underbrace{\text{the}}_{O} \; \underbrace{\text{May}}_{T} \; \underbrace{5}_{T} \; \underbrace{\text{train}}_{A} \; \underbrace{\text{collision.}}_{A}$$

In this encoding, individual mentions of an event may sometimes contain multiple tokens. For example, an accident is indicated by the phrase "*train collision*." As in the case of named entity recognition in information extraction, it is helpful to distinguish between the beginning, continuation, and end of the phrases with $\{B, C, E\}$ tags. This creates a hybrid encoding with $4 \times 3 = 12$ possible tags obtained by appending the entity type with the tag position. The resulting annotated sentence is as follows:

$$\underbrace{\text{Fifty}}_{OB} \; \underbrace{\text{passengers}}_{OE} \; \underbrace{\text{lost}}_{DB} \; \underbrace{\text{their}}_{DC} \; \underbrace{\text{lives}}_{DE} \; \underbrace{\text{in}}_{OB} \; \underbrace{\text{the}}_{OE} \; \underbrace{\text{May}}_{TB} \; \underbrace{5}_{TE} \; \underbrace{\text{train}}_{AB} \; \underbrace{\text{collision.}}_{AE}$$

For a given test segment with unmarked tokens, the goal will be to annotate the text with appropriate event markers using token level classification. It is noteworthy that this problem is identical to the token-level classification setting discussed in Section 13.2.2 of Chapter 13. Therefore, all the Markovian models discussed in Chapter 13 can be used. The method of choice is generally that of conditional random fields according to the work in [472]. Furthermore, a method for performing token-level classification with recurrent neural networks is discussed in Section 10.8.5 of Chapter 10. One can also use other features associated with tokens in combination with recurrent neural networks by using additional inputs for the features. For example, Figure 10.18 of Chapter 10 shows various types of input features for token-level classification.

The use of proper input features is helpful for accurate token-level classification. The key point in any particular application specific setting is to extract the right features for a particular event detection task. We refer the reader to [98, 340, 449, 472, 473] for a discussion of these feature extraction issues. In particular, the work in [449] provides an insightful discussion of how specific types of attributes provide indicators of events. The work also proposes the use of contextual, dictionary, part-of-speech information, and orthographic features associated with tokens. In addition, the work in [499] gathers a dictionary of event terms using *WordNet* that are shown to be effective features for event extraction. The tagging of dates is much easier than the tagging of events (because of more restricted format of dates), and it can sometimes be performed up front before event tagging with the use of a separate system like the Tempex system discussed in [366]. In such cases, the date tags become available as features, and they provide useful contextual features to the event tagging problem (which is more difficult anyway). Some part-of-speech tagging systems are also available for noisy domains like Twitter data [473].

16.4.3.2 Open Domain Event Extraction

Open domain event extraction can be viewed as a special case of open domain information extraction, and it is a more challenging setting because one is no longer restricted to extracting specific types of events. Rather, one has to use a large corpus like the Web in order to discover the different types of events in which one might be interested. There are a huge number of categories in which real-world events can be placed, and therefore one is tasked with defining these categories based on the available data from open sources like news repositories, the Web, or Twitter.

The approach in [472] is designed for Twitter data, and it mines the relevant dates and named entities associated with the events. Therefore, each event is associated with an event

Table 16.2: The list of event types discovered in [472] using open-domain event extraction

Category	Frequency	Category	Frequency
Sports	7.45%	Party	3.66%
TV	3.04%	Politics	2.92%
Celebrity	2.38%	Music	1.96%
Movie	1.92%	Food	1.87%
Concert	1.53%	Performance	1.42%
Fitness	1.11%	Interview	1.01%
ProductRelease	0.95%	Meeting	0.88%
Fashion	0.87%	Finance	0.85%
School	0.85%	AlbumRelease	0.78%
Religion	0.71%	Conflict	0.69%
Prize	0.68%	Legal	0.67%
Death	0.66%	Sale	0.66%
VideoGameRelease	0.65%	Graduation	0.63%
Racing	0.61%	Fundraiser/Drive	0.60%
Exhibit	0.60%	Celebration	0.60%
Books	0.58%	Film	0.50%
Opening/Closing	0.49%	Wedding	0.46%
Holiday	0.45%	Medical	0.42%
Wrestling	0.41%	OTHER	53.4%

type (e.g., politics or sports), a named entity, and a date. Given a stream of tweets, the work in [472] extracts the named event phrases, named entities, and dates associated with the events in the first phase. Note that the event type is not discovered at this stage. It is much easier to annotate a corpus in order to extract an event phrase, rather than an event phrase of a particular type. Therefore, the work in [472] annotates about 1000 tweets in accordance with the guidelines established in *Timebank* [448]. A conditional random field model (see previous section) is then used to annotate all the tweets with event phrases in an automated way using a supervised approach. At this point, the goal is to classify the events into different types. The first step is to use a latent variable model to cluster the event phrases into different components in which the associated entities and dates are also used for clustering. The idea is that events of similar types are more likely to share similar entities (e.g., *Michael Jordan*) and dates (e.g., September 11, 2001). In principle, one could use any type of clustering approach, although the work in [472] chooses to use LinkLDA [177] which is particularly well suited to this type of data. The automatically discovered clusters are inspected to remove the ones that are incoherent, and the remaining ones are associated with informative labels. The complete list of discovered event types that were discovered in [472] are shown in Table 16.2. Note that one is labeling clusters rather than individual tweets, and therefore the annotation process is quite fast. The process of annotating clusters automatically results in annotating the event phrases because the phrases are probabilistically associated with clusters. In principle, it is also possible to use this large type-labeled training data to extract typed events from Twitter in online fashion, although the work in [472] mentions this approach only as an avenue for future work.

16.5 Summary

This chapter discusses several text mining tasks in the sequential and temporal setting that are closely related to the topic detection and tracking effort. Many of these methods are closely related to text stream mining. In particular, we studied the tasks of text segmentation, streaming clustering, and event detection. All these tasks are closely related, and the methods for one are often used as subroutines for the other.

Text segmentation can be unsupervised or supervised. In unsupervised text segmentation, one looks for topical changes at potential segment points in a text document. In supervised text segmentation, examples of segmentation points are provided. These examples are used to predict the segmentation points in unmarked test segments.

The problems of stream clustering and event detection are closely related. It is relatively easy to adapt k-means algorithms to the problem of streaming test clustering. By identifying documents that do not naturally fit into the existing clusters, one can identify first stories on specific events. This general approach has been used in many event detection tasks. Furthermore, many segmentation methods can be used for unsupervised and supervised event detection, by treating potential event points in a stream of documents as potential segmentation points in a large document that is artificially created from the stream. Finally, the use of information extraction methods can identify events from documents at the level of individual mentions.

16.6 Bibliographic Notes

The benefits of text segmentation for better information retrieval are explained in [242, 444, 488]. The TextTiling method is proposed in [242]. A graph-based approach to segmentation is described in [491]. The C99 method is proposed in [112], which was later enhanced with LSA-based similarity in [113]. The use of HMMs for topic segmentation was proposed in the initial TDT effort [20], and it is discussed in [597]. The use of feature extraction methods in combination with off-the-shelf classifiers are discussed in [50, 343]. The use of sequential models for supervised segmentation are proposed in [308, 372]. Several interesting methods for topical text segmentation are discussed in [60, 174, 487, 491].

The streaming text clustering method discussed in this chapter is adapted from [14]. The problem of event detection in text is closely related to outlier detection in text data [287], although the latter is designed for a non-temporal setting. Much of the initial work on unsupervised event detection was done in the context of the topic detection and tracking task [20]. Methods for event detection in the context of TDT are described in [607, 621]. The latter [621] shows how one might use probabilistic and generative models for event detection. Methods for event detection in social streams are discussed in [11, 49, 486, 500]. The use of information extraction for event detection is studied in [98, 340, 448, 449, 472, 473]. A method for using dependency parsing in event detection is proposed in [377]. The problem of open-domain event detection is discussed in [472].

16.6.1 Software Resources

For linguistic text segmentation, a significant amount of software is available from traditional resources like Stanford NLP, NLTK, and Apache OpenNLP [644, 650, 652]. Many approaches like TextTiling are also available from resources like NLTK [709]. The **MALLET** toolkit supports many of the Markovian models that can be used for text segmentation [701].

The official Website of the TDT project is available at [710], although the specific results are somewhat outdated at this point. A pointer to the work on the *Timebank* corpus may be found in [448]. This work also provides guidelines for event annotation in a document.

16.7   Exercises

1. Design an approach that transforms the problem of text segmentation to that of graph partitioning. Assume that each sentence is a node in the graph. Discuss the various ways in which one might place links between nodes and set their weights. Discuss their advantages and disadvantages. [This is an open-ended question without a uniquely correct answer]

2. Implement the C99 approach for text segmentation.

3. Discuss the similarities between named entity recognition, aspect-based opinion mining, text segmentation, and event (mention) extraction. Name one core learning method that is used in all these problems.

4. Implement the streaming text clustering method discussed in this chapter. Suppose that some of the documents are marked with labels of specific event types. How would you use this information to improve the clustering.

5. Suppose that you receive a stream of text documents in which the labels of the documents are received as a separate (delayed) stream. Discuss how you would predict the label of each incoming document by modifying the streaming text clustering algorithm and combining with a centroid classification method.

Correction to: Machine Learning for Text

Correction to:
C. C. Aggarwal, *Machine Learning for Text*,
https://doi.org/10.1007/978-3-030-96623-2

This book was inadvertently published without extra supplementary material information. It has now been corrected to reflect the required information in the front matter of the book.

The updated original version for this book can be found at
https://doi.org/10.1007/978-3-030-96623-2

Bibliography

1. C. Aggarwal. Data classification: Algorithms and applications, *CRC Press*, 2014.
2. C. Aggarwal. Data mining: The textbook. *Springer*, 2015.
3. C. Aggarwal. Recommender systems: The textbook. *Springer*, 2016.
4. C. Aggarwal. Outlier analysis. *Springer*, 2017.
5. C. Aggarwal. Neural networks and deep learning. *Springer*, 2018.
6. C. Aggarwal. On the effects of dimensionality reduction on high dimensional similarity search. *ACM PODS Conference*, pp. 256–266, 2001.
7. C. Aggarwal, S. Gates, and P. Yu. On using partial supervision for text categorization. *IEEE Transactions on Knowledge and Data Engineering*, 16(2), 245–255, 2004. [Extended version of ACM KDD 1998 paper "On the merits of building categorization systems by supervised clustering."]
8. C. Aggarwal and N. Li. On node classification in dynamic content-based networks. *SDM Conference*, pp. 355–366, 2011.
9. C. Aggarwal and C. Reddy. Data clustering: algorithms and applications, *CRC Press*, 2013.
10. C. Aggarwal and S. Sathe. Outlier ensembles: An introduction. *Springer*, 2017.
11. C. Aggarwal and K. Subbian. Event detection in social streams. *SDM Conference*, 2012.
12. C. Aggarwal, Y. Xie, and P. Yu. On Dynamic Link Inference in Heterogeneous Networks. *SDM Conference*, pp. 415–426, 2012.
13. C. Aggarwal and P. Yu. On effective conceptual indexing and similarity search in text data. *ICDM Conference*, pp. 3–10, 2001.
14. C. Aggarwal and P. Yu. On clustering massive text and categorical data streams. *Knowledge and Information Systems*, 24(2), pp. 171–196, 2010.
15. C. Aggarwal, and C. Zhai, Mining text data. *Springer*, 2012.
16. C. Aggarwal and P. Zhao. Towards graphical models for text processing. *Knowledge and Information Systems*, 36(1), pp. 1–21, 2013. [Preliminary version in *ACM SIGIR*, 2010]
17. C. Aggarwal, Y. Zhao, and P. Yu. On the use of side information for mining text data. *IEEE Transactions on Knowledge and Data Engineerin*, 26(6), pp. 1415–1429, 2014.
18. E. Agichtein and L. Gravano. Snowball: Extracting relations from large plain-text collections. *ACM Conference on Digital Libraries*, pp. 85–94, 2000.
19. O. Agarwal *et al.* Large-Scale Knowledge Graph-based synthetic corpus generation for knowledge-enhanced language model pre-training. *arXiv:2010.12688*, 2020. https://arxiv.org/abs/2010.12688
20. J. Allan, J. Carbonell, G. Doddington, J. Yamron, and Y. Yang. Topic detection and tracking pilot study final report. *CMU Technical Report*, Paper 341, 1998.
21. J. Allan, R. Papka, V. Lavrenko. Online new event detection and tracking. *ACM SIGIR Conference*, 1998.

© Springer Nature Switzerland AG 2022
C. C. Aggarwal, *Machine Learning for Text*,
https://doi.org/10.1007/978-3-030-96623-2

22. C. Alt, M. Hubner, and L. Hennig. Fine-tuning pre-trained transformer language models to distantly supervised relation extraction. *arXiv:1906.08646*, 2019. https://arxiv.org/abs/1906.08646

23. A. Andreevskaia and S. Bergler. Mining WordNet for a Fuzzy Sentiment: Sentiment Tag Extraction from WordNet Glosses. *European Chapter of the Association for Computational Linguistics*, pp. 209–216, 2006.

24. R. Angelova and S. Siersdorfer. A neighborhood-based approach for clustering of linked document collections. *ACM CIKM Conference*, pp. 778–779, 2006.

25. V. Anh, O. de Kretser, and A. Moffat. Vector-space ranking with effective early termination. *ACM SIGIR Conference*, pp. 35–42, 2001.

26. V. Anh and A. Moffat. Inverted index compression using word-aligned binary codes. *Information Retrieval*, 8(1), pp. 151–166, 2005.

27. V. Anh and A. Moffat. Pruned query evaluation using pre-computed impacts. *ACM SIGIR Conference*, pp. 372–379, 2006.

28. Ashburner *et al.* Gene ontology: tool for the unification of biology. *Nature Genetics*, 25(1), pp. 25–29, 2000.

29. V. Anh and A. Moffat. Improved word-aligned binary compression for text indexing. *IEEE Transactions on Knowledge and Data Engineering*, 18(6), pp. 857–861, 2006.

30. M. Antonie and O Zaïane. Text document categorization by term association. *IEEE ICDM Conference*, pp. 19–26, 2002.

31. C. Apte, F. Damcrau, and S. Weiss. Automated learning of decision rules for text categorization, *ACM Transactions on Information Systems*, 12(3), pp. 233–251, 1994.

32. C. Apte, F. Damerau, and S. Weiss. Text mining with decision rules and decision trees. *Conference on Automated Learning and Discovery*, Also appears as *IBM Research Report*, RC21219, 1998.

33. A. Asuncion, M. Welling, P. Smyth, and Y. Teh. On smoothing and inference for topic models. *Uncertainty in Artificial Intelligence*, pp. 27–34, 2009.

34. S. Baccianella, A. Esuli, and F. Sebastiani. SentiWordNet 3.0: An enhanced lexical resource for sentiment analysis and opinion mining. *LREC*, pp. 2200–2204, 2010.

35. R. Baeza-Yates, and B. Ribeiro-Neto. Modern information retrieval. *ACM press*, 2011.

36. R. Baeza-Yates, A. Gionis, F. Junqueira, V. Murdock, , V. Plachouras, and F. Silvestri. The impact of caching on search engines. *ACM SIGIR Conference*, pp. 183–190, 2007.

37. D. Bahdanau, K. Cho, and Y. Bengio. Neural machine translation by jointly learning to align and translate. *ICLR*, 2015. https://arxiv.org/abs/1409.0473

38. L. Baker and A. McCallum. Distributional clustering of words for text classification. *ACM SIGIR Conference*, pp. 96–103, 1998.

39. L. Ballesteros and W. B. Croft. Dictionary methods for cross-lingual information retrieval. *International Conference on Database and Expert Systems Applications*, pp. 791–801, 1996.

40. M. Banko and O. Etzioni. The tradeoffs between open and traditional relation extraction. *ACL Conference*, pp. 28–36, 2008.

41. R. Banchs. Text Mining with MATLAB. *Springer*, 2012.

42. M. Baroni, G. Dinu, and G. Kruszewski. Don't count, predict! A systematic comparison of context-counting vs. context-predicting semantic vectors. *ACL*, pp. 238–247, 2014.

43. M. Baroni and A. Lenci. Distributional memory: A general framework for corpus-based semantics. *Computational Linguistics*, 36(4), pp. 673–721, 2010.

44. R. Barzilay and M. Elhadad. Using lexical chains for text summarization. *Advances in Automatic Text Summarization*, pp. 111–121, 1999.

45. R. Barzilay, N. Elhadad, and K. McKeown. Inferring strategies for sentence ordering in multidocument news summarization. *Journal of Artificial Intelligence Research*, 17, pp. 35–55, 2002.

46. R. Barzilay and K. R. McKeown. Sentence fusion for multidocument news summarization. *Computational Linguistics*, 31(3), pp. 397–328, 2005.

47. P. Baudis and J. Sedivy. Modeling of the question answering task in the yodaqa system. *International Conference of the Cross-Language Evaluation Forum for European Languages*, 2015.

48. I. Bayer. Fastfm: a library for factorization machines. *arXiv:1505.00641*, 2015. https://arxiv.org/pdf/1505.00641v2.pdf

49. H. Becker, M. Naaman, and L. Gravano. Beyond Trending Topics: Real-World Event Identification on Twitter. *ICWSM Conference*, pp. 438–441, 2011.

50. D. Beeferman, A. Berger, and J. Lafferty. Statistical models for text segmentation. *Machine Learning*, 34(1–3), pp. 177–210, 1999.

51. O. Bender, F. Och, and H. Ney. Maximum entropy models for named entity recognition. *Conference on Natural Language Learning at HLT-NAACL 2003*, pp. 148–51, 2003.

52. F. Benevenuto, G. Magno, T. Rodrigues, and V. Almeida. Detecting spammers on twitter. *Collaboration, Electronic Messaging, Anti-abuse and Spam Conference*, 2010.

53. Y. Bengio, R. Ducharme, P. Vincent, and C. Jauvin. A neural probabilistic language model. *Journal of Machine Learning Research*, 3, pp. 1137–1155, 2003.

54. J. Berant *et al.* Semantic parsing on freebase from question-answer pairs. *Conference on Empirical Methods in Natural Language Processing.* 2013.

55. D. Bertsekas. Nonlinear programming. *Athena Scientific*, 1999.

56. D. Bikel, S. Miller, R. Schwartz, and R. Weischedel. Nymble: a high-performance learning name-finder. *Applied Natural Language Processing Conference*, pp. 194–201, 1997.

57. C. M. Bishop. Pattern recognition and machine learning. *Springer*, 2007.

58. C. M. Bishop. Neural networks for pattern recognition. *Oxford University Press*, 1995.

59. D. Blei. Probabilistic topic models. *Communications of the ACM*, 55(4), pp. 77–84, 2012.

60. D. Blei and P. Moreno. Topic segmentation with an aspect hidden Markov model. *ACM SIGIR Conference*, pp. 343–348, 2001.

61. D. Blei, A. Ng, and M. Jordan. Latent dirichlet allocation. *Journal of Machine Learning Research*, 3, pp. 993–1022, 2003.

62. D. Blei and J. Lafferty. Dynamic topic models. *ICML Conference*, pp. 113–120, 2006.

63. A. Blum, and T. Mitchell. Combining labeled and unlabeled data with co-training. *COLT*, 1998.

64. A. Blum and S. Chawla. Combining labeled and unlabeled data with graph mincuts. *ICML Conference*, 2001.

65. O. Bodenreider. The Unified Medical Language System (UMLS): Integrating biomedical terminology. *Nucleic Acids Research*, 32, pp. D267–270, 2004.

66. D. Boley, M. Gini, R. Gross, E.-H. Han, K. Hastings, G. Karypis, V. Kumar, B. Mobasher, and J. Moore. Partitioning-based clustering for Web document categorization. *Decision Support Systems*, Vol. 27, pp. 329–341, 1999.

67. A. Bordes, S. Chopra, and J. Weston. Question answering with subgraph embeddings. *arXiv preprint arXiv:1406.3676*, 2014.

68. A. Bordes, N. Usunier, S. Chopra, and J. Weston. Large-scale simple question answering with memory networks. *arXiv preprint arXiv:1506.02075*, 2015.

69. E. Breck, Y. Choi, and C. Cardie. Identifying expressions of opinion in context. *IJCAI*, pp. 2683–2688, 2007.

70. L. Breiman. Random forests. *Journal Machine Learning archive*, 45(1), pp. 5–32, 2001.

71. L. Breiman. Bagging predictors. *Machine Learning*, 24(2), pp. 123–140, 1996.

72. L. Breiman and A. Cutler. Random Forests Manual v4.0, *Technical Report, UC Berkeley*, 2003. https://www.stat.berkeley.edu/~breiman/Using_random_forests_v4.0.pdf

73. S. Brin. Extracting patterns and relations from the World Wide Web. *International Workshop on the Web and Databases*, 1998.

74. S. Brin, and L. Page. The anatomy of a large-scale hypertextual web search engine. *Computer Networks*, 30(1–7), pp. 107–117, 1998.

75. T. Brown *et al.* Language models are few-shot learners. *arXiv:2005.14165*, 2020. https://arxiv.org/abs/2005.14165

76. P. Bühlmann and B. Yu. Analyzing bagging. *Annals of Statistics*, pp. 927–961, 2002.

77. J. Bullinaria and J. Levy. Extracting semantic representations from word co-occurrence statistics: A computational study. *Behavior Research Methods*, 39(3), pp. 510–526, 2007.

78. R. Bunescu and R. Mooney. A shortest path dependency kernel for relation extraction. *Human Language Technology and Empirical Methods in Natural Language Processing*, pp. 724–731, 2005.

79. R. Bunescu and R. Mooney. Subsequence kernels for relation extraction. *NeurIPS Conference*, pp. 171–178, 2005.

80. C. Burges. A tutorial on support vector machines for pattern recognition. *Data mining and knowledge discovery*, 2(2), pp. 121–167, 1998.

81. C. Burges, T. Shaked, E. Renshaw, A. Lazier, M. Deeds, N. Hamilton, and G. Hullender. Learning to rank using gradient descent. *ICML Conference*, pp. 86–96, 2005.

82. S. Buttcher, C. Clarke, and G. V. Cormack. Information retrieval: Implementing and evaluating search engines. *The MIT Press*, 2010.

83. N. Butko and J. Movellan. I-POMDP: An infomax model of eye movement. *IEEE International Conference on Development and Learning*, pp. 139–144, 2008.

84. J. Callan. Distributed information retrieval. *Advances in Information Retrieval*, Springer, pp. 127–150, 2000.

85. M. Califf and R. Mooney. Bottom-up relational learning of pattern matching rules for information extraction. *Journal of Machine Learning Research*, 4, pp. 177-210, 2003.

86. A. Carlson, J. Betteridge, B. Kisiel, B. Settles, E. R. H. Jr, and T. M. Mitchell.Toward an Architecture for Never-Ending Language Learning. *Conference on Artificial Intelligence*, pp. 1306–1313, 2010.

87. Y. Cao, J. Xu, T. Liu, H. Li, Y. Huang, and H.-W. Hon. Adapting ranking SVM to document retrieval. *ACM SIGIR Conference*, pp. 186–193, 2006.

88. Z. Cao, T. Qin, T. Liu, M. Tsai, and H. Li. Learning to rank: from pairwise approach to listwise approach. *ICML Conference*, pp. 129–136, 2007.

89. J. Carbonell and J. Goldstein. The use of MMR, diversity-based reranking for reordering documents and producing summaries. *ACM SIGIR Conference*, pp. 335–336, 1998.

90. D. Carmel, D. Cohen, R. Fagin, E. Farchi, M. Herscovici, Y. Maarek, and A. Soffer. Static index pruning for information retrieval systems. *ACM SIGIR Conference*, pp. 43–50, 2001.

91. D. Chakrabarti and K. Punera. Event Summarization Using Tweets. *ICWSM Conference*, 11, pp. 66–73, 2011.

92. S. Chakrabarti. Mining the Web: Discovering knowledge from hypertext data. *Morgan Kaufmann*, 2003.

93. S. Chakrabarti, B. Dom. R. Agrawal, and P. Raghavan. Scalable feature selection, classification and signature generation for organizing large text databases into hierarchical topic taxonomies. *The VLDB Journal*, 7(3), pp. 163–178, 1998.

94. S. Chakrabarti, B. Dom, and P. Indyk. Enhanced hypertext categorization using hyperlinks. *ACM SIGMOD Conference*, pp. 307–318, 1998.

95. S. Chakrabarti, S. Roy, and M. Soundalgekar. Fast and accurate text classification via multiple linear discriminant projections. *The VLDB Journal*, 12(2), pp. 170–185, 2003.

96. S. Chakrabarti, M. Van den Berg, and B. Dom. Focused crawling: a new approach to topic-specific Web resource discovery. *Computer Networks*, 31(11), pp. 1623–1640, 1999.

97. Y. Chali and S. Joty. Improving the performance of the random walk model for answering complex questions. *Annual Meeting of the Association for Computational Linguistics on Human Language Technologies*, pp. 9–12, 2008.

98. N. Chambers, S. Wang, and D. Jurafsky. Classifying temporal relations between events. *Annual Meeting of the ACL on Interactive Poster and Demonstration Sessions*, pp. 173–176, 2007.

99. Y. Chan and D. Roth. Exploiting syntactico-semantic structures for relation extraction. *ACL Conference: Human Language Technologies*, pp. 551–560, 2011.

100. C. Chang and C. Lin. LIBSVM: a library for support vector machines. *ACM Transactions on Intelligent Systems and Technology*, 2(3), 27, 2011. http://www.csie.ntu.edu.tw/~cjlin/libsvm/

101. Y. Chang, C. Hsieh, K. Chang, M. Ringgaard, and C. J. Lin. Training and testing low-degree polynomial data mappings via linear SVM. *Journal of Machine Learning Research*, 11, pp. 1471–1490, 2010.

102. O. Chapelle. Training a support vector machine in the primal. *Neural Computation*, 19(5), pp. 1155–1178, 2007.

103. O. Chapelle, B. Schölkopf, and A. Zien. Semi-supervised learning. *MIT Press*, 2010.

104. P. Cheeseman and J. Stutz. Bayesian classification (AutoClass): Theory and results. *Advances in Knowledge Discovery and Data Mining*, Eds. U. Fayyad, G. Piatetsky-Shapiro, P. Smyth, and R. Uthuruswamy. AAAI Press/MIT Press, 1996.

105. D. Chen, A. Fisch, J. Weston, and A. Bordes. Reading Wikipedia to answer open-domain questions. *arXiv:1704.00051*, 2017. https://arxiv.org/abs/1704.00051

106. W. Chen *et al.* Hybridqa: A dataset of multi-hop question answering over tabular and textual data. *arXiv:2004.07347*, 2020.

107. J. Cheng and M. Lapata. Neural summarization by extracting sentences and words. *arXiv:1603.07252*, 2014. https://arxiv.org/abs/1603.07252

108. D. Chickering, D. Heckerman, and C. Meek. A Bayesian approach to learning Bayesian networks with local structure. *Uncertainty in Artificial Intelligence*, pp. 80–89, 1997.

109. R. Child *et al.* Generating long sequences with sparse transformers. *arXiv:1904.10509*, 2019.

110. J. Cho, H. Garcia-Molina, and L. Page. Efficient crawling through URL ordering. *Computer Networks*, 30(1–7), pp. 161–172, 1998.

111. K. Cho, B. Merrienboer, C. Gulcehre, F. Bougares, H. Schwenk, and Y. Bengio. Learning phrase representations using RNN encoder-decoder for statistical machine translation. *EMNLP*, 2014. https://arxiv.org/pdf/1406.1078.pdf

112. F. Choi. Advances in domain independent linear text segmentation. *North American Chapter of the Association for Computational Linguistics Conference*, pp. 26–33, 2000.

113. F. Choi, P. Wiemer-Hastings, and J. Moore. Latent semantic analysis for text segmentation. *EMNLP*, 2001.

114. Y. Choi, C. Cardie, E. Riloff, and S. Patwardhan. Identifying sources of opinions with conditional random fields and extraction patterns. *Conference on Human Language Technology and Empirical Methods in Natural Language Processing*, pp. 355–362, 2005.

115. J. Chung, C. Gulcehre, K. Cho, and Y. Bengio. Empirical evaluation of gated recurrent neural networks on sequence modeling. *arXiv:1412.3555*, 2014. https://arxiv.org/abs/1412.3555

116. K. Church and P. Hanks. Word association norms, mutual information, and lexicography. *Computational Linguistics*, 16(1), pp. 22–29, 1990.

117. F. Ciravegna. Adaptive information extraction from text by rule induction and generalisation. *International Joint Conference on Artificial Intelligence*, 17(1), pp. 1251–1256, 2001.

118. J. Clarke and M. Lapata. Models for sentence compression: A comparison across domains, training requirements and evaluation measures. *ACL Conference*, pp. 377–384, 2006.

119. W. Cohen. Fast effective rule induction. *ICML Conference*, pp. 115–123, 1995.

120. W. Cohen. Learning rules that classify e-mail. *AAAI Spring Symposium on Machine Learning in Information Access*, 1996.

121. W. Cohen. Learning with set-valued features. In *National Conference on Artificial Intelligence*, 1996.

122. W. Cohen, R. Schapire, and Y. Singer. Learning to Order Things. *Journal of Artificial Intelligence Research*, 10, pp. 243–270, 1999.

123. W. Cohen and Y. Singer. Context-sensitive learning methods for text categorization. *ACM Transactions on Information Systems*, 17(2), pp 141–173, 1999.

124. M. Collins and N. Duffy. Convolution kernels for natural language. *NeurIPS Conference*, pp. 625–632, 2001.

125. R. Collobert, J. Weston, L. Bottou, M. Karlen, K. Kavukcuoglu, and P. Kuksa. Natural language processing (almost) from scratch. *Journal of Machine Learning Research*, 12, pp. 2493–2537, 2011.

126. R. Collobert and J. Weston. A unified architecture for natural language processing: Deep neural networks with multitask learning. *ICML Conference*, pp. 160–167, 2008.

127. J. Conroy and D. O'Leary. Text summarization via hidden markov models. *ACM SIGIR Conference*, pp. 406–407, 2001.

128. J. Conroy, J. Schlessinger, D. O'Leary, and J. Goldstein. Back to basics: CLASSY 2006. *Document Understanding Conference*, 2006.

129. T. Cooke. Two variations on Fisher's linear discriminant for pattern recognition *IEEE Transactions on Pattern Analysis and Machine Intelligence*, 24(2), pp. 268–273, 2002.

130. W. Cooper. Some inconsistencies and misnomers in probabilistic information retrieval. *ACM Transactions on Information Systems*, 13(1), pp. 100–111, 1995.

131. B. O'Connor, R. Balasubramanyan, B. Routledge, and N. Smith. From tweets to polls: Linking text sentiment to public opinion time series. *ICWSM*, pp. 122–129, 2010.

132. C. Cortes and V. Vapnik. Support-vector networks. *Machine Learning*, 20(3), pp. 273–297, 1995.

133. T. Cover and P. Hart. Nearest neighbor pattern classification. *IEEE Transactions on Information Theory*, 13(1), pp. 1–27, 1967.

134. N. Cristianini, and J. Shawe-Taylor. An introduction to support vector machines and other kernel-based learning methods. *Cambridge University Press*, 2000.

135. W. B. Croft. Clustering large files of documents using the single-link method. *Journal of the American Society of Information Science*, 28, pp. 341–344, 1977.

136. W. B. Croft and D. Harper. Using probabilistic models of document retrieval without relevance information. *Journal of Documentation*, 35(4), pp. 285–295, 1979.

137. W. B. Croft, D. Metzler, and T. Strohman. Search engines: Information retrieval in practice, *Addison-Wesley Publishing Company*, 2009.

138. S. Cucerzan. Large-scale named entity disambiguation based on Wikipedia data. *EMNLP-CoNLL*, pp. 708–716, 2007.

139. A. Culotta and J. Sorensen. Dependency tree kernels for relation extraction. *ACL Conference*, 2004.

140. J. Curran and S. Clark. Language independent NER using a maximum entropy tagger. *Conference on Natural Language Learning at HLT-NAACL 2003*, pp. 164–167, 2003.

141. D. Cutting, D. Karger, J. Pedersen, and J. Tukey. Scatter/gather: A cluster-based approach to browsing large document collections. *ACM SIGIR Conference*, pp. 318–329, 1992.

142. W. Dai, Y. Chen, G. Xue, Q. Yang, and Y. Yu. Translated learning: Transfer learning across different feature spaces. *NeurIPS Conference*, pp. 353–360, 2008.

143. D. Das and A. Martins. A survey on automatic text summarization. *Literature Survey for the Language and Statistics II course at CMU*, 4, pp. 1–31, 2007.

144. S. Das and M. Chen. Yahoo! for Amazon: Extracting market sentiment from stock message boards. *Asia Pacific Finance Association Annual Conference (APFA)*, 2001.

145. J. Dean and S. Ghemawat. MapReduce: simplified data processing on large clusters. *Communications of the ACM*, 51(1), pp. 107–113, 2008.

146. G. DeJong. Prediction and substantiation: A new approach to natural language processing. *Cognitive Science*, 3(3), pp. 251–273, 1979.

147. H. Deng, B. Zhao, J. Han. Collective topic modeling for heterogeneous networks. *ACM SIGIR Conference*, pp. 1109-1110, 2011.

148. H. Deng, J. Han, B. Zhao, Y. Yu, and C. Lin. Probabilistic topic models with biased propagation on heterogeneous information networks. *ACM KDD Conference*, pp. 1271–1279, 2011.

149. J. Devlin *et al.* Bert: Pre-training of deep bidirectional transformers for language understanding. *arXiv:1810.04805*, 2018. https://arxiv.org/abs/1810.04805

150. I. Dhillon. Co-clustering documents and words using bipartite spectral graph partitioning. *ACM KDD Conference*, pp. 269–274, 2001.

151. I. Dhillon and D. Modha. Concept decompositions for large sparse text data using clustering. *Machine Learning*, 42(1–2), pp. 143–175, 2001.

152. B. Dhingra, K. Mazaitis, and W. Cohen. Quasar: Datasets for question answering by search and reading. *arXiv:1707.03904*, 2017. https://arxiv.org/abs/1707.03904

153. T. Dietterich. Machine learning for sequential data: A review. *Joint IAPR International Workshops on Statistical Techniques in Pattern Recognition (SPR) and Structural and Syntactic Pattern Recognition (SSPR)*, pp. 15–30, 2002.

154. C. Ding, X. He, and H. Simon. On the equivalence of nonnegative matrix factorization and spectral clustering. *SDM Conference*, pp. 606–610, 2005.

155. C. Ding, T. Li, and M. Jordan. Convex and semi-nonnegative matrix factorizations. *IEEE Transactions on Pattern Analysis and Machine Intelligence*, 32(1), pp. 45–55, 2010.

156. C. Ding, T. Li, and W. Peng. On the equivalence between non-negative matrix factorization and probabilistic latent semantic indexing. *Computational Statistics and Data Analysis*, 52(8), pp. 3913–3927, 2008.

157. C. Ding, T. Li, W. Peng, and H. Park. Orthogonal nonnegative matrix t-factorizations for clustering. *ACM KDD Conference*, pp. 126–135, 2006.

158. X. Ding, B. Liu, and P. S. Yu. A holistic lexicon-based approach to opinion mining. *WSDM Conference*, pp. 231–240, 2008.

159. P. Domingos and M. Pazzani. On the optimality of the simple bayesian classifier under zero-one loss. *Machine Learning*, 29(2–3), pp. 103–130, 1997.

160. X. Dong, E. Gabrilovich, G. Heitz, W. Horn, N. Lao, K. Murphy, T. Strohmann, S. Sun, and W. Zhang. Knowledge Vault: A Web-scale Approach to Probabilistic Knowledge Fusion. *ACM KDD Conference*, pp. 601–610, 2014.

161. B. Dorr, D. Zajic, and R. Schwartz. Hedge Trimmer: A parse-and-trim approach to headline generation. *HLT-NAACL Workshop on Text Summarization*, pp. 1–8, 2003.

162. N. Draper and H. Smith. Applied regression analysis. *John Wiley & Sons*, 2014.

163. H. Drucker, C. Burges, L. Kaufman, A. Smola, and V. Vapnik. Support Vector Regression Machines. *NeurIPS Conference*, 1997.

164. M. Dubey *et al.* Asknow: A framework for natural language query formalization in sparql. *European Semantic Web Conference*, 2016.

165. R. Duda, P. Hart, W. Stork. *Pattern Classification*, Wiley Interscience, 2000.

166. S. Dumais. Latent semantic indexing (LSI) and TREC-2. *Text Retrieval Conference (TREC)*, pp. 105–115, 1993.

167. S. Dumais. Latent semantic indexing (LSI): TREC-3 Report. *Text Retrieval Conference (TREC)*, pp. 219–230, 1995.

168. S. Dumais, J. Platt, D. Heckerman, and M. Sahami. Inductive learning algorithms and representations for text categorization. *ACM CIKM Conference*, pp. 148–155, 1998.

169. M. Dunn *et al.*. Searchqa: A new q&a dataset augmented with context from a search engine. *arXiv:1704.05179* , 2017. https://arxiv.org/abs/1704.05179

170. S. Deerwester, S. Dumais, G. Furnas, T. Landauer, and R. Harshman. Indexing by latent semantic analysis. *Journal of the American Society for Information Science*, 41(6), 41(6), pp. 391–407, 1990.

171. C. Eckart and G. Young. The approximation of one matrix by another of lower rank. *Psychometrika*, 1(3), pp. 211–218, 1936.

172. H. P. Edmundson. New methods in automatic extracting. *Journal of the ACM*, 16(2), pp. 264–286, 1969.

173. B. Efron, T. Hastie, I. Johnstone, and R. Tibshirani. Least angle regression. *The Annals of Statistics*, 32(2), pp. 407–499, 2004.

174. J. Eisenstein and R. Barzilay. Bayesian unsupervised topic segmentation. *Conference on Empirical Methods in Natural Language Processing*, pp. 334–343, 2008.

175. P. Elias. Universal codeword sets and representations of the integers. *IEEE Transactions on Information Theory*, 21(2), pp. 194–203, 1975.

176. G. Erkan and D. Radev. LexRank: Graph-based lexical centrality as salience in text summarization. *Journal of Artificial Intelligence Research*, 22, pp. 457–479, 2004.

177. E. Erosheva, S. Fienberg, and J. Lafferty. Mixed-membership models of scientific publications. *Proceedings of the National Academy of Sciences*, 101, pp. 5220–5227, 2004.

178. A. Esuli, and F. Sebastiani. Determining the semantic orientation of terms through gloss classification. *ACM CIKM Conference*, pp. 617–624, 2005.

179. A. Esuli and F. Sebastiani. Determining term subjectivity and term orientation for opinion mining. *European Chapter of the Association of Computational Linguistics*, 2006.

180. O. Etzioni, M. Cafarella, D. Downey, A. Popescu, T. Shaked, S. Soderland, D. Weld, and A. Yates. Unsupervised named-entity extraction from the web: An experimental study. *Artificial Intelligence*, 165(1), pp. 91–134, 2005.

181. O. Etzioni, M. Banko, S. Soderland, and D. Weld. Open information extraction from the web. *Communications of the ACM*, 51(12), pp. 68–74, 2008.

182. A. Fader, S. Soderland, and O. Etzioni. Identifying relations for open information extraction. *Conference on Empirical Methods in Natural Language Processing*, pp. 1535–1545, 2011.

183. A. Fader, L. Zettlemoyer, and O. Etzioni. Paraphrase-Driven Learning for Open Question Answering. *ACL*, pp. 1608–1618, 2013.

184. A. Fader, L. Zettlemoyer, and O. Etzioni. Open question answering over curated and extracted knowledge bases. *ACM KDD Conference*, 2014.

185. C. Faloutsos and S. Christodoulakis. Signature files: An access method for documents and its analytical performance evaluation. *ACM Transactions on Information Systems*, 2(4), pp. 267–288, 1984.

186. A. Fan *et al.* Eli5: Long form question answering. *arXiv:1907.09190*, 2019. https://arxiv.org/abs/1907.09190

187. J. Fan, D. Ferrucci, D. Gondek, and A. Kalyanpur. Prismatic: Inducing knowledge from a large scale lexicalized relation resource. *NAACL HLT 2010 First International Workshop on Formalisms and Methodology for Learning by Reading*, pp. 122-127 2010.

188. R. Fan, K. Chang, C. Hsieh, X. Wang, and C. Lin. LIBLINEAR: A library for large linear classification. *Journal of Machine Learning Research*, 9, pp. 1871–1874, 2008. http://www.csie.ntu.edu.tw/~cjlin/liblinear/

189. R. Fan, P. Chen, and C. Lin. Working set selection using second order information for training support vector machines. *Journal of Machine Learning Research*, 6, pp. 1889–1918, 2005.

190. T. Fawcett. ROC Graphs: Notes and Practical Considerations for Researchers. *Technical Report HPL-2003-4*, Palo Alto, CA, HP Laboratories, 2003.

191. R. Fisher. The use of multiple measurements in taxonomic problems. *Annals of Eugenics*, 7: pp. 179–188, 1936.

192. W. Fedus, B. Zoph, and N. Shazeer. Switch Transformers: Scaling to trillion parameter models with simple and efficient sparsity. *arXiv:2101.03961*, 2021. https://arxiv.org/abs/2101.03961

193. R. Feldman and J. Sanger. The text mining handbook: advanced approaches in analyzing unstructured data. *Cambridge University Press*, 2007.

194. M. Fernandez-Delgado, E. Cernadas, S. Barro, and D. Amorim. Do we Need Hundreds of Classifiers to Solve Real World Classification Problems? *The Journal of Machine Learning Research*, 15(1), pp. 3133–3181, 2014.

195. D. Ferrucci *et al.* Building Watson: An overview of the DeepQA project. *AI magazine*, 31(3), pp. 59–79, 2010.

196. K. Filippova and M. Strube. Sentence fusion via dependency graph compression. *Conference on Empirical Methods in Natural Language Processing*, pp. 177–185, 2008.

197. K. Bollacker, C. Evans, P. Paritosh, T. Sturge, and J. Taylor. Freebase: a collaboratively created graph database for structuring human knowledge. *ACM SIGMOD Conference*, pp. 1247–1250, 2008.

198. D. Freitag and A. McCallum. Information extraction with HMMs and shrinkage. *AAAI-99 Workshop on Machine Learning for Information Extraction*, pp. 31–36, 1999.

199. C. Freudenthaler, L. Schmidt-Thieme, and S. Rendle. Factorization machines: Factorized polynomial regression models. *German-Polish Symposium on Data Analysis and Its Applications (GPSDAA)*, 2011.
https://www.ismll.uni-hildesheim.de/pub/pdfs/FreudenthalerRendle_FactorizedPolynomialRegression.pdf

200. Y. Freund, and R. Schapire. A decision-theoretic generalization of online learning and application to boosting. *Computational Learning Theory*, pp. 23–37, 1995.

201. J. Friedman. Stochastic gradient boosting. *Computational Statistics and Data Analysis*, 38(4), pp. 367–378, 2002.

202. J. Friedman, T. Hastie, and R. Tibshirani. Additive logistic regression: a statistical view of boosting (with discussion and a rejoinder by the authors). *The Annals of Statistics*, 28(2), pp. 337–407, 2000.

203. M. Fuentes, E. Alfonseca, and H. Rodriguez. Support Vector Machines for query-focused summarization trained and evaluated on Pyramid data. *ACL Conference*, pp. 57–60, 2007.

204. G. Fung and O. Mangasarian. Proximal support vector classifiers. *ACM KDD Conference*, pp. 77–86, 2001.

205. J. Fürnkranz and G. Widmer. Incremental reduced error pruning. *ICML Conference*, pp. 70–77, 1994.

206. M. Galley and K. McKeown. Lexicalized Markov grammars for sentence compression. *Human Language Technologies: The Conference of the North American Chapter of the Association for Computational Linguistics*, pp. 180–187, 2007.

207. T. Gärtner. A survey of kernels for structured data. *ACM SIGKDD Explorations Newsletter*, 5(1), pp. 49–58, 2003.

208. Y. Goldberg. A primer on neural network models for natural language processing. *Journal of Artificial Intelligence Research (JAIR)*, 57, pp. 345–420, 2016.

209. Y. Goldberg and O. Levy. word2vec Explained: deriving Mikolov et al.'s negative-sampling word-embedding method. *arXiv:1402.3722*, 2014. https://arxiv.org/abs/1402.3722

210. I. Goodfellow, Y. Bengio, and A. Courville. Deep learning. *MIT Press*, 2016.

211. W. Greiff. A theory of term weighting based on exploratory data analysis. *ACM SIGIR Conference*, pp. 11–19, 1998.

212. E. Gaussier and C. Goutte. Relation between PLSA and NMF and implications. *ACM SIGIR Conference*, pp. 601–602, 2005.

213. L. Getoor, N. Friedman, D. Koller, and B. Taskar. Learning probabilistic models of link structure. *Journal of Machine Learning Research*, 3, pp. 679–707, 2002.

214. J. Ghosh and A. Acharya. Cluster ensembles: Theory and applications. *Data Clustering: Algorithms and Applications*, CRC Press, 2013.

215. D. Gillick, K. Riedhammer, B. Favre, and D. Hakkani-Tur. A global optimization framework for meeting summarization. *IEEE International Conference on Acoustics, Speech and Signal Processing*, pp. 4769–4772, 2009.

216. S. Gilpin, T. Eliassi-Rad, and I. Davidson. Guided learning for role discovery (glrd): framework, algorithms, and applications. *ACM KDD Conference*, pp. 113–121, 2013.

217. M. Girolami and A. Kabán. On an equivalence between PLSI and LDA. *ACM SIGIR Conference*, pp. 433–434, 2003.

218. F. Girosi and T. Poggio. Networks and the best approximation property. *Biological Cybernetics*, 63(3), pp. 169–176, 1990.

219. Y. Gong and X. Liu. Generic text summarization using relevance measure and latent semantic analysis. *ACM SIGIR Conference*, pp. 19–25, 2001.

220. K. Greff, R. K. Srivastava, J. Koutnik, B. Steunebrink, and J. Schmidhuber. LSTM: A search space odyssey. *IEEE TNNLS*, 2016.

221. R. Grishman and B. Sundheim. Message Understanding Conference-6: A Brief History. *COLING*, pp. 466–471, 1996.

222. D. Grossman and O. Frieder. Information retrieval: Algorithms and heuristics, *Springer Science and Business Media*, 2012.

223. A. Graves. Supervised sequence labelling with recurrent neural networks *Springer*, 2012.

224. A. Graves. Generating sequences with recurrent neural networks. *arXiv preprint arXiv:1308.0850*, 2013. https://arxiv.org/abs/1308.0850

225. A. Graves, A. Mohamed, and G. Hinton. Speech recognition with deep recurrent neural networks. *Acoustics, Speech and Signal Processing (ICASSP)*, pp. 6645–6649, 2013.

226. M. Gutmann and A. Hyvarinen. Noise-contrastive estimation: A new estimation principle for unnormalized statistical models. *AISTATS*, 1(2), pp. 6, 2010.

227. K. Guu *et al.* Realm: Retrieval-augmented language model pre-training. *arXiv:2002.08909*, 2020.

228. B. Hagedorn, M. Ciaramita, and J. Atserias. World knowledge in broad-coverage information filtering. *ACM SIGIR Conference*, 2007.

229. A. Haghighi and L. Vanderwende. Exploring content models for multi-document summarization. *Human Language Technologies*, pp. 362–370, 2009.

230. D. Hakkani-Tur and G. Tur. Statistical sentence extraction for information distillation. *Conference on Acoustics, Speech and Signal Processing*, 4, 2007.

231. E.-H. Han, G. Karypis, and V. Kumar. Text categorization using weighted-adjusted k-nearest neighbor classification, *PAKDD Conference*, 2001.

232. E.-H. Han and G. Karypis. Centroid-based document classification: Analysis and experimental results. *PKDD Conference*, 2000.

233. J. Han, M. Kamber, and J. Pei. Data mining: concepts and techniques. *Morgan Kaufmann*, 2011.

234. T. H. Haveliwala. Topic-sensitive pagerank. *World Wide Web Conference*, pp. 517-526, 2002.

235. T. Hastie, R. Tibshirani, and J. Friedman. The elements of statistical learning. *Springer*, 2009.

236. T. Hastie and R. Tibshirani. Discriminant adaptive nearest neighbor classification. *IEEE Transactions on Pattern Analysis and Machine Intelligence*, 18(6), pp. 607–616, 1996.

237. T. Hastie, R. Tibshirani, and M. Wainwright. Statistical learning with sparsity: the lasso and generalizations. *CRC Press*, 2015.

238. T. Hastie and R. Tibshirani. Generalized additive models. *CRC Press*, 1990.

239. V. Hatzivassiloglou, J. Klavans, M. Holcombe, R. Barzilay, M.-Y. Kan, and K. R. McKeown. SIMFINDER: A flexible clustering tool for summarization. *NAACL Workshop on Automatic Summarization*, pp. 41–49, 2001.

240. M. Hatzivassiloglou, and K. McKeown. Predicting the semantic orientation of adjectives. *European Chapter of the Association for Computational Linguistics*, pp. 174–181, 1997.

241. V. Hatzivassiloglou and J. Wiebe. Effects of adjective orientation and gradability on sentence subjectivity. *Conference on Computational Linguistics*, pp. 299–305, 2000.

242. M. Hearst. TextTiling: Segmenting text into multi-paragraph subtopic passages. *Computational Linguistics*, 23(1), pp. 33–64, 1997.

243. D. Hiemstra. A linguistically motivated probabilistic model of information retrieval. *International Conference on Theory and Practice of Digital Libraries*, pp. 569–584, 1998.

244. M. Hearst. TextTiling: Segmenting text into multi-paragraph subtopic passages. *Computational Linguistics*, 23(1), pp. 33–64, 1997.

245. S. Heinz and J. Zobel. Efficient single-pass index construction for text databases. *Journal of the American Society for Information Science and Technology*, 54(8), pp. 713–729, 2003.

246. K. Hermann *et al.* Teaching machines to read and comprehend. *NeurIPS*, pp. 1684–1692, 2015

247. G. Hinton. Connectionist learning procedures. *Artificial Intelligence*, 40(1–3), pp. 185–234, 1989.

248. G. Hinton and R. Salakhutdinov. Reducing the dimensionality of data with neural networks. *Science*, 313(5786), pp. 504–507, 2006.

249. G. Hirst and D. St-Onge. Lexical chains as representation of context for the detection and correction of malapropisms. In *WordNet: An Electronic Lexical Database and Some of its Applications*, MIT Press, 1998.

250. T. K. Ho. Random decision forests. *Third International Conference on Document Analysis and Recognition*, 1995. Extended version appears as "The random subspace method for constructing decision forests" in *IEEE Transactions on Pattern Analysis and Machine Intelligence*, 20(8), pp. 832–844, 1998.

251. T. K. Ho. Nearest neighbors in random subspaces. *Lecture Notes in Computer Science*, Vol. 1451, pp. 640–648, *Proceedings of the Joint IAPR Workshops SSPR'98 and SPR'98*, 1998.

252. A. Hoang *et al.* Efficient adaptation of pretrained transformers for abstractive summarization. *arXiv:1906.00138*, 2019. https://arxiv.org/abs/1906.00138

253. S. Hochreiter and J. Schmidhuber. Long short-term memory. *Neural Computation*, 9(8), pp. 1735–1785, 1997.

254. S. Hochreiter, Y. Bengio, P. Frasconi, and J. Schmidhuber. Gradient flow in recurrent nets: the difficulty of learning long-term dependencies, *A Field Guide to Dynamical Recurrent Neural Networks*, IEEE Press, 2001.

255. T. Hofmann. Probabilistic latent semantic indexing. *ACM SIGIR Conference*, pp. 50–57, 1999.

256. T. Hofmann. Unsupervised learning by probabilistic latent semantic analysis. *Machine learning*, 41(1–2), pp. 177–196, 2001.

257. K. Hornik and B. Grün. topicmodels: An R package for fitting topic models. *Journal of Statistical Software*, 40(13), pp. 1–30, 2011.

258. K. Hornik, M. Stinchcombe, and H. White. Multilayer feedforward networks are universal approximators. *Neural Networks*, 2(5), pp. 359–366, 1989.

259. E. Hovy and C.-Y. Lin. Automated Text Summarization in SUMMARIST. in *Advances in Automatic Text Summarization*, pp. 82–94, 1999.

260. M. Hu and B. Liu. Mining opinion features in customer reviews. *AAAI*, pp. 755–760, 2004.

261. M. Hu and B. Liu. Mining and summarizing customer reviews. *ACM KDD Conference*, pp. 168–177, 2004.

262. A. Huang. Similarity measures for text document clustering. *Sixth New Zealand Computer Science Research Student Conference*, pp. 49–56, 2008.

263. M. Iyyer, J. Boyd-Graber, L. Claudino, R. Socher, and H. Daume III. A Neural Network for Factoid Question Answering over Paragraphs. *EMNLP*, 2014.

264. N. Jakob and I. Gurevych. Extracting opinion targets in a single-and cross-domain setting with conditional random fields. *Conference on Empirical Methods in Natural Language Processing*, pp. 1035–1045, 2010.

265. Z. Jiang, F. Xu, J. Araki, and G. Neubig. How can we know what language models know? *Transactions of the Association for Computational Linguistics*, 8, pp. 423–438, 2019.

266. W. Jin, H. Ho, and R. Srihari. OpinionMiner: a novel machine learning system for Web opinion mining and extraction. *ACM KDD Conference*, pp. 1195–1204, 2009.

267. N. Jindal and B. Liu. Opinion spam and analysis. *WSDM Conference*, pp. 219–230, 2008.

268. J. Jiang. Information extraction from text. *Mining Text Data*, Springer, pp. 11–41, 2012.

269. J. Jiang and C. Zhai. A systematic exploration of the feature space for relation extraction. *HLT-NAACL*, pp. 113–120, 2007.

270. H. Jing. Sentence reduction for automatic text summarization. *Conference on Applied Natural Language Processing*, pp. 310–315, 2000.

271. H. Jing. Cut-and-paste text summarization. *PhD Thesis*, Columbia University, 2001. http://www1.cs.columbia.edu/nlp/theses/hongyan_jing.pdf

272. T. Joachims. Text categorization with support vector machines: learning with many relevant features. *ECML Conference*, 1998.

273. T. Joachims. Making Large scale SVMs practical. *Advances in Kernel Methods, Support Vector Learning*, pp. 169–184, *MIT Press*, Cambridge, 1998.

274. T. Joachims. Training Linear SVMs in Linear Time. *ACM KDD Conference*, pp. 217–226, 2006.

275. T. Joachims. A probabilistic analysis of the Rocchio algorithm with TFIDF for text categorization. *ICML Conference*, 1997.

276. T. Joachims. Optimizing search engines using clickthrough data. *ACM KDD Conference*, pp. 133–142, 2002.

277. C. Johnson. Logistic matrix factorization for implicit feedback data. *NeurIPS Conference*, 2014.

278. D. Johnson, F. Oles, T. Zhang, T. Goetz. A decision tree-based symbolic rule induction system for text categorization, *IBM Systems Journal*, 41(3), pp. 428–437, 2002.

279. I. T. Jolliffe. Principal component analysis. *John Wiley & Sons*, 2002.

280. I. T. Jolliffe. A note on the use of principal components in regression. *Applied Statistics*, 31(3), pp. 300–303, 1982.

281. M. Joshi, K. Choi, D. Weld, and L. Zettlemoyer. Triviaqa: A large scale distantly supervised challenge dataset for reading comprehension. *arXiv:1705.03551*, 2017. https://arxiv.org/abs/1705.03551

282. R. Jozefowicz, W. Zaremba, and I. Sutskever. An empirical exploration of recurrent network architectures. *ICML*, pp. 2342–2350, 2015.

283. D. Jurafsky and J. Martin. Speech and language processing. *Prentice Hall*, 2008.

284. N. Kalchbrenner and P. Blunsom. Recurrent continuous translation models. *EMNLP*, 3, 39, pp. 413, 2013.

285. N. Kambhatla, Combining lexical, syntactic and semantic features with maximum entropy models for information extraction. *ACL Conference*, pp. 178–181, 2004.

286. J. Kamps, M. Marx, R. Mokken, and M. Rijke. Using wordnet to measure semantic orientations of adjectives. *LREC*, pp. 1115–1118, 2004.

287. R. Kannan, H. Woo, C. Aggarwal, and H. Park. Outlier detection for text data. *SDM Conference*, 2017.

288. H. Kanayama and T. Nasukawa. Fully automatic lexicon expansion for domain-oriented sentiment analysis. *Conference on Empirical Methods in Natural Language Processing*, pp. 355–363, 2006.

289. A. Karatzoglou, A. Smola A, K. Hornik, and A. Zeileis. kernlab – An S4 Package for Kernel Methods in R. *Journal of Statistical Software*, 11(9), 2004.
http://epub.wu.ac.at/1048/1/document.pdf http://CRAN.R-project.org/package=kernlab

290. A. Karpathy, J. Johnson, and L. Fei-Fei. Visualizing and understanding recurrent networks. *arXiv preprint arXiv:1506.02078*, 2015. https://arxiv.org/abs/1506.02078

291. A. Karpathy. The unreasonable effectiveness of recurrent neural networks, *Blog post*, 2015. http://karpathy.github.io/2015/05/21/rnn-effectiveness/

292. V. Karpukhin *et al.* Dense passage retrieval for open-domain question answering. *arXiv:2004.04906*, 2020. https://arxiv.org/abs/2004.04906

293. G. Karypis and E.-H. Han. Fast supervised dimensionality reduction with applications to document categorization and retrieval, *ACM CIKM Conference*, pp. 12–19, 2000.

294. E. Kaufmann, A. Bernstein, and R. Zumstein. Querix: A natural language interface to query ontologies based on clarification dialogs. *International Semantic Web Conference*, pp. 980–981, 2006.

295. V. Kieuvongngam, B. Tan, and Y. Niu. Automatic text summarization of covid-19 medical research articles using bert and gpt-2. *arXiv:2006.01997*, 2020.

296. H. Kim, K. Ganesan, P. Sondhi, and C. Zhai. Comprehensive Review of Opinion Summarization. *Technical Report*, University of Illinois at Urbana-Champaign, 2011. https://www.ideals.illinois.edu/handle/2142/18702

297. S. Kim and E. Hovy. Automatic identification of pro and con reasons in online reviews. *COLING/ACL Conference*, pp. 483–490, 2006.

298. Y. Kim. Convolutional neural networks for sentence classification. *arXiv preprint arXiv:1408.5882*, 2014.

299. J. Kleinberg. Authoritative sources in a hyperlinked environment. *Journal of the ACM (JACM)*, 46(5), pp. 604–632, 1999.

300. K. Knight and D. Marcu. Summarization beyond sentence extraction: A probabilistic approach to sentence compression. *Artificial Intelligence*, 139(1), pp. 91–107, 2002.

301. R. Kohavi and D. Wolpert. Bias plus variance decomposition for zero-one loss functions. *ICML Conference*, 1996.

302. E. Kong and T. Dietterich. Error-correcting output coding corrects bias and variance. *ICML Conference*, pp. 313–321, 1995.

303. T. Kwiatkowski *et al.* Natural questions: a benchmark for question answering research. *Transactions of the Association for Computational Linguistics*, 7, pp. 453–466, 2019.

304. A. Krogh, M. Brown, I. Mian, K. Sjolander, and D. Haussler. Hidden Markov models in computational biology: Applications to protein modeling. *Journal of Molecular Biology*, 235(5), pp. 1501–1531, 1994.

305. M. Kuhn. Building predictive models in R Using the caret Package. *Journal of Statistical Software*, 28(5), pp. 1–26, 2008. https://cran.r-project.org/web/packages/caret/index.html

306. J. Kupiec. Robust part-of-speech tagging using a hidden Markov model. *Computer Speech and Language, 6(3)*, pp. 225–242, 1992.

307. J. Kupiec, J. Pedersen, and F. Chen. A trainable document summarizer. *ACM SIGIR Conference*, pp. 68–73, 1995.

308. J. Lafferty, A. McCallum, and F. Pereira. Conditional random fields: Probabilistic models for segmenting and labeling sequence data. *ICML Conference*, pp. 282–289, 2001.

309. W. Lam and C. Y. Ho. Using a generalized instance set for automatic text categorization. *ACM SIGIR Conference*, 1998.

310. A. Langville, C. Meyer, R. Albright, J. Cox, and D. Duling. Initializations for the nonnegative matrix factorization. *ACM KDD Conference*, pp. 23–26, 2006.

311. H. Larochelle and G. E. Hinton. Learning to combine foveal glimpses with a third-order Boltzmann machine. *NeurIPS*, 2010.

312. J. Lau and T. Baldwin. An empirical evaluation of doc2vec with practical insights into document embedding generation. *arXiv:1607.05368*, 2016. https://arxiv.org/abs/1607.05368

313. Q. Le. Personal communication, 2017.

314. Q. Le and T. Mikolov. Distributed representations of sentences and documents. *ICML Conference*, pp. 1188–196, 2014.

315. D. Lee and H. Seung. Algorithms for non-negative matrix factorization. *Advances in NeurIPS*, pp. 556–562, 2001.

316. D. Lee and H. Seung. Learning the parts of objects by non-negative matrix factorization. *Nature*, 401(6755), pp. 788–791, 2001.

317. K. Lee, M. Chang, and K. Toutanova. Latent retrieval for weakly supervised open domain question answering. *arXiv:1906.00300*, 2019. https://arxiv.org/abs/1906.00300

318. R. Lempel and S. Moran. Predictive caching and prefetching of query results in search engines. *World Wide Web Conference*, pp. 19–28, 2003.

319. D. Lenat. CYC: A large-scale investment in knowledge infrastructure. *Communications of the ACM*, 38(11), pp. 33–38, 1995.

320. J. Leskovec, N. Milic-Frayling, and M. Grobelnik. Impact of linguistic analysis on the semantic graph: coverage and learning of document extracts. *National Conference on Artificial Intelligence*, pp. 1069–1074, 2005.

321. J. Leskovec, A. Rajaraman, and J. Ullman. Mining of massive datasets. *Cambridge University Press*, 2012.

322. N. Lester, J. Zobel, and H. Williams. Efficient online index maintenance for contiguous inverted lists. Information Processing and Management, 42(4), pp. 916–933, 2006.

323. O. Levy and Y. Goldberg. Neural word embedding as implicit matrix factorization. *NeurIPS Conference*, pp. 2177–2185, 2014.

324. O. Levy, Y. Goldberg, and I. Dagan. Improving distributional similarity with lessons learned from word embeddings. *Transactions of the Association for Computational Linguistics*, 3, pp. 211–225, 2015.

325. O. Levy, Y. Goldberg, and I. Ramat-Gan. Linguistic regularities in sparse and explicit word representations. *CoNLL*, 2014.

326. D. Lewis. An evaluation of phrasal and clustered representations for the text categorization task. *ACM SIGIR Conference*, pp. 37–50, 1992.

327. D. Lewis. Naive (Bayes) at forty: The independence assumption in information retrieval. *ECML Conference*, pp. 4–15, 1998.

328. D. Lewis and M. Ringuette. A comparison of two learning algorithms for text categorization. *Third Annual Symposium on Document Analysis and Information Retrieval*, pp. 81–93, 1994.

329. F. Li, C. Han, M. Huang, X. Zhu, Y. Xia, S. Zhang, and H. Yu. Structure-aware review mining and summarization. *Conference on Computational Linguistics*, pp. 6563–661, 2010.

330. H. Li, and K. Yamanishi. Document classification using a finite mixture model. *ACL Conference*, pp. 39–47, 1997.

331. Y. Li and A. Jain. Classification of text documents. *The Computer Journal*, 41(8), pp. 537–546, 1998.

332. Y. Li, C. Luo, and S. Chung. Text clustering with feature selection by using statistical data. *IEEE Transactions on Knowledge and Data Engineering*, 20(5), pp. 641–652, 2008.

333. D. Liben-Nowell, and J. Kleinberg. The link-prediction problem for social networks. *Journal of the American Society for Information Science and Technology*, 58(7), pp. 1019–1031, 2007.

334. E. Lim, V. Nguyen, N. Jindal, B. Liu, and H. Lauw. Detecting product review spammers using rating behaviors. *ACM CIKM Conference*, pp. 939–948, 2010.

335. C. Lin. Projected gradient methods for nonnegative matrix factorization. *Neural Computation*, 19(10), pp. 2756–2779, 2007.

336. C.-Y. Lin. Rouge: A package for automatic evaluation of summaries. *ACL Workshop*, 2004. https://www.aclweb.org/anthology/W04-1013.pdf

337. C.-Y. Lin and E. Hovy. The automated acquisition of topic signatures for text summarization. *Conference on Computational linguistics*, pp. 495–501, 2000.

338. C.-Y. Lin and E. Hovy. From single to multi-document summarization: A prototype system and its evaluation. *ACL Conference*, pp. 457–464, 2002.

339. H. Lin and J. Bilmes. Multi-document summarization via budgeted maximization of submodular functions. *Human Language Technologies*, pp. 912–920, 2010.

340. X. Ling and D. Weld. Temporal information extraction. *AAAI*, pp. 1385–1390, 2010.

341. Z. Lipton, J. Berkowitz, and C. Elkan. A critical review of recurrent neural networks for sequence learning. *arXiv:1506.00019*, 2015. https://arxiv.org/abs/1506.00019

342. L. V. Lita, A. Ittycheriah, S. Roukos, and N. Kambhatla. Truecasing. *ACL Conference*, pp. 152–159, 2003.

343. D. Litman and R. Passonneau. Combining multiple knowledge sources for discourse segmentation. *Association for Computational Linguistics*, pp. 108–115, 1995.

344. C. Lin and Y. He. Joint sentiment/topic model for sentiment analysis. *ACM CIKM Conference*, pp. 375–384, 2009.

345. B. Liu. Web data mining: exploring hyperlinks, contents, and usage data. *Springer*, New York, 2007.

346. B. Liu. Sentiment Analysis and Subjectivity. *Handbook of Natural Language Processing*, 2, pp. 627–666, 2010.

347. B. Liu. Sentiment analysis: Mining opinions, sentiments, and emotions. *Cambridge University Press*, 2015.

348. B. Liu, W. Hsu, and Y. Ma. Integrating classification and association rule mining. *ACM KDD Conference*, pp. 80–86, 1998.

349. T.-Y. Liu. Learning to rank for information retrieval. *Foundations and Trends in Information Retrieval*, 3(3), pp. 225–231, 2009.

350. Y. Liu. Fine-tune BERT for extractive summarization. *arXiv:1903.10318*, 2019. https://arxiv.org/abs/1903.10318

351. Y. Liu and M. Lapata. Text summarization with pretrained encoders. *arXiv:1908.08345*, 2019. https://arxiv.org/abs/1908.08345

352. Y. Liu *et al.* Roberta: A robustly optimized bert pretraining approach. *arXiv:1907.11692*, 2019. https://arxiv.org/abs/1907.11692

353. H. Lodhi, C. Saunders, J. Shawe-Taylor, N. Cristianini, and C. Watkins. Text classification using string kernels. *Journal of Machine Learning Research*, 2, pp. 419–444, 2002.

354. X. Long and T. Suel. Optimized query execution in large search engines with global page ordering. *VLDB Conference*, pp. 129–140, 2003.

355. Y. Lu, M. Castellanos, U. Dayal, and C. Zhai. Automatic construction of a context-aware sentiment lexicon: an optimization approach. *World Wide Web Conference*, pp. 347–356, 2011.

356. Y. Lu and C. Zhai. Opinion integration through semi-supervised topic modeling. *World Wide Web Conference*, pp. 121–130, 2008.

357. H. P. Luhn. The automatic creation of literature abstracts. *IBM Journal of Research and Development*, 2(2), pp. 159–165, 1958.

358. K. Lund and C. Burgess. Producing high-dimensional semantic spaces from lexical co-occurrence. *Behavior Research Methods, Instruments, and Computers*, 28(2). pp. 203–208, 1996.

359. M. Luong, H. Pham, and C. Manning. Effective approaches to attention-based neural machine translation. *arXiv:1508.04025*, 2015. https://arxiv.org/abs/1508.04025

360. U. von Luxburg. A tutorial on spectral clustering. *Statistics and Computing*, 17(4), pp. 395–416, 2007.

361. A. Maas, R. Daly, P. Pham, D. Huang, A. Ng, and C. Potts. Learning word vectors for sentiment analysis. *Annual Meeting of the Association for Computational Linguistics: Human Language Technologies-Volume 1*, pp. 142–150, 2011.

362. C. Mackenzie. Coded character sets: History and development. *Addison-Wesley Longman Publishing Co., Inc.*, 1980.

363. S. Madeira and A. Oliveira. Biclustering algorithms for biological data analysis: a survey. *IEEE/ACM Transactions on Computational Biology and Bioinformatics (TCBB)*, 1(1), pp. 24–45, 2004.

364. R. Malouf. A comparison of algorithms for maximum entropy parameter estimation. *Conference on Natural Language Learning*, pp. 1–7, 2002.

365. O. Mangasarian and D. Musicant. Successive overrelaxation for support vector machines. *IEEE Transactions on Neural Networks*, 10(5), pp. 1032–1037, 1999.

366. I. Mani and G. Wilson. Robust temporal processing of news. *ACL Conference*, pp. 69–76, 2000.

367. C. Manning, P. Raghavan, and H. Schütze. Introduction to information retrieval. *Cambridge University Press*, Cambridge, 2008.

368. C. Manning and H. Schütze. Foundations of statistical natural language processing. *MIT Press*, 1999.

369. E. Marsi and E. Krahmer. Explorations in sentence fusion. *European Workshop on Natural Language Generation*, pp. 109–117, 2005.

370. J. Martens and I. Sutskever. Learning recurrent neural networks with hessian-free optimization. *ICML Conference*, pp. 1033–1040, 2011.

371. A. McCallum. Bow: A toolkit for statistical language modeling, text retrieval, classification and clustering, 1996. http://www.cs.cmu.edu/~mccallum/bow

372. A. McCallum, D. Freitag, and F. Pereira. Maximum entropy Markov models for information extraction and segmentation. *ICML Conference*, pp. 591–598, 2000.

373. A. McCallum and K. Nigam. A comparison of event models for naive Bayes text classification. *AAAI Workshop on Learning for Text Categorization*, 1998.

374. P. McCullagh and J. Nelder. Generalized linear models *CRC Press*, 1989.

375. R. McDonald, K. Hannan, T. Neylon, M. Wells, and J. Reynar. Structured models for fine-to-coarse sentiment analysis. *ACL Conference*, 2007.

376. G. McLachlan. Discriminant analysis and statistical pattern recognition *John Wiley & Sons*, 2004.

377. D. McClosky, M. Surdeanu, and C. Manning. Event extraction as dependency parsing. *Annual Meeting of the Association for Computational Linguistics: Human Language Technologies-Volume 1*, pp. 1626–1635, 2011.

378. Q. Mei, D. Cai, D. Zhang, and C. Zhai. Topic modeling with network regularization. *World Wide Web Conference*, pp. 101–110, 2008.

379. Q. Mei, X. Ling, M. Wondra, H. Su, and C. Zhai. Topic sentiment mixture: modeling facets and opinions in weblogs. In *World Wide Web Conference*, pp. 171–180, 2007.

380. S. Melink, S. Raghavan, B. Yang, and H. Garcia-Molina. Building a distributed full-text index for the web. *ACM Transactions on Information Systems*, 19(3), pp. 217–241, 2001.

381. P. Melville, W. Gryc, and R. Lawrence. Sentiment analysis of blogs by combining lexical knowledge with text classification. *ACM KDD Conference*, pp. 1275–1284, 2009.

382. A. K. Menon, and C. Elkan. Link prediction via matrix factorization. *Machine Learning and Knowledge Discovery in Databases*, pp. 437–452, 2011.

383. D. Metzler, S. Dumais, and C. Meek. Similarity measures for short segments of text. *European Conference on Information Retrieval*, pp. 16-27, 2007.

384. D. Metzler *et al.* Rethinking search: Making experts out of dilettantes. *arXiv:2105.02274*, 2021. https://arxiv.org/abs/2105.02274

385. P. Michel, O. Levy, and G. Neubig. Are sixteen heads really better than one? *arXiv:1905.10650*, 2019. https://arxiv.org/abs/1905.10650

386. L. Michelbacher, F. Laws, B. Dorow, U. Heid, and H. Schütze. Building a cross-lingual relatedness thesaurus using a graph similarity measure. *LREC*, 2010.

387. R. Mihalcea and P. Tarau. TextRank: Bringing order into texts. *Conference on Empirical Methods in Natural Language Processing*, pp. 404–411, 2004.

388. S. Mika, G. Rätsch, J. Weston, B. Schölkopf, and K. Müller. Fisher discriminant analysis with kernels. *NeurIPS Conference*, 1999.

389. T. Mikolov, K. Chen, G. Corrado, and J. Dean. Efficient estimation of word representations in vector space. *arXiv:1301.3781*, 2013. https://arxiv.org/abs/1301.3781

390. T. Mikolov, I. Sutskever, K. Chen, G. Corrado, and J. Dean. Distributed representations of words and phrases and their compositionality. *NeurIPS Conference*, pp. 3111–3119, 2013.

391. T. Mikolov, M. Karafiat, L. Burget, J. Cernocky, and S. Khudanpur. Recurrent neural network based language model. *Interspeech*, Vol 2, 2010.

392. T. Mikolov, W. Yih, and G. Zweig. Linguistic Regularities in Continuous Space Word Representations. *HLT-NAACL*, pp. 746–751, 2013.

393. T. Mikolov, Q. Le, and I. Sutskever. Exploiting similarities among languages for machine translation. *arXiv preprint arXiv:1309.4168*, 2013. https://arxiv.org/abs/1309.4168

394. A. Miller *et al.* Key-value memory networks for directly reading documents. *arXiv:1606.03126*, 2016. https://arxiv.org/abs/1606.03126

395. D. Miller, T. Leek, and R. Schwartz. A Hidden Markov Model information retrieval system. *ACM SIGIR Conference*, pp. 214–221, 1999.

396. G. Miller, R. Beckwith, C. Fellbaum, D. Gross, and K. J. Miller. Introduction to WordNet: An on-line lexical database. *International Journal of Lexicography (special issue)*, 3(4), pp. 235–312, 1990. https://wordnet.princeton.edu/

397. S. Min *et al.* Knowledge guided text retrieval and reading for open domain question answering. *arXiv:1911.03868*, 2019.

398. M. Mintz, S. Bills, R. Snow, and D. Jurafsky. Distant supervision for relation extraction without labeled data. *Annual Meeting of the Association for Computational Linguistics and the International Joint Conference on Natural Language Processing*, pp. 1003–1011, 2009.

399. T. M. Mitchell. Machine learning. *McGraw Hill International Edition*, 1997.

400. T. M. Mitchell. The role of unlabeled data in supervised learning. *International Colloquium on Cognitive Science*, pp. 2–11, 1999.

401. A. Mnih and G. Hinton. Three new graphical models for statistical language modelling. *ICML Conference*, pp. 641–648, 2007.

402. A. Mnih and K. Kavukcuoglu. Learning word embeddings efficiently with noise-contrastive estimation. *NeurIPS Conference*, pp. 2265–2273, 2013.

403. A. Mnih and Y. Teh. A fast and simple algorithm for training neural probabilistic language models. *arXiv:1206.6426*, 2012. https://arxiv.org/abs/1206.6426

404. A. Moffat and J. Zobel. Self-indexing inverted files for fast text retrieval. *ACM Transactions on Information Systems*, 14(4), pp. 14(4), 1996.

405. F. Moosmann, B. Triggs, and F. Jurie. Fast Discriminative visual codebooks using randomized clustering forests. *NeurIPS Conference*, pp. 985–992, 2006.

406. T. Mullen and N. Collier. Sentiment analysis using support vector machines with diverse information sources. *Conference on Empirical Methods in Natural Language Processing (EMNLP)*, pp. 412–418, 2004.

407. J.-C. Na, H. Sui, C. Khoo, S. Chan, and Y. Zhou. Effectiveness of simple linguistic processing in automatic sentiment classification of product reviews. *Conference of the International Society for Knowledge Organization (ISKO)*, pp. 49–54, 2004.

408. N. Nakashole, G. Weikum, and F. Suchanek. PATTY: A Taxonomy of Relational Patterns with Semantic Types. *Joint Conference on Empirical Methods in Natural Language Processing and Computational Natural Language Learning*, pp. 1135–1145, 2012.

409. N. Nakashole, M. Theobald, and G. Weikum. Scalable knowledge harvesting with high precision and high recall. *WSDM Conference*, pp. 227–236, 2011.

410. R. Nallapati, F. Zhai, and B. Zhou. Summarunner: A recurrent neural network based sequence model for extractive summarization of documents. *AAAI Conference on Artificial Intelligence*, 2017.

411. R. Nallapati, B. Zhou, C. Gulcehre, and B. Xiang. Abstractive text summarization using sequence-to-sequence rnns and beyond. *arXiv:1602.06023*, 2016.

412. G. Nemhauser, L. Wolsey, and M. Fisher. An analysis of approximations for maximizing submodular set functions–I. *Mathematical Programming*, 14(1), pp. 265–294, 1978.

413. A. Nenkova and K. McKeown. Automatic Summarization *Foundations and Trends in Information Retrieval*, 5(2–3), pp. 103–233, 2011.

414. A. Nenkova and K. McKeown. A survey of text summarization techniques. *Mining Text Data*, Springer, pp. 43–76, 2012.

415. A. Ng, M. Jordan, and Y. Weiss. On spectral clustering: Analysis and an algorithm. *NeurIPS Conference*, pp. 849–856, 2002.

416. T. Nguyen and A, Moschitti. End-to-end relation extraction using distant supervision from external semantic repositories. *ACL Conference*, pp. 277–282, 2011.

417. M. Nickel, K. Murphy, V. Tresp, and E. Gabrilovich. A review of relational machine learning for knowledge graphs. *Proceedings of the IEEE*, 104(1), pp. 11–33, 2015.

418. K. Nigam, J. Lafferty, and A. McCallum. Using maximum entropy for text classification. *IJCAI Workshop on Machine Learning for Information Filtering*, pp. 61–67, 1999.

419. K. Nigam, A. McCallum, S. Thrun, and T. Mitchell. Text classification with labeled and unlabeled data using EM. *Machine Learning*, 39(2), pp. 103–134, 2000.

420. H. Niitsuma and M. Lee. Word2Vec is a special case of kernel correspondence analysis and kernels for natural language processing, *arXiv preprint arXiv:1605.05087*, 2016. https://arxiv.org/abs/1605.05087

421. F. Niu, C. Zhang, C. Re, and J. Shavlik. Elementary: Large-scale knowledge-base construction via machine learning and statistical inference. *International Journal on Semantic Web and Information Systems (IJSWIS)*, 8(3), pp. 42–73, 2012.

422. A. Ntoulas and J. Cho. Pruning policies for two-tiered inverted index with correctness guarantee. *ACM SIGIR Conference*, pp. 191–198, 2007.

423. M. Osborne. Using maximum entropy for sentence extraction. *ACL Workshop on Automatic Summarization*, pp. 1–8, 2002.

424. E. Osuna, R. Freund, and F. Girosi. Improved training algorithm for support vector machines, *IEEE Workshop on Neural Networks and Signal Processing*, 1997.

425. M. Ott, Y. Choi, C. Cardie, and J. Hancock. Finding deceptive opinion spam by any stretch of the imagination. *Association for Computational Linguistics: Human Language Technologies-Volume 1*, pp. 309–319, 2011.

426. L. Page, S. Brin, R. Motwani, and T. Winograd. The PageRank citation engine: Bringing order to the web. *Technical Report*, 1999–0120, Computer Science Department, Stanford University, 1998.

427. C. D. Paice. Constructing literature abstracts by computer: techniques and prospects. *Information Processing and Management*, 26(1), pp. 171–186, 1990.

428. A. Pak and P. Paroubek. Twitter as a Corpus for Sentiment Analysis and Opinion Mining. *LREC*, pp. 1320–1326, 2010.

429. B. Pang and L. Lee. A sentimental education: Sentiment analysis using subjectivity summarization based on minimum cuts. *ACL Conference*, 2004.

430. B. Pang and L. Lee. Seeing stars: Exploiting class relationships for sentiment categorization with respect to rating scales. *ACL Conference*, pp. 115–124, 2005.

431. B. Pang and L. Lee. Opinion mining and sentiment analysis. *Foundations and Trends in Information Retrieval*, 2(1–2), pp. 1–135, 2008.

432. B. Pang, L. Lee, and S. Vaithyanathan. Thumbs up? Sentiment classification using machine learning techniques. *Conference on Empirical Methods in Natural Language Processing (EMNLP)*, pp. 79–86, 2002.

433. R. Pascanu, T. Mikolov, and Y. Bengio. On the difficulty of training recurrent neural networks. *ICML*, (3), 28, pp. 1310–1318, 2013. http://www.jmlr.org/proceedings/papers/v28/pascanu13.pdf

434. M. Pazzani and D. Kibler. The utility of knowledge in inductive learning. *Machine Learning*, 9(1), pp. 57–94, 1992.

435. H. Paulheim and R. Meusel. A decomposition of the outlier detection problem into a set of supervised learning problems. *Machine Learning*, 100(2–3), pp. 509–531, 2015.

436. J. Pennington, R. Socher, and C. Manning. Glove: Global Vectors for Word Representation. *EMNLP*, pp. 1532–1543, 2014.

437. F. Pereira, N. Tishby, and L. Lee. Distributional clustering of English words. *ACL Conference*, pp. 183–190, 1993.

438. M. Peters et al. Deep contextualized word representations. *arXiv:1802.05365*, 2018. https://arxiv.org/abs/1802.05365

439. F. Petroni, Fabio et al. Language models as knowledge bases? *arXiv:1909.01066*, 2019. \unhbox\voidb@x\hboxhttps://arxiv.org/abs/1909.01066

440. J. C. Platt. Sequential minimal optimization: A fast algorithm for training support vector machines. *Advances in Kernel Method: Support Vector Learning*, MIT Press, pp. 85–208, 1998.

441. J. C. Platt. Probabilistic outputs for support vector machines and comparisons to regularized likelihood methods. *Advances in Large Margin Classifiers*, 10(3), pp. 61–74, 1999.

442. L. Polanyi and A. Zaenen. Contextual valence shifters. *Computing Attitude and Affect in Text: Theory and Applications*, pp. 1–10, Springer, 2006.

443. J. Ponte and W. Croft. A language modeling approach to information retrieval. *ACM SIGIR Conference*, pp. 275–281, 1998.

444. J. Ponte and W. Croft. Text segmentation by topic. *International Conference on Theory and Practice of Digital Libraries*, pp. 113–125, 1997.

445. A. Popescu and O. Etzioni. Extracting product features and opinions from reviews. *Natural Language Processing and Text Mining*, pp. 9–28, 2007.

446. T. Plotz and S. Roth. Neural nearest neighbors networks. *NeurIPS*, pp. 1087–1098, 2018.

447. J. Pritchard, M. Stephens, and P. Donnelly. Inference of population structure using multilocus genotype data. *Genetics*, 155(2), pp. 945–959, 2000.

448. J. Pustejovsky et al. The timebank corpus. *Corpus Linguistics*, pp. 40, 2003.

449. J. Pustejovsky et al. TimeML: Robust specification of event and temporal expressions in text. *New Directions in Question Answering*, 3. pp. 28–34, 2003.

450. G. Qi, C. Aggarwal, and T. Huang. Towards semantic knowledge propagation from text corpus to web images. *WWW Conference*, pp. 297–306, 2011.

451. G. Qi, C. Aggarwal, and T. Huang. Community detection with edge content in social media networks. *ICDE Conference*, pp. 534–545, 2012.

452. L. Qian, G. Zhou, F. Kong, Q. Zhu, and P. Qian. Exploiting constituent dependencies for tree kernel-based semantic relation extraction. *International Conference on Computational Linguistics*, pp. 697–704, 2008.

453. G. Qiu, B. Liu, J. Bu, and C. Chen. Opinion word expansion and target extraction through double propagation. Computational linguistics, 37(1), pp. 9–27, 2011.

454. J. Quinlan. C4.5: programs for machine learning. *Morgan-Kaufmann Publishers*, 1993.

455. J. Quinlan. Induction of decision trees. *Machine Learning*, 1, pp. 81–106, 1986.

456. L. Rabiner. A tutorial on hidden Markov models and selected applications in speech recognition. *Proceedings of the IEEE*, 77(2), pp. 257–286, 1989.

457. D. Radev, H. Jing, and M. Budzikowska. Centroid-based summarization of multiple documents: sentence extraction, utility-based evaluation, and user studies. *NAACL-ANLP Workshop on Automatic summarization*, pp. 21–30, 2000.

458. D. Radev, H. Jing, M. Stys, and D. Tam. Centroid-based summarization of multiple documents. *Information Processing and Management*, 40(6), pp. 919–938, 2004.

459. A. Radford *et al.* Language models are unsupervised multitask learners. *OpenAI blog*, 1(8), 2019. https://cdn.openai.com/better-language-models/language_models_are_unsupervised_multitask_learners.pdf

460. C. Raffel *et al.* Exploring the limits of transfer learning with a unified text-to-text transformer. *arXiv:1910.10683*, 2019. https://arxiv.org/abs/1910.10683

461. P. Rajpurkar *et al.* Squad: 100,000+ questions for machine comprehension of text. *arXiv:1606.05250*, 2016. https://arxiv.org/abs/1606.05250

462. P. Rajpurkar, R. Jia, and P. Liang. Know what you don't know: Unanswerable questions for SQuAD. *arXiv:1806.03822*, 2018. https://arxiv.org/abs/1806.03822

463. A. Ratnaparkhi. A maximum entropy model for part-of-speech tagging. *Conference on Empirical Methods in Natural Language Processing*, pp. 133–142, 1996.

464. R. Rehurek and P. Sojka. Software framework for topic modelling with large corpora. *LREC 2010 Workshop on New Challenges for NLP Frameworks*, pp. 45–50, 2010. https://radimrehurek.com/gensim/index.html

465. M. Ren, R. Kiros, and R. Zemel. Exploring models and data for image question answering. *NeurIPS*, pp. 2953–2961, 2015.

466. X. Ren, M. Jiang, J. Shang, and J. Han. Contructing Structured Information Networks from Massive Text Corpora (Tutorial), *WWW Conference*, 2017.

467. S. Rendle. Factorization machines. *IEEE ICDM Conference*, pp. 995–100, 2010.

468. S. Rendle. Factorization machines with libfm. *ACM Transactions on Intelligent Systems and Technology*, 3(3), 57, 2012.

469. B. Ribeiro-Neto, E. Moura, M. Neubert, and N. Ziviani. Efficient distributed algorithms to build inverted files. *ACM SIGIR Conference*, pp. 105–112, 1999.

470. M. Richardson, A. Prakash, and E. Brill. Beyond PageRank: machine learning for static ranking. *World Wide Web Conference*, pp. 707–715, 2006.

471. R. Rifkin. Everything old is new again: a fresh look at historical approaches in machine learning. *Ph.D. Thesis*, Massachusetts Institute of Technology, 2002. http://cbcl.mit.edu/projects/cbcl/publications/theses/thesis-rifkin.pdf

472. A. Ritter, Mausam, O. Etzioni, and S. Clark. Open domain event extraction from twitter. *ACM KDD Conference*, pp. 1104–1102, 2012.

473. A. Ritter, S. Clark, Mausam, and O. Etzioni. Named entity recognition in tweets: an experimental study. *Conference on Empirical Methods in Natural Language Processing*, pp. 1524–1534, 2011.

474. A. Roberts, C. Raffel, and N. Shazeer. How Much Knowledge Can You Pack Into the Parameters of a Language Model? *arXiv:2002.08910*, 2020. https://arxiv.org/abs/2002.08910

475. S. Robertson. Understanding inverse document frequency: On theoretical arguments for IDF. *Journal of Documentation*, 60, pp. 503–520, 2004.

476. S. Robertson and K. Spärck Jones. Relevance weighting of search terms. *Journal of the American Society for Information Science*, 27(3), pp. 129–146, 1976.

477. S. Robertson, H. Zaragoza, and M. Taylor. Simple BM25 extension to multiple weighted fields. *ACM CIKM Conference*, pp. 42–49, 2004.

478. J. Rodríguez, L. Kuncheva, and C. Alonso. Rotation forest: A new classifier ensemble method. *IEEE Transactions on Pattern Analysis and Machine Intelligence*, 28(10), pp. 1619–1630, 2006.

479. J. Rocchio. Relevance feedback information retrieval. *The Smart Retrieval System- Experiments in Automatic Document Processing*, G. Salton, Ed. Prentice Hall, Englewood Cliffs, NJ, pp. 313–323, 1971.

480. X. Rong. word2vec parameter learning explained. *arXiv preprint arXiv:1411.2738*, 2014. https://arxiv.org/abs/1411.2738

481. B. Rosenfeld and R. Feldman. Clustering for unsupervised relation identification. *ACM CIKM Conference*, pp. 411–418, 2007.

482. S. Roweis and L. Saul. Nonlinear dimensionality reduction by locally linear embedding. *Science*, 290, no. 5500, pp. 2323–2326, 2000.

483. A. M. Rush, S. Chopra, and J. Weston. A Neural Attention Model for Abstractive Sentence Summarization. *arXiv:1509.00685*, 2015. https://arxiv.org/abs/1509.00685

484. M. Sahami and T. D. Heilman. A Web-based kernel function for measuring the similarity of short text snippets. *WWW Conference*, pp. 377–386, 2006.

485. T. Sakai and K. Spärck Jones. Generic summaries for indexing in information retrieval. *ACM SIGIR Conference*, pp. 190–198, 2001.

486. T. Sakaki, M. Okazaki, and Y. Matsuo. Earthquake shakes Twitter users: real-time event detection by social sensors. *World Wide Web Conference*, pp. 851–860, 2010.

487. G. Salton and J. Allan. Selective text utilization and text traversal. *Proceedings of ACM Hypertext*, 1993.

488. G. Salton, J. Allan, and C. Buckley. Approaches to passage retrieval in full text information systems. *ACM SIGIR Conference*, pp. 49–58, 1997.

489. G. Salton and C. Buckley. Term weighting approaches in automatic text retrieval, *Technical Report 87–881*, Cornell University, 1987.
https://ecommons.cornell.edu/bitstream/handle/1813/6721/87-881.pdf?sequence=1

490. G. Salton and M. J. McGill. Introduction to modern information retrieval. *McGraw Hill*, 1986.

491. G. Salton, A. Singhal, M. Mitra, and C. Buckley. Automatic text structuring and summarization. *Information Processing and Management*, 33(2), pp. 193–207, 1997.

492. G. Salton, A. Wong, and C. Yang. A vector space model for automatic indexing. *Communications of the ACM*, 18(11), pp. 613–620, 1975.

493. H. Samet. Foundations of multidimensional and metric data structures. *Morgan Kaufmann*, 2006.

494. R. Samworth. Optimal weighted nearest neighbour classifiers. *The Annals of Statistics*, 40(5), pp. 2733–2763, 2012.

495. P. Saraiva, E. Silva de Moura, N. Ziviani, W. Meira, R. Fonseca, and B. Riberio-Neto. Rank-preserving two-level caching for scalable search engines. *ACM SIGIR Conference*, pp. 51–58, 2001.

496. S. Sarawagi. Information extraction. *Foundations and Trends in Satabases*, 1(3), pp. 261–377, 2008.

497. S. Sarawagi and W. Cohen. Semi-markov conditional random fields for information extraction. *NeurIPS Conference*, pp. 1185–1192, 2004.

498. S. Sathe and C. Aggarwal. Similarity forests. *ACM KDD Conference*, 2017.

499. R. Sauri, R. Knippen, M. Verhagen, and J. Pustejovsky. Evita: a robust event recognizer for QA systems. *Conference on Human Language Technology and Empirical Methods in Natural Language Processing*, pp. 700–707, 2005.

500. H. Sayyadi, M. Hurst, and A. Maykov. Event detection and tracking in social streams. *ICWSM Conference*, 2009.

501. F. Scholer, H. Williams, J. Yiannis, and J. Zobel. Compression of inverted indexes for fast query evaluation. *ACM SIGIR Conference*, pp. 222–229, 2002.

502. B. Schölkopf, A. Smola, and K.-R. Müller. Nonlinear component analysis as a kernel eigenvalue problem. *Neural Computation*, 10(5), pp. 1299–1319, 1998.

503. M. Schuster and K. Paliwal. Bidirectional recurrent neural networks. *IEEE Transactions on Signal Processing*, 45(11), pp. 2673–2681, 1997.

504. H. Schütze and C. Silverstein. Projections for Efficient Document Clustering. *ACM SIGIR Conference*, pp. 74–81, 1997.

505. F. Sebastiani. Machine Learning in Automated Text Categorization. *ACM Computing Surveys*, 34(1), 2002.

506. P. Sen, G. Namata, M. Bilgic, L. Getoor, B. Galligher, and T. Eliassi-Rad. Collective classification in network data. *AI magazine*, 29(3), pp. 93, 2008.

507. G. Seni and J. Elder. Ensemble methods in data mining: Improving accuracy through combining predictions. *Synthesis Lectures in Data Mining and Knowledge Discovery, Morgan and Claypool*, 2010.

508. B. Sennrich, B. Haddow, and A. Birch. Neural machine translation of rare words with subword units. *arXiv:1508.07909*, 2015. https://arxiv.org/abs/1508.07909

509. M. Seo, A. Kembhavi, A. Farhadi, and H. Hajishirzi. Bidirectional attention flow for machine comprehension. *arXiv:1611.01603*, 2016. https://arxiv.org/abs/1611.01603

510. K. Seymore, A. McCallum, and R. Rosenfeld. Learning hidden Markov model structure for information extraction. *AAAI-99 Workshop on Machine Learning for Information Extraction*, pp. 37–42, 1999.

511. F. Shahnaz, M. Berry, V. Pauca, and R. Plemmons. Document clustering using nonnegative matrix factorization. *Information Processing and Management*, 42(2), pp. 378–386, 2006.

512. S. Shalev-Shwartz, Y. Singer, N. Srebro, and A. Cotter. Pegasos: Primal estimated subgradient solver for SVM. *Mathematical Programming*, 127(1), pp. 3–30, 2011.

513. A. Shashua. On the equivalence between the support vector machine for classification and sparsified Fisher's linear discriminant. *Neural Processing Letters*, 9(2), pp. 129–139, 1999.

514. P. Shi and J. Lin. Simple BERT Models fo Relation Extraction and Semantic Role Labeling, *arXiv:1904.05255*, 2019. https://arxiv.org/abs/1904.05255

515. Y. Shinyama and S. Sekine. Preemptive information extraction using unrestricted relation discovery. *Human Language Technology Conference of the North American Chapter of the Association of Computational Linguistics*, pp. 304–311, 2006.

516. S. Siencnik. Adapting word2vec to named entity recognition. *Nordic Conference of Computational Linguistics, NODALIDA*, 2015.

517. A. Singh and G. Gordon. A unified view of matrix factorization models. *Joint European Conference on Machine Learning and Knowledge Discovery in Databases*, pp. 358–373, 2008.

518. A. Siddharthan, A. Nenkova, and K. Mckeown. Syntactic simplification for improving content selection in multi-document summarization. *International Conference on Computational Linguistic*, pp. 896–902, 2004.

519. A. Singhal, C. Buckley, and M. Mitra. Pivoted document length normalization. *ACM SIGIR Conference*, pp. 21–29, 1996.

520. N. Slonim and N. Tishby. The power of word clusters for text classification. *European Colloquium on Information Retrieval Research (ECIR)*, 2001.

521. S. Soderland. Learning information extraction rules for semi-structured and free text. *Machine Learning*, 34(1–3), pp. 233–272, 1999.

522. K. Spärck Jones. A statistical interpretation of term specificity and its application in information retrieval. *Journal of Documentation*, 28(1), pp. 11–21, 1972.

523. K. Spärck Jones. Automatic summarizing: factors and directions. *Advances in Automatic Text Summarization*, pp. 1–12, 1998.

524. K. Spärck Jones. Automatic summarising: The state of the art. *Information Processing and Management*, 43(6), pp. 1449–1481, 2007.

525. K. Spärck Jones, S. Walker, and S. Robertson. A probabilistic model of information retrieval: development and comparative experiments: Part 2. *Information Processing and Management*, 36(6), pp. 809–840, 2000.

526. M. Stairmand. A computational analysis of lexical cohesion with applications in information retrieval. *Ph.D. Dissertation*, Center for Computational Linguistics UMIST, Manchester, 1996.
http://ethos.bl.uk/OrderDetails.do?uin=uk.bl.ethos.503546

527. J. Steinberger and K. Jezek. Using latent semantic analysis in text summarization and summary evaluation. *ISIM*, pp. 93–100, 2004.

528. J. Steinberger, M. Poesio, M. Kabadjov, and K. Jezek. Two uses of anaphora resolution in summarization. *Information Processing and Management*, 43(6), pp. 1663–1680, 2007.

529. D. Stepanova, M. H. Gad-Elrab, and V. T. Ho. Rule induction and reasoning over knowledge graphs. *Reasoning Web International Summer School*, pp. 142–172, 2018.

530. G. Strang. An introduction to linear algebra. *Wellesley Cambridge Press*, 2009.

531. A. Strehl, J. Ghosh, and R. Mooney. Impact of similarity measures on web-page clustering. *Workshop on Artificial Intelligence for Web Search*, 2000.
http://www.aaai.org/Papers/Workshops/2000/WS-00-01/WS00-01-011.pdf

532. F. M. Suchanek, G. Kasneci, and G. Weikum. Yago: A Core of Semantic Knowledge. *WWW Conference*, pp. 697–706, 2007.

533. H. Sun *et al.* Open domain question answering using early fusion of knowledge bases and text. *arXiv:1809.00782*, 2018. https://arxiv.org/abs/1809.00782

534. H. Sun, T. Bedrax-Weiss, and W. Cohen. Pullnet: Open domain question answering with iterative retrieval on knowledge bases and text. *arXiv:1904.09537*, 2019. https://arxiv.org/abs/1904.09537

535. Y. Sun, J. Han, J. Gao, and Y. Yu. itopicmodel: Information network-integrated topic modeling. *IEEE ICDM Conference*, pp. 493–502, 2011.

536. M. Sundermeyer, R. Schluter, and H. Ney. LSTM neural networks for language modeling. *Interspeech*, 2010.

537. I. Sutskever, O. Vinyals, and Q. V. Le. Sequence to sequence learning with neural networks. *NeurIPS Conference*, pp. 3104–3112, 2014.

538. C. Sutton and A. McCallum. An introduction to conditional random fields. *arXiv:1011.4088*, 2010. https://arxiv.org/abs/1011.4088

539. J. Suykens and J. Venderwalle. Least squares support vector machine classifiers. *Neural Processing Letters*, 1999.

540. M. Taboada, J. Brooke, M. Tofiloski, K. Voll, and M. Stede. Lexicon-based methods for sentiment analysis. *Computational Linguistics*, 37(2), pp. 267–307, 2011.

541. K. Takeuchi and N. Collier. Use of support vector machines in extended named entity recognition. *Conference on Natural Language Learning*, pp. 1–7, 2002.

542. A. Talmor *et al.* oLMpics — On what Language Model Pre-training Captures. *arXiv:1912.13283*, 2019. https://arxiv.org/abs/1912.13283

543. P.-N Tan, M. Steinbach, and V. Kumar. Introduction to data mining. *Addison-Wesley*, 2005.

544. D. Tang, B. Qin, and T. Liu. Document modeling with gated recurrent neural network for sentiment classification. *Conference on empirical methods in natural language processing*, pp. 1422–1432, 2015.

545. J. Tang, Y. Chang, C. Aggarwal, and H. Liu. A survey of signed network mining in social media. *ACM Computing Surveys (CSUR)*, 49(3), 42, 2016.

546. J. Tang, S. Chang, C. Aggarwal, and H. Liu. (2015, February). Negative link prediction in social media. *WSDM Conference*, pp. 87–96, 2015.

547. M. Taylor, H. Zaragoza, N. Craswell, S. Robertson, and C. Burges. Optimisation methods for ranking functions with multiple parameters. *ACM CIKM Conference*, pp. 585–593, 2006.

548. W. Taylor. Cloze procedure: A new tool for measuring readability. *Journalism Bulletin*, 30(4), pp. 415–433, 1953.

549. J. Tenenbaum, V. De Silva, and J. Langford. A global geometric framework for nonlinear dimensionality reduction. *Science*, 290 (5500), pp. 2319–2323, 2000.

550. A. Tikhonov and V. Arsenin. Solution of ill-posed problems. *Winston and Sons*, 1977.

551. M. Tsai, C. Aggarwal, and T. Huang. Ranking in heterogeneous social media. *WSDM Conference*, pp. 613–622, 2014.

552. J. Turner and E. Charniak. Supervised and unsupervised learning for sentence compression. *ACL Conference*, pp. 290–297, 2005.

553. P. Turney. Thumbs up or thumbs down?: semantic orientation applied to unsupervised classification of reviews. *ACL Conference*, pp. 417–424, 2002.

554. P. Turney and M. Littman. Measuring praise and criticism: Inference of semantic orientation from association. *ACM Transactions on Information Systems*, 21(4), pp. 314–346, 2003.

555. P. Turney and P. Pantel. From frequency to meaning: Vector space models of semantics. *Journal of Artificial Intelligence Research*, 37(1), pp. 141–188, 2010.

556. C. J. van Rijsbergen. Information retrieval. *Butterworths*, London, 1979.

557. C.J. van Rijsbergen, S.E. Robertson, and M.F. Porter. New models in probabilistic information retrieval. *London: British Library. (British Library Research and Development Report, no. 5587)*, 1980.
https://tartarus.org/martin/PorterStemmer/

558. V. Vapnik. The nature of statistical learning theory. *Springer*, 2000.

559. L. Vanderwende, H. Suzuki, C. Brockett, and A. Nenkova. Beyond SumBasic: Task-focused summarization with sentence simplification and lexical expansion. *Information Processing and Managment*, 43, pp. 1606–1618, 2007

560. A. Vaswani *et al.* Attention is All you Need. *NeurIPS*, 2017. https://arxiv.org/abs/1706.03762

561. A. Vinokourov, N. Cristianini, and J. Shawe-Taylor. Inferring a semantic representation of text via cross-language correlation analysis. *NeurIPS Conference*, pp. 1473–1480, 2002.

562. O. Vinyals, A. Toshev, S. Bengio, and D. Erhan. Show and tell: A neural image caption generator. *CVPR Conference*, pp. 3156–3164, 2015.

563. E. Voorhees. Implementing agglomerative hierarchic clustering algorithms for use in document retrieval. *Information Processing and Management*, 22(6), pp. 465–476, 1986.

564. D. Vrandecic and M. Krotzsch. Wikidata: a free collaborative knowledgebase. *Communications of the ACM*, 57(1), pp. 78–85, 2014.

565. G. Wahba. Support vector machines, reproducing kernel Hilbert spaces and the randomized GACV. *Advances in Kernel Methods-Support Vector Learning*, 6, pp. 69–87, 1999.

566. H. Wallach, D. Mimno, and A. McCallum. Rethinking LDA: Why priors matter. *NeurIPS Conference*, pp. 1973–1981, 2009.

567. B. Wang *et al.* On position embeddings in BERT. *ICLR*, 2021.

568. C. Wang *et al.* Panto: A portable natural language interface to ontologies. *European Semantic Web Conference*, 2007.

569. D. Wang, S. Zhu, T. Li, and Y. Gong. Multi-document summarization using sentence-based topic models. *ACL-IJCNLP Conference*, pp. 297–300, 2009.

570. H. Wang, H. Huang, F. Nie, and C. Ding. Cross-language Web page classification via dual knowledge transfer using nonnegative matrix tri-factorization. *ACM SIGIR Conference*, pp. 933–942, 2011.

571. W. Wang *et al.* Gated self-matching networks for reading comprehension and question answering. *ACL Conference*, pp. 189–198, 2017.

572. A. Warstadt, A. Singh, and S. Bowman. Neural network acceptability judgments. *Transactions of the Association for Computational Linguistics*, 7, pp. 625–641, 2019.

573. S. Weiss, N. Indurkhya, and T. Zhang. Fundamentals of predictive text mining. *Springer*, 2015.

574. S. Weiss, C. Apte, F. Damerau, D. Johnson, F. Oles, T. Goetz, and T. Hampp. Maximizing text-mining performance. *IEEE Intelligent Systems*, 14(4), pp. 63–69, 1999.

575. X. Wei and W. B. Croft. LDA-based document models for ad-hoc retrieval. *ACM SIGIR Conference*, pp. 178–185, 2006.

576. P. Werbos. Backpropagation through time: what it does and how to do it. *Proceedings of the IEEE*, 78(10), pp. 1550–1560, 1990.

577. J. Weston, A. Bordes, S. Chopra, A. Rush, B. van Merrienboer, A. Joulin, and T. Mikolov. Towards ai-complete question answering: A set of pre-requisite toy tasks. *arXiv:1502.05698*, 2015. https://arxiv.org/abs/1502.05698

578. J. Weston, S. Chopra, and A. Bordes. Memory networks. *ICLR*, 2015.

579. J. Weston and C. Watkins. Multi-class support vector machines. *Technical Report CSD-TR-98-04*, Department of Computer Science, Royal Holloway, University of London, May, 1998.

580. B. Widrow and M. Hoff. Adaptive switching circuits. *IRE WESCON Convention Record*, 4(1), pp. 96–104, 1960.

581. W. Wilbur and K. Sirotkin. The automatic identification of stop words. *Journal of Information Science*, 18(1), pp. 45–55, 1992.

582. J. Wiebe, R. Bruce, and T. O'Hara. Development and use of a gold-standard data set for subjectivity classifications. *Association for Computational Linguistics on Computational Linguistics*, pp. 246–253, 1999.

583. J. Wiebe and E. Riloff. Creating subjective and objective sentence classifiers from unannotated texts. *International Conference on Intelligent Text Processing and Computational Linguistics*, pp. 486–497, 2005.

584. C. Williams and M. Seeger. Using the Nyström method to speed up kernel machines. *NeurIPS Conference*, 2000.

585. H. Williams, J. Zobel, and D. Bahle. Fast phrase querying with combined indexes. *ACM Transactions on Information Systems*, 22(4), pp. 573–594, 2004.

586. T. Wilson, J. Wiebe, and P. Hoffmann. Recognizing contextual polarity in phrase-level sentiment analysis. *Human Language Technology and Empirical Methodgs in Natural Language Processing*, pp. 347–354, 2005.

587. T. Wilson, J. Wiebe, and R. Hwa. Just how mad are you? Finding strong and weak opinion clauses. *Computational Intelligence*, 22(2), pp. 73–99, 2006.

588. M. J. Witbrock and V. O. Mittal. Ultra-summarization: A statistical approach to generating highly condensed non-extractive summaries. *ACM SIGIR Conference*, pp. 315–316, 1999.

589. I. H. Witten, A. Moffat, and T. C. Bell. Managing Gigabytes: Compressing and indexing documents and images. *Morgan Kaufmann*, 1999.

590. K. Wong, M. Wu, and W. Li. Extractive summarization using supervised and semi-supervised learning. *International Conference on Computational Linguistics*, pp. 985–992, 2008.

591. W. Xu, X. Liu, and Y. Gong. Document clustering based on non-negative matrix factorization. *ACM SIGIR Conference*, pp. 267–273, 2003.

592. Y. Wu *et al.* Google's neural machine translation system: Bridging the gap between human and machine translation. *arXiv:1609.08144*, 2016. https://arxiv.org/abs/1609.08144

593. Z. Wu, C. Aggarwal, and J. Sun. The troll-trust model for ranking in signed networks. *WSDM Conference*, pp. 447–456, 2016.

594. C. Xiong, S. Merity, and R. Socher. Dynamic memory networks for visual and textual question answering. *ICML*, pp. 2397–2406, 2016.

595. W. Xiong *et al.* Improving question answering over incomplete kbs with knowledge-aware reader. *arXiv:1905.07098*, 2019. https://arxiv.org/abs/1905.07098

596. J. Xu and H. Li. Adarank: a boosting algorithm for information retrieval. *ACM SIGIR Conference*, 2007.

597. J. Yamron, I. Carp, L. Gillick, S. Lowe, and P. van Mulbregt. A hidden Markov model approach to text segmentation and event tracking. *IEEE International Conference on Acoustics, Speech and Signal Processing*, pp. 333–336, 1998.

598. J. Yang, J. McAuley, and J. Leskovec. Community detection in networks with node attributes. *IEEE ICDM Conference*, pp. 1151–1156, 2013.

599. Q. Yang, Q., Y. Chen, G. Xue, W. Dai, and T. Yu. Heterogeneous transfer learning for image clustering via the social web. *Joint Conference of the ACL and Natural Language Processing of the AFNLP*, pp. 1–9, 2009.

600. T. Yang, R. Jin, Y. Chi, and S. Zhu. Combining link and content for community detection: a discriminative approach. *ACM KDD Conference*, pp. 927–936, 2009.

601. Y. Yang. Noise reduction in a statistical approach to text categorization, *ACM SIGIR Conference*, pp. 256–263, 1995.

602. Y. Yang. An evaluation of statistical approaches to text categorization. *Information Retrieval*, 1(1–2), pp. 69–90, 1999.

603. Y. Yang. A study on thresholding strategies for text categorization. *ACM SIGIR Conference*, pp. 137–145, 2001.

604. Y. Yang and C. Chute. An application of least squares fit mapping to text information retrieval. *ACM SIGIR Conference*, pp. 281–290, 1993.

605. Y. Yang and X. Liu. A re-examination of text categorization methods. *ACM SIGIR Conference*, pp. 42–49, 1999.

606. Y. Yang and J. O. Pederson. A comparative study on feature selection in text categorization, *ACM SIGIR Conference*, pp. 412–420, 1995.

607. Y. Yang, T. Pierce, and J. Carbonell. A study of retrospective and online event detection. *ACM SIGIR Conference*, pp. 28–36, 1998.

608. Z. Yang, X. He, J. Gao, L. Deng, and A. Smola. Stacked attention networks for image question answering. *CVPR*, pp. 21–29, 2016.

609. Z. Yang *et al.* Hotpotqa: A dataset for diverse, explainable multi-hop question answering. *arXiv:1809.09600*, 2018. https://arxiv.org/abs/1809.09600

610. Y. Yue, T. Finley, F. Radlinski, and T. Joachims. A support vector method for optimizing average precision. *ACM SIGIR Conference*, pp. 271–278, 2007.

611. Z. Yang *et al.* Xlnet: Generalized autoregressive pretraining for language understanding. *arXiv:1906.08237*, 2019. https://arxiv.org/abs/1906.08237

612. M. Yahya *et al.* Natural language questions for the web of data. *Joint Conference on Empirical Methods in Natural Language Processing and Computational Natural Language Learning*, pp. 379–390, 2012.

613. D. Zajic, B. Dorr, J. Lin, and R. Schwartz. Multi-candidate reduction: Sentence compression as a tool for document summarization tasks. *Information Processing and Management*, 43(6), pp. 1549–1570, 2007.

614. M. Zaki and W. Meira Jr. Data mining and analysis: Fundamental concepts and algorithms. *Cambridge University Press*, 2014.

615. O. Zamir and O. Etzioni. Web document clustering: A feasibility demonstration. *ACM SIGIR Conference*, pp. 46–54, 1998.

616. D. Zelenko, C. Aone, snd A. Richardella. Kernel methods for relation extraction. *Journal of Machine Learning Research*, 3. pp. 1083–1106, 2003.

617. C. Zhai. Statistical language models for information retrieval. *Synthesis Lectures on Human Language Technologies*, 1(1), pp. 1–141, 2008.

618. C. Zhai and J. Lafferty. A study of smoothing methods for language models applied to information retrieval. *ACM Transactions on Information Systems*, 22(2), pp. 179–214, 2004.

619. C. Zhai and S. Massung. Text data management and mining: A practical introduction to information retrieval and text mining. *Association of Computing Machinery/Morgan and Claypool Publishers*, 2016.

620. Y. Zhai and B. Liu. Web data extraction based on partial tree alignment. *World Wide Web Conference*, pp. 76–85, 2005.

621. J. Zhang, Z. Ghahramani, and Y. Yang. A probabilistic model for online document clustering with application to novelty detection. *NeurIPS Conference*, pp. 1617–1624, 2004.

622. J. Zhang, X. Long, and T. Suel. Performance of compressed inverted list caching in search engines. *World Wide Web Conference*, pp, 387–396, 2008.

623. J. Zhang *et al.* Pegasus: Pre-training with extracted gap-sentences for abstractive summarization. *International Conference on Machine Learning*, 2020.

624. M. Zhang, J. Zhang, and J. Su. Exploring syntactic features for relation extraction using a convolution tree kernel. *Human Language Technology Conference of the North American Chapter of the Association for Computational Linguistics*, pp. 288-295, 2006.

625. M. Zhang, J. Zhang, J. Su, and G. Zhou. A composite kernel to extract relations between entities with both flat and structured features. *International Conference on Computational Linguistics and the Annual Meeting of the Association for Computational Linguistics*, pp. 825–832, 2006.

626. X. Zhang, M. Lapata, F. Wei, and M. Zhou. Neural latent extractive document summarization. *arXiv:1808.07187*, 2018. https://arxiv.org/abs/1808.07187

627. S. Zhao and R. Grishman. Extracting relations with integrated information using kernel methods. *ACL Conference*, pp. 419–426, 2005.

628. Y. Zhao, G. Karypis. Empirical and theoretical comparisons of selected criterion functions for document clustering, *Machine Learning*, 55(3), pp. 311–331, 2004.

629. S. Zhong. Efficient streaming text clustering. *Neural Networks*, Vol. 18, 5–6, 2005.

630. Y. Zhou, H. Cheng, and J. X. Yu. Graph clustering based on structural/attribute similarities. *Proceedings of the VLDB Endowment*, 2(1), pp. 718–729, 2009.

631. Z.-H. Zhou. Ensemble methods: Foundations and algorithms. *CRC Press*, 2012.

632. J. Zhu *et al.* Incorporating bert into neural machine translation. *arXiv:2002.06823*, 2020. https://arxiv.org/abs/2002.06823

633. Y. Zhu, Y. Chen, Z. Lu, S. J. Pan, G. Xue, Y. Yu, and Q. Yang. Heterogeneous transfer learning for image classification. *AAAI Conference*, 2011.

634. Y. Zhu *et al.* Aligning books and movies: Towards story-like visual explanations by watching movies and reading books. *CVPR*, 2019.

635. L. Zhuang, F. Jing, and X. Zhu. Movie review mining and summarization. *ACM CIKM Conference*, pp. 43–50, 2006.

636. J. Zobel and P. Dart. Finding approximate matches in large lexicons. *Software: Practice and Experience*, 25(3), pp. 331–345, 1995.

637. J. Zobel and P. Dart. Phonetic string matching: Lessons from information retrieval. *ACM SIGIR Conference*, pp. 166–172, 1996.

638. J. Zobel, A. Moffat, and K. Ramamohanarao. Inverted files versus signature files for text indexing. *ACM Transactions on Database Systems*, 23(4), pp. 453–490, 1998.

639. J. Zobel and A. Moffat. Inverted files for text search engines. *ACM Computing Surveys (CSUR)*, 38(2), 6, 2006.

640. H. Zou and T. Hastie. Regularization and variable selection via the elastic net. *Journal of the Royal Statistical Society: Series B (Stat. Methodology)*, 67(2), pp. 301–320, 2005.

641. https://github.com/Element-Research/rnn/blob/master/examples/

642. https://github.com/lmthang/nmt.matlab

643. http://snowballstem.org/

644. http://opennlp.apache.org/index.html

645. https://archive.ics.uci.edu/ml/datasets.html

646. http://scikit-learn.org/stable/tutorial/text_analytics/working_with_text_data.html

647. https://cran.r-project.org/web/packages/tm/

648. https://www.ibm.com/developerworks/community/blogs/nlp/entry/tokenization?lang=en

649. http://www.cs.waikato.ac.nz/ml/weka/

650. http://nlp.stanford.edu/software/

651. http://nlp.stanford.edu/links/statnlp.html

652. http://www.nltk.org/

653. https://cran.r-project.org/web/packages/lsa/index.html

654. http://scikit-learn.org/stable/modules/generated/sklearn.decomposition.TruncatedSVD.html

655. http://weka.sourceforge.net/doc.stable/weka/attributeSelection/LatentSemanticAnalysis.html

656. http://scikit-learn.org/stable/modules/generated/sklearn.decomposition.NMF.html

657. http://scikit-learn.org/stable/modules/generated/sklearn.decomposition.LatentDirichletAllocation.html

658. https://cran.r-project.org/

659. http://www.cs.princeton.edu/~blei/lda-c/

660. http://scikit-learn.org/stable/modules/manifold.html

661. https://code.google.com/archive/p/word2vec/

662. https://www.tensorflow.org/tutorials/word2vec/

663. http://www.netlib.org/svdpack

664. http://scikit-learn.org/stable/modules/kernel_approximation.html

665. http://scikit-learn.org/stable/auto_examples/text/document_clustering.html

666. https://www.mathworks.com/help/stats/cluster-analysis.html

667. https://cran.r-project.org/web/packages/RTextTools/RTextTools.pdf

668. https://cran.r-project.org/web/packages/rotationForest/index.html

669. http://trec.nist.gov/data.html

670. http://research.nii.ac.jp/ntcir/data/data-en.html

671. http://www.clef-initiative.eu/home

672. https://archive.ics.uci.edu/ml/datasets/Twenty+Newsgroups

673. https://archive.ics.uci.edu/ml/datasets/Reuters-21578+Text+Categorization+Collection

674. http://www.daviddlewis.com/resources/testcollections/rcv1/

675. http://labs.europeana.eu/data

676. http://www.icwsm.org/2009/data/

677. https://www.csie.ntu.edu.tw/~cjlin/libmf/

678. http://www.lemurproject.org

679. https://nutch.apache.org/

680. https://scrapy.org/

681. https://webarchive.jira.com/wiki/display/Heritrix

682. http://www.dataparksearch.org/

683. http://lucene.apache.org/core/

684. http://lucene.apache.org/solr/

685. http://sphinxsearch.com/

686. https://snap.stanford.edu/snap/description.html

687. https://catalog.ldc.upenn.edu/LDC93T3A

688. http://www.berouge.com/Pages/default.aspx

689. https://code.google.com/archive/p/icsisumm/

690. http://finzi.psych.upenn.edu/library/LSAfun/html/genericSummary.html

691. https://github.com/tensorflow/models/tree/master/textsum

692. http://www.summarization.com/mead/

693. http://www.scs.leeds.ac.uk/amalgam/tagsets/brown.html

694. https://www.ling.upenn.edu/courses/Fall_2003/ling001/penn_treebank_pos.html

695. http://www.itl.nist.gov/iad/mig/tests/ace

696. http://www.biocreative.org

697. http://www.signll.org/conll

698. http://reverb.cs.washington.edu/

699. http://knowitall.github.io/ollie/

700. http://nlp.stanford.edu/software/openie.html

701. http://mallet.cs.umass.edu/

702. https://github.com/dbpedia-spotlight/dbpedia-spotlight/wiki

703. http://crf.sourceforge.net/

704. https://en.wikipedia.org/wiki/ClearForest

705. http://clic.cimec.unitn.it/composes/toolkit/

706. https://github.com/stanfordnlp/GloVe
707. https://deeplearning4j.org/
708. https://www.cs.uic.edu/~liub/FBS/sentiment-analysis.html
709. http://www.nltk.org/api/nltk.tokenize.html#nltk.tokenize.texttiling.TextTilingTokenizer
710. http://www.itl.nist.gov/iad/mig/tests/tdt/
711. http://colah.github.io/posts/2015-08-Understanding-LSTMs/
712. http://deeplearning.net/tutorial/lstm.html
713. http://machinelearningmastery.com/sequence-classification-lstm-recurrent-neural-networks-python-keras/
714. https://deeplearning4j.org/lstm
715. https://github.com/karpathy/char-rnn
716. https://arxiv.org/abs/1609.08144
717. https://msturing.org/
718. https://www.tensorflow.org/tutorials/text/transformer
719. https://github.com/google-research/bert
720. https://github.com/google-research/text-to-text-transfer-transformer
721. https://github.com/tensorflow/mesh/blob/master/mesh_tensorflow/transformer/moe.py
722. https://github.com/zihangdai/xlnet
723. https://github.com/allenai/bilm-tf
724. https://github.com/nlpyang/BertSum
725. https://github.com/hpzhao/SummaRuNNer
726. https://github.com/google-research/pegasus
727. https://github.com/huggingface/transformers
728. https://github.com/google/trax
729. https://huggingface.co/datasets/cnn_dailymail
730. https://huggingface.co/transformers/custom_datasets.html
731. https://trec.nist.gov/data/qa.html
732. https://github.com/facebookresearch/DrQA
733. https://github.com/hitvoice/DrQA
734. https://github.com/danqi/acl2020-openqa-tutorial
735. https://github.com/google-research/language/tree/master/language/orqa
736. https://github.com/google-research/google-research/tree/master/t5_closed_book_qa
737. https://github.com/OceanskySun/GraftNet
738. https://github.com/allenai/bilm-tf
739. https://gluebenchmark.com/
740. https://super.gluebenchmark.com
741. https://nlp.stanford.edu/sentiment/treebank.html
742. http://nlpprogress.com/
743. http://nlpprogress.com/english/natural_language_inference.html

Index

© Springer Nature Switzerland AG 2022
C. C. Aggarwal, *Machine Learning for Text*,
https://doi.org/10.1007/978-3-030-96623-2

Printed in the United States
by Baker & Taylor Publisher Services